T0336726

Imperfections in Crystalline Solids

This textbook provides students with a complete working knowledge of the properties of imperfections in crystalline solids. Readers will learn how to apply the fundamental principles of mechanics and thermodynamics to defect properties in materials science, gaining all the knowledge and tools needed to put this into practice in their own research.

Beginning with an introduction to defects and a brief review of basic elasticity theory and statistical thermodynamics, the authors go on to guide the reader in a step-by-step way through point, line, and planar defects, with an emphasis on their structural, thermodynamic, and kinetic properties.

Numerous end-of-chapter exercises enable students to put their knowledge into practice, and with solutions for instructors and MATLAB programs available online, this is an essential text for advanced undergraduate and introductory graduate courses in crystal defects, as well as being ideal for self-study.

Wei Cai is an Associate Professor in the Department of Mechanical Engineering at Stanford University. He received the Presidential Early Career Award for Scientists and Engineers in 2004, the American Society of Engineering Education Beer and Johnston New Mechanics Educator Award in 2009, and the American Society of Mechanical Engineers Hughes Young Investigator Award in 2013.

William D. Nix has been teaching courses on imperfections in crystalline solids at Stanford for more than 50 years. His teaching awards include the Bradley Stoughton Award for Young Teachers of Metallurgy, the Albert Easton White Distinguished Teacher Award, and The Educator Award from The Metallurgical Society. He has also received many awards for his research and is a Fellow of the American Academy of Arts and Sciences and a member of both the National Academy of Engineering and the National Academy of Sciences.

"This book captures the spirit of the legendary Stanford lecture course on the subject. It combines scientific authority with linguistic elegance and is suitable for students and specialists alike – a joy to read."

Eduard Arzt, Saarland University

"This authoritative book provides a thorough and quantitative description of imperfections in crystalline materials with clear text and illustrations. Students and practitioners of materials science and engineering and other related fields will benefit immensely by applying the fundamental concepts presented in the book to scientific research problems and engineering practice."

Subra Suresh, Carnegie Mellon University

"This invaluable textbook introduces knowledge of the complicated imperfections in crystalline solids and their properties unambiguously and completely, which is extremely useful for students and professionals in the field of materials science and engineering."

K. Lu, Shenyang National Laboratory for Materials Science, Chinese Academy of Science

Imperfections in Crystalline Solids

Wei Cai
Stanford University

William D. Nix
Stanford University

CAMBRIDGE
UNIVERSITY PRESS

University Printing House, Cambridge CB2 8BS, United Kingdom

One Liberty Plaza, 20th Floor, New York, NY 10006, USA

477 Williamstown Road, Port Melbourne, VIC 3207, Australia

314-321, 3rd Floor, Plot 3, Splendor Forum, Jasola District Centre, New Delhi - 110025, India

79 Anson Road, #06-04/06, Singapore 079906

Cambridge University Press is part of the University of Cambridge.

It furthers the University's mission by disseminating knowledge in the pursuit of education, learning and research at the highest international levels of excellence.

www.cambridge.org
Information on this title: www.cambridge.org/9781107123137

First published 2016

A catalogue record for this publication is available from the British Library

ISBN 978-1-107-12313-7 Hardback

Additional resources for this publication at www.cambridge.org/cai-nix

To my parents, for their encouragement and support.
Wei Cai

To my family, for their patience with my obsessions.
William D. Nix

Contents

	Preface	*page* xi
1	**Introduction**	1
1.1	Perfect crystal structures	1
1.2	Defect-controlled properties of crystals	8
1.3	Zero-dimensional defects	11
1.4	One-dimensional defects	12
1.5	Two-dimensional defects	14
1.6	Three-dimensional defects	14
1.7	Summary	14
1.8	Exercise problems	16
	PART I THEORETICAL BACKGROUND	
2	**Stress, strain, and isotropic elasticity**	21
2.1	Stress	21
2.2	Strain	25
2.3	Isotropic elasticity	28
2.4	Elastic strain energy	33
2.5	Fundamental equations of elasticity	34
2.6	Summary	42
2.7	Exercise problems	43
3	**Statistical thermodynamics**	48
3.1	Laws of thermodynamics	48
3.2	Thermodynamic potentials	50
3.3	Boltzmann's entropy expression	57
3.4	Boltzmann's distribution	59
3.5	Summary	63
3.6	Exercise problems	64

PART II POINT DEFECTS

4 Point defect mechanics 71
4.1 Hard sphere model 71
4.2 Symmetry of distortions about solutes 73
4.3 Atomic size factors 79
4.4 Elastic fields of atomic point defects 83
4.5 Elastic field of misfitting inclusion 86
4.6 Summary 95
4.7 Exercise problems 96

5 Point defect thermodynamics 101
5.1 Equilibrium concentration of solutes 101
5.2 Equilibrium concentration of vacancies 112
5.3 Vacancy experiments 118
5.4 Point defect chemical potential 122
5.5 Summary 132
5.6 Exercise problems 134

6 Point defect equilibria 140
6.1 Vacancies and self-interstitials in Si 140
6.2 Point defects in strongly ionic solids 141
6.3 Point defects in nonstoichiometric ionic solids 148
6.4 Constitutional defects in intermetallic compounds 161
6.5 Divacancies and other vacancy complexes 164
6.6 Summary 168
6.7 Exercise problems 169

7 Point defect kinetics 175
7.1 Motion of vacancies 176
7.2 Motion of solute atoms 181
7.3 Diffusion equation 183
7.4 Diffusion under stress 189
7.5 Diffusional deformation 193
7.6 Summary 200
7.7 Exercise problems 201

PART III DISLOCATIONS

8 Dislocation geometry 209
8.1 Role of dislocations in plastic deformation 209
8.2 Examples of dislocations 214
8.3 Burgers circuit and Burgers vector 219
8.4 Dislocation motion and slip 225

8.5 Dislocation sources 230
8.6 Summary 235
8.7 Exercise problems 237

9 Dislocation mechanics 244
9.1 Elastic fields of isolated dislocations 244
9.2 Dislocation line energy 257
9.3 Dislocation line tension 262
9.4 Forces on dislocations 267
9.5 Summary 275
9.6 Exercise problems 276

10 Dislocation interactions and applications 283
10.1 Interactions between two dislocations 284
10.2 Dislocation arrays 296
10.3 Strengthening mechanisms 302
10.4 Dislocation kinetics and plastic flow 306
10.5 Formation of dislocations at interfaces 312
10.6 Elastic fields of dislocations near interfaces 320
10.7 Summary 336
10.8 Exercise problems 338

11 Partial and extended dislocations 348
11.1 Partial dislocations in FCC metals 349
11.2 Dislocations in HCP metals 375
11.3 Partial dislocations in $CrCl_3$ 380
11.4 Superdislocations in ordered Ni_3Al 382
11.5 Summary 389
11.6 Exercise problems 390

12 Dislocation core structure 395
12.1 Peierls–Nabarro model 395
12.2 Dislocations in FCC metals 403
12.3 Dislocations in diamond cubic structures 406
12.4 Dislocations in BCC metals 410
12.5 Dislocation–point defect interactions 418
12.6 Summary 426
12.7 Exercise problems 427

PART IV GRAIN BOUNDARIES

13 Grain boundary geometry 433
13.1 Grain boundary orientation variables 433
13.2 Coincidence site lattice 437

13.3 Displacement shift complete lattice 446
13.4 Summary 451
13.5 Exercise problems 452

14 Grain boundary mechanics 455
14.1 Low angle tilt boundaries 456
14.2 Low angle twist boundaries 463
14.3 Dislocation content of arbitrary low angle grain boundaries 468
14.4 Grain boundary edge dislocations 470
14.5 Grain boundary screw dislocations 473
14.6 Disconnections and disclinations 474
14.7 Summary 482
14.8 Exercise problems 483

APPENDICES

A King table for solid solutions 488

B Thermoelastic properties of common crystalline solids 497

C Thermodynamic and kinetic properties of vacancies 499

D Diffusion coefficients in common crystals 501

Bibliography 507
Index 515

Preface

This book is mainly written for senior undergraduate and junior graduate students wanting to gain an understanding of the behavior of defects in crystalline materials using the fundamental principles of mechanics and thermodynamics. We choose the word *imperfections* to emphasize that the crystalline materials in which these defects are found are nearly perfect. In other words, the densities of these defects are usually very low. Yet, they can greatly alter and even control the properties of the host crystal. It can be said that the main purpose of the entire field of materials science is to modify the properties of materials through the control of their defects.

The book is written based on a set of lecture notes of a course (MSE206 Imperfections in Crystalline Solids) that one of us (WDN) taught in the Materials Science and Engineering Department of Stanford University for more than 50 years. This course is now taught in the Mechanical Engineering Department (as ME209 by WC). We wanted to turn these lecture notes into a textbook so that it can be used by students and instructors in other universities who are interested in learning/teaching this subject.

The scope of this book has significant overlap with two important books in this area: *Introduction to Dislocations* by Hull and Bacon, and *Theory of Dislocations* by Hirth and Lothe. The book by Hull and Bacon provides a clear introduction for junior undergraduate students to the field of defects in crystals, while the book by Hirth and Lothe is a monograph and a valuable reference to experienced researchers in this field. It has long been recognized by the community that what we lack is a textbook that bridges the gap between these two books, a textbook that can be used in the teaching of core senior undergraduate/junior graduate courses on defects in crystals.

To this end, we can share the personal experience of one of us (WC) while he was a graduate student himself (at MIT). Realizing that his PhD thesis research would deal with the modeling of crystalline defects, he first read through *Introduction to Dislocations* by Hull and Bacon. The book provided a very useful introduction, but after reading it, he still did not feel quite "ready" for his research tasks. So he realized it was necessary to delve into *Theory of Dislocations* by Hirth and Lothe. But that proved to be a significant challenge, especially for a junior graduate student. Looking back, it appears that a significant part of his needs as a starting graduate student could have been met if he had had the opportunity to take the MSE206 course, being offered at the other end of the continent. The possibility of helping those students in a similar situation provided a strong personal motivation for him to help turn the lecture notes of MSE206 into this book.

The ultimate goal of this book is to develop confidence in students, so that they not only *know* how defects behave in crystalline materials, but also can *apply* a set of fundamental principles to explain these behaviors, to make some (albeit approximate) predictions, and to analyze similar situations they may encounter in the future. In other words, we hope that students will develop sufficient confidence to do something with the material they have learned. To achieve this goal, most of the analyses in this book are described at a slow pace, in a step-by-step fashion, with the intention that the student can follow the steps by him/herself and will be able to apply the same approach to similar situations. The exercise problems at the end of each chapter then provide the practice necessary for mastery. The mathematics used in the analysis is kept at a level accessible to a senior undergraduate student in science or engineering.

An obvious challenge faced by this approach is that, if we describe each topic in such great detail, we cannot possibly cover many topics in a reasonably sized book. Consequently, we chose to focus on a smaller set of topics than what might appear in a broad book on defects in crystals. The motto we adopted for this book project was: we want to be *tutorial rather than encyclopedic.* When deciding on which topics to include, preference was placed on the ones that could be analyzed using the fundamental principles introduced in previous chapters.

Much emphasis in this book is placed on developing intuition. We believe it to be an important skill, which also brings a great deal of satisfaction, to be able to develop approximate, but simple, estimates of various quantities of interest, such as the fraction of vacancies in a crystal near its melting temperature. Intuition is developed (and confidence strengthened) when we have a habit of making these rough estimates and comparing them with experimental data. It is also a great way to integrate the various subjects discussed in this book and see how they fit together.

Among the existing books on this topic, this book is somewhat special in terms of the number of illustrations (300+) and number of exercise problems (300+). The more than 50 years of teaching MSE206 have resulted in the wealth of materials available for this book. We believe illustrations and diagrams are very important in conveying the essence of the subject. Therefore, we have made all the figures in this book freely available as PDF files on the book website (http://ics-book.stanford.edu), so that the reader can view them on a computer screen at arbitrary magnification.

The book consists of 14 chapters, which are grouped into four parts. Part I introduces the fundamental mechanics and statistical-thermodynamic principles that will be used in the subsequent parts of the book. In the teaching of a course, these subjects are usually covered by previous courses in the curriculum. They are included in this book to serve as a review and a quick reference for the readers. Parts II, III, and IV of the book then apply these fundamental principles to point, line, and planar defects, respectively, in crystalline materials. A summary is included at the end of each chapter. The reader can use the summary to check whether he/she has received the main messages we wish to deliver in each chapter. We emphasize that working on the exercise problems at the end of each chapter is essential for a full mastery of the subject.

While the content of this book is based on a one quarter, first year graduate level course that has been given at Stanford for many years, not all of the topics found in the book were ever presented in a given quarter. Two or more quarters would be needed to cover everything. But we anticipate that not everyone will be equally interested in all aspects of defects. Teachers will naturally want to spend more time on some topics and less, or none at all, on others. In that case, parts of the book can serve as the text for the course, with the unused parts of the books serving as a reference for the book owners if they subsequently wish to delve into different areas of crystal defects. Also, teachers might even wish to cover only elementary fundamentals in some areas, usually at the beginning of the chapters, while going to the advanced fundamentals near the ends of the chapters on topics of greater interest.

We want to thank the thousands of students who have either taken MSE206 at Stanford University or used the course notes over the past 50+ years, and have helped to shape the lecture materials that ultimately led to this book. We also thank a group of dedicated former and current students, Wendelin Wright (Bucknell University), Gang Feng (Villanova University), Seok-Woo Lee (University of Connecticut), Ill Ryu (Brown University), William Kuykendall, Ryan Sills, and Yanming Wang, who devoted their time to read the nearly final version of this book and provided valuable suggestions. WC wishes to thank Sidney Yip of MIT and Vasily V. Bulatov of LLNL for introducing him to this field, and for instilling in him a writing style of plain English and clarity. WDN wishes to thank Robert A. Huggins of Stanford for introducing him to materials science more than 55 years ago and for encouraging him in this field, and Oleg D. Sherby of Stanford for introducing the field of dislocations to him and giving him the chance to learn to teach this exciting subject.

Finally, we wish to thank the following individuals for granting us permission to use various images for the illustrations on the cover of the book: Yayu Wang of Tsinghua University for the image of point defects in $Ca_2CuO_2Cl_2$; Sang Chul Lee of Stanford University for the TEM image of a dislocation at the interface of a film of Sm doped Ceria on a Yttrium Stabilized Zirconia substrate; Wayne King of the Lawrence Livermore National Laboratory for the image of a $\Sigma 11$ grain boundary in aluminum; and James LeBeau of North Carolina State University for the image of twin boundaries in NiFeCrCo.

1 Introduction

The perfect crystal structure is an idealization of the atomic arrangements in real crystalline materials. After a brief introduction of several common perfect crystal structures, we start our study of imperfections in crystals with some remarks about why so much attention is focused on these defects. The central reason is that perfect crystals, without imperfections, would be relatively uninteresting materials, without most of the useful properties with which we are all familiar. We consider some of the physical properties that crystals would have or not have if they were perfect. Through this thought experiment, we show that most of the useful engineering properties of crystalline materials are defect controlled and thus depend on the properties and behavior of imperfections.

1.1 Perfect crystal structures

1.1.1 Single crystals and polycrystals

The word "crystal" usually brings to mind large mineral (e.g. quartz) blocks on display in museums, or the shiny diamond on a wedding ring. Their faceted surfaces and often distinct geometric shape give rise to a sense of beauty not found in other more "common" materials. As an example, Fig. 1.1a shows a photograph of a ruby crystal. However, crystalline materials are easily found in our everyday life. In fact, most engineering materials are crystalline. Metals, semiconductors, and ceramics are all crystalline materials, even though they may not have faceted surfaces.

The distinction between a large ruby crystal and an engineering metallic alloy is that the former is a single crystal and the latter is usually a polycrystal. A polycrystal is an aggregate of many small single crystals (called grains), each with a different orientation. As an example, Fig. 1.1b shows a micrograph of a nickel-based superalloy (where the word "super" refers to its superior mechanical properties). The size of each single crystal grain in this superalloy is on the order of 10 to 100 micrometers (μm), too small to be seen by the naked eye. That is why the shape of a piece of metal does not seem faceted to the eye; the facets can be observed with the aid of a microscope.

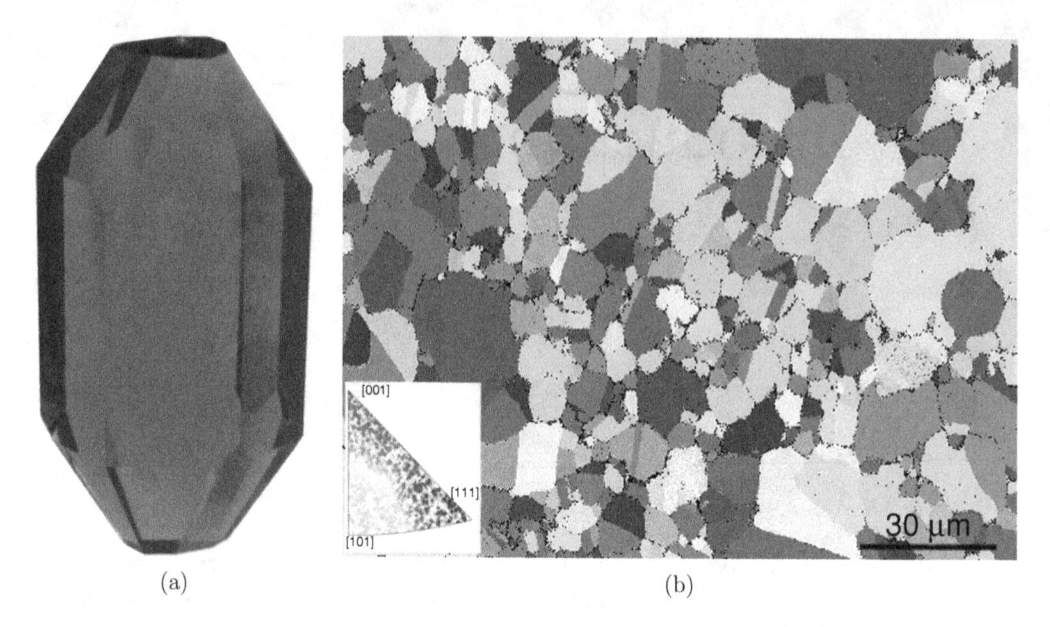

(a) (b)

Figure 1.1. (a) A ruby single crystal (Al_2O_3 doped with Cr^{3+} ions). (b) Electron back-scatter diffraction map of a polycrystalline Ni-based superalloy, with a stereographic triangle showing the orientations of the different grains. Courtesy of Michael D. Uchic, Materials and Manufacturing Directorate, Air Force Research Laboratory, and used with permission. For clarity see color figures on the book website.

If we examine crystalline materials at a length scale smaller than individual grains, then a superalloy and a ruby crystal have something in common. The atoms inside each single crystal grain are arranged in an impressively ordered, periodic array. The distance between each atom and its nearest neighbors are nearly the same everywhere in the crystal. Furthermore, there is long-range order in each crystal grain, meaning that if we know the local orientation of a cluster of atoms at one side of the grain, we can accurately predict the local orientation of another cluster of atoms at the other side of the grain, even if the two clusters may be millions of atomic distances apart. Such long-range order is unique in crystalline solids, and is not present in non-crystalline (i.e. amorphous) solids such as glass or typical polymers (plastics). It is certainly lacking in fluids (except liquid crystals).

The mathematical idealization to describe the long-range order in crystals is the perfect crystal structure. In the following, we describe a few common perfect crystal structures. The description here is meant to be elementary and readers knowledgeable of perfect crystal structures and Miller indices notation may skip on to Section 1.2. However, it should be emphasized that real crystals deviate from this idealization due to the presence of various types of defects, to be introduced later in this chapter. For more discussion of the lattice and crystal structures see [1].

1.1.2 Cubic crystals with simple basis

Simple cubic crystals

The simplest way to construct a perfect crystal structure is to first imagine a simple cubic lattice and then place the same type of atom at every lattice point. The result is a simple cubic crystal structure, as shown in Fig. 1.2.

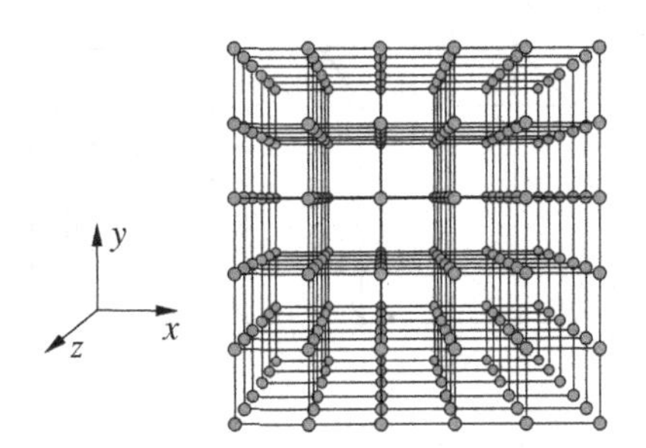

Figure 1.2. A perfect simple cubic crystal structure.

The location of all lattice points can be described by the following expression,

$$\vec{R} = a(u\hat{e}_x + v\hat{e}_y + w\hat{e}_z), \tag{1.1}$$

where a is the *lattice constant*, u, v, w are arbitrary integers, and $\hat{e}_x$, $\hat{e}_y$, $\hat{e}_z$ are unit vectors along the x, y, and z axes. Therefore, any vector or line direction connecting lattice points (i.e. crystallographic directions) can be specified by three integers.[1] In the *Miller indices* notation, the three numbers, u, v, w, are surrounded by square brackets. For example, $[2\,\bar{3}\,4]$ (reads "two three-bar four") represents the vector $(2\hat{e}_x - 3\hat{e}_y + 4\hat{e}_z)$, where it is customary to write the minus sign as a bar on top of the number. Therefore, every point in the simple cubic lattice can be represented by Miller indices as

$$\vec{R} = a[u \quad v \quad w]. \tag{1.2}$$

Angular brackets are used to represent a family of directions that have the same length and are related by symmetry (e.g. reflection). For example, $\langle 2\,3\,4 \rangle$ represents the family of directions: $[2\,3\,4]$, $[\bar{2}\,3\,4]$, $[2\,\bar{3}\,4]$, $[2\,3\,\bar{4}]$, $[\bar{2}\,\bar{3}\,4]$, ...

Miller indices can also be used to represent the crystallographic planes. These indices are found by first determining the intercepts that the planes make with the crystallographic axes, dividing each by the lattice constant along that axis, taking the reciprocals and, if necessary, multiplying all of the resulting quantities by the smallest integer to remove any fractions that may have been created. This creates a set of integer Miller indices, h, k, l, for the plane in question, commonly surrounded by parenthesis $(h\,k\,l)$. With this notation any crystallographic plane in a cubic solid can be described by the linear equation $hx + ky + lz = a$. The normal vector extending from the origin to the plane in question is then $a[h\,k\,l]/(h^2 + k^2 + l^2)$; thus the crystallographic direction $[u\,v\,w]$ is perpendicular to the plane $(h\,k\,l)$ if $u = h$, $v = k$, $w = l$. That is, crystallographic directions and planes with the same indices are perpendicular to each other in the cubic crystal system. In addition, it can be shown that the perpendicular distance from the origin to the plane can be expressed as

$$d_{hkl} = \frac{a}{\sqrt{h^2 + k^2 + l^2}}, \tag{1.3}$$

which is the spacing between such crystallographic planes. For crystallographic planes, braces are used to represent a family of planes related by symmetry. For example, $\{2\,3\,4\}$ represents the family of planes: $(2\,3\,4)$, $(\bar{2}\,3\,4)$, $(2\,\bar{3}\,4)$, $(2\,3\,\bar{4})$, $(\bar{2}\,\bar{3}\,4)$, ...

1 These numbers may also be fractional numbers for lattices other than the simple cubic lattice.

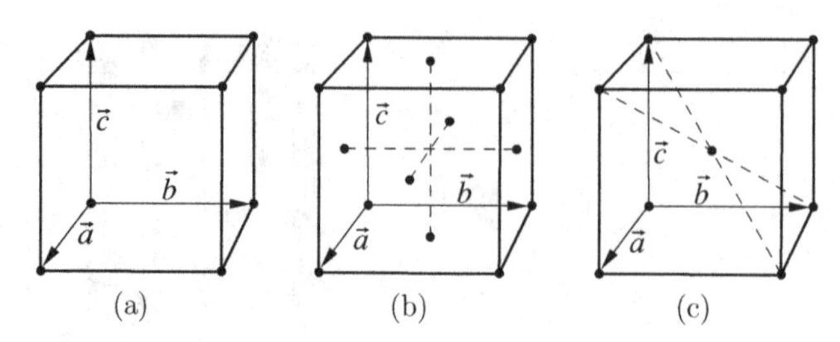

Figure 1.3. (a) The unit cell of a simple cubic lattice. (b) The unit cell of a face-centered cubic (FCC) lattice. (c) The unit cell of a body-centered cubic (BCC) lattice [2].

It is common to define three lattice vectors $\vec{a} = a\hat{e}_x$, $\vec{b} = a\hat{e}_y$, $\vec{c} = a\hat{e}_z$, which are the vectors connecting a lattice point to its neighbors. For the simple cubic lattice, $\vec{a}$, $\vec{b}$, $\vec{c}$ are orthogonal to each other, and all have the same length. These three vectors can be used to construct a cell, which is a cube for the simple cubic lattice, as shown in Fig. 1.3a. In every three-dimensional lattice, we can find such a representative cell that, when repeated, spans all space with no gaps. The smallest possible cell of this type is called the *primitive cell*; each primitive cell contains only one lattice point. However, the lattice is often represented by the *unit cell*, which may have a larger volume than the primitive cell, and hence may contain more than one lattice point. The unit cell is chosen to show the symmetry of the lattice more clearly. For the simple cubic lattice, the unit cell is the same as the primitive cell, which is a cube with lattice points occupying the corners of the cube, as shown in Fig. 1.3a.

It is useful to compute the average volume occupied by each lattice point in the simple cubic lattice, which is the same as the average volume occupied by each atom in the simple cubic crystal structure. Figure 1.3a shows that each unit cell is associated with eight lattice points. However, each lattice point is also shared by eight neighboring unit cells. Therefore, each unit cell contains $8 \times \frac{1}{8} = 1$ lattice point, and the average volume per lattice point is

$$\Omega_{\text{simple cubic}} = \frac{a^3}{8 \times \frac{1}{8}} = a^3. \tag{1.4}$$

Face-centered cubic crystals

While a simple cubic crystal structure is easy to imagine, it is rare to find a crystal in nature with that structure. The most common crystal structures in engineering materials are based on the face-centered cubic (FCC) lattice and the body-centered cubic (BCC) lattice.

The FCC lattice can be visualized by its unit cell, shown in Fig. 1.3b. In addition to placing a point at each of the eight corners of the cube, we also place a point at the center of each of the six faces of the cube. Mathematically, every lattice point in an FCC lattice is given by one of the following expressions in terms of Miller indices:

$$\begin{aligned} \vec{R} &= a[u \quad v \quad w] \\ \vec{R} &= a[u+0.5 \quad v+0.5 \quad w] \\ \vec{R} &= a[u+0.5 \quad v \quad w+0.5] \\ \vec{R} &= a[u \quad v+0.5 \quad w+0.5]. \end{aligned} \tag{1.5}$$

Gold (Au), silver (Ag), aluminum (Al), and copper (Cu) are all FCC metals.

Figure 1.3b shows that each FCC unit cell is associated with eight lattice points at the corners and six lattice points at the face centers. However, each corner point is shared by eight neighboring unit cells and each face center point is shared by two neighboring unit cells. Therefore, each unit cell contains $8 \times \frac{1}{8} + 6 \times \frac{1}{2} = 4$ lattice points, and the average volume per lattice point is

$$\Omega_{\rm FCC} = \frac{a^3}{8 \times \frac{1}{8} + 6 \times \frac{1}{2}} = \frac{a^3}{4}. \tag{1.6}$$

This is also the average volume occupied by each atom in the FCC crystal structure. Because the unit cell of the FCC lattice contains more than one lattice point, it is not the primitive cell. The primitive cell of the FCC lattice is a parallelpiped with edges formed by vectors: $\frac{a}{2}[1\ 1\ 0]$, $\frac{a}{2}[1\ 0\ 1]$, $\frac{a}{2}[0\ 1\ 1]$. The cubic symmetry of the FCC lattice is not immediately apparent from its primitive cell.

Body-centered cubic crystals

The unit cell of the BCC lattice is shown in Fig. 1.3c, where a point is placed at the center of the cube, in addition to the eight corners of the cube. Mathematically, the lattice points in a BCC lattice are given by one of the following relations:

$$\begin{aligned} \vec{R} &= a[u \quad v \quad w] \\ \vec{R} &= a[u + 0.5 \quad v + 0.5 \quad w + 0.5]. \end{aligned} \tag{1.7}$$

Tungsten (W), molybdenum (Mo), and tantalum (Ta) are BCC metals. Pure iron (Fe) is also a BCC metal at room temperature.[2]

Figure 1.3(c) shows that each BCC unit cell is associated with eight lattice points at the corners and one lattice point at the body center. While each corner point is shared by eight neighboring unit cells, the point at the body center is not shared with any other unit cells. Therefore, each unit cell contains $8 \times \frac{1}{8} + 1 = 2$ lattice points, and the average volume per lattice point is

$$\Omega_{\rm BCC} = \frac{a^3}{8 \times \frac{1}{8} + 1} = \frac{a^3}{2}. \tag{1.8}$$

This is also the average volume occupied by each atom in the BCC crystal structure. Because the unit cell of the BCC lattice contains more than one lattice point, it is not the primitive cell. The primitive cell of the BCC lattice is a parallelpiped with edges formed by vectors: $a[1\ 1\ \bar{1}]/2$, $a[1\ \bar{1}\ 1]/2$, $a[\bar{1}\ 1\ 1]/2$. The cubic symmetry of the BCC lattice is not immediately apparent from its primitive cell.

The SC, FCC, and BCC lattices, as specified by Eq. (1.2), Eq. (1.5), and Eq. (1.7), respectively, are three types of lattices with different symmetries. The symmetries of infinite, periodic lattices were studied by Auguste Bravais, and these lattices are called *Bravais* lattices. In three dimensions, there are only 14 distinct types of Bravais lattices, which are classified into seven lattice systems [3]. The three lattices (SC, FCC, BCC) introduced here form the cubic lattice system. The hexagonal lattice to be introduced in Section 1.1.4 forms its own (hexagonal) lattice system.

2 However, Fe changes to the FCC structure at high temperature, and common stainless steel (an iron–nickel–chromium alloy) also has an FCC structure.

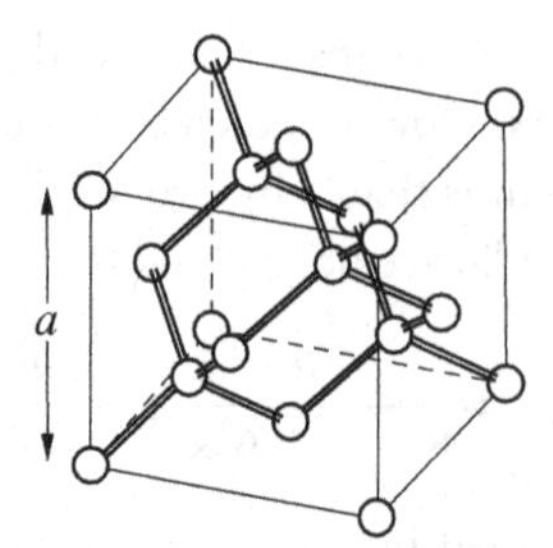

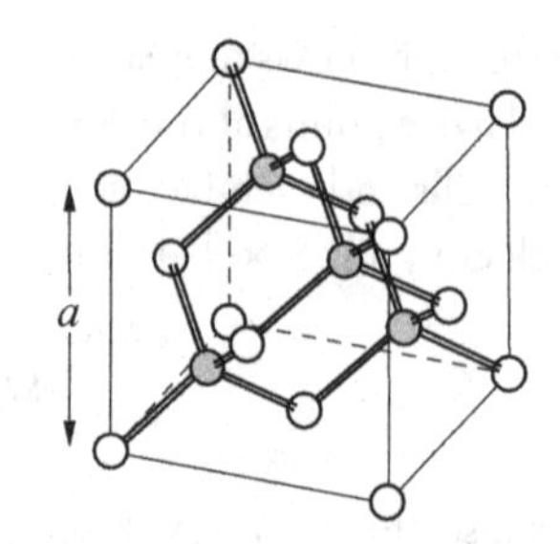

Figure 1.4. (a) The unit cell of the diamond cubic crystal structure. (b) The unit cell of the Zincblende crystal structure. The light and dark spheres represent atoms of two different chemical species. In both (a) and (b), the separation between the two atoms in the basis is $\frac{a}{4}[1\,1\,1]$.

1.1.3 Cubic crystals with complex basis

The crystal structures we have considered so far can be obtained by placing one and the same type of atom at every point of a given lattice. A multi-atom "motif" or "tiling" can be introduced at every lattice point to create more complex crystal structures. Therefore, even though there are only 14 different types of 3D Bravais lattices, the number of possible crystal structures is infinite.

The atomic "motif" placed at each lattice point is called the *basis* of the crystal structure. The basis can be arbitrarily chosen, but exactly the same basis must be placed on every lattice point (no rotation or changing of chemical species is allowed).

For example, a diamond (on a wedding ring) has the diamond cubic (DC) structure, which is created by placing a two-carbon basis at each lattice point of the FCC lattice. Elemental semiconductors such as silicon (Si) and germanium (Ge) have the DC structure. The unit cell of the DC structure is shown in Fig. 1.4a.

Because the DC structure has a two-atom basis, the average volume occupied by each atom in the DC structure is half of that in the FCC structure of the same lattice constant, i.e.

$$\Omega_{\rm DC} = \frac{\Omega_{\rm FCC}}{2} = \frac{a^3}{8}. \tag{1.9}$$

Closely related to the DC structure is the Zincblende structure, which is the structure of many Group III–V and II–VI compound semiconductors, such as GaAs, InP, ZnSe, CdTe, etc. The Zincblende structure is formed by placing a basis consisting of two different atoms, such as Ga and As, at every lattice point of the FCC lattice. The unit cell of the Zincblende structure is shown in Fig. 1.4b. Because FCC, DC, and Zincblende crystal structures are all based on the FCC lattice, the types of dislocations (line defects) in these types of crystals are very similar (see Chapter 12).

The sum of the average volume of the two atomic species in the Zincblende structure, e.g. Ga and As, equals the average volume occupied by each lattice point in the FCC lattice, i.e.

$$\Omega_{\rm Ga} + \Omega_{\rm As} = \Omega_{\rm FCC} = \frac{a^3}{4}. \tag{1.10}$$

However, the individual contribution of $\Omega_{\rm Ga}$ and $\Omega_{\rm As}$ can only be determined by the volume change of the crystal when the mole fraction of Ga and As deviates from the perfect value (50%), i.e. when defects are introduced to the crystal. This will be discussed in Chapter 6.

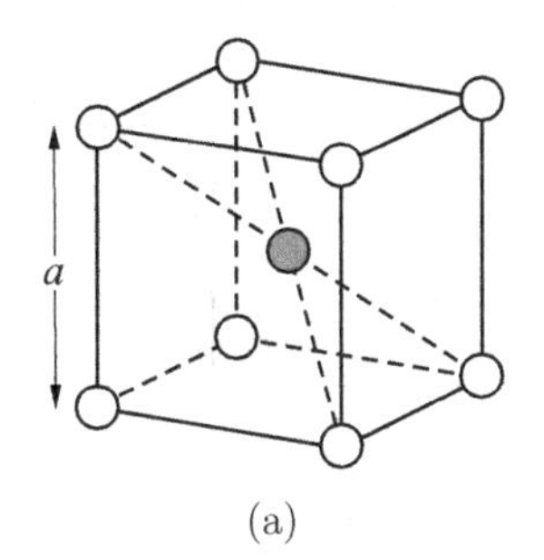

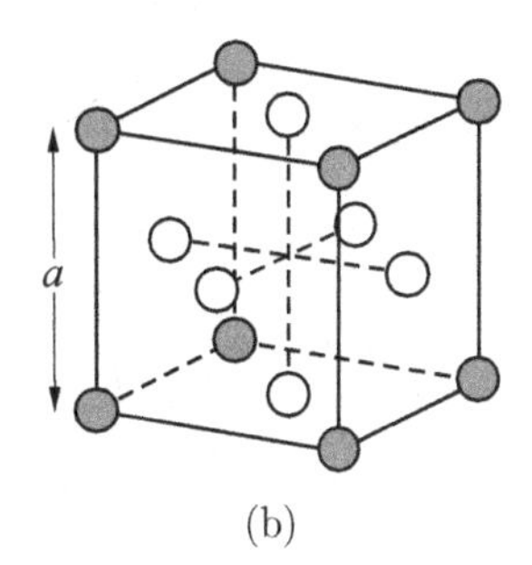

Figure 1.5. (a) The unit cell of the B2 (CsCl) crystal structure. (b) The unit cell of the $L1_2$ (Cu_3Au) crystal structure.

Intermetallic compounds are other types of crystals with multi-atom bases. For example, NiAl has the so-called B2 structure, which is also the structure of CsCl and FeAl. It is obtained by placing a Ni–Al basis at each lattice point of the simple cubic lattice. One atom (say Ni) is placed on the corner of the cube, and the other atom (say Al) is placed at the cube center. The unit cell of the B2 structure is shown in Fig. 1.5a. Note that the B2 structure does not have the BCC lattice, because we are not allowed to place different types of atoms on equivalent lattice points. The average volumes occupied by each Ni atom and Al atom in the NiAl (B2) crystal structure satisfy the relation

$$\Omega_{\mathrm{Ni}} + \Omega_{\mathrm{Al}} = \Omega_{\mathrm{SC}} = a^3. \tag{1.11}$$

Ni_3Al has the so-called $L1_2$ structure, which is also the structure of Cu_3Au and Fe_3Al. It is obtained by placing a four-atom basis (three Ni and one Al) at each lattice point of the simple cubic lattice. The three Ni atoms are placed at the face centers of the cube, and the Al atom is placed at the corner of the cube. The unit cell of the $L1_2$ structure is shown in Fig. 1.5b. The $L1_2$ structure does not have the FCC lattice, for the same reason that the B2 structure does not have the BCC lattice. The average volumes occupied by each Ni atom and Al atom in the Ni_3Al ($L1_2$) crystal structure satisfy the relation

$$3\Omega_{\mathrm{Ni}} + \Omega_{\mathrm{Al}} = \Omega_{\mathrm{SC}} = a^3. \tag{1.12}$$

1.1.4 Hexagonal close-packed structure

The hexagonal close-packed (HCP) crystal structure also has a multi-atom basis. Furthermore, it has a *hexagonal lattice*, instead of a cubic lattice. The unit cell of the hexagonal lattice is shown in Fig. 1.6a.

The lattice points of the hexagonal lattice can be described by the relation

$$\vec{R} = i\vec{a} + j\vec{b} + k\vec{c}, \tag{1.13}$$

where i, j, k are arbitrary integers. This is similar to the expression for the simple cubic lattice. The difference here is that the lattice repeat vectors $\vec{a}$ and $\vec{b}$ are at 120° relative to each other, and $\vec{c}$ does not have the same length as $\vec{a}$ and $\vec{b}$. Vectors $\vec{a}$ and $\vec{b}$ still have the same length, and vector $\vec{c}$ is perpendicular to both $\vec{a}$ and $\vec{b}$. The basis of the HCP crystal structure contains two atoms of the same chemical species. The unit cell of the HCP crystal structure is shown in

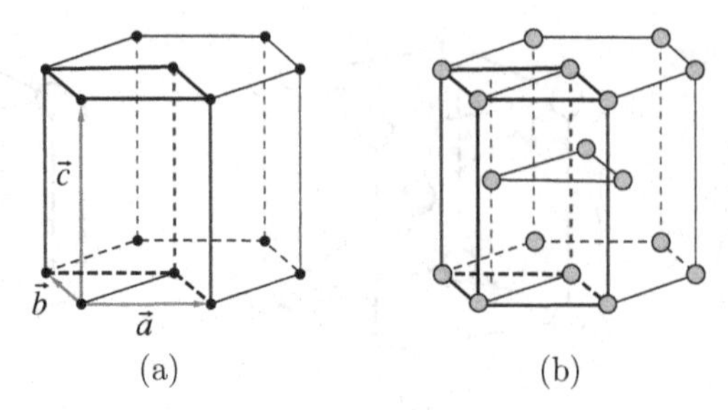

Figure 1.6. (a) The hexagonal lattice, whose unit cell is outlined by thick lines. (b) The hexagonal close-packed (HCP) crystal structure, whose unit cell is outlined by thick lines. One of the two atoms in the basis occupies the lattice point. The other atom in the basis lies immediately above the center of the triangle formed by three lattice points below.

Fig. 1.6b. Magnesium (Mg), titanium (Ti), zinc (Zn), and cobalt (Co) are some of the HCP metals.

The average volume occupied by each lattice point in the hexagonal lattice is

$$\Omega_{\text{hex. latt.}} = \frac{\sqrt{3}}{2} a^2 c, \tag{1.14}$$

where $a = |\vec{a}| = |\vec{b}|$, and $c = |\vec{c}|$. The average volume occupied by each atom in the HCP crystal structure is half of that,

$$\Omega_{\text{HCP}} = \frac{\Omega_{\text{hex. latt.}}}{2} = \frac{\sqrt{3}}{4} a^2 c. \tag{1.15}$$

The HCP crystal structure is closely related to the FCC crystal structure. Both FCC and HCP crystal structures are close-packed structures, meaning that they provide the maximum packing density if we imagine packing hard spheres of uniform radius into a pile. When an FCC crystal is viewed along the cubic diagonal direction $\hat{n} = \frac{1}{\sqrt{3}}[1\ 1\ 1]$, it can be considered as three types of triangular lattices A, B, C, stacked on top of each other in the sequence of $ABCABC\ldots$ (see Section 11.1). If the stacking sequence is changed to $ABABAB\ldots$, we get the HCP crystal structure instead (see Section 11.2.2). Therefore the FCC and HCP crystal structures can be transformed to each other by shear deformation.

1.2 Defect-controlled properties of crystals

While the perfect crystal structure captures the long-range order that exists in crystals, it would be a mistake to think of a real crystal as a collection of atoms located exactly according to the perfect crystal structure. Deviation from the perfect structure, i.e. imperfections, are vital to the physical, chemical, and electronic properties of the crystal. In order to emphasize the imperfection of real crystals, J. Frenkel in the opening chapter of his classic book, *Kinetic Theory of Liquids*, argues that crystalline solids actually have a lot in common with liquids (except for the long-range order, of course). In this section, we imagine how a crystal would behave if all atoms were located at their perfect crystal positions.

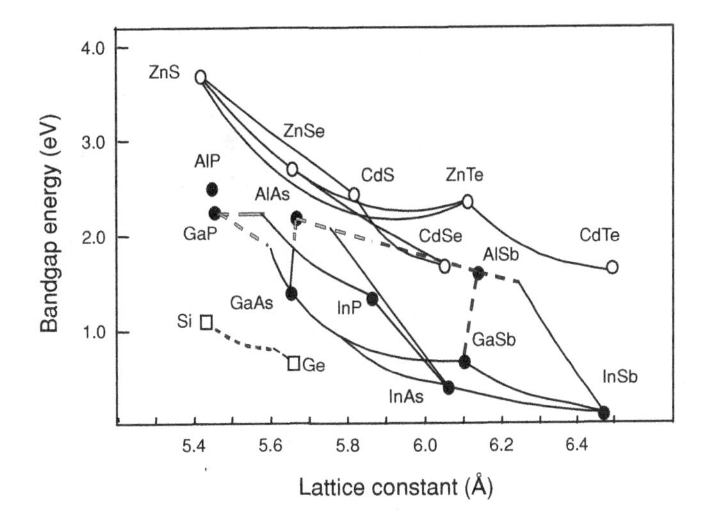

Figure 1.7. Band gap engineering – controlling the electronic band gaps in semiconductors by controlling the composition (point defect concentration). Band gap energy and lattice constant of various III–V semiconductors at room temperature. Figure adapted from one kindly provided by Professor James S. Harris of Stanford University and used with permission.

Consider an absolutely perfect crystal with all of the atoms sitting exactly on their respective lattice points, such as the one shown in Fig. 1.2, and all of the electrons in their lowest energy states. In such a perfect crystal there would be no lattice vibrations or phonons and thus no thermal conductivity. Nor would there be any heat capacity, as the constituent atoms would not be able to store thermal energy in their vibrations.

In addition, for such a perfect crystal, there would be no way to accomplish compositional changes by diffusion because there would be no atomic point defects. This would prevent us from making transistors by diffusion of dopants into silicon or do any other kind of alloying by solid-state diffusion.

Solid solution alloys are also needed for band gap engineering, which leads to semiconductor compounds with different electronic band gaps. As shown in Fig. 1.7, various Group III–V semiconductors (with Zincblende crystal structure, see Section 1.1.3) have different lattice constants as well as different band gaps. The band gap is the energy difference of the electrons between the top of the (filled) valance band and the bottom of the (empty) conduction band. The band gap is directly related to the wavelength of the light that the semiconductor can absorb or emit. By mixing two different semiconductors, e.g. GaAs and InAs, thus creating an imperfect crystal, the lattice constant of the crystal can be tuned between the two limits. The elastic strain on the lattice is coupled to the electronic band structure so that the band gap can also be tuned. The lattice constant and band gap of the impure crystal formed by mixing GaAs and InAs follow the curve connecting GaAs and InAs in Fig. 1.7. Band gap engineering enables, for example, the development of lasers at specific wavelengths for which the absorption in an optical fiber or in air is particularly low. Band gap engineering is made possible only by imperfect crystals.

If we only have perfect crystals to work with, it would not be possible to plastically deform a metal, preventing metal forming, and making it all but impossible to make such things as trains, tractors, airplanes, automobiles, and so on. Such perfect materials would be extremely strong but

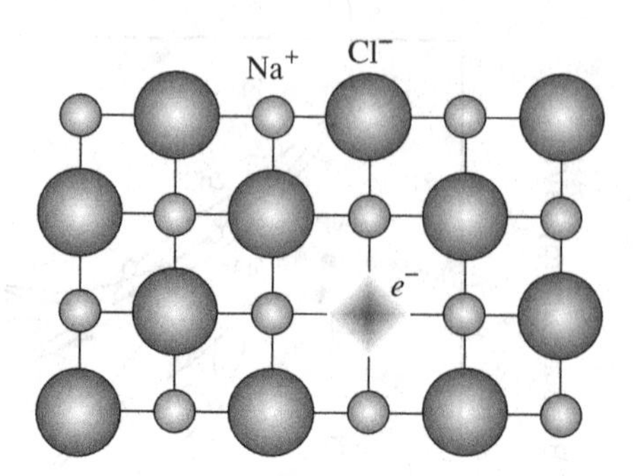

Figure 1.8. F-center in a NaCl crystal.

you could never change their shape by plastic deformation because crystal dislocations would not exist.

You would also not have transformers for changing voltage and current in everyday applications. Transformer cores depend on easy magnetization and demagnetization by the movements of defects in the magnetic structure of the crystals in the cores.

These are but a few examples of the useless properties that crystalline materials would have if they were composed of perfect crystals. Virtually all of the interesting properties that crystalline materials display arise from imperfections.

The control of imperfections in crystalline materials lies at the heart of materials science. Indeed, the very definition of materials science can be stated as: *the synthesis of useful engineering materials and the control of their properties through the control of composition and microstructure.* For crystalline materials the control of microstructure involves the control of crystal imperfections.

It is also worth noting that the most technologically important properties of crystalline materials are defect controlled. While elasticity does not involve defects, strength and plasticity do and so do most of the other engineering properties of interest.

Although one does not usually think of color as a property that can be controlled, for ceramic crystals color is very much defect controlled. Consider corundum or Al_2O_3, which is colorless in its pure state. We call this sapphire. If just 1% of the Al atoms in the crystal lattice are replaced by Cr atoms we have a red crystal called ruby. Alternatively, if 1% of the Al atoms are replaced by Ti or Fe atoms we have a blue crystal called blue sapphire.

Another example involves the color of alkali halide crystals, like KCl and NaCl. When these crystals are pure and relatively perfect they are colorless. By shining X-rays onto these crystals, lattice defects called F-centers,[3] which are anion (Cl^-) vacancies containing a trapped electron, are created (see Fig. 1.8). Such defects cause KCl to turn blue and NaCl to turn yellow. The color is controlled by the energy level of the electron that is trapped in the vacancy. Thus the color of ceramic crystals can be controlled by the control of atomic point defects.

Even though defects control most of the important properties of crystalline materials, typically only a very tiny fraction of all of the atoms in the crystal are involved in defects. Furthermore, defects must be well separated in the crystal lattice in order to have individual properties

3 The F stands for Farbe (German for color), so that F-center is also called color center.

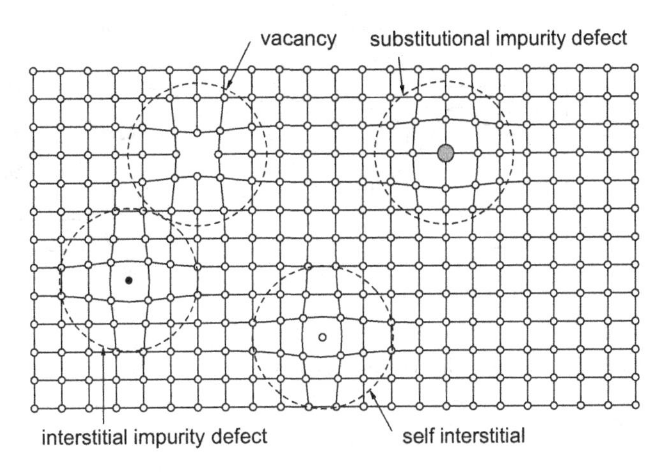

Figure 1.9. Some point defects – localized to atomic dimensions. The dashed circles indicate imperfect regions in the crystal.

of their own. Therefore, it is usually the case that

$$\text{defect spacing} \gg \text{atomic spacing}.$$

This is the *principle of dilute concentration*, with well-separated discrete defects having properties that do not depend on their concentrations. If this were not so, it would not be possible to characterize the separate properties of defects. This was explicitly recognized many decades ago in the title of the book *Imperfections in Nearly Perfect Crystals*, edited by W. Shockley *et al.* [4]. If the defects were so concentrated that their fields overlapped extensively, the material would be amorphous, instead of crystalline with defects.

In the following, we classify crystal defects according to their dimensionality and give an overview of the defects to be studied in more detail later.

1.3 Zero-dimensional defects

Zero-dimensional defects, or point defects, are lattice defects that are localized or confined to atomic dimensions in all three directions in a crystal.

Our focus is on the atomic point defects, for example: impurities or solute atoms (either substitutional or interstitial), vacancies or vacancy clusters, self-interstitials, point defect clusters, and, in some cases, small particles. As shown in Fig. 1.9, the distortions associated with these defects extend only a few atomic distances away from the defects in any direction, so they are called point defects.

Electronic point defects also exist in crystals. In a semiconductor (or insulator), these correspond to free electrons in the conduction band and free electronic holes in the valence band. Electronic point defects will not be discussed extensively in this book. In Part II of this book (Chapters 4, 5, 6, and 7), we will focus most of our attention on the distortions associated with atomic point defects and on the thermodynamics and kinetics of these defects.

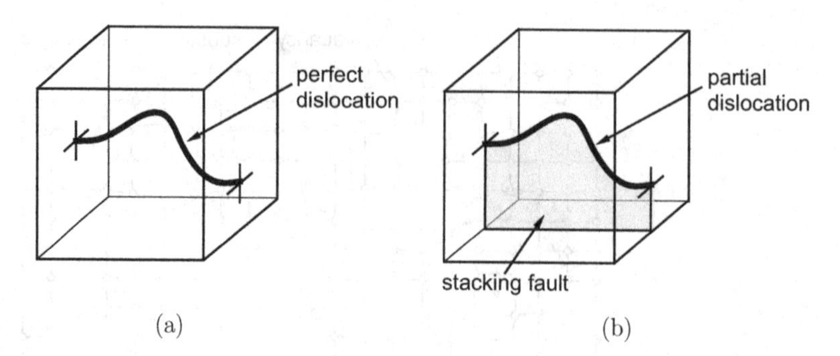

Figure 1.10. (a) Perfect dislocation as a line defect. (b) Partial dislocation attached to a stacking fault.

1.4 One-dimensional defects

1.4.1 Dislocations

The most important line defects, by far, are dislocations. They are imperfections that can extend indefinitely along a line in the crystal [5].

Perfect dislocations are line defects that are surrounded by perfect crystal, as illustrated in Fig. 1.10a. We will see in Chapter 8 that the Burgers vector, which characterizes the atomic displacements around a dislocation, must be a complete lattice translation vector in order for the perfect lattice to be maintained all around the dislocation line.

Imperfect or *partial dislocations* are line defects that lie at the edge of a stacking fault, a two-dimensional defect, as illustrated in Fig. 1.10b. Here, we will see later, that the Burgers vector is not equal to a lattice translation vector, so a fault in the crystal must extend from the partial dislocation line.

In Part III of this book (Chapters 8 through 12), we will study the structure and properties of *edge* and *screw* dislocations in detail. Our principal focus will be on: geometry of dislocations, mechanics of dislocations (stresses, forces, and energies), and atomic structure of the dislocation core (e.g. partial dislocations).

Figure 1.11 illustrates the atomic displacements around straight edge and screw dislocations. Under an electron microscope, dislocations usually appear as curved lines at the micrometer length scale. They usually exist in as-grown crystals, but large quantities of dislocations can be generated by plastic deformation. Some dislocation microstructures are shown in Fig. 1.12a, b. The stacking fault attached to one side of partial dislocations usually shows up as fringes that can be easily identified under the electron microscope, as shown in Fig. 1.12b.

1.4.2 Disclinations

Disclinations are line defects in lattices where rotational symmetry is violated. One type of disclination can be created in a crystal by removing a wedge of material and joining the two surfaces together (see Section 14.6.1). For specific angles of the wedge, it is possible for the atoms on the two surfaces of the wedge to bond as if in a perfect crystal. The atomic bonds are distorted only by a smoothly varying elastic strain field everywhere in the crystal, except at the tip of the wedge, which is the disclination line. The disclination mentioned above is the termination of a tilt grain boundary (see Chapter 14). If the wedge angle is small, it is also

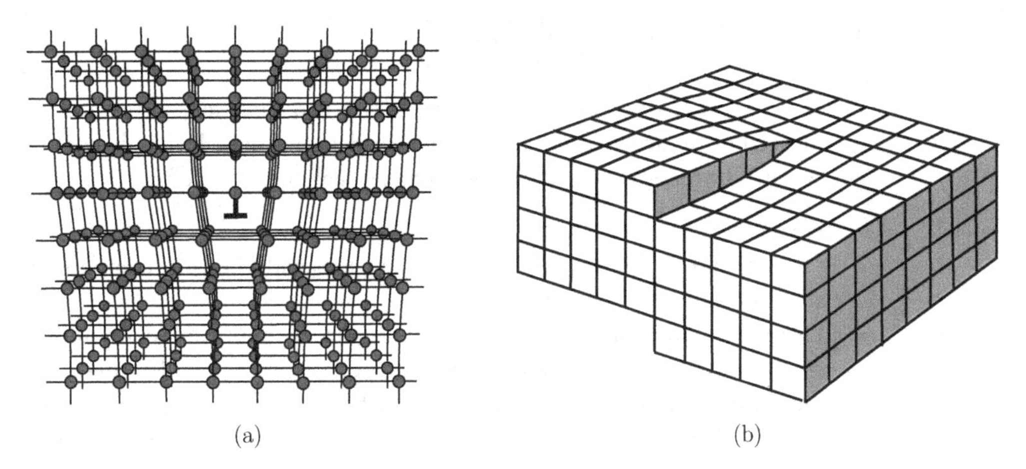

Figure 1.11. (a) Edge dislocation structure in a simple cubic crystal. (b) Screw dislocation structure in a simple cubic crystal. Atoms (not shown) are located at the corners of every cube [6].

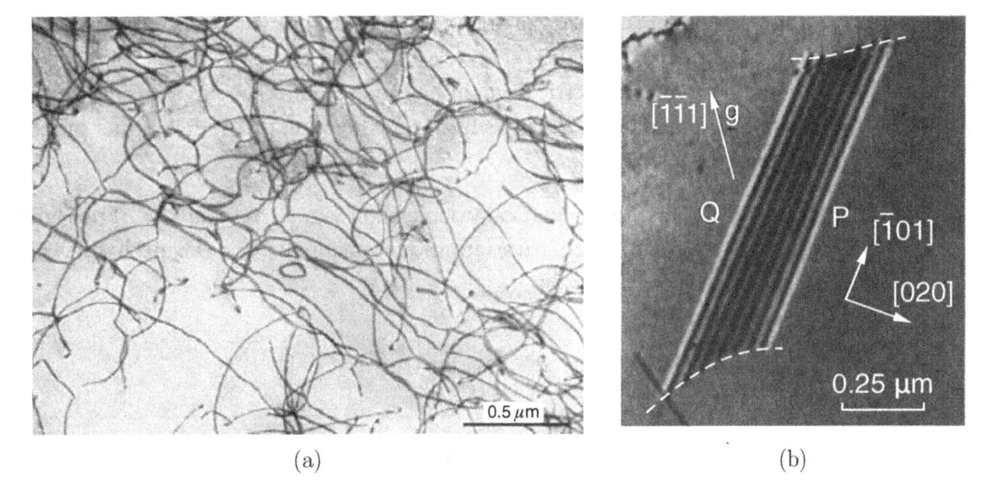

Figure 1.12. (a) Dislocation tangles in a Fe–35% Ni–20% Cr alloy, creep tested at 700 °C [7, 8]. Permission granted by R. Lagneborg. (b) Bright-field image of a stacking fault in Cu+7%Al alloy [9]. Permission granted by P. B. Hirsch. The dashed lines indicate partial dislocations. P and Q indicate the two surfaces of the thin film specimen.

equivalent to a semi-infinite array of edge dislocations. In solid crystals, a true disclination is difficult to find. Therefore, we will not have much discussion on disclinations in this book.

We mention in passing that disclinations appear frequently in some liquid crystals, in which the order of the crystal is indicated by the alignment of the rigid rod-like molecules. These alignments can be disrupted at certain places, which are disclinations. Far away from the disclinations, the orientations of rod-like molecules are similar to the orientation of their neighbors. But this becomes more and more difficult to arrange as we move closer to the disclination, and the alignment completely breaks down at the disclination. The disclination line is somewhat like a dislocation but with a displacement vector that varies with distance from the defect. Alternatively, we can say that the dislocation is a defect in positional order, while the disclination is a defect in orientational order.

1.5 Two-dimensional defects

Imperfections that can extend indefinitely along two directions in a solid are called two-dimensional or surface defects. They are planar defects although the plane of the defect need not be flat. The following is a list of two-dimensional or planar defects.

- Free surfaces.
- Grain boundaries: interfaces that separate regions with the same composition and crystal structure but with different orientations.
- Phase boundaries: interfaces that separate regions with different compositions and/or crystal structures. The crystals on both sides of the phase boundary may or may not have the same orientation. For example, the interface between a heteroepitaxial film of (001) Ge on a (001) Si substrate is a phase boundary that separates crystals with different compositions but the same structure and orientation.
- Stacking fault: a mistake in the perfect stacking order of a crystal.
- Antiphase boundary: an interface separating regions of a crystal with the same composition, crystal structure, and orientation, but with a mistake in the ordering pattern at the boundary.
- Magnetic and ferroelectric domain boundaries: interfaces in a crystal where the magnetic or electric polarization switches from one direction to another, either gradually (magnetic) or abruptly (electric).

These planar defects are schematically depicted in Fig. 1.13. In Part IV of this book (Chapters 13 and 14), we will study the structure (atomic arrangements) and the mechanics (stress and energy) of grain boundaries.

1.6 Three-dimensional defects

All of our attention will be focused on zero-, one-, and two-dimensional defects, as outlined above. Nevertheless, it is important to know that in certain industries the term "defects" usually refers to more macroscopic, three-dimensional defects. Here is a short list of this type of defects.

- Porosity (voids) and shrinkage defects in castings.
- Cracks.
- Massive inclusions of foreign materials.
- Dust particles (semiconductor processing).

1.7 Summary

Imperfections in crystalline solids are, by definition, deviations from the ideal state of perfect crystals. The study of crystal defects then requires an examination of the perfect crystal lattices that serve as references for the defects themselves. The atomic arrangements in perfect crystals can be described with Bravais lattices wherein identical atom groups are located at each of the lattice points in these lattices. This provides a useful way to describe not only the simple cubic

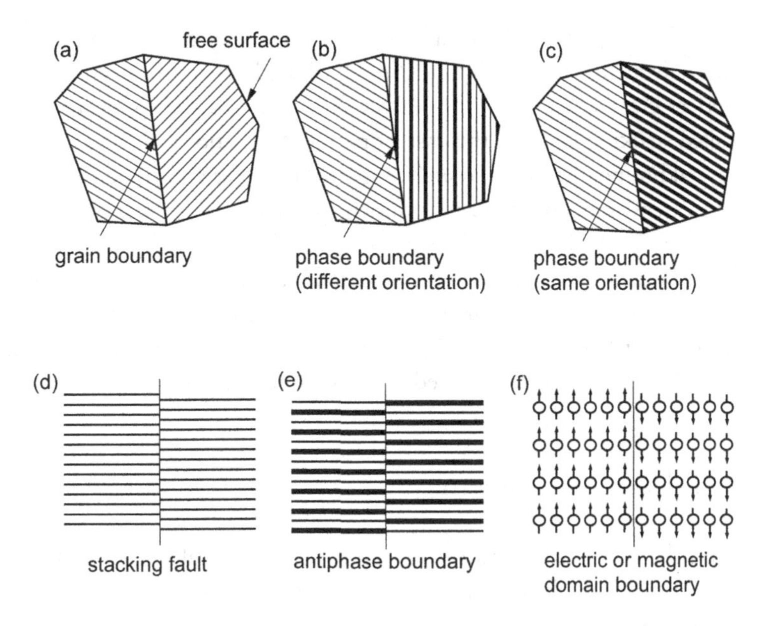

Figure 1.13. Two-dimensional (surface) defects. (a) The stripes indicate that the two grains have different orientations. (b) The thickness difference of the stripes in the two grains indicates two different materials. (c) The alignment of the stripes in both grains indicates the two crystals are aligned even though they are made of different materials. (d) The two crystals are made of the same material and have the same orientation, but have a mismatch in the stacking order. (e) The thin and thick lines indicate two types of chemical species in the crystals. (f) The arrows indicate electric or magnetic polarization on the atoms.

structures often used to illustrate basic concepts in defect theory, but also the most common crystal structures found in engineering materials: face-centered cubic (FCC), body-centered cubic (BCC), diamond cubic (DC) and other related cubic structures, and also hexagonal crystal structures. The description of perfect crystal structures includes the necessary mathematical methods for describing the positions of atoms in the crystals as well as nomenclature (e.g. Miller indices) for describing the directions and planes in crystals. The geometrical description of crystal structures also allows for the calculation of atomic volumes that play an important role in understanding imperfections.

The control of imperfections in crystalline materials lies at the heart of materials science. Indeed, the very definition of materials science can be stated as: *the synthesis of useful engineering materials and the control of their properties through the control of composition and microstructure.* For crystalline materials the control of microstructure involves the control of crystal imperfections. The hypothetical properties of perfect crystals, without defects, are shown to be mostly uninteresting. Important material properties as diverse as alloying and diffusion, strength and plasticity, magnetic permeability of ferromagnetic crystals, electronic properties of compound semiconductors, and color of ceramic crystals, all depend critically on the presence and behavior of crystal imperfections.

Imperfections can be classified according to their dimensionality. This scheme describes atomic point defects, such as solute atoms and lattice vacancies, as zero-dimensional defects with lattice distortions confined to within a few atomic dimensions in any direction. Atomic point defects are critically important in alloying and diffusion in crystalline solids. Dislocations, which are responsible for both plastic flow and strengthening of metals, are one-dimensional or line defects that can extend indefinitely along a line in the crystal. They may be surrounded by perfect crystal in the case of perfect dislocations or be connected to two-dimensional defects called stacking faults in the case of partial dislocations. In addition to stacking faults, grain or crystal boundaries and free surfaces comprise other two-dimensional defects wherein the defect can extend indefinitely in two dimensions but with distortions that are confined to within a few atomic dimensions of the plane of the defect. Other two-dimensional defects include the boundaries separating differently ordered states of the crystal, such as positional ordering of different kinds of atoms or magnetic order. Three-dimensional defects, such as voids, cracks, gas bubbles, or unwanted second phase particles, though very important in technology, are usually so massive as to not be counted as crystal imperfections.

1.8 Exercise problems

1.1 Find the distance d between nearest lattice points along the $\langle 1\,0\,0\rangle$, $\langle 1\,1\,0\rangle$, $\langle 1\,1\,1\rangle$, $\langle 1\,1\,2\rangle$, and $\langle 1\,2\,3\rangle$ lines for the simple cubic lattice with lattice constant a. Rank the line directions in increasing order of d.

1.2 Repeat the analysis in Exercise problem 1.1 for the FCC lattice.

1.3 Repeat the analysis in Exercise problem 1.1 for the BCC lattice.

1.4 Find the distance h between nearest parallel planes of lattice points perpendicular to directions of the type $\langle 1\,0\,0\rangle$, $\langle 1\,1\,0\rangle$, $\langle 1\,1\,1\rangle$, $\langle 1\,1\,2\rangle$, and $\langle 1\,2\,3\rangle$ for the simple cubic lattice with lattice constant a. Rank the directions in decreasing order of h.

1.5 Repeat the analysis in Exercise problem 1.4 for the FCC lattice.

1.6 Repeat the analysis in Exercise problem 1.4 for the BCC lattice.

1.7 Consider an atom in a simple cubic crystal and located at the origin. Find the distance between this atom and its first, second, and third nearest neighbors in terms of the lattice constant a. Express the vectors from this atom to its first, second, and third nearest neighbors in Miller indices. How many first, second, and third nearest neighbors does this atom have?

1.8 Repeat the analysis in Exercise problem 1.7 for a FCC crystal with lattice constant a.

1.9 Repeat the analysis in Exercise problem 1.7 for a BCC crystal with lattice constant a.

1.10 The NaCl (table salt) crystal has a cubic structure. Its unit cell is shown in Fig. 1.14a. The Na atoms (white spheres) form an FCC sub-lattice. The Cl atoms (gray spheres) also form an FCC sub-lattice that is offset from the Na sub-lattice by $a/2$ in the $[1\,0\,0]$ direction. Identify the lattice and basis for the crystal structure of NaCl. Express the average atomic volume in terms of the lattice constant a. The NaCl structure is also called the B1 structure, which is also the crystal structure of AgCl, MgO, and CaO.

1.11 The CaF_2 (fluorite) crystal has a cubic structure. Its unit cell is shown in Fig. 1.14b. The Ca atoms (white spheres) form an FCC sub-lattice. To specify the position of F atoms, imagine that the unit cell cube is divided into eight smaller cubes of equal sizes. The F atoms are located at the centers of these smaller cubes. Identify the lattice and basis for the crystal structure of CaF_2. Express the average atomic volume in terms of the lattice constant a. The CaF_2 structure is also called the C1 structure, which is also the crystal structure of $NiSi_2$ and cubic ZrO_2.

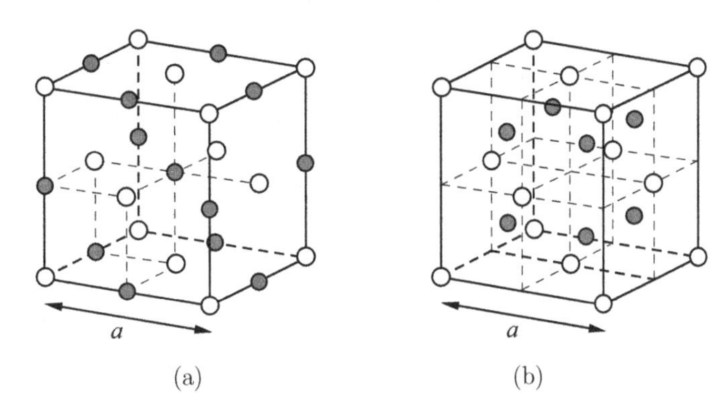

Figure 1.14. (a) The unit cell of the B1 (NaCl) crystal structure. (b) The unit cell of the C1 (CaF_2) crystal structure.

1.12 List three useful defect-controlled properties of crystalline solids that impact your daily life. For each property, identify whether the controlling defect is a point, line, or planar defect.

PART I

THEORETICAL BACKGROUND

2 Stress, strain, and isotropic elasticity

Here we give a brief review of the principles of stress, strain, and isotropic elasticity that will be needed in the study of defects. Readers familiar with this elementary material can skip on to the next chapter. The review is given here as a reference that will be used from time to time in the remainder of this book. A more thorough treatment of this subject can be found in many elasticity textbooks, such as [10, 11].

2.1 Stress

Stress is a measure of the intensity of force transmitted through a surface separating different parts of a body. The basic definition of stress is force per unit area. There are two types of stress: axial and shear, as shown in Fig. 2.1a, b, respectively. The axial stress is $\sigma = P/A$, where the force P is perpendicular to the surface area A. In other words, the force P acts along the surface normal. Hence the axial stress is also called the normal stress. The shear stress is $\tau = P/A$, where the shear force P is parallel to the surface area A. The dimension of stress is f/l^2, where f is the dimension of force and l is the dimension of length. The unit of stress is $N/m^2 = Pa$ (pascal).

2.1.1 Stress as a second-rank tensor

To completely specify the stress state at a point, we need to consider a small cube around this point and specify the traction forces per unit area on all faces of this cube. The edges of the cube are chosen to be parallel to the axes of a given coordinate system. The positive faces of the cube are defined as the three faces whose outward normal vectors are along the positive x, y, and z axes, respectively. Because the size of the cube is vanishingly small, it suffices to specify the forces on the positive faces of the cube. The forces on the negative faces must be opposite to the forces on the corresponding positive faces.

The force directions of various stress components σ_{ij} are indicated on the positive faces of the cube in Fig. 2.2. The arrows indicate the force directions for the stress components if their values are positive. According to this sign convention, a positive axial stress, e.g. σ_{xx}, corresponds to tensile stress, e.g. along the x axis. A negative axial stress corresponds to compression.

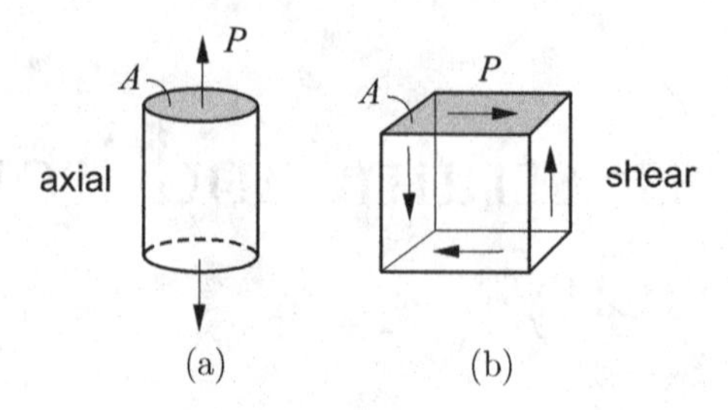

Figure 2.1. (a) Axial stress. (b) Shear stress.

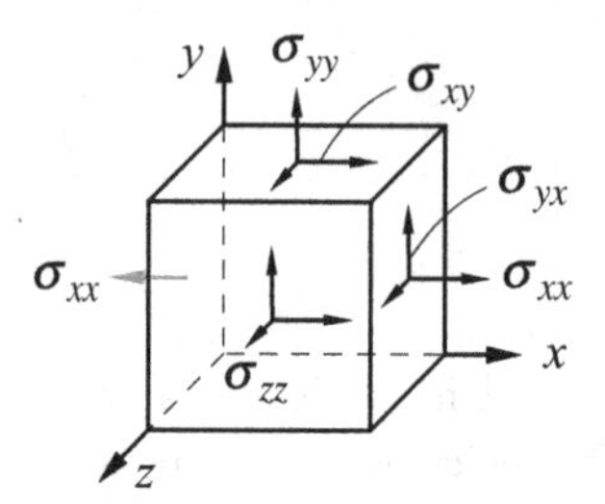

Figure 2.2. Stress at a point (infinitesimal volume). The positive notation is shown. Positive components of stress involve positive forces acting on faces with positive normals, or equivalently, negative forces on faces with normals pointing in the negative directions (as shown in gray for the component σ_{xx}).

Nine quantities are needed to completely define a state of stress (although we will show that these will be reduced to six): σ_{ij} $(i, j = x, y, z)$. Equivalently $i, j = 1, 2, 3$ are also commonly used. For example, σ_{xy} and σ_{12} mean the same thing if the Cartesian (xyz) coordinate system is implied. Here i specifies the direction of the force direction and j specifies the direction of the plane normal on which the force acts. The three force directions times the three plane normals make for nine components of a second-rank tensor. In general, a tensor of rank R in n-dimensions has $N = n^R$ components.

It is convenient to write the second-rank stress tensor as a 3×3 matrix. In Cartesian coordinates,

$$\sigma_{ij} = \begin{pmatrix} \sigma_{xx} & \sigma_{xy} & \sigma_{xz} \\ \sigma_{yx} & \sigma_{yy} & \sigma_{yz} \\ \sigma_{zx} & \sigma_{zy} & \sigma_{zz} \end{pmatrix}. \tag{2.1}$$

A similar notation is used for other orthogonal coordinate systems. The directions of the stress components in cylindrical coordinates (r, θ, z) are shown in Fig. 2.3a. The stress tensor can be written in the following matrix form

$$\sigma_{ij} = \begin{pmatrix} \sigma_{rr} & \sigma_{r\theta} & \sigma_{rz} \\ \sigma_{\theta r} & \sigma_{\theta\theta} & \sigma_{\theta z} \\ \sigma_{zr} & \sigma_{z\theta} & \sigma_{zz} \end{pmatrix}. \tag{2.2}$$

The directions of the stress components in spherical coordinates (r, θ, φ) are shown in Fig. 2.3b. The stress tensor can be written in the following matrix form

$$\sigma_{ij} = \begin{pmatrix} \sigma_{rr} & \sigma_{r\theta} & \sigma_{r\varphi} \\ \sigma_{\theta r} & \sigma_{\theta\theta} & \sigma_{\theta\varphi} \\ \sigma_{\varphi r} & \sigma_{\varphi\theta} & \sigma_{\varphi\varphi} \end{pmatrix}. \tag{2.3}$$

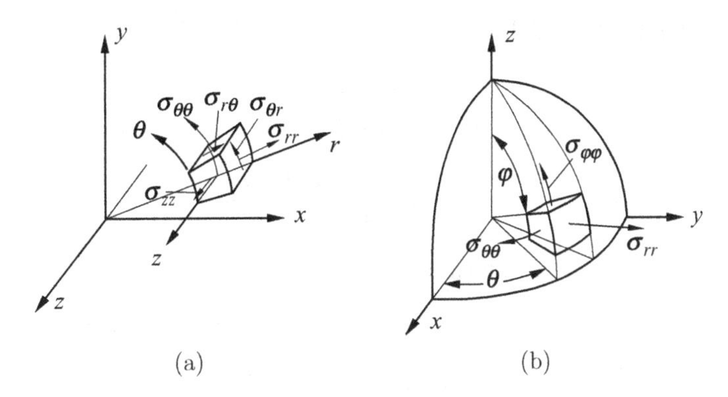

Figure 2.3. (a) Stress components in cylindrical coordinates. (b) Stress components in spherical coordinates.

2.1.2 Symmetry of stress tensor

The stress tensor is a symmetric tensor, wherein $\sigma_{ij} = \sigma_{ji}$. This can be shown from a simple moment equilibrium analysis below.

Suppose the dimension of the cube is Δx, Δy, and Δz, along the x, y, z axes, respectively. The sum of the moments about the z axis should be zero if no inertial moments exist (i.e. no dynamics).[1] Thus

$$\begin{aligned} \sum M_z &= (\sigma_{xy}\Delta x)\Delta y - (\sigma_{yx}\Delta y)\Delta x = 0, \\ \sigma_{xy} &= \sigma_{yx}. \end{aligned} \tag{2.4}$$

The vanishing moments about the other axes lead to $\sigma_{yz} = \sigma_{zy}$ and $\sigma_{xz} = \sigma_{zx}$. Therefore there are only six independent stress components. It is sometimes convenient to rename these components as follows:

$$\sigma_{ij} = \begin{pmatrix} \sigma_{xx} & \sigma_{xy} & \sigma_{xz} \\ \sigma_{yx} & \sigma_{yy} & \sigma_{yz} \\ \sigma_{zx} & \sigma_{zy} & \sigma_{zz} \end{pmatrix} = \begin{pmatrix} \sigma_1 & \sigma_6 & \sigma_5 \\ \bullet & \sigma_2 & \sigma_4 \\ \bullet & \bullet & \sigma_3 \end{pmatrix}. \tag{2.5}$$

In the so-called *Voigt notation*, the stress components are represented by a column vector,

$$\sigma_I = \begin{pmatrix} \sigma_1 \\ \sigma_2 \\ \sigma_3 \\ \sigma_4 \\ \sigma_5 \\ \sigma_6 \end{pmatrix}, \tag{2.6}$$

where $I = 1, 2, \ldots, 6$. For Cartesian coordinates,

$$\begin{aligned} &\sigma_1 = \sigma_{xx}, \quad \sigma_2 = \sigma_{yy}, \quad \sigma_3 = \sigma_{zz}, \\ &\sigma_4 = \sigma_{yz} = \sigma_{zy}, \\ &\sigma_5 = \sigma_{xz} = \sigma_{zx}, \\ &\sigma_6 = \sigma_{xy} = \sigma_{yx}. \end{aligned} \tag{2.7}$$

1 In fact, one can show that, due to the vanishingly small size of the cube, this conclusion remains valid even if there are inertial moments (i.e. dynamics).

For cylindrical coordinates,

$$\begin{aligned} &\sigma_1 = \sigma_{rr}, \quad \sigma_2 = \sigma_{\theta\theta}, \quad \sigma_3 = \sigma_{zz}, \\ &\sigma_4 = \sigma_{\theta z} = \sigma_{z\theta}, \\ &\sigma_5 = \sigma_{rz} = \sigma_{zr}, \\ &\sigma_6 = \sigma_{r\theta} = \sigma_{\theta r}, \end{aligned} \tag{2.8}$$

and for spherical coordinates,

$$\begin{aligned} &\sigma_1 = \sigma_{rr}, \quad \sigma_2 = \sigma_{\theta\theta}, \quad \sigma_3 = \sigma_{\varphi\varphi}, \\ &\sigma_4 = \sigma_{\theta\varphi} = \sigma_{\varphi\theta}, \\ &\sigma_5 = \sigma_{r\varphi} = \sigma_{\varphi r}, \\ &\sigma_6 = \sigma_{r\theta} = \sigma_{\theta r}. \end{aligned} \tag{2.9}$$

2.1.3 Stress transformation

Transformation of stresses from one coordinate system to another is needed, for example, when you are given stresses in the x, y, z system and you need stresses in the r, θ, z system, or vice versa.

In general, suppose we are given two local coordinate systems at a point p: coordinate system A with unit vectors $\hat{e}_1$, $\hat{e}_2$, $\hat{e}_3$, and coordinate system A′ with unit vectors $\hat{e}'_1$, $\hat{e}'_2$, $\hat{e}'_3$. The stress tensor at point p can be expressed by six independent components of σ_{ij} in the coordinate system A (i.e. using a cube with edges parallel to $\hat{e}_1$, $\hat{e}_2$, $\hat{e}_3$). The same stress tensor can be expressed by six independent components of σ'_{ij} in the coordinate system A′. Since they correspond to the same stress state, σ_{ij} and σ'_{ij} must be related to each other. Their relationship can be expressed as

$$\sigma'_{ij} = Q_{ik} Q_{jl} \sigma_{kl}, \tag{2.10}$$

where we have used the Einstein notation, in which all repeated indices (k and l here) are summed from 1 to 3. The matrix Q contains the dot products between the unit vectors in the A and A′ coordinate systems. For example,

$$Q_{ik} \equiv (\hat{e}'_i \cdot \hat{e}_k), \tag{2.11}$$

which equals the cosine of the angle between $\hat{e}'_i$ and $\hat{e}_k$. Equation (2.10) can be rewriten in matrix form as

$$\sigma' = Q \cdot \sigma \cdot Q^{\mathrm{T}}, \tag{2.12}$$

where Q^{T} is the transpose of matrix Q.

As an example, suppose we are given all of the stress components $\sigma_{kl}^{r\theta z}$ in the cylindrical (r, θ, z) coordinate system and we want each of the stress components in the Cartesian (x, y, z) coordinate system, σ_{ij}^{xyz}. Then coordinate system A is the cylindrical (r, θ, z) coordinate system and coordinate system A′ is the Cartesian (x, y, z) coordinate system. The matrix Q is

$$Q_{ik} = \left(\hat{e}_i^{xyz} \cdot \hat{e}_k^{r\theta z}\right) = \begin{bmatrix} (\hat{e}_x \cdot \hat{e}_r) & (\hat{e}_x \cdot \hat{e}_\theta) & (\hat{e}_x \cdot \hat{e}_z) \\ (\hat{e}_y \cdot \hat{e}_r) & (\hat{e}_y \cdot \hat{e}_\theta) & (\hat{e}_y \cdot \hat{e}_z) \\ (\hat{e}_z \cdot \hat{e}_r) & (\hat{e}_z \cdot \hat{e}_\theta) & (\hat{e}_z \cdot \hat{e}_z) \end{bmatrix}. \tag{2.13}$$

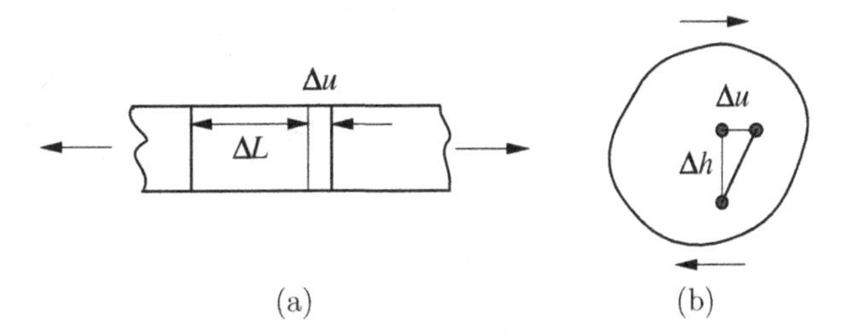

Figure 2.4. (a) Axial strain. (b) Shear strain.

Given that

$$\begin{aligned}\hat{e}_r &= \hat{e}_x \cos\theta + \hat{e}_y \sin\theta \\ \hat{e}_\theta &= -\hat{e}_x \sin\theta + \hat{e}_y \cos\theta\end{aligned} \tag{2.14}$$

we have,

$$Q_{ik} = \begin{bmatrix} Q_{xr} & Q_{x\theta} & Q_{xz} \\ Q_{yr} & Q_{y\theta} & Q_{yz} \\ Q_{zr} & Q_{z\theta} & Q_{zz} \end{bmatrix} = \begin{bmatrix} \cos\theta & -\sin\theta & 0 \\ \sin\theta & \cos\theta & 0 \\ 0 & 0 & 1 \end{bmatrix}. \tag{2.15}$$

Consider a simple scenario in which $\sigma_{\theta z} = \sigma_{z\theta}$ is the only non-zero stress component in the cylindrical coordinate system. Using the transformation law, the stress components in the Cartesian coordinate system are

$$\begin{aligned}
\sigma_{xx} &= Q_{xz} Q_{x\theta} \sigma_{z\theta} + Q_{x\theta} Q_{xz} \sigma_{\theta z} = 0, \\
\sigma_{yy} &= Q_{yz} Q_{y\theta} \sigma_{z\theta} + Q_{y\theta} Q_{yz} \sigma_{\theta z} = 0, \\
\sigma_{zz} &= Q_{zz} Q_{z\theta} \sigma_{z\theta} + Q_{z\theta} Q_{zz} \sigma_{\theta z} = 0, \\
\sigma_{xy} &= Q_{xz} Q_{y\theta} \sigma_{z\theta} + Q_{x\theta} Q_{yz} \sigma_{\theta z} = 0, \\
\sigma_{yz} &= Q_{yz} Q_{z\theta} \sigma_{z\theta} + Q_{y\theta} Q_{zz} \sigma_{\theta z} = \cos\theta \sigma_{\theta z}, \\
\sigma_{xz} &= Q_{xz} Q_{z\theta} \sigma_{z\theta} + Q_{x\theta} Q_{zz} \sigma_{\theta z} = -\sin\theta \sigma_{\theta z}.
\end{aligned}$$

Therefore, the final result for this simple case is

$$\begin{aligned}\sigma_{yz} &= \cos\theta \sigma_{\theta z}, \\ \sigma_{xz} &= -\sin\theta \sigma_{\theta z}.\end{aligned} \tag{2.16}$$

It is tempting to treat stresses as vectors and use simple vector relations in converting from one coordinate system to another, but this is dangerous and should not be done.

2.2 Strain

Strain is a measure of shape change of the body under deformation. The basic definition of strain is displacement per unit length. Similar to stress, there are also two types of strain: axial and shear, as shown in Fig. 2.4a, b, respectively. To specify strain, we consider a short segment in the undeformed body, and measure the relative displacement of the two ends of the segment in the deformed body. The axial strain is $\varepsilon = \frac{\Delta u}{\Delta L}$, where the end-to-end displacement Δu is in the same direction of the original segment length ΔL. It is also called normal strain. The

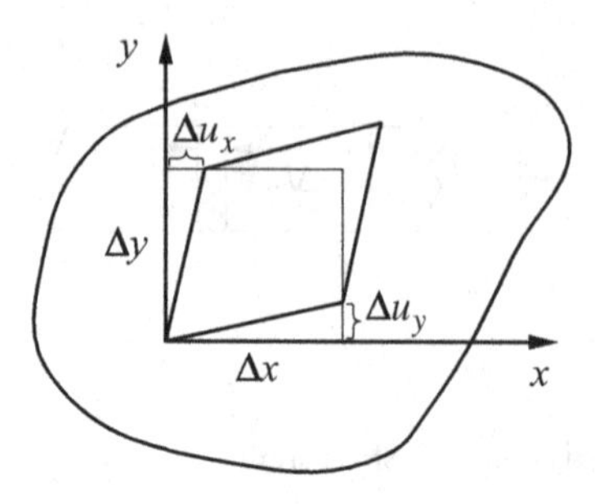

Figure 2.5. Engineering shear strain γ_{xy}.

(engineering) shear strain is $\gamma = \frac{\Delta u}{\Delta h}$, where the end-to-end displacement Δu is perpendicular to the original segment length Δh. Strain is dimensionless and has no unit.

2.2.1 Engineering strains and strain tensor

To completely specify the strain state at a point, we need to consider three infinitesimal segments in the undeformed body parallel to the axes of a given coordinate system, and specify their relative end-to-end displacements in the deformed body. There are six engineering strains. Three are normal strains and the other three are shear strains.

Let $u_i = u_x, u_y, u_z$ be the displacement field. They are functions of (x, y, z), which are the coordinates of a point in the undeformed body. The position of the same point in the deformed body is $(x + u_x, y + u_y, z + u_z)$. The normal strain along the x axis corresponds to the end-to-end displacement Δu in the x direction when the original segment Δx is also along x:

$$\varepsilon_{xx} = \lim_{\Delta x \to 0} \frac{\Delta u_x}{\Delta x} = \frac{\partial u_x}{\partial x}. \tag{2.17}$$

Therefore the three normal strains are

$$\varepsilon_{xx} = \frac{\partial u_x}{\partial x}, \quad \varepsilon_{yy} = \frac{\partial u_y}{\partial y}, \quad \varepsilon_{zz} = \frac{\partial u_z}{\partial z}. \tag{2.18}$$

The shear (engineering) strain γ_{xy} measures the combined effect of Δu along the y direction for a segment Δx along x and the effect of Δu along the x direction for a segment Δy along y, as shown in Fig. 2.5,

$$\gamma_{xy} = \lim_{\Delta y \to 0} \frac{\Delta u_x}{\Delta y} + \lim_{\Delta x \to 0} \frac{\Delta u_y}{\Delta x} = \frac{\partial u_x}{\partial y} + \frac{\partial u_y}{\partial x}. \tag{2.19}$$

So the three shear (engineering) strains are

$$\begin{aligned} \gamma_{xy} = \gamma_{yx} &= \frac{\partial u_x}{\partial y} + \frac{\partial u_y}{\partial x}, \\ \gamma_{xz} = \gamma_{zx} &= \frac{\partial u_x}{\partial z} + \frac{\partial u_z}{\partial x}, \\ \gamma_{yz} = \gamma_{zy} &= \frac{\partial u_y}{\partial z} + \frac{\partial u_z}{\partial y}. \end{aligned} \tag{2.20}$$

Unfortunately, the six engineering strains defined above do not form a second-rank tensor. If we need to transform the strains from one coordinate system to another, it is more convenient to use the strain tensor, which is defined as

$$\varepsilon_{ij} = \frac{1}{2}\left(\frac{\partial u_i}{\partial x_j} + \frac{\partial u_j}{\partial x_i}\right). \tag{2.21}$$

In component form,

$$\begin{aligned}
\varepsilon_{xx} &= \frac{\partial u_x}{\partial x}, \quad \varepsilon_{yy} = \frac{\partial u_y}{\partial y}, \quad \varepsilon_{zz} = \frac{\partial u_z}{\partial z}, \\
\varepsilon_{xy} = \varepsilon_{yx} &= \frac{1}{2}\left(\frac{\partial u_x}{\partial y} + \frac{\partial u_y}{\partial x}\right) = \frac{1}{2}\gamma_{xy}, \\
\varepsilon_{yz} = \varepsilon_{zy} &= \frac{1}{2}\left(\frac{\partial u_z}{\partial y} + \frac{\partial u_y}{\partial z}\right) = \frac{1}{2}\gamma_{yz}, \\
\varepsilon_{zx} = \varepsilon_{xz} &= \frac{1}{2}\left(\frac{\partial u_z}{\partial x} + \frac{\partial u_x}{\partial z}\right) = \frac{1}{2}\gamma_{zx}.
\end{aligned} \tag{2.22}$$

The extra factor of $\frac{1}{2}$ is needed for all the strains to transform together according to the transformation law of a tensor.

The strain tensor follows the same transformation law as the stress tensor. In general, suppose we are given two local coordinate systems at a point p: coordinate system A with unit vectors $\hat{e}_1$, $\hat{e}_2$, $\hat{e}_3$, and coordinate system A′ with unit vectors $\hat{e}'_1$, $\hat{e}'_2$, $\hat{e}'_3$. The strain tensor at point p can be described by six independent components of ε_{ij} in the coordinate system A. The same strain tensor can be expressed by six independent components of ε'_{ij} in the coordinate system A′. Since they correspond to the same strain state (shape change), ε_{ij} and ε'_{ij} must be related to each other. Their relationship can be expressed as

$$\varepsilon'_{ij} = Q_{ik} Q_{jl} \varepsilon_{kl}, \tag{2.23}$$

where we have used the Einstein notation again and the matrix Q is the same as that defined in Eq. (2.11). Equation (2.23) can be rewritten in matrix form as

$$\varepsilon' = Q \cdot \varepsilon \cdot Q^{\mathrm{T}}. \tag{2.24}$$

In the *Voigt notation*, the strain components are represented by a column vector,

$$\varepsilon_I = \begin{pmatrix} \varepsilon_1 \\ \varepsilon_2 \\ \varepsilon_3 \\ \varepsilon_4 \\ \varepsilon_5 \\ \varepsilon_6 \end{pmatrix}, \tag{2.25}$$

where $I = 1, 2, \ldots, 6$. For Cartesian coordinates,

$$\varepsilon_I = \begin{pmatrix} \varepsilon_1 \\ \varepsilon_2 \\ \varepsilon_3 \\ \varepsilon_4 \\ \varepsilon_5 \\ \varepsilon_6 \end{pmatrix} = \begin{pmatrix} \varepsilon_{xx} \\ \varepsilon_{yy} \\ \varepsilon_{zz} \\ \gamma_{yz} \\ \gamma_{xz} \\ \gamma_{xy} \end{pmatrix} = \begin{pmatrix} \varepsilon_{xx} \\ \varepsilon_{yy} \\ \varepsilon_{zz} \\ 2\varepsilon_{yz} \\ 2\varepsilon_{xz} \\ 2\varepsilon_{xy} \end{pmatrix}. \tag{2.26}$$

Notice that the shear strain in the Voigt notation is the engineering strain instead of the shear component of the strain tensor. Therefore, the strains in the Voigt notation do not transform as a tensor (nor as a vector)! This may seem confusing. Indeed the Voigt notation should be used with care. However, we shall see in Section 2.4 that, using the Voigt notation, the strain energy density can be written as a dot product between two column vectors.

2.2.2 Strains in other coordinate systems

Given the strain tensor at a point p defined in the Cartesian coordinates, Eq. (2.22), we can use the transformation law, Eq. (2.23), to obtain the components of the strain tensor in cylindrical and spherical coordinate systems. However, the result is still expressed in terms of the derivatives of the displacement (u_x, u_y, u_z) with respect to (x, y, z), which are quantities in the Cartesian coordinate system. It is often useful to express the strains entirely in terms of quantities in the cylindrical (or spherical) coordinate system.

In cylindrical coordinates, the strain components are

$$\begin{aligned}
\varepsilon_{rr} &= \frac{\partial u_r}{\partial r}, \quad \varepsilon_{\theta\theta} = \frac{1}{r}\frac{\partial u_\theta}{\partial \theta} + \frac{u_r}{r}, \quad \varepsilon_{zz} = \frac{\partial u_z}{\partial z}, \\
\varepsilon_{r\theta} &= \frac{1}{2}\left(\frac{\partial u_\theta}{\partial r} - \frac{u_\theta}{r} + \frac{1}{r}\frac{\partial u_r}{\partial \theta}\right), \\
\varepsilon_{rz} &= \frac{1}{2}\left(\frac{\partial u_r}{\partial z} + \frac{\partial u_z}{\partial r}\right), \\
\varepsilon_{\theta z} &= \frac{1}{2}\left(\frac{1}{r}\frac{\partial u_z}{\partial \theta} + \frac{\partial u_\theta}{\partial z}\right).
\end{aligned} \tag{2.27}$$

In spherical coordinates, the strain components are

$$\begin{aligned}
\varepsilon_{rr} &= \frac{\partial u_r}{\partial r}, \quad \varepsilon_{\varphi\varphi} = \frac{1}{r}\frac{\partial u_\varphi}{\partial \varphi} + \frac{u_r}{r}, \\
\varepsilon_{\theta\theta} &= \frac{1}{r\sin\varphi}\frac{\partial u_\theta}{\partial \theta} + \frac{u_r}{r} + \frac{u_\varphi \cot\varphi}{r}, \\
\varepsilon_{r\theta} &= \frac{1}{2}\left(\frac{1}{r\sin\varphi}\frac{\partial u_r}{\partial \theta} - \frac{u_\theta}{r} + \frac{\partial u_\theta}{\partial r}\right), \\
\varepsilon_{r\varphi} &= \frac{1}{2}\left(\frac{1}{r}\frac{\partial u_r}{\partial \varphi} - \frac{u_\varphi}{r} + \frac{\partial u_\varphi}{\partial r}\right), \\
\varepsilon_{\theta\varphi} &= \frac{1}{2}\left(\frac{1}{r\sin\varphi}\frac{\partial u_\varphi}{\partial \theta} - \frac{u_\theta \cot\varphi}{r} + \frac{1}{r}\frac{\partial u_\theta}{\partial \varphi}\right).
\end{aligned} \tag{2.28}$$

These expressions are more complex than those in the Cartesian coordinate system because the unit vectors depend on the coordinates in curvilinear cordinate systems.

2.3 Isotropic elasticity

We now give a brief review of the theory of isotropic linear elasticity, which relates the stress and strain tensors described above through the generalized Hooke's law. Isotropic elasticity is a

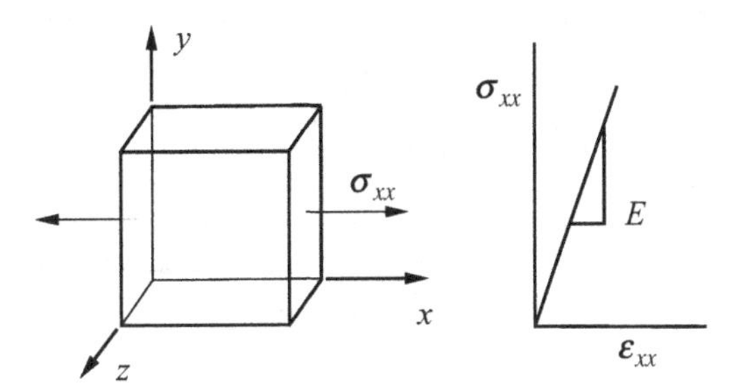

Figure 2.6. Uniaxial tension test.

very useful starting point for most defect problems, despite the fact that most single crystals are elastically anisotropic.

2.3.1 Tension and shear loading

We first consider the stress and strain relationship for a material under uniaxial loading. As shown in Fig. 2.6, the only non-zero stress component is σ_{xx} (which is positive for tension). The leads to a positive axial strain ε_{xx}, which, for a linear elastic material, is proportional to σ_{xx}.

Young's modulus E is defined as the ratio of the stress and strain in the uniaxial test,

$$E = \frac{\sigma_{xx}}{\varepsilon_{xx}}. \tag{2.29}$$

The axial stress σ_{xx} also leads to transverse strains, which are equal in the y and z directions for an isotropic material,

$$\varepsilon_{yy} = \varepsilon_{zz} = -\nu\varepsilon_{xx}, \tag{2.30}$$

where ν is called *Poisson's ratio.* For isotropic materials the limits of Poisson's ratio are $-1 < \nu < 1/2$. For most engineering materials, $\nu > 0$ and ν is often around 0.3. Materials with negative ν do exist, and they are called *auxetics.* An auxetic material has the surprising property that it swells in the transverse direction when you pull on it.

However, if $\nu > 1/2$, then a solid would increase its volume when subjected to hydrostatic pressure (i.e. having a negative bulk modulus, see Section 2.3.3) and do spontaneous work on the surroundings, thus violating the second law of thermodynamics (see Chapter 3) and common sense. If $\nu < -1$, then a solid would have a negative shear modulus (see below) and when subjected to shear stress would do spontaneous work on the surroundings, thus also violating the second law of thermodynamics (and common sense).

We now consider the stress–strain relationship for a material under shear loading. Suppose $\sigma_{xy} = \sigma_{yx}$ is the only non-zero stress component. Then for an isotropic elastic material, $\gamma_{xy} = \gamma_{yx}$ is the only non-zero strain component, as shown in Fig. 2.7.

The shear modulus μ is defined as the ratio of the stress and strain in the shear test,

$$\mu = \frac{\sigma_{xy}}{\gamma_{xy}} = \frac{\sigma_{xy}}{2\varepsilon_{xy}}. \tag{2.31}$$

For isotropic elasticity there are only two independent elastic constants (see below). This means the three elastic constants E, μ, ν introduced so far are related to each other as

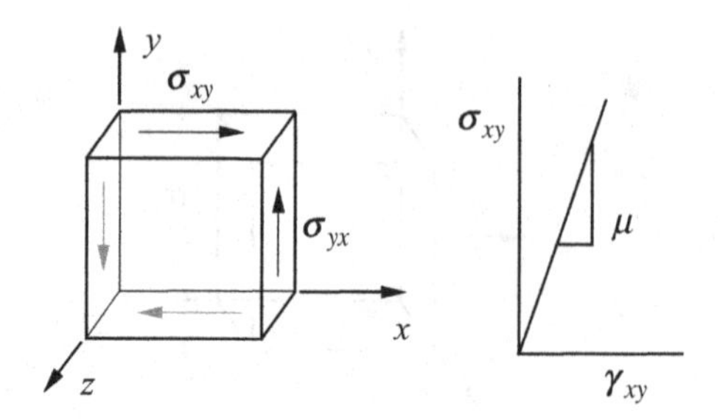

Figure 2.7. Shear deformation.

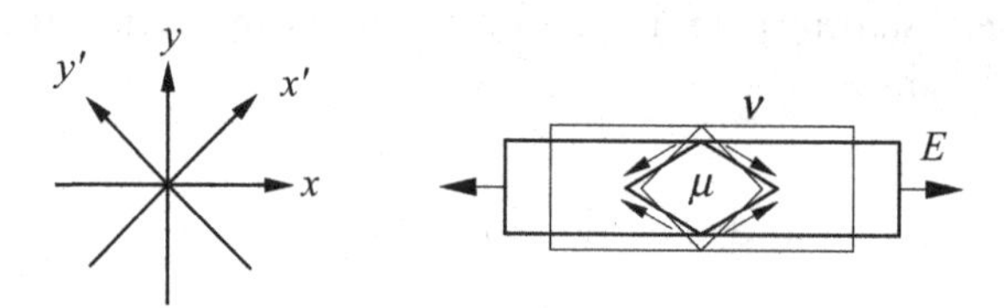

Figure 2.8. Stretching of a plate demonstrating the relation between elastic constants E, μ, and ν.

follows.

$$\boxed{\mu = \frac{E}{2(1+\nu)}} \tag{2.32}$$

Note that if $\nu < -1$, the shear modulus would be negative (assuming E is positive[2]) and the second law of thermodynamics would be violated, as mentioned above.

It is easy to see why the elastic constants should be related. Consider the stretching of a plate, as shown in Fig. 2.8. What happens to the square in the plate can be described as normal stress σ_{xx} and normal strains, ε_{xx} and ε_{yy}. However, if we choose a coordinate system x'–y' that is rotated 45° from the x–y coordinate system, then shear stress σ'_{xy} and shear strain ε'_{xy} are non-zero. (Notice the shape change of the square into a rhombus.) Thus the elastic constants E, ν, and μ have to be related. By using the transformation laws of stress and strain and the stress–strain relations provided above, you should be able to derive the relation (2.32).

The dimension of the elastic moduli E and μ is the same as the dimension of stress, i.e. f/l^2. The unit of the elastic moduli E and μ is the same as the unit of stress, i.e. $\text{N/m}^2 = \text{Pa}$. Poisson's ratio ν is dimensionless and unitless.

2.3.2 Generalized Hooke's law for an isotropic solid

Consider a solid subjected to a general state of stress (Fig. 2.9). We can calculate the resulting strains using the basic definitions above. The resulting axial strain ε_{xx} now has contributions

2 If Young's modulus E were negative, then a bar would contract when you pull on it, again violating the second law of thermodynamics.

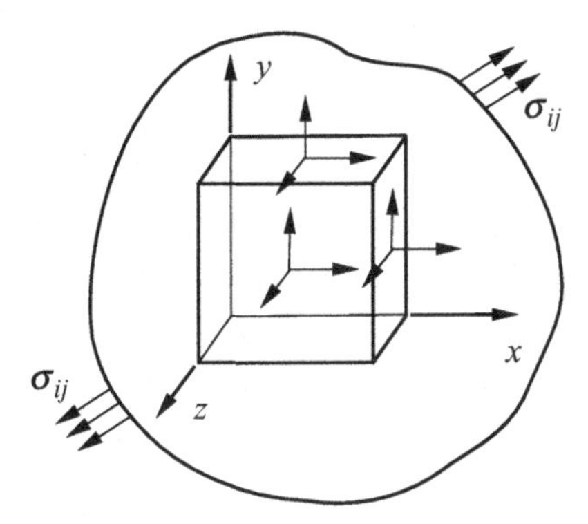

Figure 2.9. An isotropic linear elastic solid in a general stress state.

from all axial stress components, i.e.

$$\begin{aligned}\varepsilon_{xx} &= \frac{\sigma_{xx}}{E} - \nu\frac{\sigma_{yy}}{E} - \nu\frac{\sigma_{zz}}{E} \\ &= \frac{1}{E}[\sigma_{xx} - \nu(\sigma_{yy} + \sigma_{zz})].\end{aligned}$$

Therefore, all the axial strains can be written as follows.

$$\boxed{\begin{aligned}\varepsilon_{xx} &= \frac{1}{E}[\sigma_{xx} - \nu(\sigma_{yy} + \sigma_{zz})] \\ \varepsilon_{yy} &= \frac{1}{E}[\sigma_{yy} - \nu(\sigma_{xx} + \sigma_{zz})] \\ \varepsilon_{zz} &= \frac{1}{E}[\sigma_{zz} - \nu(\sigma_{xx} + \sigma_{yy})]\end{aligned}} \tag{2.33}$$

For an isotropic elastic solid, each shear strain is only coupled to the individual shear stress component.

$$\boxed{\begin{aligned}\gamma_{xy} &= \gamma_{yx} = \frac{\sigma_{xy}}{\mu} \\ \gamma_{xz} &= \gamma_{zx} = \frac{\sigma_{xz}}{\mu} \\ \gamma_{yz} &= \gamma_{zy} = \frac{\sigma_{yz}}{\mu}\end{aligned}} \tag{2.34}$$

Equations (2.33) and (2.34) are a form of the generalized Hooke's law that is most useful when the stresses are given and we wish to compute the strains. Alternatively, if the strains are given, then these equations need to be solved simultaneously (or a matrix needs to be inverted) to determine the stresses. Such a process leads to the following alternative form for the generalized Hooke's law.

$$\boxed{\begin{aligned}\sigma_{xx} &= 2\mu\varepsilon_{xx} + \lambda(\varepsilon_{xx} + \varepsilon_{yy} + \varepsilon_{zz}) \\ \sigma_{yy} &= 2\mu\varepsilon_{yy} + \lambda(\varepsilon_{xx} + \varepsilon_{yy} + \varepsilon_{zz}) \\ \sigma_{zz} &= 2\mu\varepsilon_{zz} + \lambda(\varepsilon_{xx} + \varepsilon_{yy} + \varepsilon_{zz})\end{aligned}} \tag{2.35}$$

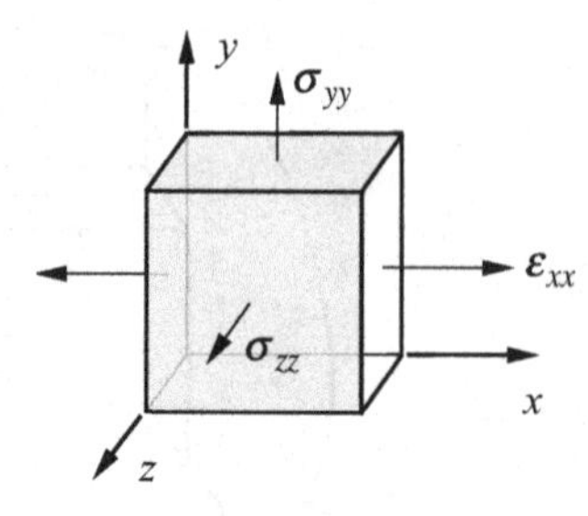

Figure 2.10. Uniaxial strain leading to transverse stresses.

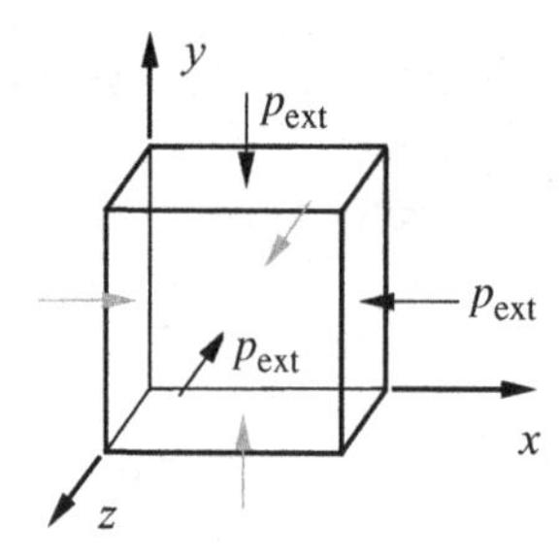

Figure 2.11. Hydrostatic pressure.

Here λ is called the Lamé coefficient, which is related to the other elastic constants as follows:

$$\lambda = \frac{\nu E}{(1+\nu)(1-2\nu)} = \frac{2\nu\mu}{1-2\nu}. \tag{2.36}$$

The Lamé coefficient describes the relation between the transverse stresses that develop and the imposed axial strain when a solid is subjected to a state of uniaxial strain (not uniaxial stress). In Fig. 2.10, the shaded faces (with normals along y and z axes) are rigidly fixed while the body is elastically strained in the x direction. Transverse stresses, $\sigma_{yy} = \sigma_{zz}$ naturally develop. The Lamé coefficient is then defined in this loading condition as

$$\lambda \equiv \frac{\sigma_{yy}}{\varepsilon_{xx}} = \frac{\sigma_{zz}}{\varepsilon_{xx}} \tag{2.37}$$

and is given by Eq. (2.36).

The shear stress–strain relations in Eq. (2.34) are easily inverted.

$$\boxed{\begin{aligned} \sigma_{yz} &= \sigma_{zy} = \mu\gamma_{yz} = 2\mu\varepsilon_{yz} \\ \sigma_{xz} &= \sigma_{zx} = \mu\gamma_{xz} = 2\mu\varepsilon_{xz} \\ \sigma_{xy} &= \sigma_{yx} = \mu\gamma_{xy} = 2\mu\varepsilon_{xy} \end{aligned}} \tag{2.38}$$

2.3.3 Hydrostatic stress and strain

We now consider a solid subjected to a hydrostatic pressure, p_{ext}, as shown in Fig. 2.11. Given our sign convention that a positive stress value corresponds to tension, the stresses for such hydrostatic loading can be written as

$$\sigma_{xx} = \sigma_{yy} = \sigma_{zz} = -p_{\text{ext}}. \tag{2.39}$$

A positive pressure leads to compressive, or negative, normal stresses. We can then calculate the resulting strains using the generalized Hooke's law

$$\varepsilon_{xx} = \varepsilon_{yy} = \varepsilon_{zz} = \frac{1}{E}[\sigma_{xx} - \nu(\sigma_{yy} + \sigma_{zz})] = -\frac{1-2\nu}{E} p_{\text{ext}}. \tag{2.40}$$

Define the volumetric strain as

$$\left(\frac{\Delta V}{V}\right)_{\text{elastic}} \equiv \varepsilon_{xx} + \varepsilon_{yy} + \varepsilon_{zz} = -\frac{3(1-2\nu)}{E} p_{\text{ext}}. \tag{2.41}$$

The bulk modulus B is defined through the relationship between the volumetric strain and the applied pressure, i.e.

$$\left(\frac{\Delta V}{V}\right)_{\text{elastic}} = -\frac{p_{\text{ext}}}{B}. \tag{2.42}$$

So by comparing the last two relations we have

$$B = \frac{E}{3(1-2\nu)}. \tag{2.43}$$

Given Eqs. (2.32) and (2.36), the bulk modulus B can also be written as

$$B = \frac{2\mu(1+\nu)}{3(1-2\nu)} = \frac{3\lambda + 2\mu}{3}. \tag{2.44}$$

Note that if $\nu > 1/2$ the bulk modulus would be negative and the second law of thermodynamics would be violated, as mentioned in Section 2.3.1.

2.4 Elastic strain energy

The elastic energy stored in a solid can be computed from the work required to produce the stress and strain state in question. Consider uniaxial loading along the x axis of an isotropic solid, as shown in Fig. 2.12a. The axial stress–strain (σ–ε) relation is shown in Fig. 2.12b. Assume the left face of the cube is fixed and the right face is displaced by distance δ due to the applied force P. The total strain energy stored in the sample equals the work done by the applied force P,

$$\Delta W = \int_0^{\delta_1} P\, d\delta = \int_0^{\varepsilon_1} (\sigma A) L\, d\varepsilon = AL \int_0^{\varepsilon_1} \sigma\, d\varepsilon. \tag{2.45}$$

Using Hooke's law, $\sigma = E\varepsilon$, we have

$$\Delta W = AL \int_0^{\varepsilon_1} E\varepsilon\, d\varepsilon = ALE\frac{\varepsilon_1^2}{2} = AL\frac{1}{2}\sigma_1\varepsilon_1. \tag{2.46}$$

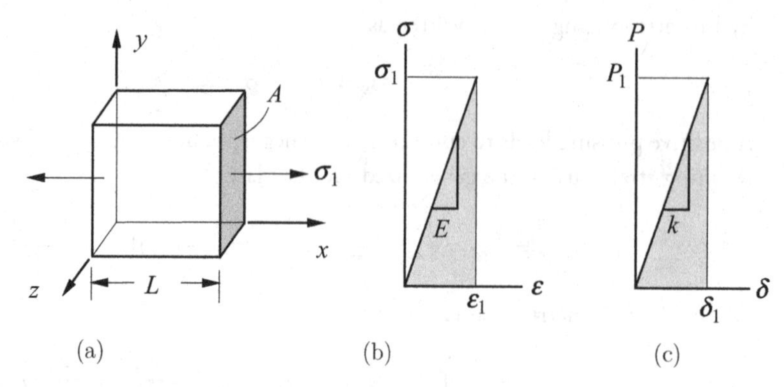

Figure 2.12. (a) An isotropic solid subjected to uniaxial tensile loading. (b) The stress–strain relation. (c) The force displacement relation.

The strain energy density (strain energy per unit volume) is then

$$\Delta w = \frac{\Delta W}{V} = \frac{\Delta W}{AL} = \frac{1}{2}\sigma_1 \varepsilon_1. \tag{2.47}$$

When this derivation is generalized for a general state of stress we have

$$\Delta w = \frac{1}{2}\sigma_{xx}\varepsilon_{xx} + \frac{1}{2}\sigma_{yy}\varepsilon_{yy} + \frac{1}{2}\sigma_{zz}\varepsilon_{zz} + \sigma_{yz}\varepsilon_{yz} + \sigma_{xz}\varepsilon_{xz} + \sigma_{xy}\varepsilon_{xy} \tag{2.48}$$

or in the Voigt notation

$$\begin{aligned}\Delta w &= \frac{1}{2}(\sigma_1\varepsilon_1 + \sigma_2\varepsilon_2 + \sigma_3\varepsilon_3 + \sigma_4\varepsilon_4 + \sigma_5\varepsilon_5 + \sigma_6\varepsilon_6) \\ &= \frac{1}{2}\sigma_I\varepsilon_I,\end{aligned} \tag{2.49}$$

where I is summed from 1 to 6. The stress and strain definitions in the Voigt notation are given in Eqs. (2.7) and (2.26).

2.5 Fundamental equations of elasticity

The purpose of the theory of elasticity is to determine the stress σ_{ij}, strain ε_{ij}, and displacement u_i fields in an elastic medium subjected to internal and external loads. By fields, we mean that in general σ_{ij}, ε_{ij}, and u_i vary from point to point in the elastic medium, i.e. they are functions of location $\vec{r} = (x, y, z)$. The fields in an elastic body satisfy a set of partial differential equations (PDEs). These fields are found by solving these PDEs subjected to suitable boundary conditions. In this section, we give a brief account of the fundamental PDEs in the theory of elasticity.

2.5.1 Equilibrium condition for stress

Throughout this book we will limit our discussions to solids at equilibrium, which means that the total force and moment exerted on any sub-regions of the solid must be zero. This leads to

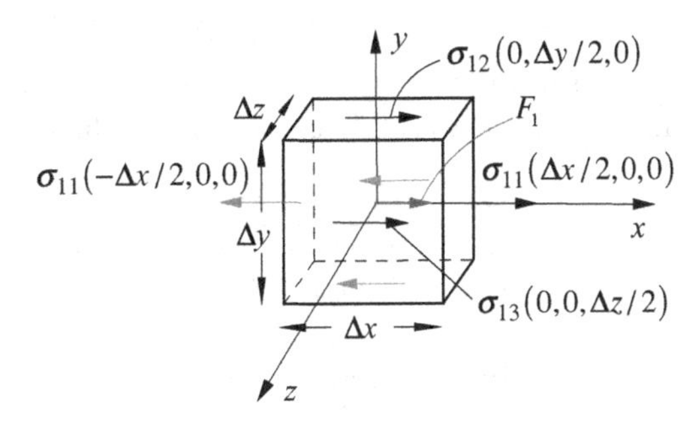

Figure 2.13. Force balance in the x direction for a small cube in the material.

a set of PDEs for the stress field. To be general, consider a solid which is not only loaded on its external surface but also subjected to a body force (e.g. gravity) distribution F_j, which has the dimension of force per unit volume.

Consider a small material cube centered at the origin with size Δx, Δy, and Δz, as shown in Fig. 2.13. The total force in the x direction has contributions from the traction forces on the six faces of the cube, as well as from the body force. At equilibrium, the total force must vanish, leading to the following equation:

$$\begin{aligned} &\left[\sigma_{11}\left(\frac{\Delta x}{2},0,0\right)-\sigma_{11}\left(-\frac{\Delta x}{2},0,0\right)\right]\cdot\Delta y\Delta z \\ &+\left[\sigma_{12}\left(0,\frac{\Delta y}{2},0\right)-\sigma_{12}\left(0,-\frac{\Delta y}{2},0\right)\right]\cdot\Delta x\Delta z \\ &+\left[\sigma_{13}\left(0,0,\frac{\Delta z}{2}\right)-\sigma_{13}\left(0,0,-\frac{\Delta z}{2}\right)\right]\cdot\Delta x\Delta y \\ &+F_1(0,0,0)\cdot\Delta x\Delta y\Delta z=0, \end{aligned} \tag{2.50}$$

where the indices 1, 2, 3 indicate x, y, z directions, respectively. In the limit of Δx, Δy, $\Delta z \to 0$, this leads to the PDE

$$\sigma_{11,1}+\sigma_{12,2}+\sigma_{13,3}+F_1=0, \tag{2.51}$$

where an index j after the comma means taking a derivative with respect to x_j, and x_1, x_2, x_3 are equivalent to x, y, z, respectively. For example, $\sigma_{13,3}=\partial\sigma_{13}/\partial x_3=\partial\sigma_{xz}/\partial z$. Considering the force balance in the y and z directions, we have

$$\sigma_{21,1}+\sigma_{22,2}+\sigma_{23,3}+F_2=0, \tag{2.52}$$

$$\sigma_{31,1}+\sigma_{32,2}+\sigma_{33,3}+F_3=0. \tag{2.53}$$

Equations (2.51)–(2.53) can be written in a concise form by using the Einstein notation,

$$\sigma_{ij,j}+F_i=0, \tag{2.54}$$

where it is implied that the repeated indices (j) are summed from 1 to 3. The non-repeated index, i, is a free index, and can vary from 1 to 3, corresponding to Eqs. (2.51)–(2.53). If there are no body forces (as if often the case in this book), then the equilibrium condition becomes

$$\sigma_{ij,j}=0. \tag{2.55}$$

It is sometimes more convenient to use cylindrical or spherical coordinates to solve certain elasticity problems due to their symmetry. It is then necessary to express the equilibrium conditions in these coordinates [10]. In cylindrical coordinates, the equilibrium conditions are

$$\begin{aligned}
&\frac{\partial \sigma_{rr}}{\partial r} + \frac{1}{r}\frac{\partial \sigma_{r\theta}}{\partial \theta} + \frac{\partial \sigma_{rz}}{\partial z} + \frac{1}{r}(\sigma_{rr} - \sigma_{\theta\theta}) + F_r = 0, \\
&\frac{\partial \sigma_{r\theta}}{\partial r} + \frac{1}{r}\frac{\partial \sigma_{\theta\theta}}{\partial \theta} + \frac{\partial \sigma_{\theta z}}{\partial z} + \frac{2}{r}\sigma_{r\theta} + F_\theta = 0, \\
&\frac{\partial \sigma_{rz}}{\partial r} + \frac{1}{r}\frac{\partial \sigma_{\theta z}}{\partial \theta} + \frac{\partial \sigma_{zz}}{\partial z} + \frac{1}{r}\sigma_{rz} + F_z = 0.
\end{aligned} \tag{2.56}$$

In spherical coordinates, the equilibrium conditions are

$$\begin{aligned}
&\frac{\partial \sigma_{rr}}{\partial r} + \frac{1}{r}\frac{\partial \sigma_{r\varphi}}{\partial \varphi} + \frac{1}{r\sin\varphi}\frac{\partial \sigma_{r\theta}}{\partial \theta} + \frac{1}{r}(2\sigma_{rr} - \sigma_{\varphi\varphi} - \sigma_{\theta\theta} + \sigma_{r\varphi}\cot\varphi) + F_r = 0, \\
&\frac{\partial \sigma_{r\varphi}}{\partial r} + \frac{1}{r}\frac{\partial \sigma_{\varphi\varphi}}{\partial \varphi} + \frac{1}{r\sin\varphi}\frac{\partial \sigma_{\varphi\theta}}{\partial \theta} + \frac{1}{r}[(\sigma_{\varphi\varphi} - \sigma_{\theta\theta})\cot\varphi + 3\sigma_{r\varphi}] + F_\varphi = 0, \\
&\frac{\partial \sigma_{r\theta}}{\partial r} + \frac{1}{r}\frac{\partial \sigma_{\varphi\theta}}{\partial \varphi} + \frac{1}{r\sin\varphi}\frac{\partial \sigma_{\theta\theta}}{\partial \theta} + \frac{1}{r}(2\sigma_{\varphi\theta}\cot\varphi + 3\sigma_{r\theta}) + F_\theta = 0.
\end{aligned} \tag{2.57}$$

2.5.2 Compatibility condition for strain

Given the displacement field $u_i(\vec{r})$, we can always find the strain field by taking spatial derivatives. We rewrite Eq. (2.21) in the following form

$$\varepsilon_{ij} = \frac{1}{2}(u_{i,j} + u_{j,i}). \tag{2.58}$$

However, we note that the displacement vector u_i has three degrees of freedom at every point $\vec{r}$, while the strain tensor ε_{ij} has six degrees of freedom. This means that if we are given an arbitrary strain field $\varepsilon_{ij}(\vec{r})$, we will not always be able to find a displacement field $u_i(\vec{r})$, which satisfies Eq. (2.58). Graphically, it means that if we cut the elastic medium into many pieces, and deform the pieces according to an arbitrary strain field, in general we will not be able to put the deformed pieces back together into a continuous body. Instead, gaps or material overlaps will be present, meaning that the strain field is not *compatible*, as shown in Fig. 2.14.

A compatible strain field must satisfy the following (second-order) PDE.

$$\boxed{\varepsilon_{ij,kl} + \varepsilon_{kl,ij} - \varepsilon_{ik,jl} - \varepsilon_{jl,ik} = 0} \tag{2.59}$$

This is called the *compatibility condition*, where i, j, k, l are all free indices that vary from 1 to 3. Therefore, Eq. (2.59) corresponds to many different equations for different choices of i, j, k, l; the compatibility condition requires all of these equations to be satisfied. Given a strain field, we can find the displacement field satisfying Eq. (2.58), if and only if a strain field satisfies the compatibility condition. Even though the compatibility condition looks rather complicated, it is greatly simplified in *plane strain* conditions (to be discussed below). The compatibility condition is automatically satisfied if we are given a displacement field $u_i(\vec{r})$ and obtain the strain field from Eq. (2.58).

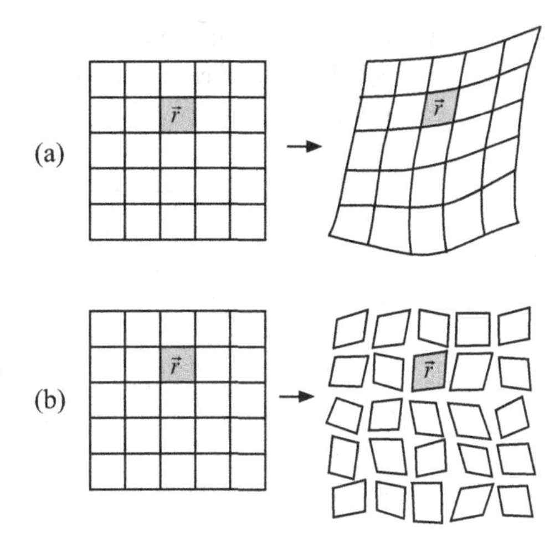

Figure 2.14. (a) A compatible strain field. (b) An incompatible strain field.

2.5.3 Solution strategies for elasticity problems

To solve an elasticity problem means that we find the stress, strain, and displacement fields that satisfy the equilibrium condition, Eq. (2.54), the compatibility condition, Eq. (2.59), and the generalized Hooke's law, Eqs. (2.33)–(2.34). For general three-dimensional problems, analytic solutions are usually difficult to find, and numerical methods such as the Finite Element Methods [12] are usually used for these purposes. However, the problem is greatly simplified when the symmetries of the problem allow us to reduce the PDEs to two-dimensions or even one-dimension. In these cases, analytic solutions can be found and they are very useful in the discussion of the elastic fields of defects in crystals. In the following, we describe the general solution strategies that take advantages of the symmetries of problem of interest.

Anti-plane problems

If by symmetry we know that the elastic fields depend only on x and y, and that the only non-zero displacement is in the z direction, then we have the so-called *anti-plane* problem. An example of the anti-plane problem is a straight screw dislocation along the z axis, to be discussed in Chapter 9. The solution strategy of an anti-plane problem is to first find the displacement field $u_z(x, y)$, from which the strain field and the stress field can be easily obtained.

Since $u_z(x, y)$ is the only non-zero displacement component, the only non-zero strain fields are

$$\begin{aligned} \gamma_{xz}(x, y) &= \frac{\partial}{\partial x} u_z(x, y) \\ \gamma_{yz}(x, y) &= \frac{\partial}{\partial y} u_z(x, y), \end{aligned} \tag{2.60}$$

and the only non-zero stress fields are

$$\begin{aligned} \sigma_{xz}(x, y) &= \mu \gamma_{xz}(x, y) \\ \sigma_{yz}(x, y) &= \mu \gamma_{yz}(x, y). \end{aligned} \tag{2.61}$$

In the absence of body forces, it can be shown that the equilibrium condition, the compatibility condition, and the generalized Hooke's law, can all be satisfied as long as the displacement field u_z satisfies Laplace's equation,

$$\nabla^2 u_z \equiv \frac{\partial^2 u_z}{\partial x^2} + \frac{\partial^2 u_z}{\partial y^2} = 0. \tag{2.62}$$

The ∇^2 symbol is called the Laplacian (operator). We note that a function that satisfies Laplace's equation is called the *harmonic function*. In cylindrical coordinates, Laplace's equation is written as

$$\left(\frac{\partial^2}{\partial r^2} + \frac{1}{r}\frac{\partial}{\partial r} + \frac{1}{r^2}\frac{\partial^2}{\partial \theta^2} \right) u_z(r, \theta) = 0. \tag{2.63}$$

This solution strategy will be used to obtain the elastic fields of straight screw dislocations in Section 9.1.1.

Plane strain problems

If by symmetry we know that the elastic fields depend only on x and y, and that the only non-zero displacements are in the x and y directions, then we have the so-called *plane strain* problem. An example of the plane strain problem is a straight edge dislocation along the z axis, to be discussed in Chapter 9. The solution strategy of a plane strain problem is to first find the so-called Airy stress function $\phi(x, y)$, from which the stress field, and then the strain field and the displacement field can be obtained.

Since $u_x(x, y)$ and $u_y(x, y)$ are the only non-zero displacement components, the only non-zero strain fields are

$$\begin{aligned}
\varepsilon_{xx}(x, y) &= \frac{\partial}{\partial x} u_x(x, y), \\
\varepsilon_{yy}(x, y) &= \frac{\partial}{\partial y} u_y(x, y), \\
\varepsilon_{xy}(x, y) &= \frac{1}{2}\left[\frac{\partial}{\partial x} u_y(x, y) + \frac{\partial}{\partial y} u_x(x, y) \right],
\end{aligned} \tag{2.64}$$

and the only non-zero stress fields are

$$\begin{aligned}
\sigma_{xx} &= (\lambda + 2\mu)\varepsilon_{xx} + \lambda \varepsilon_{yy}, \\
\sigma_{yy} &= (\lambda + 2\mu)\varepsilon_{yy} + \lambda \varepsilon_{xx}, \\
\sigma_{zz} &= \lambda(\varepsilon_{xx} + \varepsilon_{yy}) = \nu(\sigma_{xx} + \sigma_{yy}), \\
\sigma_{xy} &= 2\mu \varepsilon_{xy}.
\end{aligned} \tag{2.65}$$

In the absence of body forces, the equilibrium condition can be automatically satisfied if we introduce the *Airy stress function* $\phi(x, y)$, in terms of which the stress fields are expressed as

$$\sigma_{xx} = \frac{\partial^2 \phi}{\partial y^2}, \quad \sigma_{yy} = \frac{\partial^2 \phi}{\partial x^2}, \quad \sigma_{xy} = -\frac{\partial^2 \phi}{\partial x \partial y}. \tag{2.66}$$

In plane strain, the only non-trivial equation for the compatibility condition, Eq. (2.59) is,

$$\varepsilon_{xx,yy} + \varepsilon_{yy,xx} - 2\varepsilon_{xy,xy} = 0. \tag{2.67}$$

The compatibility condition and the generalized Hooke's law require the stress function to satisfy the *biharmonic equation*,

$$\nabla^4\phi \equiv \nabla^2(\nabla^2\phi) = \left(\frac{\partial^2}{\partial x^2} + \frac{\partial^2}{\partial y^2}\right)\left(\frac{\partial^2}{\partial x^2} + \frac{\partial^2}{\partial y^2}\right)\phi(x,y) = 0. \tag{2.68}$$

The solution to Eq. (2.68) is called a *biharmonic function*, because $\nabla^2\phi$ is a harmonic function. A harmonic function must be a biharmonic function, but the reverse is not necessarily true. In cylindrical coordinates, the biharmonic equation is written as

$$\left(\frac{\partial^2}{\partial r^2} + \frac{1}{r}\frac{\partial}{\partial r} + \frac{1}{r^2}\frac{\partial^2}{\partial \theta^2}\right)\left(\frac{\partial^2}{\partial r^2} + \frac{1}{r}\frac{\partial}{\partial r} + \frac{1}{r^2}\frac{\partial^2}{\partial \theta^2}\right)\phi(r,\theta) = 0. \tag{2.69}$$

The stress fields in the cylindrical coordinates can be expressed in terms of the stress function as follows:

$$\begin{aligned}
\sigma_{rr} &= \left(\frac{1}{r}\frac{\partial}{\partial r} + \frac{1}{r^2}\frac{\partial^2}{\partial \theta^2}\right)\phi(r,\theta),\\
\sigma_{\theta\theta} &= \frac{\partial^2}{\partial r^2}\phi(r,\theta),\\
\sigma_{r\theta} &= -\frac{\partial}{\partial r}\left(\frac{1}{r}\frac{\partial}{\partial \theta}\phi(r,\theta)\right).
\end{aligned} \tag{2.70}$$

This solution strategy will be used to obtain the elastic fields of straight edge dislocations in Section 9.1.3.

2.5.4 Elastic Green's function

The solution strategies for elasticity problems in three dimensions are different from those in two dimensions. Because the compatibility condition is very complex in three dimensions, it is often avoided by expressing the stress and strain fields in terms of the displacement field, so that the compatibility condition is automatically satisfied. Hence a common strategy for solving three-dimensional elasticity problems is to seek the displacement field that satisfies the equilibrium condition.

A solution of fundamental importance in three-dimensional elasticity is the displacement field caused by a concentrated force acting on a single point inside an infinite elastic medium, as shown in Fig. 2.15a. The solution is called the elastic Green's function and this problem is also known as Kelvin's problem. Green's function is important because, once it is known, the displacement field caused by an arbitrary distribution of body forces in an infinite medium can be obtained relatively easily by superposition. In the following, we give the analytic expression of the elastic Green's function in isotropic elastic medium and show how to use it to construct displacements caused by force dipoles.

r-dependence of stress and displacement in Kelvin's problem

Without knowning the analytic expression of the elastic Green's function, we can show that the stress field in Kelvin's problem must scale as $1/r^2$, and that Green's function (i.e. the displacement field) scales as $1/r$, where $r = \sqrt{x^2 + y^2 + z^2}$. Imagine a sphere of radius r centered at the origin, where the unit force is applied. The free-body diagram of this sphere contains the unit

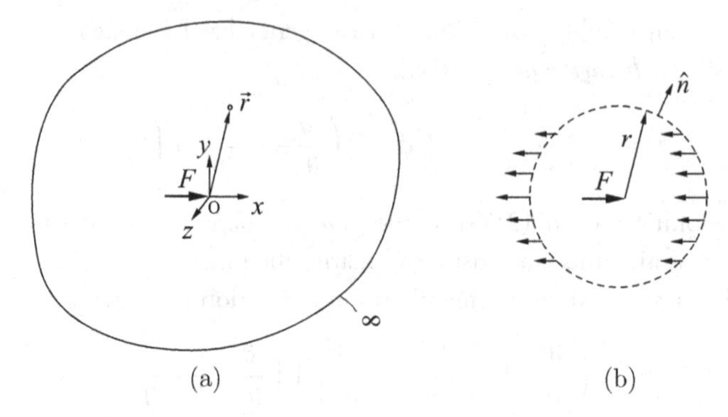

Figure 2.15. Kelvin's problem. (a) A point force is applied at the origin of an infinite medium. (b) Free body diagram of a sphere of radius r centered at the origin. The traction force on the spherical surface due to the internal stress field balances the applied force.

force applied at the center, balanced by the traction force on its surface, as shown in Fig. 2.15b. Owing to the symmetry of this problem, we expect the stress field of this problem to have the following form when expressed in spherical coordinates, r, θ, and φ:

$$\sigma_{ij}(\vec{r}) = f(r)g_{ij}(\theta, \varphi). \tag{2.71}$$

This expression means that the orientation dependence of the stress field is independent of the distance r from the origin.

The free-body diagram shows that, regardless of the radius r, the total traction force must balance the unit force applied at the center. Because the total surface area of the sphere, $4\pi r^2$, increases with increasing r, the intensity of the traction force (and stress field) must decrease with increasing r, specifically,

$$\begin{aligned} f(r)4\pi r^2 &= \text{constant}, \\ f(r) &\propto \frac{1}{r^2}. \end{aligned} \tag{2.72}$$

This proves that the stress field due to a point force scales as $1/r^2$. Consequently, the strain field also scales as $1/r^2$. Because the strain field is proportional to the spatial derivatives of the displacement field, the displacement field in Kelvin's problem must scale as $1/r$.

Analytic expression of the elastic Green's function

When the concentrated force points in the x direction and has unit magnitude, the solution of Kelvin's problem is

$$\begin{aligned} u_x &= \frac{1}{16\pi\mu(1-\nu)}\left(\frac{3-4\nu}{r} + \frac{x^2}{r^3}\right), \\ u_y &= \frac{1}{16\pi\mu(1-\nu)}\frac{xy}{r^3}, \\ u_z &= \frac{1}{16\pi\mu(1-\nu)}\frac{xz}{r^3}, \end{aligned} \tag{2.73}$$

where μ is the shear modulus and ν is Poisson's ratio of the isotropic elastic medium, and $r = \sqrt{x^2 + y^2 + z^2}$. Because the unit force is applied in the x direction, the displacement field

u_x has a different form than u_y and u_z, but an obvious symmetry can be observed between u_y and u_z.

If the unit force is applied in the y direction instead, the corresponding displacement field can be easily obtained from Eq. (2.73) through a change of coordinate system (i.e. permutation of x, y, z), and the result is

$$\begin{aligned} u_x &= \frac{1}{16\pi\mu(1-\nu)} \frac{xy}{r^3}, \\ u_y &= \frac{1}{16\pi\mu(1-\nu)} \left(\frac{3-4\nu}{r} + \frac{y^2}{r^3} \right), \\ u_z &= \frac{1}{16\pi\mu(1-\nu)} \frac{yz}{r^3}. \end{aligned} \tag{2.74}$$

The displacement field when the force is applied in the z direction can be similarly obtained.

The displacement field when the unit force is applied in x, y, or z direction can be concisely expressed using the elastic Green's function

$$G_{ij}(\vec{r}) = \frac{1}{16\pi\mu(1-\nu)} \frac{1}{r} \left[(3-4\nu)\delta_{ij} + \frac{x_i x_j}{r^2} \right], \tag{2.75}$$

where $j = x, y$, or z represents the direction of the force, and $i = x, y$, or z represents the component of the displacement field. $G_{ij}(\vec{r})$ is the displacement at point $\vec{r}$ when the unit force is applied at the origin; $r = |\vec{r}| = \sqrt{x^2 + y^2 + z^2}$; δ_{ij} is the Kronecker delta; $\delta_{ij} = 1$ if $i = j$ and $\delta_{ij} = 0$ if $i \neq j$. It can be easily verified that Eq. (2.75) reduces to Eq. (2.73) when $j = x$, and to Eq. (2.74) when $j = y$.

If the unit force is applied at point $\vec{r}'$ (instead of at the origin), then the displacement at point $\vec{r}$ is given by $G_{ij}(\vec{r} - \vec{r}')$. This is true only for an infinite medium whose elastic constants are homogeneous (i.e. do not vary in space).

From Eqs. (2.73)–(2.75) we see that the magnitude of the displacement scales as $1/r$ as we move away from the location where the force is applied, consistent with the previous expectation based on the free-body diagram.

Displacement fields of force dipoles

In Section 4.4 we shall see that the elastic distortion produced by certain point defects can be modeled as force dipoles. In the following, we show how the displacement field of force dipoles can be obtained from the elastic Green's function.

Consider, for example, a force dipole of unit strength in which the forces point in the $\pm z$ directions. The force dipole can be considered as the superposition of two concentrated forces, one acting in the $+z$ direction at point $(0, 0, \Delta z/2)$, and the other acting in the $-z$ direction at point $(0, 0, -\Delta z/2)$. Let P be the magnitude of the two forces. Because the strength of the dipole is $P\Delta z = 1$, we have $P = 1/\Delta z$. An idealized dipole is obtained in the limit of $\Delta z \to 0$.

Given the elastic Green's function, the displacement field due to the force dipole can be written as

$$\begin{aligned} u_i^{\mathrm{dp}}(\vec{r}) &= PG_{iz}(x, y, z - \Delta z/2) - PG_{iz}(x, y, z + \Delta z/2) \\ &= -\frac{G_{iz}(x, y, z + \Delta z/2) - G_{iz}(x, y, z - \Delta z/2)}{\Delta z}. \end{aligned} \tag{2.76}$$

In the limit of $\Delta z \to 0$, we obtain

$$u_i^{\text{dp}}(\vec{r}) = -\frac{\partial}{\partial z} G_{iz}(\vec{r}). \tag{2.77}$$

As another example, let us consider three mutually orthogonal force dipoles of unit strength acting at the origin. The resulting displacement field must be

$$u_i^{3\text{dp}}(\vec{r}) = -\left[\frac{\partial}{\partial x} G_{ix}(\vec{r}) + \frac{\partial}{\partial y} G_{iy}(\vec{r}) + \frac{\partial}{\partial z} G_{iz}(\vec{r})\right]. \tag{2.78}$$

Using Eq. (2.75), we can show that

$$u_i^{3\text{dp}}(\vec{r}) = \frac{1-2\nu}{8\pi\mu(1-\nu)} \frac{x_i}{r^3} = \frac{1+\nu}{12\pi(1-\nu)B} \frac{x_i}{r^3}, \tag{2.79}$$

where B is the bulk modulus. The result can be expressed more concisely in spherical coordinates

$$u_r^{3\text{dp}}(r) = \sqrt{\left(u_x^{3\text{dp}}\right)^2 + \left(u_y^{3\text{dp}}\right)^2 + \left(u_z^{3\text{dp}}\right)^2} = \frac{1+\nu}{12\pi(1-\nu)B} \frac{1}{r^2}. \tag{2.80}$$

2.6 Summary

Atomic point defects, dislocations, and even some two-dimensional defects impose elastic displacements on their surroundings and thus have elastic fields. The energies associated with these defects and, especially, interactions between the defects, are governed by these elastic fields. Thus, understanding the principles of stress, strain, and elasticity is needed for the study of the elastic properties of defects. While most crystals are elastically anisotropic, almost all defect properties can be described with sufficient fidelity by assuming elastic isotropy. Indeed, all of the fundamental relations that are used to describe defects and their interactions can be derived using isotropic elasticity.

Stress, describing the intensity of forces on and within materials, is a second-rank tensor having three axial and six shear components acting on the three positive faces of the unit cell in the coordinate system being used (Cartesian, cylindrical, spherical). The components of the stress tensor are described as σ_{ij}, where $i = 1, 2$, or 3 denotes the direction of the force and $j = 1, 2$, or 3 denotes the plane normal for the stress component of interest. Consideration of moment equilibrium shows that $\sigma_{ij} = \sigma_{ji}$, so that the six shear stresses reduce to three independent shear components. With this reduction it is often convenient to use the Voigt notation and describe the stress as a column vector, σ_I with $I = 1, 2, \ldots, 6$. Stress, like other second-rank tensors, can be transformed from one coordinate system to another using dot products of unit vectors in the different coordinate systems.

Strain is a measure of shape change of a body or an element within a body under stress. The basic definition of strain is displacement per unit length, or equivalently, a derivative of displacement with respect to position. Similar to stress, there are also two types of strain: axial and shear. The nine tensor strain components are then expressed as $\varepsilon_{ij} = \frac{1}{2}(\partial u_i/\partial x_j + \partial u_j/\partial x_i)$, where u_i represents a component of displacement and x_j represents a coordinate position. Like stress, the strain is a second-rank tensor and can be transformed from one coordinate system to another using dot products of unit vectors in the two coordinate systems. The symmetry of the strain tensor makes it obvious that $\varepsilon_{ij} = \varepsilon_{ji}$ so that there are only three independent shear strain components, making a total of only six independent components of strain. Thus, as in the case of stress, it is sometimes convenient to express strain with the Voigt notation wherein strain is expressed as a column vector, ε_I with $I = 1, 2, \ldots, 6$. But the strain components in the Voigt notation do not transform as a tensor (nor as a vector).

Isotropic linear elasticity relates the stresses and strains through the generalized Hooke's law. The relation between axial stresses and strains leads to Young's modulus, E, while the relation between shear stresses and strains leads to the shear modulus, μ. In addition, it is sometimes useful to define the transverse strain associated with a given axial strain under uniaxial loading in terms of Poisson's ratio, ν, an elastic constant that can be derived from E and μ. Other useful elastic constants can be defined in terms of these two independent elastic constants for isotropic solids. Integration of the stress and strain during elastic loading leads to expressions for elastic strain energy and strain energy density that are useful in the study of imperfections.

The distributions of stresses and strains within an elastic solid are governed by the fundamental equations of elasticity, which include the requirement that material elements be in mechanical equilibrium and that the strains at different points in the solid be compatible with each other. These fundamental equations lead to strategies for solving for the distributions of stresses and strains subject to boundary conditions imposed by the defect problems of interest. The elastic Green's function is the displacement field in an infinite homogeneous medium due to a concentrated force. Green's function can be used to construct the elastic fields around point defects modeled as force dipoles.

2.7 Exercise problems

2.1 A strip of metal is subjected to a uniform tension stress $\sigma_{yy} = \sigma$, as shown in Fig. 2.16.

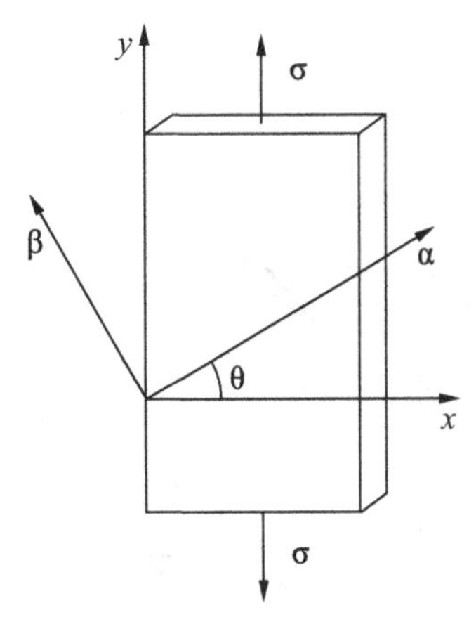

Figure 2.16. A strip of metal subjected to tensile stress.

(a) Calculate the shear stress $\sigma_{\alpha\beta}$ acting on a plane that is 45° to the horizontal axis.

(b) Write a general expression for the shear stress $\sigma_{\alpha\beta}$ on a plane making an arbitrary angle θ with the horizontal axis.

(c) Show that $\sigma_{\alpha\beta}$ is maximum at $\theta = 45°$.
(d) Calculate the hydrostatic pressure p in the strip.

2.2 An FCC single crystal thin film with the orientation shown in Fig. 2.17 is subjected to an equal-biaxial tensile stress σ. Plastic deformation of this crystal will occur by slip on the $(1\,\bar{1}\,1)$ plane (shaded) and along the [0 1 1] direction as shown. Write an expression for the shear stress acting on the slip plane in the slip direction.

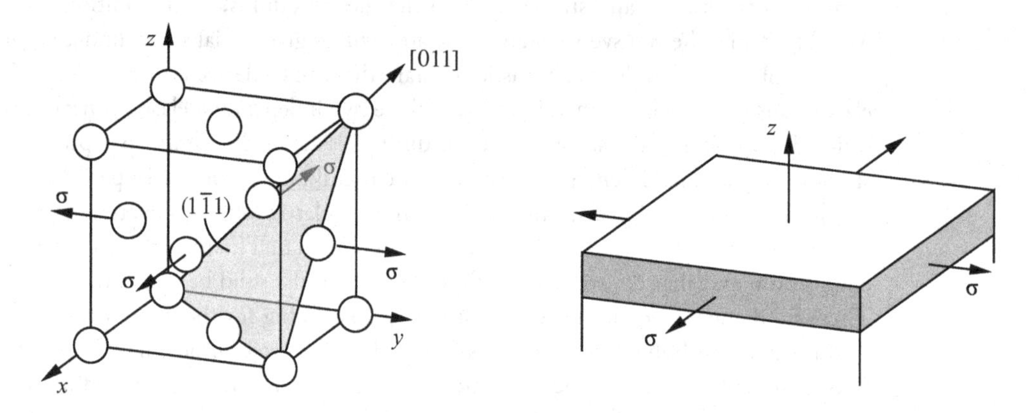

Figure 2.17. An FCC single crystal thin film under biaxial tension.

2.3 A single crystal of BCC iron is pulled in uniaxial tension along a cube direction, [1 0 0]. We assume slip occurs on the $(\bar{2}\,1\,1)$ plane and in the [1 1 1] direction as shown in Fig. 2.18a. The critical resolved shear stress (CRSS) for slip to occur is 10 MPa. Calculate the tensile stress at which plastic yielding begins.

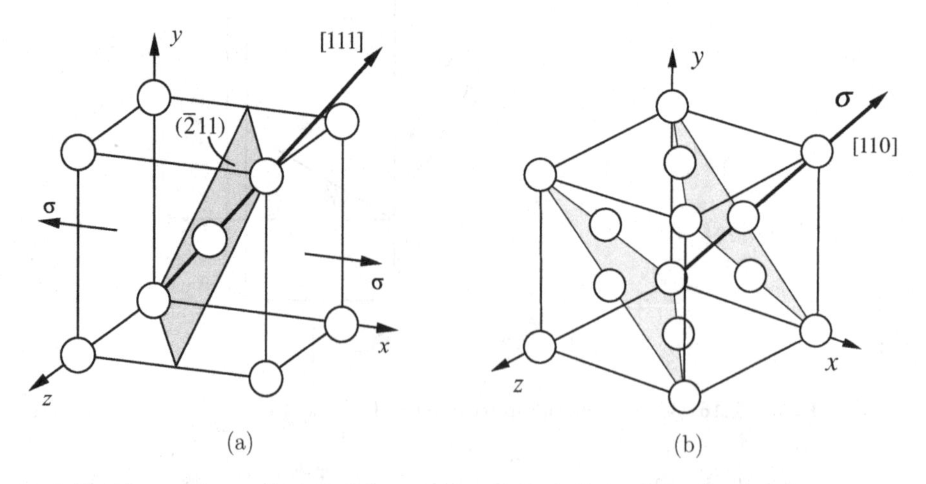

Figure 2.18. (a) A single crystal of BCC iron subjected to tension along the [1 0 0] direction. (b) An FCC single crystal subjected to tension along the [1 1 0] direction. Two $(1\,1\,\bar{1})$ planes are shaded to help show the atomic positions.

2.4 An FCC single crystal is subjected to a uniform tensile stress of magnitude σ in the [1 1 0] direction, as shown in Fig. 2.18b. Calculate all the non-zero stress components in the cubic (xyz) axes.

2.5 Consider a single crystal rod subjected to a uniaxial tensile stress σ, as shown in Fig. 2.19. Plastic deformation would occur if the resolved shear stress τ on a certain crystallographic plane (with normal vector $\hat{n}$) and along a certain crystallographic direction $\hat{m}$ exceeds a threshold value. Express the resolved shear stress τ in terms of the tensile stress σ and the angles θ and ϕ that vectors $\hat{n}$ and $\hat{m}$ make with the tensile axis. The ratio $S = \tau/\sigma$ is called the *Schmid factor.*

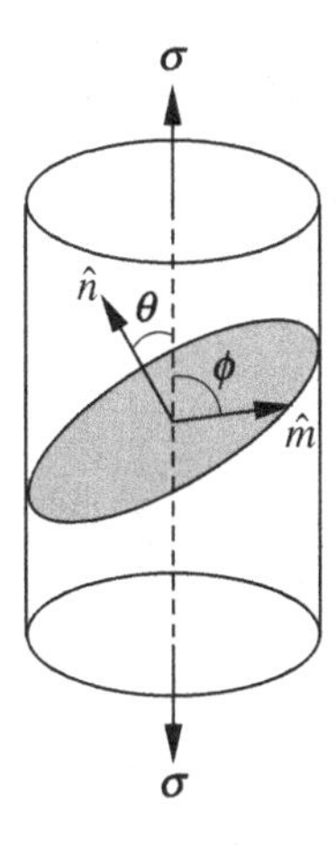

Figure 2.19. Single crystal subjected to uniaxial tension.

2.6 The stress field of an infinite isotropic elastic medium containing a pressurized cylindrical hole (of radius r_0) has the following form:

$$\begin{aligned} \sigma_{rr} &= -\frac{A}{r^2}, \\ \sigma_{\theta\theta} &= \frac{A}{r^2}, \\ \sigma_{r\theta} &= 0, \end{aligned} \tag{2.81}$$

for $r \geq r_0$. Transform the stress field into Cartesian components: σ_{xx}, σ_{yy}, σ_{xy}. Make contour plots of σ_{rr}, $\sigma_{r\theta}$, $\sigma_{\theta\theta}$ and σ_{xx}, σ_{yy}, σ_{xy}.

2.7 A strain gauge measures tensile strain along a given direction through the change of electrical resistance of thin wires aligned in that direction. Consider a solid subjected to uniaxial tensile stress, which induces a normal strain ε in the tensile direction, and a normal strain $-\nu\varepsilon$ in the perpendicular directions, as shown in Fig. 2.20a. Express the strain measured by a strain gauge oriented at 60° from the tensile axis.

2.8 Express the strain measured by a strain gauge oriented at 45° from the vertical axis of a solid subjected to an engineering shear strain of γ, as shown in Fig. 2.20b.

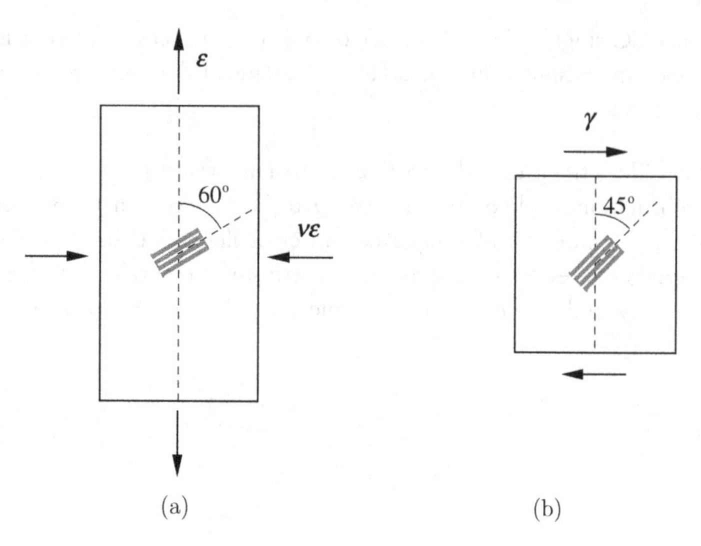

Figure 2.20. Strain gauge.

2.9 A thin film is subjected to strain $\varepsilon_{xx} = \varepsilon_1$, $\varepsilon_{yy} = \varepsilon_2$, $\varepsilon_{zz} = \varepsilon_3$. Write an expression for the axial strain ε in the arbitrary direction shown in Fig. 2.21.

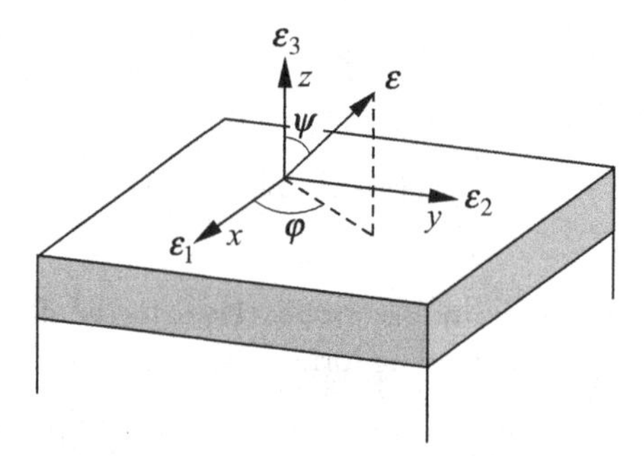

Figure 2.21. A thin film subjected to strain ε_1, ε_2, ε_3 in the x, y, z directions, respectively.

2.10 A solid cube with shear modulus μ and Poisson's ratio ν is subjected to a compressive stress along z, i.e. $\sigma_{zz} = -p$, but is not allowed to expand in the x direction due to constraints imposed by rigid plates. The solid is allowed to expand freely in the y direction. Obtain the normal stress and strain of the solid in x, y, and z directions.

2.11 Repeat the analysis in Exercise problem 2.10 but for a solid constrained (i.e. unable to expand) in both x and y directions.

2.12 Estimate the compressive stress that arises in a block of aluminum when the temperature increases from 0 °C to 30 °C if its volume is not allowed to expand. Use the thermoelastic properties of aluminum given in Table B.1. If the block were allowed to expand freely, it would experience a thermal strain of $\varepsilon^{\mathrm{T}} = \alpha \Delta T$ in all three direction with zero stress, where α is the thermal expansion coefficient and ΔT is the temperature increase.

2.13 Consider an infinite elastic solid containing a cylindrical hole of radius r_0, which is filled with a gas with pressure p. The stress field in the solid is given in Eq. (2.81). Express the coefficient A in terms of the pressure p inside the hole and the hole radius r_0. Verify that this stress field satisfies the equilibrium condition, Eq. (2.56).

2.14 Given the stress field expression, Eq. (2.81), for an infinite elastic solid containing a pressurized cylindrical hole,

(a) find the corresponding strain field, and express the strain field in Cartesian (xyz) components;
(b) verify that the strain field satisfies the compatibility condition Eq. (2.67);
(c) verify that the displacement field $u_r = C/r$, $u_\theta = 0$ is consistent with the strain field, and express the coefficient C in terms of hole radius r_0 and pressure p.

Assume the solid is a linear isotropic material with shear modulus μ and Poisson's ratio ν.

2.15 Given the stress and strain expressions in Exercise problems 2.13 and 2.14 for an infinite solid containing a pressurized cylindrical hole, express the total elastic energy per unit length along the hole in terms of the pressure p and hole radius r_0 using the following two approaches and show that they agree with each other.

(a) Integrate the strain energy density Δw over the entire volume of the solid.
(b) Integrate the work done on the interior surface of the cylindrical hole as the pressure gradually increases from 0 to p.

2.16 Find the expressions for the displacement field for Kelvin's problem, in which a unit force in the z direction is applied at the origin in an infinite elastic medium. Express the displacement field in cylindrical coordinates. Which strain and stress components are non-zero in cylindrical coordinates?

3 Statistical thermodynamics

Here we give a basic summary of thermodynamics and statistical mechanics, which are needed to derive the equilibrium distribution of point defects in a solid under external or internal stresses. A more thorough treatment of this subject can be found in many textbooks, such as [13, 14].

3.1 Laws of thermodynamics

Thermodynamics is a branch of science connected with the nature of heat and its conversion to mechanical, electrical, and chemical energy. Historically, it grew out of efforts to construct more efficient heat engines – devices for extracting useful work from expanding hot gases. The lessons people learned, through a large amount of trial and error, on how efficient a heat engine can ever be made, ultimately led to the establishment of the laws of thermodynamics. These laws were later found to be universally true for a wide range of physical and chemical processes.

3.1.1 The first law

The first law of thermodynamics establishes the total *energy* as a state variable, and also establishes the conservation of energy. A state variable is a variable that takes a unique value for a given thermodynamic state, regardless of how the state is reached. The first law of thermodynamics can be stated as follows.

You can change the total energy of the system by doing work $đW$ to it, and giving it heat $đQ$. The change of total energy is

$$dE = đW + đQ, \tag{3.1}$$

where dE is a complete differential (path independent), while $đW$ and $đQ$ are not complete differentials (they are path dependent). This difference is explained below.

For example, suppose our system is a gas, whose state can be represented by the pressure p and volume V. Then the work done for a differential change of volume is

$$đW = -p\,dV. \tag{3.2}$$

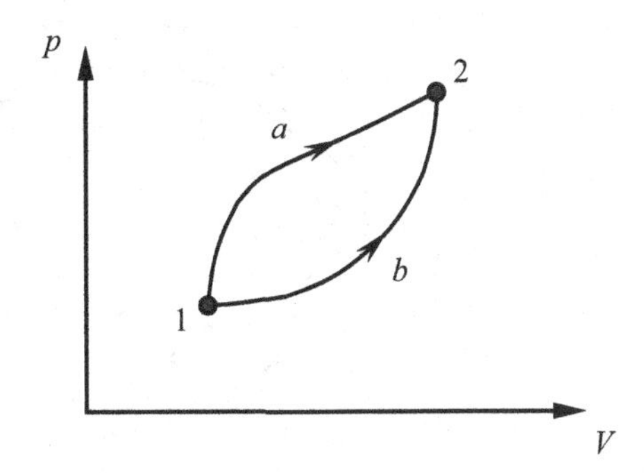

Figure 3.1. Transforming the state of gas in a container from State 1 to State 2 along two possible paths. For a given amount of gas, the temperatures in State 1 and State 2 are not the same.

Now suppose we transform the gas from State 1 to State 2. There are an infinite number of paths (ways) to do this, two of which are shown in Fig. 3.1. The total work done along path a is, in general, different from the total work done along path b, i.e.

$$\Delta W_a = \int_a đW = \int_a -p\,dV \neq \Delta W_b = \int_b đW = \int_b -p\,dV. \tag{3.3}$$

Note that $-\Delta W_a$ equals the area under curve a, and $-\Delta W_b$ equals the area under curve b. Similarly, the total heat deposited to the gas is also different along path a and along path b, i.e.

$$\Delta Q_a = \int_a đQ \neq \Delta Q_b = \int_b đQ. \tag{3.4}$$

Yet, the first law states that total change of energy is independent of the path, because it is just the energy of State 2 minus the energy of State 1, i.e.

$$E_2 - E_1 = \Delta E = \int_a dE = \int_b dE. \tag{3.5}$$

3.1.2 The second law

The second law of thermodynamics establishes the *entropy* as a state variable. Entropy is perhaps the most profound concept in the entire field of thermodynamics and statistical mechanics. Within classical thermodynamics, entropy S can be defined through the heat deposited into the system along a reversible path. A reversible path is a trajectory the system takes that is so slow that the system can be considered to be in equilibrium along the entire path, and the path can be reversed to bring the system to its original state.

As an example, we can consider the two paths shown in Fig. 3.1 and now assume that they are reversible paths, so that we will use $đQ_{\text{rev}}$ to represent the heat deposited to the system. We have seen that the total amount of heat deposited is path dependent,

$$\Delta Q_a = \int_a đQ_{\text{rev}} \neq \Delta Q_b = \int_b đQ_{\text{rev}}. \tag{3.6}$$

The second law states that when the integral is weighted by the inverse of the temperature T (measured from absolute zero), it becomes path independent, because it is simply the entropy

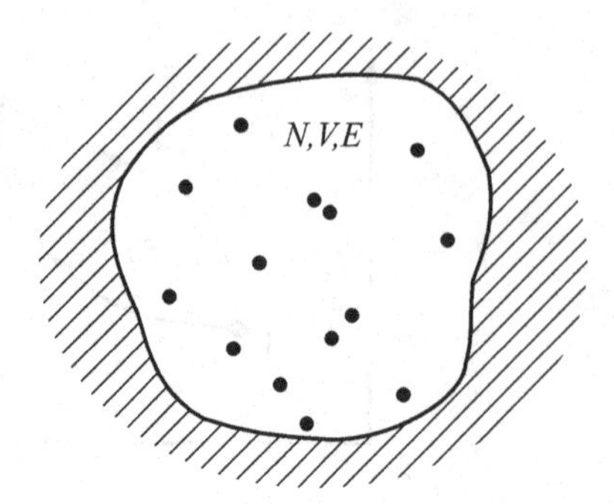

Figure 3.2. An isolated system containing N gas molecules, occupying volume V and with energy E.

difference between State 1 and State 2,

$$S_2 - S_1 = \Delta S = \int_a dS = \int_a \frac{đQ_{\text{rev}}}{T} = \int_b dS = \int_b \frac{đQ_{\text{rev}}}{T}. \tag{3.7}$$

The physical meaning of entropy is not clear in classical thermodynamics; it is a mysterious quantity that is possessed by each equilibrium state. It was not until the fundamental contribution of Boltzmann (in statistical mechanics) that people realized that the entropy S is a measure of microscopic disorder in each macroscopic state (see Section 3.3).

The second law also states that the entropy of an isolated system never decreases. An isolated system is one to which we cannot do any work or exchange heat (picture a gas in a thermally insulated container whose volume remain fixed, as shown in Fig. 3.2). As the system settles into thermal equilibrium, its entropy will reach a maximum. After that, the system will not move away from this equilibrium state because doing so would decrease its entropy, which is prohibited by the second law. Therefore, the increase of entropy provides an arrow of time that indicates the direction an isolated system should evolve towards thermal equilibrium.

3.1.3 The third law

The third law of thermodynamics is an axiom of nature regarding entropy and the impossibility of reaching absolute zero of temperature. It states that: As a system approaches absolute zero temperature, all processes cease and the entropy of the system approaches a minimum value (which can be defined as zero).

This is also known as Nernst's theorem. We will see that it is a natural consequence of Boltzmann's entropy expression (see Section 3.3).

3.2 Thermodynamic potentials

3.2.1 Equation of state

In addition to energy E, in thermodynamics we also work with other state variables, such as enthalpy H, Helmholtz free energy F, and Gibbs free energy G. They are all called *thermodynamic potentials* and are introduced below.

For simplicity, we consider again an isolated gas tank of volume V, containing N gas molecules, and let the total energy of the gas be E, as shown in Fig. 3.2. If the gas tank is left alone for a time, it should go to a thermal equilibrium state, and N, V, E stay constant. Intuitively, we expect that three variables (N, V, E) are sufficient to completely specify the state of the gas. Yet, the second law of thermodynamics states that there exists another state variable, the entropy S. Therefore, the four state variables S, N, V, E must be related, e.g. through a function $E(S, V, N)$. This function is specific to the type of material (e.g. gas) that we are studying and is called the *equation of state* of the material.[1]

Because E, S, V, N are all state functions, dE, dS, dV, dN are all complete differentials, and they are related by

$$dE = \left(\frac{\partial E}{\partial S}\right)_{V,N} dS + \left(\frac{\partial E}{\partial V}\right)_{S,N} dV + \left(\frac{\partial E}{\partial N}\right)_{S,V} dN. \tag{3.8}$$

The subscripts next to the brackets emphasize which variables are kept fixed when taking the partial derivatives. Define the first partial derivative on the right hand side to be the temperature, i.e.

$$T \equiv \left(\frac{\partial E}{\partial S}\right)_{V,N}. \tag{3.9}$$

We see that the first term on the right hand side of Eq. (3.8) corresponds to the heat deposited to the system (during a reversible process)

$$đQ_{\text{rev}} = T dS. \tag{3.10}$$

This is consistent with the relation between entropy and heat in Eq. (3.7). The second partial derivative on the right hand side of Eq. (3.8) is identified with the negative pressure

$$p \equiv -\left(\frac{\partial E}{\partial V}\right)_{S,N}, \tag{3.11}$$

so that the second term on the right hand side of Eq. (3.8) is the mechanical work done to the system

$$đW = -p dV. \tag{3.12}$$

This is consistent with Eq. (3.2). Similarly, the third partial derivative on the right hand side of Eq. (3.8) is identified with the chemical potential

$$\mu \equiv \left(\frac{\partial E}{\partial N}\right)_{S,V} \tag{3.13}$$

so that the third term on the right hand side of Eq. (3.8) is the "chemical work" done to the system if we change the number of molecules, $\mu\, dN$.

1 The term "equation of state" is often used to describe the pressure–volume–temperature relation, which can be obtained from the more fundamental equation of state defined here.

Using these new definitions, we can write the following.

$$\boxed{dE = T\,dS - p\,dV + \mu\,dN} \tag{3.14}$$

3.2.2 Gibbs–Duhem relation

E, S, V, N are all *extensive* quantities, meaning that they all scale linearly with the amount of material in the system. On the other hand, T, p, μ are all *intensive* quantities, which do not depend on the amount of material. For example, consider collecting λ number of gas tanks, each having identical E, S, V, N, and hence T, p, μ, and joining them to form a larger gas tank. The new system will have energy λE, entropy λS, volume λV, and λN number of molecules, yet the temperature T, pressure p, and chemical potential μ will stay the same as before. Therefore, the equation of state $E(S, V, N)$ is a "homogeneous function of 1st order", which means that if all the arguments of the function are multiplied by the same factor λ, the return value of the function will be multiplied by the same factor λ.

Euler studied the mathematical properties of homogeneous functions and proved an important theorem linking a homogeneous function with all of its derivatives. Applying Euler's theorem to the function $E(S, V, N)$, it says,

$$E(S, V, N) = \left(\frac{\partial E}{\partial S}\right)_{V,N} S + \left(\frac{\partial E}{\partial V}\right)_{S,N} V + \left(\frac{\partial E}{\partial N}\right)_{S,V} N \tag{3.15}$$

In other words, we can write the following.

$$\boxed{E = TS - pV + \mu N} \tag{3.16}$$

Consequently,

$$dE = T\,dS + S\,dT - p\,dV - V\,dp + \mu\,dN + N\,d\mu. \tag{3.17}$$

Combining this with Eq. (3.14)

$$dE = T\,dS - p\,dV + \mu\,dN$$

we arrive at the *Gibbs–Duhem relation*,

$$S\,dT - V\,dp + N\,d\mu = 0. \tag{3.18}$$

This relation means that the three intensive variables, T, p, μ, cannot be changed independently of each other for a system in thermal equilibrium.

3.2.3 Helmholtz free energy

In practice, the equation of state $E(S, V, N)$ is not very easy to work with, because entropy is not easy to measure and control. Instead, most experiments are performed under the condition of constant temperature T. The *Legendre transform* is used to convert function $E(S, V, N)$ into a more useful form. The result is the Helmholtz free energy F, which is a function of T, V, N.

The Helmholtz free energy F is defined as

$$F(T, V, N) \equiv E - TS. \tag{3.19}$$

To show that F is a function of T, V, and N, we only need to compute the differential of F,

$$\begin{aligned} dF &= dE - T\,dS - S\,dT \\ &= T\,dS - p\,dV + \mu\,dN - T\,dS - S\,dT \\ &= -S\,dT - p\,dV + \mu\,dN. \end{aligned} \tag{3.20}$$

Therefore S, p, and μ can also be identified with the partial derivatives of F.

$$S \equiv -\left(\frac{\partial F}{\partial T}\right)_{V,N}, \quad p \equiv -\left(\frac{\partial F}{\partial V}\right)_{T,N}, \quad \mu \equiv \left(\frac{\partial F}{\partial N}\right)_{T,V}. \tag{3.21}$$

Note that T is fixed when taking the partial derivative with respect to V and N, which is a more convenient condition to use than to keep S fixed.

Recall that the second law states that, for an isolated system, its entropy can increase with time only until it reaches thermal equilibrium, at which point the entropy is at the maximum. It can be shown that for a system kept at constant temperature T, so that it can exchange heat with its surroundings, but is unable to exchange (mechanical or chemical) work with its surroundings, the second law states that its Helmholtz free energy F can only decrease with time. F reaches the minimum at thermal equilibrium.

3.2.4 Enthalpy

When we work with solids, it is very difficult to keep the volume V of the material fixed. For example, the thermal expansion effect can induce a tremendous stress if the temperature of a solid is increased without allowing its volume to expand (see Exercise problem 2.12). Therefore, it is of interest to transform the equation of state $E(S, V, N)$ to a form that uses the pressure p as an input variable. For example, most experiments are conducted under ambient conditions, for which p equals 1 atmosphere of pressure ($\approx 10^5$ Pa).

For this purpose, the enthalpy H is introduced as another Legendre transform of E,

$$H \equiv E + pV. \tag{3.22}$$

We can easily show that H is a function of S, p, and N, by computing the differential of H,

$$\begin{aligned} dH &= dE + p\,dV + V\,dp \\ &= T\,dS - p\,dV + \mu\,dN + p\,dV + V\,dp \\ &= T\,dS + V\,dp + \mu\,dN. \end{aligned} \tag{3.23}$$

Therefore T, V, and μ can also be identified with the partial derivatives of H as

$$T \equiv \left(\frac{\partial H}{\partial S}\right)_{p,N}, \quad V \equiv \left(\frac{\partial H}{\partial p}\right)_{S,N}, \quad \mu \equiv \left(\frac{\partial H}{\partial N}\right)_{S,p}. \tag{3.24}$$

Note that p is fixed when taking the partial derivative with respect to S and N, which is sometimes a more convenient condition to use than to keep V fixed. Equation (3.22) is the proper definition of enthalpy for a solid only when the applied stress is purely hydrostatic.[2]

3.2.5 Gibbs free energy

In practice, most experiments are conducted with both pressure p and temperature T kept constant. Therefore, we wish to transform the equation of state $E(S, V, N)$ to a form that uses both p and T as input variables. This is accomplished by yet another Legendre transform,

$$G(T, p, N) \equiv E - TS + pV = F + pV = H - TS. \tag{3.25}$$

We can show that G is a function of T, p, and N, by computing the differential of G:

$$\begin{aligned} dG &= dE - T\,dS - S\,dT + p\,dV + V\,dp \\ &= T\,dS - p\,dV + \mu\,dN - T\,dS - S\,dT + p\,dV + V\,dp \\ &= -S\,dT + V\,dp + \mu\,dN. \end{aligned} \tag{3.26}$$

Therefore S, V, and μ can also be identified with the partial derivatives of G,

$$S \equiv -\left(\frac{\partial G}{\partial T}\right)_{p,N}, \quad V \equiv \left(\frac{\partial G}{\partial p}\right)_{T,N}, \quad \mu \equiv \left(\frac{\partial G}{\partial N}\right)_{T,p}. \tag{3.27}$$

Because it is a function of T and p, the Gibbs free energy is perhaps the most important among all the thermodynamic potentials. Recalling Eq. (3.16),

$$E = TS - pV + \mu N,$$

we have

$$G = E - TS + pV = \mu N. \tag{3.28}$$

In other words, we have the following,

$$\boxed{\mu = \frac{G}{N}} \tag{3.29}$$

This equation shows that the chemical potential is also the Gibbs free energy per molecule in the system.

It can be shown that for a system kept at constant temperature T and pressure p, so that it can exchange heat and mechanical work with its surroundings, the second law states that its Gibbs free energy G can only decrease with time. G reaches the minimum at thermal equilibrium.

2 Unlike fluids, a solid is able to withstand shear stresses, whose work effects need to be subtracted from the energy in the definition of enthalpy. For example, if a solid is subjected to both a hydrostatic pressure p and a shear stress τ, then $H \equiv E + pV - \tau\gamma V_0$, where γ is the shear strain and V_0 is the solid volume at zero stress. The work effects also need to be subtracted off in the definition of Gibbs free energy, which is related to the enthalpy as $G \equiv H - TS$.

3.2.6 Chemical potential of a gas

As an example of how to use the thermodynamic relations introduced so far, we derive a simple expression for the chemical potential of a gas. Consider a gas consisting of N molecules, with pressure p, volume V, and temperature T. We shall assume that the gas satisfies the ideal gas law

$$pV = Nk_B T, \tag{3.30}$$

where $k_B = 1.3806 \times 10^{-23}\ \mathrm{J \cdot K^{-1}} = 8.6173 \times 10^{-5}\ \mathrm{eV \cdot K^{-1}}$ is Boltzmann's constant.[3] In fact, this is a good approximation for real gases under most conditions. We now consider various states of the gas, all under the same temperature T, but with different pressure (and volume). We wish to find out how the chemical potential of the gas molecules is related to the pressure.

Let μ_0 be the chemical potential of the gas molecules under pressure p_0, and let μ be the chemical potential at pressure p. From Eq. (3.26), and letting $dT = 0$ and $dN = 0$, we have

$$dG = -V\,dp. \tag{3.31}$$

Using the ideal gas law, Eq. (3.30), we have

$$dG = -Nk_B T \frac{1}{p}\,dp. \tag{3.32}$$

Hence

$$\begin{aligned} G(T, p, N) - G(T, p_0, N) &= -Nk_B T \int_{p_0}^{p} \frac{1}{p}\,dp \\ &= Nk_B T \ln \frac{p}{p_0}. \end{aligned} \tag{3.33}$$

Therefore,

$$\begin{aligned} \mu - \mu_0 &= \frac{G(T, p, N)}{N} - \frac{G(T, p_0, N)}{N} = k_B T \ln \frac{p}{p_0} \\ \mu &= \mu_0 + k_B T \ln \frac{p}{p_0}. \end{aligned} \tag{3.34}$$

This result will be used for obtaining the equilibrium concentration of point defects in Sections 5.4.1 and 6.3.

3.2.7 Chemical potential of a solid under pressure

To give another example of how to use the thermodynamic relations, we derive a simple expression for the chemical potential of atoms in a pure solid as a function of pressure. Consider a solid containing N atoms kept at a constant temperature T. Define State 0 (the reference state) as the state at which the solid is subjected to zero pressure, i.e. $p = 0$, with volume V_0. Let the Helmholtz free energy of State 0 be F_0. From Eq. (3.25), the Gibbs free energy of State 0 is

$$G_0 \equiv G(T, p = 0, N) = F(T, V_0, N) = F_0. \tag{3.35}$$

3 The gas constant $R = 8.3145\ \mathrm{J \cdot mol^{-1} \cdot K^{-1}}$ equals $N_a k_B$, where $N_a = 6.0221 \times 10^{23}$ is Avogadro's number.

From Eq. (3.29), the chemical potential of State 0 is

$$\mu_0 = \frac{G_0}{N}. \tag{3.36}$$

We now consider State 1 where the solid is subjected to a non-zero pressure $p_1 > 0$, with volume $V_1 < V_0$. Our goal is to derive the chemical potential μ_1 of State 1. As an illustration of the self-consistency of the thermodynamic relationships, we will derive μ_1 using two different approaches.

Approach 1

We can start by finding the Helmholtz free energy of State 1. It can be obtained from the work done to the solid when compressing it from State 0 to State 1, i.e. by integrating Eq. (3.20),

$$F_1 = -\int_{V_0}^{V_1} p\,dV + F_0. \tag{3.37}$$

For simplicity, let us assume that at temperature T, the solid has a constant (i.e. volume independent) bulk modulus B, so that

$$p = -B\frac{V - V_0}{V_0}. \tag{3.38}$$

Therefore,

$$\begin{aligned} F_1 &= \frac{B}{V_0}\int_{V_0}^{V_1} (V - V_0)\,dV + F_0 \\ &= \frac{B(V_1 - V_0)^2}{2V_0} + F_0 \\ &= -\frac{1}{2}p_1(V_1 - V_0) + F_0 \\ &= \frac{1}{2}p_1(V_0 - V_1) + F_0. \end{aligned} \tag{3.39}$$

Define $W_1 \equiv \frac{1}{2}p_1(V_0 - V_1)$ as the strain energy in State 1. Then,

$$F_1 = F_0 + W_1. \tag{3.40}$$

The Gibbs free energy is

$$G_1 = F_1 + p_1 V_1 = F_0 + W_1 + p_1 V_1. \tag{3.41}$$

Define $w_1 \equiv W_1/N$ as the strain energy per atom, and $v_1 \equiv V_1/N$ as the volume per atom in State 1. The chemical potential of State 1 is

$$\mu_1 = \frac{G_1}{N} = \mu_0 + w_1 + p_1 v_1. \tag{3.42}$$

Approach 2

Alternatively, we can directly find the Gibbs free energy of State 1 by integrating Eq. (3.26),

$$G_1 = \int_0^{p_1} V\,dp + G_0$$
$$= \frac{1}{2}(V_0 + V_1)p_1 + G_0. \tag{3.43}$$

The chemical potential is

$$\mu_1 = \frac{G_1}{N} = \mu_0 + \frac{p_1(V_0 + V_1)}{2N} = \mu_0 + \frac{p_1(V_0 - V_1)}{2N} + \frac{p_1 V_1}{N}$$
$$= \mu_0 + w_1 + p_1 v_1, \tag{3.44}$$

which is the equivalent to Eq. (3.42). Note that

$$V_0 - V_1 = \frac{V_0}{B} p_1. \tag{3.45}$$

Hence,

$$W_1 = \frac{1}{2} p(V_0 - V_1) = \frac{V_0}{2B} p_1^2, \tag{3.46}$$

$$w_1 = \frac{V_0}{2BN} p_1^2, \tag{3.47}$$

$$\mu_1 = \mu_0 + \frac{V_0}{2BN} p_1^2 + p_1 v_1. \tag{3.48}$$

Therefore, the chemical potential contains both an $\mathcal{O}(p_1^2)$ term from the strain energy per atom, and an $\mathcal{O}(p_1)$ term due to the difference between the Gibbs free energy and Helmholtz free energy.

3.3 Boltzmann's entropy expression

According to the second law of thermodynamics, if an isolated system is left alone for a long time, it will reach the state of thermal equilibrium with a well-defined entropy S, which is a function of N, V, and E. Yet, the laws of thermodynamics do not specify how to obtain the function $S(N, V, E)$; they only state that this function exists for every material (or macroscopic system). As such, the physical meaning of the entropy remains hidden.

This problem is resolved by the fundamental contribution of Boltzmann, which states that

$$S(N, V, E) = k_B \ln \mathcal{W}, \tag{3.49}$$

where k_B is Boltzmann's constant, defined in Section 3.2.6, and $\mathcal{W}$ is the number of microscopic states consistent with the macroscopic state (which is at thermodynamic equilibrium). $\mathcal{W}$ is usually called the *partition function* in statistical mechanics [13, 14].

To understand the meaning of $\mathcal{W}$, we need to realize that, while at the macroscopic scale the system is in an equilibrium state, e.g. the temperature and pressure have ceased to change, the atoms are constantly moving around. This phenomenon is commonly referred to as thermal fluctuation. In other words, at the microscopic scale, i.e. in terms of the positions and velocities of all atoms, the system is constantly moving from one state to the next. Yet all of these microscopic states are consistent with the macroscopic state characterized by a few thermodynamic variables, such as N, V, and E. For example, all atoms in a crystal vibrate around their equilibrium positions, with energy flowing incessantly between neighboring atoms, keeping the total energy constant. The impurity atoms in a crystal can also be arranged in many different ways, all of which are consistent with the same average concentration. $\mathcal{W}$ is the total number of all these microscopic states.

The fundamental assumption in statistical mechanics is that, for an isolated system having reached thermal equilibrium, *the probability of finding the system in any one of these microscopic states is the same*, i.e.

$$p_i = \frac{1}{\mathcal{W}}, \qquad i = 1, 2, \ldots, \mathcal{W}, \tag{3.50}$$

as long as the system is not subjected to any other constraints beyond constant N, V, and E. Given the uniform probability distribution, Eq. (3.50), Boltzmann's entropy expression can be regarded as a special case of Shannon's information entropy expression

$$S = -k_{\mathrm{B}} \sum_{i=1}^{\mathcal{W}} p_i \ln p_i. \tag{3.51}$$

The value of $\mathcal{W}$ is often ridiculously large, i.e. larger than any number we usually deal with in everyday life and work. (For example, $\mathcal{W}$ can easily exceed the total number of protons in the entire universe.) However, we do not need to deal with $\mathcal{W}$ directly. For example, the entropy is proportional to $\ln \mathcal{W}$, which has reasonable values.

One of the reasons for the extremely large value of $\mathcal{W}$ is that $\mathcal{W}$ often involves the factorial $N!$ of a large number N. Therefore, in calculating the entropy, the *Stirling's formula* is often very useful:

$$\ln N! = N \ln N - N + \frac{1}{2} \ln(2\pi N) + \mathcal{O}\left(\frac{1}{N}\right). \tag{3.52}$$

The last term means that it decreases to zero in proportion to $1/N$ with increasing N for large N. In practice, the following approximation is often sufficient:

$$\ln N! \approx N \ln N - N. \tag{3.53}$$

We can see that the error made in this approximation is on the order of $\ln N$, which is much smaller than N for large N.

With decreasing temperature, the thermal fluctuation of the atoms is reduced. In the zero temperature limit, the system only occupies one microscopic state that has the lowest energy (ground state), i.e. $\mathcal{W} \approx 1$. In this limit, $S \approx 0$, consistent with the third law of thermodynamics (see Section 3.1.3).

We will use Boltzmann's entropy expression to derive the equilibrium concentration of point defects in crystals in Section 5.1.

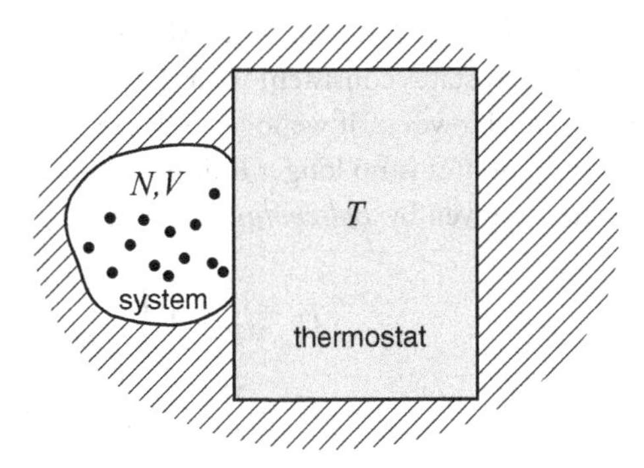

Figure 3.3. A system in thermal contact with a very large thermostat of temperature T.

3.4 Boltzmann's distribution

The material in this section is more advanced than that in previous sections. The main purpose is to show how two key results that will be used in later chapters can be obtained from the fundamental principles of statistical mechanics. The first result is the entropy of a harmonic oscillator, as given in Eq. (3.65). The second result is that the probability of finding the system in some configurational state is proportional to Boltzmann's factor containing the Gibbs free energy of that state, as given in Eq. (3.76). The reader could skip this section if he/she is willing to accept these results. For more discussions on this subject, the reader is referred to textbooks on statistical mechanics [13, 14].

3.4.1 System in contact with thermostat

As noted in Section 3.2.3, the equation of state $S(N, V, E)$ for an isolated system is not very easy to work with, because entropy is not easy to measure and control in practice. In addition, Boltzmann's entropy expression, Eq. (3.49), should only be used for systems with a large number of particles. In fact, the temperature T becomes ill defined for an isolated system with a small N. To overcome these difficulties, it is often useful to study a system in thermal contact with a very large thermostat, as shown in Fig. 3.3. Because the system can exchange heat with the thermostat, its energy E is no longer conserved. But its volume V and number of particles N still remain constant. Such a system is called a *mechanically isolated* system.

The number of particles (N_s) in the thermostat is very large so that its temperature is well defined,

$$T \equiv \left(\frac{\partial E_s}{\partial S_s}\right)_{N_s, V_s}, \tag{3.54}$$

where E_s, V_s, and S_s are the energy, volume, and entropy of the thermostat, respectively. Because the thermostat is so much larger than the system, the temperature T of the thermostat remains unchanged after it establishes thermal contact with the system. After the system has settled down into thermal equilibrium, it adopts the temperature T of the thermostat. Because the system can now exchange heat with the thermostat, the energy E of the system is no longer a constant. However, we can assume that the system and the thermostat together are isolated from the rest of the universe, i.e. $E + E_s$ stays constant. When we consider the system and the thermostat together, then all microstates consistent with the constraint on $E + E_s$ have equal probability at thermal equilibrium. However, if we focus only on the system itself, the probability of finding it in its various microstates is no longer uniform. Instead, the probability of finding the system in any microstate i is given by *Boltzmann's distribution*,

$$p_i = \frac{1}{Z} \exp\left(-\frac{E_i}{k_B T}\right), \tag{3.55}$$

$$Z \equiv \sum_i \exp\left(-\frac{E_i}{k_B T}\right), \tag{3.56}$$

where Z is also called the *partition function*. It ensures that the probability satisfies the normalization condition $\sum_i p_i = 1$, where the summation is over all possible microstates of the system. The exponential term $\exp(-\frac{E_i}{k_B T})$ is called the *Boltzmann factor*.

Because the system can exchange heat with the thermostat, its energy is no longer conserved. Hence the system's energy measured experimentally corresponds to an average of the energies among all of its possible microstates, i.e.

$$E = \sum_i p_i E_i = \frac{1}{Z} \sum_i \exp\left(-\frac{E_i}{k_B T}\right) E_i. \tag{3.57}$$

Because the probability p_i is no longer uniform, we can no longer use Boltzmann's entropy expression, Eq. (3.49), but we can still use Shannon's entropy expression, Eq. (3.51), i.e.

$$\begin{aligned} S &= -k_B \sum_i p_i \ln p_i \\ &= -k_B \sum_i \frac{1}{Z} \exp\left(-\frac{E_i}{k_B T}\right) \left(\ln \frac{1}{Z} - \frac{E_i}{k_B T}\right) \\ &= k_B \ln Z + \frac{E}{T}. \end{aligned} \tag{3.58}$$

Hence the Helmholtz free energy of the system is

$$F(T, V, N) = E - TS = -k_B T \ln Z. \tag{3.59}$$

Equation (3.59) provides the equation of state for a system maintained at constant T, V, and N, similar to Eq. (3.49) for a system maintained at constant E, V, and N.

3.4.2 Harmonic oscillator

As an example, we derive the Helmholtz free energy of a one-dimensional harmonic oscillator at temperature T. Consider a particle of mass m connected to a spring of stiffness k, as shown in Fig. 3.4.

The energy of the system depends on the position x and momentum p (not to be confused with pressure)

$$E(x, p) = \frac{1}{2} k x^2 + \frac{p^2}{2m}. \tag{3.60}$$

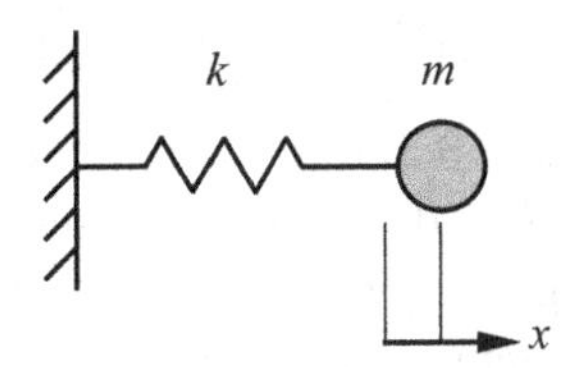

Figure 3.4. A harmonic oscillator.

To obtain the Helmholtz free energy, we need to first compute the partition function Z, which requires a summation over all possible microstates. Because the particle position x and momentum p are continuous variables, it may appear that there are an infinite number of possible microstates. However, *Heisenberg's uncertainty principle* in quantum mechanics tells us that one cannot distinguish two microstates whose differences in position and momentum satisfies

$$\Delta x \Delta p < h, \tag{3.61}$$

where h is Planck's constant. Therefore, the summation $\sum_i$ over microstates can be replaced by $\frac{1}{h}$ times the integral over x and p, i.e.

$$\begin{aligned} Z &= \frac{1}{h}\int_{-\infty}^{\infty} dx \int_{-\infty}^{\infty} dp \exp\left(-\frac{kx^2}{2k_B T} - \frac{p^2}{2mk_B T}\right) \\ &= \frac{1}{h}\sqrt{\frac{2\pi k_B T}{k}}\sqrt{2\pi m k_B T} \\ &= \frac{k_B T}{h\nu}, \end{aligned} \tag{3.62}$$

where $\nu \equiv \frac{1}{2\pi}\sqrt{k/m}$ is the vibrational frequency of the oscillator. Therefore, the Helmholtz free energy of the harmonic oscillator is

$$F = -k_B T \ln Z = -k_B T \ln \frac{k_B T}{h\nu}. \tag{3.63}$$

It can be shown that the (average) energy of the harmonic oscillator is

$$E = k_B T. \tag{3.64}$$

Therefore, the entropy of the harmonic oscillator is

$$S = \frac{E - F}{T} = k_B \left(1 + \ln \frac{k_B T}{h\nu}\right). \tag{3.65}$$

If the temperature is sufficiently high ($k_B T \gg h\nu$), then

$$S \approx k_B \ln \frac{k_B T}{h\nu}. \tag{3.66}$$

We will use this result in Section 5.2.1 to estimate the formation entropy of vacancies.

3.4.3 System in contact with thermostat and barostat

As mentioned in Section 3.2.5, it is often convenient to consider a system that is kept at a constant temperature T and pressure p. In statistical mechanics, this corresponds to allowing the system to exchange heat with a thermostat (with temperature T) and also allowing the system to exchange volume with a barostat (with pressure p). Because the number of particles N still remains constant, such a system is called a *closed* system. It can be shown that, at thermal equilibrium, the probability of finding the system in any microstate i is given by

$$p_i = \frac{1}{\Xi} \exp\left(-\frac{E_i + pV_i}{k_B T}\right), \tag{3.67}$$

$$\Xi \equiv \sum_i \exp\left(-\frac{E_i + pV_i}{k_B T}\right), \tag{3.68}$$

where V_i is the volume of microstate i, and Ξ is another *partition function*. The Gibbs free energy of the system is

$$G(T, p, N) = -k_B T \ln \Xi. \tag{3.69}$$

Equation (3.69) provides the equation of state for a system maintained at constant T, p, and N, similar to Eq. (3.59) for a system maintained at constant T, V, and N.

3.4.4 Free energy of configurational states

It is often the case that to describe the thermodynamic state of a system under consideration requires more variables than N, V, T (or N, p, T). For example, the system may be a crystal containing defects that can take several different configurations (such as those shown in Fig. 7.3). For simplicity, let us use $I = 1, 2, \ldots$ to represent the various configurations that the defects can take. We shall call the states labeled by index I *configurational states*, to distinguish from the microstates labeled by index i. The total number of configurational states (labeled by I) is much less than the total number of microstates (labeled by i), because many microstates may correspond to the same configurational state. For clarity, let us define Ω_I as the set of all microstates that is consistent with configurational state I.

We would like to find the probability p_I of finding the defect in certain configuration I at thermal equilibrium. Suppose the crystal is kept at constant volume V and constant temperature T and constant number of atoms N (i.e. a mechanically isolated system), then the probability of finding the system at any microstate i is given by Boltzmann's distribution, Eq. (3.55). The probability of finding the system at configurational state I is thus the sum of the probability of all states that are consistent with the configurational state I, i.e.

$$p_I = \sum_{i \in \Omega_I} p_i = \frac{1}{Z} \sum_{i \in \Omega_I} \exp\left(-\frac{E_i}{k_B T}\right). \tag{3.70}$$

Similar to Eq. (3.56), we define the partition function of configurational state I as the summation of the Boltzmann factor over Ω_I,

$$Z_I \equiv \sum_{i \in \Omega_I} \exp\left(-\frac{E_i}{k_B T}\right), \tag{3.71}$$

and the Helmholtz free energy of configurational state I similar to Eq. (3.59),

$$F_I \equiv -k_B T \ln Z_I. \tag{3.72}$$

Then the probability of finding the system in configurational state I is

$$p_I = \frac{1}{Z} \exp\left(-\frac{F_I}{k_B T}\right). \tag{3.73}$$

Equation (3.73) for configurational states is very similar to Boltzmann's distribution for microstates, Eq. (3.55), except that the energy E_i of microstate i is replaced by the Helmholtz free energy of configurational state I.

Analogously, if the system is kept at constant pressure p and constant temperature T (i.e. a closed system), then we can define the partition function of configurational state I as

$$\Xi_I \equiv \sum_{i \in \Omega_I} \exp\left(-\frac{E_i + pV_i}{k_B T}\right), \tag{3.74}$$

and the Gibbs free energy of configurational state I as

$$G_I \equiv -k_B T \ln \Xi_I. \tag{3.75}$$

Then the probability of finding the system in configurational state I is

$$p_I = \frac{1}{\Xi} \exp\left(-\frac{G_I}{k_B T}\right). \tag{3.76}$$

Equation (3.76) is very similar to Eq. (3.73), only with the Helmholtz free energy F_I of configurational state I replaced by its Gibbs free energy G_I. We will use Eq. (3.76) in Section 7.1 to estimate the rate at which a vacancy jumps to a neighboring lattice site.

3.5 Summary

The principles of thermodynamics and statistical mechanics are needed to derive the equilibrium distribution of point defects in a solid under external or internal stresses. While the laws of thermodynamics are often associated with the behavior of gases, they are much more general and govern the behaviors of atoms in solids as well.

The first law governs the various forms of energy that flow in and out of solids and, in particular, defines the change in internal energy, E, a state variable, as a sum of the work and heat entering that solid. For many problems the work in question involves volume changes under an external pressure. The second law introduces a second state variable, the entropy, S, which may only increase in isolated systems and reaches a maximum at equilibrium. The entropy plays a key role in defining the disordered state of solids at finite temperatures. While the entropy can be described by considering heat flow into a solid, its physical meaning was obscure until Boltzmann founded statistical mechanics and provided a statistical interpretation of entropy. With this interpretation, entropy became recognized as a measure of the microscopic disorder in macroscopic states of matter and is given by Boltzmann's famous entropy expression

$$S = k_B \ln \mathcal{W},$$

where k_B is Boltzmann's constant and W is the number of microscopic states consistent with the equilibrium macroscopic state. In addition to internal energy and entropy, both the number of atoms in the solid, N, and the volume, V, are also state variables. The thermodynamic state of a solid may then be expressed as the relation between these state variables $E(S, V, N)$, called an equation of state. According to this relation once three of the state variables are specified the fourth is completely determined.

Although $E(S, V, N)$ completely defines the thermodynamic state of a solid, it is not a very convenient form because the entropy cannot be easily measured or controlled. Instead, most experiments are performed under the condition of constant temperature T. For isothermal conditions the Helmholtz free energy, F, which is a unique function of temperature, volume, and number of atoms, can be defined as $F(T, V, N)$ and used instead of $E(S, V, N)$. The second law requires that the Helmholtz free energy F of a mechanically isolated system can only decrease with time and reaches the minimum at thermal equilibrium.

It is very difficult to keep the volume of a solid fixed when the temperature changes because very large stresses would be required to counteract the effect of thermal expansion. So it is helpful to transform the equation of state $E(S, V, N)$ to a form involving the enthalpy, $H(S, p, N)$, wherein the volume as a state variable is replaced by the pressure, which can be controlled more easily. This more convenient form of the equation of state can then be used in place of $E(S, V, N)$.

In practice, most experiments are conducted with both pressure and temperature kept constant. Therefore, it is still more convenient to transform the equation of state $E(S, V, N)$ to a form involving the Gibbs free energy, G, which is a function of pressure, temperature, and number of atoms, $G(T, p, N)$. The second law then states that under conditions of constant temperature and pressure the Gibbs free energy G of a closed system can only decrease with time and reaches the minimum at thermal equilibrium. For solids, this is the most convenient form of the equation of state. With this form, variations in the Gibbs free energy in moving from one equilibrium state to another can be expressed as

$$dG = -S\,dT + V\,dp + \mu\,dN$$

or

$$dG = \left(\frac{\partial G}{\partial T}\right)_{p,N} dT + \left(\frac{\partial G}{\partial p}\right)_{T,N} dp + \left(\frac{\partial G}{\partial N}\right)_{T,p} dN,$$

where μ is the chemical potential.

3.6 Exercise problems

3.1 What is the entropy change of 1 g of copper when the temperature changes from 0 °C to 100 °C? Assume that the specific heat of copper, $c_p = 0.386\ \text{J} \cdot \text{g}^{-1} \cdot \text{K}^{-1}$, stays constant.

3.2 The pressure–volume–temperature relation $p(V, T)$ of copper has been measured up to 8 GPa and 1100 K [15], where V is the volume of a unit cell of the FCC crystal. The data at

$T_0 = 300$ K can be described by the following (modified Birch–Murnaghan) relation

$$p(V, T_0) = \frac{3B_0}{2}\left[\left(\frac{V_0}{V}\right)^{7/3} - \left(\frac{V_0}{V}\right)^{5/3}\right], \tag{3.77}$$

where $V_0 = 47.1$ Å^3 is the volume of the unit cell under zero pressure at T_0. $B_0 = 140$ GPa is the bulk modulus at zero pressure.

(a) Find the expression for the Helmholtz free energy F of one unit cell of copper as a function of volume V, $F(V, T_0)$, relative to $F(V_0, T_0)$.
(b) Find the chemical potential μ of copper at $p = 8$ GPa at T_0 relative to the chemical potential μ_0 at $p = 0$ and T_0. Plot $\mu(p, T_0)$ for p from 0 to 8 GPa.

3.3 A collection of spherical gold nanoparticles with diameter $d = 10$ nm are dispersed in water in a container that is $h = 10$ cm deep. Find the ratio of the concentration of the nanoparticles at the bottom of the container over the concentration near the top surface of the water at $T = 300$ K, assuming that the distribution of nanoparticles has reached thermal equilibrium. The density of gold is 19.3 g $\cdot$ cm^{-3}, and the density of water is 1 g $\cdot$ cm^{-3}. How does the answer change for nanoparticles with diameter $d = 15$ nm?

3.4 A dislocation in a semiconductor has two possible core structures: A and B. Core structure A has a higher energy than core structure B by $E_A - E_B = 0.05$ eV (per repeat distance along the dislocation). Core structure A also has a higher (vibrational) entropy than core structure B by $S_A - S_B = 5\ k_B$. What is the relative probability of observing core structure A compared to core structure B at 300 K and 1000 K?

3.5 The vibrational entropy of a crystalline solid with N atoms can be approximated by that of $3N$ harmonic oscillators. The free energy of one harmonic oscillator depends on its oscillation frequency through Eq. (3.63). The oscillation frequency is given by $\nu = \frac{1}{2\pi}\sqrt{k/m}$, where k is the stiffness of the spring and m is the mass of the oscillator. Even though there are many vibrational modes in a solid, corresponding to a spectrum of vibrational frequencies, to a good approximation the effective stiffness k of these vibrational modes can be considered to be proportional to the elastic modulus of the solid times its lattice parameter.

Estimate the difference in the vibrational entropy per atom between FCC copper (Cu) and FCC aluminum (Al) at 300 K. Their Young's moduli are $E_{Cu} = 129$ GPa, $E_{Al} = 70$ GPa; their lattice parameters are $a_{Cu} = 3.61$ Å, $a_{Al} = 4.05$ Å; and their atomic masses are $m_{Cu} = 63.5$ g/mol, $m_{Al} = 27.0$ g/mol.

3.6 The vibrational frequencies ν_i in a solid change with the volume V of the solid. Owing to thermal expansion, an increasing temperature can cause a change in the vibrational frequencies and hence change the vibrational entropy as well. The Grüneisen parameter of an individual vibrational mode i is defined as

$$\gamma_i = -\frac{V}{\nu_i}\frac{\partial \nu_i}{\partial V}. \tag{3.78}$$

For simplicity, here we assume that γ_i are the same for all vibrational modes and equal to the Grüneisen parameter γ measured experimentally.

Estimate the change of vibrational entropy for one mole of gold as the temperature changes from 300 K to 400 K. Assume a constant Grüneisen parameter $\gamma = 2.6$ [16] and a constant linear thermal expansion coefficient $\alpha = 14 \times 10^{-6}\ \mathrm{K}^{-1}$.

3.7 Equation (3.63) gives the free energy of a harmonic oscillator in classical mechanics. In quantum mechanics, the energy levels of a harmonic oscillator with frequency ν are quantized and can only take the following values [17]: $E_i = (i + \frac{1}{2})h\nu$, for $i = 0, 1, 2, \ldots$.

(a) Obtain the partition function of a quantum harmonic oscillator, $Z \equiv \sum_{i=0}^{\infty} \mathrm{e}^{-\beta E_i}$, where $\beta \equiv 1/(k_B T)$.
(b) Obtain the Helmholtz free energy F, (average) energy E, and entropy S for the quantum harmonic oscillator.
(c) Plot F, E, S as a function of temperature, keeping the results from classical mechanics and quantum mechanics in the same figure.

3.8 Consider two particles with mass m connected to three springs with stiffness k, as shown in Fig. 3.5. The energy of the system can be written as

$$E(x_1, x_2, p_1, p_2) = \frac{1}{2}kx_1^2 + \frac{1}{2}k(x_2 - x_1)^2 + \frac{1}{2}kx_2^2 + \frac{p_1^2}{2m} + \frac{p_2^2}{2m}, \tag{3.79}$$

where x_1 and x_2 are displacements of the two particles from their equilibrium positions, and p_1 and p_2 are their momenta. Find the partition function Z and the Helmholtz free energy F of the system in classical mechanics. Note that the integral in the partition function can be obtained by performing a change of variables $\tilde{x}_1 \equiv (x_1 + x_2)/\sqrt{2}$, $\tilde{x}_2 \equiv (x_1 - x_2)/\sqrt{2}$.

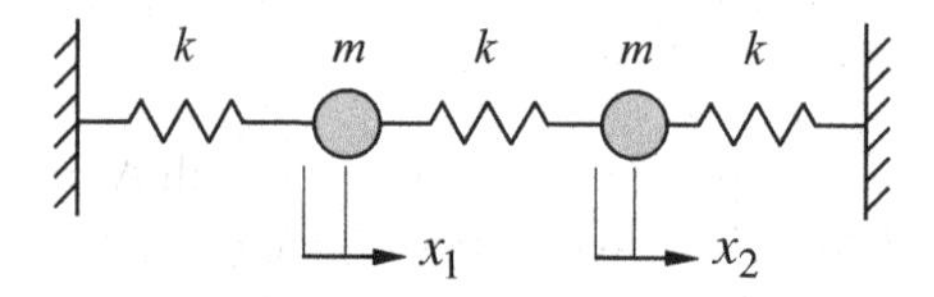

Figure 3.5. Two harmonic oscillators coupled together.

3.9 The surface stress can induce an internal stress within a nanoparticle or nanowire and this phenomenon is often called the *Gibbs–Thomson effect*. To estimate the magnitude of this effect, consider a long nanowire with a square cross section with size d, as shown in Fig. 3.6.

The Helmholtz free energy F of the nanowire includes the strain energy of the nanowire interior, $\frac{1}{2}\sigma\varepsilon L_0 d^2$, where $\sigma = E\varepsilon$ is the normal stress, $\varepsilon = (L - L_0)/L_0$ is the normal strain, L_0 is the equilibrium length of the nanowire if surface stress were not present, and E is Young's

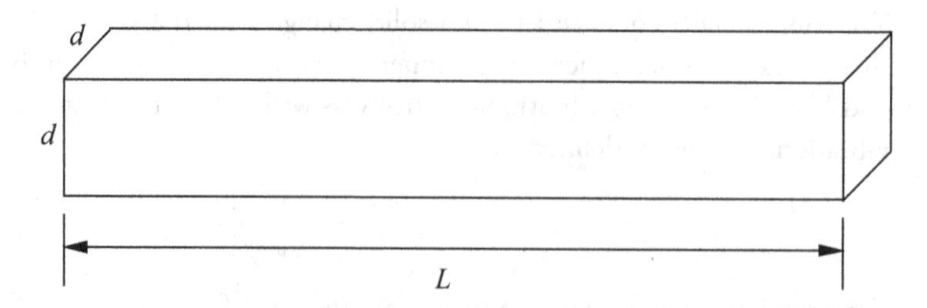

Figure 3.6. A single crystal nanowire with a square cross section.

modulus. We approximate the effect of the surface stress f by including a fA_s term in the Helmholtz free energy F, where $A_s = 4Ld$ is the area of the side surfaces of the nanowire. (The surface stress f has the same unit as the surface energy, and their values are often of the same order of magnitude. Although in reality the surface stress f is a function of strain ε, here we will assume f is a constant for simplicity.)

(a) Express the Helmholtz free energy F of the nanowire as a function of its length L.
(b) Obtain the equilibrium length L in the presence of the surface stress by minimizing F with respect to L.
(c) Obtain the stress σ and strain ε values inside a gold nanowire of width $d = 10$ nm, using a nominal value of $f = 1\ \mathrm{J \cdot m^{-2}}$ and $E = 80$ GPa for gold.

PART II

POINT DEFECTS

Point defects are imperfections in a crystal that are confined to atomic dimensions in all three directions. Depending on the chemical species involved, point defects can be considered as either *intrinsic* or *extrinsic*. Intrinsic point defects include vacancies, i.e. missing atoms, and self-interstitials, i.e. extra atoms having the same chemical species as the host crystal. Extrinsic point defects, on the other hand, are atoms having a different chemical species from the host crystal that they enter. Such point defects are often called impurity or solute atoms. Impurities usually refer to "unwanted" foreign atoms in a crystal, while solute atoms are often intentionally introduced into the crystal to alter its properties.

Point defects have a profound effect on the properties of engineering crystalline materials, either by themselves, or through their interactions with dislocations (line defects) and grain boundaries (planar defects). An example of the former situation is the semiconductor industry, whose success hinges on their ability to control the electronic properties of silicon by selective doping, through which transistors and integrated circuits can be made. An example of the latter situation is solid solution hardening, in which the elastic distortion around point defects allow them to interact with dislocations and alter the mechanical strength of the crystalline material.

In the following four chapters, we focus on the fundamental mechanics and thermodynamic principles that are needed to understand how point defects influence material properties. In Chapter 4, we study the stress and strain fields around point defects, using the elasticity theory introduced in Chapter 2. These results lead to an estimate of the energy cost of introducing point defects, as well as how point defects interact with each other and with other types of defects (e.g. dislocations) to be introduced later. In Chapter 5, the energy cost of introducing point defects is combined with the statistical thermodynamic principles of Chapter 3, to predict the concentration of point defects in a crystal at thermal equilibrium. It will be seen that, due to the entropic gain in allowing point defects, the equilibrium concentration of point defects in a crystal at finite temperature is never zero. In other words, point defects are ubiquitous! In Chapter 6, the coexistence of multiple types of point defects in the same crystal is discussed. In Chapter 7, we discuss how point defects move, which is necessary to reach the thermal equilibrium distribution. An emphasis is placed on vacancies, for which we show that an equilibrium distribution does not exist for a crystal subjected to deviatoric stress. The flow of vacancies under such conditions gives rise to the creep deformation of crystals at intermediate-to-high temperatures.

4 Point defect mechanics

A qualitative understanding of the behaviors of point defects can be established by considering atoms as hard spheres packed together to form the crystal. Crude as the hard sphere model may seem, it can be used to explain many of the observations made about point defects. In Section 4.1, we define the hard sphere radius of an atom and show its influence on the site preference of solute atoms. In Section 4.2, we use the hard sphere model to show the type of the distortions (spherically symmetric or not) in the host crystal around a solute atom. This allows us to explain why certain solutes have a much stronger solid solution hardening effect than others.

We then need to go beyond the hard sphere model in order to be more quantitative. In Section 4.3, we define the Seitz radius, which is more useful than the hard sphere radius for keeping track of the volume occupied by atoms of different kinds in solid solutions. We will see that atoms often appear to take on a different radius as a solute atom in another crystal compared to the radius it takes in its own crystal. In Section 4.4, we apply elasticity theory to predict the elastic fields around a solute atom. For simplicity, the size of the point defect is shrunk to zero and is modeled as force dipoles acting on a point in an elastic medium. In Section 4.5, a more realistic model is developed, in which the solute atom is modeled as an elastic sphere to be inserted into a hole inside an elastic medium. Elastic fields arise because the initial size of the sphere is larger than the initial size of the hole. Even though many atomistic and electronic details concerning point defects are ignored, the models developed in this chapter are increasingly more quantitative and can be used to explain a large number of behaviors of point defects.

4.1 Hard sphere model

4.1.1 Hard sphere radius

It is common to treat atoms in a crystal as undeformable spheres and to calculate the atomic sizes from the lattice parameters (measured using X-ray diffraction). We call this the hard sphere approach. As an example, let us consider a face-centered cubic (FCC) crystal made of touching

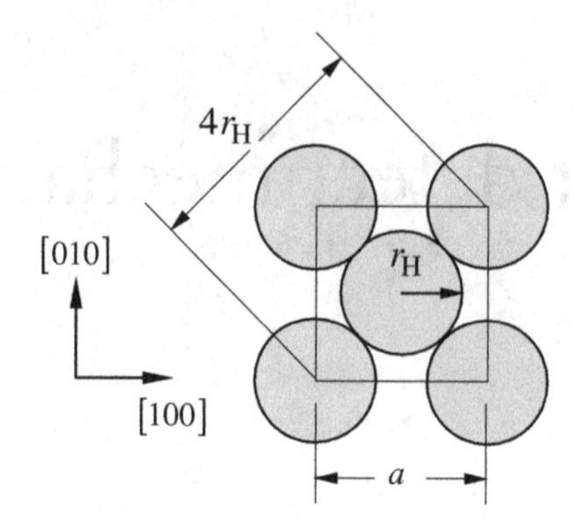

Figure 4.1. Face of an FCC unit cell showing atoms making contact along a face diagonal.

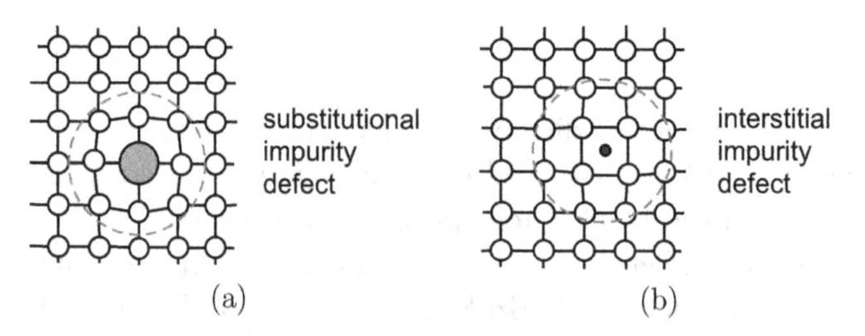

Figure 4.2. (a) Solute atom replaces solvent atom in a normal atomic site. (b) Solute atom fits into interstitial site (between the solvent atoms) if it is sufficiently small.

spheres of radius r_H. We wish to relate the hard sphere radius r_H with the lattice constant a of the FCC crystal.

Because in an FCC crystal the atoms are most closely spaced along the face diagonal, i.e. the $\langle 1\,1\,0 \rangle$ direction, the hard spheres make contact with each other along the face diagonal directions, as shown in Fig. 4.1. This leads to

$$a\sqrt{2} = 4r_H. \tag{4.1}$$

Hence the hard sphere radius is found to be

$$r_H = \frac{a\sqrt{2}}{4} = 0.354a \tag{4.2}$$

for an FCC crystal. A similar relation between the hard sphere radius and lattice constant can also be developed for a BCC crystal (see Exercise problem 4.1).

4.1.2 Substitutional and interstitial defects

It is natural to expect that atoms of different chemical species should have different radii in the hard sphere model. Therefore, when a solute atom is introduced into a crystal, the size mismatch between the atomic species leads to distortions around the solute. There are two ways a solute atom can get incorporated into a crystal: *substitutional* and *interstitial*. When solute atoms are comparable in size to the solvent atoms (the host atoms), they typically reside on substitutional sites, i.e. they replace the host atoms in the crystal lattice, as shown in Fig. 4.2a.

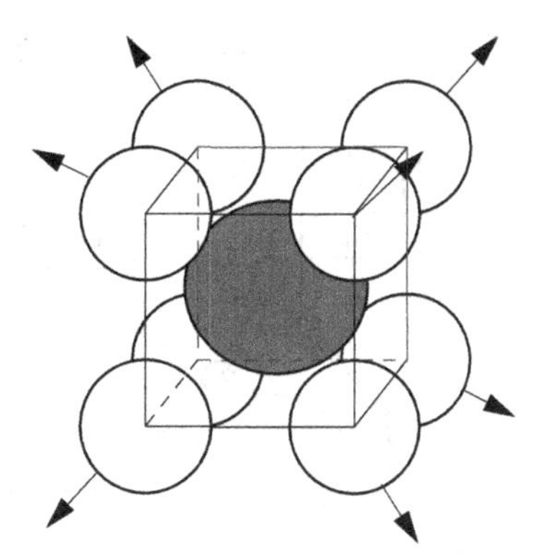

Figure 4.3. Substitutional solute defect produces spherically symmetric distortion. Here the host atom in a BCC crystal is replaced by a solute atom (gray) of a larger radius.

Roughly speaking, if the solute radius r_{solute} is 85% of the radius of the solvent (i.e. host) atoms r_{host} or greater, one can expect the solutes to reside on substitutional sites. The radius of an atom will be discussed in more detail in Section 4.3.

By comparison, if solute atoms are substantially smaller that the host (solvent) atoms then they are likely to occupy interstitial sites, i.e. in the gap between the host atoms, as shown in Fig. 4.2b. As a rule of thumb: if $r_{solute} < 0.85 r_{host}$, then interstitial sites are occupied. Small atoms such as C, N, O, B, He, and H are typically interstitial solutes in metals.

4.2 Symmetry of distortions about solutes

By considering atoms as spheres packed together to form a crystal, we are able to make several general remarks on the types of sites a solute atom can occupy in a host crystal, and the symmetry of the distortions around them.

4.2.1 Substitutional defects

In crystal structures of high symmetry (FCC, BCC, and HCP) substitutional defects create essentially spherically symmetric distortions. Consider a substitutional solute atom in a BCC lattice, shown in Fig. 4.3. The defect has eight nearest neighbors that are each displaced by the same amount, creating a spherically symmetric distortion in the far field.

For isotropic elastic solids, the displacements far from the defect are purely radial and we have a spherically symmetric distortion exactly. The explicit form of the displacement field, which is characterized by a misfit volume, will be given in Section 4.5. The far-field symmetry can be altered if the elastic properties of the surrounding crystal lattice are highly anisotropic.

4.2.2 Interstitial defects

For interstitial defects, the nature of the distortion depends on the symmetry of the site on which the defect is located. The distortion around high symmetry sites is spherically symmetric, while the distortion around low symmetry sites is not.

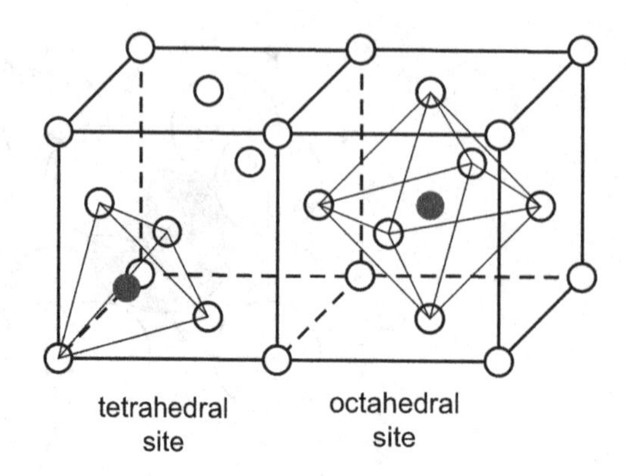

Figure 4.4. Tetrahedral and octahedral sites in an FCC crystal.

High symmetry sites

There are two types of symmetric sites for interstitial solute atoms in close-packed crystal structures (FCC and HCP): *tetrahedral sites* and *octahedral sites*, as shown in Fig. 4.4. The tetrahedral site is at the center of a regular tetrahedron (with four faces) formed by four nearest neighbor host atoms. The octahedral site is at the center of a regular octahedron (with eight faces) surrounded by six nearest neighbor host atoms. There are eight tetrahedral sites and four octahedral sites per FCC unit cell. Figure 4.4 shows the high symmetry of these sites in an FCC lattice. The far-field distortions are spherically symmetric for these symmetric sites.

We note that each tetrahedron formed by the FCC lattice points shares a (triangular) face with four octahedrons, and each octahedron shares a face with eight tetrahedrons. Furthermore, all the tetrahedrons and octahedrons formed by the FCC lattice points fill the entire space with no gaps. In other words, the entire space can be partitioned into regions of tetrahedrons and octahedrons as defined above. This geometric fact will be useful when discussing the Thompson tetrahedron and slip pyramid of FCC crystals in Chapter 11.

Using the hard sphere model, we can estimate the "free volume" at the tetrahedral and octahedral sites, i.e. the size of the solute atom that can be inserted into the tetrahedral and the octahedral site without introducing significant distortion. The critical radius for a small sphere to make touching contact with surrounding spheres for a tetrahedral site is

$$r_{\text{crit}}^{\text{tetra-FCC}} = 0.225 r_{\text{H}} \tag{4.3}$$

and the critical radius for an octahedral site is

$$r_{\text{crit}}^{\text{octa-FCC}} = 0.414 r_{\text{H}}. \tag{4.4}$$

In other words, the octahedral site has "more free volume" than the tetrahedral site in an FCC crystal, because it has a larger critical radius.

The critical radius given in Eq. (4.4) is easy to derive, as will be shown below. Consider solvent atoms as spheres of radius r_{H} and an interstitial atom as a sphere of radius r_{c} residing in a octahedral site and making contact with the surrounding atoms (spheres), as shown in Fig. 4.5. The contact between solute and host atoms is made along the cube edge direction in an FCC lattice, so that

$$a = 2r_{\text{c}} + 2r_{\text{H}}. \tag{4.5}$$

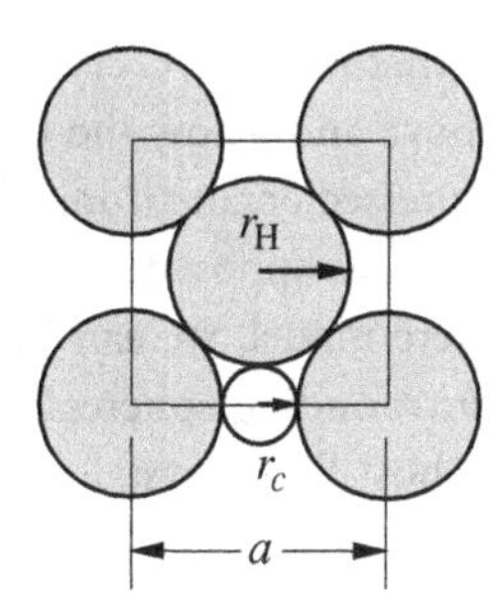

Figure 4.5. Face of an FCC unit cell showing a solute atom at the octahedral site touching the host atoms along cube edge.

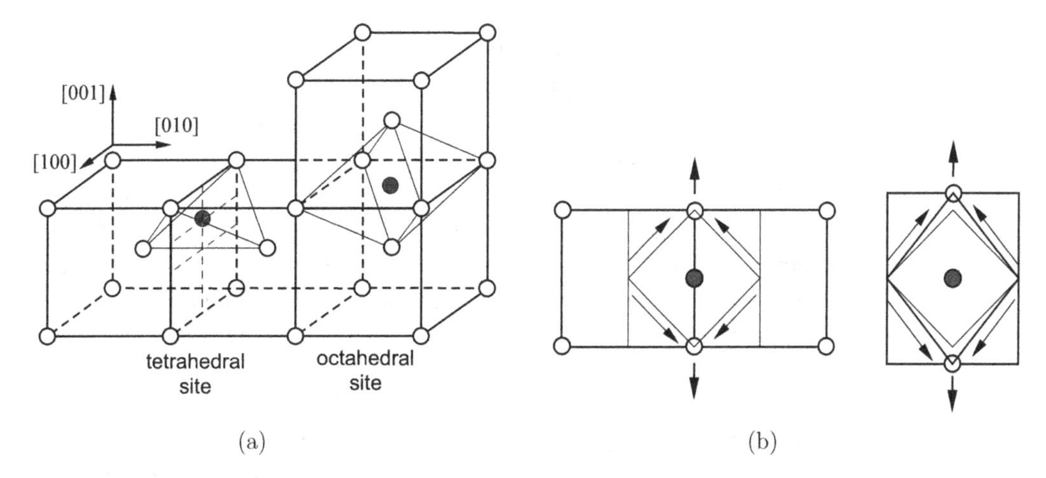

Figure 4.6. (a) Tetrahedral and octahedral sites in a BCC crystal. (b) Tetragonal (shear) distortion associated with a non-symmetric interstitial defect at the octahedral site. Left: before distortion. Right: after distortion.

Combining this relation and Eq. (4.2), we have

$$r_c = r_H\left(\sqrt{2} - 1\right) = 0.414 r_H. \tag{4.6}$$

The radii of solute atoms (see Section 4.3) are usually larger than the critical radius r_c at the interstitial sites. This leads to a distortion of the host lattice. The far-field distortion around the high symmetry sites considered above is spherically symmetric.

Low symmetry sites

Tetrahedral and octahedral sites also exist in BCC crystals, as shown in Fig. 4.6a. However, the sites shown are not a regular tetrahedron and a regular octahedron. Instead, they can be obtained from a regular tetrahedron and a regular octahedron (such as those shown in Fig. 4.4) by compressing them along the [0 0 1] direction. As a result, the tetrahedral and octahedral sites in a BCC crystal have lower symmetry than those in an FCC crystal. The far-field distortions are not spherically symmetric for these sites in a BCC crystal.

The tetrahedral site in Fig. 4.6a is at an equal distance from the four corners of the compressed tetrahedron. In other words, the tetrahedral site in a BCC crystal still has a coordination number of 4, which is the same as that in an FCC crystal. On the other hand, the octahedral site is closer to two of the corners (above and below the (0 0 1) plane) than the other four corners (on the (0 0 1) plane) of the compressed octahedron. Therefore, the octahedral site in a BCC crystal has a coordination number of 2, as opposed to 6 in an FCC crystal.

By using the hard sphere model, we can find the critical radius for a small sphere to make touching contact with surrounding spheres at tetrahedral and octahedral sites. Let r_H be the hard sphere radius of the host BCC crystal. The critical radius for a tetrahedral site is

$$r_{crit}^{tetra\text{-}BCC} = 0.291 r_H \tag{4.7}$$

and the critical radius for an octahedral site is

$$r_{crit}^{octa\text{-}BCC} = 0.155 r_H. \tag{4.8}$$

Therefore, contrary to the case of an FCC crystal, the free volume at the octahedral site is much smaller than that at the tetrahedral site.

The larger free volume in the tetrahedral site is consistent with the experimental evidence favoring the tetrahedral site occupancy of hydrogen in Fe and other BCC crystals [18]. However, interstitial atoms such as C and N in BCC metals prefer to occupy the octahedral site, despite its smaller free volume. We have seen that the octahedral site has a coordination number of 2, meaning that two host atoms (the top and bottom vertices in Fig. 4.6a) are closer to the interstitial than the other four surrounding host atoms. One reason for C atoms to prefer the octahedral site over the largest available space at the tetrahedral site is that Fe_2C is a relatively stable compound and the C atom forms a "molecule" of that composition by residing in the octahedral site.

Because the octahedral site has a coordination number of 2, the two nearest atoms are displaced more than the others and a tetragonal distortion is created. Figure 4.6b shows that such a tetragonal deformation caused by the interstitial defect is equivalent to a volumetric + shear distortion. Such interstitial defects are important because they interact more strongly with dislocations and produce a stronger solid solution hardening effect than interstitials producing a spherically symmetric strain field.

The distortional field around the tetrahedral site is also non-symmetrical, i.e. it contains both volumetric and shear distortion. However, the asymmetry is less pronounced than that around the octahedral site.

4.2.3 Interactions with dislocations

The symmetry of defects is important because it largely determines the way in which point defects interact elastically with dislocations. Spherically symmetric distortions cannot interact strongly with screw dislocations because the elastic fields of these dislocations is nearly pure shear (see Chapter 9). But tetragonal or shear distortions can interact strongly with both edge and screw components of dislocations. Figure 4.7a illustrates the interaction between the strain fields of an edge dislocation with point defects with both spherical and non-symmetric distortions. (The interaction between point defects producing spherically symmetric distortions and edge dislocations will be discussed in more detail in Section 12.5.) Figure 4.7b illustrates the interaction between the strain fields of a screw dislocation with point defects with

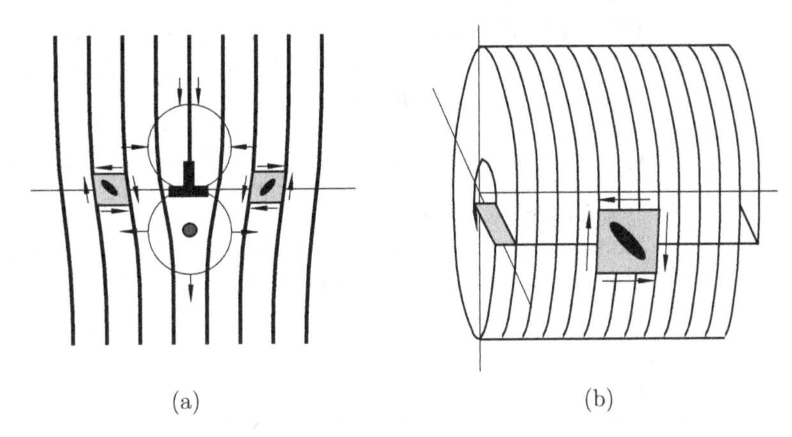

Figure 4.7. (a) Point defects interacting with the strain fields about an edge dislocation. (b) A tetragonal defect interacting with the elastic field of a screw dislocation. Illustrated are the locations and orientations preferred by the various types of point defects.

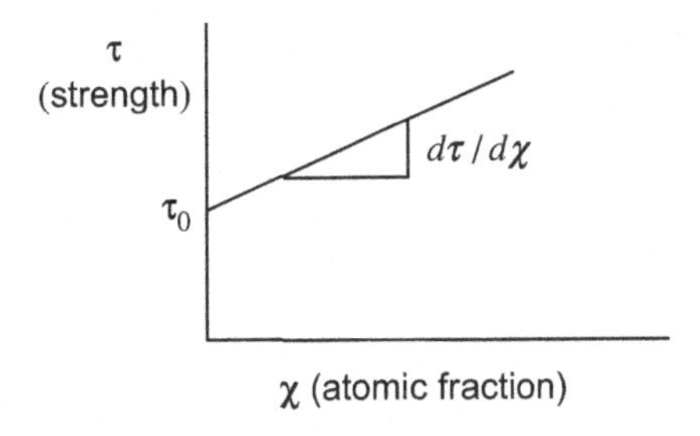

Figure 4.8. Strength versus solute concentration.

non-symmetric distortions. The consequence is that point defects with non-symmetric distortions are much more potent in solid solution hardening (see Section 4.2.4).

4.2.4 Solid solution hardening

The potency for solid solution hardening is characterized by $d\tau/d\chi$, where τ is the strength (the stress required for dislocation motion and plastic flow) and χ is the solute fraction, i.e. the ratio of the number of solute atoms over the total number of atoms (see Section 4.2.5). Figure 4.8 shows a schematic of τ as a function of χ.

Table 4.1 lists the potency of solute hardening for various host materials and point defects. It can be seen that the potency of solute hardening for non-symmetric point defects is about one-to-two orders of magnitude higher than for spherically symmetric point defects.

4.2.5 Measures of point defect density

In addition to the atomic fraction χ, there are several alternative ways to specify the density of point defects. Here we give the definitions of some of the commonly used alternatives, and discuss their relationships.

Table 4.1. Effects of different defects in hardening crystals [19]: $d\tau/d\chi$ in terms of the shear modulus μ, at a point defect fraction of 100 ppm (parts per million), i.e. $\chi = 10^{-4}$.

Host material	Defect	Typical hardening
Spherically symmetric point defects		
Aluminum	Substitutional atoms	$\mu/10$
Copper	Substitutional atoms	$\mu/20$
Iron	Substitutional atoms	$\mu/16$
Nickel	Interstitial carbon	$\mu/10$
KCl	F-centers	$\mu/2.5$
NaCl	Monovalent substitutional ions	$\mu/100$
Non-symmetric point defects		
Aluminum	Vacancy disks	2μ
Copper (irradiated)	Self-interstitials	9μ
Iron	Interstitial carbon	5μ
LiF (irradiated)	Interstitial fluorine	5μ
Niobium	Interstitial nitrogen	2μ
KCl	Interstitial chlorine	7μ
NaCl	Divalent substitutional ions	2μ

Consider N_B solute atoms of type B dissolved into a host crystal containing N_A solvent atoms of type A. The atomic fraction χ of the solute is defined as

$$\chi \equiv \frac{N_B}{N_A + N_B}. \tag{4.9}$$

The atomic fraction χ is equivalent to molar fraction and is dimensionless. Because χ is usually very small, it is often specified in units of atomic percent (at.%) or parts per million (ppm). For example, 1 at.% means $\chi = 10^{-2}$, and 1 ppm means $\chi = 10^{-6}$.

It is also common to specify the solute density by a weight fraction defined as

$$\chi_w \equiv \frac{m_B N_B}{m_A N_A + m_B N_B}, \tag{4.10}$$

where m_A and m_B are the mass of an A atom and the mass of a B atom, respectively. The weight fraction is also dimensionless and is often specified in units of weight percent (wt.%).

While the atomic fraction χ can be applied to both substitutional and interstitial solute atoms, it is sometimes useful to introduce a site fraction χ_i for interstitial solute atoms

$$\chi_i \equiv \frac{N_B}{N_s}, \tag{4.11}$$

where

$$N_s = z_i N_A \tag{4.12}$$

is the number of available interstitial sites, and z_i is the number of interstitial sites per atom in the host crystal. For example, we can consider octahedral sites in an FCC crystal. There are four octahedral sites per unit cell, and there are four host atoms per unit cell. Therefore, $z_i = 1$ for octahedral sites in an FCC crystal. However, there are eight tetrahedral sites per FCC unit cell, so that $z_i = 2$ for tetrahedral sites in an FCC crystal.

The concentration of point defects is defined as the number of defects per unit volume,

$$c \equiv \frac{N_B}{V}, \tag{4.13}$$

where V is the total volume of the crystal containing both A and B atoms. The concentration c has the dimension of inverse volume and has the unit of m^{-3} in the SI unit system. Define

$$\Omega \equiv \frac{V}{N_A + N_B} \tag{4.14}$$

as the average volume per atom for the solid solution. In Section 4.3.2, we shall see that Ω is a function of χ. Then we have

$$c = \frac{N_B}{N_A + N_B} \frac{N_A + N_B}{V} = \frac{\chi}{\Omega}. \tag{4.15}$$

In the limit of very low atomic fraction ($\chi \ll 1$), we have $c \approx \chi / \Omega_A$, where Ω_A is the average volume per atom in a pure crystal consisting of atom A.

4.3 Atomic size factors

Because the distortions in imperfect crystals depend on the relative sizes of the atoms, we need a way to determine the atomic sizes in crystals.

4.3.1 Seitz radius

The hard sphere radius r_H defined in Section 4.1 is clear and unambiguous for pure components (metals) but is not very useful for solid solutions where atoms of different sizes are present. In addition, the hard sphere radius is not very useful if we wish to keep track of the volume occupied by atoms of different kinds, because the volume of the hard spheres does not include the volume in the interstices (gaps) between the atoms.

For atomic sizes in solid solutions and for keeping track of volume it is more common to use the Seitz radius, which is defined as the radius of the hypothetical sphere that would account for all of the volume in the crystal. Start with the average atomic volume (again using the FCC lattice for illustration):

$$\Omega = \frac{\text{volume of unit cell}}{\text{atoms per unit cell}} = \frac{a^3}{4} \quad \text{(for an FCC crystal).} \tag{4.16}$$

We then imagine a sphere of radius r_s such that

$$\frac{4}{3}\pi r_s^3 = \Omega, \tag{4.17}$$

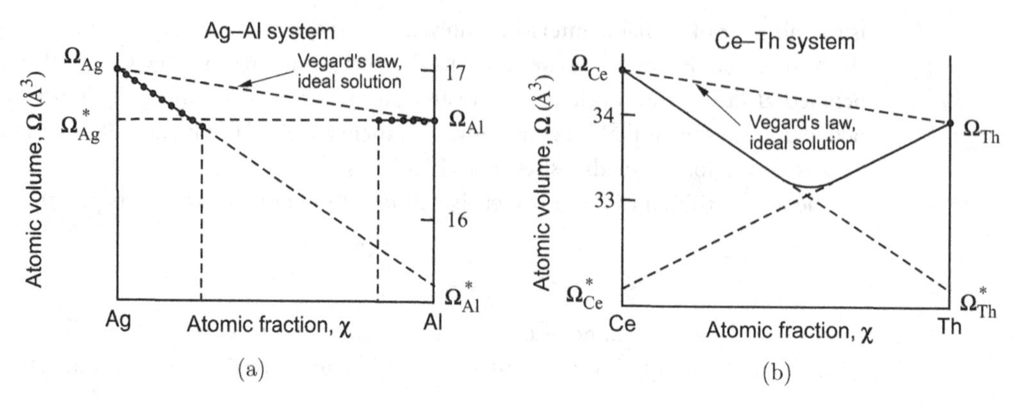

Figure 4.9. King graph for (a) Ag–Al and (b) Ce–Th alloy systems [20].

so that the Seitz radius is

$$r_s \equiv \left(\frac{3\Omega}{4\pi}\right)^{1/3} = a\left(\frac{3}{16\pi}\right)^{1/3} = 0.391a. \tag{4.18}$$

Comparing Eqs. (4.18) and (4.2), we can see that $r_s > r_H$. This is because the Seitz radius accounts for all space while the hard sphere radius does not.

The hard sphere radius is based on the lattice parameter of the pure substance. The same is true for the Seitz radius if Ω is based on the lattice parameter of the pure substance. These estimates are not very useful for solid solutions, where the atoms are of different sizes.

4.3.2 Atomic sizes in substitutional solid solutions

Consider as an example the atomic size of Si when it is dissolved in Fe (transformer core material) versus Si in single crystalline Si (semiconductor wafer material). We note that Fe is metallically bonded and has the BCC structure, while Si is covalently bonded and has the diamond cubic structure. The atomic sizes of Si in these two environments are the following

$$\begin{aligned} r_s^{Si}(\text{in Si}) &= 1.68\,\text{Å}, \\ r_s^{Si}(\text{in Fe}) &= 1.37\,\text{Å}. \end{aligned} \tag{4.19}$$

Thus the size of a Si atom in a Si crystal is quite different compared to the size of Si when it is dissolved into Fe.

The atomic radius of Si in Fe given in Eq. (4.19) is obtained using the method described by King [20]. The method is based on experimental measurements of the lattice parameters of substitutional solid solutions as a function of composition. Consider such measurements in the Ag–Al alloy system expressed as the average atomic volume Ω versus the atomic fraction χ. Here $\Omega(\chi)$ is computed from the measured lattice parameter $a(\chi)$ as $\Omega(\chi) = [a(\chi)]^3/4$, because Ag–Al has the FCC crystal structure. The data for this alloy system are shown in Fig. 4.9a, where

Ω_{Ag} = atomic volume of Ag in Ag,
Ω_{Al} = atomic volume of Al in Al,
Ω^*_{Al} = effective volume of Al when dissolved into Ag,
Ω^*_{Ag} = effective volume of Ag when dissolved into Al.

This $\Omega(\chi)$ curve has two branches not connected to each other, because the Ag–Al system is not completely miscible, so that there is a range of χ in which no data exist.

It is useful to contrast these behaviors with the predictions of Vegard's law, which holds for ideal solutions. According to this law, the average atomic volume of a substitutional solid solution of an A–B alloy follows a simple rule of mixtures,

$$\Omega(\chi) = (1 - \chi)\Omega_A + \chi\Omega_B, \tag{4.20}$$

which is a linear relation, with χ being the atomic fraction of B, as shown in Fig. 4.9. Vegard's law often holds for completely miscible systems when the electronic structure does not change with composition. For example, NaCl–KCl ionic solid solutions follow Vegard's law because the ionic structure does not change with composition. Other material systems that closely obey Vegard's law are Si–Ge, GaAs–InAs, and InP–InAs, all of which have covalent bonding that does not change with composition.

However, for most metals, particularly metallic alloys with different valences, the electronic structure changes with composition and this causes the atomic volume to change with composition in a non-ideal way. Still, an effective atomic volume of the solute (B) can be found by extrapolating the left branch of the $\Omega(\chi)$ curve (where χ is close to 0) to the limit of $\chi = 1$, even if the metals are not completely miscible. This procedure then gives the effective volume of the solute atom, Ω_B^*. As shown in Fig. 4.9a (where A = Ag and B = Al), the effective atomic volume of the solute, Ω_B^*, differs considerably from the atomic volume of that same species in its pure state, Ω_B. The average atomic volume of the solid solution, specifically the left branch of the $\Omega(\chi)$ curve, is as if Vegard's law is applied to the solvent, A with atomic volume Ω_A, and the effective solute B with atomic volume Ω_B^*. Therefore, instead of Eq. (4.20), we have

$$\Omega(\chi) = (1 - \chi)\Omega_A + \chi\Omega_B^* \qquad (\chi \text{ close to } 0). \tag{4.21}$$

When we consider the right branch of the $\Omega(\chi)$ curve, where χ is close to 1, B becomes the solvent and A becomes the solute. An effective atomic volume of the solute, Ω_A^*, can be found by extrapolating the right branch of the $\Omega(\chi)$ curve to the limit of $\chi = 0$. Then the right branch of the $\Omega(\chi)$ curve is as if Vegard's law is applied to the solvent, B, with atomic volume Ω_B, and the effective solute A with atomic volume Ω_A^*, i.e.

$$\Omega(\chi) = (1 - \chi)\Omega_A^* + \chi\Omega_B \qquad (\chi \text{ close to } 1). \tag{4.22}$$

Therefore, to the extent that Eqs. (4.21) and (4.22) are reasonable approximations, the volume–composition relation of an A–B alloy can be characterized by four parameters: Ω_A, Ω_A^*, Ω_B, and Ω_B^*. These quantities can be obtained from the King tables to be described in Section 4.3.3.

Even for completely miscible systems, like Ce–Th, Vegard's law may be violated, as shown in Fig. 4.9b. It is interesting to observe that for the Ce–Th system, Ce-rich solutions behave as a mixture of Ce atoms of volume Ω_{Ce} and Th atoms of volume Ω_{Th}^*, while for Th-rich solutions the alloy system behaves as a mixture of Ce atoms of volume Ω_{Ce}^* and Th atoms of volume Ω_{Th}. The transition from one kind of behavior to the other is surprisingly sharp.

We note that not all binary alloys satisfy the volume–composition relation discussed above. In intermetallic compounds, varying the concentration of one element may induce vacancies, which may be described as a new species with its own effective volume (see Chapter 5). The concept of effective volume can still help us understand the volume–composition relation in these materials, as long as we know what types of defects are present at each concentration.

4.3.3 Volume size factor

The King tables in Appendix A allow us to reconstruct the volume–composition curves for A–B binary alloys that satisfy Eqs. (4.21) and (4.22). First, the atomic volume of each species in its pure state is given in Table A.1, from which we can look up Ω_A and Ω_B. Second, to obtain the effective volume Ω_B^* of B as a solute atom in solvent A, we look up the so-called *volume size factor* defined as follows:

$$\Omega_{sf} \equiv \frac{\Omega_B^* - \Omega_A}{\Omega_A}. \tag{4.23}$$

The volume size factor is simply a fractional change in volume. This quantity is tabulated in Table A.2. For example, the first block in Table A.2 shows the measured values of Ω_{sf} for various elements dissolved in Ag. It can be seen that the effective volume of Al dissolved in Ag is about 9% smaller than Ag.

Given Ω_{sf}, the effective volume Ω_B^* of B as a solute atom in solvent A can be obtained from

$$\Omega_B^* = \Omega_A(1 + \Omega_{sf}). \tag{4.24}$$

From the data obtained so far, we can construct the left branch of the $\Omega(\chi)$ curve using Eq. (4.21). Table A.2 also gives the maximum atomic fraction, χ_{max}, for which Eq. (4.21) is valid.

Similarly, we can look up the volume size factor for A as a solute atom in solvent B, in order to obtain Ω_A^*, and to construct the right branch of the $\Omega(\chi)$ curve using Eq. (4.22).

In the following, we give a graphical interpretation of the volume size factor Ω_{sf}. Starting from the average atomic volume given in Eq. (4.21),

$$\Omega(\chi) = (1 - \chi)\Omega_A + \chi\Omega_B^*,$$

and taking derivatives with respect to the atomic fraction χ, we have

$$\frac{d\Omega(\chi)}{d\chi} = \Omega_B^* - \Omega_A, \tag{4.25}$$

$$\Omega_{sf} = \frac{1}{\Omega_A}\frac{d\Omega(\chi)}{d\chi}. \tag{4.26}$$

This shows how Ω_{sf} is related to the slope of the $\Omega(\chi)$ curves in Fig. 4.9.

4.3.4 Linear size factor

It is sometimes useful to define linear size factors. Because the linear dimension of an atom can be expressed as the Seitz radius, Eq. (4.18),

$$r_s \equiv \left(\frac{3\Omega}{4\pi}\right)^{1/3},$$

a linear size factor may be defined as

$$l_{sf} \equiv \frac{r_{sB}^* - r_{sA}}{r_{sA}} = \frac{r_{sB}^*}{r_{sA}} - 1 = \left(\frac{\Omega_B^*}{\Omega_A}\right)^{1/3} - 1. \tag{4.27}$$

This is, of course, related to the volume size factor as follows:

$$l_{sf} = (1+\Omega_{sf})^{1/3} - 1 = \left(1 + \frac{1}{3}\Omega_{sf} + \cdots\right) - 1 \approx \frac{1}{3}\Omega_{sf}. \tag{4.28}$$

The relation (4.28) is familiar to us in a number of ways. For example, the linear thermal expansion coefficient is also one third of the volume expansion coefficient. The volume and linear size factors for various substitutional solid solutions are listed in Table A.2.

4.3.5 Effective volume of interstitial point defects

We note that the above discussions about atomic size factors, e.g. Eq. (4.21), are meant for substitutional point defects only. If, on the other hand, B atoms enter the A-rich crystal as interstitial point defects, it is customary to consider the volume of the crystal per number of A atoms $\tilde{\Omega} \equiv V/N_A$ (as determined from lattice spacing measurements [21]). $\tilde{\Omega}$ often has a linear relationship with the number of solute atoms,

$$\tilde{\Omega} = \Omega_A + \frac{N_B}{N_A}\Omega_B^{\dagger}, \tag{4.29}$$

where $\Omega_B^{\dagger}$ can be interpreted as the effective volume of the B atoms as interstitial point defects. Therefore $\Omega_B^{\dagger}$ is the slope of the curve when $\tilde{\Omega}$ is plotted against N_B/N_A,

$$\Omega_B^{\dagger} = \frac{d\tilde{\Omega}}{d(N_B/N_A)}. \tag{4.30}$$

We note that N_B/N_A is different from the atomic fraction, which is $\chi = N_B/(N_A + N_B)$, although the difference vanishes in the small atomic fraction limit. N_B/N_A is also different from the occupancy fraction of interstitial sites, $\chi_i = N_B/N_s$, if $N_s \neq N_A$ (see Section 4.2.5).

Given the effective volume $\Omega_B^{\dagger}$ of interstitial solute atoms, a volume size factor can be defined as [22]

$$\Omega_{sf} \equiv \frac{\Omega_B^{\dagger}}{\Omega_A} = \frac{1}{\Omega_A}\frac{d\tilde{\Omega}}{d(N_B/N_A)} \tag{4.31}$$

and a linear size factor can be defined as $l_{sf} \equiv (1+\Omega_{sf})^{1/3} - 1$. The volume and linear size factors for several interstitial solid solutions are listed in Table A.3.

4.4 Elastic fields of atomic point defects

As discussed in Section 4.2, the symmetry of the site determines the nature of the displacement field in the surrounding lattice for a given point defect. To give another example, Fig. 4.10a shows a substitutional atom in a simple cubic crystal, producing a spherically symmetric displacement field. Figure 4.10(b), on the other hand, shows an interstitial atom at the cube-edge site in a simple cubic crystal. The low symmetry of this site leads to large displacement at the two nearest host atoms. The long-range displacement field of this defect has both dilatational and shear components.

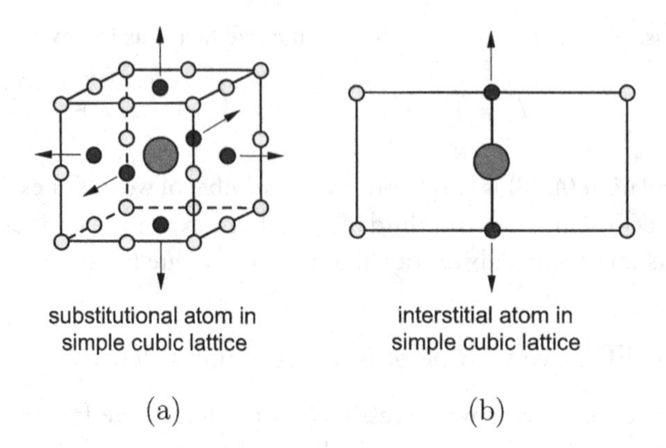

Figure 4.10. Symmetric and non-symmetric point defects in crystals.

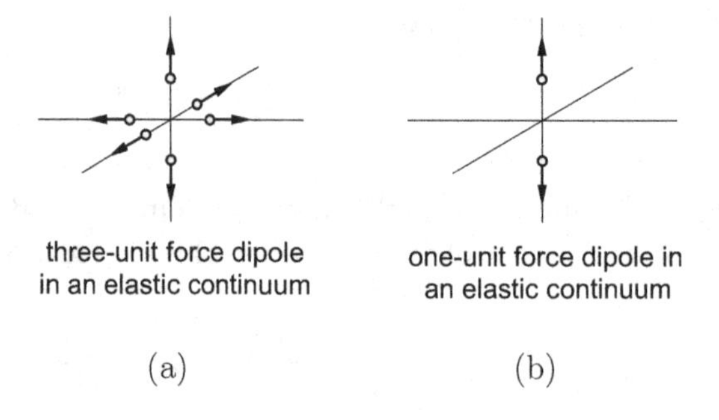

Figure 4.11. Force dipoles in an elastic continuum as models for point defects.

In this section, we apply the theory of elasticity (see Chapter 2) to obtain a more detailed description of the long-range displacement, strain, and stress fields around these defects. For simplicity, the size of the defects will be assumed to be infinitesimal, so that the predictions are valid only at distances sufficiently far away from the defect. The symmetric defect will be modeled as three mutually perpendicular force dipoles in an isotropic elastic continuum, as shown in Fig. 4.11a, and the non-symmetric or tetragonal defect will be modeled as a single force dipole, as shown in Fig. 4.11b.

4.4.1 Spherically symmetric point defect

We have argued that the displacement field of a point defect at a high symmetry site has spherical symmetry, as illustrated in Fig. 4.12. Thus we may guess that the displacement field should be of the form

$$\begin{aligned} u_r &= Ar^{-n}, \\ u_\theta &= u_\varphi = 0, \end{aligned} \tag{4.32}$$

where $n > 0$, that is, the purely radial displacements fall off with increasing radial distance from the defect and the transverse displacements are zero. But what is the value of n?

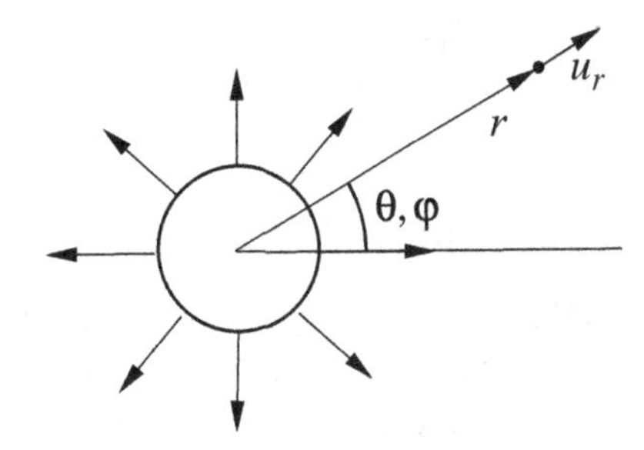

Figure 4.12. Spherical displacement field.

There are various ways to determine the value of n. Below we show that $n = 2$, hence we have the following.

$$\boxed{u_r = Ar^{-2}, \quad u_\theta = u_\varphi = 0} \tag{4.33}$$

Starting from Eq. (4.32) and using Eq. (2.28), we can show that the only non-zero strain components are

$$\begin{aligned} \varepsilon_{rr} &= \frac{\partial u_r}{\partial r} = -nAr^{-n-1}, \\ \varepsilon_{\varphi\varphi} = \varepsilon_{\theta\theta} &= \frac{u_r}{r} = Ar^{-n-1}. \end{aligned} \tag{4.34}$$

From the generalized Hooke's law, the only non-zero stress components are

$$\begin{aligned} \sigma_{rr} &= 2\mu\varepsilon_{rr} + \lambda(\varepsilon_{rr} + \varepsilon_{\varphi\varphi} + \varepsilon_{\theta\theta}) \\ &= [-2\mu n + \lambda(2 - n)]Ar^{-n-1}, \\ \sigma_{\varphi\varphi} = \sigma_{\theta\theta} &= 2\mu\varepsilon_{\varphi\varphi} + \lambda(\varepsilon_{rr} + \varepsilon_{\varphi\varphi} + \varepsilon_{\theta\theta}) \\ &= [2\mu + \lambda(2 - n)]Ar^{-n-1}. \end{aligned} \tag{4.35}$$

Given that these are the only non-zero stress components, and considering the region away from the point defect where body force is zero, the equilibrium condition in Eq. (2.57) simplifies to

$$\frac{\partial \sigma_{rr}}{\partial r} + \frac{1}{r}(2\sigma_{rr} - \sigma_{\varphi\varphi} - \sigma_{\theta\theta}) = 0. \tag{4.36}$$

Substituting Eq. (4.35) into the equilibrium condition, and after simplification, we obtain

$$(\lambda + 2\mu)(n - 2)(n + 1)Ar^{-n-2} = 0. \tag{4.37}$$

Because $n > 0$ (so that u_r decreases with increasing r), we must have $n = 2$.

The same conclusion can be reached by using the elastic Green's function of an infinite medium introduced in Section 2.5.4. According to Eq. (2.75), the displacement field due to a concentrated force acting at the origin scales with the radial distance r as r^{-1}. We have argued that a point defect at a high symmetry site can be modeled as three mutually perpendicular force dipoles. The displacement field due to a force dipole is equivalent to the spatial derivative of the displacement field due to a concentrated force. Therefore the displacement field due to a force dipole must scale as r^{-2}. When the displacement fields of three mutually orthogonal force dipoles are superimposed, the result is a purely radial displacement field that scales as r^{-2}. The displacement field due to three mutually perpendicular force dipoles of unit strength is given in Eq. (2.80).

4.4.2 Non-symmetric point defect

As shown in Fig. 4.10, the non-symmetric point defect may be modeled by a single force dipole. As an example, let us assume the forces in the dipole point in the $\pm z$ directions. The displacement field of the dipole is expressed in terms of the derivative of the elastic Green's function, in Eq. (2.77),

$$u_i^{\mathrm{dp}}(\vec{r}) = -\frac{\partial}{\partial z} G_{iz}(\vec{r}).$$

Using the analytic expression of the Green's function, Eq. (2.75), we obtain

$$\begin{aligned} u_x^{\mathrm{dp}}(\vec{r}) &= -\frac{1}{16\pi\mu(1-\nu)}\frac{x}{r^3}\left(1-\frac{3z^2}{r^2}\right), \\ u_y^{\mathrm{dp}}(\vec{r}) &= -\frac{1}{16\pi\mu(1-\nu)}\frac{y}{r^3}\left(1-\frac{3z^2}{r^2}\right), \\ u_z^{\mathrm{dp}}(\vec{r}) &= \frac{1}{16\pi\mu(1-\nu)}\frac{z}{r^3}\left(1-4\nu+\frac{3z^2}{r^2}\right). \end{aligned} \tag{4.38}$$

Note that the displacement vector is no longer spherically symmetric. However, along any given direction, the magnitude of the displacement vector still decays with r as r^{-2}.

4.5 Elastic field of misfitting inclusion

Modeling the point defect as force dipoles allows us to obtain the form of the displacement field far away from the point defect, such as Eq. (2.80), but the expression goes to infinity as r goes to zero, which is not allowed physically. As a more refined model for the solute atom embedded in a crystal, we now consider a spherical inclusion inserted into a smaller spherical hole of radius r_0 in an infinite isotropic elastic continuum [23, 24]. In addition to avoiding singularities, this model also allows us to estimate the elastic strain energy associated with creating the point defect, which will be discussed in Section 5.1.1.

Both the inclusion and the surrounding material (i.e. the matrix) are treated as linear elastic media. For simplicity, their elastic constants are assumed to be identical. We assume that the inclusion, before being inserted into the hole, is larger than the hole by a volume difference of $\Delta v > 0$. After the inclusion is inserted into the hole, the hole is expanded by volume Δv_{A}, as shown in Fig. 4.13. Here we use the lower case v to indicate volume measured near the inclusion.

The inclusion must exert forces on the surface of the hole in order to expand it. As a result, the inclusion itself is subjected to a compressive stress. Therefore, the size of the inclusion when inside the matrix is smaller than its size when outside the matrix, i.e. $\Delta v_{\mathrm{A}} < \Delta v$. While such elastic compression is reasonable for an inclusion with the size of micrometers or larger, our ultimate intention is to use the inclusion as a model of a single solute atom. Strictly speaking, an atom cannot be compressed in the same way as an inclusion. In this sense, our inclusion

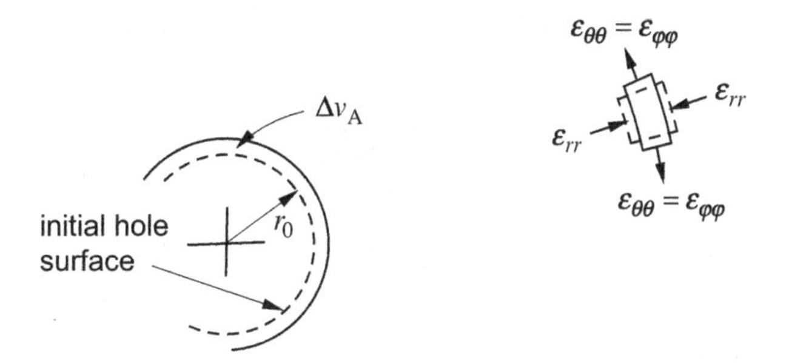

Figure 4.13. Misfit defect strain.

model is only an approximate model. Nonetheless, it does capture important physics of the solute atom.

We have shown that the displacement in the far field must be of the form of Eq. (4.33)

$$u_r = Ar^{-2}, \quad u_\theta = u_\varphi = 0,$$

by modeling the inclusion as three force dipoles. A more detailed analysis shows that this is in fact the correct form for the displacement field of the misfit strain problem everywhere in the matrix, i.e. as long as $r \geq r_0$. The constant A may be found by noting that, when Δv_A is assumed to be a thin sheet

$$4\pi r_0^2 \cdot u_r|_{r=r_0} = \Delta v_A. \tag{4.39}$$

Thus

$$\begin{aligned} u_r|_{r=r_0} &= \frac{\Delta v_A}{4\pi r_0^2} = \frac{A}{r_0^2}, \\ A &= \frac{\Delta v_A}{4\pi}. \end{aligned} \tag{4.40}$$

So the displacement field at $r \geq r_0$ is as follows.

$$\boxed{u_r = \frac{\Delta v_A}{4\pi r^2}, \quad u_\theta = u_\varphi = 0} \tag{4.41}$$

We now proceed to compute the strains and then the stresses.

4.5.1 Calculate the strains

In the spherical coordinate system the strains are computed from the displacements using Eq. (2.28), which for the simple displacement field here is

$$\begin{aligned} \varepsilon_{rr} &= \frac{\partial u_r}{\partial r} = -\frac{\Delta v_A}{2\pi r^3}, \\ \varepsilon_{\theta\theta} &= \varepsilon_{\varphi\varphi} = \frac{u_r}{r} = \frac{\Delta v_A}{4\pi r^3}, \\ \varepsilon_{r\varphi} &= \varepsilon_{r\theta} = \varepsilon_{\theta\varphi} = 0. \end{aligned} \tag{4.42}$$

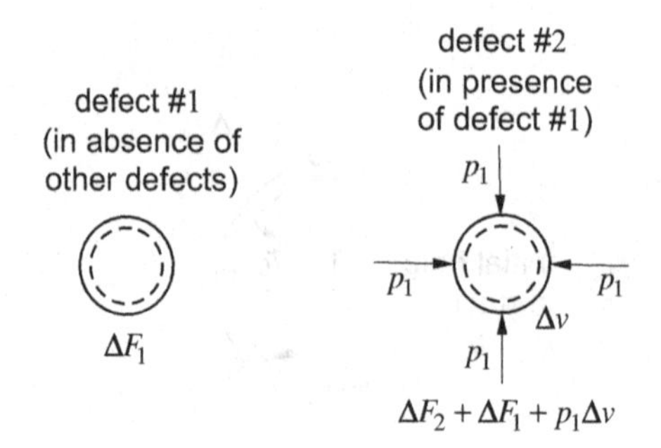

Figure 4.14. Defect interactions: Defect #1 is created first, then defect #2.

These are the strains everywhere in the infinite matrix surrounding the defect. The figure shows that for a hole expansion, the radial strains, ε_{rr}, in the surroundings are compressive (negative), as expected, but the transverse strains, $\varepsilon_{\theta\theta} = \varepsilon_{\varphi\varphi}$, are tensile (positive). We will see that a state of pure shear strain exists around the defect in an infinitely large elastic solid.

Consider the dilatation at some point in the surroundings. The dilatation is defined as

$$\begin{aligned} e &= \frac{\Delta V}{V} = \varepsilon_{rr} + \varepsilon_{\theta\theta} + \varepsilon_{\varphi\varphi} \\ &= -\frac{\Delta v_{\mathrm{A}}}{2\pi r^3} + \frac{\Delta v_{\mathrm{A}}}{4\pi r^3} + \frac{\Delta v_{\mathrm{A}}}{4\pi r^3} = 0, \end{aligned} \tag{4.43}$$

that is, there is no dilatation in the surrounding lattice. This means that every point in the surroundings is sheared but its volume is not changed by having the defect in the solid, a result that seems counterintuitive. It also means that there is no hydrostatic pressure generated in the field surrounding the defect ($r > r_0$), also a bit counterintuitive,

$$p = -Be = 0, \tag{4.44}$$

where B is the bulk modulus.

Because the dilatation is zero in the surroundings, a state of pure shear exists around each misfitting defect. This will lead to the conclusion that such defects do not elastically interact with each other. They neither attract nor repel each other, even though they are each squeezed into the lattice in the same way. Consider the mechanical energy needed to create one misfitting defect, the first defect. The elastic energy needed to stretch the hole and insert the defect (ignoring any compression of the solute atom for now) can be called ΔF_1. Since we have ignored the entropic effects in this simple model, ΔF_1 is also the formation (Helmholtz) free energy of the defect in the absence of other defects.

We now calculate the energy needed to create a second misfitting defect, ΔF_2, in the presence of the elastic field of the first one. It can be shown that $\Delta F_2 = \Delta F_1 + p_1 \Delta v$, where the second term is the pressure–volume work that must be done against any existing pressure from the first defect. But since the first defect created no hydrostatic stress in the surroundings we have $p_1 = 0$ and $\Delta F_2 = \Delta F_1$. Thus there is no interaction energy between the defects, provided the defects are far enough apart for this continuum treatment to apply (a few atomic dimensions are enough).

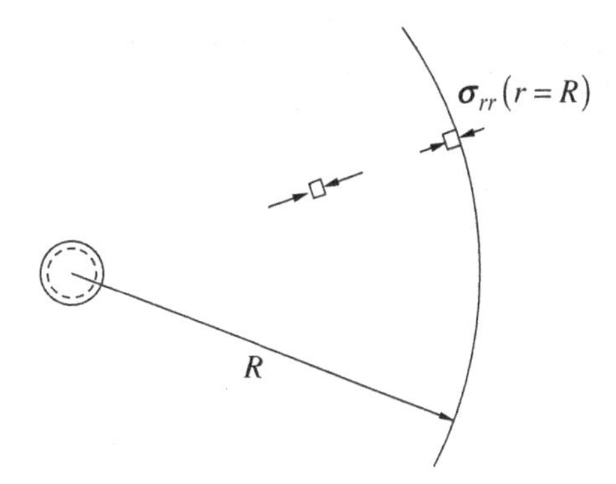

Figure 4.15. Stresses in an infinite solid.

This result also shows that $\Delta F_2 = \Delta F_1$ is independent of the relative position of the defects. Thus there is no interaction force acting between the defects – no repulsion and no attraction. This conclusion is only applicable when the distortion field of each defect is spherically symmetric and when the host crystal can be modeled as an infinitely large isotropic elastic medium.

4.5.2 Defects in a finite solid

The analysis above holds for a defect in an infinitely large solid. We now consider the corrections that are needed to describe defects in a finite solid.

Traction force on the external surface

First consider the σ_{rr} stresses at a large radial distance from the defect in an infinite solid. Using Hooke's law we can write

$$\sigma_{rr}^{\infty} = (2\mu + \lambda)\varepsilon_{rr}^{\infty} + \lambda\varepsilon_{\theta\theta}^{\infty} + \lambda\varepsilon_{\varphi\varphi}^{\infty}, \tag{4.45}$$

where μ is the shear modulus and λ is the Lamé coefficient (see Section 2.3.2). The superscript ∞ indicates the solution is for an infinitely large solid. This simplifies to

$$\begin{aligned}\sigma_{rr}^{\infty} &= 2\mu\varepsilon_{rr}^{\infty} + \lambda\varepsilon_{rr}^{\infty} + \lambda\varepsilon_{\theta\theta}^{\infty} + \lambda\varepsilon_{\varphi\varphi}^{\infty}\\ &= 2\mu\varepsilon_{rr}^{\infty} + \lambda e^{\infty}\\ &= 2\mu\varepsilon_{rr}^{\infty}\end{aligned}$$

because $e^{\infty} = 0$. We have already shown that

$$\varepsilon_{rr}^{\infty} = -\frac{\Delta v_{\mathrm{A}}}{2\pi r^3} \tag{4.46}$$

so

$$\sigma_{rr}^{\infty} = -\frac{\mu\Delta v_{\mathrm{A}}}{\pi r^3}. \tag{4.47}$$

We see that the compressive radial stress goes to zero only when $r \to \infty$. If we consider a spherical surface at $r = R$, it is not traction free; compressive stresses exist there, as illustrated in Fig. 4.15.

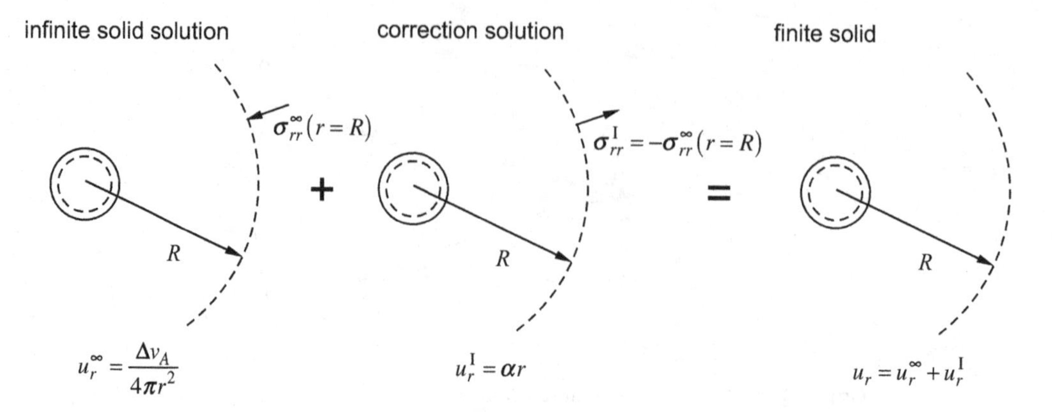

Figure 4.16. Solution for a finite solid by superposition.

Correction solution

To have a traction-free surface for a finite body we must superimpose a correcting solution onto the above solution. We do this by superimposing additional tractions σ_{rr}^{I} that cancel the ones from the infinite body solution, as shown in Fig. 4.16. Thus the correction stress is

$$\sigma_{rr}^{I}=-\sigma_{rr}^{\infty}|_{r=R}=\frac{\mu\Delta v_{A}}{\pi R^{3}}. \tag{4.48}$$

The correction stress is also called the image stress, as indicated by the superscript ^{I}. Notice that the image stress σ_{rr}^{I} is independent of r, i.e. it is a uniform field, which obviously satisfies the equation of equilibrium.

The correction solution amounts to hydrostatic tension. By inspection, the displacement field is

$$u_r^{I}=\alpha r, \tag{4.49}$$

where α is a constant to be determined. The correction strains are

$$\varepsilon_{rr}^{I}=\frac{\partial u_r^{I}}{\partial r}=\alpha,$$

$$\varepsilon_{\theta\theta}^{I}=\varepsilon_{\varphi\varphi}^{I}=\frac{u_r^{I}}{r}=\alpha.$$

The generalized Hooke's law then gives

$$\begin{aligned}\sigma_{rr}^{I}&=(2\mu+\lambda)\varepsilon_{rr}^{I}+\lambda\varepsilon_{\theta\theta}^{I}+\lambda\varepsilon_{\varphi\varphi}^{I}\\&=(2\mu+3\lambda)\alpha.\end{aligned} \tag{4.50}$$

But we know

$$\sigma_{rr}^{I}=\frac{\mu\Delta v_{A}}{\pi R^{3}}=(2\mu+3\lambda)\alpha, \tag{4.51}$$

so we can solve for α:

$$\alpha=\frac{\mu\Delta v_{A}}{\pi R^{3}(2\mu+3\lambda)}. \tag{4.52}$$

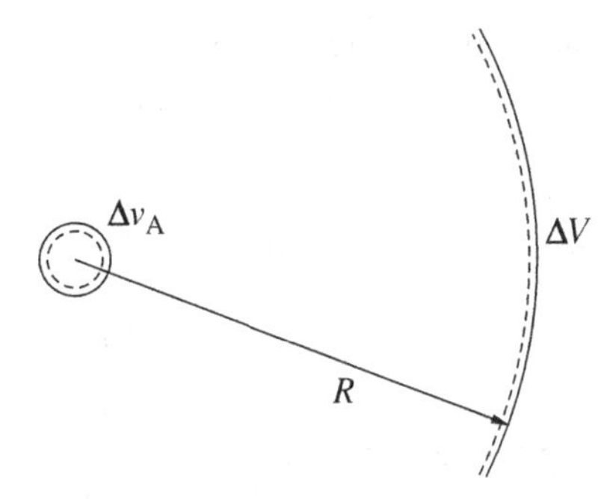

Figure 4.17. Total volume change ΔV of a finite solid of radius R containing a solute atom. The dashed lines and solid lines indicate the location of the surface before and after the solute atom is inserted, respectively.

Thus the correction displacement field must be

$$u_r^{\mathrm{I}} = \alpha r = \frac{\mu \Delta v_{\mathrm{A}}}{\pi R^3 (2\mu + 3\lambda)} r \tag{4.53}$$

and the total displacement field is

$$\begin{aligned} u_r &= u_r^{\infty} + u_r^{\mathrm{I}} \\ &= \frac{\Delta v_{\mathrm{A}}}{4\pi r^2} + \frac{\mu \Delta v_{\mathrm{A}}}{\pi R^3 (2\mu + 3\lambda)} r. \end{aligned} \tag{4.54}$$

Total volume change

Now we can calculate the total volume change associated with the introduction of the misfitting defect. The external volume change associated with the defect is

$$\Delta V = 4\pi R^2 \cdot u_r|_{r=R}. \tag{4.55}$$

Here we use capital V to indicate volume measured at the external surface of the solid. Figure 4.17 illustrates the definition of ΔV.

Using the displacement field given in Eq. (4.54) we have

$$u_r|_{r=R} = \frac{\Delta v_{\mathrm{A}}}{4\pi R^2} + \frac{\mu \Delta v_{\mathrm{A}}}{\pi R^2 (2\mu + 3\lambda)} = \frac{\Delta v_{\mathrm{A}}}{4\pi R^2} \left(\frac{6\mu + 3\lambda}{2\mu + 3\lambda} \right). \tag{4.56}$$

Now we use the identity, $3B = 2\mu + 3\lambda$, and obtain

$$u_r|_{r=R} = \frac{\Delta v_{\mathrm{A}}}{4\pi R^2} \left(1 + \frac{4\mu}{3B} \right). \tag{4.57}$$

So the external volume change is [23]

$$\begin{aligned} \Delta V &= 4\pi R^2 \frac{\Delta v_{\mathrm{A}}}{4\pi R^2} \left(1 + \frac{4\mu}{3B} \right) \\ &= \Delta v_{\mathrm{A}} \left(1 + \frac{4\mu}{3B} \right). \end{aligned} \tag{4.58}$$

Equation (4.58) shows that the total volume expansion ΔV is larger than the volume expansion right around the solute atom. The difference is caused by the tensile (image) stress that exists in the surrounding solid.

It can be shown that the excess volume of the inclusion compared to the hole at the stress-free state (i.e. before insertion) is [23, 24]

$$\Delta v = \Delta v_{\mathrm{A}} \left(1 + \frac{4\mu}{3B}\right). \tag{4.59}$$

A more general expression, with the inclusion and surroundings having different elastic properties, Eq. (4.68), is developed in Section 4.5.3. Comparing Eqs. (4.59) and (4.58), we can see that $\Delta v = \Delta V$. Therefore, considering the total volume of the inclusion and the finite matrix with a hole, the act of inserting the inclusion into the hole (thus creating an internal stress field) does not lead to any net volume change. In other words,

$$\Delta V_{\mathrm{int}} \equiv \Delta V - \Delta v = 0, \tag{4.60}$$

where ΔV_{int} is the volume change associated with the internal stress field. The reduction of volume of the inclusion due to compression is exactly compensated by the increase of volume of the surrounding solid due to the image stress. This is a useful general principle for elastically homogeneous materials – there is no volume change for internal states of stress. In general, the volume average of all components of an internal stress field vanish for a finite solid. This is known as Albenga's law [25].

When the inclusion has different elastic properties from the matrix (in which case it will be called an inhomogeneity in Eshelby's terminology), ΔV_{int} is no longer zero, as we shall see in Section 4.5.3.

Image hydrostatic stress

We found that for a misfitting defect in an infinite solid there was no hydrostatic stress in the surroundings. For the defect in a finite solid there is a hydrostatic stress, a negative pressure. The hydrostatic stress is

$$-p^{\mathrm{I}} = \frac{1}{3}\left(\sigma_{rr}^{\mathrm{I}} + \sigma_{\theta\theta}^{\mathrm{I}} + \sigma_{\varphi\varphi}^{\mathrm{I}}\right) = B\left(\varepsilon_{rr}^{\mathrm{I}} + \varepsilon_{\theta\theta}^{\mathrm{I}} + \varepsilon_{\varphi\varphi}^{\mathrm{I}}\right) = \frac{\mu \Delta v_{\mathrm{A}}}{\pi R^3}. \tag{4.61}$$

This is sometimes called the image hydrostatic stress. For a single solute at the center of a spherical matrix, the image stress is a uniform field. Eshelby has shown that, for a solid containing a uniform concentration of solutes (the distribution of solutes being locally random as long as no two solutes occupy the same site), the image stress is also a uniform field [23]. Each defect produces a hydrostatic tension stress in the matrix, but it does not depend on position. So again there are no interaction forces between the defects.

As the size R of the host crystal (containing a single misfitting defect) increases, the magnitude of the image stress decreases as R^{-3}, but its contribution to the total volume expansion remains constant. This result illustrates the importance of image effects, even in arbitrarily large solids.

Connection to experiment

Because the King tables in Appendix A are constructed by measuring the average atomic volume over the entire crystal, the data in the King tables correspond to the external volume change

ΔV, instead of the local volume change $\Delta v_{\rm A}$ just around the solute. Specifically, ΔV is related to the parameters obtainable from the King tables as follows:

$$\Delta V = \Delta v_{\rm A}\left(1 + \frac{4\mu}{3B}\right) = \Omega_{\rm B}^* - \Omega_{\rm A}, \tag{4.62}$$

where $\Omega_{\rm B}^* - \Omega_{\rm A}$ is the volume change associated with removing a host atom A and replacing it with a solute atom B in an A–B solid solution. This relation allows us to use the experimental data, $\Omega_{\rm B}^* - \Omega_{\rm A}$, to determine the parameter in the inclusion model, $\Delta v_{\rm A}$.

4.5.3 Complete misfitting sphere problem

In the analysis above, we have modeled a solute atom as a spherical misfitting inclusion inserted into a spherical hole. The inclusion is compressed by the surrounding matrix to a smaller volume than its stress-free state. This model can be challenged by arguing that continuum laws of elasticity may not be applicable down to individual atoms. An atomic-scale calculation would be needed to accurately describe what happens to the inserted atom itself.

However, the above analysis can be applied to spherical precipitate particles in a solid. Here we consider introducing a spherical precipitate, B, into a spherical hole in the matrix, A. If the ball (precipitate) is larger than the hole, then the hole is expanded and the ball is compressed in the fitting process. The size of the precipitate particle can be of the order of microns, and at this length scale, the linear elasticity theory is generally applicable. We will consider the general case in which the precipitate and the matrix have different elastic moduli, as is often the case in real crystals.

Let $\Delta v > 0$ be the volume difference between the precipitate in its stress-free state and the hole, let $\Delta v_{\rm A} > 0$ be the expansion of the hole after precipitate insertion, and let $\Delta v_{\rm B} < 0$ be the volume change of the precipitate after insertion. Since the volume difference between the hole and the ball must be accommodated by compression of the ball and expansion of the hole, we have

$$\Delta v = \Delta v_{\rm A} + |\Delta v_{\rm B}|. \tag{4.63}$$

These separate volumes can be found by requiring mechanical equilibrium at the interface between the compressed ball and the expanded hole. We compress the ball and expand the hole to the point where the interfaces match and where the tractions acting across the interfaces are the same, as shown in Fig. 4.18. Thus we write

$$\sigma_{rr}^{\rm A} = -\frac{\mu_{\rm A}\Delta v_{\rm A}}{\pi r_0^3}, \tag{4.64}$$

$$\sigma_{rr}^{\rm B} = -p_{\rm B} = +B_{\rm B}e_{\rm B} = -B_{\rm B}\frac{|\Delta v_{\rm B}|}{\frac{4}{3}\pi r_0^3}. \tag{4.65}$$

Equation (4.64) is the stress in the surroundings, evaluated at the hole–ball surface, using the shear modulus of the matrix, $\mu_{\rm A}$, and Eq. (4.65) is the hydrostatic deformation of the ball, using the bulk modulus of the ball, $B_{\rm B}$. Setting these two stresses (tractions) equal to each other leads

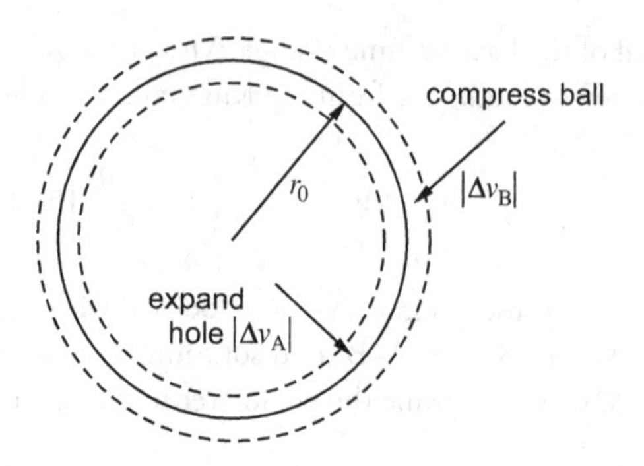

Figure 4.18. The volume difference between the precipitate and hole is accommodated elastically, with the ball being compressed and the hole being expanded.

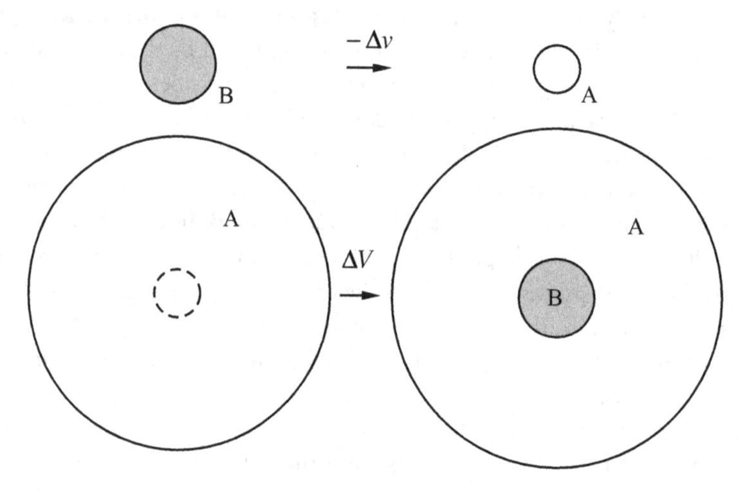

Figure 4.19. Precipitate misfitting process creating residual stresses.

to

$$\sigma_{rr}^{A} = \sigma_{rr}^{B},$$
$$-\frac{\mu_A \Delta v_A}{\pi r_0^3} = -B_B \frac{|\Delta v_B|}{\frac{4}{3}\pi r_0^3}, \tag{4.66}$$

so

$$|\Delta v_B| = \left(\frac{4\mu_A}{3B_B}\right) \Delta v_A. \tag{4.67}$$

Combining this with Eq. (4.63), we arrive at

$$\Delta v_A = \frac{\Delta v}{\left(1 + \frac{4\mu_A}{3B_B}\right)}. \tag{4.68}$$

This relation indicates how much hole expansion occurs when the ball and hole are fitted together. It is instructive to consider the overall volume change associated with the misfitting sphere problem. Consider fitting a ball of B into a sphere of A thus creating a state of residual stress (see Fig. 4.19).

The volume change associated with placing the misfitting precipitate into the smaller hole (hence creating an internal stress field) is

$$\Delta V_{\text{int}} \equiv \Delta V - \Delta v, \tag{4.69}$$

where ΔV is the volume expansion of the sphere when the precipitate is inserted and Δv is the volume change associated with replacing a B volume with an A volume (without surrounding it with a solid matrix). The expression for ΔV has been given in Eq. (4.58):

$$\Delta V = \Delta v_{\text{A}}\left(1 + \frac{4\mu_{\text{A}}}{3B_{\text{A}}}\right).$$

and Δv can be obtained from Eq. (4.68),

$$\Delta v = \Delta v_{\text{A}}\left(1 + \frac{4\mu_{\text{A}}}{3B_{\text{B}}}\right). \tag{4.70}$$

Using the relations above we can write the volume change as

$$\begin{aligned}\Delta V_{\text{int}} &= \Delta v_{\text{A}}\left(1 + \frac{4\mu_{\text{A}}}{3B_{\text{A}}}\right) - \Delta v_{\text{A}}\left(1 + \frac{4\mu_{\text{A}}}{3B_{\text{B}}}\right) \\ &= \Delta v_{\text{A}}\frac{4}{3}\mu_{\text{A}}\left(\frac{1}{B_{\text{A}}} - \frac{1}{B_{\text{B}}}\right) \\ &= \Delta v\frac{4}{3}\mu_{\text{A}}\left(\frac{1}{B_{\text{A}}} - \frac{1}{B_{\text{B}}}\right)\left(1 + \frac{4\mu_{\text{A}}}{3B_{\text{B}}}\right)^{-1}.\end{aligned} \tag{4.71}$$

In the special case where elastic properties of the precipitate and matrix are the same, $B_{\text{B}} = B_{\text{A}}$ (elastically homogeneous material), we reproduce the conclusion reached at the previous section that no volume change is associated with the act of inserting B into A, even though a state of internal stress is created. Of course, if the precipitate is much stiffer than the matrix ($B_{\text{B}} > B_{\text{A}}$), then the insertion of the precipitate into the hole in the matrix leads to a net volume expansion ($\Delta V_{\text{int}} > 0$), as expected.

4.6 Summary

Because the energies of point defects and their interactions depend in part on their elastic fields, it is necessary to describe the stresses and strains that are associated with these defects. Solute atoms comparable to the size of solvent atoms reside in substitutional sites and create spherically symmetric distortions in most lattices, while smaller solute atoms reside in interstitial sites and create distortions that depend on the symmetry of those sites. For symmetrical interstitial sites, like octahedral or tetrahedral sites in close-packed lattices, the distortions produced by interstitial defects are spherically symmetric. By comparison, for non-symmetrical interstitial sites, such as those in BCC metals, the distortions are non-symmetric or tetragonal. The nature of these distortions determines how the point defects in question interact with each other and with other imperfections.

While the sizes of solute and solvent atoms can be described by using a hard sphere picture, such a picture does not allow for a proper accounting of the volume changes associated with the insertion or removal of solutes. For this accounting, the Seitz radius, which is based on how the lattice parameter of the solid changes with the solute concentration, is more useful. Such a scheme leads to the effective volumes of solute atoms when dissolved in a given solvent matrix and to atomic size factors for describing the size mismatch.

The elastic fields associated with misfitting solute atoms or other point defects can be modeled as force dipoles imposed in an elastic continuum representing the surrounding crystal. Three orthogonal force dipoles of the same magnitude lead to a spherically symmetric distortion in an elastically isotropic solid while a single force dipole produces a tetragonal distortion. These models are useful for describing the displacement fields and the corresponding strains and stresses associated with point defects.

A more detailed model for a solute atom producing a spherically symmetric distortion is a spherical misfit inclusion inserted into a smaller hole in an isotropic elastic medium. A significant result is that spherically symmetric defects in infinite elastic solids produce pure shear stress fields in the matrix and do not elastically interact with each other. The absence of hydrostatic stresses in the field produced by each inclusion means that in an infinite elastic solid misfitting inclusions neither attract nor repel each other even when in close proximity, contrary to intuition. There is a hydrostatic image stress associated with misfitting inclusions in finite solids, but it is position independent. This means that, again, neighboring point defects do not elastically interact. These properties are of importance when considering the segregation of solutes to dislocations, to be discussed in Chapter 12.

4.7 Exercise problems

4.1 Find the hard sphere radius r_H in terms of the lattice constant a of a BCC crystal. The direction along which atoms are most closely spaced in a BCC crystal has been determined in Exercise problem 1.3.

4.2 Calculate the radius of the largest atom that could be inserted into an interstitial position in Si (diamond cubic structure) without causing any distortion. Assume that all of the atoms can be treated as touching hard spheres and take the radius of the Si atoms to be r_{Si}. Indicate where in the Si lattice such an interstitial would be located. Now calculate the radius of the largest atom that could be inserted into an interstitial position in an FCC structure with touching hard sphere atoms of the same radius, r_{Si}. This exercise should reveal the open nature of diamond cubic structures compared to close-packed FCC structures. It will help you to understand why self-interstitials are common in Si but extremely rare in close-packed metals.

4.3 In Section 4.2.2, we have seen that small atoms like C occupy interstitial positions of the type $0, 0, \frac{1}{2}$ in the BCC Fe structure. Assuming that the atoms can be treated as hard spheres, with r_{Fe} representing the radius of the host Fe atoms, determine the maximum radius of the interstitial C atom, r_C, that can be inserted into this interstitial site before making contact with the Fe atoms. This exercise will show that even very small atoms like C will cause distortions of the BCC Fe lattice.

4.4 Use the King tables in Appendix A to estimate the atomic volumes of Cu and Al, in the pure substances, in solution in each other and in solution in the following metals: Ag, Au, Co, Fe, Mg,

Mn, Ni, Pt, Zn. It is convenient to use Excel or some other spreadsheet to do this exercise. In which environments do the atoms (Cu and Al) have the largest (and smallest) atomic volumes?

4.5 Determine the atomic volume of Al in pure aluminum and compare it with the effective atomic volume of Al when it is dissolved into nickel. Briefly explain any differences you observe in light of the Ni–Al phase diagram shown in Fig. 6.12.

4.6 An alloy of 50 at.% Ni and 50 at.% Co is made by pressing powders of the pure metals together to form a fully dense compact. Then the two-phase mixture is heated to a high temperature to form a homogeneous FCC solid solution. Use the King tables to answer the following questions about this alloy.

(a) What is the effective atomic volume of Co when it is dissolved in pure Ni?
(b) What is the effective atomic volume of Ni when it is dissolved in pure Co?
(c) Estimate the fractional volume change that occurs when the two pure metals are dissolved in each other. You may assume that the atomic volume of Co is the same whether it has the HCP or FCC structure.

4.7 A fully dense alloy consisting of 95 at.% Ni (FCC) and 5 at.% Ge (diamond cubic) is to be heated to a high temperature, where the Ge dissolves completely in the FCC Ni lattice, and then cooled back to the reference temperature. Calling the atomic fraction of Ge χ_{Ge}, write an expression for and calculate the fractional volume change, $\Delta V/V$, associated with this process.

4.8 The following questions refer to a Cu–Ge alloy. The Cu–Ge phase diagram is available at the book website and a full set of King tables is available in Appendix A.

(a) Determine the atomic volume of Ge in pure germanium and compare it with the effective atomic volume of Ge when it is dissolved into copper. Briefly explain any differences you observe.
(b) A wire with a composition of 95 at.% Cu and 5 at.% Ge of unit length consists of pure Ge particles embedded in a pure Cu matrix at room temperature. The wire is heated up so that all of the Ge goes into solution and is then cooled back to room temperature. How does the length of the wire change in this process? Be as quantitative as possible.

4.9 The King tables indicate that the atomic volumes of pure Al and pure Li are: $\Omega_{\mathrm{Al}} = 0.0166\ \mathrm{nm}^3$ and $\Omega_{\mathrm{Li}} = 0.021\,61\ \mathrm{nm}^3$, that is, the atomic volume of pure Li is much greater than that of pure Al. Using what you know about atomic size effects, determine the atomic volume of Li in Al and compare it with the atomic volume of Li in pure Li. Determine the lattice parameter of an FCC Al–Li solid solution with 10 at.% Li (which is stable above 500 °C as shown in the phase diagram on the book website) and compare that with the lattice parameter of pure Al.

4.10 Molybdenum (Mo) atoms dissolve substitutionally into BCC Fe and displace the surrounding Fe atoms. We wish to estimate the positions of the first and second nearest Fe atoms that surround a Mo atom when it resides in a substitutional site in the Fe lattice. We call r_1 the distance to the first near neighbors and r_2 the distance to the second near neighbors. These near neighbors are displaced radially by δ_1 and δ_2, respectively, when a Mo atom is placed in the lattice. We wish to calculate δ_1 and δ_2 in nm. We may assume that the Fe atoms are embedded in an isotropic elastic continuum around the Mo atom.

It will be necessary to use the King tables in Appendix A to properly describe the displacements caused by the inserted Mo atom. The lattice parameter of Fe is $a = 0.286\,29$ nm and the elastic constants of Fe are available in Table B.1.

4.11 Calcium is a large atom that causes significant distortion in the Al lattice when it is dissolved substitutionally. Use your knowledge of atomic size effects and the elastic constants given in Table B.1 to write an approximate expression for the outward (radial) displacements, u_r, of the 12 Al atoms that surround a Ca atom sitting in a substitutional site in the Al lattice. For reference, the atomic volume of Al is $\Omega_{\text{Al}} = 0.0166\,\text{nm}^3$. For convenience you may call the nearest neighbor spacing between Al atoms in the Al lattice: d_{Al}.

4.12 A carbon atom is shown in an interstitial position in a BCC Fe lattice in Fig. 4.20. The symmetry of the site is such that the Fe atoms just above and just below the carbon atom are displaced along the z axis by a small amount δ in opposite directions. Derive an expression for the axial stress σ_{zz} in the Fe lattice surrounding the defect. Assume that the Fe lattice can be treated as an isotropic elastic solid.

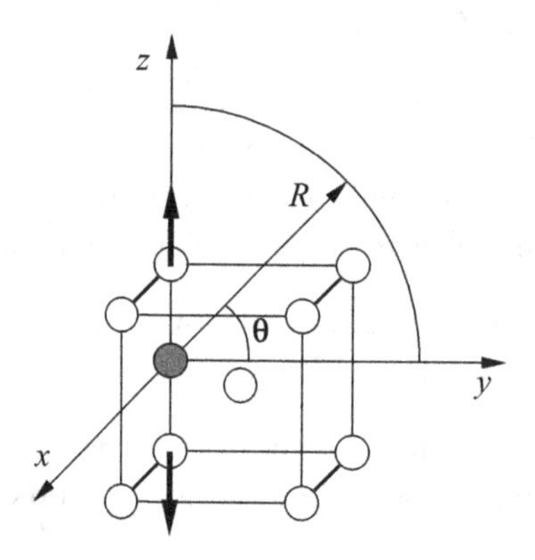

Figure 4.20. A carbon atom at an interstitial site in the Fe lattice.

4.13 Consider a misfitting cylinder in a solid. This may be regarded as a cylinder of infinite length that is inserted into an infinitely long cylindrical hole, which is slightly smaller than the cylinder. The surroundings may be regarded as an isotropic elastic solid. For this problem, it is obvious that the displacement field must be as follows:

$$u_r = Ar^{-n},$$
$$u_\theta = u_z = 0. \tag{4.72}$$

We wish to determine the value of the exponent, n, from the equilibrium equation for this cylindrical problem. This problem is analogous to the one solved in Section 4.4.1 for the spherical geometry. You should be able to take the same approach for this geometry. Once the value of n is determined, write expressions for the strains and stresses surrounding the misfitting cylinder using Eq. (2.27) and

$$\sigma_{rr} = (\lambda + 2\mu)\varepsilon_{rr} + \lambda\varepsilon_{\theta\theta} + \lambda\varepsilon_{zz},$$
$$\sigma_{\theta\theta} = (\lambda + 2\mu)\varepsilon_{\theta\theta} + \lambda\varepsilon_{rr} + \lambda\varepsilon_{zz}. \tag{4.73}$$

4.14 Al + 1wt.% Si is used as an interconnect metal in integrated circuits. The alloy is deposited onto an oxidized Si wafer prior to patterning and deposition of the passivation. We assume that the passivation is deposited at room temperature, when the interconnect metal line is in a stress-free state. All of the silicon in the alloy is in the form of Si precipitates when the passivation is applied. When the structure is heated to 500 °C, all of the Si in the alloy goes into solution (see the Al–Si phase diagram on the book website). We wish to calculate the stress in the line when the structure is quenched back to room temperature. We will assume that all of the Si remains in solution during this quenching process. We may also assume that the surroundings are perfectly rigid and do not deform.

4.15 Copper can be precipitation strengthened by cobalt. A few atomic percent Co can be dissolved in Cu at high temperatures and precipitated out as pure Co at low temperatures. The Co precipitates are spherical and perfectly coherent (lattice matching) with the FCC Cu matrix (the Co precipitates have the FCC structure even though the equilibrium structure of pure Co is HCP at room temperature). We consider a spherical particle of Co of radius R in an infinitely large pure copper matrix.

(a) Using the information and assumptions given below, describe the stress in the Co particle and write an expression that could be used to calculate that stress. All of the terms in the resulting equation should be defined so that a numerical calculation could be made.
(b) Write expressions for all non-zero components of strain in the surrounding copper matrix (use spherical coordinates) and the radial component of the stress: σ_{rr}^{∞}.

Assumptions:

(1) assume that the atomic volume of Co for the FCC structure is the same as for the HCP structure (a good assumption);
(2) assume that the Co and Cu crystals can be treated with isotropic elasticity using such quantities as μ_{Cu}, B_{Cu}, λ_{Cu}, . . ., and μ_{Co}, B_{Co}, λ_{Co}, . . ., etc. (this is not a good assumption as Cu is elastically anisotropic).

4.16 A multilayered single crystal of pure FCC Ag and pure FCC Cu with the orientation shown in Fig. 4.21a is grown by molecular beam epitaxy (MBE).

We assume that the layer dimensions are sufficiently small that the crystal layers remain perfectly coherent during growth and during subsequent annealing. After growth, the multilayered structure is heated to 700 °C and held for a sufficiently long time that diffusion of Cu into Ag and Ag into Cu occurs. The sample is then cooled to room temperature rapidly enough that no compositional changes occur. The phase diagram shown in Fig. 4.21b indicates the compositions of the layers after this annealing process. We wish to determine the misfit strain for the multilayered structure both before and after the diffusion is allowed to occur. Such information would permit us to calculate the stresses in the layers both before and after the diffusion has occurred. For this problem we will define the misfit strain as follows:

$$\varepsilon \equiv \frac{a_{(\text{Ag-phase})} - a_{(\text{Cu-phase})}}{a_{(\text{Cu-phase})}}, \tag{4.74}$$

where the lattice parameters are those for stress-free layers (obviously not the lattice parameters for the elastically strained layers). Use the King tables in Appendix A for these calculations.

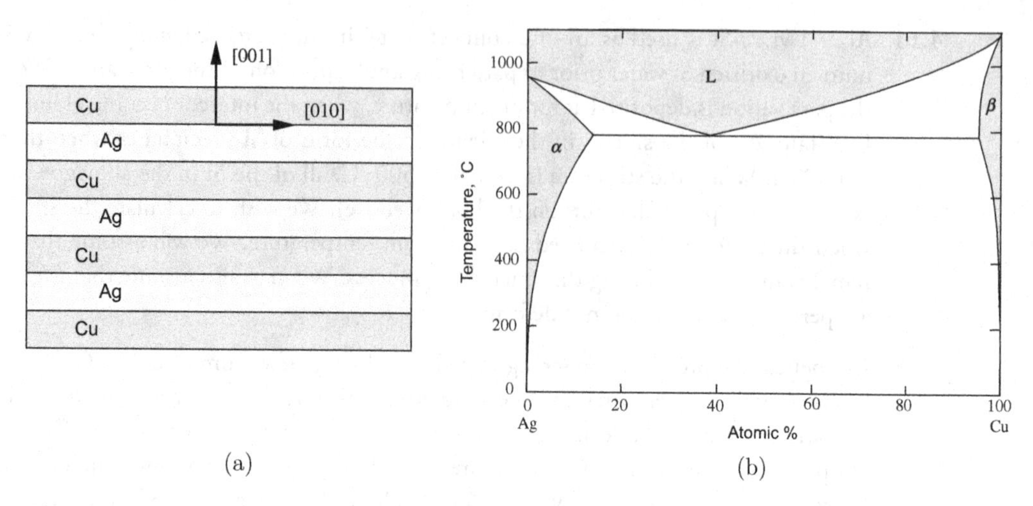

Figure 4.21. (a) Multilayered single crystal of silver and copper. (b) Silver–copper binary phase diagram. The shaded regions correspond to conditions where a single phase exists: L for the liquid phase, α for the Ag-rich solid phase, and β for the Cu-rich solid phase. The white regions correspond to conditions where the equilibrium state consists of a mixture of two phases.

4.17 We have a single crystal of pure FCC silver onto which we would like to deposit a perfectly epitaxial film of FCC aluminum. We want to have no lattice mismatch between the film and substrate. We know that the lattice parameter of Ag is $a_{Ag} = 0.4085$ nm while the lattice parameter of pure Al is $a_{Al} = 0.4049$ nm. So a pure Al film on pure Ag would have about a 1% lattice mismatch. But we know that alloying Al with Mg expands the lattice parameter of Al. Obtain an expression for the atomic fraction of Mg in Al, χ_{Mg}, that would be needed to achieve perfect lattice matching, and calculate the value of this atomic fraction.

4.18 A very large crystal (you can assume that it is an infinite body) containing an excess, non-equilibrium concentration of vacancies is subjected to a remote hydrostatic tension stress σ_H. At a later point in time some of the excess vacancies condense to form a void in the crystal of radius R. We wish to determine the complete stress field in the crystal after the void has formed. You may assume that the crystal is elastically isotropic and can be characterized by the usual elastic properties, μ, λ, B, E, ν, etc.

Derive or write equations for the following stresses, σ_{rr}, $\sigma_{\theta\theta}$, $\sigma_{\varphi\varphi}$, as a function of coordinate position in the solid. The expressions should be in the simplest form. The solution can be constructed from the stress field of a misfit inclusion using the method of superposition.

5 Point defect thermodynamics

Having discussed the elastic field around a single point defect, we now apply the thermodynamics principles (Chapter 3) to obtain the equilibrium concentration of point defects in crystals under a given temperature and pressure. The fundamental principle used repeatedly is that the Gibbs free energy of the crystal is minimized when the point defects reach the equilibrium concentration.

We start by discussing the equilibrium concentration of extrinsic point defects, i.e. substitutional and interstitial solutes, in Section 5.1. The approach is then applied, in Section 5.2, to vacancies, which are intrinsic point defects. In Section 5.3 we discuss the experimental methods to measure the equilibrium concentration and thermodynamic properties of vacancies, and compare the experimental data with theoretical estimates. The chemical potential of point defects is defined in Section 5.4.

5.1 Equilibrium concentration of solutes

We consider a dilute substitutional solution of B atoms in an A matrix. Let N_A (which is fixed) be the number of A atoms in the system and let N_B be the number of B atoms dissolved in the A-rich crystal. The total number of atomic sites in the A-rich crystal is $N = N_A + N_B$. Thus $\chi = N_B/N$ is the fraction of atomic sites where the "wrong" kind of atom is located. χ is also the molar fraction of B atoms in the crystal. We follow the regular solution/quasi-chemical approach in which the formation energy of the point defect is dominated by the energies of the chemical bonds associated with the impurity defect (quasi-chemical, Eq. (5.5)) and where the mixing entropy is that for an ideal solution (regular solution, Eq. (5.19)).

Let the A-rich crystal be in contact with a large B crystal, which acts as an infinite supply of B atoms. For simplicity, we only allow B atoms to enter the A-rich crystal as solutes, but forbid A atoms to enter the B crystal as solutes. Each time a B atom is dissolved in the lattice, the A atom it replaces takes up a site at the A/B interface and extends the A lattice by one atomic volume, as shown in Fig. 5.1a. Note that the number of A atoms N_A is conserved, while the number of B solute atoms N_B and the total number of atomic sites N for the A-rich crystal are not conserved.

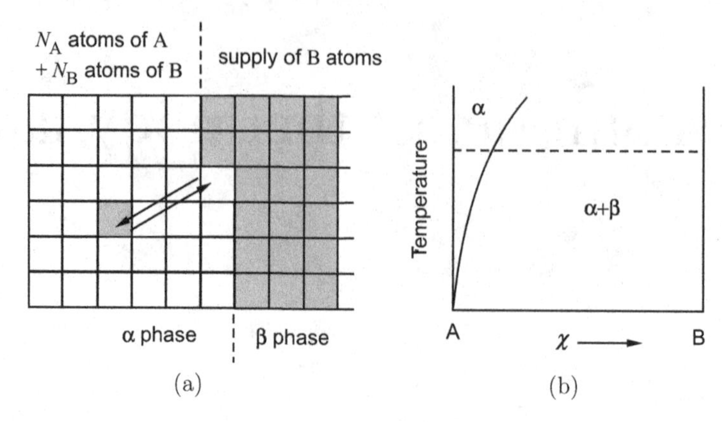

Figure 5.1. (a) Unit process of dissolving one B atom into an A lattice under the condition of constant temperature and constant external pressure. (b) Solvus line in a simple phase diagram. α indicates an A-rich crystal with B solutes. β indicates a B-rich crystal (with negligible solubility of A). $\alpha + \beta$ indicates a segregation of atoms into the A-rich α phase and B-rich β phase.

We wish to describe the equilibrium state of the system. Will any B atoms dissolve in A at equilibrium? Adding B atomic defects to the A lattice increases the internal energy of the system and that discourages the introduction of foreign, defect atoms. But the entropy of the system also increases when B atoms are introduced in the A-rich crystal and that causes some B atoms to be present at equilibrium, in spite of the increased internal energy of the system. We know this result in another context. Solutes usually dissolve spontaneously in pure components and the equilibrium concentration increases with temperature. This is shown as the solvus line on a phase diagram. A simple example is given in Fig. 5.1b. At a particular temperature, the equilibrium concentration of solute is given by the intersection of the isotherm (dashed line) with the solvus line. The solvus line indicates the maximum solubility of B in the A-rich crystal (α phase) at the given temperature. Extra B atoms will separate out from the A-rich crystal to join the B-rich crystal (β phase).

If the pressure and temperature are held constant, then the equilibrium state of the system is determined by the lowest possible Gibbs free energy, according to the second law of thermodynamics (Chapter 3). So we consider the change in Gibbs free energy that occurs when N_B B atoms are introduced into the A lattice. This may be written as

$$\Delta G = \Delta H - T\Delta S, \tag{5.1}$$

where ΔH and ΔS are the change in enthalpy and entropy, respectively, when the defects are introduced, relative to the case where no B atoms exist in the A-rich phase. We first consider the factors that make up ΔH.

5.1.1 Enthalpy

From the definition of enthalpy in Eq. (3.22),

$$H \equiv E + pV,$$

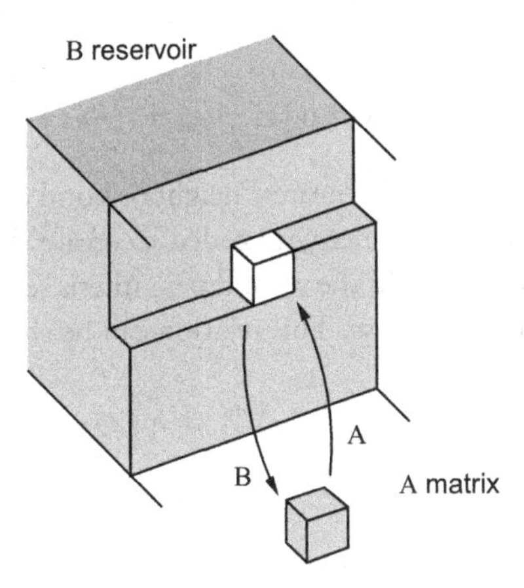

Figure 5.2. Atom exchange between the A-rich phase and the B reservoir.

the change in enthalpy can be expressed as

$$\Delta H = \Delta E + p_{\text{ext}} \Delta V, \tag{5.2}$$

where ΔE and ΔV are the change in internal energy and volume associated with creation of the defects and $p = p_{\text{ext}}$ is the external hydrostatic pressure.[1] The internal energy change can be expressed as

$$\Delta E = \chi N \Delta e_{\text{f}}, \tag{5.3}$$

where Δe_{f} is the formation internal energy of each defect, usually just called the formation energy of the defect. It, in turn, can be composed of two parts,

$$\Delta e_{\text{f}} = \Delta e_{\text{f}}^{\text{bond}} + \Delta e_{\text{f}}^{\text{strain}}, \tag{5.4}$$

where $\Delta e_{\text{f}}^{\text{bond}}$ is the internal energy change associated with making different chemical bonds at the defect and $\Delta e_{\text{f}}^{\text{strain}}$ is the elastic strain energy associated with creating the defect. We may estimate $\Delta e_{\text{f}}^{\text{bond}}$ if the energies of the nearest neighbor chemical bonds are known or can be estimated.

Bond energy

Figure 5.2 shows a process of removing an A atom from the lattice, where it is surrounded by other A atoms, and replacing it with a B atom from the phase boundary. We can estimate the change in chemical energy by counting the nearest neighbor bonds that will be created and destroyed. Each atom is surrounded by z nearest neighbors, where z is the coordination number ($z = 6$ for the simple cubic lattice shown). The net bond energy change associated with creating

1 Because Eq. (3.22) is valid only when the solid is subjected to purely hydrostatic stresses (see Section 3.2.4), so is the definition of the change of enthalpy, Eq. (5.2). However, Eq. (5.2) is a good approximation for the change of enthalpy even when the solid is subjected to non-hydrostatic stresses, as long as the defect produces a spherically symmetric strain field (see Section 4.4.1), since there is no work term coupled to shear stresses.

the defect is then

$$\Delta e_{\mathrm{f}}^{\mathrm{bond}} = \left(z\epsilon_{\mathrm{AB}} + \frac{z}{2}\epsilon_{\mathrm{AB}} + \frac{z}{2}\epsilon_{\mathrm{AA}}\right) - \left(z\epsilon_{\mathrm{AA}} + \frac{z}{2}\epsilon_{\mathrm{BB}} + \frac{z}{2}\epsilon_{\mathrm{AB}}\right), \tag{5.5}$$

where ϵ_{AB}, ϵ_{AA}, ϵ_{BB} are the nearest neighbor bond energies. The sum in the first parenthesis in Eq. (5.5) represents the energies needed to create the bonds when placing the atoms in their final places and the sum in the second parenthesis represents the energies of the bonds that must be destroyed in this process. This relation can be simplified to

$$\begin{aligned} \Delta e_{\mathrm{f}}^{\mathrm{bond}} &= z\left(\epsilon_{\mathrm{AB}} - \frac{\epsilon_{\mathrm{AA}} + \epsilon_{\mathrm{BB}}}{2}\right) \\ &= z\epsilon_{\mathrm{I}}, \end{aligned} \tag{5.6}$$

where

$$\epsilon_{\mathrm{I}} \equiv \left(\epsilon_{\mathrm{AB}} - \frac{\epsilon_{\mathrm{AA}} + \epsilon_{\mathrm{BB}}}{2}\right) \tag{5.7}$$

is sometimes called the interaction energy. Note that if the AB bond energy is the same as the average of the AA and BB bonds there is no bond energy change associated with making the defect. Typically the unlike bonds have higher energies than the like bonds so that the interaction energy is positive.

Strain energy

If the inserted atom differs in size from the host atoms then some mechanical work will be needed to create the defect. We can estimate this by thinking of inserting a spherical inclusion into a hole. Before insertion, the inclusion is larger than the hole by volume Δv. We know from Chapter 4 that the inserted inclusion will expand the hole in the matrix by Δv_{A}. The work done to the matrix during the insertion process can be computed by thinking of gradually pressurizing the hole and expanding it until it has been expanded by Δv_{A}. If p is the pressure causing the expansion of the hole and dv is the volume expansion at any arbitrary pressure then the work to expand the hole to its final state can be written as

$$\Delta e_{\mathrm{f}}^{\mathrm{strain,M}} = \int_0^{\Delta v_{\mathrm{A}}} p\,dv, \tag{5.8}$$

where M indicates the matrix. We know that when the expansion is complete, the final pressure p_{i} (i for internal) is related to the volume expansion Δv_{A} as, Eq. (4.64),

$$p_{\mathrm{i}} = \frac{\mu \Delta v_{\mathrm{A}}}{\pi r_0^3}. \tag{5.9}$$

But the pressure and volume change are linearly related during the expansion process, so that

$$p = \frac{\mu v}{\pi r_0^3}, \tag{5.10}$$

where v is the integration variable from 0 to Δv_{A}. Thus we have

$$\Delta e_{\mathrm{f}}^{\mathrm{strain}} = \int_0^{\Delta v_{\mathrm{A}}} \frac{\mu}{\pi r_0^3} v\,dv = \frac{1}{2} p_{\mathrm{i}} \Delta v_{\mathrm{A}}. \tag{5.11}$$

At the same time, work also must be done to compress the inclusion so that it eventually experiences a volume reduction of $\Delta v_{\rm A} - \Delta v$. Following the same arguments as above, we can show that the work done to the inclusion (I) is

$$\Delta e_{\rm f}^{\rm strain,I} = \frac{1}{2} p_{\rm i} (\Delta v - \Delta v_{\rm A}) \tag{5.12}$$

so that the total work done is [24]

$$\Delta e_{\rm f}^{\rm strain} = \Delta e_{\rm f}^{\rm strain,M} + \Delta e_{\rm f}^{\rm strain,I} = \frac{1}{2} p_{\rm i} \Delta v, \tag{5.13}$$

where Δv is known from Eq. (4.59),

$$\Delta v = \Delta v_{\rm A} \left(1 + \frac{4\mu}{3B} \right) = \Omega_{\rm B}^* - \Omega_{\rm A}. \tag{5.14}$$

Therefore, using Eq. (5.9), the total work done can be written as

$$\Delta e_{\rm f}^{\rm strain} = \frac{\mu (\Delta v)^2}{2\pi r_0^3 \left(1 + \frac{4\mu}{3B} \right)}. \tag{5.15}$$

Volume change

The volume change associated with creating χN defects is simply

$$\Delta V = \chi N \Delta v_{\rm f}, \tag{5.16}$$

where $\Delta v_{\rm f}$ is the formation volume of the defect. This is the volume change of the entire system shown in Fig. 5.1a when one B atom at the A/B interface exchanges with an A atom in the A-rich crystal. No atom is created or destroyed in this process. The volume occupied by the B atom being exchanged changes from $\Omega_{\rm B}$ to $\Omega_{\rm B}^*$. Therefore,

$$\Delta v_{\rm f} = \Omega_{\rm B}^* - \Omega_{\rm B}. \tag{5.17}$$

We note that $\Delta v_{\rm f}$ is different from Δv given by Eq. (5.14), which measures the local volume mismatch for a B atom in an A-rich crystal.

5.1.2 Entropy

The change in entropy associated with the creation of χN defects in N atomic sites is composed of two terms, the configurational entropy (by far the most important) and the vibrational entropy:

$$\Delta S = \Delta S_{\rm config} + \Delta S_{\rm vib}. \tag{5.18}$$

The configuration entropy is sometimes called the mixing entropy. It is used extensively in the study of phase equilibrium and is familiar to many readers.

The configurational or mixing entropy change associated with the creation of χN defects in the A-rich crystal can be written as

$$\Delta S_{\rm config} = -N k_{\rm B} [\chi \ln \chi + (1 - \chi) \ln(1 - \chi)], \tag{5.19}$$

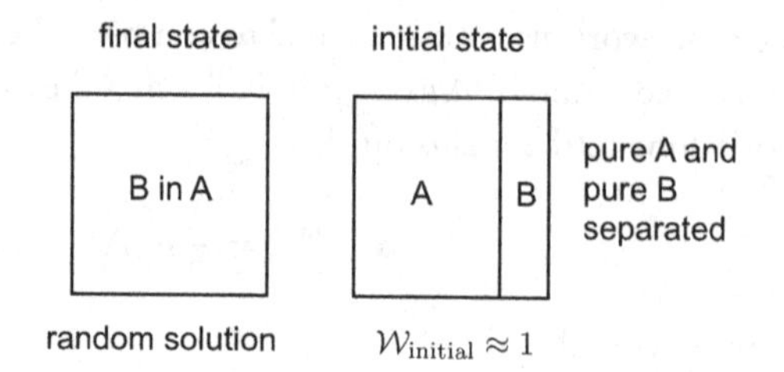

Figure 5.3. Atom configurations in the initial and final states.

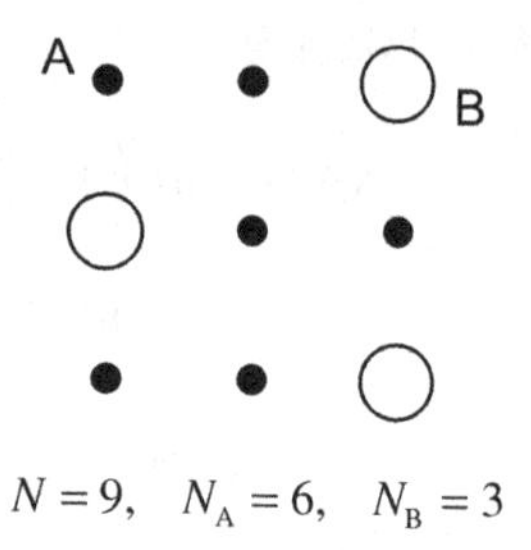

Figure 5.4. Atom configurations in the final state.

where k_B is Boltzmann's constant. For readers not familiar with the relation we derive it in the following.

Configurational entropy derivation

From statistical mechanics the basic postulate of Boltzmann is written in Eq. (3.49),

$$S = k_B \ln \mathcal{W},$$

where S is the entropy of the system, k_B is Boltzmann's constant, and $\mathcal{W}$ is the number of distinguishable configurations (or atomic arrangements) that the system can adopt. For the problem of mixing B atoms in an A-rich crystal we may express the change in entropy for defect formation as

$$\begin{aligned} \Delta S_{\text{config}} &= S_{\text{final}} - S_{\text{initial}} \\ &= k_B \ln \mathcal{W}_{\text{final}} - k_B \ln \mathcal{W}_{\text{initial}}, \end{aligned} \tag{5.20}$$

as shown in Fig. 5.3.

Since the number of distinguishable configurations in the initial state is about one (all A atoms on the left and all B atoms on the right) we may write $\mathcal{W}_{\text{initial}} \approx 1$ and $S_{\text{initial}} = k_B \ln \mathcal{W}_{\text{initial}} \approx 0$. So

$$\Delta S_{\text{config}} = k_B \ln \mathcal{W}_{\text{final}}. \tag{5.21}$$

We now calculate the total number of distinguishable atomic configurations in a random solid solution. Consider first a very simple system with nine atomic sites occupied by six A atoms and three B atoms, as shown in Fig. 5.4. It is easy to count the number of distinguishable arrangements in this system. If we think of locating the B atoms, there are nine places to put the first one, then eight places for the second one, after the first one is placed somewhere, and then seven places to put the third one, once the first two have been located. Figure 5.4 shows one particular distinguishable configuration. But there are six ways this particular configuration would have been created in the above scheme. It will have been repeated six times because for each of the three places the first B atom is located there are two places for the second and one place for the last, after the first two have been located. These six ways give the same configuration because all B atoms are identical.

Thus the total number of distinguishable arrangements of this simple system is

$$\mathcal{W}_{\text{final}} = \frac{9 \times 8 \times 7}{3 \times 2 \times 1}. \tag{5.22}$$

We may now generalize this result for N_{B} atoms on N lattice sites. The result is evidently

$$\begin{aligned}\mathcal{W}_{\text{final}} &= \frac{N(N-1)(N-2)\cdots(N-(N_{\text{B}}-1))}{N_{\text{B}}(N_{\text{B}}-1)\cdots 3 \times 2 \times 1} \\ &= \frac{N(N-1)(N-2)\cdots(N_{\text{A}}+1)}{N_{\text{B}}!} \\ &= \frac{N!}{N_{\text{B}}!N_{\text{A}}!}.\end{aligned} \tag{5.23}$$

Now we can find $\ln \mathcal{W}_{\text{final}}$. We write

$$\begin{aligned}\ln \mathcal{W}_{\text{final}} &= \ln \frac{N!}{N_{\text{B}}!N_{\text{A}}!} = \ln N! - \ln N_{\text{B}}! - \ln N_{\text{A}}! \\ &= \ln N! - \ln N_{\text{B}}! - \ln(N - N_{\text{B}})!.\end{aligned} \tag{5.24}$$

An excellent approximation for the natural logarithm of the factorial of a large number is Stirling's approximation, Eq. (3.53),

$$\ln N! \approx N \ln N - N.$$

Using the Stirling approximation we have

$$\begin{aligned}\ln \mathcal{W}_{\text{final}} &= (N \ln N - N) - (N_{\text{B}} \ln N_{\text{B}} - N_{\text{B}}) - ((N - N_{\text{B}}) \ln(N - N_{\text{B}}) - (N - N_{\text{B}})) \\ &= N \ln N - N_{\text{B}} \ln N_{\text{B}} - (N - N_{\text{B}}) \ln(N - N_{\text{B}}) \\ &= N \ln \left(\frac{N}{N - N_{\text{B}}}\right) - N_{\text{B}} \ln \left(\frac{N_{\text{B}}}{N - N_{\text{B}}}\right).\end{aligned}$$

Using $N_{\text{B}} = \chi N$ we have

$$\begin{aligned}\ln \mathcal{W}_{\text{final}} &= N \ln \left(\frac{1}{1-\chi}\right) - \chi N \ln \left(\frac{\chi}{1-\chi}\right) \\ &= -N \ln(1-\chi) - \chi N \ln \chi + \chi N \ln(1-\chi) \\ &= -N[\chi \ln \chi + (1-\chi) \ln(1-\chi)].\end{aligned}$$

Finally, using Eq. (5.21), the configurational entropy change is given by Eq. (5.19).

$$\boxed{\Delta S_{\text{config}} = -N k_{\text{B}}[\chi \ln \chi + (1-\chi) \ln(1-\chi)]}$$

Vibrational entropy derivation

The vibrational entropy arises because in a solid at finite temperature, all atoms vibrate around their equilibrium positions. As a result, there are a large number $\mathcal{W}$ of microscopic states (in terms of positions and momenta of all atoms) that correspond to a macroscopic state specified by thermodynamic variables such as T and p. At temperatures lower than half the melting point, the interaction between neighboring atoms can be approximated as harmonic springs. The vibrational entropy of the crystal can be obtained from the vibrational frequencies of the atoms (see Exercise problems 3.5–3.8). When solute atoms are introduced to the crystal, the vibrational frequencies around each solute atom are changed, leading to a change of entropy. At low solute concentration, where the average distance between two solutes is large, each solute makes the same contribution to the vibrational entropy, so that

$$\Delta S_{\text{vib}} = N_{\text{B}} \Delta s_{\text{f}} = \chi N \Delta s_{\text{f}}, \tag{5.25}$$

where Δs_{f} is the formation entropy of the point defect, and contains only the vibrational contribution. A rough estimate of the vibrational entropy of vacancies is given in Section 5.2.1.

5.1.3 Equilibrium concentration of substitutional atoms

We may now summarize all of the terms in the expression for the free energy and compute the equilibrium concentration of solute atom defects. Using the relations above we may express the free energy change as

$$\begin{aligned} \Delta G &= \Delta H - T \Delta S \\ &= \chi N (\Delta e_{\text{f}} + p_{\text{ext}} \Delta v_{\text{f}} - T \Delta s_{\text{f}}) + N k_{\text{B}} T [\chi \ln \chi + (1 - \chi) \ln(1 - \chi)]. \end{aligned} \tag{5.26}$$

For substitutional defects, $N = N_{\text{A}} + N_{\text{B}}$ is not a constant; both N and $\chi = N_{\text{B}}/N$ depend on N_{B}. Therefore, we rewrite the above expression for ΔG in terms of N_{B}:

$$\Delta G = N_{\text{B}} (\Delta e_{\text{f}} + p_{\text{ext}} \Delta v_{\text{f}} - T \Delta s_{\text{f}}) + k_{\text{B}} T \left[N_{\text{B}} \ln \frac{N_{\text{B}}}{N_{\text{A}} + N_{\text{B}}} + N_{\text{A}} \ln \frac{N_{\text{A}}}{N_{\text{A}} + N_{\text{B}}} \right]. \tag{5.27}$$

We can find the equilibrium state of the system by minimizing the free energy, ΔG, with respect to the number of defects, N_{B}. We find this state by requiring

$$\frac{\partial \Delta G}{\partial N_{\text{B}}} = 0 \quad \text{and} \quad \frac{\partial^2 \Delta G}{\partial N_{\text{B}}^2} > 0. \tag{5.28}$$

Differentiating Eq. (5.27) gives

$$\begin{aligned} \frac{\partial \Delta G}{\partial N_{\text{B}}} &= (\Delta e_{\text{f}} + p_{\text{ext}} \Delta v_{\text{f}} - T \Delta s_{\text{f}}) + k_{\text{B}} T \ln \frac{N_{\text{B}}}{N_{\text{A}} + N_{\text{B}}} \\ &= (\Delta e_{\text{f}} + p_{\text{ext}} \Delta v_{\text{f}} - T \Delta s_{\text{f}}) + k_{\text{B}} T \ln \chi. \end{aligned} \tag{5.29}$$

Setting this derivative equal to zero gives

$$\begin{aligned} k_{\text{B}} T \ln \chi &= -(\Delta e_{\text{f}} + p_{\text{ext}} \Delta v_{\text{f}} - T \Delta s_{\text{f}}), \\ \chi &= \exp\left(- \frac{\Delta e_{\text{f}} + p_{\text{ext}} \Delta v_{\text{f}} - T \Delta s_{\text{f}}}{k_{\text{B}} T} \right). \end{aligned} \tag{5.30}$$

We may identify

$$\Delta h_f = \Delta e_f + p_{ext}\Delta v_f \tag{5.31}$$

as the enthalpy of formation of the defect[2] and

$$\Delta g_f = \Delta h_f - T\Delta s_f \tag{5.32}$$

as the (Gibbs) free energy of formation of the defect. With these identifications the equilibrium fraction of defects is given by the Boltzmann relation (also see Eq. (3.76)).

$$\boxed{\chi = \exp\left(-\frac{\Delta g_f}{k_B T}\right)} \tag{5.33}$$

It can be shown that $\partial^2 \Delta G/\partial N_B^2 > 0$ as long as $N_A > 0$ and $N_B > 0$.

5.1.4 Equilibrium concentration of interstitial atoms

To obtain the equilibrium concentration of interstitial atoms, we consider a scenario similar to that shown in Fig. 5.1, where an A-rich crystal (α phase) is next to an infinite supply of B atoms (β phase). The difference here is that, when B atoms are taken from the β phase and inserted into the α phase, they enter the interstitial sites. Consequently, no A atom needs to be placed at the A/B interface and the number of A lattice sites does not grow when an interstitial defect is introduced.

For interstitial defects, the total number of available sites for B atoms is fixed, which is,

$$N_s = N_A z_i, \tag{5.34}$$

where z_i is the number of interstitial sites per A atom. The fraction of occupied sites is defined as $\chi_i = N_B/N_s$, which is not equal to the atomic fraction χ of B atoms in the crystal (see Section 4.2.5).

Given that the total number of sites N_s is fixed, it is more convenient to define the free energy change per site,

$$\Delta g \equiv \frac{\Delta G}{N_s} = \chi_i(\Delta e_f + p_{ext}\Delta v_f - T\Delta s_f) + k_B T[\chi_i \ln \chi_i + (1-\chi_i)\ln(1-\chi_i)], \tag{5.35}$$

where Δe_f, Δv_f, Δs_f are the formation energy, volume, and entropy of the interstitial point defect, respectively. In particular,

$$\Delta v_f = \Omega_B^\dagger - \Omega_B, \tag{5.36}$$

where $\Omega_B^\dagger$ is the effective volume of the interstitial B atom defined in Eq. (4.30).

2 Similar to Eq. (5.2), Eq. (5.31) is valid only when the solid is subjected to purely hydrostatic stresses. When the solid is subjected to non-hydrostatic stresses, it is a good approximation if the strain produced by the defect is spherically symmetric.

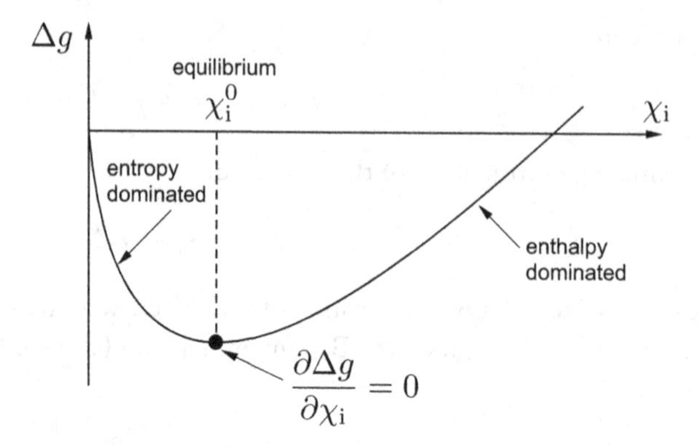

Figure 5.5. The equilibrium concentration of defects is determined by minimizing the free energy change per interstitial site.

We can find the equilibrium state of the system by minimizing the free energy Δg, with respect to the fraction of occupied sites χ_i. We find this state by requiring

$$\frac{\partial \Delta g}{\partial \chi_i} = 0 \quad \text{and} \quad \frac{\partial^2 \Delta g}{\partial \chi_i^2} > 0, \tag{5.37}$$

as illustrated by Fig. 5.5.

Differentiating Eq. (5.35) gives

$$\frac{\partial \Delta g}{\partial \chi_i} = (\Delta e_f + p_{\text{ext}} \Delta v_f - T \Delta s_f) + k_B T \ln \frac{\chi_i}{1 - \chi_i}. \tag{5.38}$$

Setting this derivative equal to zero gives

$$k_B T \ln \frac{\chi_i}{1 - \chi_i} = -(\Delta e_f + p_{\text{ext}} \Delta v_f - T \Delta s_f)$$
$$\frac{\chi_i}{1 - \chi_i} = \exp\left(-\frac{\Delta e_f + p_{\text{ext}} \Delta v_f - T \Delta s_f}{k_B T}\right) = \exp\left(-\frac{\Delta g_f}{k_B T}\right).$$

Therefore

$$\chi_i = \frac{1}{1 + \exp\left[\frac{1}{k_B T}(\Delta e_f + p_{\text{ext}} \Delta v_f - T \Delta s_f)\right]} = \frac{1}{1 + \exp\left(\frac{\Delta g_f}{k_B T}\right)}. \tag{5.39}$$

This is called the Fermi–Dirac relation. It is different from Boltzmann's relation only by the term 1 in the denominator. This term keeps $\chi_i < 1$ always, even for the case of $\Delta g_f < 0$. Physically, this means the maximum concentration of interstitials is reached if all available sites $N = N_A z_i$ are occupied.[3] It can be shown that $\partial^2 \Delta g / \partial \chi_i^2 > 0$ as long as $0 < \chi_i < 1$.

Usually $\Delta g_f \gg k_B T$, so that $\chi_i \ll 1$, i.e. we are almost always dealing with very small defect concentrations. In this limit the Fermi–Dirac relation (5.39) and the Boltzmann relation (5.33)

3 The Boltzmann relation, Eq. (5.33), is not applicable for $\Delta g_f < 0$.

have negligible difference, so that even for interstitial defects we can write the following.

$$\boxed{\chi_{\rm i} \approx \exp\left(-\frac{\Delta g_{\rm f}}{k_{\rm B}T}\right)} \tag{5.40}$$

Equation (5.39), especially the dependence of $\chi_{\rm i}$ on the pressure $p_{\rm ext}$, is valid only if the B-rich β-phase is a solid subjected to the same pressure. Interstitial atoms, such as H and N, often exist as gases outside the host crystal. In this case, the equilibrium fraction of interstitial atoms depends on the partial pressure of these atoms in the gas phase, as discussed in Section 5.4.1.

5.1.5 Factors affecting the equilibrium defect concentration

The equilibrium defect fraction can be written in various ways

$$\begin{aligned} \chi &\approx \exp\left(-\frac{\Delta g_{\rm f}}{k_{\rm B}T}\right) \\ &= \exp\left(-\frac{\Delta h_{\rm f} - T\Delta s_{\rm f}}{k_{\rm B}T}\right) \\ &= \exp\left(\frac{\Delta s_{\rm f}}{k_{\rm B}}\right)\exp\left(-\frac{\Delta h_{\rm f}}{k_{\rm B}T}\right) \\ &= \exp\left(\frac{\Delta s_{\rm f}}{k_{\rm B}}\right)\exp\left(-\frac{\Delta e_{\rm f} + p_{\rm ext}\Delta v_{\rm f}}{k_{\rm B}T}\right) \\ &= \exp\left(\frac{\Delta s_{\rm f}}{k_{\rm B}}\right)\exp\left(-\frac{\Delta e_{\rm f}}{k_{\rm B}T}\right)\exp\left(-\frac{p_{\rm ext}\Delta v_{\rm f}}{k_{\rm B}T}\right) \\ &= \chi_0 \exp\left(-\frac{p_{\rm ext}\Delta v_{\rm f}}{k_{\rm B}T}\right), \end{aligned} \tag{5.41}$$

where

$$\chi_0 = \exp\left(\frac{\Delta s_{\rm f}}{k_{\rm B}}\right)\exp\left(-\frac{\Delta e_{\rm f}}{k_{\rm B}T}\right) \tag{5.42}$$

is the equilibrium defect fraction (solubility) in the absence of any external pressure.

Temperature dependence

At atmospheric pressure, $p_{\rm ext}\Delta v_{\rm f} \ll \Delta e_{\rm f}$ (e.g. see Eq. (5.56)) so we can write

$$\chi \approx \chi_0. \tag{5.43}$$

We note that in Eq. (5.42) the term involving the formation entropy does not depend on temperature. So the temperature dependence of χ_0 is dominated by $\Delta e_{\rm f}$. The entropy term is usually close to one, so that for rough estimates it can be ignored, i.e.

$$\chi \approx \chi_0 \approx \exp\left(-\frac{\Delta e_{\rm f}}{k_{\rm B}T}\right).$$

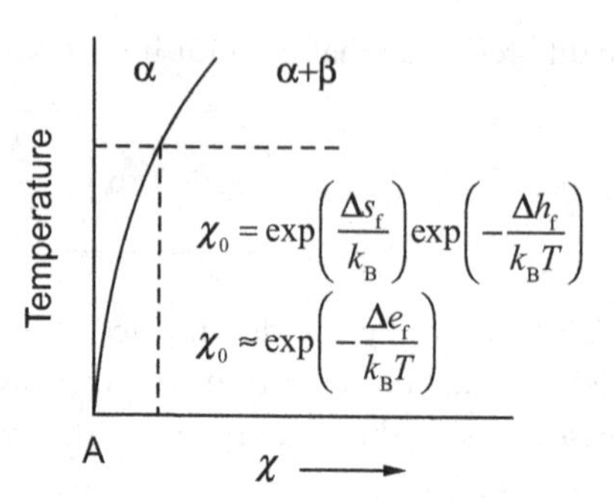

Figure 5.6. Simple model for a solvus curve.

This relation can be used as a very simple model of the solvus curve in a phase diagram, as shown in Fig. 5.6. Based on this model, knowing the solubility at one temperature allows an estimate of the solubility at another temperature, say a lower temperature where the solubility might be too small to easily measure.

Pressure dependence

The effect of pressure on the equilibrium defect fraction is given by Eq. (5.41),

$$\chi = \chi_0 \exp\left(-\frac{p_{\mathrm{ext}}\Delta v_f}{k_B T}\right),$$

where χ_0 is the equilibrium defect fraction (solubility) in the absence of any external pressure, p_{ext}, and Δv_f is the formation volume, which equals $\Omega_B^* - \Omega_B$ for a substitutional defect. Note that, for positive $\Delta v_f = \Omega_B^* - \Omega_B$, increasing pressure reduces the equilibrium concentration of solute defects. On the other hand, taking the example of Al solutes in Ag, the King tables (A.1 and A.2) indicate that $\Delta v_f = \Omega_B^* - \Omega_B$ is negative. So for this system increasing the external pressure causes the solubility to increase.

5.2 Equilibrium concentration of vacancies

We now apply the Boltzmann analysis to vacancies in a crystal lattice. The problem is the same as the one we have solved for solutes. In this case, vacancies (or vacant atomic sites) can be created by removing atoms from the crystal interior and placing them on the surface of the crystal, as shown in Fig. 5.7, much like when we moved A atoms from the lattice and replaced them with B atoms taken from the A/B interface. We may think of the vacancy as a B atom which does not occupy any volume when placed outside the A-rich crystal, i.e. $\Omega_B = 0$, but, when placed inside the A-rich crystal, takes up a finite volume somewhat less than an A atom (due to relaxation), i.e. $\Omega_B^* < \Omega_A$.

Similar to Eq. (5.27), the Gibbs free energy change for introducing N_v vacancies into a crystal containing N_A atoms can be written as

$$\Delta G = N_v(\Delta e_v + p_{\mathrm{ext}}\Delta v_v - T\Delta s_v) + k_B T\left[N_v \ln \frac{N_v}{N_A + N_v} + N_A \ln \frac{N_A}{N_A + N_v}\right], \quad (5.44)$$

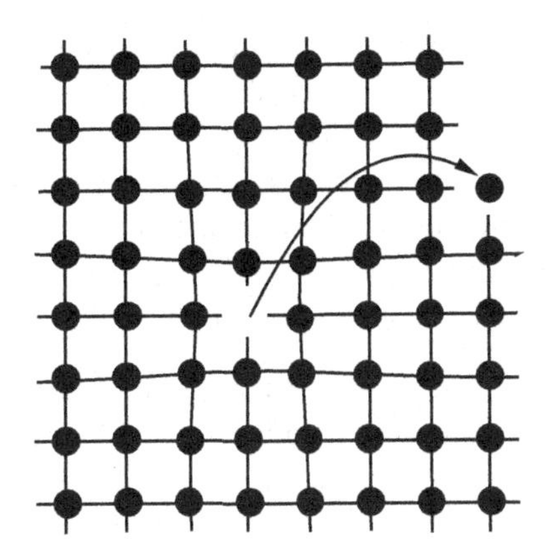

Figure 5.7. Vacancy formation.

where

Δe_v = formation energy of the vacancy,
Δv_v = formation volume of the vacancy,
Δs_v = formation entropy of the vacancy.

Minimizing ΔG with respect to N_v, we obtain the equilibrium vacancy fraction

$$\begin{aligned}\chi_\mathrm{v} \equiv \frac{N_\mathrm{v}}{N_\mathrm{A}+N_\mathrm{v}} &= \exp\left(-\frac{\Delta g_\mathrm{v}}{k_\mathrm{B}T}\right)\\ &= \exp\left(-\frac{\Delta h_\mathrm{v}-T\Delta s_\mathrm{v}}{k_\mathrm{B}T}\right)\\ &= \exp\left(\frac{\Delta s_\mathrm{v}}{k_\mathrm{B}}\right)\exp\left(-\frac{\Delta h_\mathrm{v}}{k_\mathrm{B}T}\right)\\ &= \exp\left(\frac{\Delta s_\mathrm{v}}{k_\mathrm{B}}\right)\exp\left(-\frac{\Delta e_\mathrm{v}+p_\mathrm{ext}\Delta v_\mathrm{v}}{k_\mathrm{B}T}\right). \end{aligned} \tag{5.45}$$

All the fundamental quantities, Δe_v, Δv_v, Δs_v, can be measured in various ways. For pure metals and many other crystals these quantities are well known (see Table C.1). It is instructive to make simple estimates of Δe_v, Δv_v, and Δs_v, as discussed below.

- Δe_v: We give an estimate of the formation energy of the vacancy that corresponds to the bond energy given in Eq. (5.5). Consider a simple cubic lattice with lattice parameter a, where one atom is missing. We can think of this vacancy as a cubic hole in the lattice. The cubic hole has six surfaces, each of area a^2. If these "surfaces" have the surface energy of the solid, γ_s (think of them as free surfaces), then we might estimate the formation energy of the vacancy as $\Delta e_\mathrm{v} = 6a^2\gamma_\mathrm{s}$. Taking $a = 2$ Å $= 0.2$ nm and $\gamma_\mathrm{s} = 1$ J/m^2 (a typical surface energy) leads to a vacancy formation energy of $\Delta e_\mathrm{v} = 2.4 \times 10^{-19}$ J or $\Delta e_\mathrm{v} = 1.5$ eV. Thus we see that the formation energy for the vacancy should be of the order of 1 eV.
- Δv_v: The formation volume of the vacancy can be estimated from Eq. (5.17), and the fact that $\Omega_\mathrm{B} = 0$. Therefore $\Delta v_\mathrm{v} = \Omega_\mathrm{B}^*$, which is approximately the atomic volume Ω (we have dropped the subscript A for brevity) of a perfect crystal. This corresponds to the fact

that the crystal grows by one atomic site when each vacancy is formed. Actually crystal structures are not perfectly rigid because the atoms relax inward a little when the vacancy is formed (see Fig. 5.7), so that typically $\Delta v_v \approx 0.5$–0.7Ω.

- Δs_v: The formation entropy is associated with changes in the vibrational frequencies of the atoms surrounding the vacancy. Typically $\Delta s_v \approx 2k_B$–$8k_B$. This can be understood in a semi-quantitative way by considering the changes in vibrational frequencies of the atoms next to a vacancy, as shown in Section 5.2.1.

So far in this chapter, we have assumed that the solid is subjected to a uniform and purely hydrostatic stress field, which can be characterized by a single parameter p_{ext}. Equation (5.45) is valid only when this condition is satisfied. If the stress field is either non-uniform or non-hydrostatic (or both), then the equilibrium vacancy fraction needs to be expressed in terms of the vacancy chemical potential to be introduced in Section 5.4.

In preparation for the discussion of non-hydrostatic stress conditions, it is convenient to define the relaxation volume for vacancies as

$$\Delta V_v^{rlx} = \Delta v_v - \Omega, \tag{5.46}$$

where Ω is the atomic volume. ΔV_v^{rlx} is analogous to the local volume change associated with replacing a solvent atom with a substitutional solute defect given in Eq. (5.14). For ordinary metals $\Delta v_v < \Omega$ due to relaxation so that the relaxation volume is less than zero, $\Delta V_v^{rlx} < 0$. In terms of ΔV_v^{rlx}, the equilibrium vacancy fraction under hydrostatic pressure, Eq. (5.45), can be rewritten as

$$X_v = \exp\left(\frac{\Delta s_v}{k_B}\right)\exp\left(-\frac{\Delta e_v + p_{ext}\Delta V_v^{rlx} + p_{ext}\Omega}{k_B T}\right). \tag{5.47}$$

The $p_{ext}\Delta V_v^{rlx}$ term corresponds to the work done when removing an atom from the interior of the crystal, and the $p_{ext}\Omega$ term corresponds to the work done plating the atom out on the surface. When the stress is not purely hydrostatic, the $p_{ext}\Omega$ term changes depending on the actual traction force on the surface (to be discussed in Sections 5.4.3 and 7.5.1), but the $p_{ext}\Delta V_v^{rlx}$ term stays. This illustrates the significance of the relaxation volume ΔV_v^{rlx}.

5.2.1 Rough estimate of vibrational entropy

The vibrational entropy and configurational entropy are two typical contributions to the entropy of a solid. Because the effect of configurational entropy is already accounted for in the equilibrium distribution of vacancies, leading to Eq. (5.45), only the vibrational entropy should be included in the estimate of vacancy formation entropy Δs_v.

At temperatures lower than half the melting point, the interaction between neighboring atoms can be approximated as harmonic springs, as shown in Fig. 5.8a. The entire crystal can then be considered as a large collection of coupled harmonic oscillators, whose vibration spectrum and entropy are known analytically. For a simple estimate, we can use the Einstein model, in which for each atom we assume that all of its neighbors are fixed at their perfect crystal positions. Consider a simple cubic crystal, and let each pair of nearest neighboring atoms be connected by a spring with spring constant $k_s/2$. Then in the Einstein model, each atom is considered to be held in place by three pairs of orthogonal springs, each with spring constant $k_s/2$,

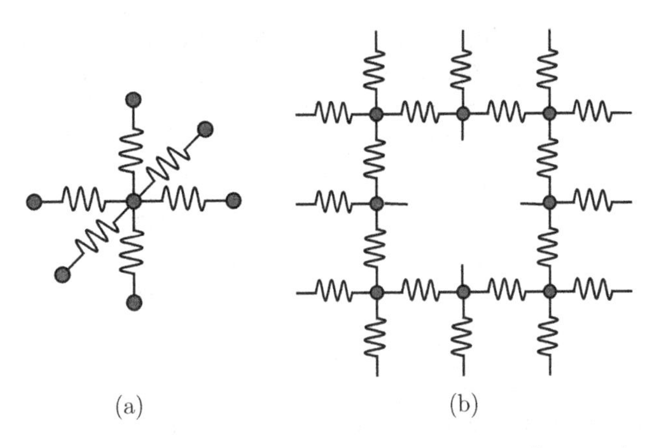

Figure 5.8. (a) Harmonic oscillators in a perfect crystal. (b) Harmonic oscillators with vacancy.

or equivalently by three orthogonal springs, each with spring constant k_s. Given the entropy expression of a harmonic oscillator at high enough temperature to neglect quantum effects, Eq. (3.66), the vibrational entropy of every atom in a perfect crystal is

$$s^{\mathrm{i}}_{\mathrm{vib}} = 3k_{\mathrm{B}} \ln \frac{k_{\mathrm{B}}T}{h\nu_{\mathrm{i}}}, \tag{5.48}$$

where i stands for the initial state (before vacancy is introduced), and

$$\nu_{\mathrm{i}} = \frac{1}{2\pi}\sqrt{\frac{k_s}{m}} \tag{5.49}$$

is the vibrational frequency, and m is the atomic mass. The factor of 3 in Eq. (5.48) accounts for the vibrational entropy in x, y, and z directions, i.e. each atom can be considered as three independent harmonic oscillators.

When a vacancy is introduced, the atoms neighboring the vacant site lose the constraining effect of a harmonic spring, so that its vibration becomes more violent. The thermal vibration along this direction only has an effective spring constant of $k_s/2$. The vibrational entropy of a neighboring atom is

$$s^{\mathrm{f}}_{\mathrm{vib}} = 2k_{\mathrm{B}} \ln \frac{k_{\mathrm{B}}T}{h\nu_{\mathrm{i}}} + k_{\mathrm{B}} \ln \frac{k_{\mathrm{B}}T}{h\nu_{\mathrm{f}}}, \tag{5.50}$$

where f stands for the final state, with

$$\nu_{\mathrm{f}} = \frac{1}{2\pi}\sqrt{\frac{k_s/2}{m}} \tag{5.51}$$

so

$$\frac{\nu_{\mathrm{i}}}{\nu_{\mathrm{f}}} = \sqrt{2}. \tag{5.52}$$

Therefore the change of the vibrational entropy associated with one atom next to the vacancy is

$$\Delta s_{\mathrm{vib}} = s^{\mathrm{f}}_{\mathrm{vib}} - s^{\mathrm{i}}_{\mathrm{vib}} = k_{\mathrm{B}} \ln \left(\frac{\nu_{\mathrm{i}}}{\nu_{\mathrm{f}}}\right) = k_{\mathrm{B}} \ln \sqrt{2}. \tag{5.53}$$

Table 5.1. Typical values of vacancy formation energy Δe_v [26] and melting temperature T_m [27] of FCC metals.

Metal	Δe_v (eV)	T_m (K)	$\Delta e_v/T_m$ (eV/K)
Au	0.93	1337	7.0×10^{-4}
Al	0.67	933	7.2×10^{-4}
Ag	1.11	1235	9.0×10^{-4}
Cu	1.28	1358	9.4×10^{-4}
Pb	0.58	601	9.7×10^{-4}

Finally, since each vacancy affects the vibration of z nearest neighbor atoms,

$$\Delta s_v = z\Delta s_{vib} = zk_B \ln\left(\frac{\nu_i}{\nu_f}\right) = zk_B \ln\sqrt{2}. \tag{5.54}$$

Here, z is called the *coordination number* of the vacancy; for a simple cubic crystal $z = 6$. Therefore the factor $\Delta s_v/k_B = z\ln\sqrt{2}$ in Eq. (5.45) leads to

$$\exp\left(\frac{\Delta s_v}{k_B}\right) = (\sqrt{2})^z = (\sqrt{2})^6 = 8. \tag{5.55}$$

Experimentally, this factor ranges between 2 and 10; thus the simple atom vibration model gives the correct order of magnitude of the pre-exponential entropy factor.

5.2.2 Estimates of the equilibrium vacancy concentration

The experimental techniques and data for the thermodynamic properties of vacancies will be discussed in more detail in Section 5.3. It will be shown that the formation energies for vacancies are of the order of 1 eV but do vary from metal to metal depending on the strength of the bonding. Table 5.1 shows that Δe_v scales approximately with the melting temperatures of the solids.

We note that the ratio $\Delta e_v/T_m$ is approximately constant for these FCC metals at about 8×10^{-4} eV/K. This means that the equilibrium vacancy fraction at the melting temperature is about the same for similarly bonded solids.

As we have shown, the equilibrium vacancy fraction is

$$\chi_v = \exp\left(\frac{\Delta s_v}{k_B}\right)\exp\left(-\frac{\Delta e_v + p_{ext}\Delta v_v}{k_B T}\right),$$

where the pre-exponential factor, $\exp(\Delta s_v/k_B)$ is about 10. Let us now compare Δe_v with $p_{ext}\Delta v_v$ when only atmospheric pressure is applied. A pressure of one atmosphere is approximately $p_{ext} = 1$ atm $\approx 10^5$ N/m^2 and the vacancy volume (for Al) is $\Delta v_v = 0.95\Omega = 1.6 \times 10^{-29}$m^3. So we may estimate

$$p_{ext}\Delta v_v \approx 10^5 \times 1.6 \times 10^{-29}\text{ J} = 1.6 \times 10^{-24}\text{ J} \approx 10^{-5}\text{ eV}. \tag{5.56}$$

Thus the term $p_{ext}\Delta v_v$ is negligible compared to Δe_v when atmospheric pressure is applied.

Now we make an estimate of the typical equilibrium vacancy fraction at the melting temperature,

$$\begin{aligned}\chi_v(T = T_m) &\approx \exp\left(\frac{\Delta s_v}{k_B}\right)\exp\left(-\frac{\Delta e_v}{k_B T_m}\right)\\ &\approx 10\exp\left(-\frac{\Delta e_v}{k_B T_m}\right)\\ &\approx 10\exp\left[-\frac{8\times 10^{-4}\text{eV/K}}{8.62\times 10^{-5}\ \text{eV/K}}\right]\\ &\approx 10^{-3}. \end{aligned} \tag{5.57}$$

So at the melting temperature about one site in 1000 is vacant at equilibrium. This does not sound like very many vacancies until one considers that this vacancy density is equivalent to that of a cubic lattice of vacancies with a lattice spacing of 10 atomic sites. The vacancies are quite close to each other at this concentration.

To illustrate the strong temperature dependence of the vacancy concentration one might estimate the equilibrium vacancy fraction at room temperature for a solid with $\Delta e_v = 1$ eV. The result is

$$\chi_v(T = 300\ \text{K}) \approx 10\exp\left[-\frac{1\ \text{eV}}{(8.62\times 10^{-5}\ \text{eV/K})(300\ \text{K})}\right] = 1.6\times 10^{-16}, \tag{5.58}$$

a very small fraction indeed. We will see later that actual vacancy concentrations at room temperature are not that low; if a solid is cooled from high temperatures, vacancies must diffuse out of the crystal or to sites of annihilation on cooling to maintain the equilibrium concentration. At low temperatures the vacancies diffuse so slowly that it takes an extremely long time for them to leave the crystal. Thus it is almost impossible to actually reach such a low concentration. This means that the vacancies are usually supersaturated in crystals at room temperature (see Sections 6.5 and 7.1.1 for more discussion).

5.2.3 Vacancy concentration in a spherical crystal

We now briefly discuss the effect of the surface energy on the equilibrium concentration of vacancies when the size of the crystal is small. Consider a crystal with a spherical shape of radius R and surface free energy γ_s (per unit area). For simplicity, we assume that the crystal is not subjected to any external pressure, i.e. $p_{ext} = 0$. The Gibbs free energy of the crystal now contains a contribution from the surface, which is $G_s = A_s\gamma_s$, where $A_s = 4\pi R^2$ is the surface area of the crystal. Creating a vacancy requires removing an atom from inside the sphere and placing it on the surface, which leads to an increase of the volume of the crystal by Δv_v. This increase of volume must be accompanied by an increase of surface area, which introduces an additional free energy cost in creating the vacancy.

To estimate the increase of surface area A_s, we assume that the crystal remains spherical after multiple vacancies have been created. Let ΔR be the average increase of the radius of the crystal due to the creation of one vacancy. Then $\Delta v_v = 4\pi R^2\Delta R$, i.e. $\Delta R = \Delta v_v/(4\pi R^2)$. The average increase in the surface area for creating one vacancy is

$$\Delta A_s = \frac{\partial A_s}{\partial R}\Delta R = 8\pi R \Delta R = \frac{2}{R}\Delta v_v. \tag{5.59}$$

Therefore the surface contribution to the free energy cost of creating a vacancy is

$$\Delta G_s = \Delta A_s \gamma_s = \frac{2\gamma_s}{R}\Delta v_v. \tag{5.60}$$

As a result, the equilibrium vacancy fraction becomes

$$\begin{aligned} \chi_v &= \exp\left[-\frac{\Delta g_v + (2\gamma_s/R)\Delta v_v}{k_B T}\right] \\ &= \exp\left(\frac{\Delta s_v}{k_B}\right)\exp\left[-\frac{\Delta e_v + (2\gamma_s/R)\Delta v_v}{k_B T}\right] \end{aligned} \tag{5.61}$$

Note that Δv_v appears in the Boltzmann factor even though $p_{ext} = 0$ (see Exercise problem 5.17). It is as if the crystal were subjected to an external pressure of the magnitude $2\gamma_s/R$. This is analogous to the capillarity pressure $p_c = 2\gamma_s/R$ induced in a liquid droplet of radius R by the surface energy.[4]

5.3 Vacancy experiments

It is instructive to review the classical experimental techniques that have been used to determine the thermodynamic properties of vacancies. We wish to measure the formation enthalpy, Δh_v, the formation entropy, Δs_v, and the formation volume, Δv_v for vacancies. It is obvious from the equilibrium expression

$$\begin{aligned} \chi_v &= \exp\left(\frac{\Delta s_v}{k_B}\right)\exp\left(-\frac{\Delta h_v}{k_B T}\right) \\ &= \exp\left(\frac{\Delta s_v}{k_B}\right)\exp\left(-\frac{\Delta e_v}{k_B T}\right)\exp\left(-\frac{p_{ext}\Delta v_v}{k_B T}\right) \end{aligned} \tag{5.62}$$

that the enthalpy, Δh_v, can be determined from the temperature dependence of χ_v, that the volume, Δv_v, can be determined from the pressure dependence of χ_v, but that a determination of the entropy, Δs_v, requires a measurement of the absolute concentration of vacancies at some temperature and pressure.

5.3.1 Positron annihilation

The absolute vacancy concentration can be measured by positron annihilation. Positrons annihilate more frequently in vacancies where free electrons tend to collect. The rate of positron annihilation in metals can be calibrated to give the fraction of vacancies in the crystals at a given temperature.

4 The internal pressure inside the crystal due to the capillarity effect is actually $p_c = 2f_s/R$, where f_s is the surface stress, which is different from the surface energy γ_s for a solid [28]. Yet it is the surface energy γ_s that enters the equilibrium vacancy fraction expression (5.61).

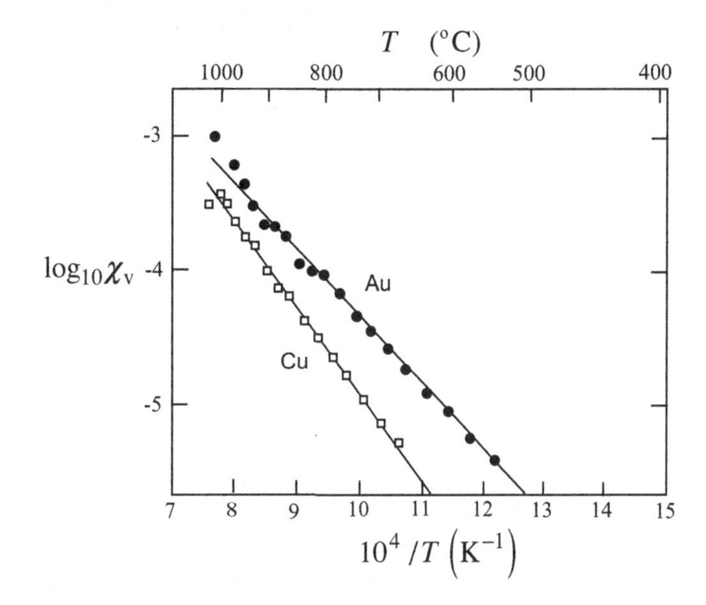

Figure 5.9. Equilibrium vacancy fractions as a function of temperature for Au (filled symbols) and Cu (open symbols) from positron annihilation experiments [26].

Figure 5.9 shows the vacancy fraction as a function of the reciprocal of the absolute temperature $1/T$, measured by the positron annihilation technique in Au (filled symbols) and Cu (open symbols). The vacancy fraction χ_v is a dimensionless number representing the fraction of atomic sites that are vacant; it is proportional to the vacancy concentration c_v, which is number of vacancies per unit volume.

The Boltzmann relation, Eq. (5.45), indicates that the equilibrium fraction of vacancies should depend exponentially on $1/T$. Thus the logarithm of equilibrium vacancy fraction should vary linearly with $1/T$, as shown in Fig. 5.9. We see that at the melting temperature the equilibrium vacancy fraction is indeed about 10^{-3} as we estimated above.

5.3.2 Measurement of Δh_v by quench-resistivity

The vacancy formation enthalpy Δh_v can be obtained from the slope of the $\ln \chi_v$ vs $1/T$ curve shown in Fig. 5.9. However, positron annihilation experiments require highly specialized apparatus. The temperature dependence of χ_v, and hence Δh_v, can be measured by other techniques that are easier to carry out in the lab.

The temperature dependence of χ_v can be found by quenching metal wires from various high temperatures and measuring the electrical resistivity to obtain information about the vacancy concentration. The experiment is as follows.

- Anneal metal wires at various high temperatures to establish various equilibrium vacancy fractions, $\chi_v(T)$.
- Cool the wires rapidly (quench) to a very low temperature, say the temperature of liquid nitrogen, where the vacancies are essentially immobile. This allows the vacancies created at

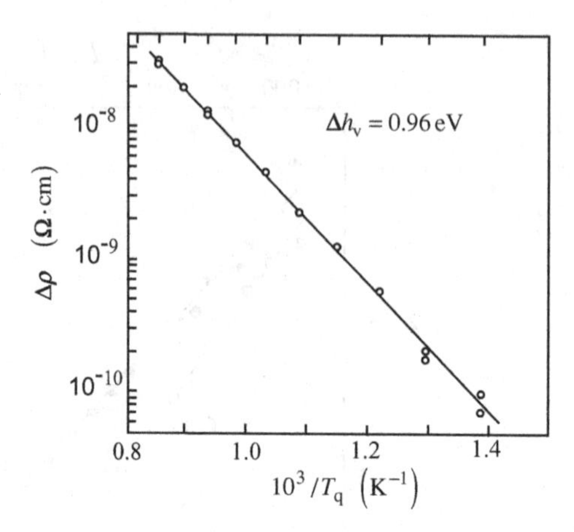

Figure 5.10. Semilogarithmic plot of quenched-in resistivity ($\Delta\rho$) versus reciprocal of the absolute quench temperature (T_q) for gold [29].

the various high temperatures to be retained in the wire. Essentially all of the vacancies remain in the wires during the quench.

- Measure the electrical resistivity of the quenched wires, $\rho(T)$. The measured resistivities ρ depend on the temperatures T from which the wires were quenched.
- Also slowly cool a wire to the same low temperature and measure the electrical resistivity, ρ_0. This is the resistivity of the wire essentially without vacancies as most of the vacancies leave the wire during the slow cool.

We note that all resistivity measurements are conducted at the same low temperature. This is to ensure all other contributions to electrical resistivity remain constant. For example, the contribution from electron scattering with phonons (i.e. lattice vibration waves) is strongly temperature dependent and can overshadow the contribution from vacancy density variations, if the resistivity were measured at different temperatures.

It is reasonable to expect the change in electrical resistivity of the wire measured in the above procedure to be proportional to the concentration of quenched-in vacancies

$$\Delta\rho \equiv \rho(T) - \rho_0 = \alpha\chi_v(T) \tag{5.63}$$

assuming that no vacancies are lost in the quenching process and that no vacancies are present in the slowly cooled wire. If so, then we may write

$$\Delta\rho = \alpha \exp\left(\frac{\Delta s_v}{k_B}\right) \exp\left(-\frac{\Delta h_v}{k_B T}\right). \tag{5.64}$$

Taking logarithms of both sides we have

$$\ln(\Delta\rho) = \ln\alpha + \frac{\Delta s_v}{k_B} - \frac{\Delta h_v}{k_B T}. \tag{5.65}$$

We then plot $\ln(\Delta\rho)$ versus $1/T$ and find the slope $d\ln(\Delta\rho)/d(1/T)$, which equals $-\Delta h_v/k_B$. The enthalpy of formation of a vacancy can be found from the slope of this plot. Figure 5.10 shows such a plot for gold where the formation enthalpy (or energy) is 0.96 eV.

The measurement of the vacancy formation volume can similarly be determined from the pressure dependence of the equilibrium vacancy concentration at some fixed temperature. Using

$$\chi_v = \chi_0 \exp\left(-\frac{p_{\text{ext}}\Delta v_v}{k_B T}\right) \tag{5.66}$$

we can write

$$\begin{aligned} \ln\left(\frac{\chi_v}{\chi_0}\right) &= \ln \chi_v - \ln \chi_0 = -\frac{p_{\text{ext}}\Delta v_v}{k_B T} \\ \Delta v_v &= -k_B T \frac{d \ln \chi_v}{d p_{\text{ext}}}. \end{aligned} \tag{5.67}$$

So one might subject wires to different pressures and quench to a low temperature to measure the pressure dependence of the electrical resistivity, $\Delta\rho(p_{\text{ext}})$, and from that determine the formation volume of a vacancy, Δv_v.

The formation enthalpy and volume can be determined from the dependence of the electrical resistivity on the temperature or pressure, respectively, without knowing the proportional constant α in Eq. (5.63), and hence without knowing the absolute vacancy concentration. However, measurement of the formation entropy requires an absolute measurement of the vacancy concentration.

5.3.3 Measurement of Δs_v by lattice and thermal expansion

As an alternative to the positron annihilation method, the absolute vacancy concentration can also be measured by combining dilatometry (crystal size expansion) and X-ray diffraction (lattice expansion) measurements. When a wire is heated to high temperatures it expands in volume or length by two processes. Ordinary thermal expansion of a perfect crystal involves an increase in the mean spacing between atoms in the crystal lattice

$$\left(\frac{\Delta V}{V_0}\right)_{\text{lattice expansion}} = 3\left(\frac{\Delta a^{\text{T}}}{a_0}\right), \tag{5.68}$$

where V_0 is the original volume of the crystal, ΔV is the change of volume, a_0 is the original lattice parameter, and Δa^{T} is the change of lattice parameter by thermal expansion alone. In addition, the volume and length increase because vacancies are formed, leading to an increased number of lattice sites, as shown in Fig. 5.11. For each vacancy formed a new lattice site is created. The volumetric change associated with the lattice growth is

$$\left(\frac{\Delta V}{V_0}\right)_{\text{lattice growth}} = \chi_v\left(\frac{\Delta v_v}{\Omega}\right), \tag{5.69}$$

where Ω is the atomic volume. This volume change may be understood as follows. If the lattice were perfectly rigid, then χ_v would be the fractional volume change associated with vacancy formation. But because typically $\Delta v_v < \Omega$, the actual volume change is less by the factor $\Delta v_v/\Omega$.

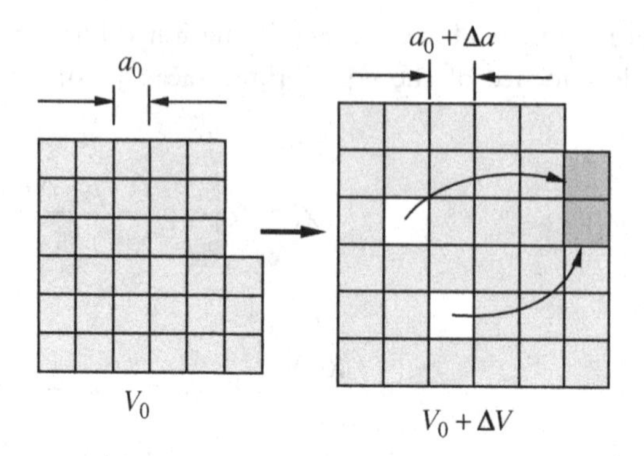

Figure 5.11. Volume change of crystal by both thermal expansion and vacancy formation.

The relaxation volume of vacancies also reduces the average lattice spacing, so that the relative change of lattice spacing measured by X-ray diffraction is

$$\frac{\Delta a}{a_0} = \frac{\Delta a^{\mathrm{T}}}{a_0} + \frac{1}{3}\chi_{\mathrm{v}}\left(\frac{\Delta V_{\mathrm{v}}^{\mathrm{rlx}}}{\Omega}\right). \tag{5.70}$$

The total change in volume, which can be measured by dilatometry (direct measurements of the length of the wire), is the sum of the contributions from thermal expansion and vacancies.

$$\begin{aligned}\left(\frac{\Delta V}{V_0}\right)_{\mathrm{total}} &= 3\left(\frac{\Delta L}{L_0}\right)_{\mathrm{total}} = \left(\frac{\Delta V}{V_0}\right)_{\mathrm{lattice\ expansion}} + \left(\frac{\Delta V}{V_0}\right)_{\mathrm{lattice\ growth}} \\ &= 3\left(\frac{\Delta a^{\mathrm{T}}}{a_0}\right) + \chi_{\mathrm{v}}\left(\frac{\Delta v_{\mathrm{v}}}{\Omega}\right) = 3\left(\frac{\Delta a}{a_0}\right) + \chi_{\mathrm{v}}\end{aligned} \tag{5.71}$$

where L_0 is the original length of the wire at low temperature and ΔL is the elongation due to temperature increase. So the vacancy fraction can then be measured by

$$\chi_{\mathrm{v}} = 3\left(\frac{\Delta L}{L_0}\right)_{\mathrm{total}} - 3\left(\frac{\Delta a}{a_0}\right). \tag{5.72}$$

This allows the determination of the absolute fraction of the vacancies χ_{v} from the difference in the fractional length change (dilatometry) and the fractional change in the lattice spacing (X-rays). As an example, the measured data for gold is shown in Fig. 5.12. If Δh_{v} and Δv_{v} are already known from previous experiments, then the formation entropy Δs_{v} can be obtained from Eq. (5.62).

Table C.1 gives a compilation of the thermodynamic properties of vacancies for pure metals, taken from [26]. Given that $p_{\mathrm{ext}}\Delta v_{\mathrm{v}}$ is negligible at atmospheric pressure, the formation enthalpy Δh_{v} and the formation energy Δe_{v} are essentially the same in this table.

5.4 Point defect chemical potential

The material in this section is more advanced than previous sections of this chapter. The main purpose is to properly define the chemical potential of the point defects we have introduced so far. It is placed here because the chemical potential is easier to define, for each type of point defect individually, before we embark on the coexistence of several types of point defects in a

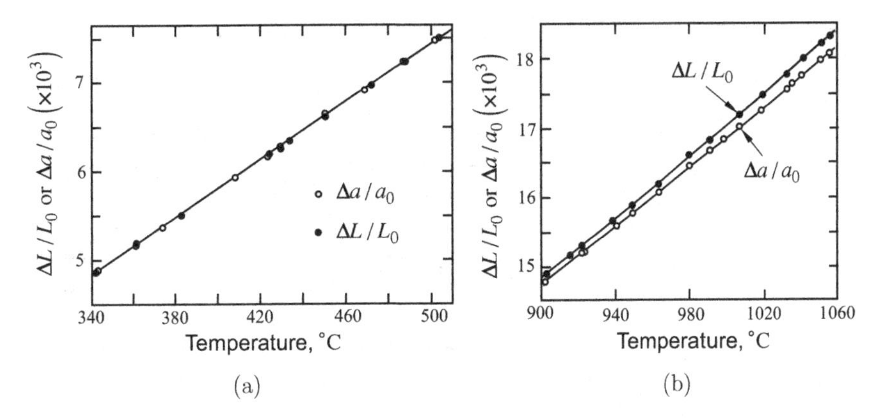

Figure 5.12. Measured length expansion $\Delta L/L_0$ and lattice parameter expansion $\Delta a/a_0$ versus temperature for gold in the (a) 340 °C–510 °C temperature interval and in the (b) 900 °C–1060 °C temperature interval [29]. In (a) $\Delta L/L_0$ and $\Delta a/a_0$ agree within experimental error, indicating negligible vacancy concentration. In (b) $\Delta L/L_0 > \Delta a/a_0$ corresponding to the thermal generation of vacancies at high temperature.

crystal. The reader could skip this section and proceed to most of the sections in Chapter 6 and Chapter 7 without a problem. However, the material introduced here will be needed in Sections 6.2, 6.3, and 7.4.

The chemical potential $\Delta\mu$ of the point defects has many similarities with the formation Gibbs free energy Δg_f of the defects, e.g. they both have enthalpic and entropic contributions, but they are not identical. The chemical potential is more convenient to use than the formation Gibbs free energy in a number of scenarios. For example, the impurity atoms in the crystal may exist in the gas phase when taken out of the crystal, such as hydrogen in nickel. The equilibrium concentration of the impurity in the crystal depends on the pressure of the gas. This dependence can be derived most conveniently using the chemical potential, as will be shown below. Another example is the diffusion of point defects driven by non-equilibrium conditions. The variation of the chemical potential in space provides the driving force for diffusion, as will be shown in Section 7.4. Point defects diffuse in a way that tends to reduce this variation. The equilibrium condition is reached if and only if the chemical potential is constant everywhere in the crystal.

5.4.1 Interstitial point defects

To properly define the chemical potential for interstitial point defects, we need to define a process in which the solute atom (B) is taken from some reference state and placed onto an unoccupied interstitial site inside the host crystal. The reference state is a hypothetical material for which the chemical potential for B atoms is defined to be zero. For simplicity, we consider the reference state as a perfect crystal formed entirely by B atoms, as shown in Fig. 5.13. The reference crystal is always under zero stress, even though the host crystal for the interstitial defects can be subjected to non-zero stress.

Let N_s be the total number of interstitial sites in the crystal, N_B be the actual number of interstitial defects, and $\chi_i = N_B/N_s$. Following Eq. (3.27), the chemical potential of interstitial atoms is defined as the derivative of the Gibbs free energy change of the entire system

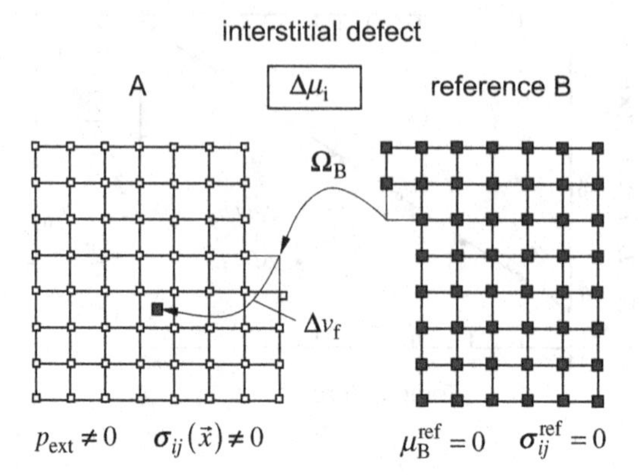

Figure 5.13. The process required in the definition of chemical potential of interstitial point defects. An atom is taken from the reference B crystal and inserted into the A crystal.

(host crystal + reference crystal) with respect to N_B,

$$\Delta\mu_i \equiv \frac{\partial \Delta G}{\partial N_B}. \tag{5.73}$$

Similar to Section 5.1.4, ΔG has contributions from a term that is proportional to N_B and another term that accounts for the configurational entropy:

$$\Delta G = \chi_i N_s(\Delta\tilde{e}_f + p_{ext}\Delta\tilde{v}_f - T\Delta\tilde{s}_f) + N_s k_B T[\chi_i \ln \chi_i + (1 - \chi_i)\ln(1 - \chi_i)]. \tag{5.74}$$

There are important conceptual differences between the quantities $\Delta\tilde{e}_f$, $\Delta\tilde{v}_f$, $\Delta\tilde{s}_f$, introduced here, and the quantities Δe_f, Δv_f, Δs_f, defined in Section 5.1, which enter the expression of the formation Gibbs free energy Δg_f. We shall see that the main difference is $\Delta\tilde{v}_f = \Delta v_f + \Omega_B$, as shown in Fig. 5.13. The differences arise because here when an atom is inserted to the interstitial site, it is taken from a reference crystal, instead of from the interface between the host crystal and the reservoir of B atoms. The differences are important when the host crystal is subjected to stress, because the reference crystal is always under zero stress, as discussed below.

- $\Delta\tilde{e}_f$: When the host crystal is under zero stress (other than those caused by the interstitials themselves), $\Delta\tilde{e}_f = \Delta e_f$. However, when the host crystal is under stress, then $\Delta\tilde{e}_f = \Delta e_f + w\Omega$, where $w > 0$ is the strain energy density that B must adopt when added to the A-rich crystal. The expression for w under hydrostatic pressure is given by Eq. (3.47). Because w is only of the second order of stress, the difference between $\Delta\tilde{e}_f$ and Δe_f is often ignored in rough estimates.
- $\Delta\tilde{v}_f$: When an interstitial atom is moved from the reference crystal to the host crystal, the volume of the host crystal expands by the effective volume $\Omega_B^\dagger$. Therefore, $\Delta\tilde{v}_f = \Omega_B^\dagger$. This is to be compared with $\Delta v_f = \Omega_B^\dagger - \Omega_B$, as given by Eq. (5.36).
- $\Delta\tilde{s}_f$: If we ignore the change of atomic vibration frequency with stress, then $\Delta\tilde{s}_f \approx \Delta s_f$. Otherwise, there may be a small difference (see Exercise problem 3.6).

Differentiating Eq. (5.74) gives the chemical potential for the interstitial defect in the host crystal,

$$\Delta\mu_{\rm i} \equiv \frac{\partial \Delta G}{\partial N_{\rm B}} = (\Delta\tilde{e}_{\rm f} + p_{\rm ext}\Delta\tilde{v}_{\rm f} - T\Delta\tilde{s}_{\rm f}) + k_{\rm B}T\ln\frac{\chi_{\rm i}}{1-\chi_{\rm i}}. \tag{5.75}$$

When the host crystal is subjected to an inhomogeneous stress field, $\sigma_{ij}(\vec{x})$, caused by either external loading or other defects (such as dislocations), the equilibrium interstitial distribution $\chi_{\rm i}^{\rm eq}(\vec{x})$ is inhomogeneous, in order to make the chemical potential $\Delta\mu_{\rm i}$ a constant throughout the crystal. The equilibrium interstitial distribution can be obtained by solving the following equation,

$$\Delta\mu_{\rm i} = (\Delta\tilde{e}_{\rm f}(\vec{x}) + p(\vec{x})\Delta\tilde{v}_{\rm f} - T\Delta\tilde{s}_{\rm f}) + k_{\rm B}T\ln\frac{\chi_{\rm i}^{\rm eq}(\vec{x})}{1-\chi_{\rm i}^{\rm eq}(\vec{x})}, \tag{5.76}$$

where we express the spatial variation of $\Delta\tilde{e}_{\rm f}$ and p explicitly and assume $\Delta\tilde{v}_{\rm f}$ and $\Delta\tilde{s}_{\rm f}$ to remain constant. In particular, $p(\vec{x}) = -\sigma_{ii}(\vec{x})/3$. The equilibrium interstitial distribution at chemical potential $\Delta\mu_{\rm i}$ is

$$\chi_{\rm i}^{\rm eq}(\vec{x}) = \frac{1}{1+\exp\left[\frac{1}{k_{\rm B}T}(\Delta\tilde{e}_{\rm f}(\vec{x}) + p(\vec{x})\Delta\tilde{v}_{\rm f} - T\Delta\tilde{s}_{\rm f} - \Delta\mu_{\rm i})\right]}. \tag{5.77}$$

This expression resembles Eq. (5.39), but there are also important differences. The most significant difference is that $\Delta v_{\rm f} = \Omega_{\rm B}^{\dagger} - \Omega_{\rm B}$ and $\Delta\tilde{v}_{\rm f} = \Omega_{\rm B}^{\dagger}$. We will use Eq. (5.77) to discuss the segregation of point defects around an edge dislocation in Section 12.5.

Equilibrium concentration of hydrogen interstitials in metals

To give an example of how to use the expressions obtained above, we consider the equilibrium concentration of hydrogen interstitials in a metal at equilibrium with hydrogen gas. In this system, the hydrogen atoms can exist either as interstitial atoms in the metal, or as diatomic molecules in the gas. Consider the process of two hydrogen interstitial atoms escaping the metal and entering the gas phase as a diatomic molecule. We may describe this process as

$$\mathrm{H_i} + \mathrm{H_i} + \Delta G \rightarrow \mathrm{H_{2(g)}}, \tag{5.78}$$

where the subscript i indicates interstitial, and ΔG is the change of Gibbs free energy of the entire system caused by this process. Specifically,

$$\Delta G = \frac{\partial G}{\partial N_{\rm H_{2(g)}}} - 2\frac{\partial G}{\partial N_{\rm H_i}} = \mu_{\rm H_{2(g)}} - 2\mu_{\rm H_i}, \tag{5.79}$$

where $N_{\rm H_{2(g)}}$ is the number of H_2 molecules in the gas phase and $N_{\rm H_i}$ is the number of hydrogen interstitial atoms in the metal; $\mu_{\rm H_{2(g)}}$ and $\mu_{\rm H_i}$ are chemical potentials of the H_2 molecules and interstitial atoms, respectively.

Let $\mu^0_{\rm H_{2(g)}}$ be the chemical potential of the H_2 molecules at the standard pressure p_0 (e.g. 1 bar). Then according to Eq. (3.34), the chemical potential of H_2 at a different pressure p (under the same temperature T) is

$$\mu_{\rm H_{2(g)}} = \mu^0_{\rm H_{2(g)}} + k_{\rm B}T\ln\frac{p}{p_0}. \tag{5.80}$$

When thermal equilibrium is reached, the change in the Gibbs free energy of the system for the process described by Eq. (5.78) must be zero, $\Delta G = 0$. Therefore,

$$\mu_{\mathrm{H_i}} = \frac{1}{2}\mu_{\mathrm{H_{2(g)}}} = \frac{1}{2}\mu^0_{\mathrm{H_{2(g)}}} + \frac{1}{2}k_B T \ln \frac{p}{p_0} \tag{5.81}$$

At the same time, the chemical potential of the hydrogen interstitial atoms is related to its fraction according to Eq. (5.75),

$$\mu_{\mathrm{H_i}} = (\Delta \tilde{e}_f + p\Delta \tilde{v}_f - T\Delta \tilde{s}_f) + k_B T \ln \frac{\chi_{\mathrm{H_i}}}{1-\chi_{\mathrm{H_i}}}. \tag{5.82}$$

For a realistic range of gas pressure p, it is reasonable to assume that the term $p\,\Delta\,\tilde{v}_f$ is negligible. If, in addition, we assume $\chi_{\mathrm{H_i}} \ll 1$, then,

$$\mu_{\mathrm{H_i}} \approx (\Delta \tilde{e}_f - T\Delta \tilde{s}_f) + k_B T \ln \chi_{\mathrm{H_i}}. \tag{5.83}$$

Combining Eqs. (5.81) and (5.83), we have

$$(\Delta \tilde{e}_f - T\Delta \tilde{s}_f) + k_B T \ln \chi_{\mathrm{H_i}} = \frac{1}{2}\mu^0_{\mathrm{H_{2(g)}}} + \frac{1}{2}k_B T \ln \frac{p}{p_0}. \tag{5.84}$$

Define $\chi^0_{\mathrm{H_i}}$ as the equilibrium fraction of hydrogen interstitials in the metal at equilibrium with H_2 gas of the standard pressure p_0. Then

$$(\Delta \tilde{e}_f - T\Delta \tilde{s}_f) + k_B T \ln \chi^0_{\mathrm{H_i}} = \frac{1}{2}\mu^0_{\mathrm{H_{2(g)}}} \tag{5.85}$$

so that

$$\begin{aligned} k_B T \ln \frac{\chi_{\mathrm{H_i}}}{\chi^0_{\mathrm{H_i}}} &= \frac{1}{2}k_B T \ln \frac{p}{p_0} \\ \ln \frac{\chi_{\mathrm{H_i}}}{\chi^0_{\mathrm{H_i}}} &= \ln \sqrt{\frac{p}{p_0}}. \end{aligned} \tag{5.86}$$

Therefore,

$$\chi_{\mathrm{H_i}} = \chi^0_{\mathrm{H_i}}\sqrt{\frac{p}{p_0}}. \tag{5.87}$$

This means that the solubility of hydrogen (or any diatomic gas which behaves as an ideal gas) in the solid is proportional to the square root of the pressure of the gas phase.[5] This is called Sievert's law.

5.4.2 Substitutional point defects

When defining chemical potentials for substitutional point defects, we need to consider a process in which an A atom is removed from the crystal and placed on a reference crystal, and a B atom

5 If the gas phase is a mixture of several components, then the solubility is proportional to the partial pressure of the gas molecules of interest.

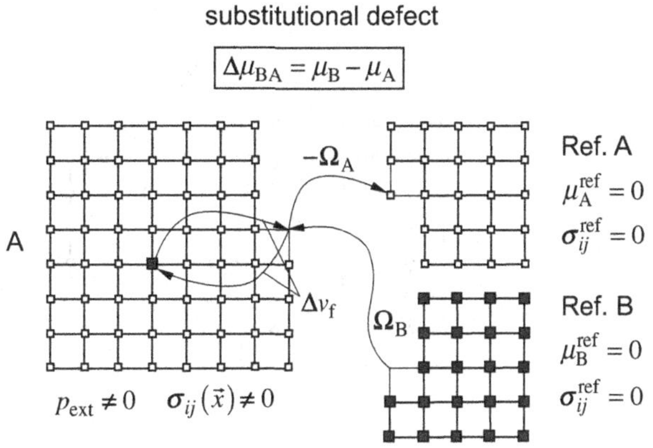

Figure 5.14. The process required in the definition of chemical potential difference between B and A atoms. An A atom is removed from the host crystal and placed on the reference A crystal, and a B atom is taken from a reference B crystal and inserted into the site just vacated by the A atom.

is taken from another reference crystal and inserted into the site just vacated by the A atom. Therefore, we now need to consider two reference crystals: one for A atoms and one for B atoms, both under zero stress, as shown in Fig. 5.14. Because the process increases the number of B atoms by one and decreases the number of A atoms by one, the required work equals the chemical potential difference between B and A atoms, i.e. $\Delta\mu_{\rm BA} \equiv \mu_{\rm B} - \mu_{\rm A}$.

Let N be the total number of atoms in the host crystal. Let $N_{\rm B}$ be the number of substitutional impurities. In the process defined here, the total number of atoms $N_{\rm A} + N_{\rm B}$ stays constant. This is different from the process depicted in Fig. 5.1.

The chemical potential difference $\Delta\mu_{\rm BA}$ associated with the substitutional defects is defined as the derivative of the Gibbs free energy change ΔG of the entire system (host crystal + two reference crystals) with respect to $N_{\rm B}$, subject to the constraint that $N_{\rm A} = N - N_{\rm B}$,

$$\Delta\mu_{\rm BA} \equiv \left.\frac{\partial \Delta G}{\partial N_{\rm B}}\right|_{N_{\rm A}=N-N_{\rm B}} = (\Delta\tilde{e}_{\rm f} + p_{\rm ext}\Delta\tilde{v}_{\rm f} - T\Delta\tilde{s}_{\rm f}) + k_{\rm B}T\ln\frac{\chi}{1-\chi}, \tag{5.88}$$

which has the same form as Eq. (5.75). $\Delta\tilde{e}_{\rm f}$, $\Delta\tilde{v}_{\rm f}$, and $\Delta\tilde{s}_{\rm f}$ are the energy, volume and entropy changes associated with introducing one substitutional point defect. The equilibrium distribution of substitutional point defects in a crystal subjected to inhomogeneous stress fields also has the same form as Eq. (5.77),

$$\chi_{\rm eq}(\vec{x}) = \frac{1}{1 + \exp\left[\frac{1}{k_{\rm B}T}(\Delta\tilde{e}_{\rm f}(\vec{x}) + p(\vec{x})\Delta\tilde{v}_{\rm f} - T\Delta\tilde{s}_{\rm f} - \Delta\mu_{\rm BA})\right]}. \tag{5.89}$$

The equilibrium distribution here has the Fermi–Dirac form, and is different from the Boltzmann form of Eq. (5.30). In addition, $\Delta\tilde{e}_{\rm f}$, $\Delta\tilde{v}_{\rm f}$, and $\Delta\tilde{s}_{\rm f}$ are different from $\Delta e_{\rm f}$, $\Delta v_{\rm f}$, and $\Delta s_{\rm f}$, as already noted in Section 5.4.1. The main difference is that $\Delta\tilde{v}_{\rm f} = \Delta v_{\rm f} + \Omega_{\rm B} - \Omega_{\rm A}$, as shown in Fig. 5.14. Recall that $\Delta v_{\rm f} = \Omega^*_{\rm B} - \Omega_{\rm B}$, so that $\Delta\tilde{v}_{\rm f} = \Omega^*_{\rm B} - \Omega_{\rm A}$.

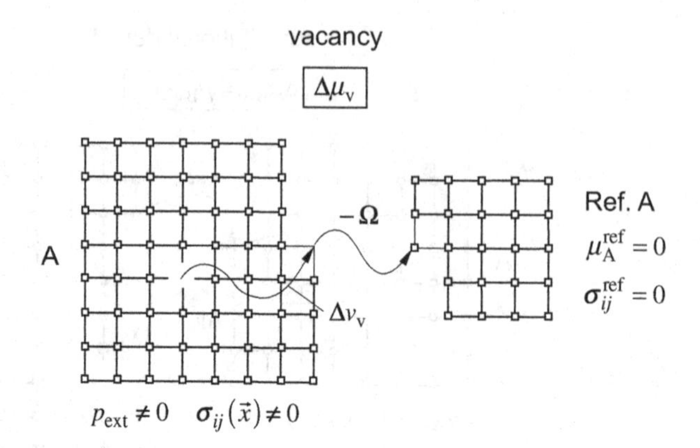

Figure 5.15. The process required in the definition of chemical potential for vacancies. An atom is removed from the host crystal and placed on the reference crystal.

5.4.3 Vacancies

To properly define the chemical potential for vacancies, we need to define a process in which an atom is removed from the crystal interior and placed on a reference crystal, as shown in Fig. 5.15. As a result, the total number of atoms in the host crystal is not conserved.

We consider a crystal containing N_s atomic sites, out of which N_v are vacant, so that the vacancy fraction is $\chi_v = N_v/N_s$. The Gibbs free energy change of the system (host crystal + reference crystal) can be written as

$$\Delta G = \chi_v N_s(\Delta\tilde{e}_v + p\Delta\tilde{v}_v - T\Delta\tilde{s}_v) + N_s k_B T[\chi_v \ln \chi_v + (1 - \chi_v) \ln (1 - \chi_v)], \quad (5.90)$$

where $\Delta\tilde{e}_v$, $\Delta\tilde{v}_v$, and $\Delta\tilde{s}_v$ are the energy, volume, and vibrational entropy changes associated with introducing one vacancy. They are different from the quantities, Δe_v, Δv_v, and Δs_v, defined in Section 5.2. The most significant difference is that $\Delta\tilde{v}_v = \Delta v_v - \Omega = \Delta V_v^{rlx}$, as shown in Fig. 5.15. These differences will be discussed further below. The chemical potential of the vacancies is defined as

$$\Delta\mu_v \equiv \frac{\partial \Delta G}{\partial N_v} = \Delta\tilde{e}_v + p\Delta\tilde{v}_v - T\Delta\tilde{s}_v + k_B T \ln \frac{\chi_v}{1 - \chi_v}. \quad (5.91)$$

Recall that even at the melting point χ_v is only about 10^{-3}, so that under most conditions $\chi_v \ll 1$, and

$$\Delta\mu_v \approx \Delta\tilde{e}_v + p\Delta\tilde{v}_v - T\Delta\tilde{s}_v + k_B T \ln \chi_v. \quad (5.92)$$

In the hypothetical case of an infinitely large crystal, subjected to an inhomogeneous stress field, and having no vacancy sources and sinks, the vacancies can reach an equilibrium distribution $\chi_v^{eq}(\vec{x})$ such that $\Delta\mu_v$ is constant everywhere,

$$\chi_v^{eq}(\vec{x}) = \frac{1}{1 + \exp\left[\frac{1}{k_B T}(\Delta\tilde{e}_v(\vec{x}) + p(\vec{x})\Delta\tilde{v}_v - T\Delta\tilde{s}_v - \Delta\mu_v)\right]}, \quad (5.93)$$

$$\approx \exp\left[-\frac{1}{k_B T}(\Delta\tilde{e}_v(\vec{x}) + p(\vec{x})\Delta\tilde{v}_v - T\Delta\tilde{s}_v - \Delta\mu_v)\right]. \quad (5.94)$$

However, this equation is not very useful because every crystal has vacancy sources and sinks (e.g. surfaces and dislocations). These sources and sinks make it generally impossible for the vacancies to reach equilibrium if the crystal is subjected to inhomogeneous stress, and/or if the stress is not purely hydrostatic, as will be further discussed in Chapter 7.

We now discuss the difference between the quantities $\Delta\tilde{e}_v$, $\Delta\tilde{v}_v$, $\Delta\tilde{s}_v$ defined here and the quantities Δe_v, Δv_v, Δs_v defined in Section 5.2. The differences arise because here the removed atom is placed on an unstressed reference crystal, while in Section 5.2 the removed atom is placed on the surface of the same (stressed) crystal.

- $\Delta\tilde{e}_v$: As a rough estimate, $\Delta\tilde{e}_v$ is similar to Δe_v, which is on the order of 1 eV. To be more precise, $\Delta\tilde{e}_v = \Delta e_v - w\Omega$, where $w > 0$ is the strain energy density and $w\Omega$ is the strain energy per atom in the stressed crystal. The negative sign in front of w appears because we remove an atom from the crystal when creating a vacancy.
- $\Delta\tilde{v}_v$: In the process considered here, the number of atomic sites N_s does not grow when an atom is removed and placed on the reference crystal. For a perfectly rigid crystal structure, $\Delta\tilde{v}_v$ is zero. Considering the inward relaxation of atoms around the vacancy, we have $\Delta\tilde{v}_v = \Delta v_v - \Omega = \Delta V_v^{\rm rlx} \approx -0.3\Omega$ to -0.5Ω, which is defined in Eq. (5.46).
- $\Delta\tilde{s}_v$: If we ignore the change of atomic vibration frequency with stress, then $\Delta\tilde{s}_v$ is approximately the same as Δs_v, which is about 2–8 k_B.

Vacancy chemical potential at zero stress

In order to build intuition about the quantity $\Delta\mu_v$, we now discuss the vacancy chemical potential in a solid under zero stress. Notice that the exact expression of the equilibrium vacancy fraction, when expressed in terms of the chemical potential, Eq. (5.93), has the Fermi–Dirac form. But the result obtained earlier by minimizing the Gibbs free energy of the crystal, Eq. (5.45), has the Boltzmann form. This may seem surprising. Here we show that both expressions are correct and there is no conflict.

When the solid is under zero stress, the relation, Eq. (5.91), between the the vacancy chemical potential and the vacancy fraction χ_v (which may or may not be at equilibrium), can be simplified to

$$\Delta\mu_v = \Delta e_v - T\Delta s_v + k_B T \ln\frac{\chi_v}{1-\chi_v}, \tag{5.95}$$

$$\chi_v = \frac{1}{1+\exp\left[\frac{1}{k_B T}\left(\Delta e_v - T\Delta s_v - \Delta\mu_v\right)\right]}. \tag{5.96}$$

At the same time, the equilibrium vacancy fraction is related to the vacancy formation energy and entropy through

$$\Delta e_v - T\Delta s_v = -k_B T \ln \chi_v^{\rm eq}. \tag{5.97}$$

Equation (5.97) states that the vacancy chemical potential in a solid at equilibrium under zero stress has a positive value, and is approximately the product of $k_B T$ and the equilibrium vacancy fraction $\chi_v^{\rm eq}$.

To understand this result, we recall the Gibbs free energy change to the crystal, ΔG, caused by introducing vacancies, as given in Eq. (5.44). The equilibrium vacancy fraction

χ_v^{eq} is obtained by minimizing ΔG with respect to N_v. If we now substitute the equilibrium value of N_v into Eq. (5.44), we obtain the Gibbs free energy of the crystal at thermal equilibrium

$$\Delta G^{eq} = N_v(\Delta e_v - T\Delta s_v) + k_B T\left[N_v \ln \chi_v^{eq} + N_A \ln(1 - \chi_v^{eq})\right]. \tag{5.98}$$

Because

$$k_B T \ln \chi_v^{eq} = -(\Delta e_v - T\Delta s_v) \tag{5.99}$$

we have

$$\Delta G^{eq} = N_A k_B T \ln(1 - \chi_v^{eq}). \tag{5.100}$$

This means that the chemical potential for the A atoms in the crystal with an equilibrium concentration of vacancies is

$$\Delta\mu_A = \frac{\partial \Delta G^{eq}}{\partial N_A} = k_B T \ln(1 - \chi_v^{eq}). \tag{5.101}$$

In the definition of chemical potential, creating a vacancy is equivalent to removing an atom. Therefore, we should expect $\Delta\mu_v = -\Delta\mu_A$, which is consistent with Eq. (5.97).

Vacancy chemical potential under uniform hydrostatic stress

To further develop our intuition, we now consider a solid subjected to a uniform, purely hydrostatic stress. In this case, the vacancy formation free energy Δg_v remains well defined. We can use this case to discuss the difference between the vacancy chemical potential $\Delta\mu_v$ and the formation free energy Δg_v. Here the stress field can be characterized by the pressure p, which is a constant. From Eq. (5.92) and Fig. 5.15, the vacancy chemical potential is

$$\Delta\mu_v \approx \Delta\tilde{e}_v + p\Delta\tilde{v}_v - T\Delta\tilde{s}_v + k_B T \ln \chi_v,$$

where

$$\begin{aligned} \Delta\tilde{e}_v &= \Delta e_v - w\Omega = \Delta e_v - \frac{p^2\Omega}{2B}, \\ \Delta\tilde{v}_v &= \Delta v_v - \Omega = \Delta V_v^{rlx}, \\ \Delta\tilde{s}_v &\approx \Delta s_v, \end{aligned} \tag{5.102}$$

and w is the strain energy density in the solid, and B is the bulk modulus. Therefore

$$\Delta\mu_v \approx \Delta g_v - \frac{p^2\Omega}{2B} - p\Omega + k_B T \ln \chi_v, \tag{5.103}$$

where $p^2\Omega/(2B)$ represents the strain energy difference per atom in the uniformly stressed crystal and the stress-free reference crystal, and $p\Omega$ represents the work done against the external

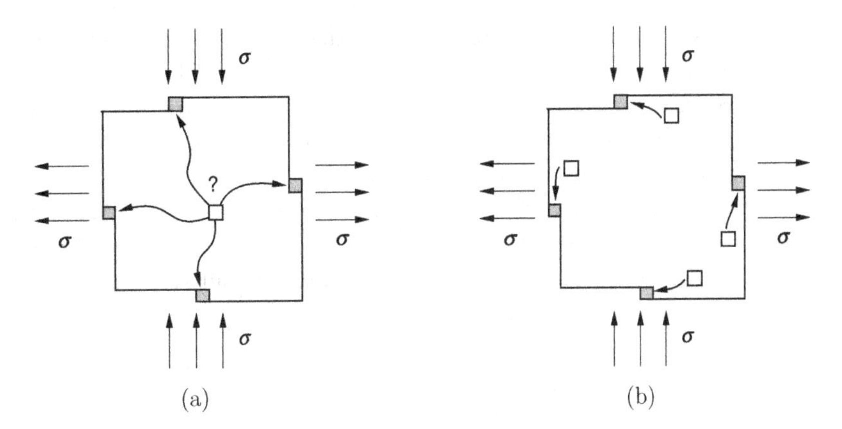

Figure 5.16. Vacancy creation under shear stress. (a) When an atom inside the crystal is removed, it is not clear on which surface it should be placed, leading to ambiguity of the vacancy formation free energy. (b) The ambiguity is removed if only atoms in the vicinity of the surface can be removed to create vacancies.

pressure when the removed atom is placed on the surface of the stressed crystal. Equation (5.103) shows that $\Delta\mu_v$ includes the configurational entropy effect, while Δg_v does not.

To fully express the dependence of $\Delta\mu_v$ on pressure p, we note that Δg_v also depends on p. The most significant effect is the work done by the external pressure $p\Delta v_v$ related to the formation volume of the vacancy, Δv_v. In addition, if we model the vacancy as a spherical hole of volume Ω in an infinite isotropic elastic medium, then an elastic energy of $3p^2\Omega/(8\mu)$ will be relaxed in the surrounding matrix when the spherical hole is created. Therefore,

$$\Delta g_v \approx \Delta g_v^0 - \frac{3p^2\Omega}{8\mu} + p\Delta v_v, \tag{5.104}$$

where Δg_v^0 is the vacancy formation free energy at zero stress. Therefore[6]

$$\Delta\mu_v \approx \Delta g_v^0 - \frac{p^2\Omega}{2B}\left(1 + \frac{3B}{4\mu}\right) + p\Delta V_v^{\text{rlx}} + k_B T \ln \chi_v. \tag{5.105}$$

Vacancy chemical potential under uniform non-hydrostatic stress

We now discuss the complications that arise when the solid is subjected to a uniform but non-hydrostatic stress. We use this example to further illustrate the differences between the vacancy chemical potential $\Delta\mu_v$ and the formation free energy Δg_v.

For simplicity, we consider a square grain subjected to a uniform shear stress, as shown in Fig. 5.16. The vacancy formation free energy Δg_v cannot be uniquely defined here. The definition of Δg_v involves a process in which an atom is removed from the crystal and placed on the crystal surface. When an atom inside the crystal is removed, it is not clear on which surface it should be placed, as shown in Fig. 5.16a, and different choices lead to different amounts of work required to create the vacancy, making the vacancy formation free energy ambiguous. If, based

6 Note that in most cases, $p \ll B$, so that the term proportional to p^2 is usually much smaller than the term linear with p, and is often ignored.

on physical intuition, we impose the constraint that only atoms very close to the surface can be removed when creating the vacancy, then there is no ambiguity in where the atoms should be placed, as shown in Fig. 5.16b. However, even in this case, the vacancy formation free energy can be defined only in the region very close to the surfaces. The vacancy formation free energy deep inside the crystal remains undefined.

On the other hand, there is no ambiguity in defining the chemical potential of vacancies everywhere in the crystal using Eq. (5.92), because atoms removed from the crystal interior are supposed to be placed on a reference crystal that remains stress free. Specifically,

$$\Delta\mu_v \approx \Delta\bar{h}_v - T\Delta\bar{s}_v + k_B T \ln \chi_v,$$

where

$$\begin{aligned} \Delta\bar{h}_v &\approx \Delta h_v^0 - \frac{\sigma^2\Omega}{2\mu}\frac{15(1-\nu)}{7-5\nu}, \\ \Delta\bar{s}_v &\approx \Delta s_v^0, \end{aligned} \tag{5.106}$$

and μ is the shear modulus.[7] Here, Δh_v^0 and Δs_v^0 are the vacancy formation enthalpy and entropy, respectively, when the shear stress is zero. Even though $p = 0$, there is a difference between $\Delta\bar{h}_v$ and Δh_v^0 because of the shear stress. The expression for $\Delta\bar{h}_v$ is obtained assuming the vacancy is a spherical hole of volume Ω in an infinite isotropic elastic medium [30]. Equation (5.106) includes the recovery of elastic strain energy both inside the spherical volume and in the surrounding matrix. Therefore

$$\Delta\mu_v \approx \Delta g_v^0 - \frac{\sigma^2\Omega}{2\mu}\frac{15(1-\nu)}{7-5\nu} + k_B T \ln \chi_v, \tag{5.107}$$

where Δg_v^0 is the vacancy formation free energy when the crystal is subjected to zero stress. Since both Δg_v^0 and σ are independent of position, the spatial gradient of the chemical potential is only dependent on the gradient of the vacancy fraction χ_v.

Equation (5.107) shows that the shear stress σ alters the vacancy chemical potential only through a term proportional to σ^2, i.e. there is no contribution to the first order in σ. This is in contrast to Eq. (5.105), in which the pressure p makes a first-order contribution to the vacancy chemical potential. Given that $\sigma \ll \mu$ and $p \ll B$, the second-order terms are usually much smaller than the first-order term and are often ignored in the analysis.

5.5 Summary

Although point defects increase the internal energy and enthalpy of the crystals in which they reside, they also increase the entropy, causing a reduction of the Gibbs free energy and leading to an equilibrium concentration of such defects. A consequence is that crystals at equilibrium are invariably impure and imperfect. The equilibrium concentration of solute atoms in a crystal at a given temperature and pressure can be found by minimizing the Gibbs free energy with respect to the number of solute atoms added. This analysis involves accounting for the increase

7 Not to be confused with the chemical potential $\Delta\mu_v$.

in internal energy and enthalpy as well as the increase in entropy, especially the configurational entropy. For substitutional solutes, with a growing number of possible defect sites created as defects are added, the resulting equilibrium fraction of defects is given by the Boltzmann relation

$$\chi = \exp\left(-\frac{\Delta g_{\mathrm{f}}}{k_{\mathrm{B}}T}\right),$$

where $\Delta g_{\mathrm{f}} = \Delta e_{\mathrm{f}} - T\Delta s_{\mathrm{f}} + p_{\mathrm{ext}}\Delta v_{\mathrm{f}}$ is the formation Gibbs free energy of the solutes, and Δe_{f}, Δs_{f}, and Δv_{f} are the formation energy, formation (vibrational) entropy, and formation volume, respectively. For interstitial solutes, with a fixed number of possible defect sites, the resulting equilibrium fraction is given by

$$\chi_{\mathrm{i}} = \left[1 + \exp\left(\frac{\Delta g_{\mathrm{f}}}{k_{\mathrm{B}}T}\right)\right]^{-1},$$

an expression having the form of a Fermi–Dirac distribution. The analysis used for substitutional solutes also applies to vacancies because the process of creating vacant atomic sites involves the removal of atoms from the interior of a crystal to surface sites, thereby increasing the number of atomic sites each time this occurs. Thus the Boltzmann analysis applies to vacancies as well, predicting the equilibrium vacancy fraction

$$\chi_{\mathrm{v}} = \exp\left(-\frac{\Delta g_{\mathrm{v}}}{k_{\mathrm{B}}T}\right) = \exp\left(\frac{\Delta s_{\mathrm{v}}}{k_{\mathrm{B}}}\right)\exp\left(-\frac{\Delta e_{\mathrm{v}}}{k_{\mathrm{B}}T}\right)\exp\left(-\frac{p_{\mathrm{ext}}\Delta v_{\mathrm{v}}}{k_{\mathrm{B}}T}\right).$$

The formation energy of vacancies, Δe_{v}, can be estimated by accounting for the energies of chemical bonds that are broken or created when vacancies are created and can be found by measuring how the equilibrium vacancy concentration, or some property proportional to that concentration, varies with temperature. Similarly the formation volume of the vacancy, Δv_{v}, can be estimated from the atomic volume and can be measured from the pressure dependence of the equilibrium fraction of vacancies. The formation entropy of vacancies, Δs_{v}, can be estimated by considering the vibrational frequencies of the atoms adjacent to the vacancies. However, the measurement of Δs_{v} requires an absolute measurement of the vacancy concentration such as through a comparison of thermal expansion with the expansion of the crystal lattice on heating to a high temperature.

When the crystal is subjected to a spatially inhomogeneous stress field, either because of external loading or as a result of other defects (such as dislocations), the equilibrium concentration of point defects is also inhomogeneous. To analyze these situations, it is more convenient to define a point defect chemical potential $\Delta\mu$, which measures the Gibbs free energy change when atoms taken from (or inserted into) the host crystal are placed on (or taken from) reference crystals that are always under zero stress. If thermal equilibrium is reached, point defects distribute in such a way that the chemical potential is spatially homogeneous. The corresponding point defect distribution function has a Fermi–Dirac form, and in this expression, the term ($\Delta\tilde{v}_{\mathrm{f}}$) that couples to the local pressure is different from the formation volume (Δv_{f}). For impurity atoms that exist in the gas phase when taken out of the host crystal, an analysis based on the chemical potential can be used to obtain the relationship between their equilibrium fraction in the crystal and the vapor pressure.

5.6 Exercise problems

5.1 For the change of Gibbs free energy ΔG as a function of the number of substitutional atoms, N_B, given in Eq. (5.27), show that the second derivative, $\partial^2 \Delta G / \partial N_B^2$, is positive for $N_A > 0$ and $N_B > 0$.

5.2 For the change of Gibbs free energy per site Δg as a function of the fraction of interstitial atoms, χ_i, given in Eq. (5.35), show that the second derivative, $\partial^2 \Delta g / \partial \chi_i^2$, is positive for $0 < \chi_i < 1$.

5.3 The maximum solubility of Sn in solid aluminum is 0.1 at.% ($\chi = 10^{-3}$) and this occurs at 600 °C. This solubility limit is determined by the formation energy of Sn solute atoms in aluminum. Estimate the formation energy of the solute defect and estimate the solubility (i.e. equilibrium concentration) of Sn in aluminum at room temperature.

5.4 Pure Ni is able to dissolve a small amount of carbon at high temperatures. The carbon atoms dissolve interstitially into the FCC Ni lattice. We expect to find the C atoms in both octahedral and tetrahedral interstitial sites, as shown in Fig. 4.4. Because the octahedral interstitial sites are larger than the tetrahedral sites, the octahedral sites will have a higher fraction of occupancy than the tetrahedral sites. Let n_i^{octa} and n_i^{tetra} be the number of interstitial C atoms in the two kinds of interstitial sites in a crystal containing n_0 Ni atoms. Using the assumptions and information given below, calculate the ratio of n_i^{tetra} and n_i^{octa}, at 950 °C.
Assumptions:

(a) the fraction of octahedral sites occupied by C atoms at 950 °C is 0.01;
(b) the internal energies of formation of the interstitial defects are $\Delta e_i^{tetra} = 1.1 \Delta e_i^{octa}$;
(c) the entropies of formation can be neglected;
(d) only atmospheric pressure acts on the system.

5.5 Both Si and Ge are covalent crystals having the diamond cubic structure. Ge dissolves substitutionally into the Si crystal. We wish to consider the distortions and strain energies associated with placing a Ge atom onto a substitutional site in the Si crystal using the continuum misfitting sphere model as a guide.

The King table A.1 indicates that the atomic volumes for pure Si and pure Ge are $\Omega_{Si} = 20.02\ \text{Å}^3$ and $\Omega_{Ge} = 22.64\ \text{Å}^3$, respectively. Use the continuum model to determine the effective volume of a Ge atom, Ω_{Ge}^*, when it is dissolved in Si and compare the result with the experimental value found using the King table A.2. In the continuum model, a Ge atom is considered to be a small sphere having the elastic properties of Ge and the substitutional site is considered to be a small spherical hole in a continuum having the elastic properties of Si. The elastic properties of Si and Ge are given in Table B.1.

Derive an expression for and calculate the total work needed to fit a Ge atom into the Si lattice using the misfitting sphere model. Use this result to estimate the solubility of Ge in Si at room temperature. For this analysis, we assume that the strain energy associated with fitting the Ge into the substitutional site is the only work that is required to form the defect. Compare the calculated solubility with the known solubility of Ge in Si and discuss the reasons for any differences.

5.6 The maximum solubility of Sn in solid aluminum is 0.1 at.% ($\chi = 10^{-3}$) and this occurs at 600 °C. This solubility limit is determined by the formation energy of Sn solute atoms in aluminum, which has both a chemical bonding and strain energy component. We note from the King tables in Appendix A that the effective volume of Sn is larger than that of Al so that the size misfit (or strain energy) must be at least partially responsible for the low solubility. Use the model of the misfitting sphere to estimate the expected solubility of Sn in Al in the absence of size effects (i.e. if the effective volume of Sn were the same as that of Al). To do this you may consider the sample to be a sphere 2 mm in diameter. Also, you may assume that the volume expansion associated with each Sn atom is as if the atom were located at the center of the sphere. The elastic properties of Al at 600 °C are $\mu = 20$ GPa and $B = 60$ GPa. When estimating the strain energy, treat the Sn atom as an elastic inclusion with the same elastic properties as those of Al.

5.7 The Al–Mg equilibrium phase diagram (available on book website) indicates that at 450 °C Al can dissolve up to 17 at.% Mg in solid solution while at 437 °C Mg can dissolve up to 11 at.% Al. These are the maximum solubilities at atmospheric pressure for these metals. We wish to determine how the application of hydrostatic pressure would affect these solubilities.

(a) Using the King tables given in the Appendix A, find the atomic volumes of Al and Mg in the pure metals, Ω_{Al} and Ω_{Mg}.
(b) Calculate the effective volumes that these atoms have when they are dissolved in each other, Al in Mg, Ω^*_{Al} and Mg in Al, Ω^*_{Mg}.
(c) Write an equation showing how these solubilities, Al dissolved in Mg: $\chi_{\text{Al in Mg}}$ and Mg dissolved in Al: $\chi_{\text{Mg in Al}}$, should depend on pressure, p_{ext}, at the temperatures of maximum solubility.
(d) Determine which of these solubilities ($\chi_{\text{Al in Mg}}$ or $\chi_{\text{Mg in Al}}$) would be most affected by the application of hydrostatic pressure.
(e) Would the pressure increase or decrease that solubility?

5.8 The Ag–Ge phase diagram (available on book website) shows that Ge can dissolve into Ag, up to about 9 at.% at the eutectic temperature (the lowest melting temperature of the alloy as a function of composition), but that Ag is essentially insoluble in Ge at all temperatures.

(a) Assume that pure Ag is brought into contact with pure Ge just below the eutectic temperature. How does the lattice parameter of Ag change as Ge dissolves into the Ag lattice (increase or decrease)? Show how you reach this conclusion.
(b) Use the King tables in Appendix A to determine, qualitatively, how pressure would affect the solubility of Ge in Ag, i.e. whether the solubility increases or decreases with pressure. Show how you reach this conclusion.

5.9 In an integrated circuit, aluminum conductors are in contact with and may be in equilibrium with essentially pure silicon. They are also covered with a very rigid material such as silicon nitride.

During processing, residual stresses can be built up in such thin films as a result of differences in thermal expansion. It is not uncommon for such films to be subjected to hydrostatic tension in these instances. For the present problem we may consider the aluminum line to be encased in perfectly rigid surroundings. For this approximation the stress state in the aluminum line is perfectly hydrostatic and the aluminum line can be considered to be in the shape of a sphere. We suppose that the aluminum is first deposited onto Si, then heated to 500 °C, covered with silicon nitride and then cooled to room temperature. We wish to estimate the maximum hydrostatic tension stress that might be imposed on the aluminum as a result of cooling from 500 °C and then determine the effect of the stress state on the equilibrium concentration of Si in Al.

(a) Consult the phase diagram for this system (available on book website) and determine or estimate the equilibrium atomic fraction of Si in Al at room temperature (Si dissolves substitutionally in Al) under zero stress.
(b) Calculate the hydrostatic stress in the Al line at room temperature. Use the thermal expansion coefficient for Al given in Table B.1.
(c) Determine the effect of the hydrostatic stress on the equilibrium concentration of Si in Al at room temperature.

5.10 The results of positron annihilation experiments for gold and copper are shown in Fig. 5.9. Determine the enthalpy (energy) and entropy of formation for vacancies in Au from these data. Assume that divacancies and other vacancy complexes can be ignored.

5.11 The results of quenching–resistivity experiments and lattice parameter–thermal expansion measurements for gold are shown in Figs. 5.10 and 5.12. Determine the entropy of formation for vacancies from these data. Assume that divacancies and other vacancy complexes can be ignored. Use the measured vacancy volume given in Table C.1.

5.12 Using information contained in Table C.1, calculate the equilibrium fraction of vacant atomic sites in pure indium at 100 °C. Briefly explain why the vacancy fraction is so high at such a low temperature.

5.13 A long wire of pure Cd is held at room temperature $T = 20\,^\circ\mathrm{C}$ for a very long time so that equilibrium is established. The length of the wire in this state is L_0. Then the wire is heated to T_1 and held for a sufficiently long time to establish equilibrium at that temperature, before being quenched, instantaneously, to room temperature. The length of the wire in the quenched state will be called $L_q(T_1)$.

(a) Write an expression for the change of length of the wire in the quenched state relative to the initial state, normalized by the initial length of the wire, $(L_q - L_0)/L_0 = \Delta L/L_0$, as a function of T_1, and the thermodynamic properties of vacancies in Cd.
(b) Calculate the relative change in the length of the wire, $(L_q - L_0)/L_0$, after annealing at $T_1 = 300\,^\circ\mathrm{C}$ and quenching from that temperature to room temperature.

5.14 We consider a spherical, single crystal nanoparticle of Au with a radius, r, under no external pressure. For this problem, we also ignore the capillarity pressure caused by the surface free energy. We wish to determine the temperature at which only one vacancy would be found in the nanoparticle under equilibrium conditions.

(a) First, write an expression for the critical temperature, T_c, at which just one vacancy would be found in the nanoparticle, using the usual thermodynamic properties of the vacancy (Δe_v, Δv_v, etc.) and calling the atomic volume Ω.

(b) Calculate the critical temperature for a nanoparticle of diameter $d = 20$ nm, using the data from Table C.1.

5.15 A sample of FCC platinum is held at $T = 2000$ K, just below its melting temperature for a long period of time so that vacancy equilibrium is established.

(a) Calculate the equilibrium concentration of vacancies (in units of vacancies per m^3) when no external pressure is acting on the solid.
(b) Now consider the effects of imposing an external pressure, p_{ext}, onto the solid at the temperature of 2000 K. Calculate the pressure needed to reduce the vacancy concentration to 1% of the value found in (a).

The lattice and vacancy quantities needed for these calculations are given in Tables A.1 and C.1.

5.16 We wish to conduct an experiment on a sample of a pure metal as a function of temperature, under the condition that the equilibrium vacancy concentration be held constant at exactly $\chi_v = 10^{-6}$, i.e. 1 part per million vacancies. In order to have the vacancy concentration be constant as a function of temperature, we plan to vary the external pressure on the sample as a function of temperature.

(a) Using the following quantities for vacancies in the metal: Δe_v, Δs_v, and Δv_v, write an explicit expression for the external pressure p_{ext} which would be needed to maintain the vacancy fraction at $\chi_v = 10^{-6}$, as a function of the absolute temperature.
(b) Make a sketch of the computed p_{ext} as a function of temperature, from half the melting temperature to the melting temperature, T_m. The sketch should indicate where the pressure is positive and where it is negative.
(c) Write an expression for the critical temperature, T_c, where the required pressure to hold the vacancy concentration at $\chi_v = 10^{-6}$ would be zero.

5.17 Consider a spherical particle of Pt of radius R held at 2000 K, near its melting temperature. The surface free energy of the particle modifies the equilibrium vacancy fraction according to Eq. (5.61).

(a) Write an expression for and calculate the equilibrium concentration of vacancies, c_0 (in units of vacancies per m^3) for $R \to \infty$, i.e. when the surface effect can be ignored.
(b) Write an expression for the vacancy concentration in a nanoparticle of radius R, in terms of c_0.
(c) What is the largest nanoparticle size (R) in which one would expect to find less than one vacancy in the particle at 2000 K?

The lattice and vacancy quantities needed for these calculations can be obtained from Tables A.1 and C.1. Assume the surface free energy of Pt is $\gamma_s = 2\ \mathrm{J/m^2}$.

5.18 A single crystal of FCC gold is held at $T = 1000\,^\circ$C for a long time so that vacancy equilibrium is established. We now consider the effects of imposing an external pressure, p_{ext}, onto the crystal at the temperature of $1000\,^\circ$C. Take the bulk elastic modulus of gold at $1000\,^\circ$C to be represented by B.

(a) Write a general expression for the dilatational strain, $e = (\Delta V/V)$, as a function of the external pressure. The dilatational strain is composed of both the elastic strain due to the pressure and the volumetric strain associated with removal of vacancies in the system. Assume that the pressure is applied slowly so that vacancy equilibrium is maintained at all times. We also assume that the surface of the crystal is a good source and sink for vacancies.

(b) Because vacancy formation and annihilation processes contribute to the dilatation, one can imagine that the static bulk modulus would be changed from the purely elastic value mentioned above. Write an expression for the effective bulk modulus (including these vacancy processes) in the limit of very small pressures. If the elastic bulk modulus at 1000 °C is $B = 145$ GPa, what is the value of the effective modulus?

5.19 A free standing sphere of Au is heated to 1000 °C and allowed to reach equilibrium with respect to the formation of vacancies. Then the sphere is encapsulated in (or surrounded by) a perfectly rigid material and the entire composite is cooled to 800 °C and held for a long time. The rigid surrounding material prevents the Au sphere from shrinking when the composite is cooled to 800 °C. Using the information given below, do the following.

(a) Calculate the equilibrium vacancy fraction at 1000 °C (before the sphere is encapsulated into the rigid material).

(b) Derive an expression for the equilibrium vacancy fraction in the Au at any temperature less than 1000 °C. The expression need not be in explicit form.

(c) Use the relation derived in (b) to calculate (approximate calculation is acceptable) the equilibrium vacancy fraction at 800 °C under the constrained conditions described above.

The various properties of Au needed for this problem are listed in Tables C.1 and B.1. Assume the bulk modulus B of Au is 145 GPa in the temperature range of 800 °C–1000 °C.

5.20 We consider vacancies in a sphere of pure Al at the melting temperature, 933 K, under two different conditions.

(a) First, calculate the equilibrium fraction of vacant atomic sites in the solid at the melting temperature if the sphere is completely free to expand when heated.

(b) We now imagine that a vacancy-free sphere, initially at 0 K, is heated to the melting temperature under the constraint that its volume cannot change at all. We wish to find the pressure that would develop in the sphere under these conditions and the equilibrium fraction of vacancies. Assume that the thermal expansion coefficient is independent of temperature.

(i) Write an equation, the solution of which would give the resulting pressure. You need not solve this equation explicitly but the equation should contain known quantities except for the pressure.

(ii) Then write an equation for the equilibrium fraction of vacancies at the melting temperature in this constrained state.

5.21 The equilibrium concentration of hydrogen in Pd surrounded by H_2 gas at 10^3 Pa is about 0.5 at.%. What is the equilibrium concentration of hydrogen in Pd when the hydrogen gas pressure is increased to 10^4 Pa?

5.22 Consider a Si single crystal slab of thickness $h = 1$ cm kept at $T = 1000$ K for a long time so that the vacancy concentration has reached the equilibrium value. A very thin layer of SiC is then grown epitaxially onto the surfaces of the slab. The SiC layer prevents the creation or destruction of vacancies at the surfaces, and there are no vacancy sources or sinks inside the Si slab. We now apply a bending moment to the Si slab such that the top surface ($y = h/2$) is subjected to a tensile stress of 300 MPa, and the bottom surface ($y = -h/2$) is subjected to a compressive stress of 300 MPa.

(a) What are the equilibrium vacancy fractions χ_v^{eq} near the top and bottom surfaces? Obtain an expression for χ_v^{eq} as a function of y and plot it.
(b) Obtain an expression for χ_v^{eq} if the slab is cooled to 300 K while subjected to the same bending moment.

Refer to Eq. (6.2) for equilibrium vacancy fraction at zero stress and assume that the vacancy relaxation volume is $\Delta V_v^{rlx} = -0.3\,\Omega$.

6 Point defect equilibria

In our discussion of the equilibrium concentration of point defects in Chapter 5, we have assumed that for a given crystal there is only one type of defect with non-negligible concentration. We now discuss the scenario in which multiple types of point defects coexist in the same crystal. In Section 6.1, we show that the open crystal structure of silicon allows both vacancies and self-interstitials to exist with appreciable concentrations at thermal equilibrium. In Section 6.2, we discuss strongly ionic solids, in which all point defects are charged, and the charge neutrality condition requires point defects to be created in pairs. The equilibrium point defect concentration can be obtained by minimizing the Gibbs free energy of the crystal, or, alternatively, by the method of chemical equilibrium, each subjected to the charge neutrality condition. In Section 6.3, we discuss nonstoichiometric ionic solids in which atomic point defects can exist in both charged or neutral states, and electronic defects can also be present. The concentration of various atomic and electronic defects can be controlled by the partial pressures in the vapor phase.

In Section 6.4, we show that the type of the dominant point defect in intermetallic compounds, such as Ni–Al alloys, can change with composition. In Section 6.5, we discuss the formation of vacancy clusters. When a crystal is quenched from a high temperature to a low temperature, there is not sufficient time for vacancies to reach the true equilibrium concentration at the low temperature, and they tend to form clusters instead. We show that the thermodynamic principles can be applied to predict the concentration of vacancy clusters in the state where the total number of vacant sites is constrained.

6.1 Vacancies and self-interstitials in Si

Most of our discussion so far has concerned close-packed metals where vacancies can form but where self-interstitials are so energetic that they are essentially never present (except for metals irradiated by energetic particles like neutrons). On the other hand, the open crystal structure of Si (see Chapter 1) allows self-interstitials to form relatively easily. There is enough room in some interstitial positions to allow Si atoms (treated as spheres) to fit into the interstitial sites with no distortion at all (see Exercise problem 4.2). So for Si we think of the equilibrium state

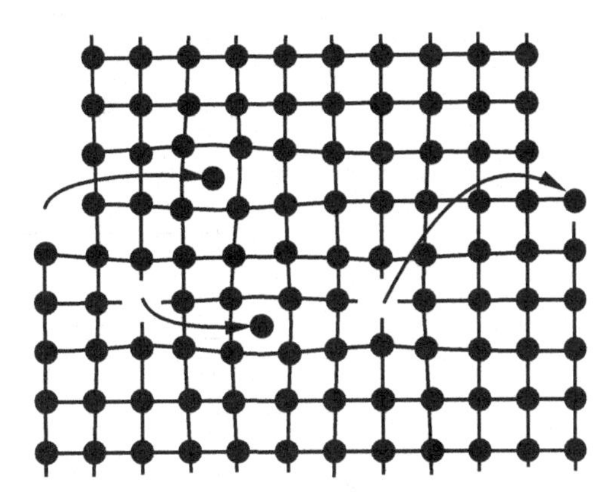

Figure 6.1. Vacancy and self-interstitial formation in Si.

of the crystal as involving both vacancies and self-interstitials. Various unit formation processes might be envisioned, as shown in Fig. 6.1.

Estimates of the fraction of self-interstitials and vacancies in Si have been made based on impurity diffusion models and positron annihilation data [31]:[1]

$$\chi_{\mathrm{i}}^{\mathrm{Si}} \approx 2 \times 10^4 \exp\left(-\frac{3.8\,\mathrm{eV}}{k_{\mathrm{B}}T}\right), \tag{6.1}$$

$$\chi_{\mathrm{v}}^{\mathrm{Si}} \approx 20 \exp\left(-\frac{2.6\,\mathrm{eV}}{k_{\mathrm{B}}T}\right). \tag{6.2}$$

We see that the formation energy of the self-interstitial is larger than that for vacancies, as expected. We also note a huge pre-exponential factor for self-interstitials, suggesting a large formation entropy.

It is instructive to compute the equilibrium defect concentrations at the melting temperature of Si. At the melting temperature of $T_{\mathrm{m}} = 1683$ K, the defect fractions are

$$\chi_{\mathrm{i}}^{\mathrm{Si}}(T = T_{\mathrm{m}}) \approx 8 \times 10^{-8},$$
$$\chi_{\mathrm{v}}^{\mathrm{Si}}(T = T_{\mathrm{m}}) \approx 3 \times 10^{-7}.$$

We note that the equilibrium vacancy concentration in Si is about three orders of magnitude lower than for metals at the melting temperature, because of the strong covalent bonding in Si. However, the equilibrium concentration of self-interstitials in Si is huge when compared to metals. At the melting temperature the concentration of self-interstitials in Si is only about four times smaller than that for vacancies. For metals the equilibrium self-interstitial concentration is essentially zero, even at the melting temperature.

6.2 Point defects in strongly ionic solids

Our discussion of point defects so far has mainly focused on single component systems. For strongly ionic compounds like the alkali halides (NaCl, KCl, LiF, ...) and many oxides (MgO,

1 Equation (6.1) is obtained from the measured concentration of self-interstitials, $c_{\mathrm{i}}^{\mathrm{Si}} \approx 10^{27} \exp[-3.8\,\mathrm{eV}/(k_{\mathrm{B}}T)]\,\mathrm{cm}^{-3}$. Equation (6.2) is obtained from the knowledge of the difference in the formation entropy of the interstitials and that of the vacancies, $\Delta s_{\mathrm{i}} - \Delta s_{\mathrm{v}} = 7k_{\mathrm{B}}$. See Exercise problem 6.1.

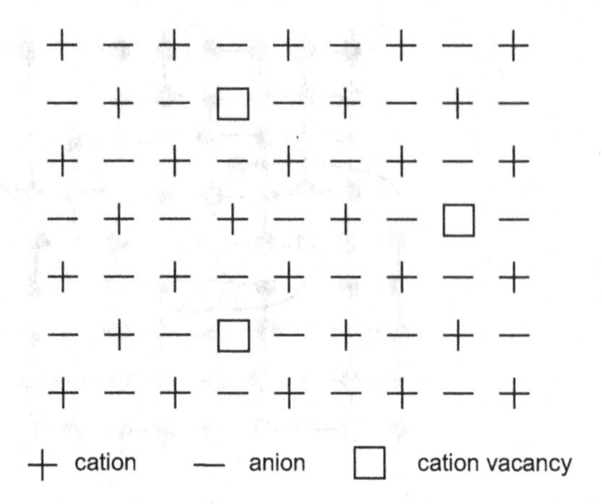

Figure 6.2. Hypothetical ionic crystal with cation vacancies only, leading to a net negative charge.

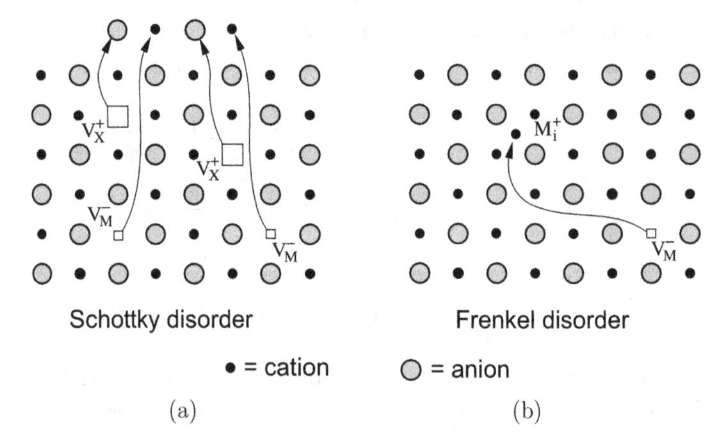

Figure 6.3. (a) Schottky disorder. Displacement of anion and cation to surface leaves a pair of vacancies. (b) Frenkel disorder. Ion leaving normal site forms an intersitial ion and leaves a vacancy.

FeO, ZnO,...) the electric charge associated with vacancies and interstitials needs to be taken into account.

If one had only cation vacancies in an ionic lattice and no other defects, then the crystal would have a huge negative electric charge, as shown in Fig. 6.2, the energy of which would be so large as to essentially prohibit cation vacancies from forming. The energies associated with such electric charging completely prohibit an excess concentration of defects of a particular charge. For this reason, point defects in strongly ionic crystals always come in pairs, each pair being charge neutral. Two types of defect pairs are usually discussed in ionic solids, as shown in Fig. 6.3:

- Schottky defect: a cation vacancy and an anion vacancy forming a pair;
- Frenkel defect: a vacancy and an interstitial forming a pair on either the cation or the anion sub-lattice.

6.2.1 Schottky defects

Consider an MX ionic crystal, where M represents the electropositive element (i.e. cations, such as Na) and X represents the electronegative element (i.e. anions, such as Cl) [32]. For simplicity, we assume the cations are monovalent in this section. Removing a cation from the crystal would result in a net negative charge relative to the perfect crystal structure, and we shall represent a cation vacancy as V_M^-. Similarly, an anion vacancy will be represented as V_X^+. The cation and anion interstitials will be represented as M_i^+ and X_i^-, respectively.

Suppose there are N cation sites containing $n_{V_M^-}$ vacancies and N anion sites containing $n_{V_X^+}$ vacancies. Because equal numbers of vacancies must be created on the cation and anion sub-lattices to avoid charging the crystal, we have $n_v \equiv n_{V_M^-} = n_{V_X^+}$. The Gibbs free energy of the crystal relative to the perfect state is then

$$\Delta G = n_{V_M^-} \Delta g_{V_M^-} + n_{V_X^+} \Delta g_{V_X^+} - T \Delta S_{\text{mix}}, \tag{6.3}$$

where $\Delta g_{V_M^-}$ is the formation free energy of a cation vacancy and $\Delta g_{V_X^+}$ is the formation free energy of an anion vacancy and ΔS_{mix} is the mixing (i.e. configurational) entropy. Following methods established earlier we write

$$\Delta S_{\text{mix}} = k_B \ln \mathcal{W}, \tag{6.4}$$

where $\mathcal{W}$ is the number of distinguishable arrangements of the system. For mixing cation vacancies and anion vacancies on their respective lattices we can write

$$\mathcal{W} = \mathcal{W}_{V_M^-} \mathcal{W}_{V_X^+}, \tag{6.5}$$

where $\mathcal{W}_{V_M^-}$ and $\mathcal{W}_{V_X^+}$ are the distinguishable arrangements for each separate mixing problem. The total number of distinguishable arrangements $\mathcal{W}$ equals the product of $\mathcal{W}_{V_M^-}$ and $\mathcal{W}_{V_X^+}$ because for every arrangement of cation vacancies we can have all possible arrangements of anion vacancies, and vice versa. Given that

$$\mathcal{W}_{V_M^-} = \frac{N!}{n_{V_M^-}!(N - n_{V_M^-})!}$$
$$\mathcal{W}_{V_X^+} = \frac{N!}{n_{V_X^+}!(N - n_{V_X^+})!},$$

and using $n_{V_M^-} = n_{V_X^+} = n_v$ we can then write

$$\mathcal{W} = \mathcal{W}_{V_M^-} \mathcal{W}_{V_X^+} = \left[\frac{N!}{n_v!(N - n_v)!} \right]^2. \tag{6.6}$$

Finally we define $\chi \equiv n_v/N$ as the fraction of Schottky pairs and write

$$\begin{aligned} \Delta G &= n_v \Delta g_{V_M^-} + n_v \Delta g_{V_X^+} - T \Delta S_{\text{mix}} \\ &= n_v \left(\Delta g_{V_M^-} + \Delta g_{V_X^+} \right) + 2N k_B T \left[\chi \ln \chi + (1 - \chi) \ln(1 - \chi) \right]. \end{aligned} \tag{6.7}$$

Again using

$$\frac{\partial \Delta G}{\partial n_v} = 0, \tag{6.8}$$

subjected to the condition that the number of cations (or anions), $N - n_v$, remains constant, we can set up an equation for the fraction of Schottky pairs at thermal equilibrium, $\chi_{eq}^{Schottky}$,

$$\Delta g_{V_M^-} + \Delta g_{V_X^+} + 2k_B T \ln \chi_{eq}^{Schottky} = 0. \tag{6.9}$$

$$\boxed{\chi_{eq}^{Schottky} = \exp\left(-\frac{\Delta g_{V_M^-} + \Delta g_{V_X^+}}{2k_B T}\right)} \tag{6.10}$$

We see that the equilibrium fraction of cation vacancies depends on the formation free energies of both cation and anion vacancies (because they come in pairs) and also that a factor of two appears in the denominator of the exponential term (due to the two mixing problems). Equation (6.10) predicts the fraction of cation vacancies (which equals the fraction of anion vacancies), when the Schottky pairs are the only point defects in the crystal, i.e. when there are no Frenkel pairs.

6.2.2 Frenkel defects

Following a similar analysis to Section 6.2.1, we can show that, if Frenkel pairs (i.e. vacancies and interstitials) on the cation sub-lattice are the only point defects in the crystal, then their equilibrium fraction is as follows.

$$\boxed{\chi_{eq}^{FrenkelM} = \exp\left(-\frac{\Delta g_{M_i^+} + \Delta g_{V_M^-}}{2k_B T}\right)} \tag{6.11}$$

Here, $\Delta g_{M_i^+}$ is the formation free energy of a cation interstitial, and $\Delta g_{V_M^-}$ is the formation free energy of a cation vacancy. The Boltzmann form of Eq. (6.11) is an approximation, which is valid only in the small concentration limit, i.e. if $\chi_{eq}^{FrenkelM} \ll 1$. Furthermore, there cannot be Schottky pairs, or Frenkel pairs on the anion sub-lattice. It also assumes that the number of cation interstitial sites equal the number of cations.

Similarly, if Frenkel pairs on the anion sub-lattice are the only point defects in the crystal, then their equilibrium fraction is as follows.

$$\boxed{\chi_{eq}^{FrenkelX} = \exp\left(-\frac{\Delta g_{X_i^-} + \Delta g_{V_X^+}}{2k_B T}\right)} \tag{6.12}$$

Here, $\Delta g_{X_i^-}$ is the formation free energy of an anion interstitial, and $\Delta g_{V_X^+}$ is the formation free energy of an anion vacancy. For Eq. (6.12) to apply, there cannot be Schottky pairs, or Frenkel pairs on the cation sub-lattice. It also assumes that the number of anion interstitial sites equal the number of anions.

6.2.3 Coexistence of Schottky and Frenkel defects

In principle, all three types of defect pairs can exist simultaneously in an ionic crystal. Yet, it is often the case that only one type of the defect pairs has a non-negligible concentration, because the formation free energies of the other defect pairs are much higher. When this is the case, then one of the three equations, (6.10), (6.11), or (6.12) can be used to predict the point defect fraction. However, when this condition is not satisfied, we can no longer use these equations directly. Instead, a more complex analysis is needed to predict the equilibrium fraction of point defects.

As an example, consider the case in which both Schottky pairs and cation Frenkel pairs are present with non-negligible fractions. We now need to determine the equilibrium fraction of three types of point defects: cation vacancy $\chi_{\mathrm{V_M^-}}$, anion vacancy $\chi_{\mathrm{V_X^+}}$, and cation interstitial $\chi_{\mathrm{M_i^+}}$. This can be accomplished by minimizing the Gibbs free energy $\Delta G(\chi_{\mathrm{V_M^-}}, \chi_{\mathrm{V_X^+}}, \chi_{\mathrm{M_i^+}})$,

$$\begin{aligned}\Delta G = {} & n_{\mathrm{V_M^-}} \Delta g_{\mathrm{V_M^-}} + N k_{\mathrm{B}} T \left[\chi_{\mathrm{V_M^-}} \ln \chi_{\mathrm{V_M^-}} + \left(1 - \chi_{\mathrm{V_M^-}}\right) \ln \left(1 - \chi_{\mathrm{V_M^-}}\right)\right] \\ & + n_{\mathrm{V_X^+}} \Delta g_{\mathrm{V_X^+}} + N k_{\mathrm{B}} T \left[\chi_{\mathrm{V_X^+}} \ln \chi_{\mathrm{V_X^+}} + \left(1 - \chi_{\mathrm{V_X^+}}\right) \ln \left(1 - \chi_{\mathrm{V_X^+}}\right)\right] \\ & + n_{\mathrm{M_i^+}} \Delta g_{\mathrm{M_i^+}} + N k_{\mathrm{B}} T \left[\chi_{\mathrm{M_i^+}} \ln \chi_{\mathrm{M_i^+}} + \left(1 - \chi_{\mathrm{M_i^+}}\right) \ln \left(1 - \chi_{\mathrm{M_i^+}}\right)\right]\end{aligned} \tag{6.13}$$

subjected to the charge neutrality constraint,

$$\chi_{\mathrm{V_M^-}} = \chi_{\mathrm{V_X^+}} + \chi_{\mathrm{M_i^+}}. \tag{6.14}$$

We shall limit the discussion in the small concentration limit, $\chi_{\mathrm{V_M^-}}, \chi_{\mathrm{V_X^+}}, \chi_{\mathrm{M_i^+}} \ll 1$, and ignore the increase of lattice sites when Schottky pairs are created. This means that N will be treated as a constant.

Method of Lagrange multiplier

A standard method for constrained minimization is to introduce a Lagrange multiplier λ and to minimize the modified function

$$\Delta \tilde{G}(\chi_{\mathrm{V_M^-}}, \chi_{\mathrm{V_X^+}}, \chi_{\mathrm{M_i^+}}) = \Delta G(\chi_{\mathrm{V_M^-}}, \chi_{\mathrm{V_X^+}}, \chi_{\mathrm{M_i^+}}) - N\lambda(\chi_{\mathrm{V_M^-}} - \chi_{\mathrm{V_X^+}} - \chi_{\mathrm{M_i^+}}) \tag{6.15}$$

without any constraint. Afterward, the constant λ is removed (or solved for) by the imposed constraint. Minimization of the modified function $\Delta \tilde{G}$ leads to

$$\begin{aligned}\frac{\chi_{\mathrm{V_M^-}}}{1 - \chi_{\mathrm{V_M^-}}} &= \exp\left(-\frac{\Delta g_{\mathrm{V_M^-}} - \lambda}{k_{\mathrm{B}} T}\right), \\ \frac{\chi_{\mathrm{V_X^+}}}{1 - \chi_{\mathrm{V_X^+}}} &= \exp\left(-\frac{\Delta g_{\mathrm{V_X^+}} + \lambda}{k_{\mathrm{B}} T}\right), \\ \frac{\chi_{\mathrm{M_i^+}}}{1 - \chi_{\mathrm{M_i^+}}} &= \exp\left(-\frac{\Delta g_{\mathrm{M_i^+}} + \lambda}{k_{\mathrm{B}} T}\right).\end{aligned} \tag{6.16}$$

Since we are already in the limit of $\chi_{\mathrm{V_M^-}}, \chi_{\mathrm{V_X^+}}, \chi_{\mathrm{M_i^+}} \ll 1$, we have

$$\begin{aligned}\chi_{\mathrm{V_M^-}} &= \exp\left(-\frac{\Delta g_{\mathrm{V_M^-}} - \lambda}{k_{\mathrm{B}} T}\right), \\ \chi_{\mathrm{V_X^+}} &= \exp\left(-\frac{\Delta g_{\mathrm{V_X^+}} + \lambda}{k_{\mathrm{B}} T}\right), \\ \chi_{\mathrm{M_i^+}} &= \exp\left(-\frac{\Delta g_{\mathrm{M_i^+}} + \lambda}{k_{\mathrm{B}} T}\right).\end{aligned} \tag{6.17}$$

The Lagrange multiplier λ can be eliminated by taking products,

$$\chi_{V_M^-} \cdot \chi_{V_X^+} = \exp\left(-\frac{\Delta g_{V_M^-} + \Delta g_{V_X^+}}{k_B T}\right),$$
$$\chi_{V_M^-} \cdot \chi_{M_i^+} = \exp\left(-\frac{\Delta g_{V_M^-} + \Delta g_{M_i^+}}{k_B T}\right). \tag{6.18}$$

Combining Eq. (6.18) with the charge neutrality equation (6.14), we have

$$\begin{aligned}\chi_{V_M^-} &= \left[\exp\left(-\frac{\Delta g_{V_M^-} + \Delta g_{V_X^+}}{k_B T}\right) + \exp\left(-\frac{\Delta g_{V_M^-} + \Delta g_{M_i^+}}{k_B T}\right)\right]^{1/2} \\ &= \exp\left(-\frac{\Delta g_{V_M^-}}{2k_B T}\right)\left[\exp\left(-\frac{\Delta g_{V_X^+}}{k_B T}\right) + \exp\left(-\frac{\Delta g_{M_i^+}}{k_B T}\right)\right]^{1/2}.\end{aligned} \tag{6.19}$$

The equilibrium fraction of the other two point defects can be obtained by substituting Eq. (6.19) into Eq. (6.18):

$$\chi_{V_X^+} = \exp\left(-\frac{\Delta g_{V_M^-}}{2k_B T} - \frac{\Delta g_{V_X^+}}{k_B T}\right)\left[\exp\left(-\frac{\Delta g_{V_X^+}}{k_B T}\right) + \exp\left(-\frac{\Delta g_{M_i^+}}{k_B T}\right)\right]^{-1/2}, \tag{6.20}$$

$$\chi_{M_i^+} = \exp\left(-\frac{\Delta g_{V_M^-}}{2k_B T} - \frac{\Delta g_{M_i^+}}{k_B T}\right)\left[\exp\left(-\frac{\Delta g_{V_X^+}}{k_B T}\right) + \exp\left(-\frac{\Delta g_{M_i^+}}{k_B T}\right)\right]^{-1/2}. \tag{6.21}$$

In the limit of $\Delta g_{M_i^+} \gg \Delta g_{V_X^+}$, we have

$$\chi_{V_M^-} \approx \chi_{V_X^+} \approx \exp\left(-\frac{\Delta g_{V_M^-} + \Delta g_{V_X^+}}{2k_B T}\right) \gg \chi_{M_i^+}, \tag{6.22}$$

which is consistent with Eq. (6.10). This is the limit where the fraction of Frenkel pairs is negligible compared with Schottky pairs.

Method of chemical equilibrium

An easier way to obtain the same result is to consider the creation of Schottky pairs and cation Frenkel pairs as two chemical reactions,

$$\text{N.O.} + \Delta G_S \rightarrow V_M^- + V_X^+ \tag{6.23}$$
$$\text{N.O.} + \Delta G_{FM} \rightarrow V_M^- + M_i^+, \tag{6.24}$$

where N.O. represents "normal occupancy" (i.e. no defect). ΔG_S and ΔG_{FM} are the change of the total Gibbs free energy of the system caused by these two reactions.

Define $[V_M^-]$, $[V_X^+]$, and $[M_i^+]$ as concentrations (numbers per unit volume) of the point defects. It can be shown that (see below), when equilibrium is reached, for each chemical reaction, the multiplication of the concentrations of the products divided by the concentrations of the reactants equals a constant specific to that reaction. For the two reactions given above, we have

$$K'^{\,\text{Schottky}}_{\text{eq}} = [V_M^-] \cdot [V_X^+] \tag{6.25}$$
$$K'^{\,\text{FrenkelM}}_{\text{eq}} = [V_M^-] \cdot [M_i^+], \tag{6.26}$$

where $K'^{\,\mathrm{Schottky}}_{\mathrm{eq}}$ and $K'^{\,\mathrm{FrenkelM}}_{\mathrm{eq}}$ are Chemical equilibrium (or reaction rate) constants. Since the vacancy concentrations are proportional to the vacancy fractions, we may write

$$K'^{\,\mathrm{Schottky}}_{\mathrm{eq}} = \left[\mathrm{V}_{\mathrm{M}}^{-}\right] \cdot \left[\mathrm{V}_{\mathrm{X}}^{+}\right] \propto \chi_{\mathrm{V}_{\mathrm{M}}^{-}} \cdot \chi_{\mathrm{V}_{\mathrm{X}}^{+}} = K^{\mathrm{Schottky}}_{\mathrm{eq}} \tag{6.27}$$

$$K'^{\,\mathrm{FrenkelM}}_{\mathrm{eq}} = \left[\mathrm{V}_{\mathrm{M}}^{-}\right] \cdot \left[\mathrm{M}_{\mathrm{i}}^{+}\right] \propto \chi_{\mathrm{V}_{\mathrm{M}}^{-}} \cdot \chi_{\mathrm{M}_{\mathrm{i}}^{+}} = K^{\mathrm{FrenkelM}}_{\mathrm{eq}}. \tag{6.28}$$

The constants with the prime, i.e. $K'^{\,\mathrm{Schottky}}_{\mathrm{eq}}$ and $K'^{\,\mathrm{FrenkelM}}_{\mathrm{eq}}$ have the dimension of concentration squared, while the constants without the prime, i.e. $K^{\mathrm{Schottky}}_{\mathrm{eq}}$ and $K^{\mathrm{FrenkelM}}_{\mathrm{eq}}$, are dimensionless. Because $K^{\mathrm{Schottky}}_{\mathrm{eq}}$ stays constant, regardless of the specific values of $\chi_{\mathrm{V}_{\mathrm{M}}^{-}}$ or $\chi_{\mathrm{V}_{\mathrm{X}}^{+}}$, its value can be determined by considering a special case in which $\chi_{\mathrm{V}_{\mathrm{M}}^{-}}$ and $\chi_{\mathrm{V}_{\mathrm{X}}^{+}}$ are known. In Section 6.2.1 we have seen that, if the concentration of Frenkel defects are negligible, then, from Eq. (6.10), we have

$$\chi_{\mathrm{V}_{\mathrm{M}}^{-}} = \chi_{\mathrm{V}_{\mathrm{X}}^{+}} = \exp\left(-\frac{\Delta g_{\mathrm{V}_{\mathrm{M}}^{-}} + \Delta g_{\mathrm{V}_{\mathrm{X}}^{+}}}{2k_{\mathrm{B}}T}\right). \tag{6.29}$$

Therefore, the constant $K^{\mathrm{Schottky}}_{\mathrm{eq}}$ is

$$K^{\mathrm{Schottky}}_{\mathrm{eq}} = \chi_{\mathrm{V}_{\mathrm{M}}^{-}} \cdot \chi_{\mathrm{V}_{\mathrm{X}}^{+}} = \exp\left(-\frac{\Delta g_{\mathrm{V}_{\mathrm{M}}^{-}} + \Delta g_{\mathrm{V}_{\mathrm{X}}^{+}}}{k_{\mathrm{B}}T}\right). \tag{6.30}$$

We emphasize that Eq. (6.29) is valid only in the special case where only Schottky defects exist, while Eq. (6.30) remains valid even if Frenkel defects coexist with Schottky defects.

Similarly, by considering the special case where only cation Frenkel defects exist, we can determine the constant $K^{\mathrm{FrenkelM}}_{\mathrm{eq}}$ from Eq. (6.11),

$$K^{\mathrm{FrenkelM}}_{\mathrm{eq}} = \chi_{\mathrm{V}_{\mathrm{M}}^{-}} \cdot \chi_{\mathrm{M}_{\mathrm{i}}^{+}} = \exp\left(-\frac{\Delta g_{\mathrm{V}_{\mathrm{M}}^{-}} + \Delta g_{\mathrm{M}_{\mathrm{i}}^{+}}}{k_{\mathrm{B}}T}\right). \tag{6.31}$$

Combined with the charge neutrality equation (6.14), Eqs. (6.30) and (6.31) will lead to the same predictions for the point defect fraction as given in Eqs. (6.19) to (6.21). Therefore, the method of Lagrange multiplier is mathematically equivalent to the method of chemical equilibrium considered here.

Derivation of chemical equilibrium constants

Here we give a brief derivation of Eq. (6.27) to show how it follows from the thermodynamic principles introduced previously. Following the approach described in Section 5.4.1 and Eq. (5.79), the Gibbs free energy change for the Schottky pair formation, Eq. (6.23), may be expressed as

$$\Delta G_{\mathrm{S}} = \mu_{\mathrm{V}_{\mathrm{M}}^{-}} + \mu_{\mathrm{V}_{\mathrm{X}}^{+}} - \mu_{\mathrm{N.O.}}, \tag{6.32}$$

where $\mu_{\mathrm{V}_{\mathrm{M}}^{-}}$ and $\mu_{\mathrm{V}_{\mathrm{X}}^{+}}$ are the chemical potentials of the vacancies and the chemical potential of the perfect lattice is taken to be zero, $\mu_{\mathrm{N.O.}} = 0$. Similar to Eq. (5.83), we may express the chemical potentials of the vacancies as

$$\begin{aligned} \mu_{\mathrm{V}_{\mathrm{M}}^{-}} &\approx \left(\Delta\tilde{e}_{\mathrm{V}_{\mathrm{M}}^{-}} - T\Delta\tilde{s}_{\mathrm{V}_{\mathrm{M}}^{-}}\right) + k_{\mathrm{B}}T \ln \chi_{\mathrm{V}_{\mathrm{M}}^{-}} \\ \mu_{\mathrm{V}_{\mathrm{X}}^{+}} &\approx \left(\Delta\tilde{e}_{\mathrm{V}_{\mathrm{X}}^{+}} - T\Delta\tilde{s}_{\mathrm{V}_{\mathrm{X}}^{+}}\right) + k_{\mathrm{B}}T \ln \chi_{\mathrm{V}_{\mathrm{X}}^{+}}, \end{aligned} \tag{6.33}$$

where $\chi_{V_M^-}$ and $\chi_{V_X^+}$ represent the vacancy fractions and the first terms in these relations may be regarded as the chemical potentials of these defects in their hypothetical, "pure" states ($\chi_{V_M^-} = 1$ and $\chi_{V_X^+} = 1$)

$$\begin{aligned} \mu^{\circ}_{V_M^-} &\approx \Delta\tilde{e}_{V_M^-} - T\Delta\tilde{s}_{V_M^-} \\ \mu^{\circ}_{V_X^+} &\approx \Delta\tilde{e}_{V_X^+} - T\Delta\tilde{s}_{V_X^+}. \end{aligned} \tag{6.34}$$

With these indentifications, the Gibbs free energy change for the Schottky pair formation, Eq. (6.23), may be written as

$$\Delta G_S = \Delta G_S^{\circ} + k_B T \ln \chi_{V_M^-} \chi_{V_X^+}, \tag{6.35}$$

where

$$\Delta G_S^{\circ} \equiv \mu^{\circ}_{V_M^-} + \mu^{\circ}_{V_X^+}. \tag{6.36}$$

At equilibrium, where $\Delta G_S = 0$, we then have

$$\chi_{V_M^-} \chi_{V_X^+} = \exp\left(-\frac{\Delta G_S^{\circ}}{k_B T}\right) \equiv K_{eq}^{Schottky}. \tag{6.37}$$

6.3 Point defects in nonstoichiometric ionic solids

Our discussion of point defects in ionic solids thus far has assumed that only cation and anion defects are present in the crystal, as either vacancies or interstitials, and that the concentrations of electronic defects, such as free electrons in the conduction band or electronic holes in the valance band, are negligible. We have also assumed that the atomic point defects are fully ionized, and that charge neutrality conditions are met by having equal numbers of ionic defects of opposite sign. We shall see that point defects in some ionic crystals may not be fully ionized so that both neutral and ionized point defects, as well as the associated electronic defects, may be present at equilibrium. We will show that for some ionic crystals, point defect concentrations, including those for electrons and holes, may be altered by controlling the partial pressures of the components of the ionic crystal in the vapor phase. In this way the electronic conductivity of some ionic crystals can be determined by controlling the partial pressures in the vapor phase. A comprehensive treatment of point defect chemistry in ionic solids can be found in the review paper by Kröger and Vink [33]. This section of the book may be regarded as being more specialized than other sections as it relies heavily on the principles of chemical equilibrium. The reader may skip on to Section 6.4 without difficulty.

Charged and neutral point defects

In Fig. 6.2, we considered the hypothetical case in which an ionic crystal contains only cation vacancies, resulting in a net negative charge and a prohibitively high energy. The purpose of the figure is to show why defects must appear in pairs to maintain charge neutrality. However, in constructing this hypothetical case, we have chosen to neglect the electrons and holes that can exist in the crystal. For many ionic crystals, it is a reasonable approximation to make because

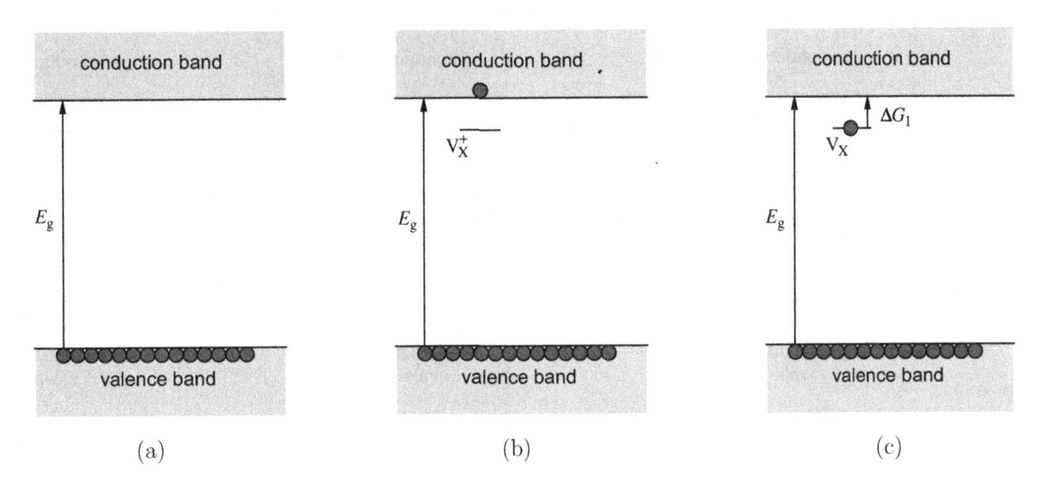

Figure 6.4. Energy band diagram in an MX crystal. (a) Perfect crystal with valence band completely filled by electrons and conduction band empty. (b) Crystal containing an ionized anion vacancy and a free electron in the conduction band. (c) Crystal containing a neutral vacancy.

the densities of electrons and holes are much smaller than those of the ionic point defects. However, we now go beyond this approximation.

Let us consider anion vacancies in an ionic crystal, while taking electrons into account. Before creating the anion vacancies, the electronic energy states of the perfect ionic crystal can be described by the band diagram shown in Fig. 6.4a. There is an energy gap E_g between the valence band and the conduction band. At zero temperature, all the states in the valence band are filled by electrons (as indicated by filled circles) and all the states in the conduction band are empty. The resulting electronic conductivity is zero. We now introduce one anion vacancy, V_X^+, into the crystal. Charge balance can be restored if we also insert one electron into the crystal, as shown in Fig. 6.4b. The result is a free electron, e^-, in the conduction band, which would impart electronic conductivity to the crystal.

Because the anion vacancy V_X^+ and the electron e^- have opposite charges, they attract each other and can lower the energy by forming a bound state. The result is a vacancy V_X that is charge neutral, and no free electron exists in the conduction band, as shown in Fig. 6.4c. In this hypothetical case where the ionic crystal contains a single vacancy, the lowest energy state is the neutral vacancy, V_X. At finite temperature, the electron trapped at the vacancy can be excited into the conduction band, i.e. going from Fig. 6.4c to Fig. 6.4b. By this process, the neutral vacancy V_X becomes ionized and contributes a free electron to the conduction band. In this sense, the anion vacancy acts as a n-type dopant.

Similarly, we may consider a hypothetical case in which the ionic crystal contains a single cation vacancy V_M^- and no other atomic point defects. An electron would have to be removed from the crystal to maintain charge neutrality. This leads to a free electronic hole, e^+, in the valence band, as shown in Fig. 6.5a. Because the cation vacancy V_M^- and the electronic hole e^+ have opposite effective charges, they attract each other and can lower the energy by forming a bound state. The result is a neutral vacancy V_M as shown in Fig. 6.5b. In this hypothetical case where the ionic crystal contains a single vacancy, the lowest energy state is the neutral vacancy, V_M. At finite temperature, an electron in the valence band can be excited into the localized region of the vacancy, i.e. going from Fig. 6.5b to Fig. 6.5a. By this process, the neutral vacancy

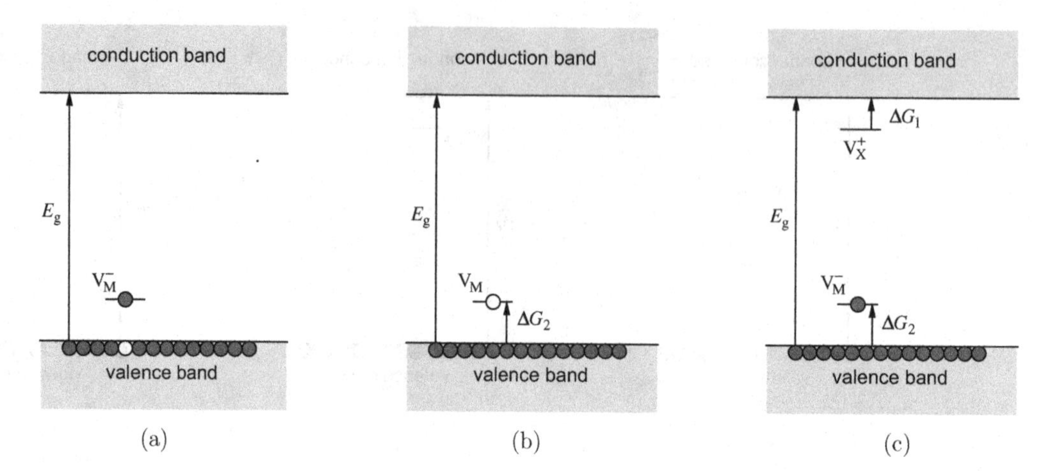

Figure 6.5. Energy band diagram in an MX crystal. (a) Crystal containing an ionized cation vacancy and a free hole in the valence band. (b) Crystal containing a neutral vacancy. (c) Crystal containing a Schottky pair and no free electrons or holes.

V_M becomes ionized and contributes a free hole to the valence band. In this sense, the cation vacancy acts as a *p*-type dopant.

For ionic crystals with a large band gap, an even lower energy state can be achieved when a cation vacancy and an anion vacancy are both created. In this case, both vacancies are ionized, i.e. V_X^+ and V_M^-, as shown in Fig. 6.5c. There is neither a free electron in the conduction band, nor a free hole in the valence band. This is the Schottky pair configuration considered in Section 6.2.1. The hypothetical scenario considered here shows that in an ionic crystal when atomic point defects can come in pairs, they tend to be fully ionized and no free electrons or holes need to be considered. However, if there is an excess of atomic point defects of one type, then neutral point defects, free electrons, or holes are created. We shall see that the concentration of these defects depends on the vapor pressure.

Relation between vapor pressures

We imagine that the MX crystal is at equilibrium with the vapor phase, which includes both metal atoms, $M_{(g)}$, and electronegative molecules, $X_{2(g)}$, in the gas phase. Often the electronegative element in the gas phase is a diatomic molecule, such as O_2 or S_2. At thermal equilibrium, there is a definite relationship between the (partial) vapor pressure of the two vapors, $M_{(g)}$ and $X_{2(g)}$, in the gas phase. Here we use the method of chemical equilibrium to obtain this relationship.

Figure 6.6 illustrates the process through which a crystal would be in equilibrium with the gas phase. This process can be represented as a chemical reaction

$$\mathrm{MX} + \Delta G_{\mathrm{MX}} \rightarrow \mathrm{M}_{(g)} + \frac{1}{2}\mathrm{X}_{2(g)}, \tag{6.38}$$

where ΔG_{MX} is the free energy change associated with the reaction. Similar to Eq. (6.35), and using Eq. (3.34), ΔG_{MX} may be expressed as

$$\Delta G_{\mathrm{MX}} = \Delta G^{\circ}_{\mathrm{MX}} + k_B T \ln \frac{p_{\mathrm{M}}}{p^{\circ}_{\mathrm{M}}} + \frac{1}{2} k_B T \ln \frac{p_{\mathrm{X}_2}}{p^{\circ}_{\mathrm{X}_2}} - k_B T \ln \chi_{\mathrm{MX}}, \tag{6.39}$$

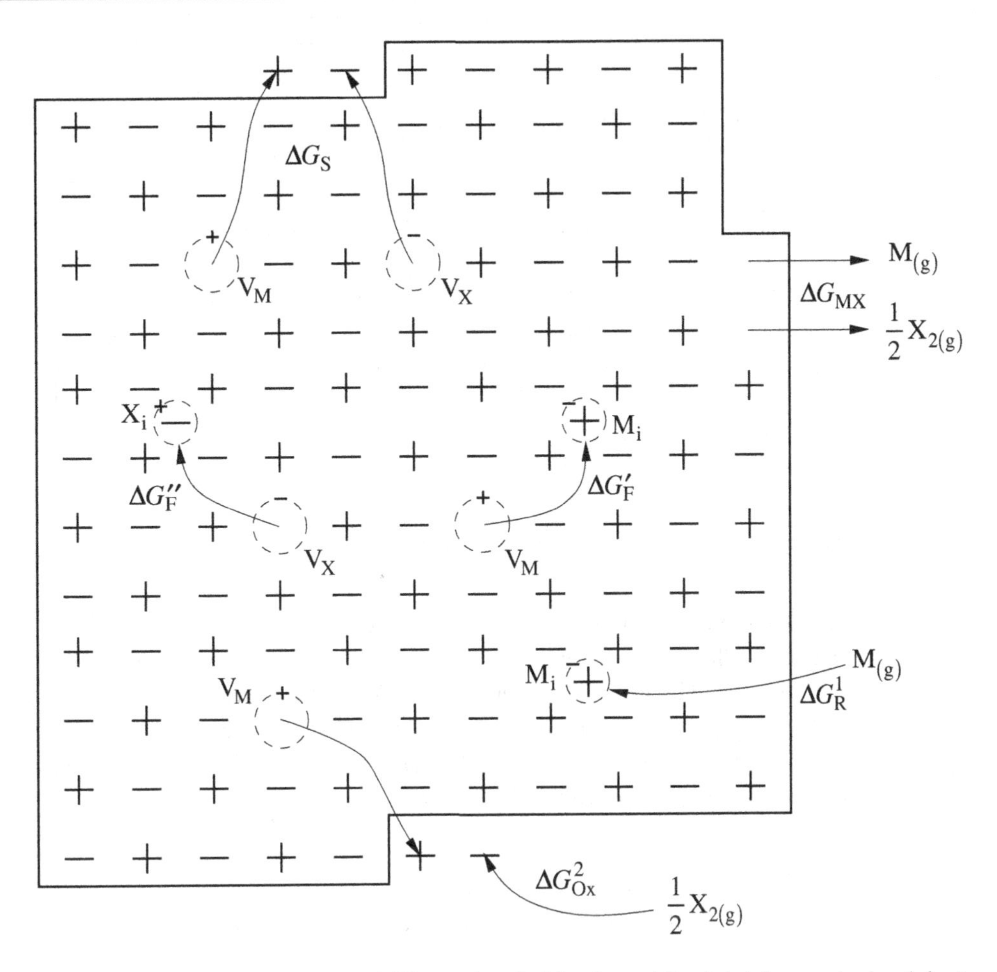

Figure 6.6. Chemical reactions in an MX crystal with Schottky and Frenkel defects on both sub-lattices, leading to equilibrium with the vapor phase.

where ΔG°_{MX} is the free energy change under the standard condition, in which the partial pressure for the M and X_2 gases are p°_{M} and $p^{\circ}_{X_2}$, respectively, and the MX crystal is perfect. Because the defect fraction in the crystal is very small, we can assume $\chi_{MX} \approx 1$. Then,

$$\Delta G_{MX} = \Delta G^{\circ}_{MX} + k_B T \ln \frac{p_M \sqrt{p_{X_2}}}{p^{\circ}_M \sqrt{p^{\circ}_{X_2}}}. \tag{6.40}$$

At equilibrium $\Delta G_{MX} = 0$, so that

$$p_M \sqrt{p_{X_2}} = p^{\circ}_M \sqrt{p^{\circ}_{X_2}} \exp\left(-\frac{\Delta G^{\circ}_{MX}}{k_B T}\right) \equiv K'_{MX}, \tag{6.41}$$

where K'_{MX} is the equilibrium constant for the reaction. In the following, we will further apply the method of chemical equilibrium to describe the formation of point defects as well as their ionization.

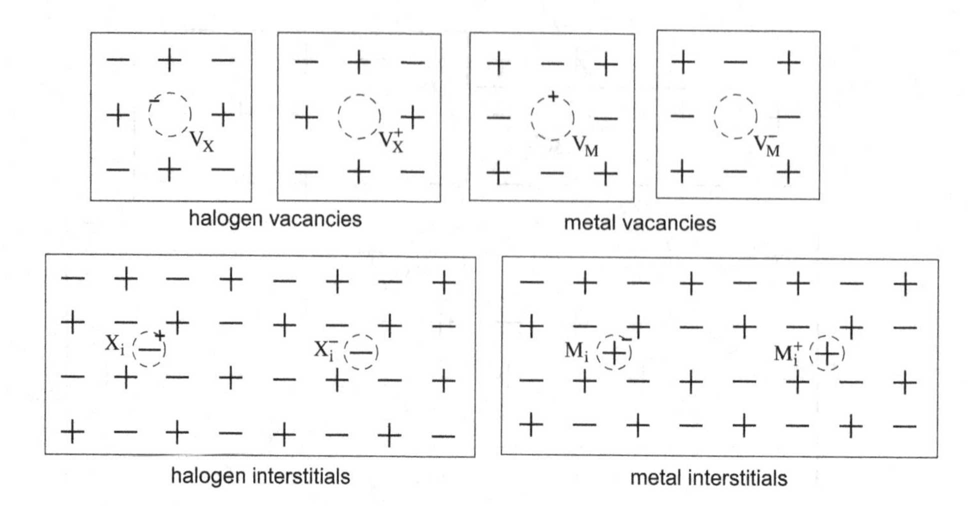

Figure 6.7. Charged and neutral vacancies and interstitials in an MX crystal.

Application of chemical equilibria to defects

Figure 6.7 illustrates both neutral and charged Schottky defects that might exist in an alkali halide crystal. The neutral halogen vacancy, V_X, can be created by removing the halogen ion and placing an electron in the vacant site. Sufficient thermal energy could cause the electron present in the vacancy to escape to the conduction band, leaving a charged halogen vacancy V_X^+, as described by the following reaction,

$$V_X + \Delta G_1 \rightarrow V_X^+ + e^-. \tag{6.42}$$

Applying the method of chemical equilibrium to this ionization gives

$$K_1' = \frac{[V_X^+]\,n}{[V_X]}, \tag{6.43}$$

where n is the concentration of electrons in the conduction band and the brackets, [ ], indicate the concentration (number per unit volume) of the lattice defects.

Similarly, a neutral metal vacancy V_M can be ionized, resulting in a negatively charged metal vacancy V_M^- and a free hole in the valence band, e^+. The chemical reaction describing this defect ionization and the corresponding equilibrium constant are both shown in Table 6.1, where the electronic hole concentration is represented by p.

Although the illustrations in Fig. 6.7 are specifically for alkali halide crystals, such as NaCl, where the individual ions are singly charged, the defect equilibria described here are also valid for other ionic crystals, such as ZnO or PbS, where the individual ions are multiply charged. For a neutral oxygen vacancy in ZnO, for example, two electrons would be present in the ionic cavity, so that the natural charge distribution of the crystal would be unchanged. Removal of one of those electrons to the conduction band would produce a positively charged halogen vacancy, V_X^+, just as for an alkali halide crystal. Removal of the second electron to the conduction band would create a doubly charged halogen vacancy, V_X^{++}. Thus V_X^+ and V_M^- stand for singly charged point defects, regardless of the charges on the ions that make up the crystal.

Table 6.1. Chemical reactions and equilibrium constants for ionization reactions in an MX crystal with vacancies and interstitials on either sub-lattice.

Defect reaction	Chemical reaction	Equilibrium constant
Ionization of halogen vacancy	$V_X + \Delta G_1 \rightarrow V_X^+ + e^-$	$K_1' = [V_X^+]\, n / [V_X]$
Ionization of metal vacancy	$V_M + \Delta G_2 \rightarrow V_M^- + e^+$	$K_2' = [V_M^-]\, p / [V_M]$
Ionization of metal interstitial	$M_i + \Delta G_3 \rightarrow M_i^+ + e^-$	$K_3' = [M_i^+]\, n / [M_i]$
Ionization of halogen interstitial	$X_i + \Delta G_4 \rightarrow X_i^- + e^+$	$K_4' = [X_i^-]\, p / [X_i]$
Electron–hole equilibrium	$\text{N.O.} + E_g \rightarrow e^- + e^+$	$K_g' = [e^-][e^+] = np$

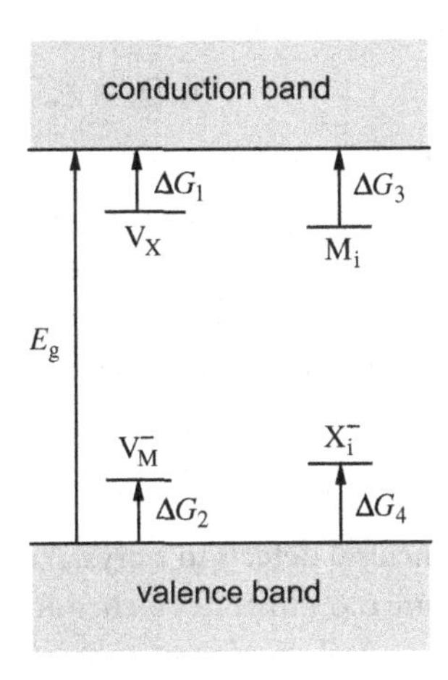

Figure 6.8. Energy band diagram showing the energy levels of the valence and conduction band, as well as the energy levels of vacancies and interstitials in an MX crystal. V_X and M_i represent the state of the electron when it is bound to the halogen vacancy and metal interstitial, respectively. The jumping of the electron from these states to the conduction band leads to ionization of the point defects. V_M^- and X_i^- represent the state of the electron after it has jumped from the valance band to ionize the metal vacancy and halogen interstitial, respectively, leaving a hole in the valance band.

Interstitials, both neutral and ionized, may also exist in MX crystals at equilibrium. These are also illustrated in Fig. 6.7 where a neutral metal interstitial, M_i, and a neutral halogen interstitial, X_i, are shown. The neutral metal interstitial is shown as a cation with an electron attached, making it a neutral atom. With sufficient thermal energy the neutral metal interstitial, M_i, may be ionized by removing the electron to the conduction band, as described by the chemical reaction and equilibrium constant shown in Table 6.1. Similarly the halogen interstitial may be ionized by creating a hole in the valence band, as described by the reaction and equilibrium constant shown in Table 6.1.

The energies associated with the defect ionizations described here may be represented with an energy diagram as shown in Fig. 6.8, where the conduction and valence band are separated by an energy gap, E_g, and where the energy levels for the various point defects described are also shown. We note the energies needed to ionize these point defects are expected to be smaller than the energy required to create an electron–hole pair in this ionic solid and this is reflected in the diagram. The creation of free electrons in the conduction band and holes in the valance

Table 6.2. Chemical reactions and equilibrium constants describing the formation of Schottky and Frenkel defects in an MX crystal.

Defect reaction	Chemical reaction	Equilibrium constant
Schottky defect formation	$\mathrm{N.O.} + \Delta G_S \rightarrow V_M + V_X$	$K'_S = [V_M][V_X]$
Frenkel defect formation on metal sub-lattice	$\mathrm{N.O.} + \Delta G_{FM} \rightarrow M_i + V_M$	$K'_{FM} = [M_i][V_M]$
Frenkel defect formation on halogen sub-lattice	$\mathrm{N.O.} + \Delta G_{FX} \rightarrow X_i + V_X$	$K'_{FX} = [X_i][V_X]$

band may also be described using the method of chemical equilibrium as

$$\mathrm{N.O.} + E_g \rightarrow e^- + e^+, \tag{6.44}$$

where N.O. stands for "normal occupancy," with all electrons in their ground state (valence band fully occupied and conduction band empty). This is the reference or perfect state whose Gibbs free energy is set to zero. The band gap energy, E_g, represents the energy needed to create an electron–hole pair by moving an electron from the valence band to the conduction band. The equilibrium constant for electron-hole equilibrium is expressed as

$$K'_g = [e^-]\left[e^+\right] = np. \tag{6.45}$$

This too is included in Table 6.1.

The formation of neutral defects in a crystal can also be described using the method of chemical equilibrium. Figure 6.6 illustrates a chemical reaction wherein neutral Schottky defects are created by removing neutral atoms from the crystal interior and placing them on surface sites. Such a reaction is shown in Table 6.2 along with the corresponding equilibrium constant. The reactions and equilibrium constants for the creation of Frenkel defects on either of the sub-lattices are also given in Table 6.2.

Finally we consider how the presence of defects in the crystal may affect chemical equilibrium with the vapor phase. In Eq. (6.41) we showed that chemical equilibrium between a stoichiometric MX crystal and its vapor leads to an equilibrium constant, K'_{MX}, involving the partial pressures of the two components in the vapor phase. For a nonstoichiometric crystal with defects the chemical equilibrium with the vapor phase may result in the creation or annihilation of defects. As illustrated in Fig. 6.6, a metal atom in the vapor phase might join the crystal as an interstitial; this would be described by the chemical reaction

$$M_{(g)} + \Delta G^1_R \rightarrow M_i, \tag{6.46}$$

with an equilibrium constant $K'_{Ri} = [M_i]/p_M$, where the subscript R indicates a *reduction* reaction wherein an electropositive element (M) combines with the MX crystal (in this case entering the crystal as an interstitial). This reaction and equilibrium constant is shown in Table 6.3. Alternatively, if metal vacancies are present, equilibrium with the vapor phase might be established by allowing a metal atom in the vapor phase to join the MX crystal by annihilating a metal vacancy, as described in Table 6.3.[2]

2 The chemical reactions and equilibrium constants in Tables 6.2 and 6.3 are related to each other. For example, it can be easily shown that $\Delta G_{FM} = \Delta G^1_R + \Delta G^2_R$ and $K'_{Ri}K'_{RV} = K'_{FM}$.

Table 6.3. Chemical reactions and equilibrium constants describing possible reduction and oxidation reactions between an MX crystal with defects and its vapor.

Defect reaction	Chemical reaction	Equilibrium constant
Reduction reaction through the formation of metal interstitials	$M_{(g)} + \Delta G_R^1 \rightarrow M_i$	$K'_{Ri} = [M_i]/p_M$
Reduction reaction through the annihilation of metal vacancies	$M_{(g)} + V_M + \Delta G_R^2 \rightarrow \text{N.O.}$	$K'_{RV} = \{[V_M]p_M\}^{-1}$
Oxidation reaction through the annihilation of metal interstitials	$\frac{1}{2}X_{2(g)} + M_i + \Delta G_{Ox}^1 \rightarrow \text{N.O.}$	$K'_{Oxi} = \{[M_i]\sqrt{p_{X_2}}\}^{-1}$
Oxidation reaction through the creation of metal vacancies	$\frac{1}{2}X_{2(g)} + \Delta G_{Ox}^2 \rightarrow \text{N.O.} + V_M$	$K'_{OxV} = [V_M]/\sqrt{p_{X_2}}$

Since the equilibrium partial pressures of the two components are related through the equilibrium constant, K'_{MX}, as shown in Eq. (6.41), equilibrium relations between the vapor and the nonstoichiometric crystal with defects can alternately be expressed in terms of the partial pressure of the electronegative element, $X_{2(g)}$, through oxidation reactions. The chemical reactions and equilibrium constants describing equilibrium with the vapor through the annihilation of metal interstitials or the creation of metal vacancies are also shown in Table 6.3, where the subscript Ox denotes an oxidation reaction with an electronegative element. The oxidation reaction leading to the formation of metal vacancies is also illustrated in Fig. 6.6.

Charge neutrality

The chemical reactions described above, together with their equilibrium constants, define the relations between the defect concentrations in nonstoichiometric crystals and the partial pressures of the components in the vapor phase, at a given high temperature. For a given set of possible defects, the equilibrium defect concentrations can be determined by solving the various equilibrium relations simultaneously. One additional condition is needed to solve this system of equations, namely: charge neutrality. For example, if both charged and neutral Frenkel defects on the metal sub-lattice are present in the crystal, along with the electrons and holes associated with their ionizations, then the charge neutrality condition would be

$$n + \left[V_M^-\right] = p + \left[M_i^+\right], \tag{6.47}$$

that is, the number of negative charges in the crystal must be balanced by the number of positive charges. Similarly, if Schottky defects are present on both sub-lattices, then the charge neutrality condition would read

$$n + \left[V_M^-\right] = p + \left[V_X^+\right]. \tag{6.48}$$

The equilibrium state of the crystal is then finally found by solving the various equilibrium relations, subject to the constraint of a charge neutrality condition.

Example 1. Equilibrium defect concentrations in a PbS crystal with Schottky defects

We illustrate the principles we have described by considering defect equilibrium in nonstoichiometric PbS crystals at high temperatures in a partial pressure of sulfur, a system that has been well studied. Extensive previous research has shown that Schottky defects on the two sub-lattices are the dominant point defects in PbS. Using the principles we have described, the defect equilibrium with the vapor phase may be described by

$$V_S + \frac{1}{2}S_{2(g)} + \Delta G_{Ox} \rightarrow \text{N.O.}, \tag{6.49}$$

with an equilibrium constant

$$K'_{Ox} = \frac{1}{[V_S]\sqrt{p_{S_2}}}, \tag{6.50}$$

or

$$[V_S] = \frac{1}{K'_{Ox}\sqrt{p_{S_2}}}. \tag{6.51}$$

This relation establishes the concentration of neutral vacancies on the sulfur sub-lattice for a given partial pressure of sulfur, assuming the equilibrium constant is known.

Schottky pair formation is governed by a chemical reaction like that shown in Eq. (6.23) and this leads to

$$K'_S = [V_{Pb}][V_S]. \tag{6.52}$$

The Schottky defects on the two sub-lattices may be neutral or charged so that the chemical equilibria associated with ionization of these defects, as described in Table 6.1, are required and this leads to the following equilibrium constants

$$K'_1 = \frac{[V_S^+]\,n}{[V_S]}, \tag{6.53}$$

and

$$K'_2 = \frac{[V_{Pb}^-]\,p}{[V_{Pb}]}. \tag{6.54}$$

In addition, electron–hole equilibrium, Eq. (6.44), leads to the relation

$$K'_g = np. \tag{6.55}$$

Finally, the charge neutrality condition for Schottky defects reads

$$n + [V_{Pb}^-] = p + [V_S^+]. \tag{6.56}$$

Equations (6.51)–(6.56) constitute a system of six independent relations which, if solved simultaneously, would give the six unknown defect concentrations, $[V_S]$, $[V_S^+]$, $[V_{Pb}]$, $[V_{Pb}^-]$, n, p, at a given sulfur partial pressure. As noted above, the equilibrium constants are assumed to be known at a given high temperature. We illustrate the solution of these relations for a given set of known equilibrium constants. Specifically, we take

$$K'_S = 10^{29} \quad K'_1 = K'_2 = 10^{18} \quad K'_g = 10^{32}, \tag{6.57}$$

where the numbers are chosen to describe defect concentrations in units of cm^{-3} (following the examples given in Kröger and Vink [33]). Using these constants, the six equations, (6.51)–(6.56), are most conveniently expressed in logarithmic form:

$$\begin{aligned}
\log[\mathrm{V_S}] &= -\frac{1}{2}\log\left(K'^2_{\mathrm{Ox}} p_{\mathrm{S_2}}\right), \\
\log[\mathrm{V_{Pb}}] + \log[\mathrm{V_S}] &= 29, \\
\log\left[\mathrm{V_S^+}\right] + \log n - \log[\mathrm{V_S}] &= 18, \\
\log\left[\mathrm{V_{Pb}^-}\right] + \log p - \log[\mathrm{V_{Pb}}] &= 18, \\
\log n + \log p &= 32, \\
\log\left\{n + \left[\mathrm{V_{Pb}^-}\right]\right\} &= \log\left\{p + \left[\mathrm{V_S^+}\right]\right\}.
\end{aligned} \tag{6.58}$$

We see that this is a system of linear equations, except for the neutrality condition. Because the defect concentrations vary by many orders of magnitude as the partial pressure of sulfur is varied, it is convenient to approximate the charge neutrality condition at different partial pressures to make it linear also. This is called the *Brouwer approximation*. For example, at very low sulfur partial pressures one expects a relatively high concentration of sulfur vacancies, so the dominant charged defects would be e^- and $\mathrm{V_S^+}$. In this regime of low partial pressure of sulfur the approximate charge neutrality conditions would then be

$$\log n = \log\left[\mathrm{V_S^+}\right]. \tag{6.59}$$

With this approximation the six equations, Eqs. (6.58), can be solved simultaneously to read

$$\begin{aligned}
\log[\mathrm{V_S}] &= -\frac{1}{2}\log\left(K'^2_{\mathrm{Ox}} p_{\mathrm{S_2}}\right), \\
\log[\mathrm{V_{Pb}}] &= 29 + \frac{1}{2}\log\left(K'^2_{\mathrm{Ox}} p_{\mathrm{S_2}}\right), \\
\log n &= 9 - \frac{1}{4}\log\left(K'^2_{\mathrm{Ox}} p_{\mathrm{S_2}}\right), \\
\log p &= 23 + \frac{1}{4}\log\left(K'^2_{\mathrm{Ox}} p_{\mathrm{S_2}}\right), \\
\log\left[\mathrm{V_S^+}\right] &= 9 - \frac{1}{4}\log\left(K'^2_{\mathrm{Ox}} p_{\mathrm{S_2}}\right), \\
\log\left[\mathrm{V_{Pb}^-}\right] &= 24 + \frac{1}{4}\log\left(K'^2_{\mathrm{Ox}} p_{\mathrm{S_2}}\right).
\end{aligned} \tag{6.60}$$

We see that for the regime of low partial pressure of sulfur, $\log n = \log[\mathrm{V_S^+}]$, which both decrease with increasing sulfur partial pressure. In the same regime, $\log[\mathrm{V_{Pb}^-}] > \log p$, and both increase with increasing $p_{\mathrm{S_2}}$. So as the partial pressure of sulfur increases, $\log[\mathrm{V_{Pb}^-}]$ will exceed $\log n$ before $\log p$ exceeds $\log[\mathrm{V_S^+}]$. Thus at intermediate sulfur partial pressures the

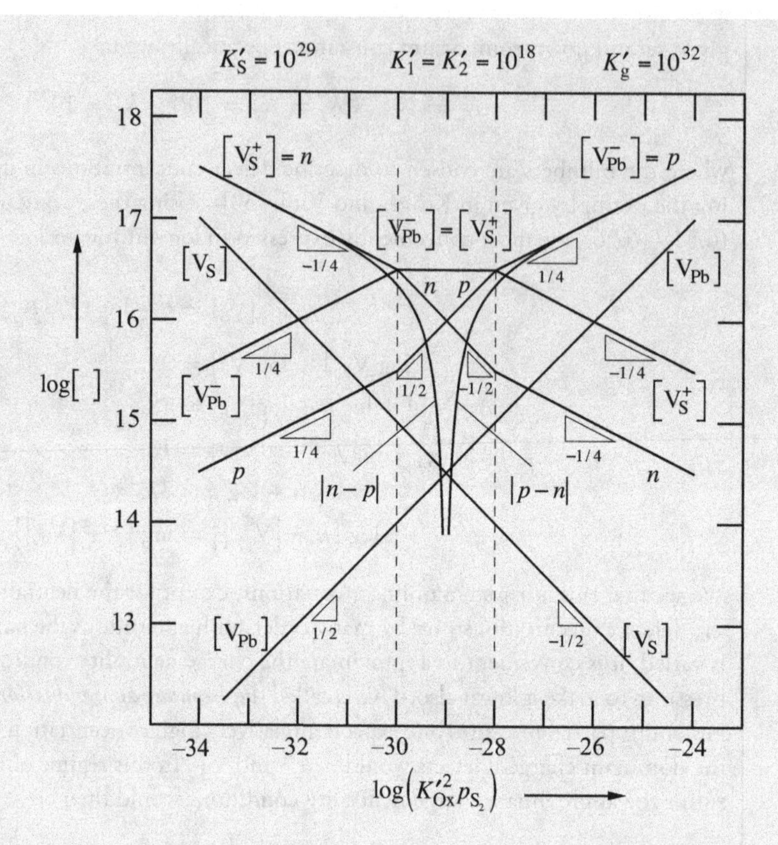

Figure 6.9. Brouwer diagram for PbS showing how the concentrations of both neutral and charged Schottky defects and electron and hole concentrations depends on the partial pressure of sulfur. The equilibrium constants leading to the diagram are shown in the figure.

Brouwer approximation becomes $\log[V_{Pb}^-] = \log[V_S^+]$ and a new set of linear equations, like those shown in Eqs. (6.60), would be produced.

Finally, at high partial pressures of sulfur, where Pb vacancies are expected to be dominant, the approximate neutrality condition becomes $\log[V_{Pb}^-] = \log p$ and a new set of linear equations describe the defect concentrations. These relations can be shown graphically in a log–log plot of defect concentration vs sulfur partial pressure $(K_{Ox}'^2 p_{S_2})$, as shown in Fig. 6.9. This is called a *Brouwer diagram*; for the defect model we have described and for the equilibrium constants we have chosen, it shows how the concentrations of the different point defects are expected to vary with sulfur partial pressure.

One of the predictions of this Brouwer diagram is that electrons are the dominant electronic carriers at low partial pressures of sulfur and that holes dominate at high sulfur partial pressures. Specifically, the logarithm of the net concentration of electronic charges, $\log|n - p|$, is predicted to decline with increasing $\log(K_{Ox}'^2 p_{S_2})$ at low sulfur partial pressures, first with a slope of $-1/4$ and then with a greater slope of $-1/2$ and higher before increasing with increasing sulfur partial pressure at higher partial pressures, eventually with a slope of $+1/2$ and finally with a slope of $+1/4$. In the mid 1950s J. Bloem *et al.* [34, 35] measured

$|n-p|$ vs p_{S_2} for PbS using the Hall effect and confirmed the predictions of this Brouwer diagram. Figure 6.10 shows their measurements at two different temperatures with the same slopes as those in the Brouwer diagram.

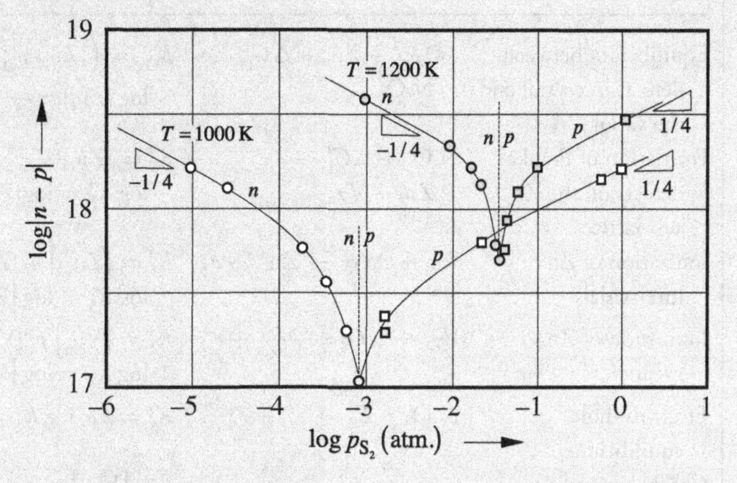

Figure 6.10. Measurements of the net electronic carrier concentration in PbS as a function of partial pressure of sulfur at two high temperatures (data taken from J. Bloem *et al.* [34]).

Example 2. Equilibrium defect concentrations in a ZnO crystal with Frenkel defects on the metal sub-lattice

ZnO is another system for which point defect concentrations and electronic conductivity at high temperatures have been studied as a function of partial pressures in the vapor phase. Frenkel defects on the Zn sub-lattice are expected because Zn atoms are small enough to fit into interstitial positions. In a reducing atmosphere of high partial pressure of Zn (or, equivalently, a low partial pressure of oxygen), Zn interstitials are expected to be the dominant point defects. We can construct a Brouwer diagram to describe the defect properties of ZnO by considering the reactions and equilibrium constants in Table 6.4.

A solution to these relations can be found by constructing a Brouwer diagram using the following equilibrium constants: $K_F' = 10^{30}$, $K_3' = K_2' = 10^{18}$, $K_g' = 10^{32}$, again using constants chosen to describe possible defect concentrations per cubic centimeter. Starting with the Brouwer approximation $\log n = \log[\mathrm{Zn}_i^+]$ at low oxygen partial pressures, the linear equations can be solved for different regimes of oxygen partial pressure, as described above, and this leads to the Brouwer diagram shown in Fig. 6.11. We see that, at low partial pressures of oxygen (or equivalently at high Zn partial pressures), the ZnO crystal is predicted to be an n-type conductor with a carrier concentration and conductivity that varies with oxygen partial pressure as $\sigma \propto n \propto p_{O_2}^{-1/4}$, consistent with experiment. At higher partial pressures of oxygen the conductivity is predicted, first, to decrease more rapidly than $p_{O_2}^{-1/4}$, and then increase rapidly before increasing as $\sigma \propto p_{O_2}^{1/4}$ after the crystal becomes a p-type conductor.

Table 6.4. Chemical reactions and equilibrium constants describing ZnO crystal with Frenkel defects on the metal sub-lattice.

Defect reaction	Chemical reaction	Equilibrium constant and constraint
Equilibrium between defects in crystal and the vapor	$\frac{1}{2}O_{2(g)} + Zn_i + \Delta G^1_{Ox} \rightarrow$ N.O.	$K'_{Ox} = \left([Zn_i]\, p^{1/2}_{O_2}\right)^{-1}$, $\log[Zn_i] = -\frac{1}{2}\log\left(K'^2_{O_2} p_{O_2}\right)$
Formation of Frenkel defects on the Zn sub-lattice	N.O. $+ \Delta G'_F \rightarrow Zn_i + V_{Zn}$	$K'_F = [Zn_i][V_{Zn}]$, $\log K'_F = \log[Zn_i] + \log[V_{Zn}]$
Ionization of Zn interstitials	$Zn_i + \Delta G_3 \rightarrow Zn_i^+ + e^-$	$K'_3 = [Zn_i^+]\, n/[Zn_i]$, $\log K'_3 = \log[Zn_i^+] + \log n - \log[Zn_i]$
Ionization of Zn vacancies	$V_{Zn} + \Delta G_2 \rightarrow V^-_{Zn} + e^+$	$K'_2 = [V^-_{Zn}]\, p/[V_{Zn}]$, $\log K'_2 = \log[V^-_{Zn}] + \log p - \log[V_{Zn}]$
Electron–hole equilibrium	N.O. $+ E_g \rightarrow e^- + e^+$	$K'_g = np$, $\log K'_g = \log n + \log p$
Charge neutrality		$n + [V^-_{Zn}] = p + [Zn_i^+]$, $\log\{n + [V^-_{Zn}]\} = \log\{p + [Zn_i^+]\}$

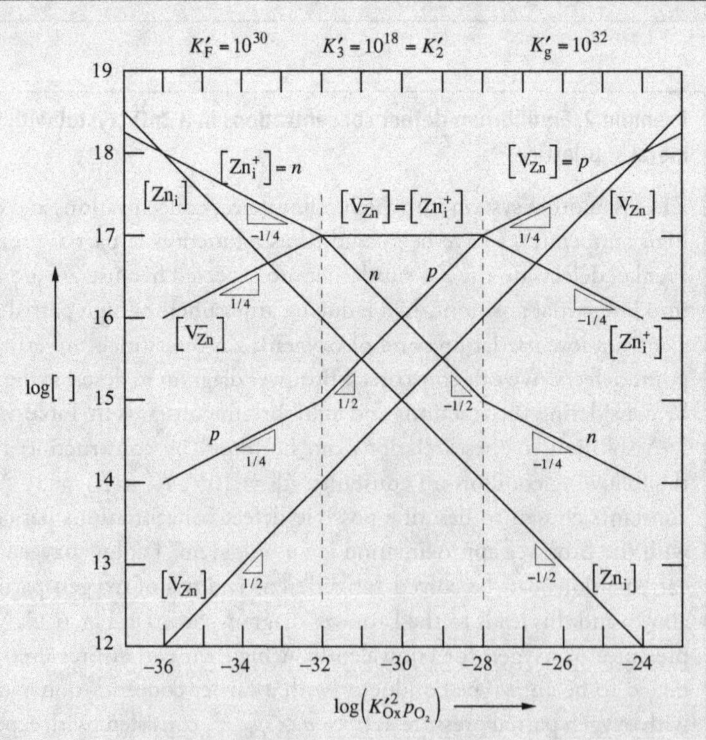

Figure 6.11. Brouwer diagram for ZnO showing how the concentrations of both neutral and charged cation defects and electron and hole concentrations depend on the partial pressure of oxygen. The equilibrium constants leading to the diagram are shown in the figure.

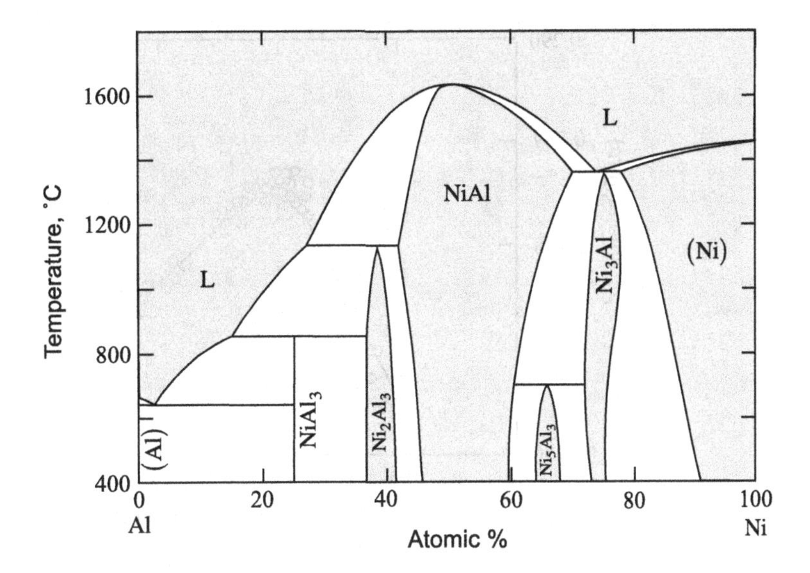

Figure 6.12. Ni–Al phase diagram [36].

6.4 Constitutional defects in intermetallic compounds

Many intermetallic alloys are not line compounds; rather they show a wide range of composition, even if they are still described using the molecular compound terminology of A_xB_y, as if the material were always stoichiometric. When the compositions deviate from the stoichiometric composition they do so by the introduction of the so-called constitutional defects, that is, defects that accommodate constitutional or compositional changes. Consider the simple example of NiAl, an intermetallic alloy having the CsCl or B2 crystal structure, shown in Fig. 1.5a.

As shown in the phase diagram in Fig. 6.12, the NiAl phase can exist over a wide range of compositions from about 45 at.% Ni to 60 at.% Ni (at 400 °C). The deviation from stoichiometry considered here is orders of magnitude larger than that for ionic solids discussed in Section 6.3. We may ask how such compositional changes can occur in a compound with a definite structure. The answer is that constitutional defects are responsible for the different compositions. Two different types of constitutional defects exist in the NiAl crystal depending on the composition.

- In Ni-rich alloys, there are Ni atoms residing on the Al sites. These are called *anti-site* defects or *anti-structure* defects.
- In Ni-poor (Al-rich) alloys, Ni vacancies are present on the Ni sub-lattice.

Much research on the structural and physical properties of this phase has led to this knowledge. There is no way to know, *a priori*, that these are the defects that account for the compositional changes.

The single most important evidence suggesting this particular defect model involves the composition dependence of the lattice parameter of this phase. As shown in Fig. 6.13, lattice parameter measurements indicate that the stoichiometric composition has the largest lattice parameter, a_0. With either increasing or decreasing nickel content the lattice parameter shrinks linearly

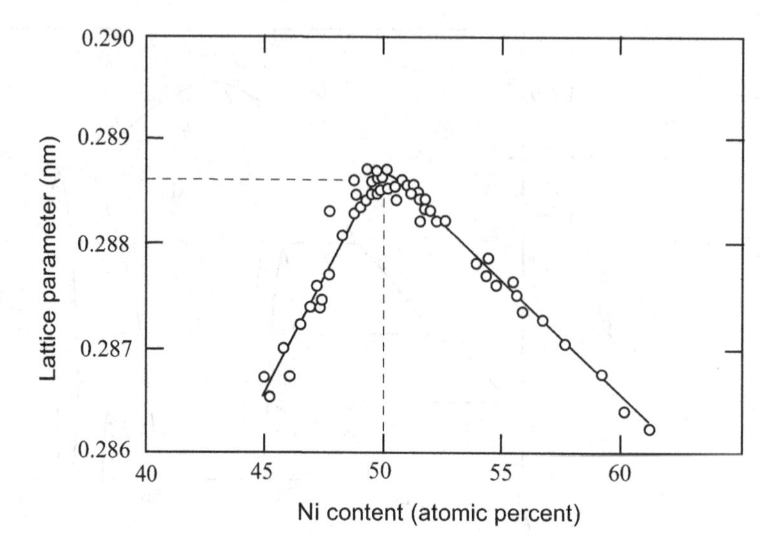

Figure 6.13. Lattice parameter measurements for Ni–Al alloys in the composition range where the NiAl phase is stable [37].

with composition. From the dependence of the lattice parameter on composition we can determine the separate sizes of the Ni and Al atoms, Ω_{Ni} and Ω_{Al}, respectively, as well as the size of the vacancies on the Ni sub-lattice, $\Omega_{\mathrm{v}}^{\mathrm{Ni}}$.

If the defect model we have described is responsible for the deviation from stoichiometry, then it is apparent that $\Omega_{\mathrm{Ni}} < \Omega_{\mathrm{Al}}$, that is, nickel atoms must be smaller than Al atoms in this structure to account for the decreasing lattice parameter with increasing Ni content. Similarly it must be true that the Ni vacancies in Ni-poor alloys must also be smaller than the Ni atoms they replace because the lattice parameter decreases with decreasing Ni content.

Consider a nickel-rich phase with N unit cells having n_{Ni} anti-structure defects (Ni atoms residing on Al sites). The volume of such a sample would be Na^3, where a is the composition-dependent lattice parameter. In a perfect NiAl crystal with stoichiometric composition, there are N Ni atoms and N Al atoms. Using the defect model of Ni anti-structure defects on Al sites, we may say that the total number of Ni atoms in the sample is $N + n_{\mathrm{Ni}}$, since all the Ni sites are occupied, plus n_{Ni} Ni atoms on the Al sites. The total number of Al atoms must be $N - n_{\mathrm{Ni}}$, because some of the Al sites are now occupied by Ni atoms. Thus we may write the total volume of this sample to be

$$\begin{aligned} V &= Na^3 = (N + n_{\mathrm{Ni}})\Omega_{\mathrm{Ni}} + (N - n_{\mathrm{Ni}})\Omega_{\mathrm{Al}}, \\ a^3 &= \left(1 + \frac{n_{\mathrm{Ni}}}{N}\right)\Omega_{\mathrm{Ni}} + \left(1 - \frac{n_{\mathrm{Ni}}}{N}\right)\Omega_{\mathrm{Al}} \\ &= \frac{n_{\mathrm{Ni}}}{N}(\Omega_{\mathrm{Ni}} - \Omega_{\mathrm{Al}}) + (\Omega_{\mathrm{Ni}} + \Omega_{\mathrm{Al}}). \end{aligned} \tag{6.61}$$

The fraction χ_{Ni} of Ni in the alloy is related to n_{Ni}/N through the following relation,

$$\chi_{\mathrm{Ni}} = \frac{\text{number of Ni atoms}}{\text{number of Ni and Al atoms}} = \frac{N + n_{\mathrm{Ni}}}{2N}, \tag{6.62}$$

so

$$\frac{n_{\mathrm{Ni}}}{N} = 2\chi_{\mathrm{Ni}} - 1. \tag{6.63}$$

Thus the lattice parameter can be expressed as

$$a^3 = (2\chi_{\mathrm{Ni}} - 1)(\Omega_{\mathrm{Ni}} - \Omega_{\mathrm{Al}}) + (\Omega_{\mathrm{Ni}} + \Omega_{\mathrm{Al}}). \tag{6.64}$$

Now if we differentiate this expression with respect to χ_{Ni} we have

$$3a^2 \frac{da}{d\chi_{\mathrm{Ni}}} = 2(\Omega_{\mathrm{Ni}} - \Omega_{\mathrm{Al}}) \tag{6.65}$$

or

$$\begin{aligned} \Omega_{\mathrm{Ni}} &= \frac{3}{2} a^2 \frac{da}{d\chi_{\mathrm{Ni}}} + \Omega_{\mathrm{Al}} \\ &\approx \frac{3}{2} a_0^2 \frac{da}{d\chi_{\mathrm{Ni}}} + \Omega_{\mathrm{Al}}, \end{aligned} \tag{6.66}$$

where the lattice parameter at any composition, a, has been replaced by the lattice parameter at the stoichiometric composition, a_0. This is acceptable since a changes only very slightly with composition. We also know that for the stoichiometric composition

$$a_0^3 = \Omega_{\mathrm{Ni}} + \Omega_{\mathrm{Al}}. \tag{6.67}$$

Solving Eqs. (6.66) and (6.67) simultaneously we can find

$$\Omega_{\mathrm{Ni}} = \frac{a_0^3}{2} + \frac{3}{4} a_0^2 \frac{da}{d\chi_{\mathrm{Ni}}}, \tag{6.68}$$

$$\Omega_{\mathrm{Al}} = \frac{a_0^3}{2} - \frac{3}{4} a_0^2 \frac{da}{d\chi_{\mathrm{Ni}}}. \tag{6.69}$$

By taking the slope of the lattice parameter versus composition plot in the Ni-rich regime we can determine $da/d\chi_{\mathrm{Ni}}$ experimentally. This, together with a_0, allows us to find

$$\begin{aligned} \Omega_{\mathrm{Ni}} &= 10.66\,\text{Å}^3, \\ \Omega_{\mathrm{Al}} &= 13.41\,\text{Å}^3. \end{aligned}$$

These values can be compared with the atomic values of Al and Ni in pure metals. From the King table A.1 we can find that $\Omega_{\mathrm{Ni}} = 10.94\,\text{Å}^3$ in pure Ni and $\Omega_{\mathrm{Al}} = 16.6\,\text{Å}^3$ in pure Al. Therefore, both Ni and Al atoms become smaller when they form the intermetallic compound NiAl. It means that Ni and Al bond with each other more strongly than they bond with themselves. This is consistent with the fact that the melting point of NiAl is higher than that of pure Ni and pure Al (see Fig. 6.12).

A similar analysis would allow us to determine the volume of vacancies on the Ni sub-lattice, where the lattice parameter decreases with decreasing Ni content. The result is

$$\Delta v_{\mathrm{v}}^{\mathrm{Ni}} = 7.9\,\text{Å}^3.$$

Thus we find that a Ni vacancy occupies a smaller volume than a Ni atom in NiAl.

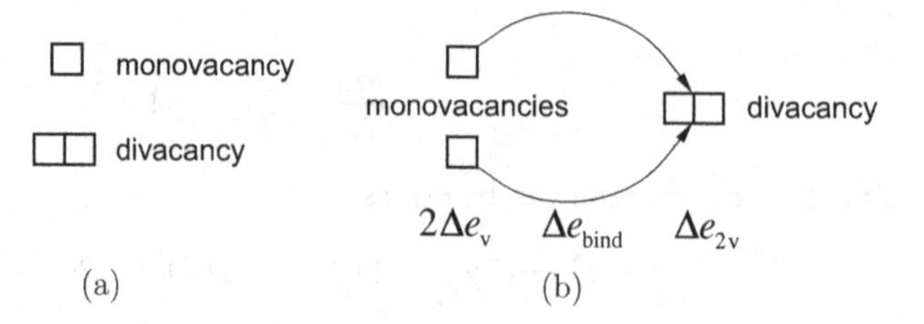

Figure 6.14. (a) Monovacancies and divacancy. (b) Creating divacancy by combining two vacancies.

Other intermetallic alloys have different constitutional defects accommodating the compositional changes. Each compound must be studied in detail, in terms of structure and physical properties, to determine the defects in each case.

6.5 Divacancies and other vacancy complexes

6.5.1 Equilibrium concentration

Just as single vacancies, or monovacancies, exist in crystals at equilibrium, so do divancancies, trivacancies, quadravacancies, and other point defect clusters. Considering monovacancies and divacancies, we may expect the equilibrium state of the crystal to be determined by

$$\frac{\partial \Delta G}{\partial n_v} = 0 \quad \text{and} \quad \frac{\partial \Delta G}{\partial n_{2v}} = 0, \tag{6.70}$$

where n_v and n_{2v} are the number of monovacancies and divacancies, respectively, in a crystal with N atomic sites. The results are

$$\chi_v = \exp\left(-\frac{\Delta g_v}{k_B T}\right) \quad \text{and} \quad \chi_{2v} = \exp\left(-\frac{\Delta g_{2v}}{k_B T}\right), \tag{6.71}$$

where χ_{2v} is the fraction of divacancy sites occupied by divacancies and Δg_{2v} is the Gibbs free energy of formation of the divacancy. We may symbolize these lattice defects as in Fig. 6.14a.

The fraction of divacancies is defined as

$$\chi_{2v} \equiv \frac{n_{2v}}{N_{2v}^{\text{sites}}}, \tag{6.72}$$

where N_{2v}^{sites} is the number of divacancy sites in the crystal having N atomic sites. Because there are N atomic sites and z nearest neighbors for each atomic site (coordination number) then

$$N_{2v}^{\text{sites}} = \frac{Nz}{2}, \tag{6.73}$$

where the factor of 2 is needed to avoid double counting indistinguishable divacancies. Thus the number of divacancies in a crystal is then

$$\begin{aligned} n_{2v} &= \frac{Nz}{2}\exp\left(-\frac{\Delta g_{2v}}{k_B T}\right) \\ &= \frac{Nz}{2}\exp\left(\frac{\Delta s_{2v}}{k_B}\right)\exp\left(-\frac{\Delta e_{2v}}{k_B T}\right)\exp\left(-\frac{p_{\text{ext}}\Delta v_{2v}}{k_B T}\right), \end{aligned} \tag{6.74}$$

where Δe_{2v}, Δv_{2v}, and Δs_{2v} are the formation energy, volume, and entropy of divacancies, respectively. For comparison with monovacancies it is convenient to express the formation energy of the divacancy as

$$\Delta e_{2v} = 2\Delta e_v + \Delta e_{bind}, \tag{6.75}$$

where Δe_{bind} is the binding energy for the process of creating the divacacy as envisioned in Fig. 6.14b. The divacancy is created by first creating two separate monovacancies that are then bound together to make the divacancy. For this comparison we can approximate the less important terms in the divacancy formation free energy as

$$\Delta s_{2v} = 2\Delta s_v \quad \text{and} \quad \Delta v_{2v} = 2\Delta v_v. \tag{6.76}$$

With this approximation we can then write

$$n_{2v} = \frac{Nz}{2} \exp\left(\frac{2\Delta s_v}{k_B}\right) \exp\left(-\frac{2\Delta e_v + \Delta e_{bind}}{k_B T}\right) \exp\left(-\frac{2p_{ext}\Delta v_v}{k_B T}\right).$$

This may be written as

$$\begin{aligned}
n_{2v} &= \frac{Nz}{2} \chi_v^2 \exp\left(-\frac{\Delta e_{bind}}{k_B T}\right), \\
\frac{n_{2v}}{N} &= \frac{z}{2}\left(\frac{n_v}{N}\right)^2 \exp\left(-\frac{\Delta e_{bind}}{k_B T}\right), \\
\chi_{2v} &= \left(\frac{n_v}{N}\right)^2 \exp\left(-\frac{\Delta e_{bind}}{k_B T}\right),
\end{aligned} \tag{6.77}$$

using the definition in Eq. (6.72). We see that the equilibrium concentration of divacancies varies as the square of the monovacancy concentration. This means that the divacancy concentration is typically much less than the monovacancy concentration, except under certain conditions, as described below. If the binding energy were zero, then this expression would simply give the concentration of divacancies that would be present by the chance occurrence of monovacancies residing on adjacent sites. The binding energy, which is typically negative, tends to increase the divacancy concentration above this chance value. The binding energy will be further discussed in Section 6.5.2 and an estimate will be given in Eq. (6.85).

6.5.2 Method of chemical equilibrium

We may think of Eq. (6.77) as describing a "chemical" equilibrium between monovacancies and divacancies. We can think of these defects as chemical species that can react to create or destroy each other, as illustrated in Fig. 6.15. Under this interpretation, the binding energy Δe_{bind} must be related to the chemical equilibrium constant K'_{eq}, which is what we will show in the following.

Divacancies

The chemical reaction can be analyzed as

$$\mathrm{v} + \mathrm{v} \Leftrightarrow 2\mathrm{v}, \tag{6.78}$$

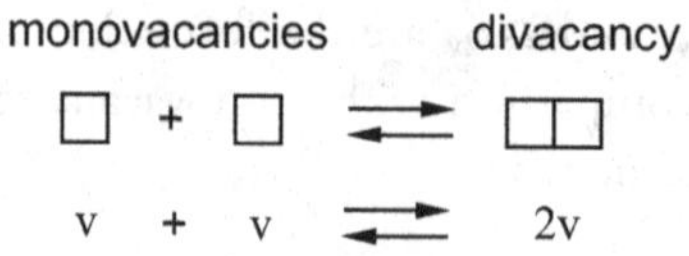

monovacancies divacancy

$$\square + \square \rightleftharpoons \square\square$$

$$\mathrm{v} + \mathrm{v} \rightleftharpoons 2\mathrm{v}$$

Figure 6.15. Monovacancies and divacancy equilibrium.

with a chemical equilibrium constant

$$K'_{\mathrm{eq}} = \frac{[2\mathrm{v}]}{[\mathrm{v}]^2}, \tag{6.79}$$

$$[2\mathrm{v}] = [\mathrm{v}]^2 K'_{\mathrm{eq}}, \tag{6.80}$$

where $[2\mathrm{v}] = n_{2\mathrm{v}}/V$ and $[\mathrm{v}] = n_{\mathrm{v}}/V$ are the concentrations of divacancies and vacancies and V is the volume of the crystal. Then from Eq. (6.77), and noticing $\Omega = V/N$, we have

$$[2\mathrm{v}] = \frac{z}{2}[\mathrm{v}]^2 \Omega \exp\left(-\frac{\Delta e_{\mathrm{bind}}}{k_{\mathrm{B}}T}\right), \tag{6.81}$$

which has the same form as Eq. (6.80), leading to the identification

$$K'_{\mathrm{eq}} = \frac{z}{2}\Omega \exp\left(-\frac{\Delta e_{\mathrm{bind}}}{k_{\mathrm{B}}T}\right). \tag{6.82}$$

This identification means that we may indeed think of a chemical equilibrium between monovacancies and divacancies with an equilibrium constant related to the binding energy.

Earlier in Section 5.2 we made a rough estimate of the formation energy of the vacancy as

$$\Delta e_{\mathrm{v}} = 6a^2\gamma_{\mathrm{s}}, \tag{6.83}$$

where the surface of the vacancy (treated as a cubic hole) is regarded as a free surface with energy γ_{s}. A similar argument for a divacancy (side by side cubic holes) leads to

$$\Delta e_{2\mathrm{v}} = 10a^2\gamma_{\mathrm{s}}. \tag{6.84}$$

A comparison of these estimates allows us to estimate the binding energy as

$$\Delta e_{\mathrm{bind}} = \Delta e_{2\mathrm{v}} - 2\Delta e_{\mathrm{v}} = -2a^2\gamma_{\mathrm{s}} = -\frac{1}{3}\Delta e_{\mathrm{v}}. \tag{6.85}$$

This is a useful estimate showing the order of magnitude of the binding energy and also that the binding energy is negative, i.e. the vacancies tend to stick together. The reasoning we have used here can be extended to other vacancy complexes.

Trivacancies

Considering the creation of trivacancies as the combination of a divacancy with another vacancy, we have the following estimate of the trivacancy fraction,

$$\chi_{3\mathrm{v}} \approx \left(\frac{n_{\mathrm{v}}}{N}\right)^3 \exp\left(-\frac{\Delta e^{(1)}_{\mathrm{bind}}}{k_{\mathrm{B}}T}\right) \exp\left(-\frac{\Delta e^{(2)}_{\mathrm{bind}}}{k_{\mathrm{B}}T}\right), \tag{6.86}$$

where $\Delta e^{(1)}_{\text{bind}} = \Delta e_{\text{bind}}$ is the binding energy for the first two vacancies and $\Delta e^{(2)}_{\text{bind}}$ is the binding energy of the third vacancy to the divacancy. There are several different types of trivacancies – three in a line, three at an angle, ... Hence several values of $\Delta e^{(2)}_{\text{bind}}$ can be defined, and Eq. (6.86) is only an approximation. Another approximation made in Eq. (6.86) is the neglect of formation entropy changes as vacancies coalesce.

Quadravacancies

Extending the arguments above to a cluster of four vacancies, we have the following estimate of the quadravacancy fraction:

$$\chi_{4\text{v}} \approx \left(\frac{n_{\text{v}}}{N}\right)^4 \exp\left(-\frac{\Delta e^{(1)}_{\text{bind}}}{k_{\text{B}}T}\right) \exp\left(-\frac{\Delta e^{(2)}_{\text{bind}}}{k_{\text{B}}T}\right) \exp\left(-\frac{\Delta e^{(3)}_{\text{bind}}}{k_{\text{B}}T}\right). \tag{6.87}$$

These relations show that the equilibrium concentration of defect clusters decreases very rapidly with increasing number of monovacancies in the cluster. For all intents and purposes, the equilibrium concentration of these higher-order clusters is nearly zero, even at the melting temperature.

6.5.3 Constrained vacancy equilibrium

There are certain conditions where the concentration of defect clusters, like divacancies, can be high, even higher that the monovacancy concentration. Consider a crystal with N atomic sites first held at a high temperature of T_1 and then rapidly cooled to a lower temperature of T_2 where the vacancies and divacancies are allowed to come to equilibrium with respect to each other but that the total number of vacant atomic sites is not permitted to change. The monovacancies and divacancies do not have enough time to diffuse to sinks and leave the crystal (see Section 7.1.1). So we have a constrained equilibrium – the constraint being that the total number of vacant atomic sites remain fixed at the value established at the high temperature.

At the high temperature we have

$$\left(\frac{n_{\text{v}}}{N}\right)_{T_1} \approx \exp\left(-\frac{\Delta e_{\text{v}}}{k_{\text{B}}T_1}\right) \tag{6.88}$$

and

$$\left(\frac{n_{2\text{v}}}{N}\right)_{T_1} = \frac{z}{2}\left(\frac{n_{\text{v}}}{N}\right)^2_{T_1} \exp\left(-\frac{\Delta e_{\text{bind}}}{k_{\text{B}}T_1}\right). \tag{6.89}$$

After rapid cooling to the lower temperature, we assume the total number of vacant atomic sites remain unchanged, so that

$$\left(\frac{n_{\text{v}}}{N}\right)_{T_2} + \left(\frac{2n_{2\text{v}}}{N}\right)_{T_2} = \left(\frac{n_{\text{v}}}{N}\right)_{T_1} + \left(\frac{2n_{2\text{v}}}{N}\right)_{T_1} \equiv C, \tag{6.90}$$

where C is the total fraction of vacant atomic sites in the system. Define

$$\alpha \equiv \left(\frac{n_{2\text{v}}}{N}\right)_{T_2}. \tag{6.91}$$

The monovacancy–divacancy equilibrium, Eq. (6.77), at the lower temperature requires

$$\alpha = \left(\frac{n_{2v}}{N}\right)_{T_2} = \frac{z}{2}\left(\frac{n_v}{N}\right)_{T_2}^2 \exp\left(-\frac{\Delta e_{\text{bind}}}{k_B T_2}\right). \tag{6.92}$$

Therefore

$$\left(\frac{n_v}{N}\right)_{T_2} = \left[\frac{2}{z}\exp\left(+\frac{\Delta e_{\text{bind}}}{k_B T_2}\right)\right]^{1/2} \alpha^{1/2}. \tag{6.93}$$

Plugging Eqs. (6.91) and (6.93) into Eq. (6.90), we have

$$C = \left[\frac{2}{z}\exp\left(+\frac{\Delta e_{\text{bind}}}{k_B T_2}\right)\right]^{1/2} \alpha^{1/2} + 2\alpha,$$

$$2\alpha + \left[\frac{2}{z}\exp\left(+\frac{\Delta e_{\text{bind}}}{k_B T_2}\right)\right]^{1/2} \alpha^{1/2} - C = 0. \tag{6.94}$$

Now we may solve for $\sqrt{\alpha}$ as

$$\sqrt{\alpha} = \frac{1}{4}\left\{-\left[\frac{2}{z}\exp\left(+\frac{\Delta e_{\text{bind}}}{k_B T_2}\right)\right]^{1/2} \pm \sqrt{\frac{2}{z}\exp\left(+\frac{\Delta e_{\text{bind}}}{k_B T_2}\right) + 8C}\right\}, \tag{6.95}$$

where the + should be taken for the ± sign, so that $\sqrt{\alpha} > 0$. As a result

$$\left(\frac{n_{2v}}{N}\right)_{T_2} = \frac{1}{16}\left\{-\left[\frac{2}{z}\exp\left(+\frac{\Delta e_{\text{bind}}}{k_B T_2}\right)\right]^{1/2} + \sqrt{\frac{2}{z}\exp\left(+\frac{\Delta e_{\text{bind}}}{k_B T_2}\right) + 8C}\right\}^2. \tag{6.96}$$

Consider the following typical values: $\Delta e_v = 1$ eV, $\Delta e_{\text{bind}} = -\frac{1}{3}$ eV. The above analysis gives the following estimates.
At $T_1 = 1000$ K:

$$\left(\frac{n_v}{N}\right)_{T_1} = 9.2 \times 10^{-6} \quad \text{and} \quad \left(\frac{n_{2v}}{N}\right)_{T_1} = 2.4 \times 10^{-8}, \tag{6.97}$$

that is, the number of divacancies is far less than the number of monovacancies, as expected. But under the quench conditions,
at $T_2 = 300$ K:

$$\left(\frac{n_v}{N}\right)_{T_1} = 1.3 \times 10^{-6} \quad \text{and} \quad \left(\frac{n_{2v}}{N}\right)_{T_1} = 4 \times 10^{-6}. \tag{6.98}$$

We see that in the quenched condition the divacancy concentration exceeds the monovacancy concentration, that is, the vacancies tend to cluster together. Such clustering also leads to the formation of trivacancies, quadravacancies, and other vacancy clusters and eventually to the formation of dislocation loops, which will be discussed in Chapter 7, see Fig. 7.2.

6.6 Summary

In semiconductors, ionic solids, and intermetallic compounds, vacancies co-exist with other intrinsic point defects, including interstitials and anti-site defects. In strongly ionic solids, point defects exist in the form of Schottky pairs or Frenkel pairs, in order to satisfy the charge neutrality condition. An analysis of how the Gibbs free energy changes with the introduction of any of these point defects still results in Boltzmann-like expressions for the equilibrium concentrations. The charge neutrality condition can be imposed when minimizing the Gibbs free energy using the method of Lagrange multipliers. Alternatively, the method of chemical equilibrium can also be used to obtain the equilibrium concentrations.

In nonstoichiometric ionic solids, the point defects can be present in both neutral and ionized states, and free electrons and holes also need to be considered. The concentrations of various point defects depend on the equilibrium partial pressures of the components of the ionic crystal in the vapor phase. This dependence can be obtained by applying the method of chemical equilibrium subjected to the charge neutrality condition. A log–log plot of the resulting relations between the concentrations and partial pressure is called a Brouwer diagram.

For intermetallic compounds with anti-site defects an analysis of how the lattice parameter changes with composition can lead to estimates of the sizes of the different atoms in these compounds.

When a crystal is quenched from a high temperature to a low temperature, there is not sufficient time for vacancies to reach the true equilibrium concentration at the low temperature. A consequence is that vacancies aggregate to form divacancies and other vacancy clusters that may reach even higher concentrations than monovacancies under such conditions. Thermodynamic principles can still be used to predict the concentration of vacancy clusters in the constrained equilibrium states where the total number of vacant sites is fixed.

6.7 Exercise problems

6.1 Diffusion experiments have indicated that the equilibrium fractions of self-interstitials and vacancies in Si can be expressed as Eqs. (6.1) and (6.2), respectively. In the following we may assume that the self-interstitials will be found only in the largest, hard sphere interstices in the diamond structure. The lattice parameter of Si may be taken to be 0.543 nm.

(a) Determine the number of self-interstitial sites per unit cell in the diamond cubic structure (see Exercise problem 4.2).

(b) Determine the entropy of formation of both self-interstitials, Δs_i, and vacancies, Δs_v, using the information provided in Eqs. (6.1) and (6.2).

(c) Determine the formation free energies of both self-interstitials, Δg_i, and vacancies, Δg_v, at 1000 K.

(d) Calculate the equilibrium concentration of vacancies and self-interstitials in Si (defects per cubic meter) at 1000 K.

(e) Assuming that the vacancy volume relaxation is $\Delta V_v^{rlx} = -0.3\Omega$, and assuming also that the Si self-interstitial causes no distortion of the lattice, develop an expression for the critical pressure, p_{ext}^c, at which the concentration of vacancies and self-interstitials would be expected to be the same, at an arbitrary temperature, T. Write this expression using the usual notation where the thermodynamic properties of the defects are Δs_i, Δe_i, Δs_v, Δe_v, etc., and the atomic volume is Ω.

(f) Calculate the critical pressure at which the vacancy and self-interstitial concentrations are the same at 1000 K.

6.2 A nearly perfect (001) surface of a simple cubic solid with N surface lattice sites is shown in Fig. 6.16. A few of the surface atoms have been displaced from the plane of the surface and placed on top of the surface, creating holes in the surface. We call n^+ the number of isolated atoms sitting on the top of the surface and n^- the number of "holes" in the surface from which those atoms were removed. We treat the case in which there is one isolated surface atom for every "hole". We assume that the energy needed to remove an atom from the surface and place it on top, in the manner shown, is Δe_f. This would be the formation energy of the defect pair. Write an expression for the fraction of surface lattice sites occupied by isolated surface atoms, n^+/N, by minimizing the Gibbs free energy of the entire system. You may assume that the concentration of surface defects is very small. You may also assume that there is no external pressure and that the vibrational frequencies of the atoms do not change when the defects are formed.

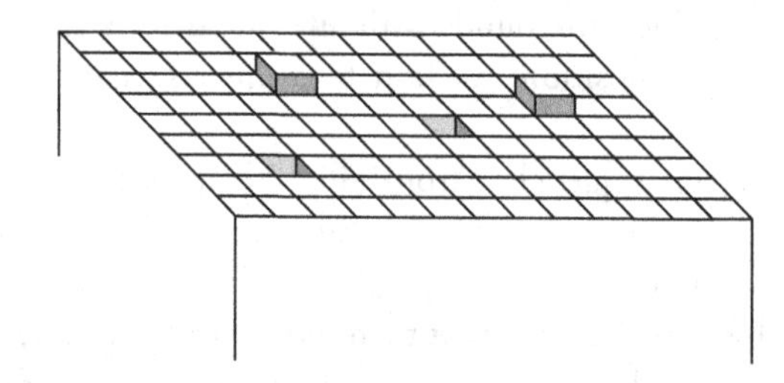

Figure 6.16. A nearly perfect (0 0 1) surface of a simple cubic solid.

6.3 The dominant point defects in NaCl are Schottky pairs with a formation enthalpy around 2.4 eV. Estimate the densities of cation and anion vacancies (number per unit volume) in NaCl at its melting temperature $T_m = 1074$ K. NaCl has the B1 structure, as shown in Fig. 1.14a, with a lattice constant $a = 5.64$ Å.

6.4 In TiO_2, which is an MX_2 ionic crystal, each cation has a charge of $+4e$ and each anion has a charge of $-2e$, so that a Schottky defect consists of one cation vacancy and two anion vacancies to maintain charge neutrality. Assume that Schottky defects are the dominant defects in the crystal. Then the number of cation vacancies n_v^{4-} and the number of anion vacancies n_v^{2+} must satisfy the condition, $n_v^{4-} = 2n_v^{2+}$.

(a) Derive an expression for the fraction of cation (and anion) vacancies by minimizing the Gibbs free energy of the entire crystal subjected to the charge neutrality constraint.
(b) Estimate the fraction of vacancies in TiO_2 at its melting temperature $T_m = 2116$ K assuming that the formation enthalpy of the Schottky defect is 5.2 eV [38].

6.5 Cubic ZrO_2 has the C1 structure shown in Fig. 1.14b in which the cations form an FCC sub-lattice. Assume that the formation enthalpy of a Schottky defect (one cation vacancy and two anion vacancies) is 5.9 eV, and the formation energy of an anion Frenkel defect (one anion vacancy and one anion interstitial) is 5.4 eV [39]. Assume that the anion interstitials occupy the octahedral sites of the FCC sub-lattice.

(a) Derive the expressions for the fraction of cation vacancies, anion vacancies, and anion interstitials by minimizing the Gibbs free energy of the entire crystal subjected to the charge neutrality constraint.

(b) Estimate the fraction of point defects in ZrO_2 at its melting temperature $T_m = 2988$ K. Are the Schottky defects or the Frenkel defects the dominant defects in this crystal?

6.6 Yttria-stabilized zirconia (YSZ) is made by doping ZrO_2 with Y_2O_3, with yttrium appearing as substitutional defects on the cation sub-lattice. (The ZrO_2–Y_2O_3 phase diagram is available on the book website.) Because an yttrium ion has a charge of $+3e$, for every two yttrium ions there must be one oxygen (anion) vacancy, to maintain charge balance. Suppose the fraction of yttrium ions (number of yttrium divided by total number of yttrium and zirconium ions) is 1%. Use the Schottky and Frenkel formation energies given in Exercise problem 6.5 to estimate the fraction of cation vacancies, anion vacancies, and anion interstitials in the crystal at 2000 K.

6.7 An MX crystal with Frenkel disorder on the metal sub-lattice is characterized by the following equilibrium constants: Frenkel defect formation, $K_F' = 10^{26}$; vacancy ionization, $K_2' = 10^{18}$; interstitial ionization, $K_3' = 10^{19}$; electron–hole equilibrium: $K_g' = 10^{32}$. Construct a Brouwer diagram for this model showing how the concentrations of various point defects depend on the partial pressure of the metal in the vapor phase. Determine the dominant carriers in different regimes of partial pressure and show how the conductivity would be expected to vary with metal partial pressure over a wide range of partial pressures.

6.8 High-temperature lattice parameter and density measurements for Fe-deficient FeO equilibrated at different oxygen partial pressures have indicated that the lattice expands and the density increases as the composition of the crystal approaches the stoichiometric composition, FeO. This suggests that cation vacancies are responsible for the deviation from stoichiometry. Also, because neutral Fe vacancies include two electron holes in the ionic cavity, there is reason to expect that each Fe vacancy would contribute two holes to the valence band when fully ionized. Write a chemical reaction for the double ionization of neutral Fe vacancies and show how the partial pressure of oxygen would be expected to affect the concentration of holes in the valence band. How would the electronic conductivity vary with increasing oxygen partial pressure?

6.9 The electronic conductivity of Cu_2O has been measured at high temperatures as a function of the partial pressure of oxygen. The results suggest that singly ionized Cu vacancies are responsible for the observed properties. Write a chemical reaction showing how equilibrium with the vapor phase can produce Cu vacancies in Cu_2O and then show how the ionization of these vacancies leads to a relation between the carrier concentration and oxygen partial pressure. Show how the conductivity would be expected to vary with oxygen partial pressure according to this model.

6.10 Construct a Brouwer diagram for an MX crystal with Schottky defects that are almost completely ionized. In Section 6.2.1 we described Schottky defects in stoichiometric ionic crystals as being fully ionized, so that equal concentrations of ionic vacancies on the two sub-lattices are required for charge neutrality. Here we show how the method of Brouwer diagrams can be used to recover this result. In the context of our present treatment of defect chemistry, vacancies would be almost fully ionized if the ionization energies are very small relative to the other energies involved or if the equilibrium constants are large relative to the other equilibrium constants. Under these conditions the vacancies would be ionized so easily that essentially all of the vacancies would be ionized. We can model this situation by selecting the following equilibrium constants: $K_S' = 10^{29}$, $K_1' = K_2' = 10^{20}$, and $K_g' = 10^{32}$. Construct the Brouwer diagram for this crystal showing the concentrations of defects as functions of partial pressure P_{X_2}.

6.11 Consider the equilibrium point defects in stoichiometric NiAl. Based on the defect structures in nonstoichiometric NiAl, we may assume that the thermal vacancies in stoichiometric NiAl are created by also creating anti-structure Ni defects on the Al sub-lattice. Figure 6.17 shows how these two types of defects are created when a pair of Ni and Al atoms are removed to the surface. The net effect is to create two Ni vacancies and one anti-structure defect (a Ni atom on an Al site). Let N be the total number of unit cells in the system and let n be the number of anti-site defects (Ni atoms on Al sites), so that the fraction of Al sites occupied by Ni atoms is $\chi = n/N$. Also let Δg_v be the formation free energy of Ni vacancies and Δg_{Al} be the formation energy of the anti-site defect (Ni on the Al lattice).

(a) Write an expression for the change in free energy (per unit cell) when n anti-structure defects (and $2n$ Ni vacancies) are introduced into the crystal. Write this expression in terms of χ.
(b) Derive or write an expression for the equilibrium fraction of anti-site defects.

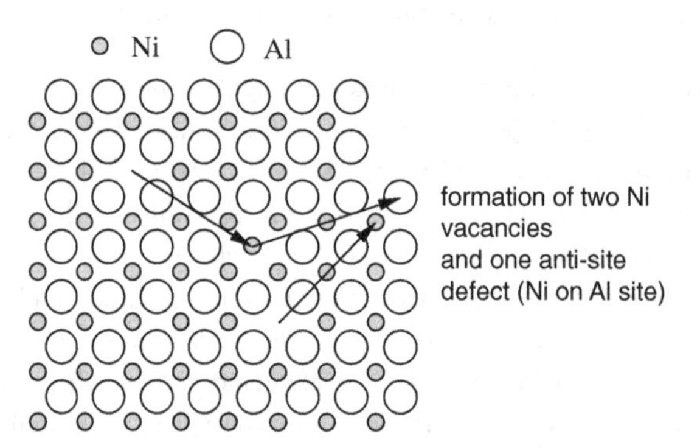

Figure 6.17. Vacancies and anti-site defect in NiAl.

6.12 A single crystal of FCC gold is held at 1000 °C for a long time so that point defect equilibrium is established. We wish to calculate the equilibrium concentrations of monovacancies, divacancies, and triangular trivacancies. These defects are illustrated in Fig. 6.18, which shows the positions of the missing atoms in a close-packed plane in the FCC lattice. Atoms in the close-packed planes above and below the one in the diagram are not shown.

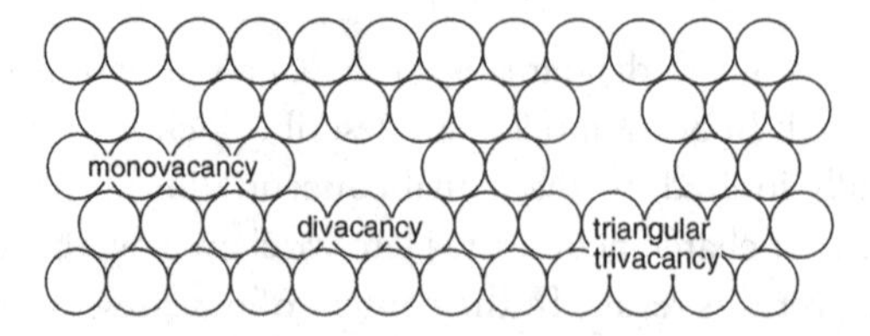

Figure 6.18. Monovacancy, divacancy, and trivacancy.

The lattice and monovacancy quantities needed for these calculations are given in Tables A.1 and C.1. Because the formation free energies of divacancies and trivacancies are not well known, we will use a broken bond model to estimate these quantities. We will assume that the formation energy for each of these defects is directly proportional to the number of broken gold bonds found on the surface of the defect. According to this model, the formation energy of the monovacancy (which is known) is equal to 12 times the energy of one of the broken bonds. Such an approach permits an estimate of the formation energies of divacancies and trivacancies. For simplicity we will assume that the formation entropies and volumes of these defects are as follows: $\Delta s_{2v} = 2\Delta s_v$, $\Omega_{2v} = 2\Omega_v$, $\Delta s_{3v} = 3\Delta s_v$, $\Omega_{3v} = 3\Omega_v$. Use the broken bond model to obtain the equilibrium concentration (in units of defects per m^3) of monovacancies, divacancies, and triangular trivacancies at 1000 °C.

6.13 An FCC metal crystal with N atomic sites contains $\chi_s N$ substitutional solute atoms (χ_s is the atomic fraction of solute in the alloy). At equilibrium, the crystal also contains vacancies, as shown in Fig. 6.19. Two different types of vacancies are to be distinguished, those that are bound to the solutes (on adjacent sites) and those that are free of the solutes. Let n_0^{bound} and n_0^{free} represent the total number of possible sites for bound and free vacancies, respectively. The formation free energy of the free vacancy is Δg_v and the binding energy of the vacancy to the solute atom is Δg_b (this is a negative quantity).

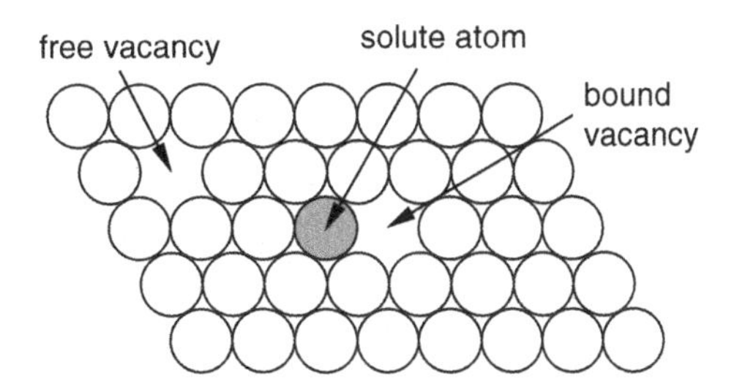

Figure 6.19. A free vacancy and a vacancy bound to a solute atom.

(a) Assuming a dilute solid solution ($\chi_s \ll 1$), write expressions for n_0^{bound} and n_0^{free}.
(b) Write an expression for the ratio of the number of bound vacancies to the number of free vacancies, $n_v^{\text{bound}}/n_v^{\text{free}}$, at equilibrium.
(c) Taking $\Delta g_v = 1$ eV, $\Delta g_b = -\frac{1}{3}$ eV, and $\chi_s = 10^{-3}$, find the temperature for which the number of bound and free vacancies is the same.
(d) We now hold the solid solution at a high temperature T_0 for a long time and then suddenly quench it to a much lower temperature T. At the lower temperature, the vacancies can move from site to site but are not able to leave the crystal (the vacancy sinks are too far away). Write the equations that would need to be solved to determine the equilibrium numbers of vacancies on the two kinds of sites under the constrained equilibrium condition.

6.14 A single crystal alloy contains n impurity atoms that are distributed between N_L lattice sites and N_S surface sites. Call n_L the number of impurity atoms found on lattice sites and n_S the number of impurity atoms found on surface sites. The formation free energy of an impurity atom on a lattice site is Δg_L and the formation free energy for an impurity on a surface site is Δg_S. We consider the case of surface segregation of impurities, where $\Delta g_S < \Delta g_L$. We wish to consider the equilibrium state of this system at any arbitrary temperature.

(a) Write an expression for the free energy change of the system when n_L impurity atoms are added to the lattice and n_S impurity atoms are added to the surface. (There is no need to derive this from scratch – use derived relations in this chapter.)
(b) Derive an expression for the equilibrium concentration (fraction) of impurities at the surface in terms of the equilibrium concentration in the bulk.
(c) Briefly explain how the surface concentration would vary with temperature, assuming a fixed number, n, of impurity atoms in solution.
(d) How would the impurity concentration at the surface change if precipitates of the impurity atoms form in the solid at low temperatures?

7 Point defect kinetics

In the previous two chapters we have studied the concentration of point defects in crystals after thermal equilibrium has been reached. We now discuss the motion of point defects, which is necessary for the equilibrium concentration to be reached in the first place. In fact, the motion of individual point defects never stops even after thermal equilibrium is reached. It is only when we take a coarse-grained view, do we find that the continuum concentration field of the point defects stops changing once equilibrium is reached.

In this chapter, we consider the motion of point defects both at the individual (discrete) level and at the collective (continuum) level. In Section 7.1, we consider a single vacancy and discuss the mechanism of its motion. We use the principles of statistical mechanics to show that the rate at which a vacancy jumps to a neighboring site is determined by a migration free energy (barrier) through a Boltzmann factor, and hence is strongly sensitive to temperature. In Section 7.2, we extend this result to the motion of interstitial and substitutional solute atoms. Because a neighboring vacancy is often required for a substitutional solute atom to jump, the jump rate of a substitutional solute atom is determined by a Boltzmann factor containing the sum of the migration free energy and the vacancy formation free energy.

In Section 7.3, we consider a large collection of point defects, each making random jumps to neighboring sites at a constant probability rate, and show that the evolution of their concentration field with time can be described by the diffusion equation. In Section 7.4, we show that if the crystal is subjected to an inhomogeneous stress field, then the equilibrium point defect concentration, if it exists, is not necessarily uniform. In this case, it would be necessary to use a generalized diffusion equation using the chemical potential defined in Section 5.4.

Under certain boundary conditions, no equilibrium solution exists for the diffusion equation, although a steady-state solution may exist. This is particularly common for vacancies, which can be constantly created at vacancy sources, travel across the crystal lattice, and be annihilated at vacancy sinks. As a simple model for this scenario, in Section 7.5 we consider vacancies in crystals subjected to a uniform deviatoric stress, to understand diffusional creep of crystals at high temperatures.

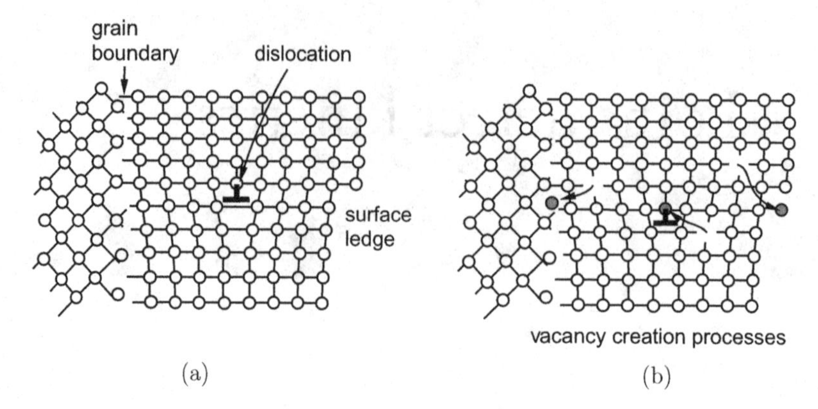

Figure 7.1. (a) Crystalline solid with a grain boundary, dislocation, and surface ledge. (b) Unit processes for vacancy creation.

7.1 Motion of vacancies

7.1.1 Mechanisms of vacancy concentration changes

We have shown that the equilibrium vacancy concentration increases dramatically with increasing temperature. We now consider the kinetics and mechanisms by which changes in the vacancy concentration can occur. Consider an imperfect solid with grain boundaries, dislocations, and free surfaces, as shown in Fig. 7.1a. At very low temperatures the equilibrium concentration of vacancies is very low so that there might not be any vacancies in the considered volume. We wish to describe what happens when the solid is heated and vacancies are created.

When a crystal is heated, vacancies are created when atoms move from their lattice positions to the defects. The vacancy creation process is equivalent to an atom plating process. For every vacancy created, an atom is plated out somewhere in the structure, as shown in Fig. 7.1b. In this case, the defects act as vacancy sources. Once the vacancies are created in this way, they can diffuse into the bulk of the crystal by diffusion (exchanging positions with neighboring atoms). Similarly, when a crystal is cooled slowly, vacancies leave the crystal by diffusing to these defects and annihilating there. Vacancy annihilation involves atoms at the defects filling the vacant lattice sites. In this case, the defects act as vacancy sinks.

When a crystal is rapidly cooled to a low temperature there is often not enough time for the vacancies to diffuse to the defects where they can annihilate. In that case the huge vacancy supersaturation created by quenching causes another mode of vacancy annihilation to occur. As shown in Fig. 7.2a, under these conditions vacancies can agglomerate within the crystal and arrange themselves into plate-like structures. (See Section 6.5 for discussions on vacancy clusters.) Once the vacancies have agglomerated in this way the plate of vacancies, one atomic dimension in thickness, is unstable with respect to lattice collapse. The result is that vacancies can precipitate out of the lattice to form dislocation loops or rings, as shown in Fig. 7.2a. Over 50 years ago Hirsch and Westmacott, and others, showed that when Al is quenched from a high temperature the excess vacancies condense to form dislocation loops [40] (also see Section 11.1.12). Figure 7.2b is taken from that classic work. Each loop is the result of vacancy condensation and lattice collapse.

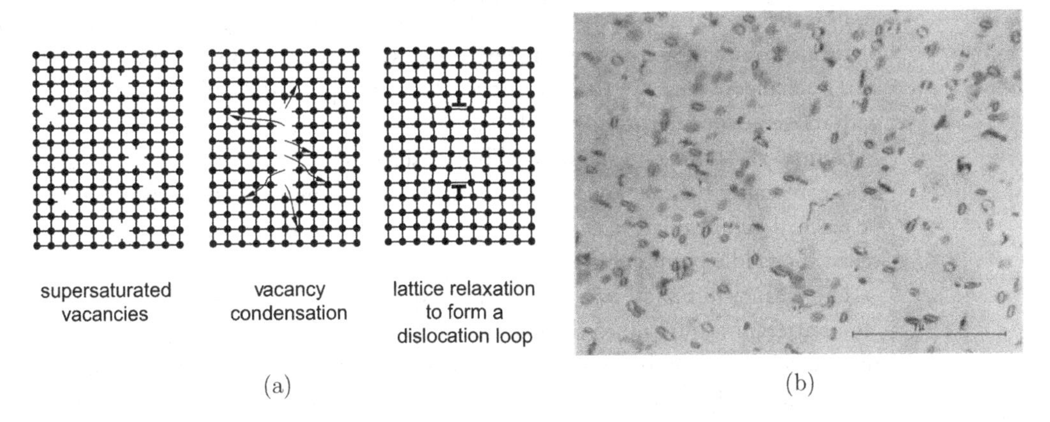

Figure 7.2. (a) Vacancy condensation into dislocation loop. (b) Vacancy loops in quenched aluminum from 560 °C to iced brine [40]. Permission granted by P. B. Hirsch.

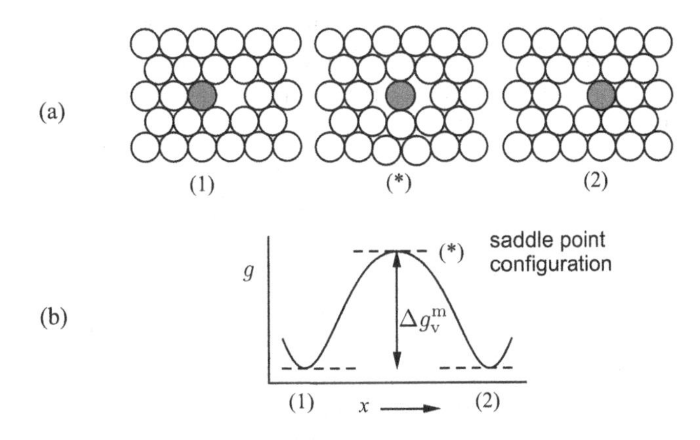

Figure 7.3. Vacancy motion. (a) A neighboring atom (filled circle) exchanges place with the vacancy. (b) The change of free energy of the crystal during the process shown in (a).

7.1.2 Theory of vacancy jump rate

Here we consider the kinetic aspects of vacancies, which are not only necessary for vacancies to reach their equilibrium concentration, but are also important for diffusion and all processes controlled by diffusion, such as recovery, diffusional deformation, and, in some cases, phase changes. We consider the application of reaction rate theory to the movement of vacancies.

Consider a vacancy in a close-packed crystal structure shown in Fig. 7.3a. The thermal energy of the solid is manifested in the vibrations of the atoms. Using the Einstein model of the solid, each atom can be regarded as three harmonic oscillators, corresponding to vibrations along three orthogonal axes (see Fig. 5.8). The mass of the atom and the spring constant holding it in place (related to the elastic properties of the solid) lead to a natural vibration frequency, typically about $\nu = 10^{12}\ \mathrm{s}^{-1}$. The Boltzmann distribution of energies in the solid will allow some of the atoms adjacent to vacancies to make successful jumps into the neighboring vacancy.

Alternatively we can think of the vacancies as vibrating and making attempts to exchange positions with one of the neighboring atoms. As shown in Fig. 7.3a, when a neighboring atom and vacancy exchange places, the neighboring atoms are displaced and the coordination of some of the atoms and the corresponding bond angles change. As a consequence, the free energy of the system changes when the vacancy moves. We can represent the free energy changes associated with movement of the atom into the vacancy as shown in Fig. 7.3b.

Figure 7.3a shows two stable equilibrium states, (1) and (2), which are the states of an atom before and after the exchange with the vacancy. Also shown is the activated state (*) that must be reached before an atom can make the transition into the vacancy. The activated state is also called the saddle point configuration.[1]

The vacancy jump rate r_{jump} can now be expressed as the product of the frequency with which atoms attempt to jump into the vacancy, ν, the number of atoms adjacent to each vacancy, i.e. the coordination number, z (each of the near neighbor atoms is free to jump into the vacancy) and the probability that a given neighbor atom has sufficient energy to move to the activated state, $p(\Delta g_{\mathrm{v}}^{\mathrm{m}})$,

$$r_{\mathrm{jump}} \approx \nu z p(\Delta g_{\mathrm{v}}^{\mathrm{m}}), \tag{7.1}$$

where $\Delta g_{\mathrm{v}}^{\mathrm{m}}$ is the Gibbs free energy of the activated state (*) relative to the initial state (1), as shown in Fig. 7.3. $\Delta g_{\mathrm{v}}^{\mathrm{m}}$ is also called the activation (Gibbs) free energy for vacancy migration. Since the probability of finding the system at different states is given by the Boltzmann relation (see Eq. (3.76)),

$$p(\Delta g_{\mathrm{v}}^{\mathrm{m}}) = \exp\left(-\frac{\Delta g_{\mathrm{v}}^{\mathrm{m}}}{k_{\mathrm{B}} T}\right), \tag{7.2}$$

the vacancy jump rate is

$$r_{\mathrm{jump}} \approx \nu z \exp\left(-\frac{\Delta g_{\mathrm{v}}^{\mathrm{m}}}{k_{\mathrm{B}} T}\right). \tag{7.3}$$

More accurate predictions of the jump rate are provided by the classical nucleation theory of Becker and Döring [41] and the transition state theory [42, 43].

The activation free energy consists of several contributions,

$$\Delta g_{\mathrm{v}}^{\mathrm{m}} = \Delta h_{\mathrm{v}}^{\mathrm{m}} - T\Delta s_{\mathrm{v}}^{\mathrm{m}} = \Delta e_{\mathrm{v}}^{\mathrm{m}} + p_{\mathrm{ext}}\Delta v_{\mathrm{v}}^{\mathrm{m}} - T\Delta s_{\mathrm{v}}^{\mathrm{m}}, \tag{7.4}$$

where $\Delta h_{\mathrm{v}}^{\mathrm{m}}$, $\Delta e_{\mathrm{v}}^{\mathrm{m}}$, $\Delta v_{\mathrm{v}}^{\mathrm{m}}$, and $\Delta s_{\mathrm{v}}^{\mathrm{m}}$ are activation enthalpy, energy, volume, and entropy, respectively. Since the entropy factor and pressure dependence are both usually small, we have $\Delta g_{\mathrm{v}}^{\mathrm{m}} \approx \Delta h_{\mathrm{v}}^{\mathrm{m}} \approx \Delta e_{\mathrm{v}}^{\mathrm{m}}$. Hence we conclude that

$$r_{\mathrm{jump}}^{\mathrm{vacancy}} \approx \nu z \exp\left(-\frac{\Delta e_{\mathrm{v}}^{\mathrm{m}}}{k_{\mathrm{B}} T}\right). \tag{7.5}$$

The activation energy $\Delta e_{\mathrm{v}}^{\mathrm{m}}$ is, by far, the most important factor in determining the rate at which vacancies move about in a crystal. The activation enthalpies for vacancy motion are given in the Wollenberger table C.1 in Appendix C. The vacancy jump rate is, of course, very strongly

1 At the activated state the energy is at a local maximum when the migrating atom is constrained to move along the x direction defined in Fig. 7.3. But if the migrating atom moves along a perpendicular direction, the activated state is at a local energy minimum. So the energy landscape has the shape of a saddle near the activated state.

temperature dependent. Using Eq. (7.5) we can show that the vacancy jump rate in copper ($\Delta e_{\mathrm{v}}^{\mathrm{m}} \approx 0.7$ eV) is about $3 \times 10^{10}\,\mathrm{s}^{-1}$ at the melting temperature (1358 K) and only $20\,\mathrm{s}^{-1}$ at room temperature (300 K).

7.1.3 Measurement of vacancy migration barrier

The classic experimental technique for determining $\Delta e_{\mathrm{v}}^{\mathrm{m}}$ not only shows how this important kinetic parameter can be determined but also what happens to vacancies in quenched solids. The experiment can be described as follows.

- Quench wires from a fixed high temperature T_{H} to freeze in a supersaturation of vacancies and then measure the electrical resistivity in the quenched state. We quench to a sufficiently low temperature T_{L} (e.g. liquid nitrogen temperature) so that the vacancies are essentially immobile. This leaves the wire with a vacancy concentration c_i and an electrical resistivity of ρ_i.
- As a reference, we also slowly cool a wire to the same low temperature T_{L} so that the vacancy concentration is essentially zero and then measure the electrical resistivity of the wire, ρ_0.
- Next, the wires that have been quenched to T_{L} are heated to an intermediate temperature T where the vacancies become mobile and can migrate and annihilate at sinks. The equilibrium vacancy concentration at T is also close to zero. After holding the wires at this intermediate temperature, T, for some period of time, t, the wires are again quickly cooled to the low temperature T_{L} and the resistivity $\rho(T, t)$ is measured, which is a function of the intermediate temperature T and the holding period t.

Assuming that the electrical resistivity in excess of ρ_0 is proportional to the vacancy concentration, the resistivity ratio

$$\frac{\rho(T, t) - \rho_0}{\rho_i - \rho_0} = \frac{\Delta\rho}{\Delta\rho_i} \tag{7.6}$$

then measures the remaining vacancy concentration divided by the initial concentration,

$$\frac{\Delta\rho}{\Delta\rho_i} = \frac{c}{c_i}. \tag{7.7}$$

In the as-quenched state, $c/c_i = 1$; c/c_i tends toward zero as vacancies move and annihilate at the annealing temperature. The rate of loss of vacancies at the annealing temperature can be described by first-order kinetics, wherein the rate of vacancy loss is proportional to the current vacancy concentration through a time constant, τ:

$$\frac{dc}{dt} = -\frac{c}{\tau}, \tag{7.8}$$

where the time constant depends on temperature through the inverse of the Boltzmann factor

$$\tau = \tau_0 \exp\left(\frac{\Delta e_{\mathrm{v}}^{\mathrm{m}}}{k_{\mathrm{B}} T}\right). \tag{7.9}$$

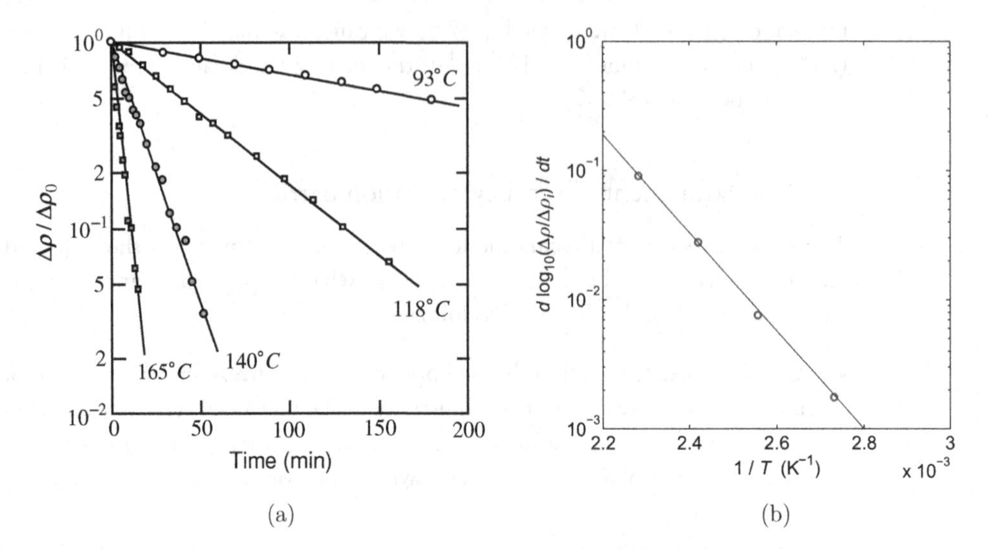

Figure 7.4. (a) Isothermal annealing curves obtained on 0.04-mm pure gold wires after quenching from 700 °C [29]. (b) Arrhenius plot of $[d \log_{10}(\Delta\rho/\Delta\rho_i)]/dt$ based on the data from (a).

By integrating the first-order kinetic law we have

$$\int_{c_i}^{c} \frac{dc}{c} = -\int_0^t \frac{dt}{\tau},$$

$$\ln\left(\frac{c}{c_i}\right) = -\frac{t}{\tau},$$

$$\log_{10}\left(\frac{\Delta\rho}{\Delta\rho_i}\right) = \log_{10}\left(\frac{c}{c_i}\right) = -\frac{t}{2.303\tau}. \tag{7.10}$$

This latter expression suggests that if $\log_{10}(\Delta\rho/\Delta\rho_i)$ is plotted against t, the slope of the plot would equal $-1/(2.303\tau)$, i.e.

$$\frac{d \log_{10}(\Delta\rho/\Delta\rho_i)}{dt} = -\frac{1}{2.303\tau_0} \exp\left(-\frac{\Delta e_{\mathrm{v}}^{\mathrm{m}}}{k_{\mathrm{B}} T}\right). \tag{7.11}$$

Figure 7.4a shows that $\log_{10}(\Delta\rho/\Delta\rho_i)$ is indeed linear with time, and the slope of the curve depends on temperature. The activation energy for vacancy migration can be obtained from these data. One would need to extract the slopes of the curves in Fig. 7.4a and make an Arrhenius plot, i.e. the logarithm of $[d \log_{10}(\Delta\rho/\Delta\rho_i)]/dt$ as a function of the reciprocal of the absolute temperature, as in Fig. 7.4b. Specifically,

$$\log_{10}\left[-\frac{d \log_{10}(\Delta\rho/\Delta\rho_i)}{dt}\right] = \log_{10}\left(\frac{1}{2.303\tau_0}\right) - \left(\frac{\Delta e_{\mathrm{v}}^{\mathrm{m}}}{2.303 k_{\mathrm{B}}}\right) \cdot \frac{1}{T}. \tag{7.12}$$

The slope of the Arrhenius plot gives $-\Delta e_{\mathrm{v}}^{\mathrm{m}}/(2.303 k_{\mathrm{B}})$. From Fig. 7.4b, we find that $\Delta e_{\mathrm{v}}^{\mathrm{m}}/k_{\mathrm{B}}$ is approximately 8.7×10^3 K, which gives $\Delta e_{\mathrm{v}}^{\mathrm{m}} \approx 0.75$ eV. This is consistent with the value $\Delta h_{\mathrm{v}}^{\mathrm{m}} = 0.71 \pm 0.05$ eV in Table C.1 for Au. Similar approaches have been used to determine the activation energies for motion of other point defects.

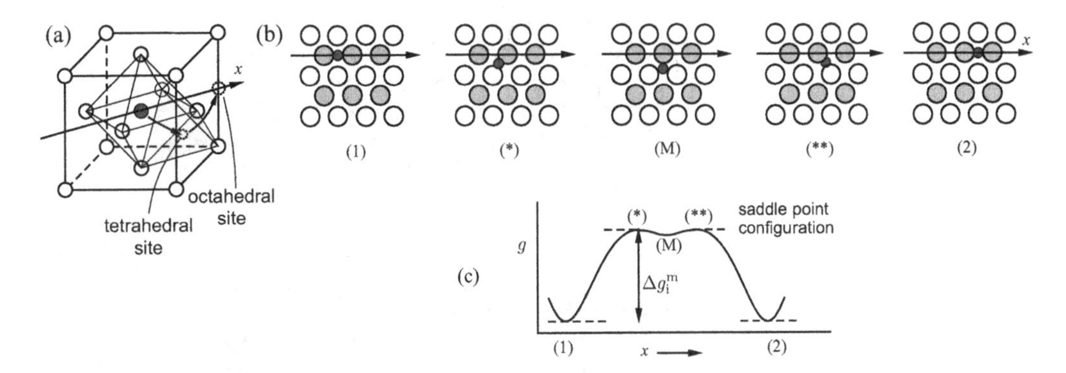

Figure 7.5. (a) A solute atom (dark circle) in an octahedral site of an FCC crystal. (b) The solute atom jumps to a neighboring octahedral site via a tetrahedral site. The host atoms that are beneath the plane of the page by $a/\sqrt{2}$ are shaded in light gray. (c) The change of free energy of the crystal during the process shown in (b).

7.2 Motion of solute atoms

We now consider the motion of extrinsic point defects, i.e. interstitial and substitutional solute atoms, in a crystal. For simplicity, we only discuss the most common mechanisms of point defect motion. The reader is referred to [44–47] for a more comprehensive discussion of the various mechanisms of point defect migration.

7.2.1 Interstitial solutes

Consider an interstitial solute atom occupying an octahedral site in an FCC crystal. Figure 7.5a shows the solute atom and a neighboring octahedral site into which it can potentially jump. The $\langle 1\,1\,0 \rangle$ line direction that connects these two octahedral sites is defined as the x axis here.

Considering atoms as hard spheres, we can see that the solute atom is unlikely to follow the straight path along the x axis. Doing so would require the solute atom to squeeze through the narrow region between two nearest neighbor atoms of the host crystal. Instead, the solute atom is more likely to first go to a tetrahedral site and then go to the next octahedral site. There are two tetrahedral sites adjacent to both octahedral sites; one above and the other below the x axis. The lower one is shown in Fig. 7.5a.

Figure 7.5b sketches the various states the solute atom goes through during this jump. In the two stable equilibrium states, (1) and (2), the solute atom is at the initial and final octahedral sites, respectively. The state in which the solute atom occupies the intermediate tetrahedral site is marked (M), for metastable. Here we have two activated states, or saddle point configurations, (*) and (**), that lie between state (1) and state (M), and between state (M) and state (2), respectively. By symmetry, the two activated states have the same Gibbs free energy. The free energy difference between the activated states and the equilibrium states is the activation (Gibbs) free energy for interstitial migration, Δg_i^m, as shown in Fig. 7.5c.

Again, the activation free energy may be written as,

$$\Delta g_i^m = \Delta h_i^m - T\Delta s_i^m = \Delta e_i^m + p_{ext}\Delta v_i^m - T\Delta s_i^m, \tag{7.13}$$

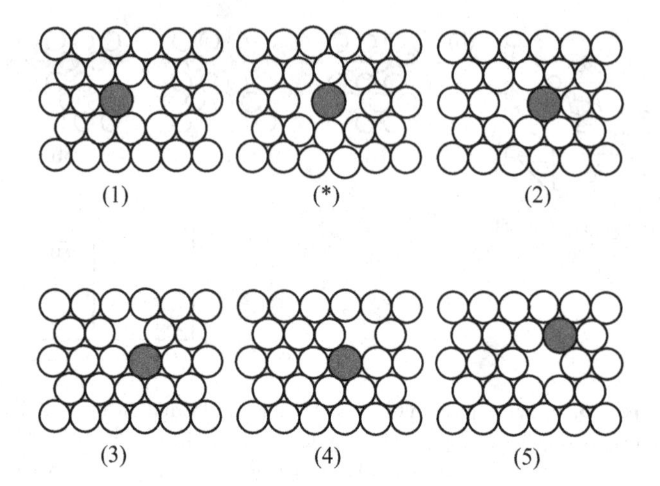

Figure 7.6. A solute atom (filled circle) moving in a crystal with the assistance of a vacancy.

where Δh_i^m, Δe_i^m, Δv_i^m, Δs_i^m are activation enthalpy, energy, volume, and entropy, respectively, and Δe_i^m is often the dominant term. In practice, the free energy of state (M) is often very close to that of the activated state (*). Therefore, the energy difference between the two states in which the solute occupies octahedral and tetrahedral sites, respectively, can be used as an estimate of the activation free energy for migration.

Similar to the vacancy jump rate, the jump rate of the interstitial solute atom can be approximated as

$$r_{\text{jump}}^{\text{interstitial}} \approx \nu z \exp\left(-\frac{\Delta g_i^m}{k_B T}\right) \approx \nu z \exp\left(-\frac{\Delta e_i^m}{k_B T}\right). \tag{7.14}$$

7.2.2 Substitutional solutes

Substitutional defects are larger than interstitial defects and are usually very difficult to move without assistance from other defects in the crystal. Because vacancies usually exist in much higher concentrations than self-interstitials, substitutional solutes often move by the vacancy mechanism described below.

Figure 7.6 shows various states as a substitutional solute atom moves through the crystal assisted by a nearby vacancy. The exchange of the solute with the vacancy, i.e. the transition from state (1) to state (2) through the activated state (*), is similar to the process shown in Fig. 7.3a. Define Δg_s^m as the Gibbs free energy difference between state (*) and state (1), i.e. the migration free energy barrier, then the rate of transition between state (1) and state (2) is

$$r_{1\to 2} \approx \nu \exp\left(-\frac{\Delta g_s^m}{k_B T}\right). \tag{7.15}$$

However, if the vacancy switches places only with the (same) solute atom, then the solute atom will return to its original position in the next step, and will only go back and forth between states (1) and (2) and never move far away.

For the solute atom to move to new locations, the vacancy must sometimes exchange location with the neighboring host atoms. For example, state (2) can transition into state (3) and then into state (4), when the vacancy first exchanges location with a host atom to its upper right, and then with another host atom to its right. Starting from state (4), if the vacancy then exchanges location with the solute atom, the result is state (5), in which the solute atom is at a new location that is different from those in states (1) and (2).

In the vacancy mechanism described here, the solute atom can jump to neighboring sites only if a vacancy is nearby. The probability that a given solute atom has a vacancy nearby is proportional to the vacancy fraction in the host crystal, $\chi_v = \exp(-\Delta g_v/(k_B T))$, where Δg_v is the vacancy formation free energy defined in Section 5.2. For simplicity, we ignore the possible binding between solute atoms and vacancies (see Exercise problem 6.13), so that the jump rate for a substitutional solute atom can be written as

$$\begin{aligned} r_{\text{jump}}^{\text{subs.}} &\approx \nu z \exp\left(-\frac{\Delta g_s^m}{k_B T}\right) \exp\left(-\frac{\Delta g_v}{k_B T}\right) \\ &= \nu z \exp\left(-\frac{\Delta g_s^m + \Delta g_v}{k_B T}\right). \end{aligned} \tag{7.16}$$

Given that the Boltzmann factor in Eq. (7.16) contains a sum of solute migration and vacancy formation free energies, we can expect that the jump rate of substitutional solute atoms should, in general, be lower than that of interstitial solute atoms, and have a stronger temperature dependence.

The above analysis can also be applied to the self-diffusion of atoms in a pure crystal by the vacancy mechanism. In this case, the shaded atom in Fig. 7.6 is of the same type as all the neighboring host atoms. The jump rate of any host atom by the vacancy mechanism is

$$r_{\text{jump}}^{\text{host atom}} \approx \nu z \exp\left(-\frac{\Delta g_v^m + \Delta g_v}{k_B T}\right), \tag{7.17}$$

where Δg_v^m and Δg_v are vacancy migration and formation free energies, respectively. The self-diffusion of most (FCC, HCP, BCC) metals usually occurs by the vacancy mechanism.

7.3 Diffusion equation

7.3.1 Fick's laws

So far we have considered individual point defects making random jumps in the host crystal. We now consider a large collection of point defects (of the same type) and explore the consequences when every defect is making such jumps. The discussion here is not restricted to a particular type of point defect, so we will not make a distinction between vacancies, interstitials, and substitutional solutes.

We will take a coarse-grained view, in which we are not concerned about the exact location of any individual point defect. Instead, we describe the point defect distribution by a density function, $c(x, y, z)$. To define $c(x, y, z)$, we consider a cube with side length Δx and centered at point (x, y, z). Then

$$c(x, y, z) \equiv \frac{\text{number of defects in the cube}}{(\Delta x)^3}. \tag{7.18}$$

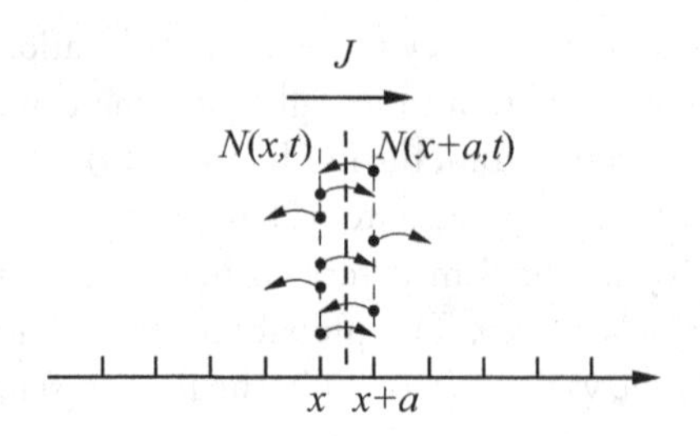

Figure 7.7. A simplified one-dimensional model of diffusion. At time t, the number of point defects per unit area on plane x is $N(x, t)$, and that on plane $x + a$ is $N(x + a, t)$. The curved arrows indicate their jump directions at time $t + \tau$.

For a concentration function to be meaningful, the cube size Δx has to be much greater than the lattice parameter, so that the number of point defects in the cube is sufficiently large (say, >100). At the same time Δx needs to be much smaller than the sample size, so that it can be considered infinitesimal in the diffusion equation that we will set up. Roughly speaking, we may consider Δx to be in the range 0.1–10 μm, depending on the order of magnitude of the point defect concentration.

Imagine that the point defects are distributed according to some non-equilibrium concentration function as an initial condition. We now construct the equations that describe how the concentration function evolves with time. For simplicity, we consider the one-dimensional case, in which the concentration function (as well as every other aspect of the problem) is independent of y and z, so that we can simply write $c(x)$.

Fick's first law

Imagine a plane normal to the x axis, shown as the thick dashed line in Fig. 7.7. Because of the random motion of the point defects, there are always point defects jumping across the plane in both directions. At thermal equilibrium, the flux of point defects from left to right and that from right to left exactly balance each other; but this is not the case in a general, non-equilibrium state. Define J as the net flux of point defects in the positive x direction per unit time per unit area across this plane.

In order to link the net flux J with the jump rate of individual point defects, we consider a highly simplified one-dimensional model, in which point defects can exist only at sites with $x = 0, \pm a, \pm 2a, \ldots$. Furthermore, we assume that every point defect jumps at discrete times: $t = \tau, 2\tau, 3\tau, \ldots$, with $\tau = 1/r_{\text{jump}}$. This is obviously a highly idealized situation, but it simplifies our discussion without changing the main conclusions.

Suppose at some time t, the number of point defects on plane x per unit area is $N(x, t)$, and that on plane $x + a$ is $N(x + a, t)$. The number per unit area is related to the volume density through $c(x, t) = N(x, t)/a$. Because the random motion of point defects is not biased in any direction, at time $t + \tau$, about half of the $N(x, t)$ point defects jump from x to $x + a$, i.e. in the positive x direction. Similarly, about half of the $N(x + a)$ point defects jump from $x + a$ to x, i.e. in the negative x direction. Therefore, on the average, the net flux across the plane at

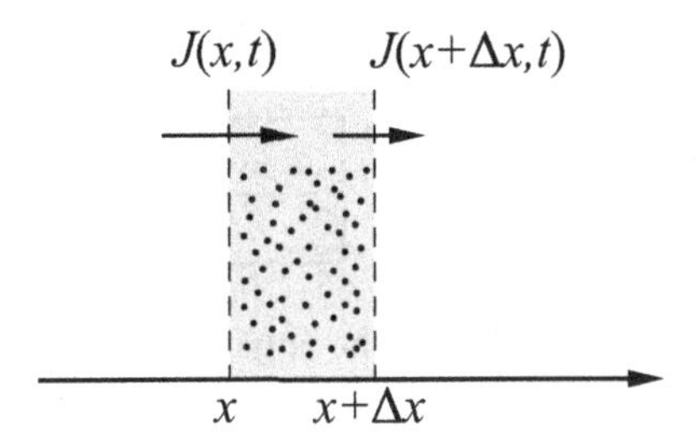

Figure 7.8. Accumulation of point defects in the domain of $[x, x+\Delta x]$ if the in-flux $J(x,t)$ exceeds the out-flux $J(x+\Delta x, t)$.

$x + a/2$ is

$$\begin{aligned} J &\approx \frac{N(x,t)/2 - N(x+a,t)/2}{\tau} \\ &= -\frac{a}{2\tau}\frac{N(x+a,t) - N(x,t)}{a} \\ &= -\frac{a^2}{2\tau}\frac{c(x+a,t) - c(x,t)}{a} \\ &\approx -\frac{a^2}{2\tau}\frac{\partial c(x,t)}{\partial x}. \end{aligned} \tag{7.19}$$

If we identify

$$D = \frac{a^2}{2\tau} = \frac{a^2 r_{\text{jump}}}{2}, \tag{7.20}$$

then we arrive at Fick's first law of diffusion.

$$\boxed{J(x,t) = -D\frac{\partial c(x,t)}{\partial x}} \tag{7.21}$$

Here, D is called the *diffusion coefficient*. Equation (7.21) means that the net flux is proportional to and points in the opposite direction of the concentration gradient.

Conservation of mass

We now consider how the diffusional flux J can lead to local changes of the defect concentration with time. We focus on the situation in which the point defects are neither created nor destroyed. While this is rather obvious for solute atoms, when applying it to vacancies it means that we assume there are no vacancy sources or sinks in the region of interest.

Consider a domain $[x, x+\Delta x]$ as shown in Fig. 7.8. Because the net flux is defined to be positive when pointing in the positive x direction, $J(x,t)$ represents the flux into this domain, and $J(x+\Delta x, t)$ represents the flux out of this domain. If $J(x,t)$ exceeds $J(x+\Delta x, t)$, then there will be an accumulation of point defects in this domain. Specifically,

$$\frac{\partial c(x,t)}{\partial t} \approx \frac{J(x,t) - J(x+\Delta x, t)}{\Delta x}. \tag{7.22}$$

In the limit of $\Delta x \to 0$, we have the following.

$$\boxed{\frac{\partial c(x,t)}{\partial t} = -\frac{\partial J(x,t)}{\partial x}} \tag{7.23}$$

This is commonly referred to as the equation for the conservation of mass.

Fick's second law

Combining Eqs. (7.21) and (7.23), we arrive at the diffusion equation in one dimension.

$$\boxed{\frac{\partial c}{\partial t} = D\frac{\partial^2 c}{\partial x^2}} \tag{7.24}$$

This is also called Fick's second law. The diffusion equation in three dimensions is a generalization of Eq. (7.24),

$$\frac{\partial c}{\partial t} = D\nabla^2 c, \tag{7.25}$$

where $\nabla^2 \equiv \frac{\partial^2}{\partial x^2} + \frac{\partial^2}{\partial y^2} + \frac{\partial^2}{\partial z^2}$ is the Laplacian operator in three dimensions.

7.3.2 Diffusion coefficients in crystals

Vacancies

The diffusion coefficient D is linked to the jump rate of individual point defects, as given in Eq. (7.20). As an example, we give an estimate for the diffusion coefficient of vacancies:

$$\begin{aligned} D_{\mathrm{v}} &\approx r_{\mathrm{jump}} \cdot \frac{a^2}{2} \\ &\approx \nu z \frac{a^2}{2} \exp\left(-\frac{\Delta h_{\mathrm{v}}^{\mathrm{m}} - T\Delta s_{\mathrm{v}}^{\mathrm{m}}}{k_{\mathrm{B}}T}\right) \\ &= \nu z \frac{a^2}{2} \exp\left(\frac{\Delta s_{\mathrm{v}}^{\mathrm{m}}}{k_{\mathrm{B}}}\right) \exp\left(-\frac{\Delta h_{\mathrm{v}}^{\mathrm{m}}}{k_{\mathrm{B}}T}\right) \\ &\approx \nu z \frac{a^2}{2} \exp\left(-\frac{\Delta e_{\mathrm{v}}^{\mathrm{m}}}{k_{\mathrm{B}}T}\right), \end{aligned} \tag{7.26}$$

where a is the distance of one vacancy jump, which is the nearest neighbor distance between atoms in the perfect crystal. Therefore the diffusion coefficient is strongly temperature dependent.

The experimentally measured diffusivity data are often expressed in the form of

$$D = D_0 \exp\left(-\frac{Q}{k_{\mathrm{B}}T}\right), \tag{7.27}$$

where D_0 and Q are constants (in the temperature range where the Arrhenius plot of D is a straight line). Comparing Eqs. (7.26) and (7.27), we can see that, for vacancies,

$$D_0 \approx \nu z \frac{a^2}{2} \exp\left(\frac{\Delta s_v^m}{k_B}\right), \quad Q = \Delta h_v^m. \tag{7.28}$$

Therefore, a larger D_0 value usually indicates a higher migration entropy Δs_v^m.

Interstitial solutes

Similarly, for interstitial solute atoms,

$$D_0 \approx \nu z \frac{a^2}{2} \exp\left(\frac{\Delta s_i^m}{k_B}\right), \quad Q = \Delta h_i^m. \tag{7.29}$$

Substitutional solutes

However, for substitutional solute atoms, given Eq. (7.16), we have

$$\begin{aligned} D_s &\approx \nu z \frac{a^2}{2} \exp\left(-\frac{\Delta h_s^m - T\Delta s_s^m}{k_B T}\right) \exp\left(-\frac{\Delta h_v - T\Delta s_v}{k_B T}\right) \\ &= \nu z \frac{a^2}{2} \exp\left(\frac{\Delta s_s^m + \Delta s_v}{k_B}\right) \exp\left(-\frac{\Delta h_s^m + \Delta h_v}{k_B T}\right). \end{aligned} \tag{7.30}$$

Therefore,

$$D_0 \approx \nu z \frac{a^2}{2} \exp\left(\frac{\Delta s_s^m + \Delta s_v}{k_B}\right), \quad Q = \Delta h_s^m + \Delta h_v. \tag{7.31}$$

Self-diffusivity

Similar expressions exist for the self-diffusivity D_L of host atoms, where L stands for lattice diffusivity. In particular,

$$D_L = \chi_v D_v \approx \nu z \frac{a^2}{2} \exp\left(\frac{\Delta s_v^m + \Delta s_v}{k_B}\right) \exp\left(-\frac{\Delta h_v^m + \Delta h_v}{k_B T}\right). \tag{7.32}$$

Therefore,

$$D_0 \approx \nu z \frac{a^2}{2} \exp\left(\frac{\Delta s_v^m + \Delta s_v}{k_B}\right), \quad Q = \Delta h_v^m + \Delta h_v. \tag{7.33}$$

When self-diffusivity of pure metals is measured over an extended range of temperature, it is often found that the Arrhenius plot of D_L is not a perfectly straight line. In these cases, the diffusivity data are often expressed as

$$D = D_{01} \exp\left(-\frac{Q_1}{k_B T}\right) + D_{02} \exp\left(-\frac{Q_2}{k_B T}\right), \tag{7.34}$$

with $Q_1 < Q_2$. This behavior can be understood as other mechanisms, such as divacancy mechanisms, participating in diffusion at higher temperature. Hence we would expect Q_1 to correspond to the monovacancy mechanism discussed here (with $Q_1 = \Delta h_v^m + \Delta h_v$) and Q_2 to correspond to the other diffusion mechanism.

The diffusion coefficient data for various species in common crystals are listed in Tables D.1, D.2, and D.3.

7.3.3 Application of diffusion equation

As an example, we now use the diffusion equation to study how the vacancy concentration in a crystal reaches a new equilibrium as the temperature is changed. Consider a polycrystal that is chemically homogeneous and subjected to a uniformly applied, purely hydrostatic stress (with no internal stress). The crystal has been maintained at temperature T_0 for a long time, so that its vacancies have reached the equilibrium distribution. Suppose the temperature of the crystal is suddenly increased to a higher value T_1. As a result, more vacancies will be created near the sources, leading to a non-uniform concentration of vacancies. These vacancies need to diffuse away from the sources until a new equilibrium distribution is reached, in which the vacancy concentration is again uniform.

For simplicity, let us consider a one-dimensional model and assume that the grain boundaries are the only vacancy sources and sinks. Consider a grain of size $2l$, occupying the domain $-l \le x \le l$, with the grain boundaries located at $x = \pm l$. Suppose that at time $t = 0$, the vacancy concentration is the equilibrium value at temperature T_0,

$$c_v(x, t = 0) = \frac{1}{\Omega} \exp\left(-\frac{\Delta g_v}{k_B T_0}\right) \equiv c_0 \qquad \text{for } -l < x < l. \tag{7.35}$$

This is the initial condition of our diffusion equation. Let us assume that, at $t = 0$, the temperature is suddenly changed to a higher value T_1. The grain boundaries at $x = \pm l$ will immediately start to create more vacancies so that the local vacancy concentration becomes higher. We shall model this effect by requiring that

$$c_v(x = \pm l, t) = \frac{1}{\Omega} \exp\left(-\frac{\Delta g_v}{k_B T_1}\right) \equiv c_1 \qquad \text{for } t > 0. \tag{7.36}$$

This is the boundary condition for the diffusion equation. To obtain the evolution of the vacancy concentration profile, we need to solve the diffusion equation for vacancies,

$$\frac{\partial c_v(x, t)}{\partial t} = D_v \frac{\partial^2 c_v(x, t)}{\partial x^2}, \tag{7.37}$$

subjected to the initial and boundary conditions specified above.

Such a solution can be obtained semi-analytically [48] or numerically (see Exercise problem 7.14). The result is plotted in Fig. 7.9 in dimensionless form. The solution is symmetric around $x = 0$. Figure 7.9 shows that, for small t, the vacancy concentration is significantly different from c_0 only in the region close to the grain boundaries at $x = \pm l$. However, for $Dt/l^2 \ge 1$, the change in the vacancy concentration $(c - c_0)$ has reached about 90% or a higher fraction

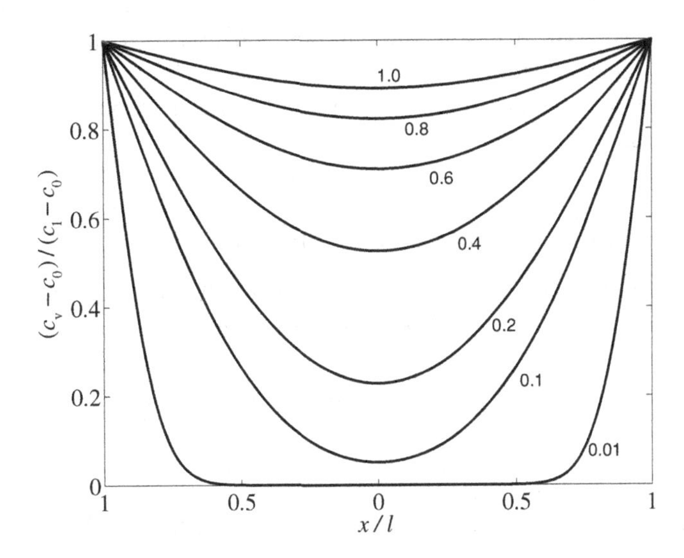

Figure 7.9. Solution of the one-dimensional diffusion equation under the initial condition Eq. (7.35) and boundary condition Eq. (7.36). Numbers on curves are values of Dt/l^2 [48].

of the change needed to reach the new equilibrium $(c_1 - c_0)$. Therefore, we can define the characteristic time necessary for vacancies to reach the new equilibrium concentration

$$t_c \equiv \frac{l^2}{D}. \tag{7.38}$$

Even though, strictly speaking, the equilibrium (uniform) concentration is reached only at $t \to \infty$, the concentration profile is very close to the equilibrium value for $t > t_c$.

Because the diffusion coefficient D is strongly temperature dependent, so is the characteristic equilibration time t_c. Combining Eqs. (7.38) and (7.26), we have

$$t_c \approx \frac{l^2}{\nu z a^2} \exp\left(\frac{\Delta e_v^m}{k_B T}\right). \tag{7.39}$$

Note that the characteristic time t_c is proportional to the square of the grain size $(2l)$. This is the hallmark of diffusional processes. Using the parameters for copper (migration energy $\Delta e_v^m \approx 0.7$ eV, nearest neighbor distance $a = 2.56$ Å), we find that for a grain size of 2 μm ($l = 1$ μm), t_c is about 5×10^{-4} s at the melting temperature (1358 K) and 7×10^5 s (about 8 days) at room temperature (300 K). For a grain size of 20 μm, t_c is about 5×10^{-2} s at the melting temperature and 7×10^7 s (about 800 days) at room temperature.

7.4 Diffusion under stress

So far we have only considered the diffusion equation in solids under no external or internal stresses. In order to discuss point defect diffusion in the presence of a non-uniform stress field, we need to rewrite the diffusion equation in terms of the chemical potential of the point defects defined in Section 5.4. To be specific, we will limit our discussions to vacancies, even though most of the results can be easily generalized to interstitials and substitutional solutes.

7.4.1 Vacancy diffusion under uniform hydrostatic stress

For simplicity, we first consider the case where the solid is subjected to a uniform, purely hydrostatic stress. The vacancy chemical potential under this condition is given by Eq. (5.105)

$$\Delta\mu_{\rm v} \approx \Delta g_{\rm v}^0 - \frac{p^2\Omega}{2B}\left(1+\frac{3B}{4\mu}\right) + p\Delta V_{\rm v}^{\rm rlx} + k_{\rm B}T\ln\chi_{\rm v},$$

where $\Delta g_{\rm v}^0$ is the vacancy formation free energy at zero stress, p is the pressure, and $\Delta V_{\rm v}^{\rm rlx} = \Delta v_{\rm v} - \Omega$ is the vacancy relaxation volume. Since both $\Delta g_{\rm v}^0$ and p are independent of position, the spatial gradient of the chemical potential is

$$\frac{\partial\Delta\mu_{\rm v}}{\partial x} = k_{\rm B}T\frac{1}{\chi_{\rm v}}\frac{\partial\chi_{\rm v}}{\partial x} = k_{\rm B}T\frac{1}{c_{\rm v}}\frac{\partial c_{\rm v}}{\partial x}. \tag{7.40}$$

Therefore, Fick's first law, Eq. (7.21), can be rewritten as follows.

$$\boxed{J_{\rm v} = -\frac{D_{\rm v}c_{\rm v}}{k_{\rm B}T}\frac{\partial\Delta\mu_{\rm v}}{\partial x}} \tag{7.41}$$

The diffusion equation (7.37) can be rewritten as follows (assuming $D_{\rm v}$ and T are constants).

$$\boxed{\frac{\partial c_{\rm v}}{\partial t} = \frac{D_{\rm v}}{k_{\rm B}T}\frac{\partial}{\partial x}\left(c_{\rm v}\frac{\partial\Delta\mu_{\rm v}}{\partial x}\right)} \tag{7.42}$$

What we have shown is that, when the stress is uniform, the two forms of Fick's first law, Eq. (7.21) and Eq. (7.41), are equivalent to each other. In fact, Eq. (7.41) is the fundamentally correct form, which reduces to Eq. (7.21) when the stress is uniform. When the stress is no longer uniform, we need to use Eqs. (7.41) and (7.42) as the generalized Fick's first law and diffusion equation, respectively.

The physical meaning of Eqs. (7.41) is that vacancies flow in the direction of the negative chemical potential gradient. As a result of the vacancy flux, the Gibbs free energy of the entire crystal is spontaneously reduced, as required by the second law of thermodynamics (see Section 3.2.5).

7.4.2 Vacancy diffusion under inhomogeneous stress

If a solid contains either chemical inhomogeneity or an inhomogeneous internal stress field, then the equilibrium vacancy concentration (if it exists) is generally non-uniform, as shown

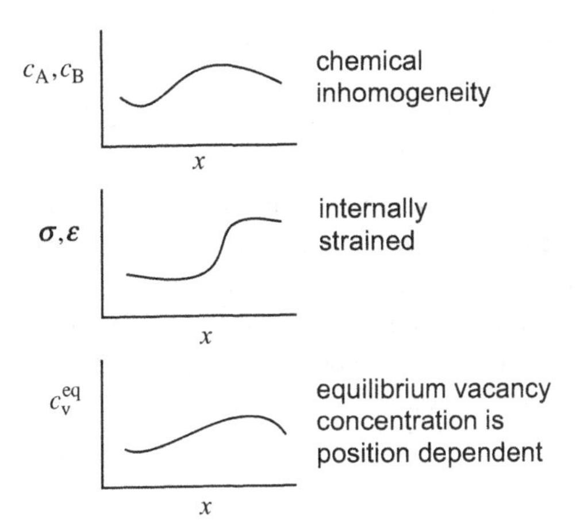

Figure 7.10. Inhomogeneous equilibrium vacancy concentration and local stress.

in Fig. 7.10. In this section, we consider the effect of internal stress on vacancy diffusion, and we ignore vacancy sources and sinks for now. Since surfaces and grain boundaries are vacancy sources and sinks, here we are discussing either the hypothetical case of an infinite single crystal free of internal sources and sinks, or only the diffusion behavior in the interior of such a single crystal away from the surface.

For simplicity, we ignore contributions to the vacancy chemical potential that are second order in stress. Based on Eqs. (5.92), (5.105), and (5.107), the vacancy chemical potential is

$$\Delta\mu_v(x) \approx \Delta g_v^0 + p(x)\Delta V_v^{rlx} + k_B T \ln \chi_v(x), \tag{7.43}$$

where $p(x)$ is the inhomogeneous pressure field.

Equilibrium vacancy concentration

If the boundary condition of the diffusion equation is such that an equilibrium state can be reached, then at the equilibrium state, $J_v = 0$, and

$$\Delta\mu_v = \text{constant}. \tag{7.44}$$

The equilibrium vacancy fraction χ_v^{eq} must satisfy the relation

$$p(x)\Delta V_v^{rlx} + k_B T \ln \chi_v^{eq} = \text{constant}, \tag{7.45}$$

so that

$$\chi_v^{eq}(x) = (\text{constant}) \cdot \exp\left[-\frac{p(x)\Delta V_v^{rlx}}{k_B T}\right]. \tag{7.46}$$

The constant in Eq. (7.46) can be determined by defining χ_0 as the value of $\chi_v^{eq}(x)$ at locations where $p(x) = 0$. Then,

$$\chi_v^{eq}(x) = \chi_0 \exp\left[-\frac{p(x)\Delta V_v^{rlx}}{k_B T}\right]. \tag{7.47}$$

The equilibrium vacancy concentration is

$$c_{\rm v}^{\rm eq}(x) = c_0 \exp\left[-\frac{p(x)\Delta V_{\rm v}^{\rm rlx}}{k_{\rm B}T}\right], \tag{7.48}$$

where $c_0 = \chi_0/\Omega$ is the vacancy concentration at locations of zero pressure.

Because $\Delta V_{\rm v}^{\rm rlx} < 0$, $\chi_{\rm v}^{\rm eq}(x)$ is enhanced by a positive local pressure $p(x)$. This may seem counterintuitive since a positive pressure should make it more difficult to create vacancies. But here we have assumed that there are no vacancy sources and sinks, and the equilibrium distribution is only the result of the vacancy diffusion process.[2]

When the vacancy concentration field has not reached the equilibrium distribution defined in Eq. (7.48), the chemical potential can be obtained from Eq. (7.43)

$$\begin{aligned}\Delta\mu_{\rm v} &= \Delta g_{\rm v}^0 - k_{\rm B}T\ln\frac{\chi_{\rm v}^{\rm eq}(x)}{\chi_0} + k_{\rm B}T\chi_{\rm v}(x)\\ &= k_{\rm B}T\ln\frac{\chi_{\rm v}(x)}{\chi_{\rm v}^{\rm eq}(x)} + \left(\Delta g_{\rm v}^0 + k_{\rm B}T\ln\chi_0\right).\end{aligned} \tag{7.49}$$

Therefore,

$$\Delta\mu_{\rm v} = k_{\rm B}T\ln\frac{c_{\rm v}(x)}{c_{\rm v}^{\rm eq}(x)} + \text{constant}, \tag{7.50}$$

where the constant may be non-zero if $\chi_0 \neq \exp(-\Delta g_{\rm v}^0/(k_{\rm B}T))$, which occurs if the total number of vacancies in the crystal does not have sufficient time to reach the equilibrium value at temperature T.

Vacancy flux

Now consider again the one-dimensional flux equation (7.41),

$$J_{\rm v} = -\frac{D_{\rm v}c_{\rm v}}{k_{\rm B}T}\frac{\partial\Delta\mu_{\rm v}}{\partial x}.$$

Given Eq. (7.50), we have

$$\frac{\partial\Delta\mu_{\rm v}}{\partial x} = k_{\rm B}T\left[\frac{1}{c_{\rm v}}\frac{\partial c_{\rm v}}{\partial x} - \frac{1}{c_{\rm v}^{\rm eq}}\frac{\partial c_{\rm v}^{\rm eq}}{\partial x}\right]. \tag{7.51}$$

Thus the vacancy flux is

$$J_{\rm v} = -D_{\rm v}\left[\frac{\partial c_{\rm v}}{\partial x} - \frac{c_{\rm v}}{c_{\rm v}^{\rm eq}}\frac{\partial c_{\rm v}^{\rm eq}}{\partial x}\right]. \tag{7.52}$$

If $c_{\rm v}^{\rm eq}$ is a constant (independent of x), we reproduce Eq. (7.21), which is Fick's first law

$$J_{\rm v} = -D_{\rm v}\frac{\partial c_{\rm v}}{\partial x}.$$

2 If there were vacancy sources and sinks, then the equilibrium vacancy distribution may be unreachable, because the vacancies can continuously be created at the sources, diffuse to the sinks and get annihilated there.

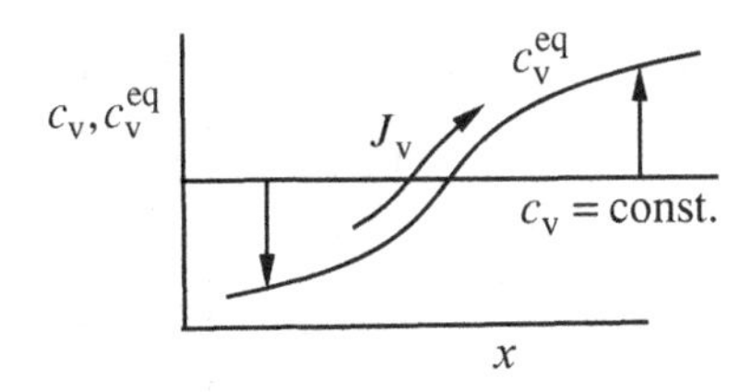

Figure 7.11. Vacancy flux when the equilibrium concentration is not uniform.

However, suppose that c_v is a constant but c_v^{eq} is not a constant, then

$$J_v = +D_v \left[\frac{c_v}{c_v^{eq}} \frac{\partial c_v^{eq}}{\partial x} \right]. \tag{7.53}$$

This means that the vacancy flux will be determined by the gradient of the equilibrium vacancy concentration. In the example shown in Fig. 7.11, the vacancy flux is positive, $J_v > 0$ where $\partial c_v^{eq}/\partial x > 0$, so that vacancies are depleted from the left side and accumulate on the right side. Eventually, an equilibrium state is reached wherein $c_v = c_v^{eq}$ everywhere.

7.5 Diffusional deformation

Usually the external pressure or applied stress has a negligible effect on the vacancy concentration, as described in Eq. (5.56). But when polycrystalline solids are subjected to shear stresses, differently signed tractions develop on different grain boundaries and that leads to vacancy concentration differences that can have profound effects. We will see that, at very high temperatures, where vacancies are mobile, polycrystalline solids can deform by stress-driven atomic diffusion. This is called diffusional creep. As we will see, the creep response is the result of vacancy concentration gradients that are set up by the stress. So even though the applied stress has a very small effect on the absolute vacancy concentration, important vacancy concentration differences are established within the solid.

7.5.1 Nabarro–Herring creep

Imagine a polycrystalline solid (modeled as a square array of grains as shown in Fig. 7.12) at a high temperature, T, and under no stress. The equilibrium vacancy fraction is

$$\chi_0 = \exp\left(\frac{\Delta s_v^0}{k_B}\right) \exp\left(-\frac{\Delta e_v^0}{k_B T}\right) = \exp\left(-\frac{\Delta f_v^0}{k_B T}\right), \tag{7.54}$$

where

$$\Delta f_v^0 = \Delta e_v^0 - T \Delta s_v^0 \tag{7.55}$$

is the Helmholtz free energy of formation of the vacancy under zero stress. Δf_v also equals the Gibbs free energy of formation because the stress is zero. The equilibrium vacancy concentration

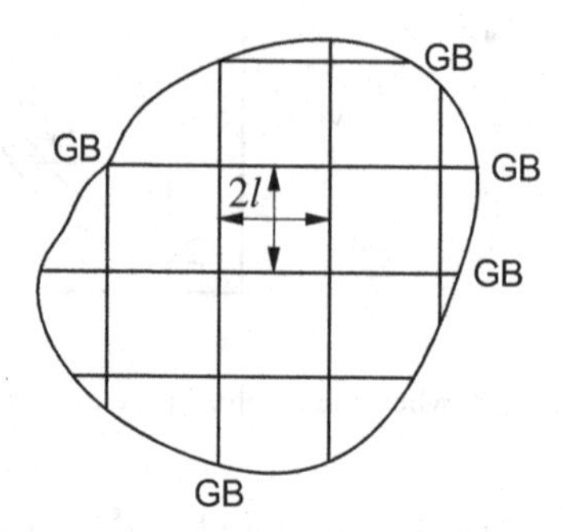

Figure 7.12. Array of square grains with size $L = 2l$. GB means grain boundary.

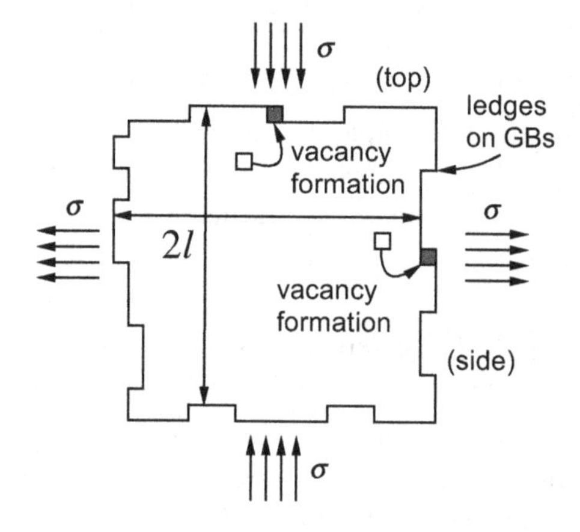

Figure 7.13. Vacancy formation under non-hydrostatic stress.

is

$$c_0 = \frac{\chi_0}{\Omega} = \frac{1}{\Omega} \exp\left(-\frac{\Delta f_v^0}{k_B T}\right). \tag{7.56}$$

Local vacancy concentration near grain boundary

Now consider the application of a uniform shear stress, σ, i.e. $\sigma_{xx} = \sigma, \sigma_{yy} = -\sigma$. As shown in Fig. 7.13, each grain in the polycrystalline aggregate is subjected to compression in the vertical direction and tension in the transverse direction. We now wish to determine the local equilibrium vacancy concentrations at the top (or bottom) and side surfaces of the crystal.

If we consider the formation of a vacancy near the top surface (i.e. grain boundary) of the square grain, an atom will be moved to that surface and additional work must be done against the applied stress in the vacancy formation process. Thus the formation Gibbs free energy of the vacancy here is

$$\Delta g_v^{top} = \Delta f_v^0 - \Delta W, \tag{7.57}$$

where ΔW is the work done by the compressive tractions acting on the top surface of the grain. From Fig. 7.14a, we have

$$\Delta W = -(\sigma a^2)a = -\sigma a^3 = -\sigma\Omega. \tag{7.58}$$

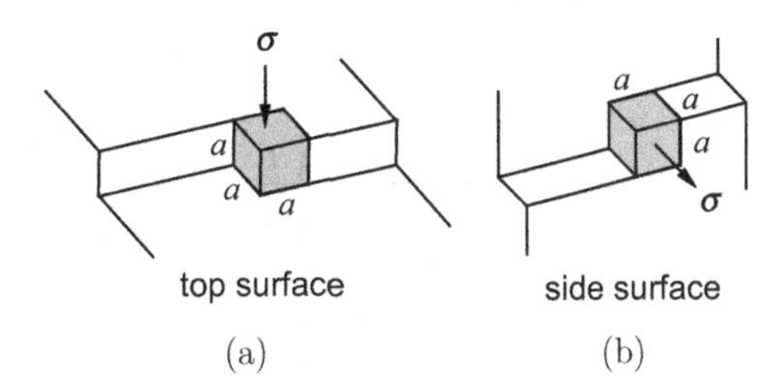

Figure 7.14. (a) Vacancy formation on top surface. (b) Vacancy formation on side surface.

So

$$\Delta g_v^{top} = \Delta f_v^0 + \sigma\Omega. \tag{7.59}$$

The increased formation free energy means that the local equilibrium vacancy concentration (enforced by the grain boundary) is depressed by the compressive tractions on the top surface,

$$c_v^{top} = c_0 \exp\left(-\frac{\sigma\Omega}{k_B T}\right). \tag{7.60}$$

Now consider an equivalent process on the side of the grain where tensile tractions are acting. Here the formation free energy of the vacancy is

$$\Delta g_v^{side} = \Delta f_v^0 - \Delta W, \tag{7.61}$$

where ΔW is the work done by the existing (tensile) tractions (acting on the side surface of the grain), as shown in Fig. 7.14b,

$$\Delta W = (\sigma a^2)a = \sigma a^3 = \sigma\Omega, \tag{7.62}$$

thus reducing the Gibbs free energy of formation to

$$\Delta g_v^{side} = \Delta f_v^0 - \sigma\Omega \tag{7.63}$$

and leading to an increased local equilibrium vacancy concentration

$$c_v^{side} = c_0 \exp\left(\frac{\sigma\Omega}{k_B T}\right). \tag{7.64}$$

Diffusional flux

Because the stress is uniform, the generalized Fick's first law, Eqs. (7.41), is equivalent to the simple form, Eq. (7.21), i.e. the vacancy flux is proportional to the concentration gradient. The difference between c_v^{top} and c_v^{side} causes vacancies to drift or diffuse down the concentration gradient from the sides of the crystal, where the concentration is high, to the top and bottom surfaces, where the concentration is low. But the tractions that act on the boundaries continually maintain the local vacancy concentration at the values we have calculated. So the concentration difference remains even though the vacancies are diffusing from the sides to the top and bottom. As vacancies arrive at the top surface by diffusion they annihilate there so that the local

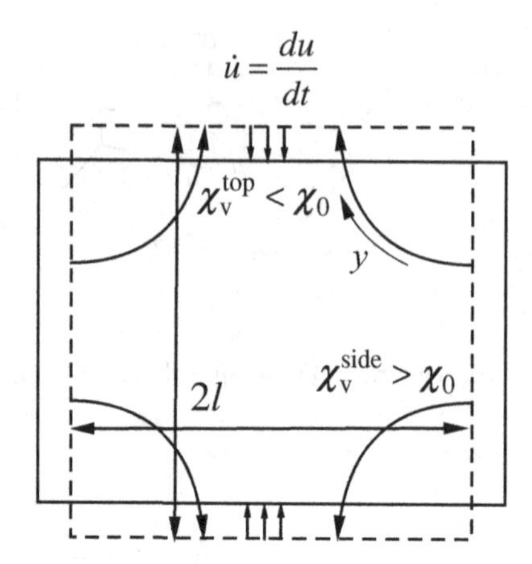

Figure 7.15. Nabarro–Herring creep.

equilibrium is maintained. Similarly, as vacancies diffuse away from the side surface more are created to replace them.

Therefore, we should regard c_v^{top} and c_v^{side} as boundary conditions to the diffusion equation (7.25). This boundary condition is imposed by the grain boundaries, assumed here to act as perfect sinks and sources of vacancies, so that the local vacancy concentration near the grain boundaries are maintained at values different from c_0. The vacancy concentration at locations away from the grain boundary can only be influenced by the grain boundary through diffusion. Therefore, the vacancy concentration in the grain interior should be obtained by solving the diffusion equation (7.25) using c_v^{top} and c_v^{side} as boundary conditions.

To provide a rough estimate, we shall not solve the three-dimensional (or two-dimensional) diffusion equation, but consider the path y from the side of the grain to the top, as shown in Fig. 7.15, and apply the one-dimensional diffusion equation along this curved path. Using Fick's first law we can express the vacancy flux as

$$J_v = -D_v \frac{dc_v}{dy}, \tag{7.65}$$

Now we may make a simple estimate of the vacancy concentration gradient

$$\frac{dc_v}{dy} \approx \frac{c_v^{top} - c_v^{side}}{l} = \frac{1}{l\Omega}\left(\chi_v^{top} - \chi_v^{side}\right), \tag{7.66}$$

where we have approximated the length of the diffusion distance to be $l = L/2$, i.e. half the size of the square grain (see Fig. 7.12). We will see later that this simple approximation gives a result very close to the exact result based on an account of the full diffusion field.

Now the concentration gradient can be evaluated as

$$\begin{aligned}\frac{dc_v}{dy} &\approx \frac{1}{l}\left[c_0 \exp\left(-\frac{\sigma\Omega}{k_B T}\right) - c_0 \exp\left(\frac{\sigma\Omega}{k_B T}\right)\right] \\ &\approx -\frac{2c_0}{l}\sinh\left(\frac{\sigma\Omega}{k_B T}\right).\end{aligned}$$

At high temperatures and low stresses, where diffusional deformation is dominant, $\sigma\Omega/k_B T \ll 1$ so that

$$\frac{dc_v}{dy} \approx -\frac{2c_0}{l} \cdot \left(\frac{\sigma\Omega}{k_B T}\right). \tag{7.67}$$

Thus the flux of vacancies is

$$J_v = -D_v \frac{dc_v}{dy} \approx \frac{2D_v c_0}{l} \cdot \left(\frac{\sigma\Omega}{k_B T}\right). \tag{7.68}$$

Creep rate

Figure 7.15 shows that the arrival of vacancies at the top surface and their annihilation there causes the top surface of the grain to be eaten away at a displacement rate $\dot{u}$. It is convenient to multiply this displacement rate by a unit area on the top surface and then divide by that same unit area

$$\frac{\dot{u} \cdot \text{area}}{\text{area}} = \frac{\text{volume rate}}{\text{area}} = J_v\Omega,$$
$$\dot{u} = J_v\Omega. \tag{7.69}$$

With this consideration the displacement rate at the top surface is

$$\dot{u} \approx \frac{2D_v c_0 \Omega}{l} \cdot \left(\frac{\sigma\Omega}{k_B T}\right). \tag{7.70}$$

Finally the axial compressive strain rate would be

$$\dot{\varepsilon} \approx \frac{\dot{u}}{l} = \frac{2D_v c_0 \Omega}{l^2} \cdot \left(\frac{\sigma\Omega}{k_B T}\right) = \frac{2D_v \chi_0}{l^2} \cdot \left(\frac{\sigma\Omega}{k_B T}\right). \tag{7.71}$$

We note that the lattice diffusivity is related to the vacancy diffusivity through Eq. (7.32),

$$D_L = \chi_0 D_v,$$

and noting that the grain size is $L = 2l$, we have the following.

$$\boxed{\dot{\varepsilon}_{N\text{-}H} = \frac{8D_L}{L^2} \cdot \left(\frac{\sigma\Omega}{k_B T}\right)}$$

This is called the Nabarro–Herring creep law. It describes lattice diffusion controlled diffusional deformation in polycrystalline solids. Each grain in the polycrystalline aggregate changes shape as indicated in the diagram above and this causes the entire solid to change shape in the same way. We note that the creep rate depends directly on the lattice diffusivity (which in turn depends exponentially on the temperature). The rate of creep also depends linearly on the applied stress and inversely on the square of the grain size. Thus, fine-grained solids at high temperatures are likely to deform in this way. It should be remembered that all of this deformation was brought about by the fact that the local equilibrium vacancy concentrations near the differently oriented grain boundaries were slightly different. An exact treatment of this problem with a two-dimensional [49] diffusional field leads to a result of the same form except that the numerical factor is 6 instead of the approximate 8.

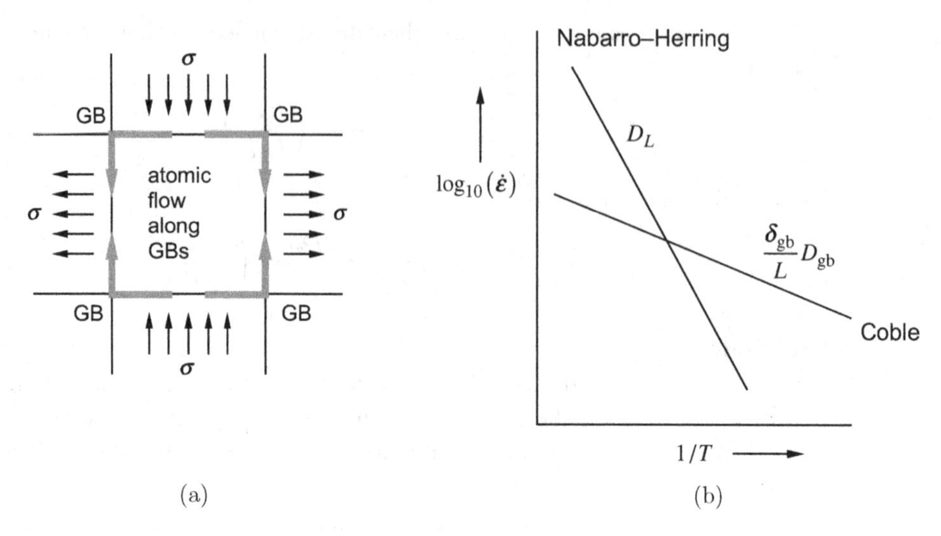

Figure 7.16. (a) Diffusion currents in Coble creep. (b) Arrhenius plot of creep rates.

7.5.2 Coble creep

Atomic diffusion may also occur along the grain boundaries connecting the tops (and bottoms) of the grains with the sides. This leads to atom diffusivity in the grain boundary D_{gb} that has a different temperature dependence from the lattice diffusivity D_L, as in

$$D_L = D_L^0 \exp\left(-\frac{Q_L}{k_B T}\right) \tag{7.72}$$

$$D_{gb} = D_{gb}^0 \exp\left(-\frac{Q_{gb}}{k_B T}\right) \tag{7.73}$$

where Q_L and Q_{gb} are the activation energies for lattice and grain boundary diffusion, respectively, and we often have $Q_L > Q_{gb}$. However, grain boundary diffusion is only limited to the region within a width δ_{gb} of the grain boundary. The creep law resulting from grain boundary diffusion is as follows.

$$\boxed{\dot{\varepsilon}_C = \frac{8\delta_{gb}D_{gb}}{L^3}\left(\frac{\sigma\Omega}{k_B T}\right)} \tag{7.74}$$

This is called Coble creep, as shown in Fig. 7.16. In the Coble creep rate expression, an extra factor of δ_{gb}/L appears, causing the creep rate to depend on the reciprocal of the cube of the grain size. In general, both grain boundary and lattice paths are open to the transport of atoms from the sides to the tops and bottoms of the grains. So we consider the creep rate to be the sum of the creep rates for these two separate paths

$$\begin{aligned}\dot{\varepsilon} &= \dot{\varepsilon}_{N\text{-}H} + \dot{\varepsilon}_C \\ &= \frac{8}{L^2}\left(\frac{\sigma\Omega}{k_B T}\right)\left(D_L + \frac{\delta_{gb}D_{gb}}{L}\right).\end{aligned} \tag{7.75}$$

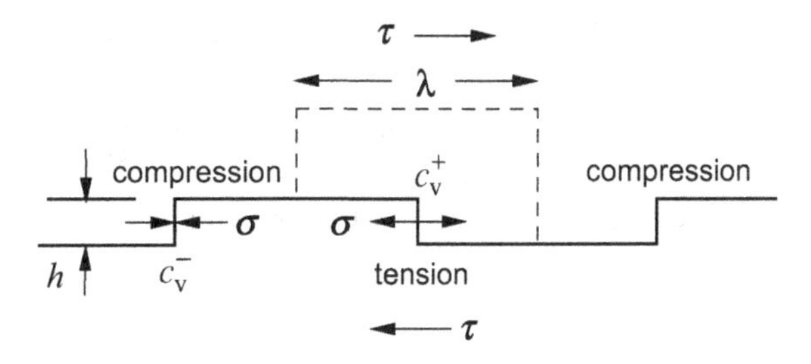

Figure 7.17. Grain boundary sliding controlled by diffusion.

The temperature dependence of the creep rates depends almost completely on the temperature dependence of the diffusivities, which are known to depend exponentially on temperature according to Eqs. (7.72) and (7.73).

Thus, as shown in Fig. 7.16b, Nabarro–Herring creep tends to be dominant (faster) for higher temperatures and coarser grains while Coble creep dominates at lower temperatures and smaller grains.

7.5.3 Grain boundary sliding controlled by diffusion

The sliding of crystals along their grain boundaries can also be controlled by stress-driven diffusional processes. Consider a non-flat grain boundary subjected to a shear stress at very high temperatures. If the flat parts of the boundary are free to slide easily, then normal tractions will be built up at the steps in the boundary to support the applied shear stresses.

A simple free-body force analysis of the element shown in dotted lines in Fig 7.17 gives

$$\tau\lambda = \sigma h \tag{7.76}$$

(where λ is the average spacing between the steps and h is the step height) so that the normal stresses acting on the stepped grain boundary segments are

$$\sigma = \frac{\tau\lambda}{h}. \tag{7.77}$$

These normal stresses cause the local vacancy concentrations to be enhanced and depressed, in tensile and compressive regions, respectively, as shown in Fig. 7.17. This, in turn, causes vacancies to flow from the region of high concentration to the region of low concentration (or atomic flow in the opposite direction), just as in the case of diffusional creep. As vacancies arrive and leave the different stepped parts of the grain boundary, a material gap is created that must be accommodated by the sliding of the grain boundary. If the process is controlled by lattice diffusion, one can think of vacancies diffusing out from (or into) the stepped source (or sink) through a cylinder centered on the source (sink). One then thinks of the vacancy concentration at the midpoint between the sources and sinks to be the equilibrium vacancy concentration in the absence of stress, c_0. Following this approach the rate of sliding, controlled by lattice diffusion, is (see Exercise problem 7.2)

$$\dot{u}^{\mathrm{L}}_{\mathrm{sliding}} = \frac{2\pi D_{\mathrm{L}}}{\ln(\lambda/h)} \frac{\tau\Omega}{k_{\mathrm{B}}T} \frac{\lambda}{h^2}. \tag{7.78}$$

If diffusion is controlled by atomic flow along the grain boundary then the result is [50]

$$\dot{u}^{\mathrm{gb}}_{\mathrm{sliding}} = \frac{4\delta_{\mathrm{gb}} D_{\mathrm{gb}}}{h^2} \frac{\tau \Omega}{k_{\mathrm{B}} T}, \tag{7.79}$$

which is, interestingly, independent of the spacing between the steps, λ.

7.6 Summary

The equilibrium vacancy concentration increases dramatically with increasing temperature. It also decreases with increasing pressure in a solid subjected to a uniform pressure. For these changes to occur vacancies either need to be created at sources and move into the surrounding lattice or move to sinks where they may be annihilated. Surfaces, grain boundaries, dislocations, and other point defect clusters may serve as sources and sinks for these processes. Under some conditions an excess vacancy concentration may be lessened by the coalescence of vacancies to form dislocation loops, without migrating to existing sinks.

The rate at which vacancy concentrations change depends on the rate at which vacancies can jump from one lattice site to the next. This is a thermally activated process governed by a Boltzmann relation involving the Gibbs free energy change, $\Delta g_{\mathrm{v}}^{\mathrm{m}}$, associated with an atom squeezing between neighboring atoms as it moves into the vacancy, thereby moving the vacancy from one site to the next. Hence the vacancy jump rate can be written as a product of the frequency for the neighboring atoms attempting to jump into the vacancy, the number of neighboring atoms, and the probability that each attempt is successful as given by the Boltzmann relation.

The activation energy for vacancy motion, $\Delta e_{\mathrm{v}}^{\mathrm{m}}$, which is the principal term in $\Delta g_{\mathrm{v}}^{\mathrm{m}}$, has been measured using a quench/resistivity technique. In this technique wires are first quenched from a high temperature to establish a vacancy supersaturation at a low temperature and then heated to various moderate temperatures to allow the excess vacancies to diffuse to and annihilate at sinks. The electrical resistivity is used to monitor the concentration of vacancies.

The jump rate of an interstitial solute atom in a host crystal is similarly governed by a Boltzmann relation involving the activation free energy of migration $\Delta g_{\mathrm{i}}^{\mathrm{m}}$, although the interstital atom may not follow a straight path when jumping between two equivalent sites. The jump of a substitutional solute atom, on the other hand, often requires the assistance of a neighboring vacancy. As a result, the jump rate of a substitutional solute atom is governed by a Boltzmann relation involving the sum of the activation free energy of migration and the free energy of vacancy formation.

The evolution of the concentration field of a large number of point defects, each making random jumps to neighboring sites, can be described by the diffusion equation. In crystals with no internal stress nor chemical inhomogeneity (and ignoring any interaction between the point defects), the flux of point defects is proportional to the negative of the concentration gradient. The proportionality constant is the diffusion coefficient, which is directly related to the jump rate of the individual point defects. For crystals subjected to non-uniform internal stresses, it is necessary, instead, to write the point defect flux in terms of the gradient of the chemical potential. As a result, in the absence of vacancy sources and sinks, the local equilibrium vacancy concentration is enhanced at locations of positive pressure. This pressure effect is opposite to that in a solid containing vacancy sources and sinks and subjected to a uniform pressure.

In the case of crystals without internal stresses, but subjected to a uniform non-hydrostatic applied shear stresses, different local vacancy concentrations are established at different surfaces (or grain boundaries) of the stressed crystal. The resulting vacancy concentration gradients lead to vacancy flow and diffusional deformation. At high temperatures and for large crystals self-diffusion occurs primarily through the crystal lattice and this leads to Nabarro–Herring creep, whereas at low temperatures and for smaller crystals diffusion occurs primarily through the grain boundaries and we have Coble creep. Grain boundary sliding may also be controlled by similar diffusive processes.

7.7 Exercise problems

7.1 A long wire of pure Au is held at room temperature for a very long time so that equilibrium is established. The length of the wire in this state is L_0. Then the wire is heated to $T_1 = 900\,^\circ\text{C}$ and held for a sufficiently long time to establish equilibrium at that temperature, before being quenched instantaneously to room temperature. The length of the wire in the quenched state is L_q.

(a) Write an expression for L_q/L_0 as a function of T_1 and the thermodynamic properties of vacancies in Au, which are given in Table C.1.

(b) The quenched wires are heated to a slightly elevated temperature, T, and annealed for a time, t, before being cooled again to room temperature where the length of the wire, L, is again measured. Write an expression showing how the length of the wire would be expected to change with increasing annealing time. You may assume that the equilibrium vacancy concentration at the annealing temperature T is so small that it can be ignored.

7.2 We wish to derive the rate of the grain boundary sliding controlled by diffusion, Eq. (7.78). Consider a cylindrical region with radius $\lambda/2$ centered at each grain boundary step, as shown in Fig. 7.18. For simplicity, assume that at the boundary of each cylindrical region, the vacancy concentration has already reached the far-field value, c_v^0. The vacancy concentration at the center of each cylinder is either c_v^+ or c_v^-, depending on whether the local stress is tensile or compressive.

(a) Obtain the steady-state solution of the vacancy concentration field inside each cylinder, $c_v(r, \theta, z)$, where the center of the cylinder is chosen as the origin of the coordinate

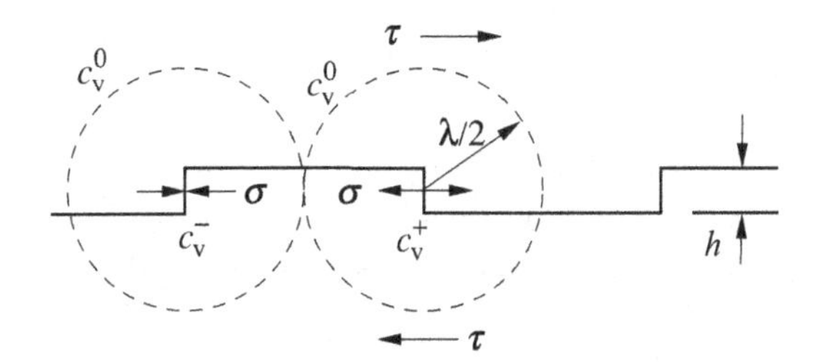

Figure 7.18. Grain boundary sliding controlled by diffusion.

system. The Laplacian operator ∇^2 in cylindrical coordinates is defined as

$$\nabla^2 c_v = \frac{1}{r}\frac{\partial}{\partial r}\left(r\frac{\partial c_v}{\partial r}\right) + \frac{1}{r^2}\frac{\partial^2 c_v}{\partial \theta^2} + \frac{\partial^2 c_v}{\partial z^2}. \tag{7.80}$$

By symmetry, we expect c_v to be independent of θ and z, so that

$$\nabla^2 c_v(r) = \frac{1}{r}\frac{\partial}{\partial r}\left(r\frac{\partial c_v(r)}{\partial r}\right). \tag{7.81}$$

(b) Obtain the steady-state vacancy flux $J_v(r) = -D_v \partial c_v(r)/\partial r$ in the radial direction and determine the integrated vacancy flux out of the cylindrical surface.

(c) Obtain the grain boundary sliding rate assuming that the flat parts of the boundary are free to slide easily.

7.3 A very large crystal is filled with a high concentration of nanovoids, each of radius R. The surfaces of the voids have a surface energy γ_s.

(a) Develop or write an expression for the equilibrium vacancy fraction in this crystal assuming the equilibrium vacancy fraction in the absence of voids is χ_0. You may wish to first determine the work needed to create a single vacancy in the presence of the void by removing at atom from the crystal and placing it on the void surface. This is similar to the problem that arises in diffusional creep.

(b) Based on your result, indicate what would happen to if two voids of different size, R_1 and $R_2 > R_1$ are located near each other, assuming that vacancies are mobile.

7.4 We wish to estimate the time that would be needed for a small spherical void of radius a, in a much larger spherical particle of radius b, to shrink and be completely eliminated by vacancy diffusion from the surface of the cavity to the surface of the particle. The cavity surface and the surface of the particle have a surface free energy γ_s so that local equilibrium vacancy fraction is modified according to Eq. (5.61). Specifically, the equilibrium vacancy concentrations at the two surfaces are

$$c^+ = c_0 \exp\left[-p_c^{(a)}\Omega/(k_B T)\right] \approx c_0\left[1 - p_c^{(a)}\Omega/(k_B T)\right] \tag{7.82}$$

and

$$c^- = c_0 \exp\left[-p_c^{(b)}\Omega/(k_B T)\right] \approx c_0\left[1 - p_c^{(b)}\Omega/(k_B T)\right], \tag{7.83}$$

where $p_c^{(a)} = -2\gamma_s/a$ and $p_c^{(b)} = 2\gamma_s/b$.

(a) Assuming that movements of the interfaces can be ignored and that steady-state diffusion between the two surfaces provides an adequate solution to the problem, derive an expression for the time needed for the cavity to be eliminated if the particle is held at some temperature, T.

(b) Compute the time needed for complete shrinkage of a cavity of radius $a = 1\ \mu\text{m}$ in a large gold particle ($b \to \infty$) held at $T = 1000$ K, again assuming that a steady-state solution is adequate. The lattice parameter of FCC gold is 0.407 nm and the surface energy of gold is about 1 J/m^2.

(c) Repeat the calculation in (b) for a temperature of $T = 500$ K . The calculations show how difficult it is to eliminate porosity in metals by annealing alone.

7.5 The quench-resistivity technique for measuring the thermodynamic properties of vacancies in metals requires that wires be quenched sufficiently quickly that most of the vacancies present in the wire at high temperature are retained at low temperatures. Here we wish to estimate the cooling rates needed to conduct such experiments. We can make a rough estimate of the required cooling rate in the following way. Imagine that a wire, initially equilibrated at some high initial temperature, T_0, is cooled to a series of temperatures, $T_i = T_0 - i\Delta T, i = 1, 2, \ldots$. At each step, the wire temperature is maintained at temperature T_i for a short time period Δt. The cooling rate is therefore $\theta = \Delta T / \Delta t$.

(a) Let N_0 be the equilibrium number of vacancies in the wire at T_0. Let $N_{eq}(T_i)$ be the equilibrium number of vacancies at temperature T_i, if the wire were kept at temperature T_i for a very long time. Express $N_{eq}(T_i)$ in terms of N_0 and the thermodynamic properties of the vacancies.

(b) We shall assume that the cooling rate is sufficiently fast, so that the number of vacancies in the wire at the beginning of each temperature period is very close to N_0. Therefore, $N_0 - N_{eq}(T_i)$ is the excess number of vacancies at temperature T_i. Assuming that the only vacancy sink is the surface of the wire, whose radius is R, write down the expression for the rate of vacancy loss $f(T_i)$ at temperature T_i.

(c) For gold wires 400 μm in diameter and quenched from $T_0 = 900\,^\circ\text{C}$, find the critical temperature T_c at which the loss of vacancies is most rapid.

(d) The total number of vacancies lost during the quenching process can be approximated as $N_{loss} = \sum_i f(T_i)\Delta t = \sum_i f(T_i)\Delta T/\theta$. In the limit of $\Delta T \to 0$, $N_{loss} = \int_0^{T_0} f(T) dT/\theta$. If we require that less than 1% of the vacancies are lost during the quenching process, then the cooling rate must exceed the critical value $\theta_c = \int_0^{T_0} f(T) dT/(0.01 N_0)$. Find the critical cooling rate for gold wires specified in (c). You may perform the integral by numerical methods.

7.6 Assuming that self-diffusion occurs by the vacancy mechanism, predict the activation energy Q for self-diffusion, as defined in Eq. (7.27), for Ag, Al, Au, Cu, Fe, Mo, Ni, Ta, and W, using the vacancy properties given in Table C.1. Compare the predictions with Table D.1. What conclusions can be drawn from this comparison?

7.7 Estimate the activation enthalpy Δh_s^m for substitutional Al, Au, Cu, Pt, V, and W atoms to jump into a neighboring vacancy in a Ni crystal, using the diffusivity data in Table D.2 and vacancy properties in Table C.1. Can the ranking of Δh_s^m be explained in terms of the atomic size factors of substitutional impurities given in King's table A.2?

7.8 Mark all tetrahedral and octahedral sites in a BCC unit cell. Sketch a likely pathway for an interstitial atom to jump between two neighboring tetrahedral sites.

7.9 Consider Al impurity atoms in a Au crystal at 300 K.

(a) Estimate the diffusion coefficient of Al impurity atoms, assuming that the diffusion data in Table D.2 can be extrapolated to 300 K.

(b) In (a) we have assumed that the vacancies in the Au crystal have reached their equilibrium concentration at 300 K, which is usually difficult to achieve in practice. Assume that the Au crystal has been kept at 800 K for a long time and then rapidly quenched to 300 K.

Estimate the diffusion coefficient of Al impurity atoms in the quenched Au crystal, using the vacancy properties given in Table C.1.

7.10 Consider a Cu polycrystal with an average grain size of 2 μm and subjected to a shear stress of 50 MPa.

(a) Estimate the strain rate by Nabarro–Herring creep at 500 K, assuming that the solid has attained the equilibrium vacancy concentration at this temperature.
(b) If the crystal has been kept at 800 K for a long time and then rapidly quenched to 500 K, what is the creep rate for the quenched polycrystalline solid? Use the diffusion data in Table D.1 and vacancy properties in Table C.1.

7.11 Consider the charge transport in an ionic solid, yttria-stabilized zirconia (YSZ, see Exercise problem 6.6), by the diffusion of oxygen vacancies, V_O^{++}. Assume the crystal is subjected to zero stress but under a uniform electric field $\mathcal{E}$ along the x direction. The vacancy chemical potential under this condition can be written as

$$\Delta\mu_v = k_B T \ln \chi_v - 2e\mathcal{E}x + \text{constant}, \tag{7.84}$$

where $e = 1.602 \times 10^{-19}$ coulombs is the magnitude of the charge of one electron.

(a) Use the generalized Fick's first law, Eq. (7.41), to obtain the vacancy flux J_v when the vacancy concentration c_v is uniformly distributed in the crystal.
(b) The electrical conductivity σ is defined as the current density divided by the electric field. Express the electrical conductivity of the crystal in terms of the self-diffusivity of oxygen anions and temperature.
(c) Assume that the oxygen diffusivity in YSZ containing 10 mol-% Y_2O_3 is $D_O = 5 \times 10^{-12} \text{m}^2\,\text{s}^{-1}$ at 1000 K and that the lattice constant is $a = 5.14$ Å. Estimate the electrical conductivity of the YSZ crystal at 1000 K.

7.12 An important solution of the one-dimensional diffusion equation, Eq. (7.24), in the infinite domain ($-\infty < x < \infty$) is

$$c(x,t) = \frac{1}{\sqrt{4\pi Dt}} e^{-\frac{x^2}{4Dt}}. \tag{7.85}$$

This solution corresponds to the initial condition of $c(x,0) = \delta(x)$, where $\delta(x)$ is the Dirac delta function. Physically this corresponds to the scenario in which all diffusing particles are located on the plane $x = 0$ at time $t = 0$. Equation (7.85) is called the Green's function of the diffusion equation.

(a) Verify that Eq. (7.85) satisfies the diffusion equation.
(b) Plot the corresponding flux $J(x,t)$ as a function of time at $x = 0.01$ m, assuming $D = 10^{-8}\,\text{m}^2\,\text{s}^{-1}$.

7.13 One way to measure the diffusion coefficient of an impurity species in a solid is to place a reservoir of the impurity atoms next to the solid surface and watch the concentration profile of the impurity atoms in the solid evolve as a function of time. We shall assume the solid surface corresponds to the plane $x = 0$ and the solid occupies the half space $x \geq 0$. We assume that initially the impurity atoms have a uniform concentration c_0 in the solid, corresponding to the initial condition of $c(x,0) = c_0$ for $x > 0$. We also assume the reservoir maintains a constant concentration c_1 on the solid surface, corresponding to the boundary condition of $c(0,t) = c_1$ for $t \geq 0$. The solution of the diffusion equation subjected to such initial and boundary conditions is

$$c(x,t) = c_1 - (c_1 - c_0)\,\mathrm{erf}\left(\frac{x}{2\sqrt{Dt}}\right), \tag{7.86}$$

where the erf function is defined as

$$\mathrm{erf}(z) \equiv \frac{2}{\sqrt{\pi}} \int_0^z e^{-t^2}\, dt. \tag{7.87}$$

(a) Find the location x_h where the concentration is halfway between c_0 and c_1, as a function of time, i.e. $c(x_h, t) = (c_0 + c_1)/2$.
(b) If the measurement shows that $x_h = 1$ cm at $t = 1$ hour, what is the diffusivity D?
(c) Plot the concentration profile at $t = 1$ hour assuming $c_0 = 10^{20}\ \mathrm{m}^{-3}$ and $c_1 = 10^{22}\ \mathrm{m}^{-3}$.

7.14 We now use a computer to solve the one-dimensional diffusion problem described in Section 7.3.3. We first represent the concentration function on a uniform grid with spacing $\Delta x = 2l/N$. The grid points are located at $x_i = -l + (i-1)\Delta x$, $i = 1, \ldots, N+1$. The second spatial derivative of the concentration function can be approximated as

$$\frac{\partial^2 c(x_i,t)}{\partial x^2} \approx \frac{c(x_i + \Delta x, t) + c(x_i - \Delta x, t) - 2c(x_i, t)}{(\Delta x)^2} \tag{7.88}$$

and the time derivative can be approximated as

$$\frac{\partial c(x_i,t)}{\partial t} \approx \frac{c(x_i, t+\Delta t) - c(x_i, t)}{\Delta t}, \tag{7.89}$$

where Δt is the time step of the numerical method. Given these approximations, the diffusion equation (7.24) can be rewritten as

$$c(x_i, t+\Delta t) = c(x_i, t) + \frac{D\Delta t}{(\Delta x)^2}[c(x_i + \Delta x, t) + c(x_i - \Delta x, t) - 2c(x_i, t)]. \tag{7.90}$$

This equation provides a method to obtain the concentration values on the grid points at time $t + \Delta t$ based on the information at time t (the previous time step).

(a) Download and run the Matlab program `P_7_14_diffusion.m` on the book website.
(b) Modify the Matlab program to plot $(c - c_0)/(c_1 - c_0)$ for $Dt/l^2 = 0.01, 0.1, 0.2, 0.4, 0.6, 0.8, 1.0$, using $N = 500$ and $D\Delta t/l^2 = 5 \times 10^{-6}$. Compare your results with Fig. 7.9.

7.15 Consider a modification to the diffusion process in Exercise problem 7.14 in which the concentration at the right boundary is maintained at half the value at the left boundary. In other words, the boundary conditions are now: $c(-l,t) = c_1$, $c(l,t) = c_2$, where $c_2 = c_1/2$.

(a) Modify the Matlab program in Exercise problem 7.14 to solve the diffusion equation under the new boundary condition, with $c_0 = 10^{-3}$ and $c_1 = 10^{-2}$ (in arbitrary units).
(b) Plot the flux J across the $x = 0$ plane as a function of time.

7.16 We now apply numerical methods to solve the generalized diffusion equation (7.42) for vacancies in a solid subjected to an inhomogeneous pressure field $p(x)$. Consider a crystal in the domain $-l \le x \le l$ containing an initially uniform concentration of vacancies c_0. The vacancy concentrations are also kept at c_0 at the two boundaries. However, the solid is subjected to an internal stress such that $-p(x)\Delta V_{\rm v}^{\rm rlx}/(k_{\rm B}T) = 1 - (x/l)^2$. According to Eq. (7.48), the equilibrium vacancy concentration would be non-uniform and given by $c_{\rm v}^{\rm eq}(x) = c_0 \exp[1 - (x/l)^2]$. From Eq. (7.50), we have

$$\frac{\partial \Delta\mu_{\rm v}}{\partial x} = k_{\rm B}T\left(\frac{1}{c_{\rm v}}\frac{\partial c_{\rm v}}{\partial x} - \frac{\partial \ln c_{\rm v}^{\rm eq}}{\partial x}\right), \tag{7.91}$$

so that the diffusion equation (7.42) becomes

$$\begin{aligned}\frac{\partial c_{\rm v}}{\partial t} &= D_{\rm v}\frac{\partial}{\partial x}\left(\frac{\partial c_{\rm v}}{\partial x} - c_{\rm v}\frac{\partial \ln c_{\rm v}^{\rm eq}}{\partial x}\right)\\ &= D_{\rm v}\left(\frac{\partial^2 c_{\rm v}}{\partial x^2} - \frac{\partial c_{\rm v}}{\partial x}\frac{\partial \ln c_{\rm v}^{\rm eq}}{\partial x} - c_{\rm v}\frac{\partial^2 \ln c_{\rm v}^{\rm eq}}{\partial x^2}\right).\end{aligned} \tag{7.92}$$

(a) Download and run the Matlab program `P_7_16_diffusion.m` on the book website.

(b) Modify the Matlab program to plot $c_{\rm v}/c_0$ for $Dt/l^2 = 0.01, 0.1, 0.2, 0.4, 0.6, 0.8, 1.0$, using $N = 500$ and $D\Delta t/l^2 = 5\times 10^{-6}$. Assume $c_0 = 10^{-3}$ (in arbitrary units). At what time is the relative difference between $c_{\rm v}$ and $c_{\rm v}^{\rm eq}$ at $x = 0$ less than 1%?

PART III

DISLOCATIONS

The following five chapters constitute Part III of this book and focus on dislocations, the major line defects in crystals. Chapter 8 on dislocation geometry introduces the variables needed to properly define the dislocation line, specifically the Burgers vector and line sense vector. It also describes how dislocation motion produces plastic strain in the crystal. Chapter 9 on dislocation mechanics discusses the stress field and energy of dislocations. The discussion is based on the linear elasticity theory, which is an accurate description of the solid only at sufficiently small strain, i.e. in regions sufficiently far away from the dislocation center. The interaction between dislocations, and effect of interfaces on the dislocation stress field, are studied in Chapter 10, which also discusses several applications of dislocation mechanics, such as the strain relaxation between misfit layers. In Chapter 11, we examine the structure of the dislocation core in close-packed crystals, in which a perfect dislocation dissociates into partial dislocations bounding a stacking fault area. The atomistic structures of the dislocation in several non-close-packed crystal structures are discussed in Chapter 12. The dislocation core structure is the result of non-linear interatomic interactions, and strongly influences the dislocation mobility.

It is instructive to compare the layout of Part III on dislocations with Part II on point defects. For example, Chapter 9 (dislocation mechanics) is the counterpart of Chapter 4 (point defect mechanics); in these chapters the stress fields of individual defects are discussed. However, Chapters 8 (dislocation geometry), 11 (partial dislocations) and 12 (dislocation core structure) have no corresponding chapters in Part II. This is because the geometry and atomistic core structure of dislocations are much more complex than those of the point defects; the latter are mostly covered in Sections 4.1 and 4.2.

At the same time, there is no chapter on "dislocation thermodynamics" that corresponds to Chapter 5 on point defect thermodynamics. This is because the energy of a dislocation line is so large that they should not exist if the crystal were in a truly thermodynamic equilibrium state. In comparison, a finite concentration of point defects should always exist at thermal equilibrium. Therefore, when discussing dislocations, we are necessarily dealing with non-equilibrium (perhaps meta-stable) states. For example, dislocations are usually generated in great quantities during plastic deformation, which is a highly dissipative, non-equilibrium process.

8 Dislocation geometry

The main purpose of this chapter is to introduce the geometrical properties of dislocations, the rules governing dislocation reactions, and the directions of dislocation motion in response to applied stress. The goal is to develop an intuitive understanding of the basic behaviors of dislocations without obtaining their stress field (which is the subject of the next chapter).

We start with Section 8.1 on why dislocations are necessary for plastic deformation of crystals. In Section 8.2, we introduce Volterra dislocations in an elastic continuum, and then describe the differences between them and dislocations in a crystal. In Section 8.3, we define the Burgers vector of a dislocation, and describe the geometric rule for Burgers vectors that must be satisfied when dislocations react. Section 8.4 shows which direction a dislocation should move on its glide plane under an applied stress. It also introduces cross-slip and climb, as alternative modes of dislocation motion. Section 8.5 describes where crystal dislocations come from. Several mechanisms are presented in which the motion of existing dislocations can lead to multiplication, i.e. an increase of total dislocation length.

8.1 Role of dislocations in plastic deformation

We begin our study of dislocations by first thinking about plastic deformation in crystals – a problem that first led to the concept of crystal dislocations in the early 1930s. Although one's common experience with plastic deformation usually involves the continuous bending or stretching of a soft metal wire, the fundamental mechanism of plastic deformation is a shear process, as shown in Fig. 8.1.

The crystal, represented as a rectangular box, is plastically deformed in tension by sequential slip on various crystal planes. The bold lines indicate the active slip plane for that particular strain increment. Notice that the cumulative effect of these events is to make the crystal permanently longer and narrower. So the macroscopic shape change associated with ordinary tensile deformation is actually the cumulative effect of a large number of shear events. This can be confirmed by observing the surface of a plastically deformed metal crystal under an optical microscope, which reveals lots of surface steps, called slip traces. These are the intersection lines between slip planes and the sample surface. Based on this understanding, it is tempting to relate the yield stress of

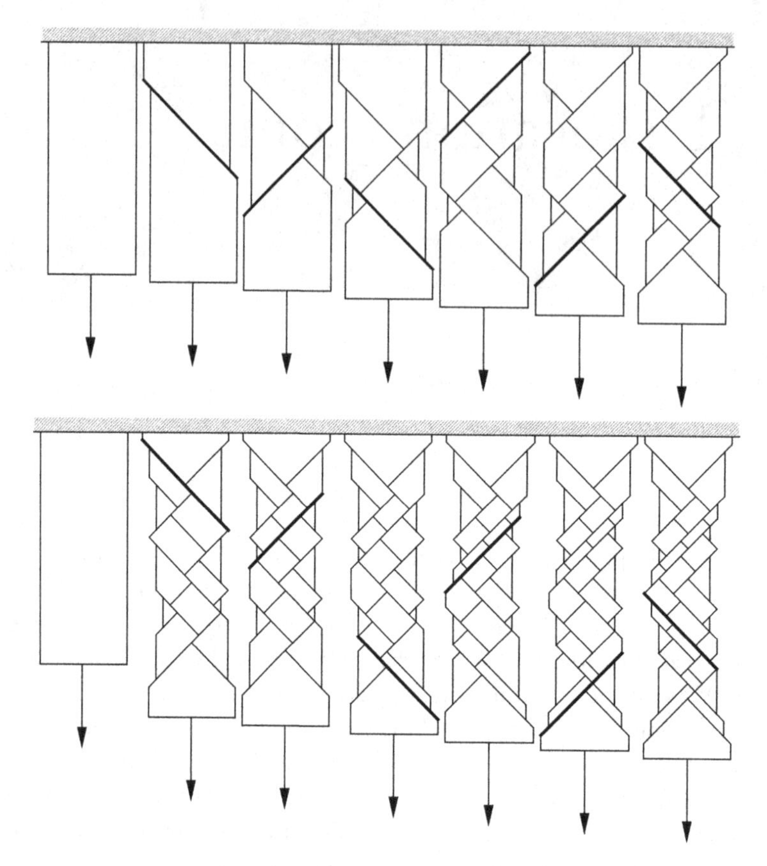

Figure 8.1. Through a series of shear events on two sets of slip planes, the original sample gradually becomes longer and thinner. The new slip plane in each snapshot is emphasized as a thick line.

a metal, at which slip initiates between adjacent crystallographic planes, and the interatomic forces between the atomic planes. In the 1920s Frenkel did this analysis and was the first to show that the shear stress needed to cause slip to occur along a slip plane in a perfect crystal was much, much greater than the experimentally observed yield stress. This discrepancy eventually led to the discovery of the crystal dislocation. The idea of a crystal dislocation was introduced in 1934, almost simultaneously, by three different people: G. I. Taylor [51], E. Orowan [52], and M. Polanyi [53]. It was suggested as a way of explaining why ordinary metal crystals were so weak in shear. The first direct observation of dislocations in a crystal was not made until the late 1950s [54] after the invention of the transmission electron microscope (TEM).

8.1.1 Frenkel model of slip in perfect crystal

Consider the atoms on either side of a crystallographic slip plane, as shown in Fig. 8.2. We wish to estimate the shear stress that would be needed to cause the plane of atoms on the top to slide past the plane of atoms below. One may think of the lower plane of atoms as being fixed to the crystal below and held in place, while the upper plane of atoms is fixed to the crystal above and is free to slide. We expect the shear stress τ, which holds the upper half in equilibrium at

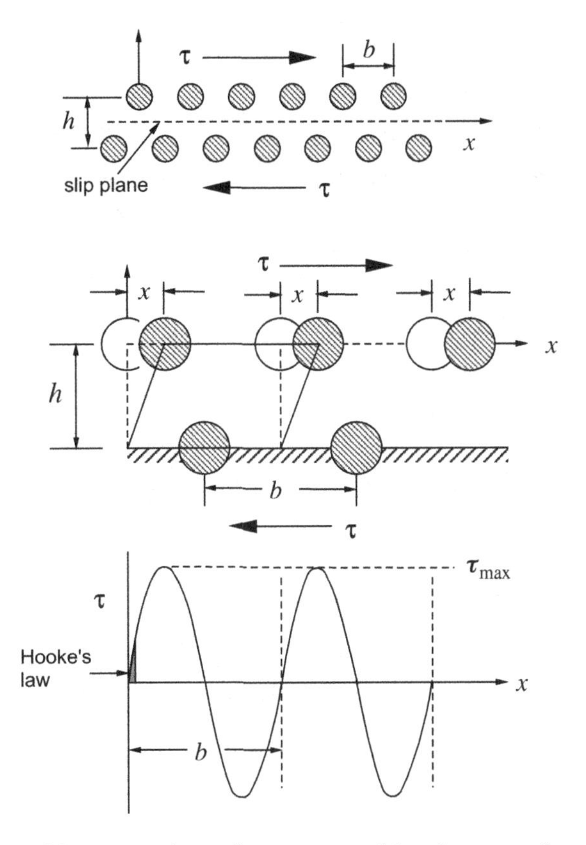

Figure 8.2. The process of shearing one layer of atoms on top of the other in a perfect crystal. The maximum shear stress required is called the ideal shear strength, τ_{max}.

any arbitrary shear displacement, x, to be a periodic function of x with periodicity b. This is because the perfect crystal arrangement is restored after a shearing displacement $x = b$. As a simple estimate, we can assume that the shear stress τ varies approximately sinusoidally with x, as shown. Notice that when the layer of atoms has been displaced by an amount $x = b/2$, then the upper plane of atoms will be balanced right on top of the lower layer. No shear stress is needed to maintain this state but the system is, of course, unstable. Negative shear stresses are needed to hold the upper layer in place when $x > b/2$ because the atomic layer is seeking its new equilibrium position at $x = b$. Let τ_{max} be the maximum shear stress that could be supported by the adjacent atomic planes. Then,

$$\tau = \tau_{max} \sin\left(\frac{2\pi x}{b}\right). \tag{8.1}$$

In this simple model, τ_{max} is reached at $x = b/4$.

When x is very small, Eq. (8.1) can be approximated as a linear function of x:

$$\tau \approx \tau_{max} \frac{2\pi x}{b}. \tag{8.2}$$

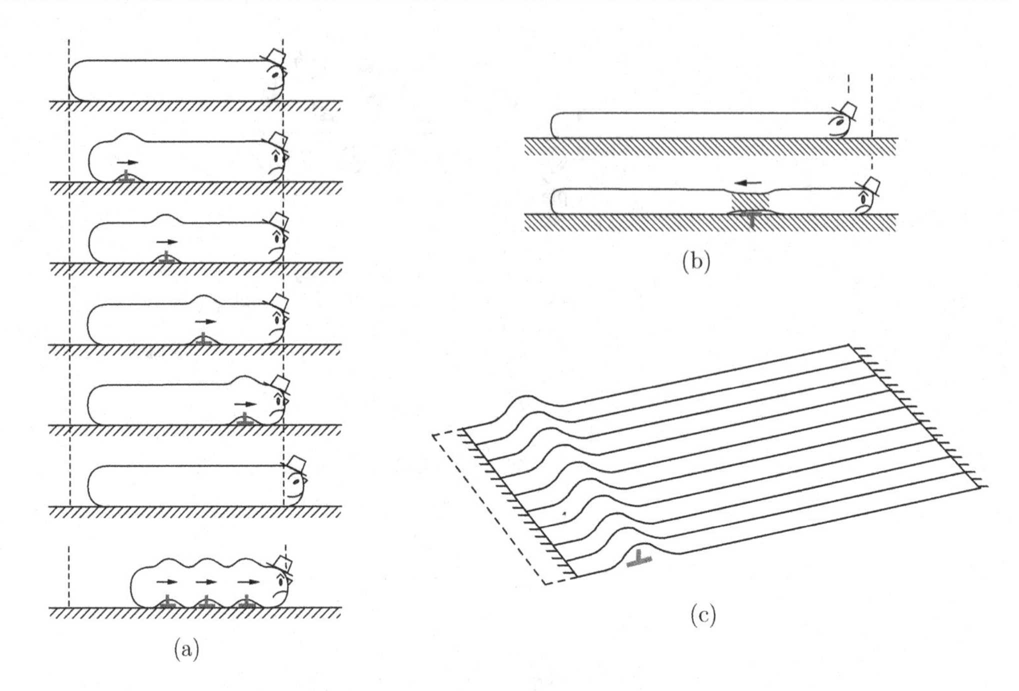

Figure 8.3. (a) Movement of a caterpillar by "dislocation-like" kink motion. The speed of the caterpillar is determined by the kink velocities and the number of kinks at any one time. (b) Movement of a worm by the propagation of a tensile defect. (c) Movement of a rug by dislocation-like motion. The $\perp$ sign is the symbol typically used to indicate an edge dislocation.

At the same time, a small shear displacement x corresponds to an elastic shear strain, $\gamma = x/h$, which leads to a shear stress τ through the elastic shear modulus μ, i.e.

$$\tau = \mu\gamma = \mu\frac{x}{h}. \tag{8.3}$$

Combining Eqs. (8.2) and (8.3), we have

$$\tau_{\max} \approx \frac{\mu}{2\pi}\cdot\frac{b}{h} \approx \frac{\mu}{2\pi} \approx \frac{\mu}{10}. \tag{8.4}$$

This is Frenkel's estimate of the theoretical shear strength of a perfect crystal. The prediction is many orders of magnitude greater than the critical shear stress needed to cause plastic flow in a typical metal crystal. For example, the shear modulus of a copper alloy (Red Brass C83400) is $\mu = 37$ GPa, while its yield strength is $\sigma_Y = 70$ MPa, so that its critical shear stress τ_c at yield is about 35 MPa. Therefore $\tau_c \approx \mu/1000$ for this alloy, which is two orders of magnitude lower than the prediction of $\tau_{\max}$ in Eq. (8.4). For pure copper, the yield stress is so low ($\sim$1 MPa) that τ_c is about four orders of magnitude smaller than the $\tau_{\max}$ predicted by Eq. (8.4). This huge discrepancy led to the idea of crystal dislocations.

8.1.2 Slip by dislocation movement

If we think more about it, the notion that slip in a crystal occurs all at once, as required in the Frenkel calculation, is actually at odds with common experience. Anyone who has watched an ordinary caterpillar move along a surface knows this. As shown in Fig. 8.3a, a caterpillar

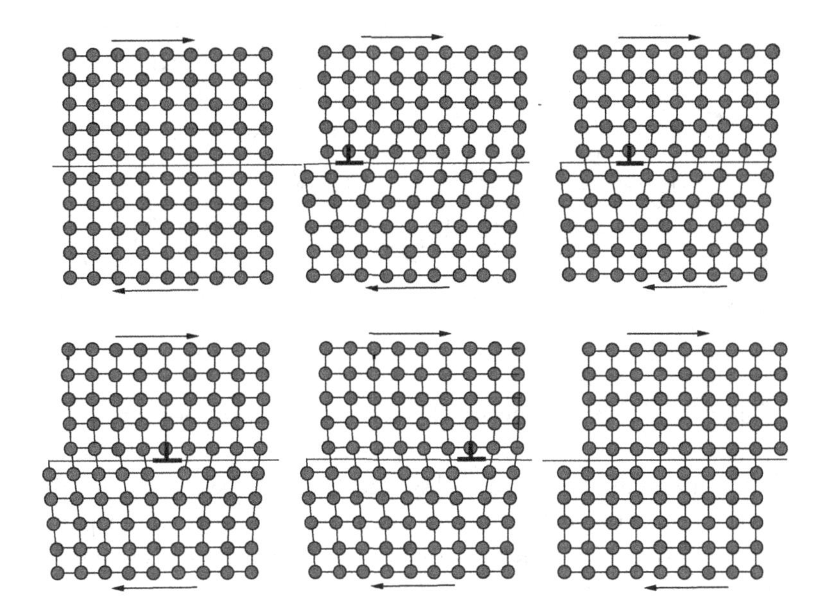

Figure 8.4. Slip in a crystal by the motion of an edge dislocation (⊥).

moves by inching its rear forward slightly, thereby creating a kink in the caterpillar. As the kink gradually moves forward, the whole caterpillar is advanced by the amount determined by the kink. Eventually the kink pops out the front and the caterpillar has moved a unit distance forward. Normally caterpillars do not wait for one kink to move the entire length of the caterpillar; instead several kinks are initiated before the first one pops out the front.

Worms move in a similar way, though they do so by stretching out their neck, creating a tensile defect that travels to the rear of the worm and allows forward movement. This is illustrated in Fig. 8.3b.

Still a better piece of ordinary experience involves the movement of a rug on a floor. If a rug on a floor is to be moved a small distance it is unwise to try to move it all at once. If you tug on one end of the rug it will be hard to move and will likely move an uncontrolled distance once it starts to slide. Instead we all know that the way to do it is to pick up one edge of the rug and move it a predetermined distance by creating a kink in the rug, as shown in Fig. 8.3c. Then by standing on the moved part of the rug and giving the kink a slight nudge, the kink will move easily to the other edge of the rug, causing the entire rug to move by the predetermined distance. The kink in the rug is perfectly analogous to the line defect we call an edge dislocation. The rug represents the upper half of a crystal while the floor represents the lower half.

These simple examples show how shear takes place gradually and not all at once. These ideas must have led Taylor, Orowan, and Polanyi to apply the same reasoning to slip in a crystal. Figure 8.4 illustrates show the shear between two planes of atoms in a simple cubic crystal can be achieved without having the slip occurring all at once. In this illustration, an edge dislocation forms near the left surface, so that slip has occurred between the atomic layers to the left of the dislocation (⊥), but the atomic layers to the right of the dislocation have not slipped. It takes a much smaller stress than τ_{max} to move the edge dislocation, because only a few atomic bonds need to be broken and re-formed. As the dislocations moves to the right, more area on the atomic planes experiences slip. When the dislocation escapes from the right surface, the entire

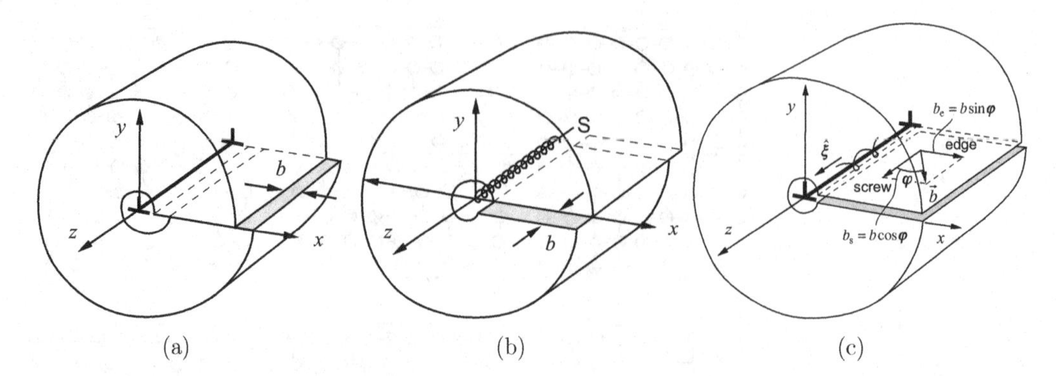

Figure 8.5. Volterra dislocations. (a) Edge dislocation (⊥). (b) Screw dislocation (S). (c) Mixed dislocation.

upper half of the crystal is now displaced by one lattice vector to the right relative to the bottom half. The end result is the same as one shear step illustrated in Fig. 8.2. However, the required shear stress τ_c is much less than $\tau_{\max} \approx \mu/10$ because the slip does not occur simultaneously across the entire plane.

8.2 Examples of dislocations

8.2.1 Dislocations in a continuum

Long before crystal dislocations were postulated or observed, the concept of a dislocation in a continuum body was well established as a certain kind of distortion in an elastic medium. These are called Volterra dislocations after the Italian mathematician Vito Volterra who first explored their properties. A brief description of Volterra dislocations is useful because it leads to a better understanding of some of the properties of crystal dislocations.

Consider an elastic medium having a cylindrical shape, with the cylindrical axis along the z direction, as shown in Fig. 8.5. A cut is made on the plane connecting the cylindrical axis and the cylindrical surface. The two surfaces created by the cut are displaced by a vector $\vec{b}$ relative to each other, and then glued together. The result is an edge dislocation if $\vec{b}$ is perpendicular to the cylindrical axis, as shown in Fig. 8.5a. A screw dislocation is created if $\vec{b}$ is parallel to the cylindrical axis, as shown in Fig. 8.5b. When the slip vector $\vec{b}$ between the two surfaces is neither parallel nor perpendicular to the dislocation line, the dislocation is called *mixed*, as shown in Fig. 8.5c. Let φ be the angle between $\vec{b}$ and the dislocation line; φ is called the *character angle* of the dislocation. A mixed dislocation has both edge component ($b_e = b \sin \varphi$) and screw component ($b_s = b \cos \varphi$). For an edge dislocation $\varphi = 90°$ and for a screw dislocation $\varphi = 0°$.

We note that before the dislocation is introduced, a small cylindrical hole is created at the center of the elastic medium. The subsequently created dislocation line is located inside the cylindrical hole. The small cylindrical hole not only makes it easier to draw the displacement field associated with the dislocation, it also avoids a more fundamental conceptual problem. As we will see in Chapter 8, the Volterra dislocation has a singular elastic field, with the stress and strain varying as r^{-1}, where r is the distance from the center of the dislocation line. By creating the small cylindrical hole around the dislocation line, the singularity on the dislocation line

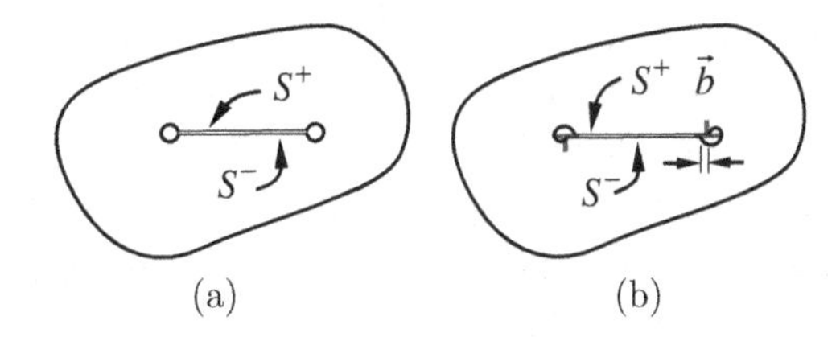

Figure 8.6. Formation of a Volterra dislocation dipole in a two-dimensional elastic medium. (a) A cut is introduced in the medium. (b) The upper surface S^+ of the cut is displaced by $\vec{b}$ relative to the lower surface S^-.

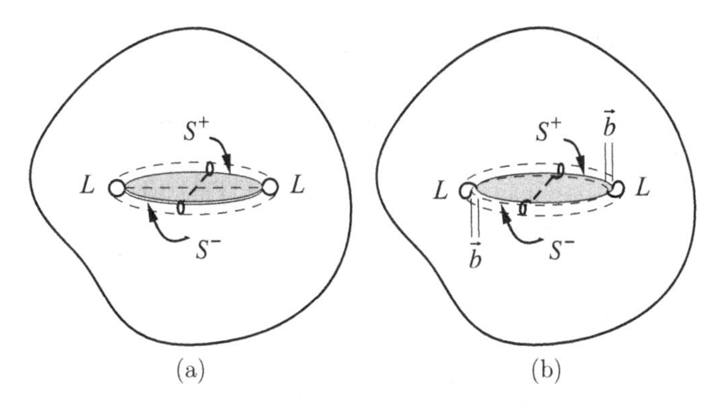

Figure 8.7. Formation of a Volterra dislocation loop in a three-dimensional elastic medium.

is removed, allowing us to discuss the properties of the Volterra dislocation without having to deal with infinities. We will continue to use the Volterra model when discussing the stress–strain fields around the dislocations in Chapter 9. The Volterra model provides an accurate description of the elastic field of a dislocation in a crystal far away from the dislocation line ($r \gg b$).

Let us return to the discussion of creating dislocations in an elastic medium through cuts and displacements. If the elastic medium is very large and the cuts do not extend all the way to the external surface, then the result is a dislocation dipole (in two dimensions) or a dislocation loop (in three dimensions). Figure 8.6 shows the creation of an edge dislocation dipole by the "cut-and-shift" operation inside a two-dimensional elastic medium. The upper and lower free surfaces created by the cut are labeled S^+ and S^-, respectively, as shown in Fig. 8.6a. The upper surface is then displaced relative to the lower surface to the right by $\vec{b}$, and the two surfaces are then glued together, as shown in Fig. 8.6b. The end result is a Volterra dislocation dipole, consisting of two oppositely signed edge dislocations.

Figure 8.7 shows the creation of a dislocation loop by the "cut-and-shift" operation inside a three-dimensional elastic medium. In this example, the internal surface S has a disc shape and has a perimeter L. After the cut is introduced at S, two surfaces, S^+ and S^- are created. The upper surface S^+ is then rigidly displaced relative to the lower surface S^- by a small amount, $\vec{b}$, and the two surfaces are then glued together. The result is a Volterra dislocation loop L at the perimeter of the surface S. In this example, the displacement vector $\vec{b}$ is parallel to the surface S, so that no material needs to be inserted or removed before S^+ and S^- can be glued back together. The resulting dislocation loop is called a glide loop. If, on the other hand, $\vec{b}$ is not parallel to the surface S, then material needs to be inserted or removed to fix any gap or material overlap that may have been created when the two surfaces were displaced relative to each other.

For the dislocation loop shown in Fig. 8.7b, where the local line direction is perpendicular to $\vec{b}$, the dislocation is of the edge type. At the locations where the local line direction is parallel (or antiparallel) to $\vec{b}$, the dislocation is of the screw type. However, at most places along the dislocation line, the line direction is neither perpendicular nor parallel to $\vec{b}$. The dislocation line is of the mixed type at these locations.

From the examples above, we see that the dislocation line is the boundary between slipped and unslipped areas. The definition of the Volterra dislocation given here leads to certain rules about the properties of dislocations:

(1) If we call the displacement vector $\vec{b}$, the Burgers vector[1] (after J. M. Burgers of screw dislocation fame), then we see that the Burgers vector must be the same for all parts of the dislocation line; that is, the same displacement discontinuity between S^+ and S^- applies to all parts of the line L.
(2) A dislocation may not end within a solid. The dislocation line lies along the perimeter of a surface, which must be continuous, so it may not end.

8.2.2 Dislocations in a crystal

Dislocations in a crystal are similar to Volterra dislocations in the sense that they are both boundary lines between slipped and un-slipped areas. However, there are several important differences between the two.

While the Volterra dislocations in a continuum are associated with arbitrary slip vectors, crystal dislocations are associated with slip vectors that take only discrete values associated with the crystal lattice. Except for a possible sign difference, the slip vector is the same as the dislocation's Burgers vector, which will be properly defined in the next section.

For *perfect dislocations*, as sketched in Fig. 1.10a, the Burgers vector is a multiple of a lattice translation vector. That is, if the crystal is to be perfect (except for small elastic distortions) all around the dislocation, then the relative displacement discontinuity between S^+ and S^- must be a multiple of a lattice translation vector. If the displacement is not an integral number of lattice translations, then a fault would be created in the crystal. The result is a *partial dislocation*, as sketched in Fig. 1.10b. We will see, in Chapter 9, that the elastic energy of the dislocation scales with the square of the length of the Burgers vector, so that it is energetically favorable for dislocations to have the shortest possible Burgers vector. For perfect dislocations, the Burgers vector prefers to be the shortest lattice translation vector. For partial dislocations, the Burgers vector can be even shorter.

The atomic arrangements around an edge and a screw dislocation are illustrated in Fig. 1.11a, b, respectively. Such pictures are convenient to illustrate the atomic arrangements of straight dislocations. For curved dislocation lines, it is useful to look at the atomic arrangements immediately above and below the slip plane, as shown in Fig. 8.8. Starting from the bottom of the figure and going upward, the dislocation gradually turns from the screw orientation to the edge

1 The Burgers vector will be properly defined in Section 8.3. The Burgers vector of the dislocation shown in Fig. 8.7 is either $\vec{b}$ or $-\vec{b}$ depending on the direction of the line sense vector.

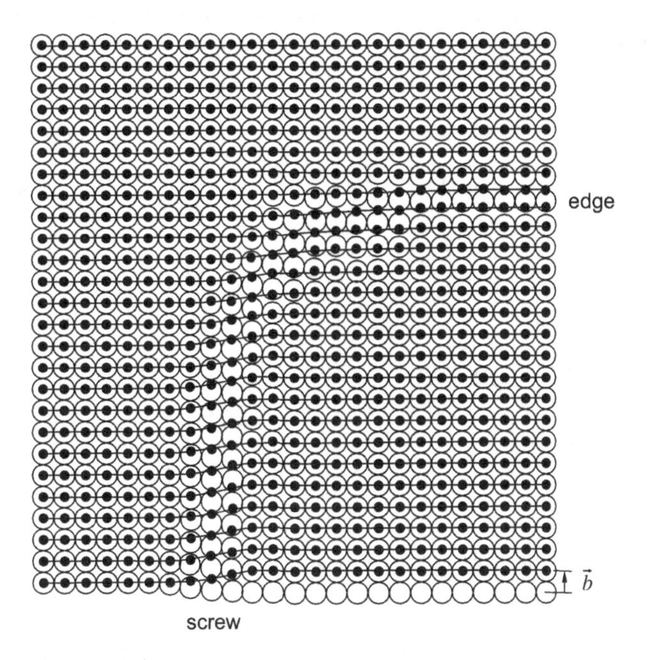

Figure 8.8. Atomic arrangements in a curved dislocation in a simple-cubic crystal. The character of the dislocation changes gradually from a right-handed screw (RHS) at the bottom of the figure to a pure edge dislocation at the right side. Dots represent atoms above the slip plane, and unfilled circles represent atoms below the slip plane.

orientation. The dislocation is of the mixed type where the dislocation line is neither parallel nor perpendicular to the Burgers vector $\vec{b}$.

We have mentioned that the Volterra dislocation is a line singularity in an elastic continuum and the infinity is often avoided by making a cylindrical hole around the dislocation line. On the other hand, dislocations in a real crystal do not contain any singularity, nor do they require a hollow core.[2] Instead, linear elasticity theory breaks down in the region close to the center of a crystal dislocation ($r \approx b$). The non-linear interaction between the atoms leads to a dislocation core structure that depends on the types of both the material and the dislocation. This will be discussed in Chapter 12.

Dislocations are called line defects because they extend along lines (not necessarily straight) in the crystal. Imaging by transmission electron microscopy (TEM) is perhaps the best way to illustrate the line nature of dislocations. Figure 8.9 shows TEM micrographs of dislocations in NiAlTi crystals. What is observed in these TEM micrographs are actually the elastic distortions around the dislocation. The crystal is oriented to suitable orientations so that electrons can easily pass through the sample. However, near the dislocation line, the atomic planes are significantly rotated away from their original orientation, as can be seen from Fig. 8.4. These atomic planes strongly diffract the incoming electron beam, leaving a dark contrast on the observing screen [7, 9].

2 Dislocations in some crystals with large lattice constants do contain a hollow core [55] as sketched in Fig. 8.5 (also see Exercise problem 9.4).

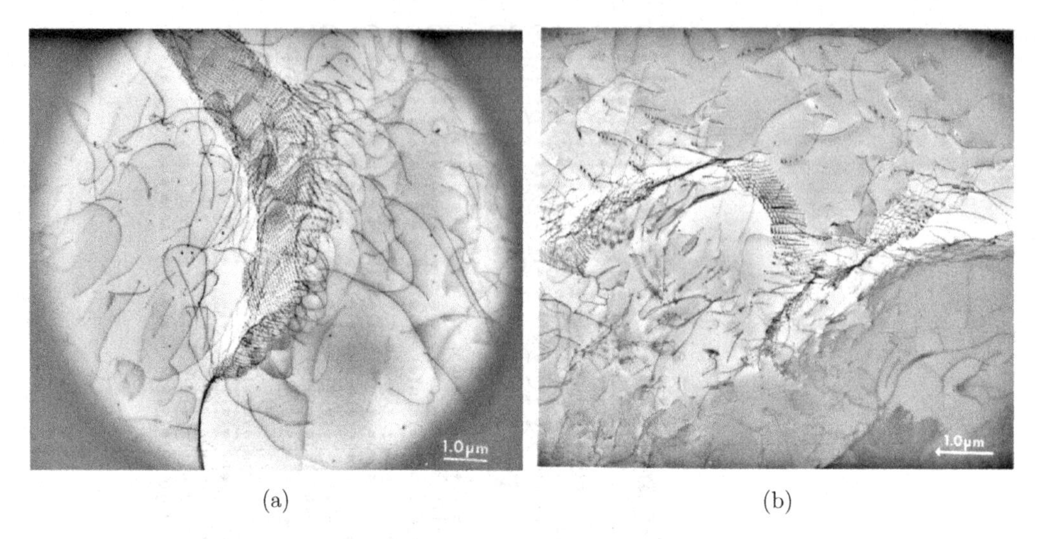

(a) (b)

Figure 8.9. Bright field TEM micrograph of a single crystal of NiAl–2.5Ti deformed under creep conditions at 1200 °C. The lines are the imaged dislocations. They terminate at the top and bottom surfaces of the foil through which the imaging electrons have passed. The grids are organized aggregates of dislocations that constitute low angle grain boundaries [56]. Images provided by Dr. Anita Garg, NASA Glenn Research Center, and used with permission.

Most of the dislocations in these micrographs are short line segments terminating at the two free surfaces of the metal foil, which must be thin enough to be transparent to the electron beam. Figure 8.9 also shows that dislocations sometimes self-organize into ordered arrays; these arrays are equivalent to low angle grain boundaries, and will be discussed in Chapter 14.

8.2.3 Dislocation density

Because dislocations are line objects, the dislocation density is defined as the total dislocation line length L divided by the volume V of interest,

$$\rho = \frac{L}{V}. \tag{8.5}$$

Consequently, the dislocation density ρ has the dimension of inverse area. (In comparison, the concentration of point defects has the dimension of inverse volume.) The dimension of the dislocation density can be illustrated by considering a randomly oriented plane in a crystal and the intersection points between this plane and the dislocation lines. The dislocation density ρ is proportional to the areal density of the intersection points on the plane, which has the dimension of inverse area.

Unlike the density of point defects, the dislocation density in a crystal almost always greatly exceeds the value to be expected at thermal equilibrium. Dislocations are usually created during the initial formation of the crystal, though for silicon and other semiconductors very large crystals are commonly grown with no dislocations at all. When dislocations are present they are very difficult to remove by annealing and can easily multiply by several orders of magnitude if the crystal is plastically deformed, especially for metals that deform easily (see Section 8.5).

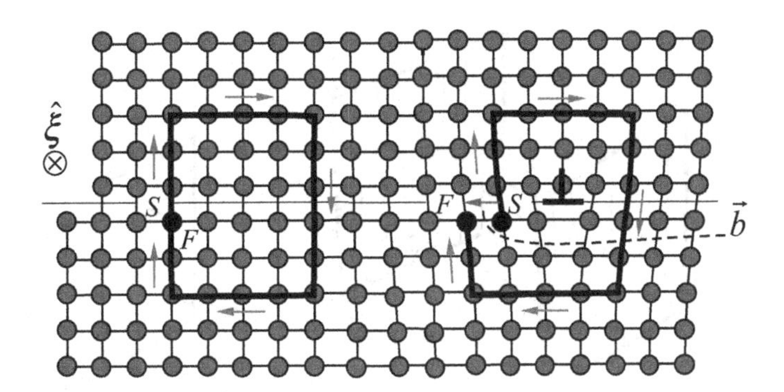

Figure 8.10. Burgers circuit and Burgers vector using the RH/SF convention.

The dislocation density in a well-annealed crystal is typically in the range of 10^9–10^{10} m^{-2}. In a cold worked metal, the dislocation density can get in the range of 10^{14}–10^{16} m^{-2} [57, 58].

8.3 Burgers circuit and Burgers vector

Two vectors are needed to completely specify the type (and sign) of a dislocation: one is called the *sense vector*, $\hat{\xi}$, a unit vector parallel to the dislocation line at a particular point, and the other is called the *Burgers vector*, $\vec{b}$, which is defined using the *Burgers circuit*.

8.3.1 Burgers circuit

There are serval conventions in the literature to draw the Burgers circuit and to define the Burgers vector of a dislocation. Any one may be used as long as it is used consistently everywhere in the analysis. In this book, we use the so-called *right hand, start–finish* (RH/SF) convention, to be explained below.

Consider a crystal shown in Fig. 8.10, and choose the direction of the sense vector to point into the page. Looking at the crystal in the direction of the sense vector, we first create a Burgers circuit in the perfect crystal region. The Burgers circuit is a sequence of discrete steps from one lattice point to the next in a clockwise direction that closes to form a complete loop when drawn on a perfect crystal. Here we have followed the right hand (RH) convention because when the thumb of the right hand points along the sense vector, the curved fingers points along the direction of the Burgers circuit. In Fig. 8.10, the Burgers circuit involves discrete steps from one lattice point to the next, first by three steps north, then by four steps east, then by five steps south, and then by four steps west and, finally, by two steps north.[3] We denote the start of the circuit as S and the finish as F, which are located at the same lattice point in a perfect lattice. When the same circuit is drawn around the dislocation, a closure failure is created, where the finish of the circuit does not coincide with the start. By convention we draw the Burgers vector $\vec{b}$ from the starting point S to the finishing point F. Hence we have followed the right hand, start–finish (RH/SF) convention in the above definition of the Burgers vector $\vec{b}$.

3 Here we have used the notation commonly used in cartography (i.e. maps), where north is up, east is right, south is down, and west is left.

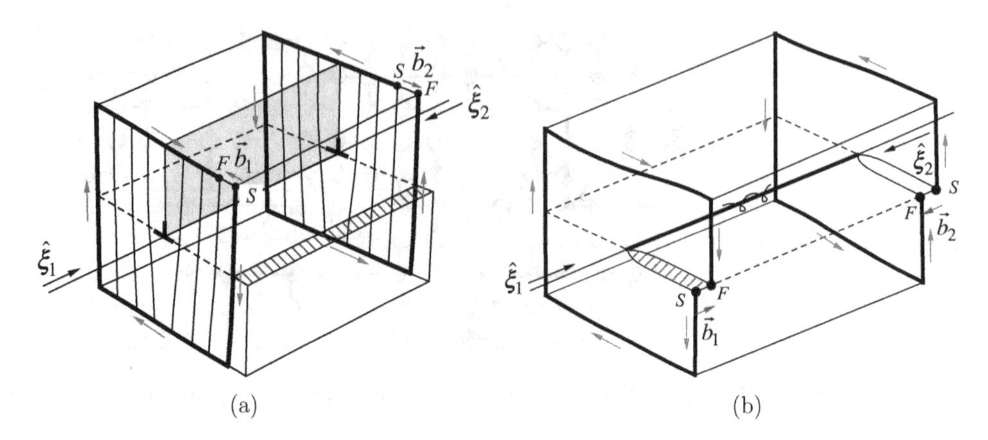

Figure 8.11. Sense vector and Burgers vector for (a) an edge dislocation and (b) a right-handed screw dislocation, using the RH/SF convention.

8.3.2 Sign of edge and screw dislocations

The character of the dislocation is defined unambiguously by the sense vector and the Burgers vector together, regardless of which way the sense vector points. This is shown in Fig. 8.11 for both pure edge and pure screw dislocations. Notice that when the sense vector is reversed, the resulting Burgers vector, using the RH/SF convention, is also reversed. Therefore, whichever way the sense vector is assigned, the cross product of the sense vector and the Burgers vector always points to the extra half-plane of atoms that defines the edge dislocation:

$$\hat{\xi} \times \vec{b} \quad \Rightarrow \quad \text{extra half-plane.}$$

Note also that for a pure edge dislocation the sense vector and Burgers vector are perpendicular to each other so that their dot product is zero: $\hat{\xi} \cdot \vec{b} = 0$.

The plane that contains both vectors $\hat{\xi}$ and $\vec{b}$ is called the *glide plane* of the dislocation (see Section 8.4). By convention, we can choose one side of the glide plane (e.g. the up-side) as the "positive" side. The edge dislocation with the extra half-plane on the positive side of the glide plane is called the positive edge dislocation. The edge dislocation with the extra half-plane on the other side of the glide plane is called the negative edge dislocation. For example, the edge dislocation shown in Fig. 8.8 has the extra half-plane coming out of the plane of the diagram. If we define the direction pointing out of the plane of the diagram as positive, then this dislocation is a positive edge dislocation.

The character of a pure screw dislocation is, of course, also independent of the choice of the direction of the sense vector. Figure 8.11b shows the Burgers circuits around a right-handed screw (RHS) dislocation. Whichever way the sense vector $\hat{\xi}$ is chosen, the Burgers vector $\vec{b}$ always points in the same direction as $\hat{\xi}$, i.e. $\hat{\xi} \cdot \vec{b} > 0$. This dislocation is called RHS because when the four fingers of the right hand curve around the Burgers circuit, the thumb points in the direction of the Burgers vector. The screw dislocation in Fig. 8.8 is a RHS. A left-handed screw (LHS) dislocation would be characterized by $\hat{\xi} \cdot \vec{b} < 0$.

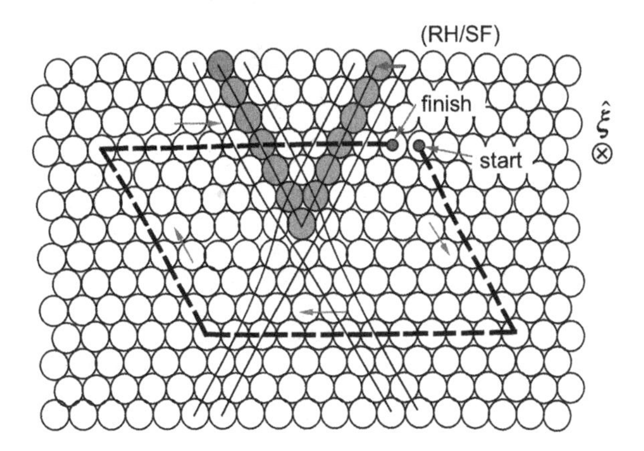

Figure 8.12. Edge dislocation in a close-packed crystal showing a Burgers circuit and resulting Burgers vector using the RH/SF convention. The shaded atoms indicate two possible "extra half-planes."

The designation of an edge dislocation as positive or negative depends on the choice of a coordinate system relative to the crystal, i.e. on the direction at which we look at the dislocation. Suppose we are examining a crystal containing a positive edge dislocation but accidently dropped the crystal. When we pick up the crystal to examine it again, we happen to have rotated the crystal around the dislocation line by 180°. Then the same edge dislocation becomes a negative edge dislocation to us. On the contrary, the designation of screw dislocations as RHS or LHS has no such ambiguity. As long as we follow the RH/SF convention, the RHS is characterized by $\hat{\xi} \cdot \vec{b} > 0$, while the LHS is characterized by $\hat{\xi} \cdot \vec{b} < 0$. This means that even if we dropped the crystal and lost its original orientation, when we examine the crystal again, the screw dislocation will still be designated by RHS or LHS in the same way as before.

While it is tempting to always think of the edge dislocation as the termination of a definite half-plane of atoms, Fig. 8.12 shows that this is too restrictive. In a close-packed crystal structure shown here, it is not clear which of the two shaded half-planes is the "extra half-plane" associated with the edge dislocation. On the other hand, the Burgers circuit analysis unambiguously shows that the Burgers vector $\vec{b}$ points to the left when the sense vector $\hat{\xi}$ points into the page. Therefore $\hat{\xi} \times \vec{b}$ points north, so that the edge dislocation in Fig. 8.12 is of the same sign as the one shown in Fig. 8.10.

8.3.3 Frank's rule

Because a dislocation is the boundary between slipped and unslipped areas, the dislocation line cannot end by itself inside the crystal. However, it can end at other dislocations. Figure 8.13 shows the creation of two dislocation loops, 1 and 2, with Burgers vectors, $\vec{b}_1$ and $\vec{b}_2$, having a common line, AB. This configuration can be created by starting from two small dislocation loops, which expand until they touch at line AB. Therefore, we may think of line AB as the place where two dislocation lines coincide. Alternatively, we may think of the dislocation extending from A to B as a different dislocation, say dislocation 3. For the choice of sense vectors shown in Fig. 8.13, the Burgers vector of dislocation 3 is $\vec{b}_3 = \vec{b}_1 + \vec{b}_2$. The points where the dislocations

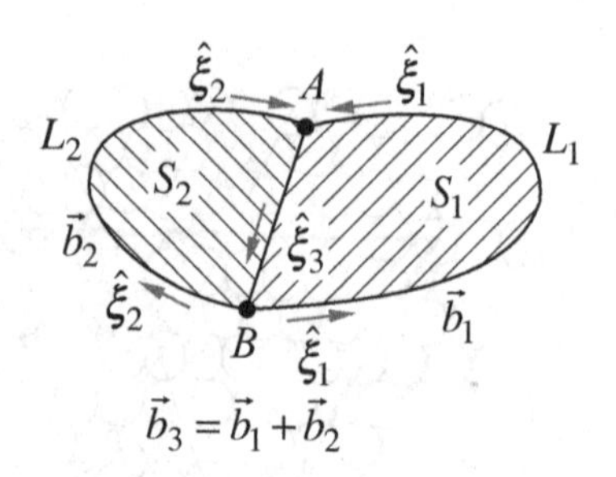

Figure 8.13. Two dislocation loops forming a junction.

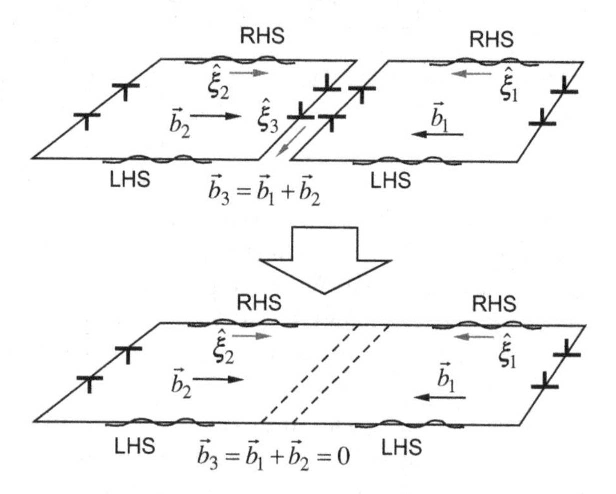

Figure 8.14. Annihilation of two oppositely signed edge dislocations.

are joined are called nodes. This shows that dislocations may end at the nodes they form with other dislocations.

Note that in the above example, the sense vectors for loop L_1 and loop L_2 are chosen to coincide along segment AB. This is the reason for $\vec{b}_3 = \vec{b}_1 + \vec{b}_2$. If the sense vector for loop L_2 is chosen to be counterclockwise, so that along segment AB, it is opposite to the sense vector of L_1, then we must have $\vec{b}_3 = \vec{b}_1 - \vec{b}_2$. In the special case where L_1 and L_2 are of the same type, i.e. $\vec{b}_1 = \vec{b}_2$ when both sense vectors are chosen to be counterclockwise, then $\vec{b}_3 = 0$. This means that when two identical dislocation loops touch each other, the contacting segments annihilate and the two loops merge into a single, larger loop, as shown in Fig. 8.14.

The relationship between the Burgers vectors of multiple dislocation lines meeting at the same node can be regarded as a conservation law for the Burgers vectors. The conservation of the Burgers vector at dislocation nodes is called Frank's rule, after F. C. Frank. It is analogous to Kirchoff's law for the conservation of charge (or current) in electric circuits. The difference here is that the conserved quantity along the dislocation lines is a vector.

In Fig. 8.13, the sense vector of the three dislocations connected to node A is such that $\hat{\xi}_3$ flows out of node A, while $\hat{\xi}_1$ and $\hat{\xi}_2$ flow into node A. For such choices of the sense vectors, the conservation law of the Burgers vector reads: $\vec{b}_3 = \vec{b}_1 + \vec{b}_2$. However, if we reverse the sense

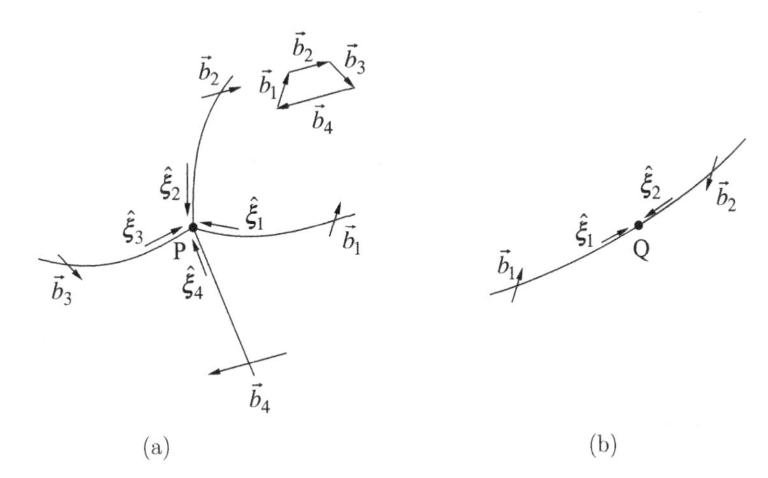

Figure 8.15. (a) Frank's rule applied to a point P where four dislocation lines merge. (b) Frank's rule applied to a point Q on the dislocation line.

vector of dislocation 3 to also flow into node A, then the direction of $\vec{b}_3$ should reverse, and the conservation law of the Burgers vector reads: $\vec{b}_1 + \vec{b}_2 + \vec{b}_3 = 0$.

In general, imagine an arbitrary number (n) of dislocation lines ending at a node. Suppose all of the sense vectors point into the node, then Franks rule can be stated as

$$\sum_{i=1}^{n} \vec{b}_i = 0, \tag{8.6}$$

that is, the sum of the Burgers vectors "flowing into the node" is zero, as illustrated in Fig. 8.15a. This is like Kirchoff's law where the total current flowing into a node is zero. This rule can be used to determine the character of a dislocation connected to a node if the characters of the other dislocations connected to that same node are known.

An almost trivial example of Frank's rule is the case of a dislocation line passing through a point Q, as shown in Fig. 8.15b. We can also consider this as two dislocation lines ending at node Q, i.e. $n = 2$. According to the convention adopted above, we can choose the sense vectors of both dislocation lines to flow into Q. In this case, Frank's rule reads: $\vec{b}_1 + \vec{b}_2 = 0$, i.e. $\vec{b}_2 = -\vec{b}_1$. This simply states the fact that the Burgers vector stays constant along the dislocation line, but reverses direction if the sense vector reverses.

As a more complex application of Frank's rule, consider the three dislocation lines connected at a node shown in Fig. 8.16a. All three dislocations make angles of 120° with each other. Supposed that two of the dislocations, (1) and (2), are right-handed screw dislocations each having Burgers vectors of magnitude b. The character of the third can be found by using Frank's rule, as shown in Fig. 8.16b. For convenience, we have chosen sense vectors of all dislocations to flow into the node. By using Frank's rule we see that the third dislocation (3) must also be a right-handed screw dislocation.

As another example, consider the three dislocation lines connected at a node shown in Fig. 8.17a. One of the dislocations, (2), is a left-handed screw dislocation and the other, (1), is a right-handed screw dislocation, each with Burgers vectors of magnitude b. The character of the third dislocation can be found by using Frank's rule, as shown in Fig. 8.17b. The third

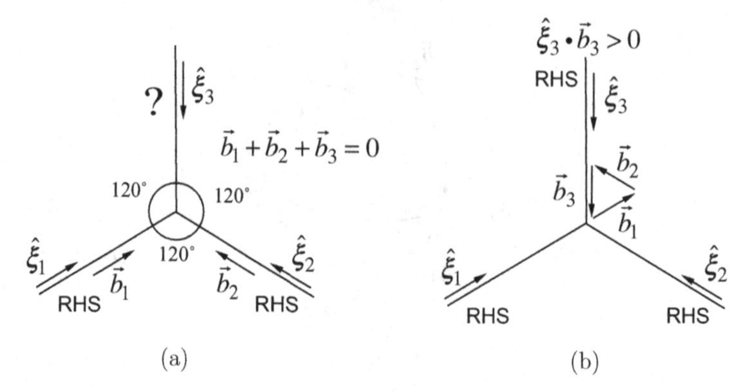

Figure 8.16. Application of Frank's rule: case 1.

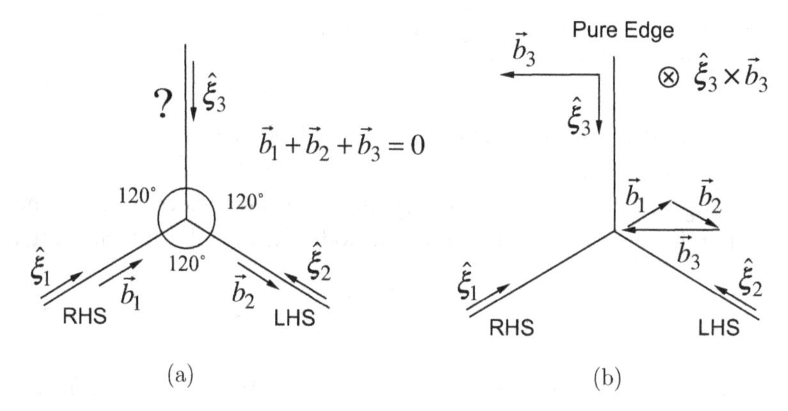

Figure 8.17. Application of Frank's rule: case 2.

dislocation, (3), turns out to be a pure edge dislocation with Burgers vector $\sqrt{3}b$. The extra-half-plane of this edge dislocation can be found from $\hat{\xi} \times \vec{b}$, which points into the plane of the diagram. Hence the extra-half-plane is beneath the plane of the diagram. These examples show how Frank's rule can be used to determine the character of unknown dislocations in a network provided the character of some of the others are known.

8.3.4 Dislocations intersecting crystal boundaries

While dislocations cannot end inside a crystal, they can end at free surfaces or at grain boundaries, as shown in Fig. 8.18. In fact, we have presented many figures where a dislocation intersects the free surface, such as Fig. 8.11.

In Fig. 8.18, the dislocation line L_2 "ends" at the surface. There is a displacement discontinuity on the surface where the slip plane S_2 intersects the free surface. When the Burgers vector of L_2 has a non-zero component along the normal of the free surface, then the displacement discontinuity can be visually identified as a surface step.[4] The surface step starts where the

4 When the Burgers vector is in the plane of the free surface, there is no surface step but the displacement discontinuity is nonetheless still present at the surface.

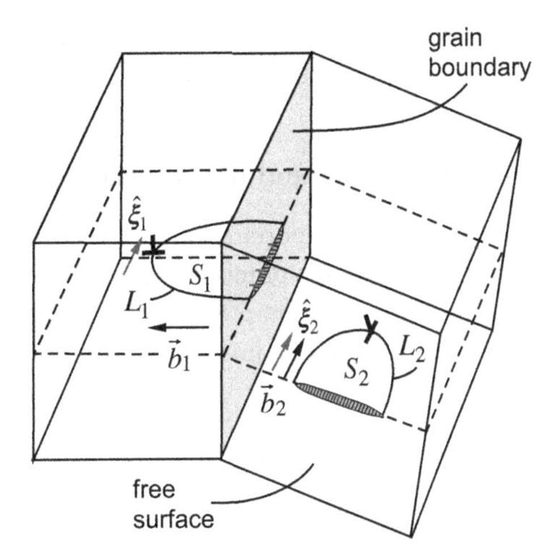

Figure 8.18. Dislocations "ending" at free surfaces and grain boundaries.

dislocation line "ends"; hence the surface step may be considered as a continuation of the dislocation line. However, we usually do not call that surface step a dislocation because there is no stress or strain singularity associated with the surface step.[5]

In Fig. 8.18, the dislocation line L_1 ends at a grain boundary. A displacement discontinuity is still present along the grain boundary where the surface S_1 intersects with the boundary. This could represent a dislocation line in a grain boundary, so that L_1 and the grain boundary dislocation form a dislocation loop. Alternatively, we may think of the dislocation line in the boundary as simply changing the structure of the boundary. If the discontinuity is confined to the line of intersection of S_1 with the boundary, then there would be a singular elastic field associated with the grain boundary dislocation. If instead the core of the grain boundary dislocation becomes more diffuse, then the singular field would be nearly lost when the dislocation is in the grain boundary. As another alternative, for special grain boundaries, the Burgers vector $\vec{b}_1$ of the crystal dislocation can dissociate into distinct but much shorter Burgers vectors of grain boundary dislocations (see Chapter 14).

8.4 Dislocation motion and slip

Dislocation motion causes plastic deformation of crystals. The most common type of dislocation motion is dislocation *glide* on its glide plane, which is called *conservative* motion because no atomic diffusion is required. All that is required is relative slip between atoms on both sides of the glide plane. Conservative glide will be discussed first in this section. Non-screw dislocations can only glide conservatively on one plane, but screw dislocations can glide conservatively from one plane to another, in a process called *cross-slip*, which will be discussed next. The *climb* motion of non-screw dislocations off their glide planes will be discussed last. Climb motion

5 In a simple continuum model, a surface step does not produce any stress field. Surface steps in real crystals can induce stresses due to reconstruction of atoms. But the associated stress field is very different from that of dislocations.

is *non-conservative* because atoms need to move either into or away from the dislocation by diffusion as the dislocation climbs.

8.4.1 Glide of edge, screw, and mixed dislocations

Dislocation glide is the most important kind of motion as it dominates plastic deformation of crystals at all but the very highest of temperatures. The glide plane (i.e. slip plane) for a gliding dislocation contains the sense vector, $\hat{\xi}$ and the Burgers vector, $\vec{b}$. Therefore, as long as the dislocation is not purely screw, the unit vector of the slip plane normal is defined by

$$\hat{n} = \frac{\hat{\xi} \times \vec{b}}{|\hat{\xi} \times \vec{b}|}. \tag{8.7}$$

We see that the slip plane for a dislocation is well defined as long as it has a non-zero edge component.

It is useful to have an understanding of how the motion of dislocations in their glide plane leads to slip deformation. We will see that the simple "rug" model of edge dislocation motion (Fig. 8.3c) is too restrictive and that we should consider a mixed dislocation moving on its slip plane. We start by considering an element of a crystal that is to be sheared along its slip plane and in the slip direction, as shown in Fig. 8.19a.

If we imagine that the shearing of the crystal element (in the direction of the applied shear stress τ) starts at the left-front surface of the element and propagates gradually over the slip plane, then when a unit shearing process is in mid-course an edge dislocation will be present in the element as shown in Fig. 8.19b. The dislocation moves from the left-front surface to the right-back surface as the two half crystals shear against each other.

It is a common mistake to associate the direction of dislocation motion with the direction of the applied shear force. Even though for the case of an edge dislocation considered here the two directions happen to coincide, we shall see shortly that this is in general not the case.

For example, if, instead, the slip of the element had started on the right-front face of the element, then progressive slip on the slip plane in the direction of shear would involve the motion of a right-handed screw dislocation, as shown in Fig. 8.19c. By starting at the front corner of the element, shearing produces a mixed dislocation, as shown in Fig. 8.19d.

The shear stress on the slip plane in the direction of the Burgers vector is the stress component that drives dislocation motion. Because the Burgers vector is invariant along the dislocation line, the same stress component drives the motion of the entire dislocation line regardless of the local line orientation. On the other hand, the dislocation motion is always in the direction locally perpendicular to the dislocation line, expanding or shrinking the slipped area. Hence the directions of the dislocation motion and shear force do not need to coincide with each other.

Figure 8.19d shows how a right-handed screw dislocation (starting from the front-left surface) gradually becomes a positive edge dislocation[6] (ending at the front-right surface) if you follow

6 Here we define the "positive" side of the glide plane to be along the positive y axis. The edge dislocation is positive because its extra-half-plane is on the positive side of the glide plane.

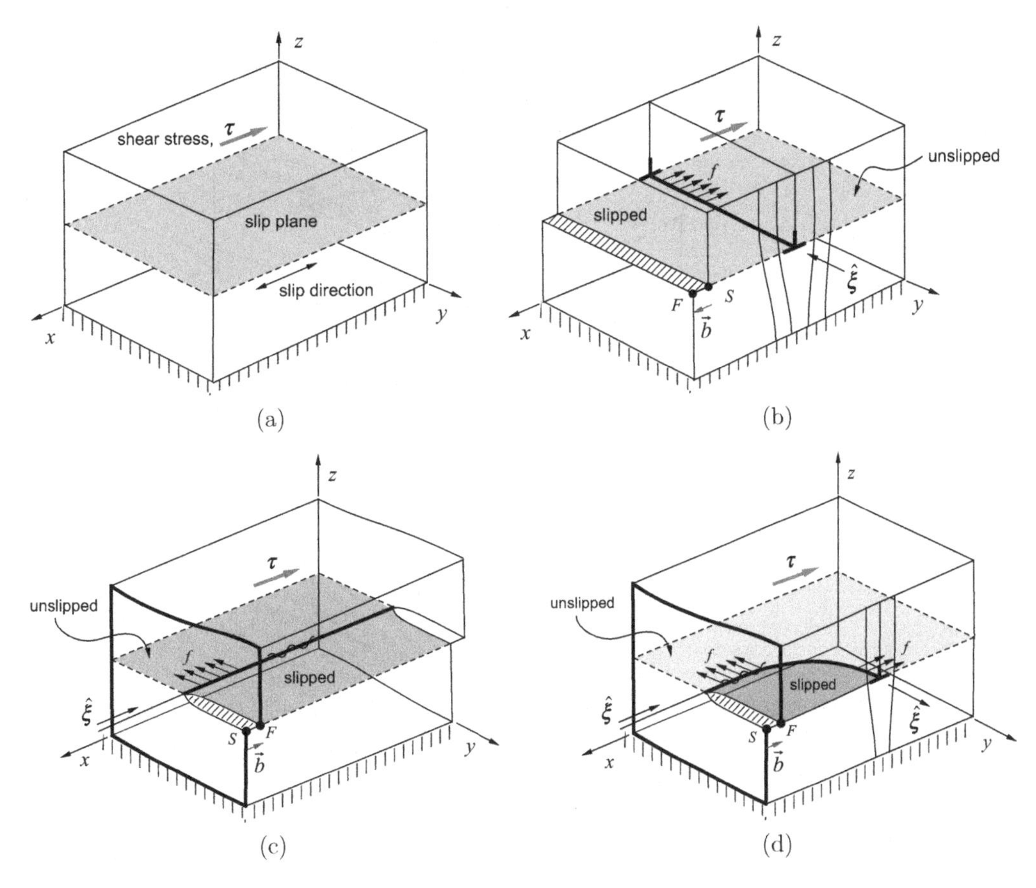

Figure 8.19. (a) Perfect single crystal element under shear. The bottom of the element is considered to be fixed. (b) Edge dislocation motion, (c) screw dislocation motion, and (d) motion of a curved dislocation line and the corresponding slip induced in the crystal. The short arrows indicate the local directions of dislocation motion.

the dislocation line from the point where the Burgers vector is parallel to the line to the point where it is perpendicular. In between these two limits, the Burgers vector is neither parallel nor perpendicular to the dislocation line, and the dislocation has a mixed character. Naturally, various kinds of mixed dislocations can be created, depending on how the element is sheared. In Fig. 8.20 we see how shearing the elements in different ways can produce all of the possible types of mixed dislocations for glide. In each of the four elements, shear initiates from a corner, leading to a curved dislocation line with edge and screw characters at the two ends and mixed dislocation in between. The top element, containing a right-handed screw (RHS) and a positive edge dislocation is the one that corresponds to Fig. 8.8.

When these elements are joined together, a standard glide loop is created, as shown in Fig. 8.21a. This standard glide loop shows how different types of dislocations move in response to the applied shear stress. It is convenient to describe the standard glide loop by viewing the loop from above, so that it is represented by a loop on a plane as shown in Fig. 8.21b. The loop expands in response to the applied shear stress.

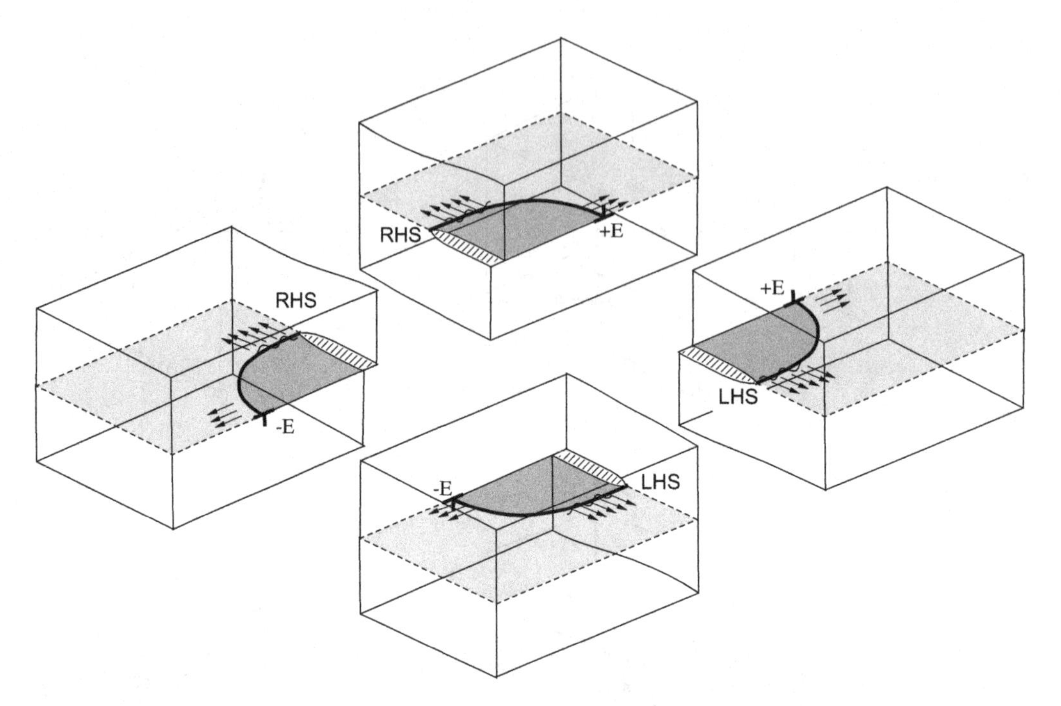

Figure 8.20. Construction of a standard glide loop from four quarter loops by shearing initiated at element corners. Despite the fact that the lines nucleate and move from different corners, they produce the same shear strain component.

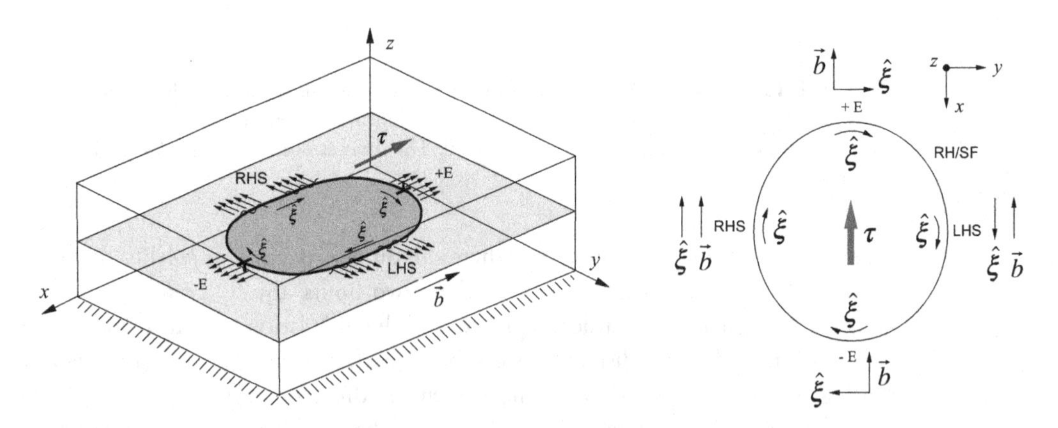

Figure 8.21. (a) Standard glide loop showing how the applied shear stress moves different kinds of dislocations. Short arrows indicate the local direction of dislocation motion driven by the shear stress τ. (b) The standard glide loop viewed from above, marked by sense and Burgers vectors using the RH/SF convention. The applied stress τ drives the loop to expand.

8.4.2 Cross-slip of screw dislocations

We now discuss the non-planar motion of screw dislocations in a crystal. Because the sense vector $\hat{\xi}$ and Burgers vector $\vec{b}$ lie along the same line for a screw dislocation ($\varphi = 0$ in Fig. 8.5c), there are an infinite number of planes that contain both of these vectors so there are an infinite

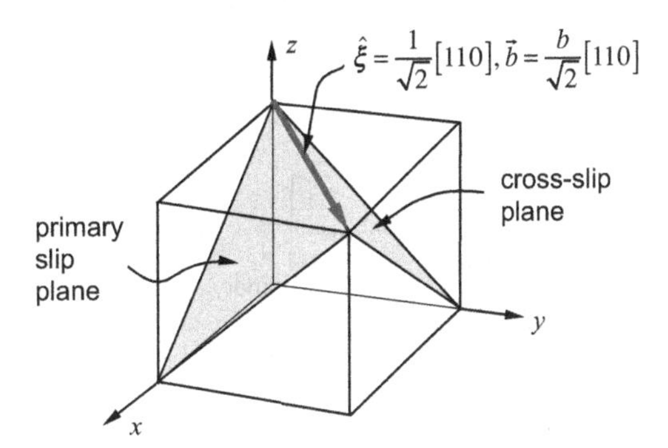

Figure 8.22. FCC cross-slip crystallography. A screw dislocation with sense vector and Burgers vector both along the [110] direction can move on two {111}-type planes.

number of possible slip planes. However, slip usually occurs only on certain crystal planes, the ones that are most widely spaced and atomically smooth. Hence screw dislocation motion is ordinarily restricted to these select planes. For an FCC crystal where the {111} planes are the most widely spaced, each screw dislocation with a Burgers vector of the type $\vec{b} = \frac{a}{2}\langle 110 \rangle$ has two {111}-type planes to move on. For example, a screw dislocation with $\vec{b} = \frac{a}{2}[110]$ can move on both $(1\bar{1}1)$ and $(\bar{1}11)$ planes, as shown in Fig. 8.22. The screw dislocation motion from one {111}-type plane to another is called *cross-slip*. We call the initial slip plane the primary slip plane and we call the plane on which the dislocation moves after cross-slip the cross-slip plane.

We shall see in Chapter 11 that perfect dislocations in FCC crystals dissociate into pairs of partial dislocations bounding an area of stacking fault on {111} planes, on which the dislocation glides. As a result, cross-slip requires the changing of the plane of dissociation from one {111} plane to another, which requires overcoming an activation barrier. Therefore, cross-slip is a rare event in FCC crystals and requires the assistance of thermal fluctuation and local stress. Several mechanisms of cross-slip are discussed in Section 11.1.10.

Figure 8.23 shows a screw dislocation that encounters an obstacle that inhibits its motion on the primary slip plane (where the resolved shear stress is highest). If the stress is sufficiently high, the dislocation might cross-slip onto the cross-slip plane to avoid the obstacle and then return to the primary slip plane (i.e. double cross-slip) once the obstacle is passed and eventually leave the crystal element.

8.4.3 Climb of non-screw dislocations

Edge dislocations, as well as any mixed dislocations with a non-zero edge component, cannot move out of their slip planes by glide. However, at sufficiently high temperatures, where diffusion can occur, edge (and mixed) dislocations can *climb* out of their slip planes. In the following, we will only discuss edge dislocations for brevity.

When an edge dislocation encounters an obstacle that inhibits its glide motion in its slip plane, it may climb up (or down) by absorbing (or emitting) vacancies from (or to) the surroundings. For up-climb of a positive edge dislocation, some of the atoms at the edge of the extra-half-plane have to be removed to the surrounding for the dislocation to climb. This can

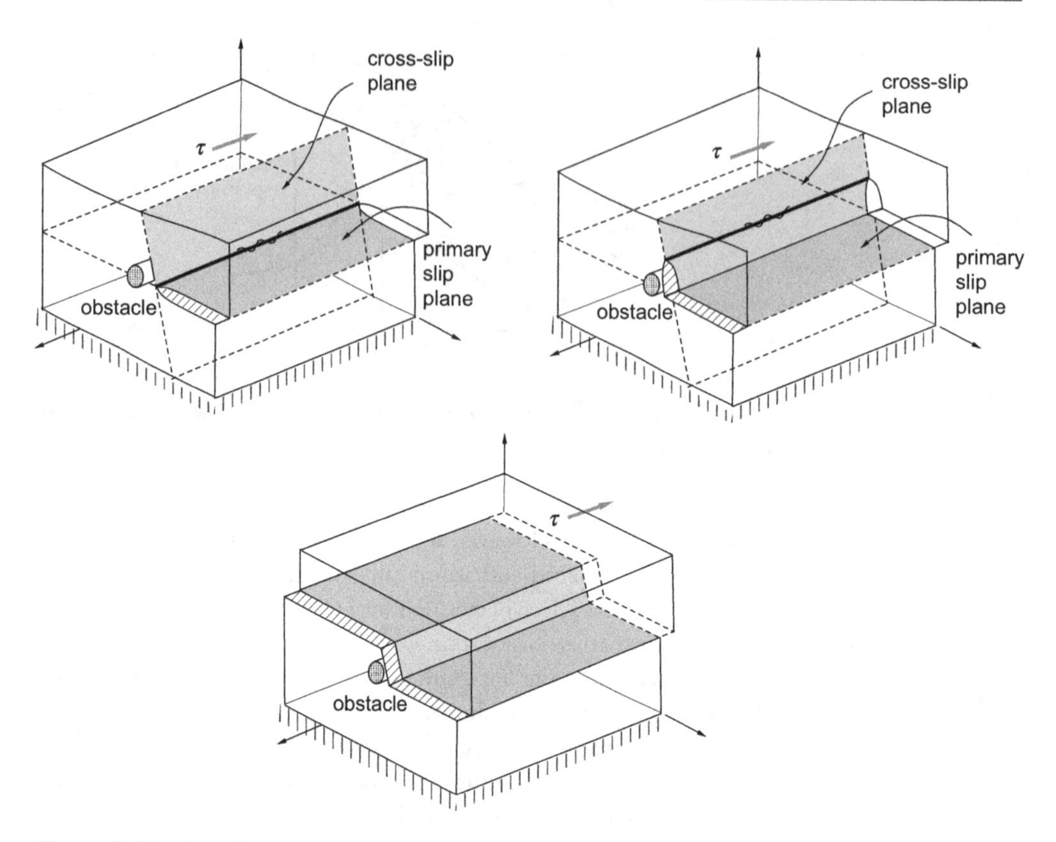

Figure 8.23. Cross-slip process to avoid an obstacle.

occur by absorbing vacancies from the surroundings. For down-climb of a positive edge dislocation, the extra-half-plane has to be extended and this requires that additional atoms be taken from the surroundings, or equivalently, that vacancies be injected into the surroundings.

The climb process illustrated in Fig. 8.24 occurs by absorbing vacancies from the surroundings. This requires that vacancies diffuse to the edge dislocation, where they can annihilate and remove atoms from the bottom of the extra-half-plane. This vacancy diffusion process can occur only at high temperatures, typically above half of the melting temperature, where high concentrations of vacancies are present and where vacancy motion can occur. Thus the rate of climb of edge dislocations is ultimately controlled by the rate at which vacancies arrive from the surroundings which is, in turn, controlled by self-diffusion (see Section 7.3). Once the dislocation has climbed a sufficient distance away from the obstacle, it can move forward by glide and produce plastic strain.

8.5 Dislocation sources

Before closing this chapter, we give a brief account of the origin of the dislocations in crystals. Unlike point defects, the presence of dislocations is not required by thermal equilibrium.

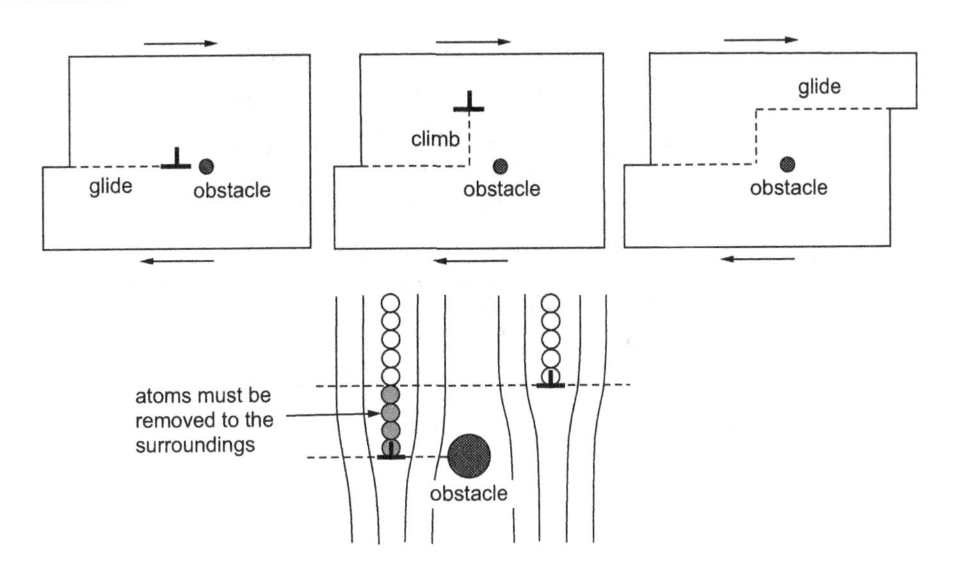

Figure 8.24. Edge dislocation climb by removal of atoms to the surroundings. This can be accomplished by absorbing vacancies from the surroundings.

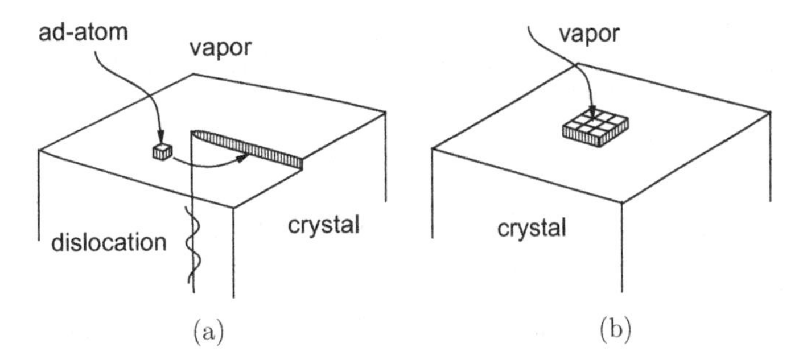

Figure 8.25. (a) F. C. Frank's picture of crystal growth from the vapor for a crystal with a screw dislocation. (b) The perfect crystal grows more slowly because new atomic layers have to be nucleated after each successive layer.

Instead, the dislocations are so energetic that they should not exist in a crystal if a fully equilibrium state is reached; they are kinetically trapped inside the crystal after being created from the various sources described below.

8.5.1 Grown-in dislocations

Most crystals have dislocations within them from the time they are grown, whether they are grown from a vapor, liquid, or solid. One of the first successes of the theory of dislocations was F. C. Frank's assertion that crystals containing dislocations would grow faster from the vapor than perfect crystals, so that imperfect crystals would dominate in crystal growth. Frank's idea is illustrated in Fig. 8.25. A crystal containing a screw dislocation can grow easily because an atomic ledge is always present to accept ad-atoms, which land on the surface from the vapor and diffuse to the ledge. A new layer of atoms does not have to be nucleated once a given layer is

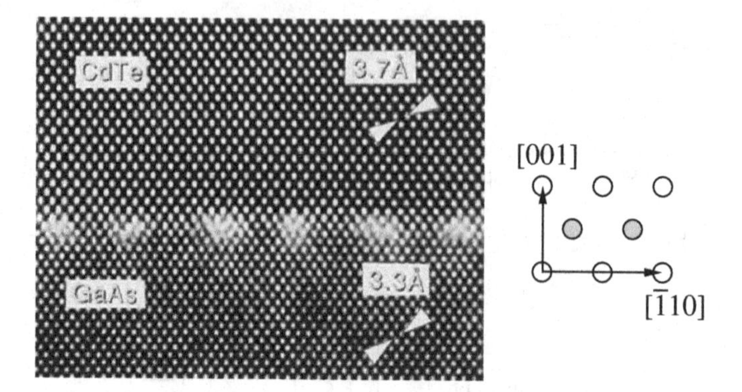

Figure 8.26. Misfit dislocations at the interface between CdTe and GaAs [59]. Permission granted by R. Sinclair.

covered, as in the case of the perfect crystal. For this reason, under some conditions, the crystal with the dislocation grows faster.

Engineering metals and semiconductors are often grown by solidification from the melt (i.e. liquid). It is almost impossible to obtain dislocation-free crystals this way,[7] because the thermal gradient in the crystals sets up internal stress fields that are sufficient to nucleate and multiply dislocations, because the crystals are very fragile at such high temperatures. At the same time, impurities tend to segregate at these grown-in dislocations (see Section 12.5 for dislocation–point defect interactions). These impurities pin the dislocations in place, making it nearly impossible to remove them from the crystal.

8.5.2 Dislocations in elastically strained films

In some cases crystals are elastically strained as they grow. The growth of heteroepitaxial films onto crystalline substrates is an example of this. In this example, the film, with its own equilibrium lattice spacing, has to be elastically strained to match the lattice spacing of the substrate.

Figure 8.26 shows an array of dislocations at the interface between CdTe and GaAs. The equilibrium lattice constants of CdTe is larger than that of GaAs (see Fig. 1.7). Therefore the CdTe film must be compressed in order to match the lattice spacings of the GaAs substrate. This creates a large elastic strain energy density. When the film becomes sufficiently thick, it becomes energetically favorable to form a periodic array of dislocations. These dislocations are called misfit dislocations because they come to relieve the misfit strain between the film and the substrate crystal. The critical thickness of the film for the misfit dislocations to be energetically stable will be discussed in Section 10.5.2.

8.5.3 Frank–Read and related sources

While dislocations are commonly created during crystal growth, most of the dislocations in ordinary metals are created by multiplication mechanisms by which the few dislocations present in well-grown crystals multiply in the course of plastic deformation. The most famous example

7 A notable exception is silicon. The ability to grow large, essentially dislocation-free Si wafers provides the foundation of the semiconductor industry.

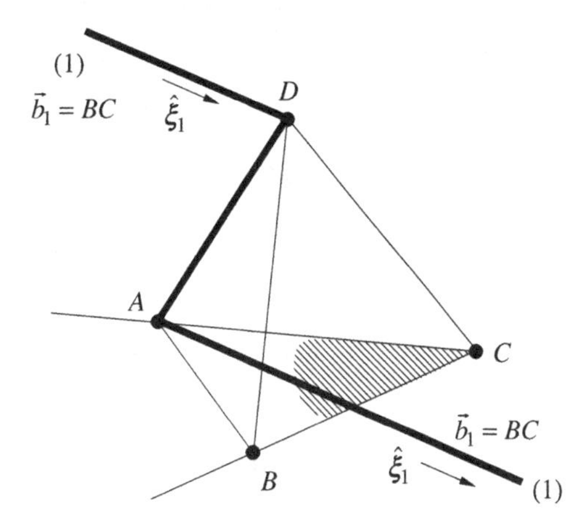

Figure 8.27. Grown-in, jogged edge dislocation relative to the slip planes and slip directions in an FCC crystal.

is the Frank–Read source, conceived independently, but at the same time, by F. C. Frank and W. T. Read [60, 61].

Consider the dislocation structure shown in Fig. 8.27. The dislocation line is shown relative to the slip planes and directions for the FCC crystal, which we will call the Thompson tetrahedron later in Section 11.1.5.

The four faces of the tetrahedra represent the four {111}-type slip planes; the four edges of the tetrahedra represent the four ⟨110⟩-type directions that are the Burgers vectors of perfect dislocations in the FCC crytal. The dislocation shown in Fig. 8.27 has a Burgers vector along the *BC* direction and can glide on planes parallel to the *ABC* plane. The glide plane for segment *AD* can be obtained by the cross product $\hat{\xi} \times \vec{b}$, and is a {001} type plane. Because dislocations in FCC crystals do not glide easily on {001} planes, the segment *AD* is not free to glide (due to its non-planar core-structure, see Section 11.1.11). Thus we may consider the points *A* and *D* to be fixed pinning points, which cannot move in the course of plastic deformation by ordinary glide. Hence the glide segment on the *ABC* plane has a pinning point at *A*. It is easy to see how a given glide segment could have pinning points on either end of the segment. In that case the two ends of the segment would be pinned. Frank and Read independently considered what would happen to such a segment if it were forced to glide by the application of a shear stress.

Figure 8.28, also called the Frank–Read diagram, shows what happens to a positive edge dislocation segment with pinning points at each end when a shear stress is applied to make the dislocation glide in its slip plane. Because the ends of the segment are fixed, the segment is forced to rotate around these points when glide occurs. The positive edge dislocation leads to a RHS element on the left and a LHS element on the right when the dislocation rotates in its slip plane. Further rotation leads to negative edge segments and finally, to new oppositely signed screw dislocation segments. The oppositely signed screw segments created by these rotations are in the same slip plane, so that annihilation occurs when they come together. When this happens one part of the dislocation returns to the original configuration while the other part is a completely new dislocation loop that did not exist before. The part of the dislocation that returns to the initial position can again bow out and again create a new loop, just as before. In this way the dislocation segment can continue to create new dislocation loops until the dislocation configuration is changed in some way. This is called the Frank–Read source.

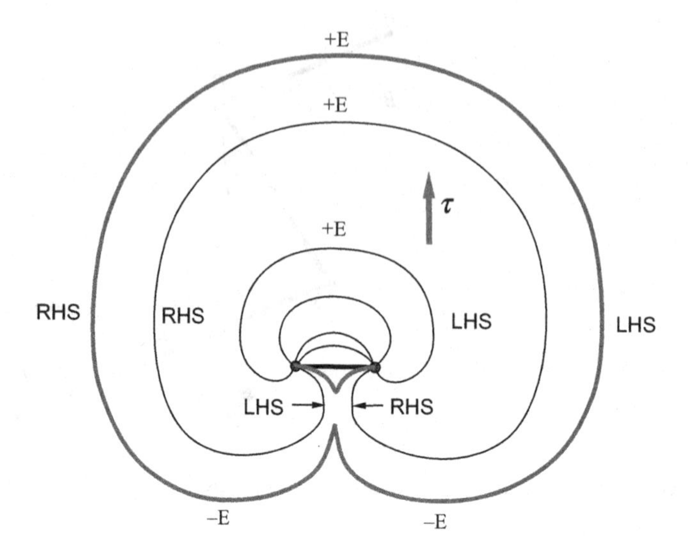

Figure 8.28. Successive configurations of a dislocation line in a Frank–Read source. A completely new dislocation loop is created in this process while the initial segment is recovered and is free to create additional loops. The original dislocation segment is shown as a bold straight line.

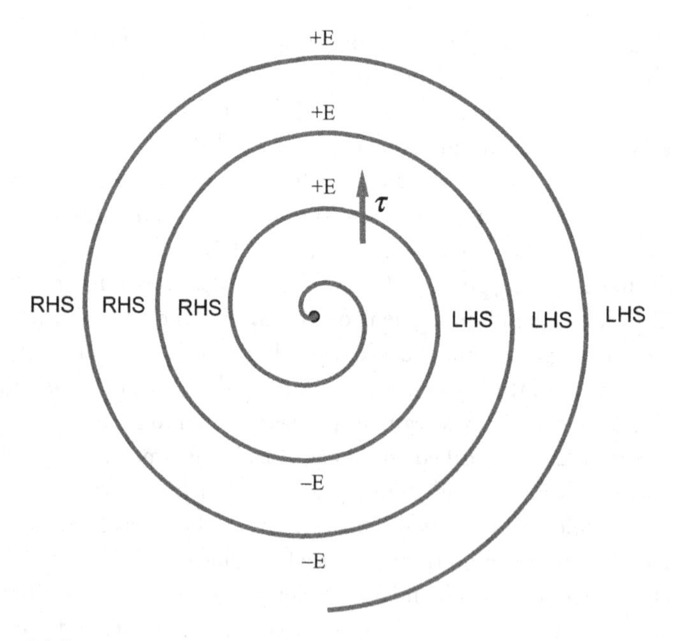

Figure 8.29. Spiral dislocation source.

The Frank–Read source is but one example of dislocation multiplication by bowing of dislocations about pinning points. Another example, usually called the spiral source, is illustrated in Fig. 8.29. There the line length of the segment gradually increases as the dislocation spirals around the pinning point. (Compare this figure with the expansion of the standard loop in Fig. 8.21.) Since the dislocation density is the dislocation line length per unit volume (see Section 8.2.3), it increases when the dislocation spirals around the pinning point, even if a

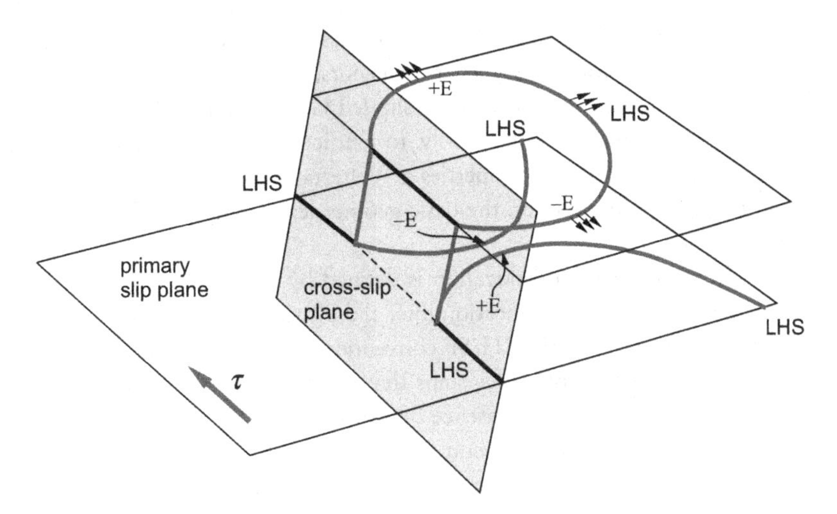

Figure 8.30. Multiplication by double cross-slip of a screw dislocation. Short arrows indicate the local direction of dislocation motion.

completely new dislocation line is not created. Indeed, this brings us to the most fundamental statement that can be made about these kinds of dislocation sources. It is that the dislocation density increases with plastic deformation because the existing dislocation lines are stretched (made longer) by the bowing process. This is true of the Frank–Read source as well. In each successive configuration of the source, the dislocation line is a little longer than before. Therefore, the dislocation density, ρ, defined as the total line length per unit volume, is gradually increasing even before the pinch-off event occurs.

The sources we have discussed assume the existence of fixed pinning points. Another kind of source involves the temporary formation of pinning points that are created at one time and destroyed later. This can occur by double cross-slip of a screw dislocation segment, as shown in Fig. 8.30. There, if a portion of a left-handed screw dislocation cross-slips onto a cross-slip plane, the jogs created in the dislocation line are not free to glide along with the screw. This means that the jogged segments would act as pinning points, about which the screw segments would rotate. The cross-slipped segment could operate like a Frank–Read source while the uncross-slipped segments would rotate in their slip planes and eventually create another Frank–Read source on the primary slip plane. If the screw segment cross-slips back to the primary slip plane later then the source will cease to operate.

Interaction between dislocations can also create temporary pinning points (e.g. from dislocation junctions) that can lead to dislocation multiplication. During plastic deformation, these pinning points can be created and destroyed at a steady rate. As long as these pinning points hold back the dislocations long enough for bowing to occur, the overall dislocation density can increase even if no pinning point is permanent.

8.6 Summary

Understanding the properties and behavior of dislocations starts with their geometry and how, through their movement, they allow crystals to change their shape when subjected to shear stresses. An intuitive understanding of these processes serves as a valuable guide to formulating the mechanics of dislocation behaviors, as discussed in Chapter 9. The ease with which metal crystals can be deformed by slip led G. I. Taylor, E. Orowan, and M. Polanyi, independently and almost simultaneously, to predict the existence of crystal dislocations in the early 1930s. Building on the properties of Volterra dislocations, which had been studied previously in continuum mechanics, the basic geometrical properties of crystal dislocations were quickly established.

The character of a dislocation is defined by two vectors: a sense vector, $\hat{\xi}$, which is a unit vector parallel to the dislocation line, and a corresponding Burgers vector, $\vec{b}$, defined using the right hand, start–finish, RH/SF convention. By looking along the sense vector and drawing a clockwise sequence of lattice steps that would close in a perfect lattice, any closure failure of the circuit indicates the presence of a dislocation. By convention the Burgers vector is drawn from the start of such a circuit to the finish: RH/SF. Edge dislocations are characterized by the Burgers vector being perpendicular to the sense vector, $\hat{\xi} \cdot \vec{b} = 0$, while screw dislocations are characterized by the condition that the Burgers vector and sense vector are parallel to each other, $\hat{\xi} \parallel \vec{b}$. Dislocations with Burgers vectors and sense vectors making arbitrary angles with each other are called mixed dislocations, having both edge and screw character.

When different dislocation lines meet at a node their Burgers vectors must sum to zero if the sense vectors are all chosen as pointing into (or out of) the node. This continuity relation is called Frank's rule. It permits the character of one of the dislocations to be determined if the character of the others is known.

Dislocation glide is the most important kind of dislocation motion because it dominates plastic deformation at all but the very highest of temperatures. The plane of glide (or the slip plane) contains both the Burgers vector, $\vec{b}$, and the sense vector, $\hat{\xi}$. The resolved shear stress acting both on the glide plane and in the direction of the Burgers vector provides the driving force for glide motion. An intuitive understanding of this relationship, as represented by an expanding glide loop subjected to a resolved shear stress, helps to determine how dislocations respond to different kinds of loading.

Screw dislocations can glide either on a primary slip plane or on another equivalent crystallographic plane called the cross-slip plane. This kind of glide motion out of the primary slip plane and onto the cross-slip plane is possible because both planes contain the sense vector and Burgers vector. Such out-of-plane glide motion is not possible for dislocations having any edge component. Edge dislocations can move out of their glide planes only by climb, a process involving the addition or subtraction of atoms at the edge of the extra-half-plane defining the edge dislocation. This process is active only at very high temperatures where vacancies can diffuse from or to the climbing dislocation to accommodate the addition or subtraction of atoms there.

While most metal crystals contain dislocations in the as-grown state, the density of such dislocations is usually far below that needed to accommodate significant plasticity. Thus dislocations need to be created or multiplied to support ordinary plastic deformation. The most prominent mechanisms of dislocation multiplication involve processes by which the dislocation line length per unit volume (the dislocation density) gradually increases through the glide motion of partially pinned dislocation segments. The Frank–Read mechanism and the closely related spiral sources and double cross-slip are prominent types of dislocation sources in metal crystals.

8.7 Exercise problems

8.1 The transmission electron micrograph given in Fig. 8.26 shows atoms (or more precisely, columns of atoms) near the interface between crystals of CdTe and GaAs, both of which have the FCC lattice. The orientation of the crystals relative to the interface is shown in the micrograph. The micrograph shows misfit dislocations at the interface between the crystals.

(a) Taking the sense vector for the dislocations to be pointing into the page, determine the Burgers vectors of the dislocations shown.
(b) Indicate the character of these dislocations and make a simple sketch to illustrate the dislocations with appropriate symbols.
(c) Assuming that the dislocations are perfect dislocations in the FCC lattice, give a complete description of the Burgers vectors for the chosen sense vectors in terms of the lattice parameter a of the FCC lattice.

8.2 Figure 8.31 shows a TEM image of the crystal lattice surrounding a dislocation in NiAl. The dislocation seen in the center of the image has two separate parts as indicated by the white arrows. Note that the plane of the picture is the (1 0 0) plane and that the sides of the diagram are parallel to directions of the type ⟨0 1 1⟩. The dislocation shown is a perfect dislocation with a Burgers vector which extends from one lattice point to another. NiAl has the B2 (CsCl) structure as shown in Fig. 1.5a.

(a) Determine the Burgers vector and character of the dislocation shown. The Burgers vector should indicate not only direction but also the magnitude. You may call the lattice parameter a_0 for this purpose.
(b) Determine the Burgers vectors and characters of the two separate parts of the dislocation indicated by the arrows.

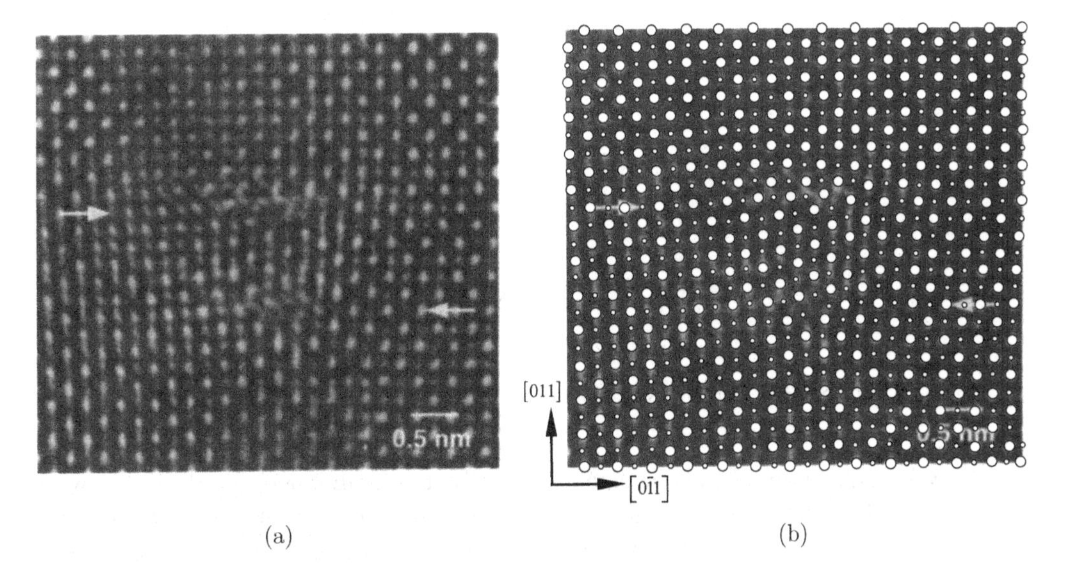

Figure 8.31. (a) A TEM image of NiAl crystal containing a dislocation [62]. Permission granted by M. J. Mills. (b) An enhanced version to simplify the Burgers circuit analysis.

8.3 A pure right-handed screw dislocation is shown in a crystal in Fig. 8.32. The dislocation is made to glide, first in plane 1, next in plane 2, then in plane 3, and finally in plane 4, such that the dislocation returns to the initial position. Describe what happens to the shape of the crystal when the dislocation traverses that path 100 times. The dislocation can take such a path by cross-slipping from one crystallographic plane to another.

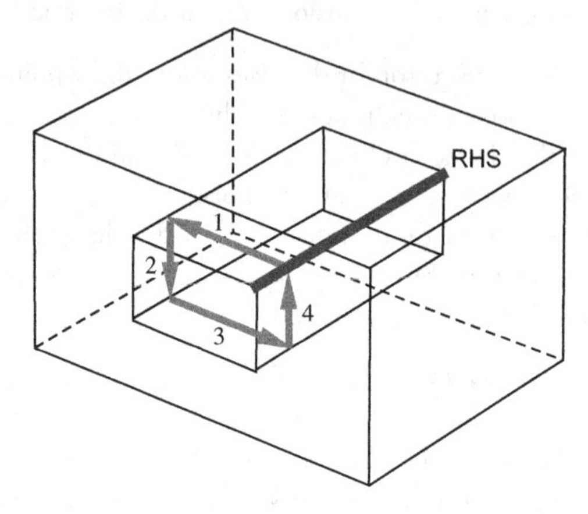

Figure 8.32. A right-handed screw dislocation traverses along a closed path.

8.4 A planar array of dislocations in the plane of the page is shown in Fig. 8.33. Dislocation segments AD, DC, and CB are all right-handed screw dislocations with Burgers vectors of length b. The angles made at the junctions of the dislocation segments are all 120°, some of which are labeled. Determine the character of dislocation segment EF. Determine the magnitude of the Burgers vector and give a clear statement of the type of dislocation found in that segment.

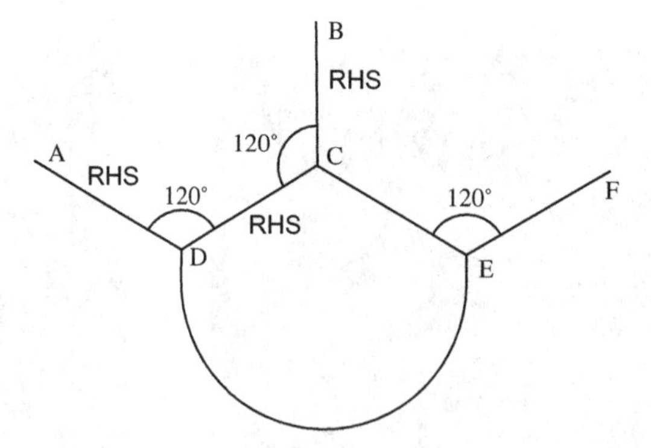

Figure 8.33. A planar dislocation network.

8.5 Dislocations often leave dipoles and trails of dislocation loops in their wake when they move. Figure 8.34 shows how dipoles can be transformed into a trail of loops. Subsequent experiments and analyses have shown that loops formed by this process eventually shrink and disintegrate into vacancies. The rate at which this happens depends on the rate of vacancy diffusion away from the dipoles (or rows of loops).

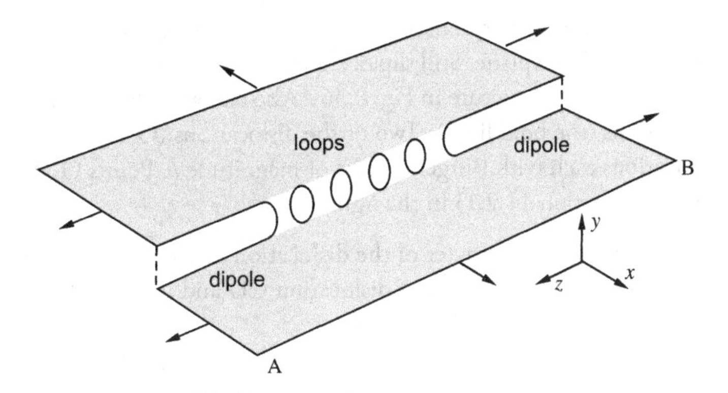

Figure 8.34. Formation of dislocation loops from a dislocation dipole.

Use the information given above and your knowledge of dislocations to determine the character (edge, screw, half-plane orientation, etc.) for the dislocation segment labeled A–B.

8.6 The solid lines in Fig. 8.35a represent dislocation segments which lie in the xy plane. The character of four of the segments is indicated. The dashed lines are dislocation segments parallel to z and the dotted segment is parallel to y. Determine the character of the dotted segment and indicate how it would move in response to a positive shear stress σ_{zx}.

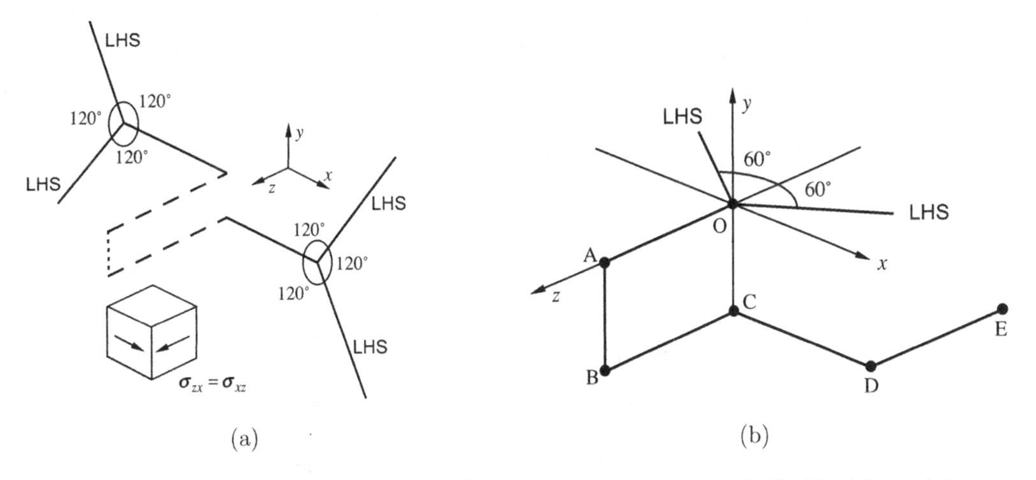

Figure 8.35. (a) Dislocation dipole connected to a screw dislocation network. (b) Two left-handed screw dislocation and a non-planar dislocation.

8.7 Two pure LHS dislocations with Burgers vectors of the same magnitude lie on the xz plane as shown in Fig. 8.35b and form a node with the dislocation segment OA. The dislocation segment is then connected to others, making a non-planar dislocation line as shown.

(a) Determine the character of all of the dislocations segments: OA, AB, BC, CD, and DE.

(b) Assuming that the screws are not allowed to move, determine which segments, if any, would move if a positive shear stress $\sigma_{zy} = \sigma_{yz}$ were imposed and determine the direction of motion.

8.8 In some HCP metal crystals slip occurs preferentially on the prism planes, rather than on the close-packed basal plane. Still slip occurs in the close-packed directions (of the type shown in the hexagonal crystal structure in Fig. 8.36). Also shown in Fig. 8.36 is a particular configuration of dislocations (the bold lines). Two of the dislocations (OA and OB) are known to be pure LHS dislocations each with Burgers vector of magnitude b. Points O, A, B, C all lie on the top basal plane (see Section 11.2.1) in the figure.

(a) Determine the character of the dislocation segment labeled DE.
(b) Determine the character of dislocation CD and indicate the plane on which it could glide.

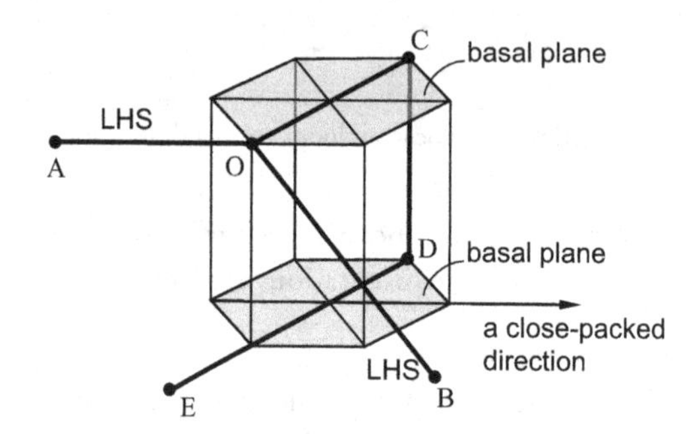

Figure 8.36. Dislocations in an HCP crystal.

8.9 A dislocation loop is created in Al by quenching from a high temperature and allowing the excess vacancies to aggregate onto a plane to form a circular loop (on the yz plane) as shown in Fig. 8.37. (Note: this is not a glide loop of the kind used to describe slip.) Another straight dislocation, parallel to the x axis, is shown near the loop. When the straight dislocation moves close to the loop, it reacts with the loop and forms the configuration shown in Fig. 8.37. Determine the character of the straight dislocation.

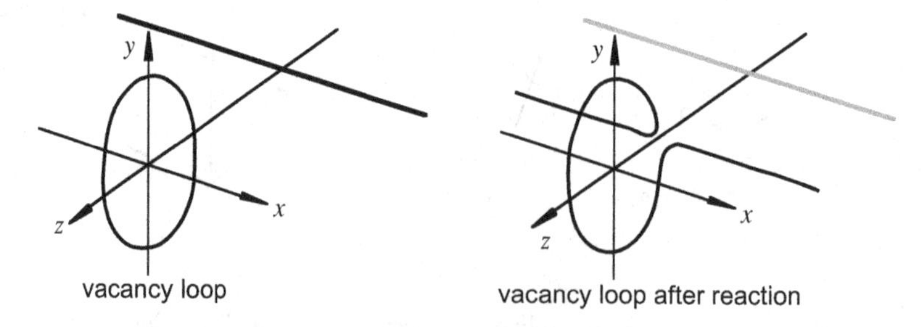

Figure 8.37. A dislocation line reacts with a circular dislocation loop formed from vacancies.

8.10 Three circular glide dislocation loops lying on a close-packed plane in an FCC crystal are shown in Fig. 8.38. The Burgers vectors of the loops, which are of the type $(a/2)\langle 1\ 1\ 0\rangle$, are indicated by the double-ended arrows (the precise Burgers vector depends on the choice of sense vector). The magnitudes of the Burgers vectors of the loops are assumed to be the same. If the loops expand by glide, they can combine with each other to produce the heavy gray dislocation structure shown.

Determine the character (sense vector, Burgers vector, half-plane, etc.) of each of the dislocation segments AO, BO, and CO.

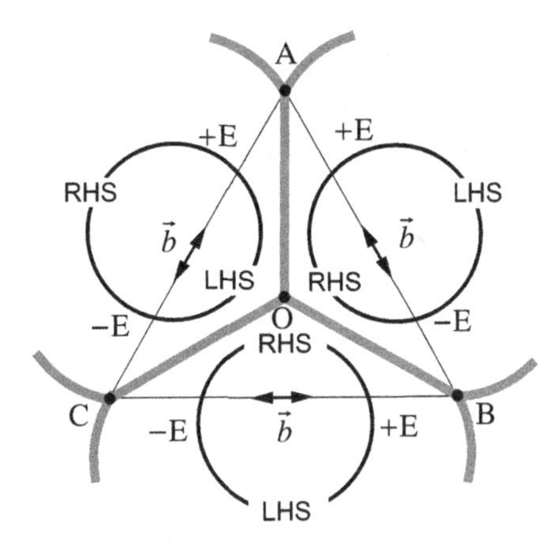

Figure 8.38. Three circular glide dislocation loops expand and combine with each other.

8.11 A non-planar dislocation loop is shown in Fig. 8.39. All of the dislocation segments are parallel or perpendicular to the coordinate axes shown. The segment labeled L–A is a positive edge dislocation.

(a) Determine the character of the segments C–D, D–E, E–F, and F–G.
(b) Assuming that all of the screw dislocation segments are locked and are not free to move, determine which of the dislocation segments will move in response to a positive shear stress σ_{xy}. Indicate the direction of motion of those segments that will move in response to this shear stress.

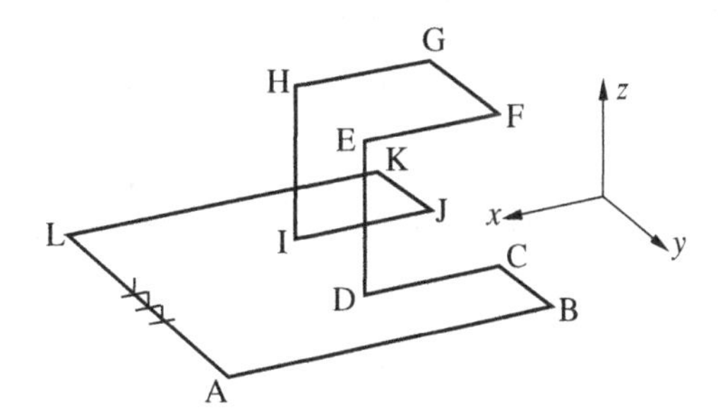

Figure 8.39. A non-planar dislocation loop.

8.12 A pure edge dislocation is shown in a cylindrical crystal subjected to pure torque or twist, as shown in Fig. 8.40a. The shaded end of the crystal is twisted in the clockwise direction, as shown.

(a) Assuming that the dislocation is free to glide, but not climb or cross-slip, show how the dislocation would move on its slip plane and where it would eventually stop. It is assumed that no other dislocations are formed by the applied torque.
(b) Determine the character of the dislocation in its final state.

8.13 A pure edge dislocation is shown in a cylindrical crystal subjected to pure torque or twist, as shown in Fig. 8.40b. The torque shown produces a positive $\sigma_{\theta z}$.

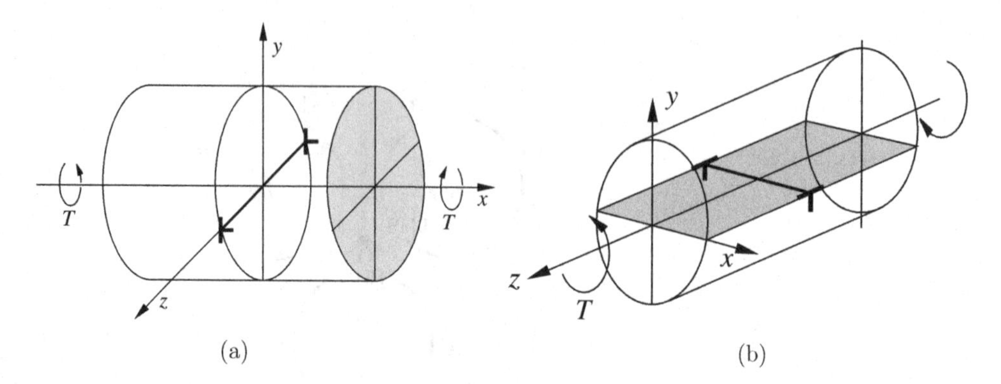

Figure 8.40. An edge dislocation in a cylinder under twist. (a) The Burgers vector is perpendicular to the cylinder axis. (b) The Burgers vector is parallel to the cylinder axis.

(a) Choose a sense vector and determine the Burgers vector for the dislocation shown.
(b) Assuming that the dislocation is free to glide on the shaded slip plane (the xz plane), but not climb or cross-slip, show how the dislocation would move on its slip plane and where it would eventually stop. It is assumed that no other dislocations are formed by the applied torque.
(c) Determine the character of the dislocation in its final state.
(d) Determine the sense vector and Burgers vector in the final state.

8.14 A hexagonal array of pure right-handed screw dislocations resides on the basal slip plane of a round, c-axis-oriented HCP nanowire of radius R, as shown in Fig. 8.41(a). The c axis of the round wire is perpendicular to the page.

(a) Show the direction of movement of all of the dislocation segments if the wire is subjected to the torsion shown.
(b) Describe what happens to the dislocations, assuming that they are all free to glide easily on the basal plane and also assuming that no new dislocations are nucleated.

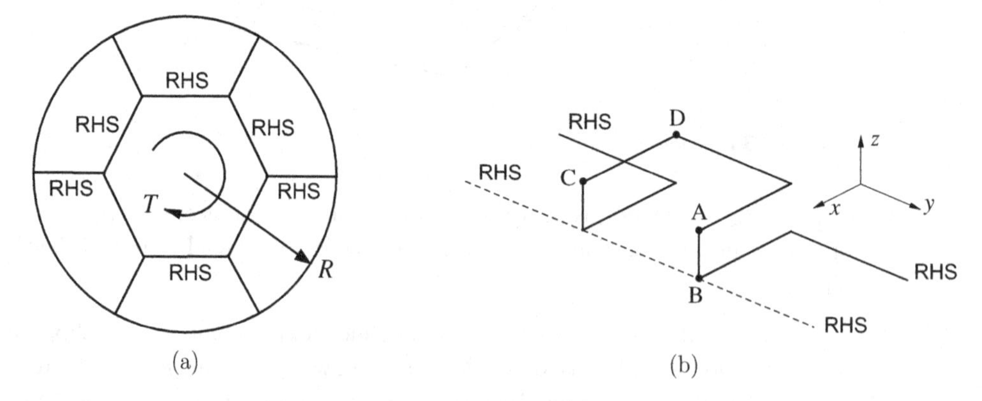

Figure 8.41. (a) A hexagonal array of right-handed screw dislocations in a nanowire of radius R. (b) A right-handed screw dislocation having undergone double cross-slip.

8.15 A right-handed screw dislocation, initially located in the dotted position, as shown in Fig. 8.41b, is forced to glide on the xy plane in the $-x$ direction in response to a positive shear stress σ_{yz}.

The central portion of the screw dislocation cross-slips twice and then glides on a plane parallel to the primary slip plane. This creates the non-planar dislocation configuration shown.

(a) Determine the character of dislocation segment AB and indicate whether it will move in response to the applied shear stress and, if so, in what direction.
(b) Determine the character of dislocation segment CD and indicate whether it will move in response to the applied shear stress and, if so, in what direction.

9 Dislocation mechanics

Our study of dislocations up to this point has focused on their geometry and their role in accommodating plastic deformation through their motion. We now turn to another important aspect of dislocations: their elastic fields. The remarkable thing about (perfect) dislocations is that while they leave a crystal internally perfect after they have passed through the crystal, they produce elastic distortions in the crystal as long as they are present. Their elastic fields determine, to a large extent, how they interact with each other and with other structural defects in the crystal.

Our study of the elastic properties of dislocations will assume that the material in which they are found is elastically isotropic. While most crystals are not elastically isotropic, the framework that emerges by assuming elastic isotropy is still quite useful and even reasonably accurate for most crystals. In this chapter, we will use the Volterra dislocation model, in which the dislocation line is a singularity that needs to be avoided. The atomistic structure of the dislocation core will be discussed in Chapter 12.

In Section 9.1 we discuss the stress, strain, and displacement fields of infinitely long, straight dislocations, and in Section 9.2 we obtain the elastic energy of these dislocations. Based on the elastic energy results, we describe in Section 9.3 the line tension model, in which a curved dislocation line is approximated as a taut string with a certain resistance to stretching. In Section 9.4 we derive the Peach–Koehler formula that determines the force per unit length on the dislocation exerted by the local stress. Given that dislocations generate their own stress fields, the Peach–Koehler formula can be used to predict the interaction between dislocations, which will be discussed in Chapter 10.

9.1 Elastic fields of isolated dislocations

The goal of this section is to obtain the stress, strain, and displacement fields of an infinitely long screw, edge, or mixed dislocation in an infinite medium, or at the center of an infinitely long cylinder of finite radius.

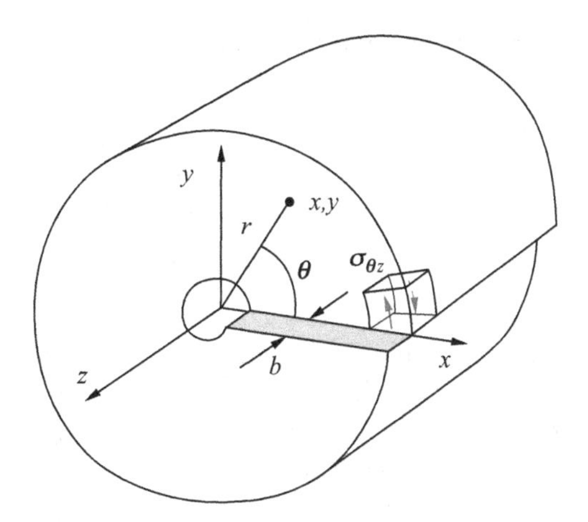

Figure 9.1. Right-handed screw dislocation in an isotropic solid.

9.1.1 Screw dislocation in an infinitely long cylinder

Figure 9.1 shows an infinitely long cylinder containing a right-handed screw dislocation running along its axis. The outside radius of the cylinder is R while the radius of the small hole at the core is r_0. The core material is assumed to be removed to avoid singular stresses and strains, as discussed in Section 8.2.1.

We choose the line sense vector $\hat{\xi}$ to be along the positive z axis, then the dislocation Burgers vector $\vec{b}$ also points to the positive z axis. The elastic field of such a screw dislocation in an isotropic elastic solid is a problem of *anti-plane strain*, as discussed in Section 2.5. The only non-zero displacement component is $u_z(x, y)$, which satisfies Laplace's equation, Eq. (2.62),

$$\nabla^2 u_z \equiv \left(\frac{\partial^2}{\partial x^2} + \frac{\partial^2}{\partial y^2} \right) u_z(x, y) = 0.$$

We need a displacement field $u_z(x, y)$ that satisfies this equation and also satisfies the Burgers condition,

$$\oint_C \left(\frac{\partial \vec{u}}{\partial s} \right) \mathrm{d}s = \vec{b}, \tag{9.1}$$

where the integral is around any loop C that goes counterclockwise and contains the z axis (i.e. the dislocation line). The Burgers condition is equivalent to the definition of Burgers vector by drawing the Burgers circuit following the RH/SF rule (see Section 8.3).

The rotational symmetry of the problem suggests to us that the cylindrical coordinate system should be used, in which Laplace's equation becomes Eq. (2.63),

$$\nabla^2 u_z = \left(\frac{\partial^2}{\partial r^2} + \frac{1}{r}\frac{\partial}{\partial r} + \frac{1}{r^2}\frac{\partial^2}{\partial \theta^2} \right) u_z(r, \theta) = 0.$$

By inspection one can guess that the correct solution is

$$u_z = \frac{b}{2\pi}\theta = \frac{b}{2\pi}\arctan\left(\frac{y}{x}\right). \tag{9.2}$$

It is easy to demonstrate that this solution satisfies Laplaces's equation:

$$\frac{\partial u_z}{\partial r} = 0$$
$$\frac{\partial u_z}{\partial \theta} = \frac{b}{2\pi}$$
$$\frac{\partial^2 u_z}{\partial \theta^2} = 0,$$

so that we obtain the result: $\nabla^2 u_z = 0$. It is important to note that $\theta = \arctan(y/x)$ is a multi-valued function on the x–y plane. To make it single-valued, it is necessary to introduce a "branch-cut" where θ jumps by 2π. For example, we can introduce the branch-cut along the positive x axis, so that the domain of θ is $[0, 2\pi)$, and θ jumps by 2π as the field point crosses the positive x axis from above. The mathematical branch-cut corresponds to the cut-plane in Fig. 9.1 needed to introduce the dislocation into the solid. Because the very definition of a dislocation requires a displacement jump by b across a cut-plane, we expect the $\theta b/(2\pi)$ term to be present in the displacement field of any straight dislocation.[1] Note that the displacement field of a straight screw dislocation is independent of the elastic constants (μ, ν) of the solid. The displacement field is also independent of the radius R of the cylinder, as long as the cylinder has infinite length (i.e. no ends).

Given the displacement field, it is a simple matter to compute the strain and stress fields. In cylindrical coordinates,

$$\gamma_{\theta z} = \frac{1}{r}\frac{\partial u_z}{\partial \theta} = \frac{b}{2\pi r} \tag{9.3}$$

$$\sigma_{\theta z} = \mu \gamma_{\theta z} = \frac{\mu b}{2\pi r}. \tag{9.4}$$

All other components are zero. Therefore, in cylindrical coordinates, the only non-zero stress component is $\sigma_{\theta z}$ (marked on Fig. 9.1), which is also independent of θ. This is a manifestation of the symmetry of the screw dislocation in an isotropic medium, as one might expect intuitively. Even though the displacement field is not entirely symmetric with respect to rotation around the z axis (due to the branch-cut of θ), the stress and strain fields are symmetric when expressed in cylindrical coordinates.

A simple way to rationalize the basic result of Eq. (9.3) is to consider a thin shell of radius r and thickness dr surrounding a unit length of dislocation, as shown in Fig. 9.2. If this shell is unwrapped and flattened out we have a rectangular sheet of height $2\pi r$ and unit width with a shear strain of $\gamma_{z\theta} = b/(2\pi r)$, consistent with Eq. (9.3).

We note that the stresses near a dislocation line depend linearly on the shear modulus and Burgers vector and inversely on the distance from the dislocation line. Thus the elastic field is singular, with a singularity of the form $\sigma_{ij} \sim r^{-1}$. It is the singular nature of this field that requires us to remove the core from consideration. Obviously, the solution breaks down in the core of the dislocation because the stress in a real solid must be bounded. Linear elasticity breaks down in the core as the interactions between atoms become non-linear. The non-linear

1 For the displacement field of a dislocation loop in 3D, $\theta/(2\pi)$ is replaced by $\Omega/(4\pi)$, where Ω is the solid angle that the loop subtends at the field point.

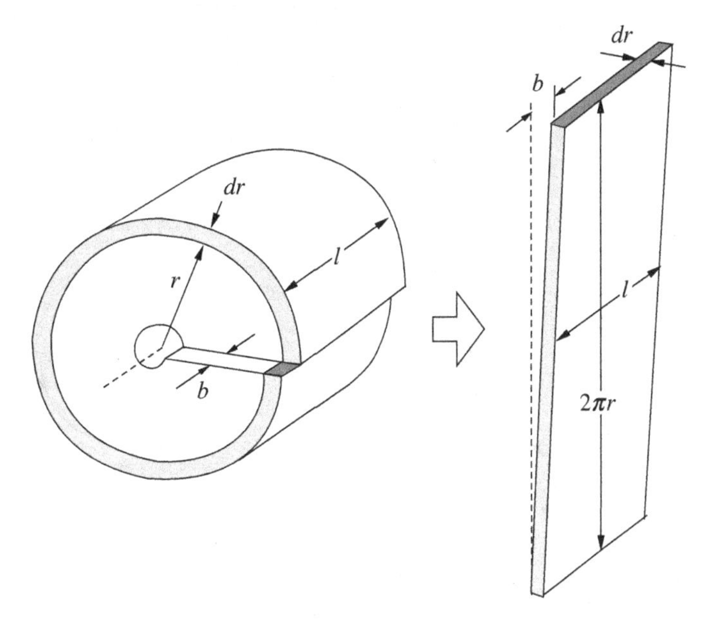

Figure 9.2. A shell of material surrounding the screw dislocation is unwrapped and flattened out, leading to a rectangular sheet with a shear strain.

interatomic interactions give rise to various dislocation core structures, which will be discussed in Chapter 12.

The stress–strain fields in Cartesian coordinates can be obtained either by coordinate transformation from cylindrical coordinates, or by differentiating the displacement field directly in Cartesian coordinates. Here we take the latter approach. The only two non-zero strain components are

$$\gamma_{xz} = \frac{\partial u_z}{\partial x} = -\frac{b}{2\pi}\frac{y}{x^2+y^2} \tag{9.5}$$

$$\gamma_{yz} = \frac{\partial u_z}{\partial y} = \frac{b}{2\pi}\frac{x}{x^2+y^2}. \tag{9.6}$$

Using Hooke's law, the only two non-zero stress components are

$$\sigma_{xz} = \mu\gamma_{xz} = -\frac{\mu b}{2\pi}\frac{y}{x^2+y^2} \tag{9.7}$$

$$\sigma_{yz} = \mu\gamma_{yz} = \frac{\mu b}{2\pi}\frac{x}{x^2+y^2}. \tag{9.8}$$

9.1.2 Screw dislocation in a finite cylinder

The above solution for the stresses about a screw dislocation is valid in an infinitely long cylinder with no ends, so that there is no need to consider the boundary conditions at the two ends. In this section, we consider cylinders of finite length L. We are interested in the regime where $L \gg R$, so that we are dealing with long cylinders instead of disks. If L is finite, the solution above is still valid if the ends of the cylinder are constrained so that it does not twist. If, instead,

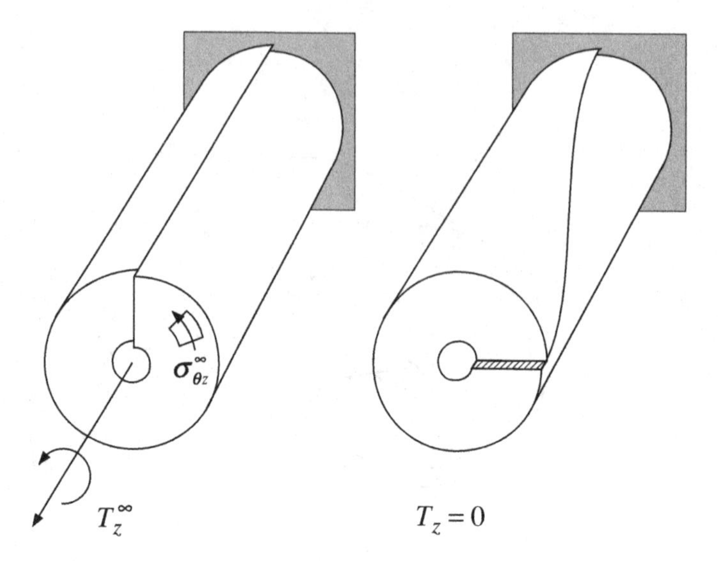

Figure 9.3. Screw dislocation in an infinitely long, constrained cylinder and in a cylinder that is unconstrained and torque-free.

the cylinder is of finite length and radius, and is torque-free, then the cylinder will twist when the screw dislocation is created. The resulting stress field will be modified from that given by Eq. (9.4). The result is no longer an anti-plane problem because twisting around the z axis leads to displacements in the x and y directions. However, the solution can be easily constructed from the anti-plane solution given above plus another image solution.

Figure 9.3 shows a dislocated cylinder in a constrained state and the twist associated with the torque-free state. The twist that occurs when the cylinder is made torque-free is called the Eshelby twist, after J. D. Eshelby who first studied these properties in the 1950s [63].

As shown in Fig. 9.3, when the cylinder is constrained from twisting, the shear stress will now be called $\sigma_{\theta z}^{\infty}$, where the superscript indicates that this is the solution valid in an infinitely long cylinder. This shear stress produces a non-zero torque on the cross section of the cylinder,

$$T_z^{\infty} = \int_{r_0}^{R} \int_0^{2\pi} \left(\sigma_{\theta z}^{\infty} \cdot r\right) r \, d\theta \, dr. \tag{9.9}$$

Given the stress field in Eq. (9.4), we have

$$T_z^{\infty} = \int_{r_0}^{R} \int_0^{2\pi} \left(\frac{\mu b}{2\pi r}\right) \cdot r \cdot r \, d\theta \, dr = \mu b \int_{r_0}^{R} r \, dr = \mu b \left(\frac{R^2}{2} - \frac{r_0^2}{2}\right). \tag{9.10}$$

In the limit of $r_0 \ll R$,

$$T_z^{\infty} = \mu b \frac{R^2}{2}. \tag{9.11}$$

This is the torque that has to be applied to prevent the cylinder from twisting. If this torque is released (the equivalent of applying an equal but oppositely signed torque) then the cylinder will twist as shown in Fig. 9.4.

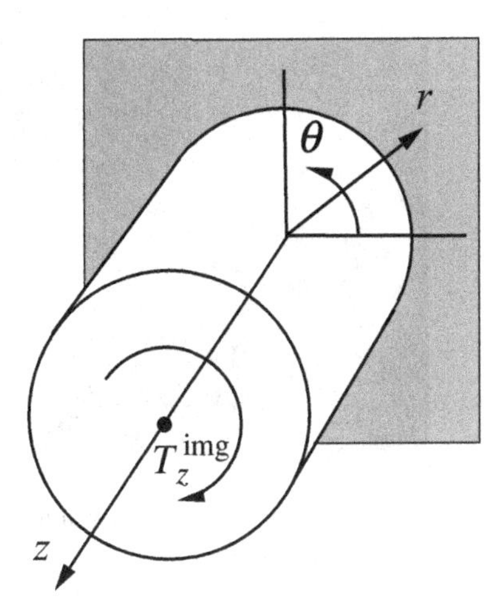

Figure 9.4. Torsion of a cylindrical bar.

To make the end condition of the cylinder torque free, the solution given in the previous section must be corrected by superimposing an "image" torque, $T_z^{\text{img}} = -T_z^{\infty} = -\mu b R^2/2$. We can calculate the additional or "image" displacements, strains, and stresses associated with the "image" torque by finding a solution for pure torsion. By inspection we can guess that the displacement field for pure torsion is

$$u_\theta^{\text{img}} = -Arz, \tag{9.12}$$

where A is a constant to be determined and the displacements are linear in both r and z. The corresponding elastic strains are

$$\gamma_{\theta z}^{\text{img}} = \frac{\partial}{\partial z} u_\theta^{\text{img}} = -Ar$$

$$\gamma_{r\theta}^{\text{img}} = \left(\frac{\partial}{\partial r} - \frac{1}{r} + \frac{1}{r}\frac{\partial}{\partial \theta} \right) u_\theta^{\text{img}} = 0$$

with all other components zero. The corresponding shear stresses are then

$$\sigma_{\theta z}^{\text{img}} = \mu \gamma_{\theta z}^{\text{img}} = -\mu A r. \tag{9.13}$$

The corresponding image torque is then

$$T_z^{\text{img}} = \int_{r_0}^{R} \int_0^{2\pi} \left(\sigma_{\theta z}^{\text{img}} \cdot r \right) r \, d\theta \, dr, = -2\pi \mu A \int_{r_0}^{R} r^3 \, dr,$$

which, in the limit of $r_0 \ll R$, is

$$T_z^{\text{img}} = -\pi \mu A \frac{R^4}{2}. \tag{9.14}$$

By setting T_z^{img} to $-T_z^{\infty} = -\mu b R^2/2$ we can find the constant A,

$$A = \frac{b}{\pi R^2}. \tag{9.15}$$

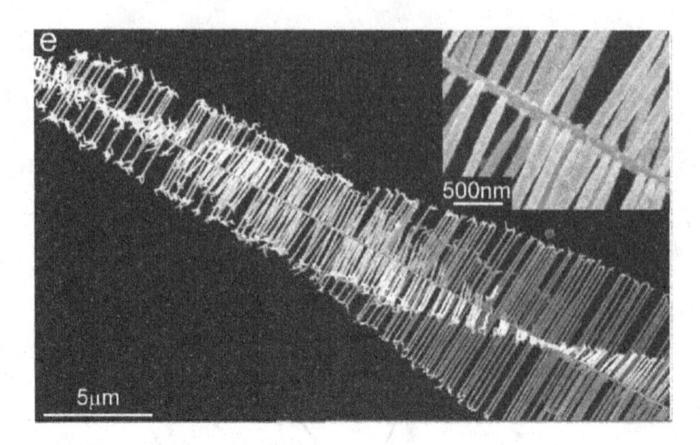

Figure 9.5. Eshelby twist in a PbSe nanowire. A screw dislocation is found along the axis of the main nanowire. The nanowire branches are free of dislocations and are not twisted [64]. Permission granted by Y. Cui.

We now have the complete solution for a screw dislocation in a cylinder of finite length and radius[2]:

$$\sigma_{\theta z} = \sigma_{\theta z}^{\infty} + \sigma_{\theta z}^{\text{img}} = \frac{\mu b}{2\pi r} - \frac{\mu b}{\pi R^2} r = \frac{\mu b}{2\pi r}\left[1 - 2\left(\frac{r}{R}\right)^2\right]. \tag{9.16}$$

We see that near the core of the dislocation, where $r \ll R$, the standard solution is recovered. But the image solution becomes more and more important far away from the core. Note that the shear stress changes sign near the outer boundary of the cylinder (at $r = R/\sqrt{2}$).

One interesting aspect of this solution is that a finite cylinder containing a screw dislocation along its core is expected to twist. Using the image displacement field, we can calculate the periodic length z_c for which one complete revolution would be expected by setting $u_\theta^{\text{img}}(r = R) = -2\pi R$:

$$u_\theta^{\text{img}}(r = R) = -\frac{b}{\pi R} z_c = -2\pi R,$$
$$z_c = \frac{2\pi^2 R^2}{b}. \tag{9.17}$$

It is remarkable that the periodic length z_c is independent of the elastic constants of the solid. We can see that as $R \to \infty$ (keeping in mind $L \gg R$), $z_c \to \infty$, meaning that the twisting becomes negligible for very thick cylinders.

This kind of twisting has recently been observed in semiconductor nanowires [64, 65]. Figure 9.5 shows the Eshelby twist of PbSe nanowires [64]. PbSe has the B1 (NaCl) structure, with an FCC lattice and a two atom basis along the edge of the unit cell. The lattice constant of PbSe is $a = 0.613$ nm [27]. Because the nanowires grow along the [0 0 1] direction, the Burgers vector of the screw dislocation along the nanowire axis must be $b = a = 0.613$ nm. The radius of these nanowires is $R = 48$ nm. Using Eq. (9.17), the periodic length for one complete revolution is predicted to be $z_c \approx 75$ μm. This is in good agreement with the experimentally measured value of $z_c \approx 73$ μm.

2 The image stress $\sigma_{\theta z}^{\text{img}}$ does not cancel $\sigma_{\theta z}^{\infty}$ point by point. Hence the traction forces on the end surfaces are not completely removed. However, the net torque is canceled by $\sigma_{\theta z}^{\text{img}}$. According to Saint-Venant's principle, $\sigma_{\theta z}^{\infty} + \sigma_{\theta z}^{\text{img}}$ is the correct solution at distances d away from the end surface, as long as $d \gg 2R$.

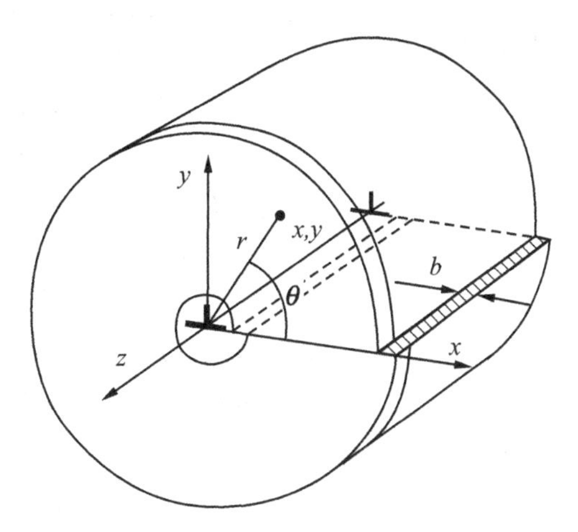

Figure 9.6. Positive edge dislocation.

9.1.3 Edge dislocation in an infinitely long cylinder

We now describe the elastic field of a positive edge dislocation in an elastically isotropic solid. Figure 9.6 shows an edge dislocation in an infinitely long cylinder of radius R with a small hole at the core of radius r_0. This edge dislocation is "positive" in the sense that the extra half-plane is along the positive y axis. When the line sense vector is chosen to be the positive z axis, the Burgers vector is along the positive x axis.

When the dislocation is created, the slab of material lying in the x–y plane remains planar when the dislocation is formed and the thickness of that slab neither increases nor decreases when the distortion is created. Therefore, all strains are in the x–y plane. This means that the elastic field of an edge dislocation in an infinitely long cylinder is a problem of *plane strain*, as discussed in Section 2.5. Plane strain problems are usually solved by first finding the Airy stress function $\phi(x, y)$ that satisfies the *biharmonic equation*, Eq. (2.68),

$$\left(\frac{\partial^2}{\partial x^2} + \frac{\partial^2}{\partial y^2}\right)\left(\frac{\partial^2}{\partial x^2} + \frac{\partial^2}{\partial y^2}\right)\phi(x, y) = 0,$$

and the appropriate boundary conditions. The stress fields can be obtained from the derivatives of the stress function, as given in Eqs. (2.66) and (2.70).

Stress function solution

The general solution to the biharmonic equation for an edge dislocation takes the following form in cylindrical coordinates

$$\begin{aligned}\phi(r, \theta) &= f(r)\sin\theta,\\ f(r) &= Br^3 - Ar\ln r,\end{aligned} \tag{9.18}$$

where B and A are constants to be determined from the boundary conditions. The $\sin\theta$ term in the stress function makes the stress field a periodic function of θ with periodicity 2π, which is obviously the case. It is easy to show that this function satisfies the biharmonic equation in the cylindrical coordinate system, Eq. (2.69). From Eq. (2.70), the stress fields corresponding to this stress function are

$$\begin{aligned}
\sigma_{rr} &= \left(-\frac{A}{r} + 2Br\right)\sin\theta, \\
\sigma_{\theta\theta} &= \left(-\frac{A}{r} + 6Br\right)\sin\theta, \\
\sigma_{r\theta} &= \left(\frac{A}{r} - 2Br\right)\cos\theta.
\end{aligned} \tag{9.19}$$

Since for a dislocation in an infinite solid the stresses are expected to go to zero at $r \to \infty$, it follows that $B = 0$ for a cylinder of infinite radius, i.e. $R \to \infty$. With $B = 0$, the stress field becomes

$$\begin{aligned}
\sigma_{rr}^{\infty} &= -\frac{A}{r}\sin\theta, \\
\sigma_{\theta\theta}^{\infty} &= -\frac{A}{r}\sin\theta, \\
\sigma_{r\theta}^{\infty} &= \frac{A}{r}\cos\theta.
\end{aligned} \tag{9.20}$$

Here the superscript indicates the limit that both the cylinder length L and radius R go to infinity,[3] i.e. the edge dislocation exists in a medium that is infinite in all three dimensions. The displacement fields corresponding to this solution are given by [11]

$$\begin{aligned}
u_r^{\infty} &= \frac{A}{\mu}(1-\nu)\theta\cos\theta - \frac{A}{2\mu}\sin\theta\left[(1-2\nu)\ln r - \frac{1}{2}\right], \\
u_\theta^{\infty} &= -\frac{A}{\mu}(1-\nu)\theta\sin\theta - \frac{A}{2\mu}\cos\theta\left[(1-2\nu)\ln r + \frac{1}{2}\right],
\end{aligned} \tag{9.21}$$

where ν is Poisson's ratio. To show this we would need to compute the strains from generalized Hooke's law, and integrate the strains to get the displacements. Fortunately, for typical stress functions like the one given in Eq. (9.18), the corresponding stress and displacement fields are available in look-up tables [11]. The displacement field in Cartesian coordinates can be obtained by vector rotation,

$$\begin{aligned}
u_x &= u_r\cos\theta - u_\theta\sin\theta, \\
u_y &= u_r\sin\theta + u_\theta\cos\theta,
\end{aligned} \tag{9.22}$$

which gives

$$\begin{aligned}
u_x^{\infty} &= \frac{A}{\mu}(1-\nu)\theta + \frac{A}{4\mu}\sin 2\theta, \\
u_y^{\infty} &= -\frac{A}{2\mu}(1-2\nu)\ln r - \frac{A}{4\mu}\cos 2\theta.
\end{aligned} \tag{9.23}$$

Similar to the screw dislocation (where $u_z = b\theta/(2\pi)$), here we expect u_x to contain the $b\theta/(2\pi)$ term in order to satisfy the jump condition at the cut plane. Hence we must select the

3 This is different from the case of screw dislocation, in which $\sigma_{\theta z}^{\infty}$ is the correct stress field as long as L goes to infinity regardless of R.

constant A as

$$A = \frac{\mu b}{2\pi(1-\nu)}. \tag{9.24}$$

Therefore, the stress function for a positive edge dislocation in an infinite solid is

$$\phi^\infty = -\frac{\mu b}{2\pi(1-\nu)}(r \ln r)\sin\theta \tag{9.25}$$

or, in Cartesian coordinates,

$$\phi^\infty = -\frac{\mu b}{2\pi(1-\nu)} y \ln\sqrt{x^2+y^2}. \tag{9.26}$$

Displacements around a positive edge dislocation

With the constant A thus determined, we can write down the explicit expressions of the displacement field around a positive edge dislocation in an infinite medium. In cylindrical coordinates, they are

$$\begin{aligned} u_r^\infty &= \frac{b}{2\pi}\theta\cos\theta - \frac{b}{4\pi(1-\nu)}\sin\theta\left[(1-2\nu)\ln r - \frac{1}{2}\right], \\ u_\theta^\infty &= -\frac{b}{2\pi}\theta\sin\theta - \frac{b}{4\pi(1-\nu)}\cos\theta\left[(1-2\nu)\ln r + \frac{1}{2}\right]. \end{aligned} \tag{9.27}$$

The displacements in Cartesian coordinate are

$$\begin{aligned} u_x^\infty &= \frac{b}{2\pi}\left[\theta + \frac{1}{4(1-\nu)}\sin 2\theta\right], \\ u_y^\infty &= -\frac{b}{2\pi}\left[\frac{1-2\nu}{2(1-\nu)}\ln r + \frac{1}{4(1-\nu)}\cos 2\theta\right]. \end{aligned} \tag{9.28}$$

Using $x = r\cos\theta$, $y = r\sin\theta$, these can be rewritten in terms of x and y as

$$\begin{aligned} u_x &= \frac{b}{2\pi}\left[\arctan\left(\frac{y}{x}\right) + \frac{1}{2(1-\nu)}\frac{xy}{x^2+y^2}\right], \\ u_y &= -\frac{b}{2\pi}\left[\frac{1-2\nu}{4(1-\nu)}\ln(x^2+y^2) + \frac{1}{4(1-\nu)}\frac{x^2-y^2}{x^2+y^2}\right]. \end{aligned} \tag{9.29}$$

Note that the displacement field of a straight edge dislocation is independent of shear modulus μ, but depends on the Poisson's ratio ν of the solid.

Stresses around a positive edge dislocation

Given $A = \mu b/[2\pi(1-\nu)]$ and $B = 0$, the stress fields in Eq. (9.19) become the following.

$$\boxed{\begin{aligned} \sigma_{rr}^\infty = \sigma_{\theta\theta}^\infty &= -\frac{\mu b}{2\pi(1-\nu)}\frac{\sin\theta}{r} \\ \sigma_{r\theta}^\infty &= \frac{\mu b}{2\pi(1-\nu)}\frac{\cos\theta}{r} \end{aligned}} \tag{9.30}$$

The stresses in Cartesian coordinates are easily obtained by differentiating the stress function given in Eq. (9.26). The results are

$$
\begin{aligned}
\sigma_{xx}^{\infty} &= \frac{\partial^2 \phi^{\infty}}{\partial y^2} = -\frac{\mu b}{2\pi(1-\nu)} \frac{y(3x^2+y^2)}{(x^2+y^2)^2}, \\
\sigma_{yy}^{\infty} &= \frac{\partial^2 \phi^{\infty}}{\partial x^2} = \frac{\mu b}{2\pi(1-\nu)} \frac{y(x^2-y^2)}{(x^2+y^2)^2}, \\
\sigma_{xy}^{\infty} &= -\frac{\partial^2 \phi^{\infty}}{\partial x \partial y} = \frac{\mu b}{2\pi(1-\nu)} \frac{x(x^2-y^2)}{(x^2+y^2)^2}.
\end{aligned}
\tag{9.31}
$$

For this plane-strain problem, we can also compute the out-of-plane stress from Eq. (2.65),

$$
\sigma_{zz}^{\infty} = \nu\left(\sigma_{xx}^{\infty} + \sigma_{yy}^{\infty}\right) = -\frac{\mu b \nu}{\pi(1-\nu)} \frac{y}{x^2+y^2}, \tag{9.32}
$$

which, of course, comes from the condition that $\varepsilon_{zz} = 0$. We may also want to know the hydrostatic pressure at any point in the solid around the dislocation,

$$
p^{\infty} = -\frac{1}{3}\left(\sigma_{xx}^{\infty} + \sigma_{yy}^{\infty} + \sigma_{zz}^{\infty}\right) = \frac{\mu b(1+\nu)}{3\pi(1-\nu)} \frac{y}{(x^2+y^2)}. \tag{9.33}
$$

Visualization of the stress field

It is helpful to develop an intuition about the stresses around a dislocation, so that one can anticipate how dislocations interact with each other and with other defects. Figure 9.7 shows the in-plane stress components about a positive edge dislocation, as described by the equations above.

We note that the signs of some of the components can be guessed from the nature of the distortion about the extra half-plane. For example, since the extra half-plane is located along the positive y axis, we expect the hydrostatic pressure to be positive in the region $y > 0$, and negative in the region $y < 0$. For the same reason, we expect the normal stress σ_{xx} to be negative in the region $y > 0$, and positive in the region $y < 0$. These are indeed the case, as shown in Fig. 9.7.

Furthermore, from the distortion of the atomic planes around the extra half-plane, we expect the shear stress, σ_{xy}, to be positive on the right hand side of the dislocation, and negative on the left hand side. This is indeed the case along the x axis, as shown in Fig. 9.7. As we will see shortly, a positive σ_{xy} produces a glide force in the positive x direction on an edge dislocation of the same Burgers vector as the one at the origin. Therefore, two edge dislocations on the same glide plane repel each other if they have the same sign (and attract each other if they have the opposite sign).

However, the shear stress σ_{xy} changes sign multiple times as a function of x along a horizontal line if $y \neq 0$. Consequently, if we consider another edge dislocation moving on a parallel glide plane above the origin, the glide force it experiences from the dislocation can be either attractive or repulsive depending on its position. This will be discussed in more detail in Chapter 10. The sign of σ_{xy} for $y \neq 0$ may not be easily guessed by intuitive thinking. Similarly, while the sign of σ_{yy} may have been guessed correctly along the y axis, its behavior may not seem so intuitive along vertical lines with $x \neq 0$.

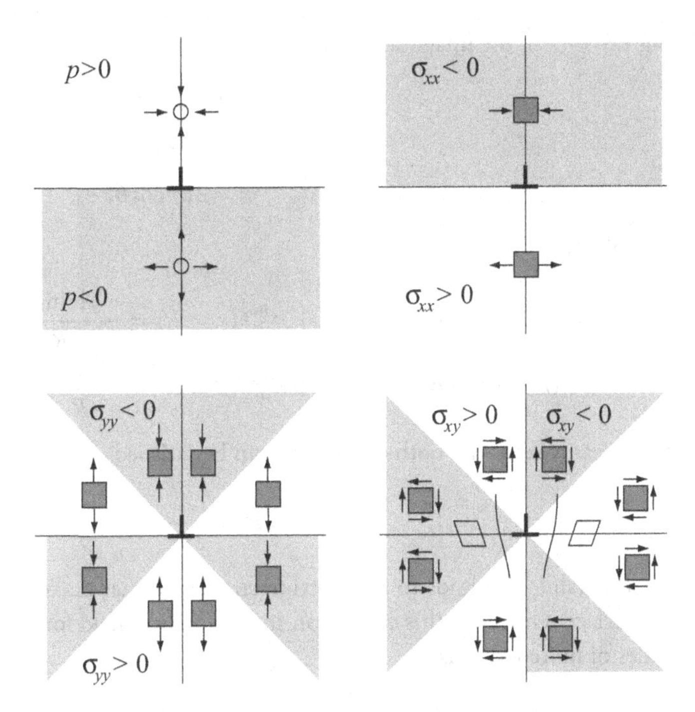

Figure 9.7. Visualization of the stress field about a positive edge dislocation. For each stress component (p, σ_{xx}, σ_{yy}, σ_{xy}), the white regions correspond to positive values and the gray regions correspond to negative values. Note that $p = -(\sigma_{xx} + \sigma_{yy} + \sigma_{zz})/3$ is the hydrostatic pressure.

Certain symmetries of the stress fields may also have been guessed intuitively without knowing the analytic solution. For example, the normal stresses (σ_{xx}, σ_{yy}, and p) are all symmetric with respect to mirror reflection against the y–z plane, and anti-symmetric with respect to mirror reflection against the x–z plane. The shear stress σ_{xy}, on the other hand, is anti-symmetric with respect to mirror reflection against the y–z plane, and symmetric to mirror reflection against the x–z plane. These symmetries are obeyed by the shading patterns shown in Fig. 9.7.

9.1.4 Edge dislocation in a finite radius cylinder

The solution we have described is valid for a positive edge dislocation in an infinite elastic solid. If we want to have a solution for an edge dislocation in an infinitely long cylinder of finite radius, then corrections need to be made to account for the traction-free boundary conditions at the cylindrical surface,

$$\sigma_{rr}(r = R) = \sigma_{\theta r}(r = R) = 0. \tag{9.34}$$

This can be accomplished by choosing a non-zero constant B in the stress function in Eq. (9.18). In other words, we have an image stress function to describe the correction field:

$$\phi^{\text{img}} = B r^3 \sin\theta. \tag{9.35}$$

Using Eq. (9.19), the image stress fields are

$$\begin{aligned}\sigma_{rr}^{\text{img}} &= 2Br\sin\theta,\\ \sigma_{\theta\theta}^{\text{img}} &= 6Br\sin\theta,\\ \sigma_{r\theta}^{\text{img}} &= -2Br\cos\theta.\end{aligned} \tag{9.36}$$

We require that

$$\sigma_{rr}(r=R) = \sigma_{rr}^{\infty}(r=R) + \sigma_{rr}^{\text{img}}(r=R) = -\frac{A\sin\theta}{R} + 2BR\sin\theta = 0,$$

$$\sigma_{r\theta}(r=R) = \sigma_{r\theta}^{\infty}(r=R) + \sigma_{r\theta}^{\text{img}}(r=R) = \frac{A\cos\theta}{R} - 2BR\cos\theta = 0.$$

It can be easily seen that both conditions can be satisfied if

$$B = \frac{A}{2R^2}. \tag{9.37}$$

Thus, this value of B leads to a correction stress field that makes all the tractions on the cylindrical surface zero. With this correction the shear stress field for a positive edge dislocation in a cylinder of finite radius is

$$\begin{aligned}\sigma_{rr} &= \sigma_{rr}^{\infty} + \sigma_{rr}^{\text{img}} = -\frac{\mu b\sin\theta}{2\pi(1-\nu)r}\left[1-\left(\frac{r}{R}\right)^2\right],\\ \sigma_{\theta\theta} &= \sigma_{\theta\theta}^{\infty} + \sigma_{\theta\theta}^{\text{img}} = -\frac{\mu b\sin\theta}{2\pi(1-\nu)r}\left[1-3\left(\frac{r}{R}\right)^2\right],\\ \sigma_{r\theta} &= \sigma_{r\theta}^{\infty} + \sigma_{r\theta}^{\text{img}} = \frac{\mu b\cos\theta}{2\pi(1-\nu)r}\left[1-\left(\frac{r}{R}\right)^2\right].\end{aligned} \tag{9.38}$$

The total stress field is similar in form to the stress field of a screw dislocation in a finite cylinder, Eq. (9.16), except that for the edge dislocation considered here, the total stresses σ_{rr} and $\sigma_{r\theta}$ go continuously to zero at the surface $r = R$ and do not change sign as a function of r.

From the condition that $\varepsilon_{zz} = 0$, the out-of-plane stress can be obtained as

$$\sigma_{zz} = \nu(\sigma_{rr} + \sigma_{\theta\theta}) = -\frac{\mu b\nu\sin\theta}{\pi(1-\nu)r}\left[1-2\left(\frac{r}{R}\right)^2\right]. \tag{9.39}$$

The hydrostatic pressure at any point in the cylinder around the dislocation is

$$p = -\frac{1}{3}(\sigma_{rr} + \sigma_{\theta\theta} + \sigma_{zz}) = \frac{\mu b(1+\nu)\sin\theta}{3\pi(1-\nu)r}\left[1-2\left(\frac{r}{R}\right)^2\right]. \tag{9.40}$$

We note that σ_{zz} and p do change sign near the outer boundary of the cylinder (at $r = R/\sqrt{2}$).

9.1.5 Stress field for a straight mixed dislocation

Using the principle of superposition of solutions for linear elastic problems, we can easily describe the stress field for a mixed dislocation by superimposing the fields of the screw and edge components of that dislocation.

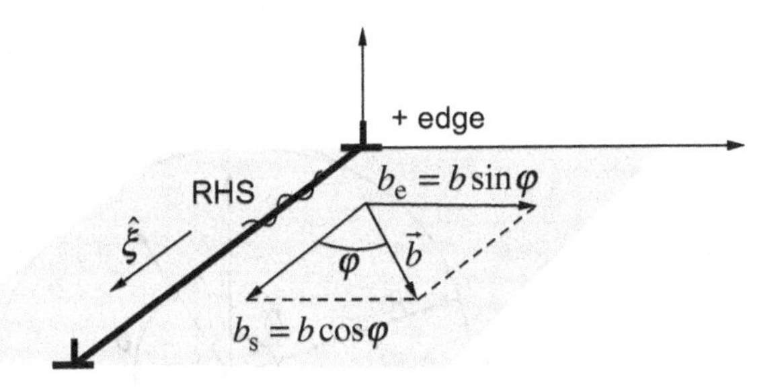

Figure 9.8. Burgers vectors for a mixed dislocation.

In the cylindrical coordinate system (r, θ, z) the stress field of a mixed dislocation in an infinite solid can be expressed as

$$\sigma_{ij}^{\infty} = \begin{bmatrix} \sigma_{rr}^{\infty} & \sigma_{r\theta}^{\infty} & 0 \\ \sigma_{\theta r}^{\infty} & \sigma_{\theta\theta}^{\infty} & \sigma_{\theta z}^{\infty} \\ 0 & \sigma_{z\theta}^{\infty} & \sigma_{zz}^{\infty} \end{bmatrix}, \tag{9.41}$$

where

$$\begin{aligned} \sigma_{rr}^{\infty} = \sigma_{\theta\theta}^{\infty} &= -\frac{\mu b_e \sin\theta}{2\pi(1-\nu)r} = -\frac{\mu b \sin\varphi \sin\theta}{2\pi(1-\nu)r}, \\ \sigma_{r\theta}^{\infty} = \sigma_{\theta r}^{\infty} &= \frac{\mu b_e \cos\theta}{2\pi(1-\nu)r} = \frac{\mu b \sin\varphi \cos\theta}{2\pi(1-\nu)r}, \\ \sigma_{z\theta}^{\infty} = \sigma_{\theta z}^{\infty} &= \frac{\mu b_s}{2\pi r} = \frac{\mu b \cos\varphi}{2\pi r}. \end{aligned} \tag{9.42}$$

The stress, strain, and displacement fields of any two dislocations can be superimposed (i.e. added) to produce the elastic fields when both dislocations are present. The above stress fields of a mixed dislocation are simply the superposition of the stress fields of an edge dislocation (with Burgers vector b_e) and a screw dislocation (with Burgers vector b_s) both located along the z axis.

However, the energy associated with different elastic fields in general cannot be added in this way, because the strain energy density depends quadratically on the stresses and strains (see Chapter 2). Nonetheless, in the following section, we shall see that the strain energy of a straight mixed dislocation is just the sum of the strain energies of its edge and screw components. This is because parallel straight edge and screw dislocations (of infinite length) produce different stress/strain components that don't interact with each other at all in an isotropic elastic medium.

9.2 Dislocation line energy

9.2.1 Screw dislocation

The stresses and strains in the elastic field about a dislocation line lead to the concept of a dislocation line energy – the energy per unit length of the dislocation. Consider a unit length of dislocation in a long, but finite, cylinder of radius R.

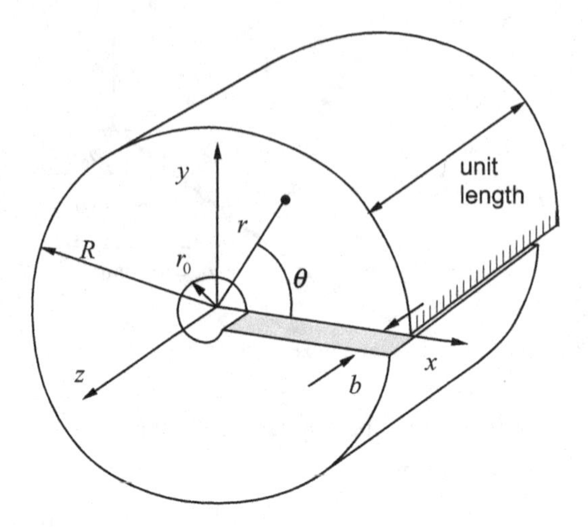

Figure 9.9. Creating a positive screw dislocation in an elastic cylinder by displacing the lower side of the cut surface in the z direction by b.

There are multiple ways to obtain the elastic strain energy stored in the medium. A direct approach is to integrate the strain energy density over the entire cylinder volume. The energy per unit length of the cylinder is

$$W_s^{\mathrm{el}} = \int_{r_0}^{R} \int_0^{2\pi} \frac{(\sigma_{\theta z})^2}{2\mu} r\, dr d\theta. \tag{9.43}$$

Given the analytic expression for $\sigma_{\theta z}$, Eq. (9.16), this integral can be carried out analytically.

However, it is instructive and often easier to obtain the elastic energy using an alterative approach, where a reversible path is envisioned to create a dislocation in an elastic cylinder. The work done along this path equals the elastic energy of the final state, because the elastic energy of the initial state is zero. As long as the path is reversible (i.e. equilibrium is maintained at every intermediate state along the path), the same energy is obtained independent of the path. This allows creativity in the selection of paths that sometimes greatly simplifies the derivation of the energy expression.

Here we consider the following path. We first make a "cut" in the cylinder on the x–z plane on one side of the centerline, $x > 0$. As shown in Fig. 9.9, the upper side of the "cut" is held fixed, while the lower side is displaced in the z direction by an amount q_z. As q_z increases from 0 to b, the original elastic cylinder is gradually transformed to the final configuration that contains the screw dislocation with Burgers vector b.

At every intermediate state along the path, the system must be at equilibrium. Thus a force must be applied to the lower side of the cut surface. Let F_z be the total applied force per unit length along the cylinder, which is a function of q_z. The work done (per unit length) along the path is simply

$$W_s^{\mathrm{el}} = \int_0^b F_z\, dq_z. \tag{9.44}$$

Because the cylinder is a linear elastic medium, F_z must be a linear function of q_z. Therefore,

$$W_s^{\text{el}} = \frac{1}{2} F_z(q_z = b) \cdot b. \tag{9.45}$$

Note that $F_z(q_z = b)$ is the force applied to the lower side of the cut surface when the dislocation is fully created, i.e.

$$F_z(q_z = b) = \int_{r_0}^{R} \sigma_{z\theta}(\theta = 0)\, dr. \tag{9.46}$$

Given Eq. (9.16), we have

$$\begin{aligned} W_s^{\text{el}} &= \frac{b}{2} \int_{r_0}^{R} \frac{\mu b}{2\pi r} \left[1 - 2\left(\frac{r}{R}\right)^2 \right] dr \\ &= \frac{\mu b^2}{4\pi} \left[\ln \frac{R}{r_0} - 1 + \left(\frac{r_0}{R}\right)^2 \right]. \end{aligned} \tag{9.47}$$

The same answer would be obtained if we carry out the double integral in Eq. (9.43). Assuming $r_0/R \ll 1$, we have

$$W_s^{\text{el}} \approx \frac{\mu b^2}{4\pi} \left[\ln \frac{R}{r_0} - 1 \right]. \tag{9.48}$$

This is the elastic energy per unit length. Of course we have omitted the energy of the dislocation core, E_s^{core} (due to non-linear interatomic interactions), in this calculation. With this added in we can express the total energy per unit length of the screw as

$$E_s = W_s^{\text{el}} + E_s^{\text{core}} = \frac{\mu b^2}{4\pi} \left[\ln \frac{R}{r_0} - 1 \right] + E_s^{\text{core}}. \tag{9.49}$$

It is common to write this result in compact form as

$$E_s = \frac{\mu b^2}{4\pi} \ln \frac{\alpha_s R}{r_0}, \tag{9.50}$$

where α_s is given by

$$\alpha_s = \exp\left(\frac{4\pi E_s^{\text{core}}}{\mu b^2} - 1 \right). \tag{9.51}$$

To determine α_s is equivalent to finding the core energy, which requires the use of an atomistic model [66]. The resulting value of α_s is usually of the order unity. We usually estimate the radius of the core of the dislocation to be approximately $r_0 \approx b$ so that

$$E_s = \frac{\mu b^2}{4\pi} \ln \frac{\alpha_s R}{b}. \tag{9.52}$$

A more approximate relation that is often good enough can be expressed by noting that

$$\frac{1}{4\pi} \ln \frac{\alpha_s R}{b} \approx \frac{1}{2}, \tag{9.53}$$

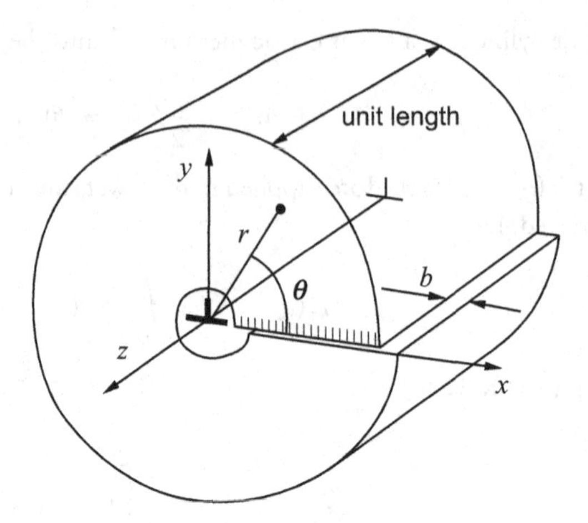

Figure 9.10. Creating a positive edge dislocation in an elastic cylinder by displacing the lower side of the cut surface in the x direction by b.

so that

$$E_s \approx \frac{\mu b^2}{2}. \tag{9.54}$$

We note that the dislocation line energy, which naturally has the dimensions of energy per unit length, depends mainly on the shear modulus and the square of the Burgers vector.

The approximation in Eq. (9.54), although very crude, may be justified by the following. Equation (9.53) is equivalent to the following choice of cylinder radius R,

$$R \approx \frac{b}{\alpha_s} \exp(2\pi) \approx 500b. \tag{9.55}$$

If we assume $b \approx 0.3$ nm, then $R \approx 150$ nm, which is a reasonable value for the mean spacing between dislocations in a crystal. Even though the size of a macroscopic crystal greatly exceeds 150 nm, a real crystal usually contains a distribution of dislocations that screen the long-range stress fields of each other. Therefore, to estimate the line energy of dislocations in a real crystal, it is more appropriate to replace R by the mean spacing between the dislocations. Furthermore, E_s only depends logarithmically on R. Hence E_s would not change appreciably even if R (i.e. mean spacing between dislocations) changes by one order of magnitude. This is why Eq. (9.54) is reasonable as a rough estimate.

9.2.2 Edge dislocation

The energy of a unit length of edge dislocation can be similarly computed by calculating the surface work needed to create the dislocation.

Again we make a "cut" in the cylinder on the x–z plane on one side of the centerline, $x > 0$. As shown in Fig. 9.10, the upper side of the "cut" in the cylinder is held fixed and the lower side is displaced in the x direction by the amount q_x. As q_x increases from 0 to b, the original elastic cylinder is gradually transformed to the final configuration that contains the edge dislocation with Burgers vector b.

At every intermediate state along the path, a force F_x (per unit length) must be applied to the lower side of the cut surface. The work done (per unit length) along the path is simply

$$W_{\rm e}^{\rm el} = \int_0^b F_x \, dq_x = \frac{1}{2} F_x(q_x = b) \cdot b. \tag{9.56}$$

Note that $F_x(q_x = b)$ is the force applied to the lower side of the cut surface when the dislocation is fully created, i.e.

$$F_x(q_z = b) = \int_{r_0}^R \sigma_{r\theta}(\theta = 0) \, dr. \tag{9.57}$$

Given Eq. (9.38), we have

$$\begin{aligned} W_{\rm e}^{\rm el} &= \frac{b}{2} \int_{r_0}^R \frac{\mu b}{2\pi(1-\nu)r} \left[1 - \left(\frac{r}{R}\right)^2\right] dr \\ &= \frac{\mu b^2}{4\pi(1-\nu)} \left[\ln \frac{R}{r_0} - \frac{1}{2} + \frac{r_0^2}{2R^2}\right], \end{aligned} \tag{9.58}$$

which with $r_0/R \ll 1$ leads to

$$W_{\rm e}^{\rm el} = \frac{\mu b^2}{4\pi(1-\nu)} \left[\ln \frac{R}{r_0} - \frac{1}{2}\right]. \tag{9.59}$$

This is the elastic energy per unit length of the edge dislocation, where, again, we have omitted the energy of the dislocation core, $E_{\rm e}^{\rm core}$. With this added in we can express the total energy per unit length of the edge as

$$E_{\rm e} = W_{\rm e}^{\rm el} + E_{\rm e}^{\rm core} = \frac{\mu b^2}{4\pi(1-\nu)} \left[\ln \frac{R}{r_0} - \frac{1}{2}\right] + E_{\rm e}^{\rm core}. \tag{9.60}$$

We can again write this in compact form as

$$E_{\rm e} = \frac{\mu b^2}{4\pi(1-\nu)} \ln \frac{\alpha_{\rm e} R}{r_0}, \tag{9.61}$$

where $\alpha_{\rm e}$ is given by

$$\alpha_{\rm e} = \exp\left[\frac{4\pi(1-\nu)E_{\rm e}^{\rm core}}{\mu b^2} - \frac{1}{2}\right] \tag{9.62}$$

and is usually considered as a constant of order unity. We usually estimate the radius of the core of the dislocation to be approximately $r_0 \approx b$ so that

$$E_{\rm e} = \frac{\mu b^2}{4\pi(1-\nu)} \ln \frac{\alpha_{\rm e} R}{b}. \tag{9.63}$$

A more approximate relation that is often good enough can be expressed by noting that

$$\frac{1}{4\pi} \ln \frac{\alpha_{\rm e} R}{b} \approx \frac{1}{2} \tag{9.64}$$

so that

$$E_{\rm e} \approx \frac{1}{(1-\nu)} \frac{\mu b^2}{2}. \tag{9.65}$$

We see that to a good approximation the energy per unit length of an edge dislocation is typically greater than that of a screw

$$E_e = \frac{1}{(1-\nu)} E_s. \tag{9.66}$$

Taking a typical Poisson's ratio value of $\nu = 1/3$, we find

$$E_e = \frac{3}{2} E_s. \tag{9.67}$$

9.2.3 Mixed dislocation

As will be clear later (using the Peach–Koehler formula), there is no interaction force between an edge dislocation and a screw dislocation if they are both infinitely long and are parallel to each other. Consequently, the total energy of an infinite solid containing these two dislocations is independent of the distance between them. Since a mixed dislocation can be regarded as the superposition of edge and screw components (see Fig. 9.8), it follows that the energy per unit length of a mixed dislocation is simply the sum of the energies of the two components. If φ is the character angle between the line direction and the Burgers vector, then the energies of each of the components are

$$E_e = \frac{\mu b_e^2}{4\pi(1-\nu)} \ln \frac{\alpha_e R}{b} = \frac{\mu (b \sin\varphi)^2}{4\pi(1-\nu)} \ln \frac{\alpha_e R}{b}$$

$$E_s = \frac{\mu b_s^2}{4\pi} \ln \frac{\alpha_s R}{b} = \frac{\mu (b \cos\varphi)^2}{4\pi} \ln \frac{\alpha_s R}{b}$$

and taking $\alpha_e = \alpha_s = \alpha'$ we have

$$E_L(\varphi) = E_e + E_s = \frac{\mu b^2}{4\pi(1-\nu)} \ln \frac{\alpha' R}{b} [\sin^2\varphi + (1-\nu)\cos^2\varphi]$$

$$= \frac{\mu b^2}{4\pi(1-\nu)} \ln \frac{\alpha' R}{b} [1 - \nu \cos^2\varphi]$$

or

$$E_L(\varphi) \approx E_e[1 - \nu \cos^2\varphi], \tag{9.68}$$

where the subscript L indicates this is the energy per unit length. Obviously, this relation recovers the edge and screw limits when $\varphi = \pi/2$ and $\varphi = 0$, respectively.

9.3 Dislocation line tension

In the previous section, we have obtained the energy per unit length of straight dislocations with arbitrary character angle φ, as given in Eq. (9.68). This allows us to estimate the energy of a curved dislocation. Let us now consider a dislocation loop C, which is the simplest curved dislocation structure that satisfies the conservation of the Burgers vector. The energy of this

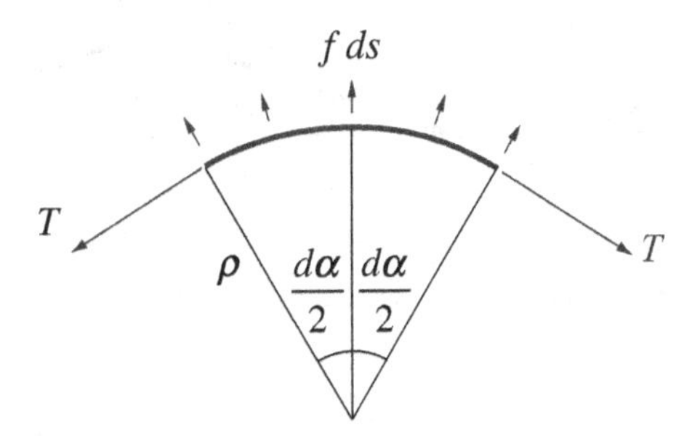

Figure 9.11. Force balance of a taut string subjected to lateral loading.

dislocation loop can be approximated as

$$E \approx \oint_C E_L(\varphi)\, dL, \tag{9.69}$$

where the line integral is carried along the dislocation loop C. Equation (9.69) is only an approximation, because it considers the dislocation as a line that carries an energy per unit length along itself, and ignores the long-range interaction between different sections of the line. The exact expressions for the dislocation energy involves a double integral around the dislocation line [5, 67], in order to account for the long-range interactions.

Because dislocations have line energy, they tend to remain straight, so as to minimize their lengths and, thus, their energies. They act as if a tension force were present along the line of the dislocation, much like the tension in a taut string. Therefore, Eq. (9.69) is called the *line tension* approximation. Despite being only approximate, the line tension expression is able to qualitatively capture certain physical behaviors of dislocations, and hence is a useful model.

9.3.1 Force balance in a string with line tension

For a taut string (such as a stretched rubber band), the line energy E_L is a constant and equals the line tension force T. Figure 9.11 shows a section of the string of length ds at equilibrium while subjected to a lateral force distribution f per unit length. Let ρ be the radius of curvature for this section of the string and $d\alpha$ be the rotation of the line orientation from one end to the other end, so that $ds = \rho\, d\alpha$. The two ends of the string are subjected to the line tension force T. In the limit of small $d\alpha$, the force balance in the vertical direction can be written as

$$\begin{aligned} f ds - 2T \sin\frac{d\alpha}{2} &\approx f ds - T\, d\alpha = 0, \\ f = T\frac{d\alpha}{ds} &= \frac{T}{\rho}, \\ \rho &= \frac{T}{f}. \end{aligned} \tag{9.70}$$

Therefore, a taut string with constant line energy (or line tension), $E_L = T$, when subjected to a constant lateral load f per unit length, will bend into a circular arc with radius $\rho = T/f$. The greater the line tension, the larger the radius, i.e. the straighter the string. This means that the line tension provides the string with a resistance against bending.

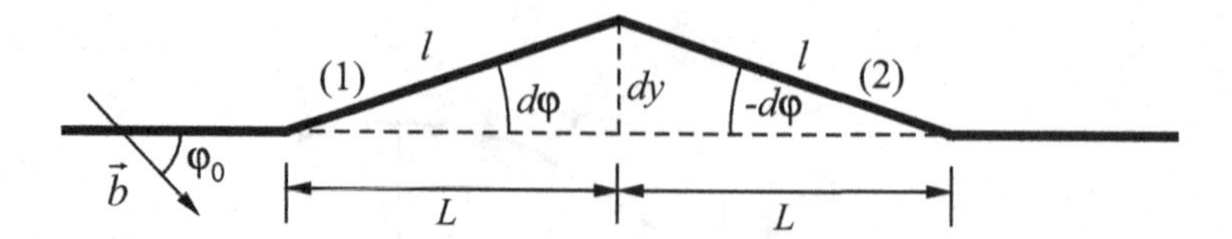

Figure 9.12. Small bow-out on a mixed dislocation line.

A string with constant line energy is a very crude approximation for a dislocation line, because the dislocation line energy depends on its orientation. The edge orientation ($\varphi = \pi/2$) has higher line energy than the screw orientation ($\varphi = 0$), for the usual case of $\nu > 0$. A better approximation is a string with an orientation-dependent line energy, $E_L(\varphi)$. However, the stretched rubber band is no longer a valid model for such a string. This is because, in order to lower its energy, a section of such a string not only has a tendency to shrink its length, it also has a tendency to rotate towards the low energy screw orientation $\varphi = 0$. It is as if each small segment of the string is also subjected to a torque, except when the local orientation is exactly screw or edge. These torques must be taken into account in the calculations of the equilibrium shape of the string subjected to lateral loading.

A full analysis of the equilibrium shape of a string with orientation-dependent line energy requires the method of calculus of variation [68], which is omitted here. Instead, we give a simpler analysis that provides an *effective line tension* T, which can be used to predict the equilibrium shape of the dislocation line, without having to deal with the torques mentioned above. The resulting effective line tension is also a function of the local line orientation φ, but it does not equal to the line energy $E_L(\varphi)$. The result from the simple analysis below is the same as what would be obtained from a full analysis using the calculus of variation.

9.3.2 Orientation-dependent line tension

Since the line tension tends to make a curved dislocation straighten out, we can define the effective line tension T of the dislocation as the change in energy associated with a small deviation from the straight line by bending and stretching. Imagine that a part of the dislocation line of length $2L$ is bowed-out as shown in Fig. 9.12, so that

$$T(\varphi_0) = \frac{dE_L(\varphi_0)}{dL}, \tag{9.71}$$

where $dE_L(\varphi_0)$ is the total energy change and dL is the total length change for the bow-out and where φ_0 is the character angle of the dislocation before the bow-out. In the following, we compute the length change dL and the energy change $dE_L(\varphi_0)$ separately.

Length change dL

The new length l for segment (1) and segment (2) is

$$l = \sqrt{L^2 + (dy)^2}, \tag{9.72}$$

where $dy = L\tan(d\varphi)$. Then in the limit of $d\varphi \to 0$, $\tan(d\varphi) \approx d\varphi$, and so

$$l \approx L[1 + (d\varphi)^2]^{1/2} = L\left[1 + \frac{(d\varphi)^2}{2} + \cdots\right] \approx L\left[1 + \frac{(d\varphi)^2}{2}\right]. \tag{9.73}$$

Hence the length change for the bow-out is (in the limit of small bow-out)

$$dL = 2l - 2L = L(d\varphi)^2. \tag{9.74}$$

Energy change $dE_{\mathrm{L}}(\varphi_0)$

The character angle between the Burgers vector and line direction increases for segment (1) and decreases for segment (2). We can express the energy per unit length of any part of the dislocation line as a series expansion

$$E_{\mathrm{L}}(\varphi) \approx E_{\mathrm{L}}(\varphi_0) + \frac{\partial E_{\mathrm{L}}(\varphi)}{\partial \varphi} d\varphi + \frac{\partial^2 E_{\mathrm{L}}(\varphi)}{\partial \varphi^2}\frac{(d\varphi)^2}{2}. \tag{9.75}$$

Within the line tension approximation, Eq. (9.69), the energy change for the bow-out, $dE_{\mathrm{L}}(\varphi_0)$, is caused by both the length change and orientation change for each of the rotated segments, i.e.

$$\begin{aligned} dE_{\mathrm{L}}(\varphi_0) &= E_{\mathrm{L}}(\varphi_0 + d\varphi)l + E_{\mathrm{L}}(\varphi_0 - d\varphi)l - E_{\mathrm{L}}(\varphi_0)(2L) \\ &\approx \left[E_{\mathrm{L}}(\varphi_0) + \frac{\partial E_{\mathrm{L}}(\varphi)}{\partial \varphi} d\varphi + \frac{\partial^2 E_{\mathrm{L}}(\varphi)}{\partial \varphi^2}\frac{(d\varphi)^2}{2}\right] l \\ &\quad + \left[E_{\mathrm{L}}(\varphi_0) - \frac{\partial E_{\mathrm{L}}(\varphi)}{\partial \varphi} d\varphi + \frac{\partial^2 E_{\mathrm{L}}(\varphi)}{\partial \varphi^2}\frac{(d\varphi)^2}{2}\right] l - E_{\mathrm{L}}(\varphi_0)(2L), \end{aligned} \tag{9.76}$$

which, since the first derivatives cancel out, is

$$dE_{\mathrm{L}}(\varphi_0) = 2\left[E_{\mathrm{L}}(\varphi_0) + \frac{\partial^2 E_{\mathrm{L}}(\varphi)}{\partial \varphi^2}\frac{(d\varphi)^2}{2}\right] l - E_{\mathrm{L}}(\varphi_0)(2L). \tag{9.77}$$

Using the expression in Eq. (9.73) for l, we have

$$\begin{aligned} dE_{\mathrm{L}}(\varphi_0) &\approx 2L\left[E_{\mathrm{L}}(\varphi_0) + \frac{\partial^2 E_{\mathrm{L}}(\varphi)}{\partial \varphi^2}\frac{(d\varphi)^2}{2}\right]\left[1 + \frac{(d\varphi)^2}{2}\right] - E_{\mathrm{L}}(\varphi_0)(2L) \\ &\approx 2L\frac{(d\varphi)^2}{2}\left[E_{\mathrm{L}}(\varphi_0) + \frac{\partial^2 E_{\mathrm{L}}(\varphi)}{\partial \varphi^2}\right], \end{aligned} \tag{9.78}$$

where we have ignored terms of higher order than $(d\varphi)^2$ in the last step.

Line tension $T(\varphi_0)$

Following the definition in Eq. (9.71), and using Eqs. (9.74) and (9.78), the line tension is as follows.

$$\boxed{T(\varphi_0) = E_{\mathrm{L}}(\varphi_0) + \frac{\partial^2 E_{\mathrm{L}}(\varphi)}{\partial \varphi^2}} \tag{9.79}$$

Taking the line energy expression in Eq. (9.68)

$$E_{\mathrm{L}}(\varphi) = E_{\mathrm{e}}(1 - \nu\cos^2\varphi),$$

Table 9.1. Comparison of edge and screw line energies and line tensions for $\nu = 1/3$.

Character	φ	Line energy	Line tension
Edge	$\pi/2$	E_e	$T_e = (1-2\nu)E_e = (1/3)E_e$
Screw	0	$E_s = (1-\nu)E_e = (2/3)E_e$	$T_s = (1+\nu)E_e = (4/3)E_e$

we have

$$T(\varphi_0) = E_e\left(1 - \nu\cos^2\varphi_0 + 2\nu\cos 2\varphi_0\right). \tag{9.80}$$

Using a typical Poisson's ratio value of $\nu = 1/3$, the line energies and line tensions of edge and screw dislocations are as shown in Table 9.1.

We see that while the energy of an edge dislocation is greater than that of a screw, $E_e > E_s$, the line tension of an edge dislocation is smaller than that of a screw, $T_e < T_s$. This apparent anomaly arises because when an edge dislocation is bowed-out, the less energetic screw segments are created, while when a screw dislocation is bowed-out, the more energetic edge segments are created. For this reason it is easier to bow an edge dislocation compared to a screw dislocation.

9.3.3 Equilibrium dislocation shape

The effective line tension given in Eq. (9.80) can be used to analyze the force balance of a section of a string as a model for the dislocation line, as shown in Fig. 9.11. The result is similar to Eq. (9.70), but depends on the local line orientation,

$$\rho(\varphi_0) = \frac{T(\varphi_0)}{f}. \tag{9.81}$$

Here $\rho(\varphi_0)$ is the local radius of curvature of a dislocation line with character angle φ_0, and f is the lateral force per unit length on the dislocation. As we shall see in Section 9.4, $f = \tau b$, where τ is the appropriate shear stress component that exerts force on the dislocation in the glide plane, and b is the Burgers vector of the dislocation.

One consequence of the differences in line tension between the edge and screw is that a dislocation segment does not take a circular shape when it is bowed out by the action of a stress. Instead the screw component bows more gently and the edge component more sharply when bowing occurs. Assume a typical value of $\nu = 1/3$, the line tension of the screw component is four times greater than for an edge component (see Table 9.1). Therefore, the local radius of curvature at the screw dislocation is four times greater than the local radius curvature at the edge dislocation, as shown in Fig. 9.13.

De Wit and Koehler [70] derived the analytic expression for the equilibrium shape of the dislocation line under constant stress τ. For an arbitrary line energy function $E_L(\varphi)$, the coordinates for a dislocation line with an equilibrium shape are described by the following equations

$$\begin{aligned} x(\varphi) &= \frac{1}{\tau b}\left[E_L(\varphi)\sin\varphi + \frac{dE_L(\varphi)}{d\varphi}\cos\varphi\right], \\ y(\varphi) &= \frac{1}{\tau b}\left[E_L(\varphi)\cos\varphi - \frac{dE_L(\varphi)}{d\varphi}\sin\varphi\right], \end{aligned} \tag{9.82}$$

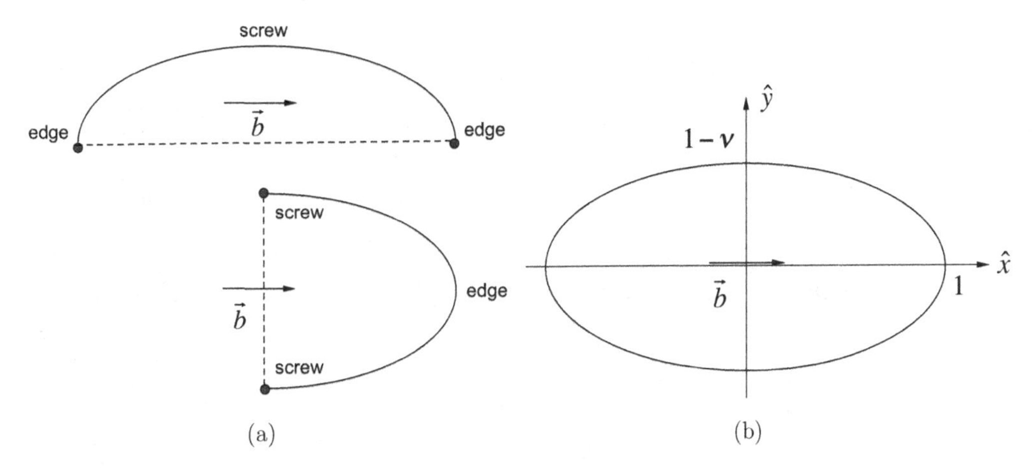

Figure 9.13. (a) Bowing of edge and screw components showing how an edge bows out more easily than a screw. The local radius of curvature is smaller at the edge dislocation than that at the screw dislocation under the same stress condition. (b) Equilibrium shape of a glide dislocation loop under stress in scaled coordinates $\hat{x}$–$\hat{y}$, where $\hat{x} = x\tau b/E_e$, $\hat{y} = y\tau b/E_e$ [69].

where φ is also used as a parameter to trace the curve in the x–y plane. For the line energy expression given in Eq. (9.68), we can obtain more explicit expressions [69]:

$$x(\varphi) = \frac{E_e}{\tau b}\left[\left(1 + \frac{\nu}{4}\right)\sin\varphi + \frac{\nu}{4}\sin 3\varphi\right],$$
$$y(\varphi) = \frac{E_e}{\tau b}\left[\left(1 - \frac{5\nu}{4}\right)\cos\varphi + \frac{\nu}{4}\cos 3\varphi\right]. \tag{9.83}$$

These expressions can be used to describe the stable equilibrium shapes of bowed-out dislocations with their ends pinned as shown in Fig. 9.13a, as well as the (unstable) equilibrium shape of a glide dislocation loop under stress, as shown in Fig. 9.13b. From Eq. (9.83), we have

$$\varphi = 0: \quad x = 0, \quad y = \frac{E_e}{\tau b}(1 - \nu),$$
$$\varphi = \frac{\pi}{2}: \quad x = \frac{E_e}{\tau b}, \quad y = 0. \tag{9.84}$$

Hence the aspect ratio of the equilibrium glide loop is $1 : (1 - \nu)$, as shown in Fig. 9.13b, reflecting the loop's tendency to have longer screw components than edge components.

9.4 Forces on dislocations

In Section 8.4 we discussed the direction of dislocation motion on its glide plane under an applied shear stress. We now consider the forces that act on the dislocation both in and out of the glide plane as a result of the general stress state in the crystal. We will derive the famous Peach–Koehler formula for computing forces on dislocations by first considering some special cases. Throughout this discussion we will note that the configurational forces acting on dislocations arise because the total energy of the system, E, changes as the position x of the dislocation is

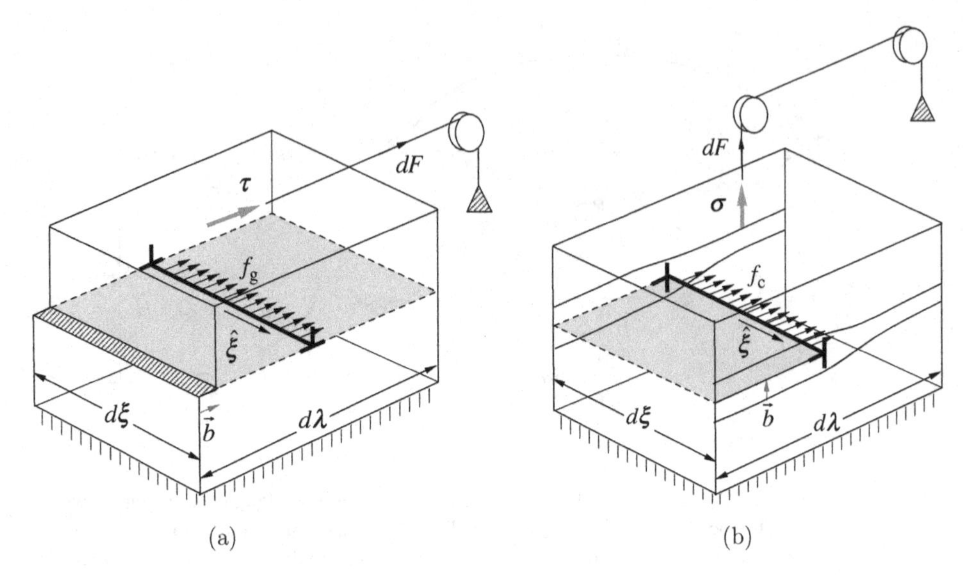

Figure 9.14. (a) Thermodynamic system for calculation of Peach–Koehler glide force on an edge dislocation. (b) Thermodynamic system for calculation of Peach–Koehler climb force on an edge dislocation.

changed. The force is then computed in the usual way by noting that $F_x = -(dE/dx)$, for example. This leads to an expression for the force on a dislocation that is proportional to the local stresses. In the examples considered in this section, the stresses are provided by an external loading mechanism. However, the Peach–Koehler relation between forces and stresses is equally applicable when the stresses are produced by internal sources, such as other defects.

9.4.1 Glide and climb forces on an edge dislocation

First consider an infinitesimal crystal element with dimensions $d\xi$ and $d\lambda$ through which an edge dislocation is passing by glide, as shown in Fig. 9.14a. The Burgers vector $\vec{b}$ and the line sense vector $\hat{\xi}$ of the dislocation are consistent with the RH/SF convention. The character of the dislocation is such that, if the dislocation moves across the crystal element on the glide plane in the direction indicated by the short arrows, then the top part of the element moves by $\vec{b}$.

The shear stress acting on the crystal element can be supplied by a weight, as shown, so that an infinitesimal force, $dF = \tau(d\xi \cdot d\lambda)$, acts on the top face of the infinitesimal element. The applied stress causes a force per unit length, f_g, to act on the dislocation, tending to push it through the element.[4] This is because, as the dislocation moves, the top part of the element moves in the direction of the applied load, allowing the weight to drop. Therefore, the shear stress created by the external weight leads to a distributed (glide) force f_g (per unit length) on the dislocation.

To determine the magnitude of f_g, we make use of a reversible work argument wherein the dislocation is allowed to move reversibly through the crystal element. We consider the energy changes associated with allowing the dislocation segment to glide all the way through the crystal

4 We have used the lower-case f to represent force per unit length and the upper-case F to represent the total force acting on a line or area element. We will use this convention throughout this book.

element. When this happens, the work done by the external loading agent is the force dF on the top part of the element times the distance the *top surface* moves in the process, b, i.e.

$$dW_\tau = b \cdot dF = \tau b(d\xi \cdot d\lambda). \tag{9.85}$$

Equivalently we can express the work as the distributed force $f_g \, d\xi$ on the dislocation times the distance the *dislocation* moves in the process, $d\lambda$, i.e.

$$dW_{\text{equiv}} = (f_g \, d\xi) \cdot d\lambda. \tag{9.86}$$

Equating these equivalent work effects we have

$$\begin{aligned} dW_\tau &= dW_{\text{equiv}} \\ \tau b \, (d\xi \cdot d\lambda) &= (f_g \, d\xi) \cdot d\lambda, \end{aligned} \tag{9.87}$$

so that

$$f_g = \tau b. \tag{9.88}$$

Note that τ is the stress acting on the plane of dislocation glide and in the direction of the Burgers vector. Equation (9.88) is an elementary form of the Peach–Koehler formula; it indicates that the distributed force per unit length on a glide dislocation is simply the product of the shear stress resolved on the slip plane and in the direction of the Burgers vector and the magnitude of the Burgers vector. This simple relation for the glide force (per unit length) holds whatever the character of the dislocation (e.g. edge, screw, mixed).

Another simple case involves the force on a climbing edge dislocation. Consider a crystal element through which an edge dislocation is climbing in response to a tensile stress, as shown in Fig. 9.14b. The character of the dislocation is such that, if the dislocation moves across the crystal element on the climbing plane in the direction indicated by the short arrows, then the top part of the element moves by $\vec{b}$. The work done by the tensile stress on the element is then

$$dW_\sigma = b \cdot dF = \sigma b(d\xi \cdot d\lambda), \tag{9.89}$$

while the equivalent work done by the distributed climb force is

$$dW_{\text{equiv}} = (f_c \, d\xi) \cdot d\lambda. \tag{9.90}$$

Equating these equivalent work effects we have

$$\begin{aligned} dW_\sigma &= dW_{\text{equiv}} \\ \sigma b(d\xi \cdot d\lambda) &= (f_c \, d\xi) \cdot d\lambda, \end{aligned} \tag{9.91}$$

so that

$$f_c = \sigma b. \tag{9.92}$$

In this case the distributed climb force is the product of the tensile stress acting in the direction of the Burgers vector times the magnitude of the Burgers vector.

Both glide force f_g and climb force f_c are thermodynamic driving forces for the given virtual displacement of the dislocation segment, indicating how much free energy reduction would occur, if the virtual displacement were to take place. However, the velocities they produce in the glide and climb directions are usually very different. Because dislocation climb requires diffusion, the climb velocity is usually much smaller than the glide velocities under driving forces with equal magnitudes. Therefore, a common approximation at room temperature is to assume that only glide motion occurs. However, this does not mean that there is no driving for climb; it only means that climb is inhibited kinetically.

9.4.2 Peach–Koehler formula

In both Eqs. (9.88) and (9.92), the stress is acting on the plane of dislocation motion (or virtual motion) being considered and in the direction of the Burgers vector. This is true regardless of the type of dislocation (e.g. screw, edge, mixed), or whether the motion is glide or climb. Therefore, it is reasonable to expect that a more general expression can be obtained that gives the correct force on arbitrary dislocations under an arbitrary stress state, with Eqs. (9.88) and (9.92) being the special cases. Indeed, the more general expression has the following form:

$$f_n = \sigma_{ij} n_j b_i, \tag{9.93}$$

where n_j are the three components of the (unit) normal vector $\hat{n}$ of the plane on which the virtual motion of the dislocation is considered. Here we have adopted the Einstein convention, in which repeated indices (both i and j) are summed over from 1 to 3. Define the dot product between the Burgers vector $\vec{b}$ and the stress tensor $\overleftrightarrow{\sigma}$ as a vector $(\vec{b} \cdot \overleftrightarrow{\sigma})$, whose components are

$$\left(\vec{b} \cdot \overleftrightarrow{\sigma}\right)_j \equiv b_i \sigma_{ij}. \tag{9.94}$$

Then $\sigma_{ij} n_j b_i$ is just the dot product between the vector $(\vec{b} \cdot \overleftrightarrow{\sigma})$ and vector $\hat{n}$, so that,

$$f_n = \left(\vec{b} \cdot \overleftrightarrow{\sigma}\right) \cdot \hat{n} = \vec{b} \cdot \left(\overleftrightarrow{\sigma} \cdot \hat{n}\right) \tag{9.95}$$

Equation (9.95) can tell us the magnitude of the force produced by the stress tensor $\overleftrightarrow{\sigma}$ on the dislocation line projected on any plane $\hat{n}$. However, it is still limited because it cannot unambiguously determine the sign of the force. For one thing, if we reverse the sense vector $\hat{\xi}$ of the dislocation, the Burgers vector $\vec{b}$ will reverse direction, and the f_n given by Eq. (9.93) will change sign. Upon more careful examination, it can be shown that the following equation uniquely specifies the force vector on the dislocation projected on any plane $\hat{n}$,

$$\vec{f}^{(n)} = \left[\left(\vec{b} \cdot \overleftrightarrow{\sigma}\right) \cdot \hat{n}\right]\left(\hat{n} \times \hat{\xi}\right). \tag{9.96}$$

The first term on the right hand side (in square brackets) is a scalar, whose absolute value is the magnitude of the force (per unit length), and the second term is a unit vector; the product of the two terms completely specifies the force vector. If we reverse the sense vector $\hat{\xi}$ and the Burgers vector $\vec{b}$ simultaneously, the force vector $\vec{f}^{(n)}$ given by Eq. (9.96) stays unchanged, as it should. Furthermore, the force vector $\vec{f}^{(n)}$ is also independent of the choice of the sign for $\hat{n}$, as it should.

We can apply Eq. (9.96) to an arbitrary mixed dislocation and determine the glide and climb forces exerted on it by a general stress tensor. Figure 9.15 shows a straight mixed dislocation with sense vector $\hat{\xi}$ and Burgers vector $\vec{b}$, with edge component b_e and screw component b_s. The glide plane, with normal vector $\hat{n}^{(g)}$, is the plane that simultaneously contains the line vector $\hat{\xi}$ and

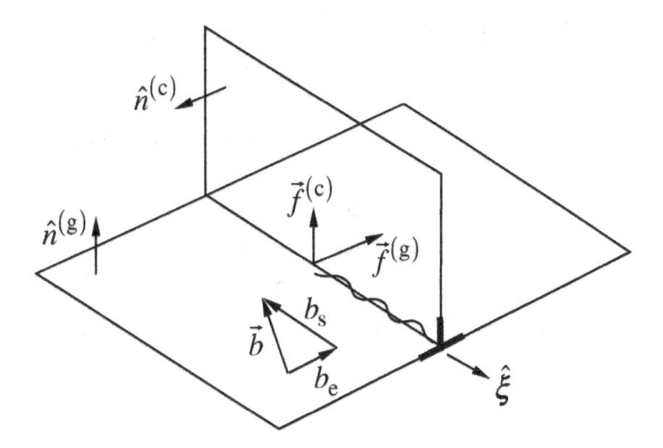

Figure 9.15. Glide and climb forces on a mixed dislocation.

the Burgers vector $\vec{b}$. The climb plane, with normal vector $\hat{n}^{(c)}$, is the plane that contains $\hat{\xi}$ and is perpendicular to the glide plane.

When we substitute $\hat{n}^{(g)}$ into Eq. (9.96), we obtain the glide force vector,

$$\vec{f}^{(g)} = \left[\left(\vec{b} \cdot \overleftrightarrow{\sigma}\right) \cdot \hat{n}^{(g)}\right]\left(\hat{n}^{(g)} \times \hat{\xi}\right). \tag{9.97}$$

This is consistent with Eq. (9.88) if the dislocation is pure edge. When we substitute $\hat{n}^{(c)}$ into Eq. (9.96), we obtain the climb force vector,

$$\vec{f}^{(c)} = \left[\left(\vec{b} \cdot \overleftrightarrow{\sigma}\right) \cdot \hat{n}^{(c)}\right]\left(\hat{n}^{(c)} \times \hat{\xi}\right). \tag{9.98}$$

This is consistent with Eq. (9.92) if the dislocation is pure edge.

Because a dislocation is a line object, motion along the line has no physical meaning. Therefore, a dislocation's velocity is only meaningful in the two directions perpendicular to the dislocation line. The same is true for the force acting on the dislocation, so that this force has only two independent components. For a mixed dislocation, the glide force vector $\vec{f}^{(g)}$ and the climb force vector $\vec{f}^{(c)}$ are two orthogonal vectors. They can be considered as the two independent components of the total force, so that the total force vector on the dislocation is

$$\begin{aligned}\vec{f} &= \vec{f}^{(g)} + \vec{f}^{(c)} \\ &= \left[\left(\vec{b} \cdot \overleftrightarrow{\sigma}\right) \cdot \hat{n}^{(g)}\right]\left(\hat{n}^{(g)} \times \hat{\xi}\right) + \left[\left(\vec{b} \cdot \overleftrightarrow{\sigma}\right) \cdot \hat{n}^{(c)}\right]\left(\hat{n}^{(c)} \times \hat{\xi}\right).\end{aligned} \tag{9.99}$$

This expression can be shown to simply reduce to the following.

$$\boxed{\vec{f} = \left(\vec{b} \cdot \overleftrightarrow{\sigma}\right) \times \hat{\xi}} \tag{9.100}$$

This is the celebrated Peach–Koehler formula.

The Peach–Koehler formula gives the total force vector (per unit length) acting on an arbitrary dislocation with Burgers vector $\vec{b}$. It can be seen from Eq. (9.100) that, if the directions of both $\hat{\xi}$ and $\vec{b}$ are reversed simultaneously, the Peach–Koehler force $\vec{f}$ remains unchanged, as it should.

Given the Peach–Koehler force vector, the glide force component can be obtained by projecting the force $\vec{f}$ onto the glide plane, i.e.

$$\vec{f}^{(g)} = \vec{f} - \left(\vec{f} \cdot \hat{n}^{(g)}\right)\hat{n}^{(g)}. \tag{9.101}$$

It can be shown that this is identical to Eq. (9.97). Obviously, if we just want to find out the glide force on the dislocation, it is easier to use Eq. (9.97) directly, instead of using the Peach–Koehler formula, Eq. (9.100), and then carrying out the orthogonalization, Eq. (9.101).

Similarly, given the Peach–Koehler force vector, the climb force component can be obtained by

$$\vec{f}^{(c)} = \vec{f} - \left(\vec{f} \cdot \hat{n}^{(c)}\right)\hat{n}^{(c)}. \tag{9.102}$$

It can be shown that this is identical to Eq. (9.98). If we just want to find out the climb force on the dislocation, it is easier to use Eq. (9.98) directly, instead of using Eqs. (9.100) and then (9.102).

9.4.3 Examples

Here we give a few examples on how to evaluate the force on a dislocation for a given stress state. We will see that sometimes it is more convenient to use the Peach–Koehler formula, Eq. (9.100), while at other times, it is more convenient to use the in-plane force expression Eq. (9.93) or Eq. (9.96).

Force on a mixed dislocation

If we want to find the force on a dislocation projected on a specific plane, it is more convenient to use Eq. (9.96), which states that the in-plane force is the unit vector $(\hat{n} \times \hat{\xi})$ times the scalar,

$$f_n = \sigma_{ij} n_j b_i.$$

When f_n is positive, the in-plane force is along the $\hat{n} \times \hat{\xi}$ direction; when f_n is negative, the in-plane force points opposite to the $\hat{n} \times \hat{\xi}$ direction.

Consider an arbitrarily oriented mixed dislocation shown in Fig. 9.16a. The projected force on an arbitrary plane $\hat{n}$ can be written out explicitly as

$$\begin{aligned} f_n = {} & \sigma_{11} n_1 b_1 + \sigma_{12} n_2 b_1 + \sigma_{13} n_3 b_1 + \sigma_{21} n_1 b_2 + \sigma_{22} n_2 b_2 + \sigma_{23} n_3 b_2 \\ & + \sigma_{31} n_1 b_3 + \sigma_{32} n_2 b_3 + \sigma_{33} n_3 b_3. \end{aligned} \tag{9.103}$$

Notice that f_n is independent of the line vector $\hat{\xi}$, as long as $\hat{\xi}$ lies within the plane $\hat{n}$, i.e. $\hat{\xi} \cdot \hat{n} = 0$.

As an example consider a dislocation segment lying in the $x_1 x_3$ plane as shown in Fig. 9.16b. The plane normal of interest is now $n_1 = 0$, $n_2 = 1$, $n_3 = 0$ so the projected force is

$$f_n = \sigma_{12} b_1 + \sigma_{22} b_2 + \sigma_{32} b_3. \tag{9.104}$$

The first term in this expression is the glide force on an edge component b_1, the second term is the climb force on an edge component b_2, and the last term is a glide force on a screw component b_3, as shown schematically in Fig. 9.16b.

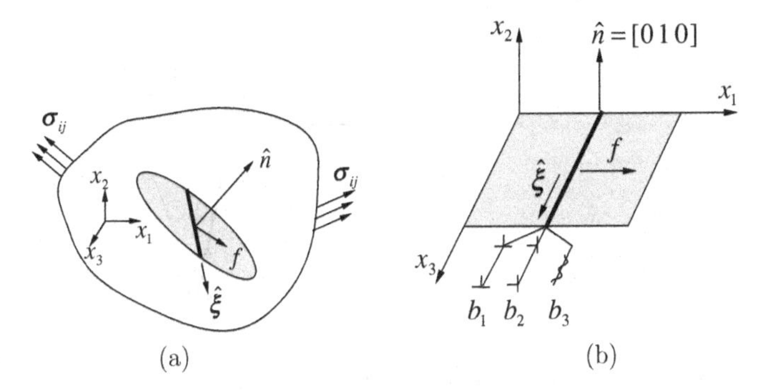

Figure 9.16. (a) Projection of Peach–Koehler force on plane $\hat{n}$ for an arbitrarily oriented mixed dislocation. (b) Projection of Peach–Koehler force when the dislocation line vector $\hat{\xi}$ is along [0 0 1] and the projection plane normal is $\hat{n} = [0\ 1\ 0]$.

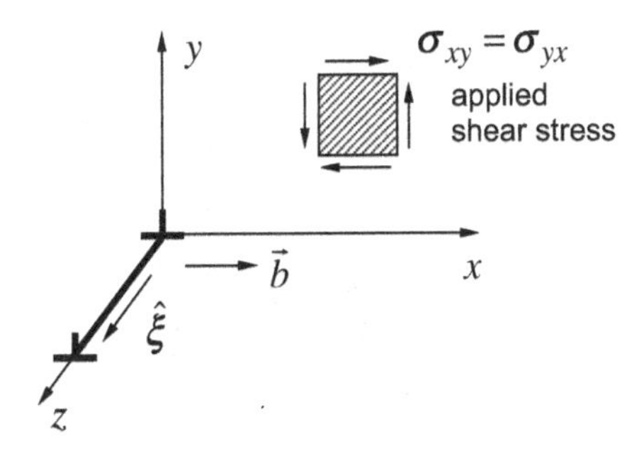

Figure 9.17. Positive edge dislocation subjected to a shear stress.

Edge dislocation under shear stress

We next consider some simple cases for which the Peach–Koehler force can be determined by using basic formulas derived earlier and our intuition. Indeed, the intuitive approach described here can be quite useful because many problems are sufficiently simple that the full tensor multiplication in the Peach–Koehler formula is not necessary.

Suppose a positive edge dislocation lies along the z axis and the crystal is subjected to a pure shear stress σ_{xy}, as shown in Fig. 9.17. In this case the shear stress σ_{xy} is the resolved shear stress τ in the slip plane of the dislocation and in the direction of the Burgers vector. So the simple Eq. (9.88), $f_g = \tau b$, can be used to write the vector force (per unit length) on the dislocation:

$$\vec{f} = [\sigma_{xy}b \quad 0 \quad 0] \tag{9.105}$$

and the full tensor multiplication is not required.

As a tutorial exercise, in the following we shall use the full Peach–Koehler formula to obtain the same result. As shown in the figure, the sense vector can be chosen to be $\hat{\xi} = [0\ 0\ 1]$ so that the Burgers vector is $\vec{b} = b[1\ 0\ 0]$. Notice that it is convenient to write the Burgers vector as the product of the magnitude of the Burgers vector, b, and a unit vector, in this case, [1 0 0]. We

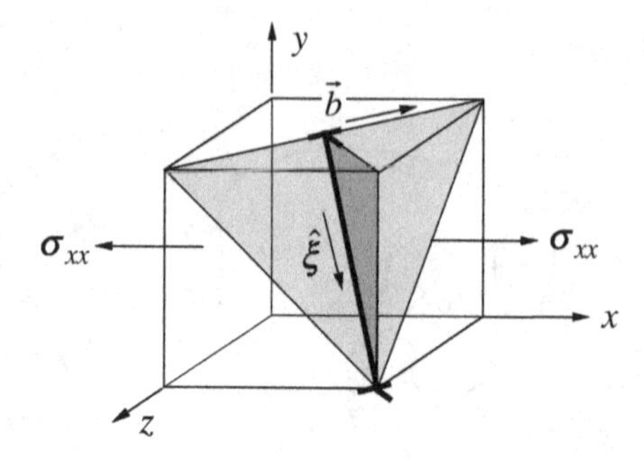

Figure 9.18. Edge dislocation subjected to tension stress.

will use this practice often in this book. The applied stress is simply

$$\overleftrightarrow{\sigma} = \begin{bmatrix} 0 & \sigma_{xy} & 0 \\ \sigma_{yx} & 0 & 0 \\ 0 & 0 & 0 \end{bmatrix}. \tag{9.106}$$

We now use the Peach–Koehler formula to obtain the vector force (per unit length) acting on the dislocation

$$\begin{aligned} \vec{f} &= (\vec{b} \cdot \overleftrightarrow{\sigma}) \times \hat{\xi} \\ &= \left\{ b\,[1\,0\,0] \cdot \begin{bmatrix} 0 & \sigma_{xy} & 0 \\ \sigma_{yx} & 0 & 0 \\ 0 & 0 & 0 \end{bmatrix} \right\} \times \begin{bmatrix} 0 \\ 0 \\ 1 \end{bmatrix} \\ &= b \begin{bmatrix} 0 \\ \sigma_{xy} \\ 0 \end{bmatrix} \times \begin{bmatrix} 0 \\ 0 \\ 1 \end{bmatrix} \\ &= [\sigma_{xy} b \quad 0 \quad 0]. \end{aligned} \tag{9.107}$$

This is, of course, identical to the result, Eq. (9.105), obtained using the simple formula and intuition.

Edge dislocation under tensile stress

We now consider a simple Peach–Koehler problem for which the solution cannot be easily guessed using the simple formulas. Figure 9.18 shows an edge dislocation on the (1 1 1) plane in an FCC crystal element subjected to a pure tensile stress, σ_{xx}.

The sense vector in Fig. 9.18 is a unit vector, $\hat{\xi} = \frac{1}{\sqrt{6}}[1\,\bar{2}\,1]$ and the corresponding Burgers vector is $\vec{b} = b\frac{1}{\sqrt{2}}[1\,0\,\bar{1}]$. As mentioned earlier, it is convenient to express the Burgers vector as the product of the magnitude of the Burgers vector, b, times a unit vector in the direction of the Burgers vector, $\frac{1}{\sqrt{2}}[1\,0\,\bar{1}]$. The applied stress is simply

$$\overleftrightarrow{\sigma} = \begin{bmatrix} \sigma_{xx} & 0 & 0 \\ 0 & 0 & 0 \\ 0 & 0 & 0 \end{bmatrix}. \tag{9.108}$$

Now using the Peach–Koehler formula we can compute the force vector (per unit length) on the dislocation line.

$$
\begin{aligned}
\vec{f} &= (\vec{b}\cdot\overleftrightarrow{\sigma}) \times \hat{\xi} \\
&= \left\{ \frac{b}{\sqrt{2}}[1 \quad 0 \quad -1] \cdot \begin{bmatrix} \sigma_{xx} & 0 & 0 \\ 0 & 0 & 0 \\ 0 & 0 & 0 \end{bmatrix} \right\} \times \frac{1}{\sqrt{6}} \begin{bmatrix} 1 \\ -2 \\ 1 \end{bmatrix} \\
&= \frac{1}{\sqrt{12}} b \begin{bmatrix} \sigma_{xx} \\ 0 \\ 0 \end{bmatrix} \times \begin{bmatrix} 1 \\ -2 \\ 1 \end{bmatrix} \\
&= -\frac{b\sigma_{xx}}{\sqrt{12}} [0 \quad 1 \quad 2].
\end{aligned} \tag{9.109}
$$

The glide force vector can be obtained by subtracting out the component normal to the plane $\hat{n}^{(g)} = \frac{1}{\sqrt{3}}[1\ 1\ 1]$,

$$
\vec{f}^{(g)} = \vec{f} - (\vec{f} \cdot \hat{n}^{(g)})\hat{n}^{(g)} = \frac{b\sigma_{xx}}{\sqrt{12}}[1 \quad 0 \quad \bar{1}]. \tag{9.110}
$$

This is in the same direction as the Burgers vector $\vec{b}$, as may be expected from intuition. The climb force vector can be obtained by

$$
\vec{f}^{(c)} = \vec{f} - \vec{f}^{(g)} = -\frac{b\sigma_{xx}}{\sqrt{12}}[1 \quad 1 \quad 1]. \tag{9.111}
$$

This is in the opposite direction of the slip plane normal, as may be expected from intuition.

9.5 Summary

When dislocations traverse a crystal during plastic deformation they carry with them elastic fields through which they interact with other dislocations and other lattice defects. Understanding these fields starts with an analysis of the displacements, strains, and stresses surrounding infinitely long, straight dislocations. A pure screw dislocation in an infinitely long cylinder is a problem of anti-plane strain, with the out-of-plane displacement being governed by Laplace's equation. The solution, together with the Burgers condition, leads to shear stresses of the form $\sigma_{r\theta} = \mu b/(2\pi r)$. This shows the characteristic r^{-1} singularity that is found for the stresses and strains for all types of dislocations. Allowing the end of the cylinder to be torque-free produces a twist that alters the field of the screw dislocation by adding a torsional elastic field. The torsional field stabilizes the dislocation at the center of the cylinder.

An edge dislocation in a long cylinder is a plane strain elasticity problem that can be solved by finding the Airy stress function that satisfies the biharmonic equation as well as the Burgers condition. The stresses are found by differentiating the Airy stress function. The strains are then found from the stresses using Hooke's law and the displacements are obtained by integrating the strains. If the cylinder has a finite radius, a correction stress function is needed to account for the traction-free cylindrical surface. This introduces corrections to the stress field that are significant only far from the center of the dislocation. The elastic fields of screw and edge dislocations can be superposed to obtain the elastic fields of mixed dislocations having both screw and edge components.

The line energy of a straight dislocation may be found either by integrating the strain energy density in the field of that dislocation or by computing the work required to create the dislocation by enforcing a Burgers displacement at a suitable cut extending away from the dislocation line. The result for a pure screw dislocation takes the approximate form $E_s \approx \mu b^2/2$, while that for an edge is $E_e \approx \mu b^2/[2(1-\nu)]$, both of which have the dimensions of energy per unit length. The line energies of mixed dislocations, found by summing the energies of their edge and screw components, leads to an expression for line energy per unit length that depends on the orientation of the dislocation in its slip plane. That, in turn, leads to an orientation-dependent line tension that controls the shape of a dislocation under stress.

The work associated with a virtual movement of a dislocation subjected to stress can be described with the Peach–Koehler formula, which gives the force per unit length, causing the dislocation line to move normal to itself, $f_n = \sigma_{ij} n_j b_i$, where σ_{ij} is the local stress at that point, n_j is the normal to the plane on which the dislocation is moving and b_i is the Burgers vector. For the simple case of a resolved shear stress τ, this gives $f_{\text{glide}} = \tau b$, whereas for the simple case of a tensile stress σ in the direction of the Burgers vector, this gives $f_{\text{climb}} = \sigma b$. In general, the force vector per unit length acting on a dislocation can be expressed as $\vec{f} = (\vec{b} \cdot \overleftrightarrow{\sigma}) \times \hat{\xi}$.

9.6 Exercise problems

In all the exercise problems below, we model the solid as an elastically isotropic medium with shear modulus μ and Poisson's ratio ν.

9.1 Show that the Airy stress function, Eq. (9.18), for an edge dislocation satisfies the biharmonic equation, Eq. (2.69).

9.2 An edge dislocation is located at the very center of an infinitely long cylinder with radius R and surfaces that are traction free. Write an expression for the σ_{xy} component of stress along the diagonal line $x = y$.

9.3 Show that an edge dislocation in a long cylinder of finite radius, R, and finite length, L, would not cause the cylinder to either twist or bend (i.e. the cylinder would not twist or bend when the dislocation is created).

9.4 Epitaxial thin films of GaN grown on sapphire are of importance in optoelectronics because they permit blue LEDs to be created (an achievement recognized by the 2014 Nobel Prize in Physics). When this material is grown by metal organic chemical vapor deposition (MOCVD) techniques, unusual defects in the form of tiny cylindrical holes are found in the films. One theory for these defects involves the concept of a "hollow screw dislocation," in which the material in the core of a screw dislocation intersecting the surface of the film is removed by surface diffusion. The high elastic energy of the material at the core of the dislocation is assumed to be the driving force for this process. Figure 9.19 shows the development of such a cylindrical hole by surface diffusion. As material is removed from the core of the dislocation, the cylindrical hole grows bigger and bigger, with the consequence that the elastic energy of the dislocation decreases. But the total surface energy of the system also increases in this process and this ultimately limits the size of the hole.

Using your knowledge of dislocation theory and assuming that the free surface energy of the solid is represented by γ_s, develop an expression for the equilibrium radius, a_{eq}, of the hollow dislocation.

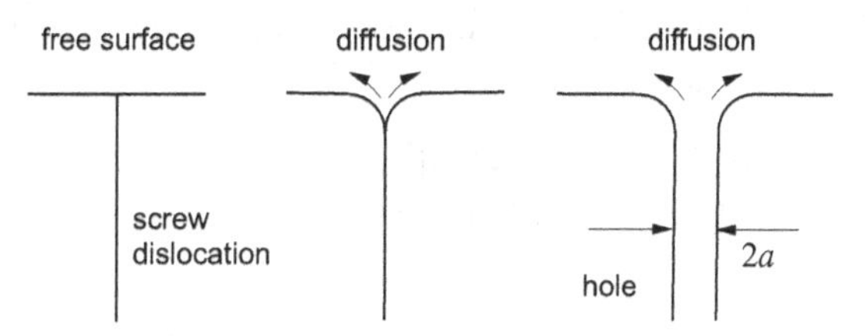

Figure 9.19. Formation of a hollow screw dislocation.

9.5 A right-handed screw dislocation is positioned a small distance $x_0 \ll a$ from the center of a long whisker or nanowire of radius a, as shown in Fig. 9.20.

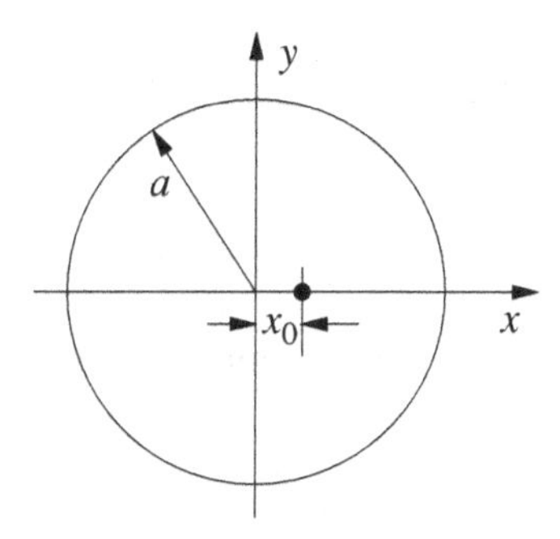

Figure 9.20. A screw dislocation offset from the center of a cylinder.

We wish to investigate the stability of such a dislocation with respect to whether it tends to leave the nanowire or return to the center. To perform the stability analysis we need to know the elastic field of a screw dislocation in a cylinder when the dislocation is not at the center. Based on the results in Section 10.6.6, in particular Eq. (10.133), the stress field for a right-handed screw dislocation in an infinitely long and clamped cylinder, is:

$$\begin{aligned}\sigma_{xz} &= -\frac{\mu b}{2\pi}\left[\frac{y}{(x-x_0)^2+y^2} - \frac{y}{(x-a^2/x_0)^2+y^2}\right],\\ \sigma_{yz} &= \frac{\mu b}{2\pi}\left[\frac{x-x_0}{(x-x_0)^2+y^2} - \frac{x-a^2/x_0}{(x-a^2/x_0)^2+y^2}\right].\end{aligned} \tag{9.112}$$

The first terms in these expressions are those expected for a dislocation in an infinite solid. The second terms are the "image" corrections that arise when the cylindrical surface is made traction-free. As noted above, these expressions apply if the cylinder is constrained at its ends and not free to twist.

(a) Using the expressions given in Eq. (9.112), determine the force acting on the dislocation positioned at $x = x_0 \ll a$.

(b) What can be said about the stability of the screw dislocation when placed in such a constrained cylinder?

(c) We now allow the cylinder to be torque-free. Assuming that the dislocation is very close to the center, at $x = x_0 \ll a$, estimate the force (per unit length) on the dislocation associated with removal of the torque (this can be done approximately using results in Section 9.1.2).

(d) Based on this analysis, comment on the stability of a screw dislocation near the center of a nanowire that is torque free.

9.6 Derive an expression for the elastic energy (per unit length) of a screw dislocation which lies along the axis of a cylinder of finite length with outside radius R and inside radius r_0 (the radius of the core).

Compare the result with the energy per unit length of a screw dislocation in a similar cylinder of infinite length. Give a qualitative argument to account for any differences found.

9.7 Determine the sign (RHS or LHS) of the screw dislocation in the central PbSe nanowire causing the twist shown in the SEM micrograph in Fig. 9.5. Explain your reasoning in terms of the elastic fields of dislocations.

9.8 A right-handed screw dislocation in a BCC crystal is shown in Fig. 9.21a. The shaded plane indicates the slip plane for the dislocation. An equal-biaxial tension stress, $\sigma_{xx} = \sigma_{yy} = \sigma$, is applied to the crystal.

(a) Write an expression for the vector force per unit length on the dislocation due to this applied stress.

(b) Indicate the direction of glide motion caused by the applied stress.

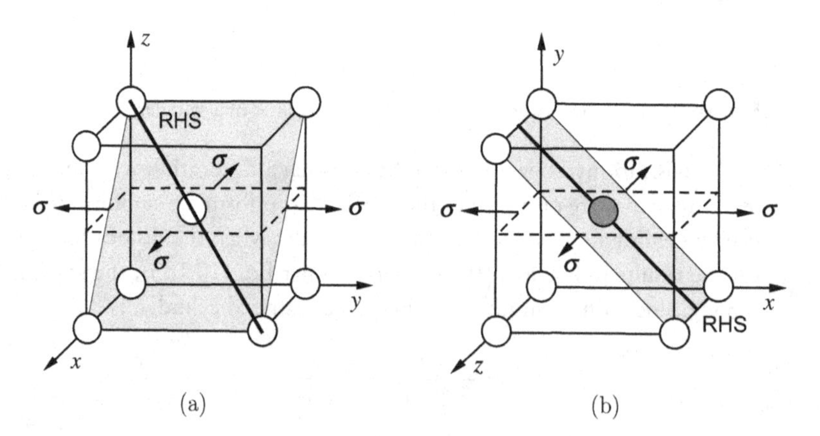

Figure 9.21. (a) A screw dislocation under a biaxial tensile stress. (b) A right-handed screw dislocation in NiAl subjected to a biaxial tensile stress.

9.9 A single crystal of NiAl with a pure right-handed screw dislocation lying on the (110) plane is subjected to a biaxial tensile stress, σ, as shown in Fig. 9.21b. Use the Peach–Koehler formula to determine the glide force (per unit length) acting on the dislocation.

9.10 A (0 0 1) epitaxial, single crystal film of Ge (DC crystal structure and FCC lattice) is deposited onto a (0 0 1) single crystal Si substrate and thus subjected to a biaxial tensile stress

$\sigma_{xx} = \sigma_{yy} = \sigma$. Figure 9.22 shows a (1 1 1) slip plane (slighted shaded) in the crystal as well as the sense vector and Burgers vector of a dislocation gliding on that plane (the bold line is the dislocation). As the dislocation segment AA moves in the film, it deposits a "misfit" dislocation at the film substrate interface. We wish to calculate the forces on the segment AA due to the applied stress.

(a) Use the Peach–Koehler formula and write an expression for the vector force (per unit length) on the dislocation segment AA.
(b) Determine the direction of the glide force on dislocation segment AA by computing the dot product of the force on the dislocation with the direction vector $\vec{V}$, as shown in Fig. 9.22.
(c) Will the biaxial stress drive the segment AA in the direction of $\vec{V}$ or in the opposite direction?

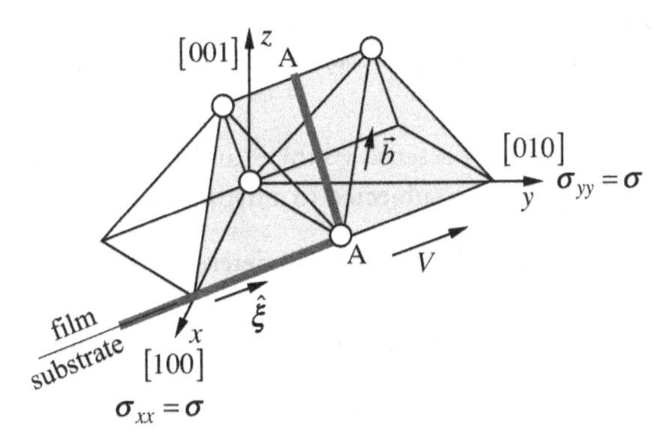

Figure 9.22. Slip pyramids in an (0 0 1) epitaxial film of Ge under biaxial tension. The spheres are the apexes of a Thompson's tetrahedron (see Section 11.1.5). The shaded plane is the slip plane for the dislocation shown as a bold line.

9.11 An FCC single crystal film with the (0 0 1) orientation is subjected to an in-plane equal-biaxial stress $\sigma_{xx} = \sigma_{yy} = \sigma$, as shown in Fig. 9.23a. A pure edge dislocation with sense vector $\hat{\xi} = \frac{1}{\sqrt{6}}[\bar{1}\,\bar{1}\,2]$ and Burgers vector $\vec{b} = \frac{b}{\sqrt{2}}[1\,\bar{1}\,0]$ is present in the crystal.

(a) Derive an expression for the force per unit length on the edge dislocation due to the applied stress. Give the force in terms of its components in the xyz coordinate system.
(b) Transform the stresses to find the applied stress components $\sigma_{\alpha\alpha}$ and $\sigma_{\alpha\beta}$.
(c) Calculate the glide force per unit length acting in the α direction and the climb force per unit length acting in the β direction.

9.12 A pure edge dislocation in an FCC crystal is shown in Fig. 9.23b. The crystal is loaded in compression along the y axis, as shown.

(a) Use the Peach–Koehler formula to develop an expression for the vector force (per unit length) on the dislocation.
(b) Write expressions for the glide and climb forces (per unit length) acting on the dislocation and indicate the direction of these forces.
(c) Make a sketch showing the direction for the glide and climb forces.

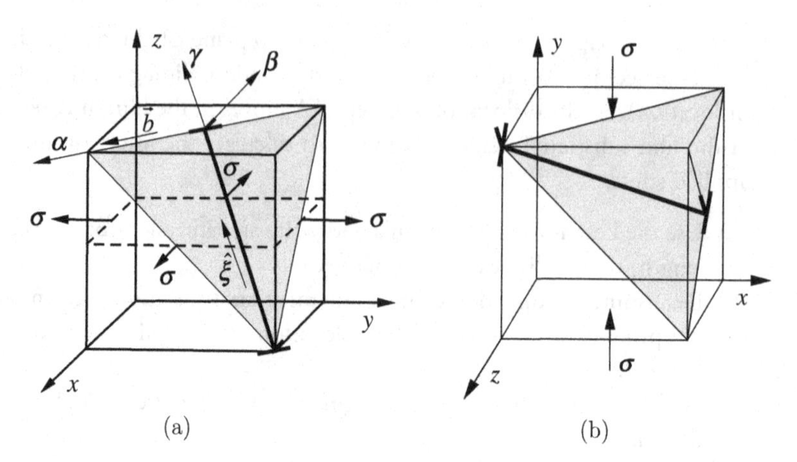

Figure 9.23. (a) An edge dislocation in an FCC single crystal subjected to a biaxial tensile stress. (b) An edge dislocation in an FCC single crystal subjected to a uniaxial compressive stress.

9.13 A 60° dislocation with a sense vector $\hat{\xi}$ and Burgers vector $\vec{b}$ in a cubic crystal is shown in Fig. 9.24a. The crystal is subjected to a hydrostatic pressure, p.

(a) Use the Peach–Koehler formula to determine the force (per unit length) on the dislocation due to the applied pressure.
(b) Show that the direction of the force and the magnitude is consistent with physical expectation.

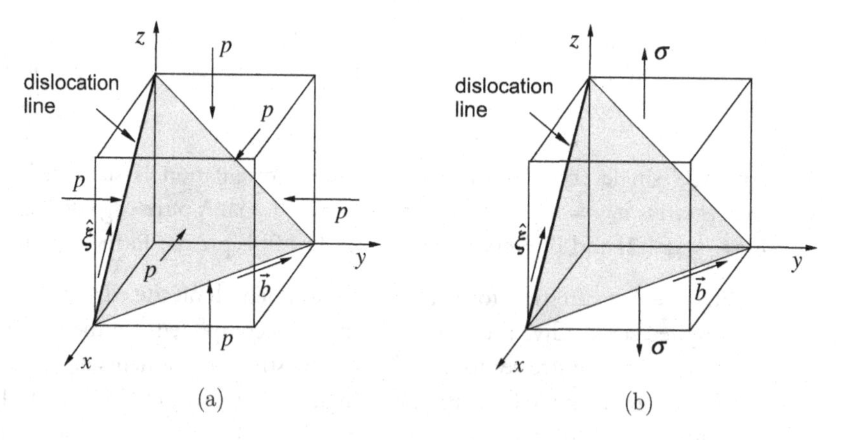

Figure 9.24. (a) A 60° dislocation in a cubic crystal under hydrostatic pressure. (b) A 60° dislocation subjected to uniaxial tension.

9.14 A 60° dislocation with a sense vector $\hat{\xi}$ and Burgers vector $\vec{b}$ in a cubic crystal is shown in Fig. 9.24b. The crystal is subjected to a tensile stress in the z direction.

(a) Use the Peach–Koehler formula to determine the force (per unit length) on the dislocation due to the applied stress.
(b) Indicate whether the calculated force is consistent with physical expectation.

9.15 A right-handed screw dislocation lies along a $[\bar{1}\,1\,1]$ direction on a $(1\,1\,0)$ slip plane in a BCC crystal as shown in Fig. 9.25a. The crystal is subjected to a tensile stress σ along the x axis as shown.

(a) Write an expression for the force (per unit length) on the dislocation due to the applied stress. This should be a vector expression with x, y, and z components specified.
(b) If the dislocation is constrained to glide on the slip plane shown, indicate the direction of glide motion due to the applied stress.
(c) Write an expression for the magnitude of the glide force per unit length of dislocation.

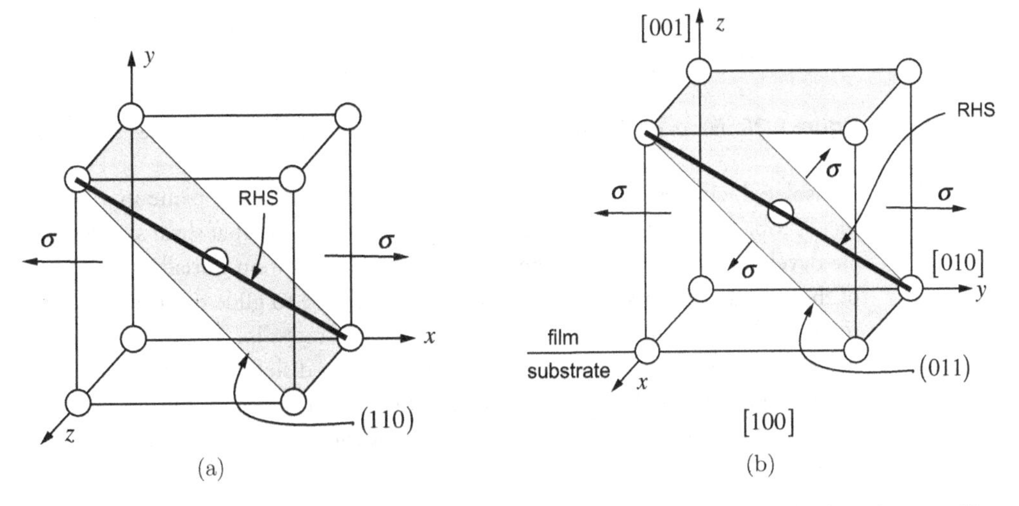

Figure 9.25. (a) A right-handed screw dislocation in a BCC crystal subjected to uniaxial tensile stress. (b) A right-handed screw dislocation in a BCC crystal subjected to biaxial tensile stress.

9.16 A pure screw dislocation in a BCC metal crystal film lies along the $[\bar{1}\,1\,\bar{1}]$ axis as shown in Fig. 9.25b. The film is subjected to an equal-biaxial (in-plane) tensile stress of σ. The normal to the surface of the film is $[0\,0\,1]$ as shown in the figure.

(a) Use the Peach–Koehler formula to determine the force per unit length acting on the dislocation due to the applied biaxial stress.
(b) Write an expression for the force per unit length on the dislocation if a compression stress $\sigma_{zz} = -\sigma$ were acting on the crystal.
(c) Briefly discuss how the answer to (b) could have been deduced from the answer to (a).

9.17 A circular edge dislocation loop is shown in Fig. 9.26a. The elastic energy of such a loop is

$$E_{\text{loop}} = 2\pi R \frac{\mu b^2}{4\pi(1-\nu)} \ln\left(\frac{\alpha R}{b}\right). \qquad (9.113)$$

(a) Describe the applied stress that would be needed to hold the dislocation loop in equilibrium.
(b) Write an expression for the magnitude of the applied stress needed to maintain equilibrium.
(c) Determine if the equilibrium state produced by applying an appropriate stress is a stable equilibrium state.

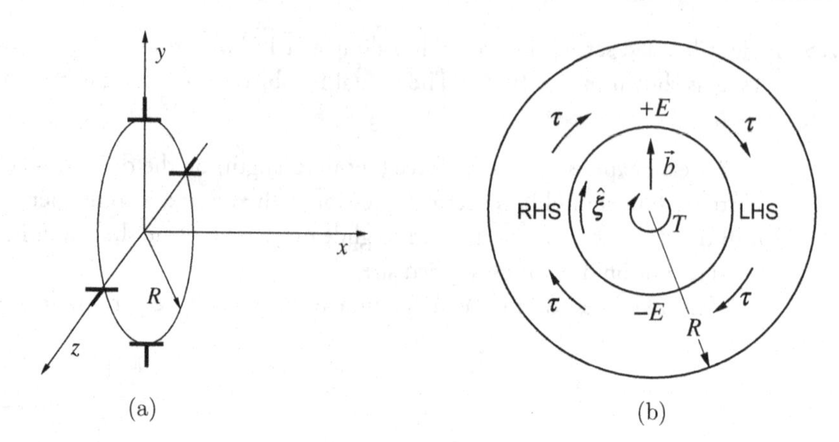

Figure 9.26. (a) An edge dislocation loop. (b) A glide dislocation loop subjected to an applied torque.

9.18 An isolated, circular glide dislocation loop is shown on a slip plane in a cylindrical crystal in Fig. 9.26b. The cylinder is subjected to a twist or torque so that shear stresses of the kind shown are developed in the crystal. The center of the circular loop initially coincides with the center of the cylinder. We assume that the dislocation is free to glide on its slip plane (in response to resolved shear stresses) but is not free to climb or cross-slip. The dislocation is free to run out of the crystal but the formation of completely new dislocations, at the surface of the crystal, for example, is completely inhibited. Describe how the dislocation moves when the torque is applied. Indicate the final position of the dislocation, where it will no longer be moved by the applied torque.

9.19 The Orowan bowing stress is the critical stress at which a dislocation bowing between two pinning points becomes unstable. Using the line tension expressions derived in this chapter, write expressions for the critical shear stresses that are needed to force edge and screw dislocations between two point obstacles separated by a distance L. Assume that the dislocations take the shape shown in Fig. 9.13 as they bow.

10 Dislocation interactions and applications

In Chapter 9, we have seen that dislocations produce stress fields in the crystal that contains them. We have also seen that stresses produce Peach–Koehler forces on dislocations. Therefore, dislocations exert forces on each other through the stress fields they produce. In this chapter, we discuss dislocation–dislocation interactions, as well as the interaction between dislocations and other defects in the crystal, and the consequence of these interactions on the strength of bulk crystalline materials. We also consider the applications of these interactions to the mechanical properties of thin films.

In Section 10.1, we consider the interaction between two dislocations in an infinite medium, starting with two infinitely long parallel screw or edge dislocations. A few examples of the interaction between two non-parallel dislocation lines are also considered. In Section 10.2, we consider arrays formed by more than two dislocations of the same sign. When the dislocation interactions are attractive, the dislocation array corresponds to low angle grain boundaries. When the dislocation interactions are repulsive, the dislocation array is called a pile-up, because it can be found in front of an obstacle which blocks the dislocation motion.

In Section 10.3, we discuss two dislocation mechanisms which increase the strength of crystals. In the Taylor hardening mechanism, the strength is controlled by the interaction between dislocations themselves. In the Orowan bowing mechanism, the strength is associated with the presence of non-shearable particles, between which the gliding dislocation must bow for plastic deformation to occur. In Section 10.4, we consider several models in which the kinetics of dislocation motion, multiplication, and annihilation are used to explain the plastic deformation behavior of single crystals, such as the phenomena of yield point drop and sigmoidal creep.

The last two sections of this chapter deal with dislocations near interfaces. In Section 10.5, we discuss the conditions under which it becomes energetically favorable for dislocations to form at the interface between two materials to relieve the misfit strain. In Section 10.6, we discuss the stress and strain fields of dislocations near free surfaces and interfaces between two materials, and the forces exerted on these dislocations by the interfaces. For straight screw dislocations parallel to the interfaces, the effect of the interfaces can be modeled by sets of image dislocations in infinite, homogeneous elastic media.

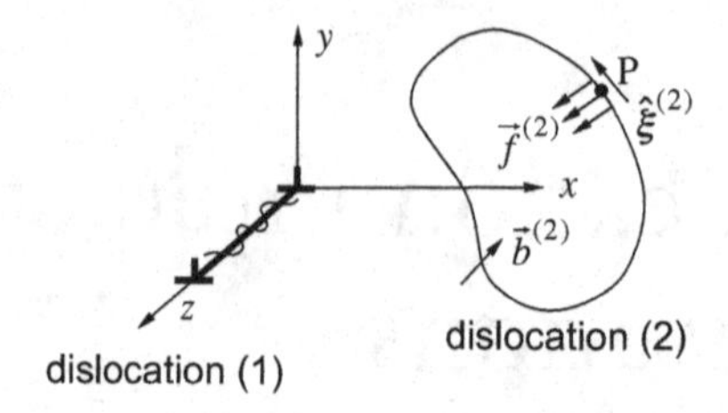

Figure 10.1. Dislocation–dislocation interaction.

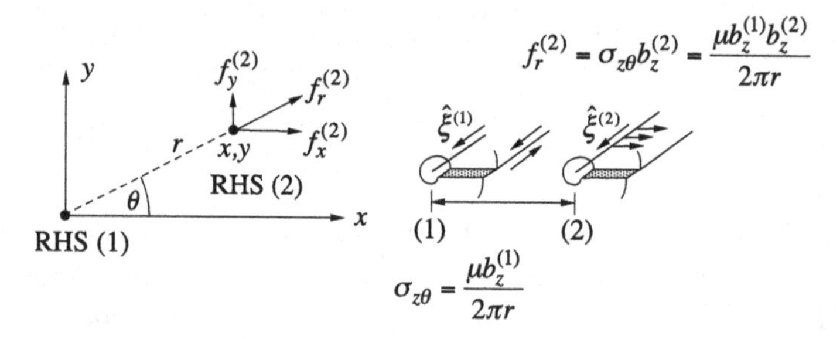

Figure 10.2. Screw–screw interaction.

10.1 Interactions between two dislocations

Forces are exerted on dislocations when the crystals that contain them are subjected to external stresses. Similarly when the stress field of one dislocation acts on another dislocation, an interaction force (per unit length) is exerted on the second dislocation. The stress field of the second dislocation also creates forces on the first dislocation.

Consider a straight dislocation (1) in the proximity of a dislocation loop (2) as shown in Fig. 10.1. Using the Peach–Koehler formula we may say that the interaction force (per unit length) acting on dislocation (2) at the point P is simply

$$\vec{f}^{(2)} = \left(\vec{b}^{(2)} \cdot \overset{\leftrightarrow}{\sigma}{}^{(1)}\right) \times \hat{\xi}^{(2)}, \tag{10.1}$$

where $\hat{\xi}^{(2)}$ is the sense vector of dislocation (2) at the point P, $\vec{b}^{(2)}$ is the Burgers vector of dislocation (2), and $\sigma_{ij}^{(1)}$ is the stress field of dislocation (1) at the point P.

In the following, we apply Eq. (10.1) to evaluate the interaction forces between dislocations in simple structures, some of which are commonly found in crystals.

10.1.1 Parallel screw dislocations

Consider two parallel right-handed screw dislocations as shown in Fig. 10.2. Dislocation (1) lies along the z axis and dislocation (2) is parallel to it at the position (r, θ) or (x, y). The sense vector for both dislocations may be chosen to be $\hat{\xi}^{(1)} = \hat{\xi}^{(2)} = [0\,0\,1]$ so that $\vec{b}^{(1)} = [0\,0\,b_z^{(1)}]$ and $\vec{b}^{(2)} = [0\,0\,b_z^{(2)}]$. The stress field of dislocation (1) is well known and can be expressed as

$$\overleftrightarrow{\sigma}^{(1)} = \begin{bmatrix} 0 & 0 & \sigma_{xz}^{(1)} \\ 0 & 0 & \sigma_{yz}^{(1)} \\ \sigma_{zx}^{(1)} & \sigma_{zy}^{(1)} & 0 \end{bmatrix}, \tag{10.2}$$

where

$$\begin{aligned} \sigma_{xz}^{(1)} = \sigma_{zx}^{(1)} &= -\frac{\mu b_z^{(1)}}{2\pi} \frac{y}{x^2+y^2} \\ \sigma_{yz}^{(1)} = \sigma_{zy}^{(1)} &= \frac{\mu b_z^{(1)}}{2\pi} \frac{x}{x^2+y^2}. \end{aligned} \tag{10.3}$$

Using the Peach–Koehler formula we have

$$\begin{aligned} \vec{f}^{(2)} &= (\vec{b}^{(2)} \cdot \overleftrightarrow{\sigma}^{(1)}) \times \hat{\xi}^{(2)} \\ &= \left\{ b_z^{(2)} [0 \quad 0 \quad 1] \cdot \begin{bmatrix} 0 & 0 & \sigma_{xz}^{(1)} \\ 0 & 0 & \sigma_{yz}^{(1)} \\ \sigma_{zx}^{(1)} & \sigma_{zy}^{(1)} & 0 \end{bmatrix} \right\} \times \begin{bmatrix} 0 \\ 0 \\ 1 \end{bmatrix} \\ &= b_z^{(2)} \begin{bmatrix} \sigma_{zx}^{(1)} \\ \sigma_{zy}^{(1)} \\ 0 \end{bmatrix} \times \begin{bmatrix} 0 \\ 0 \\ 1 \end{bmatrix} \\ &= \begin{bmatrix} \sigma_{zy}^{(1)} b_z^{(2)} & -\sigma_{zx}^{(1)} b_z^{(2)} & 0 \end{bmatrix}. \end{aligned} \tag{10.4}$$

So the components of the interaction force on dislocation (2) are

$$\begin{aligned} f_x^{(2)} &= \sigma_{zy}^{(1)} b_z^{(2)} = \frac{\mu b_z^{(1)} b_z^{(2)}}{2\pi} \frac{x}{x^2+y^2} = \frac{\mu b_z^{(1)} b_z^{(2)}}{2\pi} \frac{\cos\theta}{r} \\ f_y^{(2)} &= -\sigma_{zx}^{(1)} b_z^{(2)} = \frac{\mu b_z^{(1)} b_z^{(2)}}{2\pi} \frac{y}{x^2+y^2} = \frac{\mu b_z^{(1)} b_z^{(2)}}{2\pi} \frac{\sin\theta}{r}. \end{aligned} \tag{10.5}$$

From this the radial interaction force may be computed as

$$f_r^{(2)} = f_x^{(2)} \cos\theta + f_y^{(2)} \sin\theta = \frac{\mu b_z^{(1)} b_z^{(2)}}{2\pi r}. \tag{10.6}$$

We see that this result could have been found using the simple relations given earlier and some intuition. Since the shear stress τ about a screw dislocation is simply $\sigma_{z\theta}^{(1)} = \mu b_z^{(1)}/(2\pi r)$, the radial force on dislocation (2) can be computed using the basic relation, $f = \tau b$, to be

$$f_r^{(2)} = \sigma_{z\theta}^{(1)} b_z^{(2)} = \frac{\mu b_z^{(1)} b_z^{(2)}}{2\pi r}, \tag{10.7}$$

which is, of course, the same result, as shown in Fig. 10.2. If dislocations (1) and (2) are of the same sign, i.e. $b_z^{(1)} b_z^{(2)} > 0$, then $f_r^{(2)} > 0$. This means that two parallel screw dislocations of the same sign always repel. Conversely, if $b_z^{(1)} b_z^{(2)} < 0$, then $f_r^{(2)} < 0$, meaning that two parallel screw dislocations of opposite sign always attract.

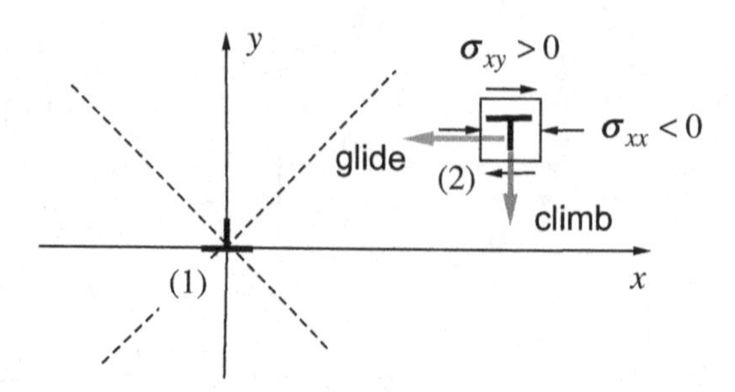

Figure 10.3. Edge–edge interaction. The inset shows how the glide and climb force can be obtained using the simple relations and intuition: $f_g = f_x^{(g)} = \sigma_{xy} b_x^{(2)}$, $f_c = f_y^{(c)} = -\sigma_{xx} b_x^{(2)}$. The arrows indicate the directions of glide and climb forces when $b_x^{(1)} > 0$ and $b_x^{(2)} < 0$.

10.1.2 Parallel edge dislocations

We now consider edge dislocations of opposite signs lying on parallel slip planes as shown in Fig. 10.3. We position dislocation (1), a positive edge dislocation, along the z axis, and the oppositely signed dislocation (2) at the position (x, y). The forces acting on dislocation (2) due to the stress field of dislocation (1) can again be found using the Peach–Koehler formula. For this we choose the sense vector for both dislocations to be $\hat{\xi}^{(1)} = \hat{\xi}^{(2)} = [0\,0\,1]$ for which the Burgers vectors are $\vec{b}^{(1)} = [b_x^{(1)}\,0\,0]$ and $\vec{b}^{(2)} = [b_x^{(2)}\,0\,0]$. Figure 10.3 corresponds to the scenario where $b_x^{(1)} > 0$ and $b_x^{(2)} < 0$. However, the expressions developed in the following are valid regardless of the signs of $b_x^{(1)}$ and $b_x^{(2)}$. The stress field of dislocation (1) is

$$\overset{\leftrightarrow}{\sigma}^{(1)} = \begin{bmatrix} \sigma_{xx}^{(1)} & \sigma_{xy}^{(1)} & 0 \\ \sigma_{yx}^{(1)} & \sigma_{yy}^{(1)} & 0 \\ 0 & 0 & \sigma_{zz}^{(1)} \end{bmatrix}. \tag{10.8}$$

Using the Peach–Koehler formula

$$\begin{aligned} \vec{f}^{(2)} &= (\vec{b}^{(2)} \cdot \overset{\leftrightarrow}{\sigma}^{(1)}) \times \hat{\xi}^{(2)} \\ &= \left\{ b_x^{(2)} [1 \quad 0 \quad 0] \cdot \begin{bmatrix} \sigma_{xx}^{(1)} & \sigma_{xy}^{(1)} & 0 \\ \sigma_{yx}^{(1)} & \sigma_{yy}^{(1)} & 0 \\ 0 & 0 & \sigma_{zz}^{(1)} \end{bmatrix} \right\} \times \begin{bmatrix} 0 \\ 0 \\ 1 \end{bmatrix} \\ &= b_x^{(2)} \begin{bmatrix} \sigma_{xx}^{(1)} \\ \sigma_{xy}^{(1)} \\ 0 \end{bmatrix} \times \begin{bmatrix} 0 \\ 0 \\ 1 \end{bmatrix} \\ &= \begin{bmatrix} \sigma_{xy}^{(1)} b_x^{(2)} & -\sigma_{xx}^{(1)} b_x^{(2)} & 0 \end{bmatrix}. \end{aligned} \tag{10.9}$$

This result could have been obtained using the simple relations and some intuition. From Eqs. (9.88) and (9.92), we find that the glide force on dislocation (2) is

$$f_x^{(g)} = \sigma_{xy}^{(1)} b_x^{(2)} \tag{10.10}$$

and that the climb force is

$$f_y^{(c)} = -\sigma_{xx}^{(1)} b_x^{(2)}. \tag{10.11}$$

These results are consistent with Eq. (10.9), which is obtained from the full Peach–Koehler analysis. In Fig. 10.3, $b_x^{(2)} < 0$, hence the glide force $f_x^{(g)}$ has the opposite sign as $\sigma_{xy}^{(1)}$, and the climb force $f_y^{(c)}$ has the same sign as $\sigma_{xx}^{(1)}$.

Using the known expressions of the stress field of a positive edge dislocation, Eq. (9.31),

$$\sigma_{xy}^{(1)} = \frac{\mu b_x^{(1)}}{2\pi(1-\nu)} \frac{x(x^2-y^2)}{(x^2+y^2)^2}$$

$$\sigma_{xx}^{(1)} = -\frac{\mu b_x^{(1)}}{2\pi(1-\nu)} \frac{y(3x^2+y^2)}{(x^2+y^2)^2},$$

the interaction forces on dislocation (2) are

$$\begin{aligned} f_x^{(g)} &= \sigma_{xy}^{(1)} b_x^{(2)} = \frac{\mu b_x^{(1)} b_x^{(2)}}{2\pi(1-\nu)} \frac{x(x^2-y^2)}{(x^2+y^2)^2} \\ f_y^{(c)} &= -\sigma_{xx}^{(1)} b_x^{(2)} = \frac{\mu b_x^{(1)} b_x^{(2)}}{2\pi(1-\nu)} \frac{y(3x^2+y^2)}{(x^2+y^2)^2}. \end{aligned} \tag{10.12}$$

The interaction forces depend not only on the signs of the Burgers vectors but also on the relative positions of the two dislocations. In Fig. 10.3, $x > y > 0$, so that $\sigma_{xy}^{(1)} > 0$ and $\sigma_{xx}^{(1)} < 0$. Consequently, $f_x^{(g)} < 0$ and $f_y^{(c)} < 0$, as indicated by the arrows.

10.1.3 Interaction forces in an edge dipole

The interaction forces between two oppositely signed edge dislocations allow us to study the properties of edge dislocation dipoles and the circumstances under which they are in stable equilibrium with respect to glide or in stable equilibrium with respect to climb. Let $b_x^{(1)} = b$ and $b_x^{(2)} = -b$, and by using Eq. (10.12), the interaction force on dislocation (2) is

$$\begin{aligned} f_x^{(g)} &= -\sigma_{xy}^{(1)} b = -\frac{\mu b^2}{2\pi(1-\nu)} \frac{x(x^2-y^2)}{(x^2+y^2)^2}, \\ f_y^{(c)} &= \sigma_{xx}^{(1)} b = -\frac{\mu b^2}{2\pi(1-\nu)} \frac{y(3x^2+y^2)}{(x^2+y^2)^2}. \end{aligned} \tag{10.13}$$

Again, the glide force (along x) has the opposite sign as $\sigma_{xy}^{(1)}$, and the climb force (along y) has the same sign as $\sigma_{xx}^{(1)}$. The domains where $\sigma_{xx}^{(1)}$ and $\sigma_{xy}^{(1)}$ are positive or negative are shown in Fig. 9.7. Therefore, the climb force points to the negative y direction if dislocation (2) is above the x axis, and points to the positive y direction if dislocation (2) is below the x axis. In other words, the climb force always tends to bring dislocation (2) closer to the x axis. The direction of the glide force, on the other hand, is more complex.

The directions of the glide and climb forces on dislocation (2) are indicated in Fig. 10.4 by arrows denoted as g and c, respectively, in each of the octants. The curved arrows indicate the path that the second dislocation might take relative to the first if it were able to both glide and climb in response to the interaction forces. As the curved arrows indicate, the ultimate fate of the second dislocation would be annihilation with the oppositely signed dislocation at the origin, if both glide and climb are allowed to occur. If the second dislocation lies anywhere along the dotted lines representing $|x| = |y|$, then the two dislocations are in stable equilibrium

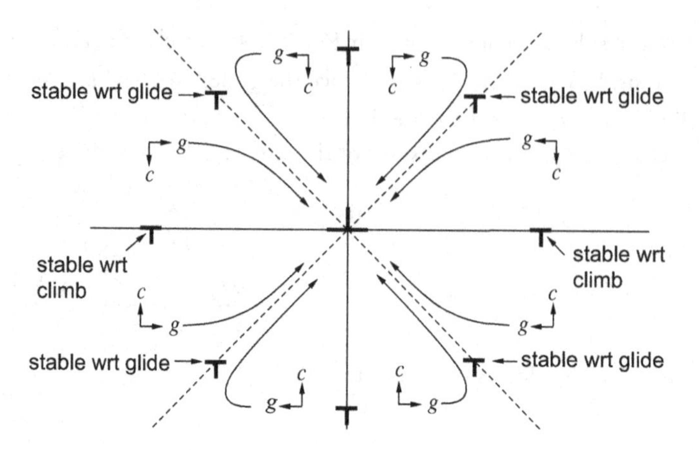

Figure 10.4. Edge dipole interaction. The arrows indicate the direction of glide (g) and climb (c) forces on the negative dislocation at various positions. Curved arrows indicate possible trajectories.

with respect to glide (if climb is not allowed), not only because the glide force is zero at these places but also because the glide forces increase on either side of these positions to restore the dislocation to the $|x| = |y|$ line. Similarly, when the second dislocation lies along $y = 0$ the dislocations are in equilibrium with respect to climb (if glide is not allowed) for similar reasons.

Of course the glide mobility is almost always much higher than the climb mobility, so for most problems only glide need be considered. In that case the second dislocation will glide to the nearest $|x| = |y|$ line and remain there in stable equilibrium (with respect to glide). Only at high temperatures, where dislocations can climb by diffusional processes, could annihilation of the two dislocations occur.

10.1.4 Critical passing stress for edge dislocations

With the interaction forces we have calculated we can now determine the critical stress needed to cause two oppositely signed edge dislocations on parallel slip planes to glide past each other. This is a fundamental problem in plasticity and strain hardening of crystals.

We let dislocation (1) be fixed at the origin and seek the critical applied shear stress, τ_c, that would be needed to cause the second dislocation, initially at $x \gg h$, to move past the first dislocation. We can analyze this problem by first determining the applied shear stress, τ, that would be needed to hold the second dislocation at any point in its slip plane (Fig. 10.5). The interaction glide force on dislocation (2) due to the stress field of dislocation (1) is

$$f_x^{\text{int}} = \sigma_{xy}^{(1)} b_x^{(2)} = \frac{\mu b_x^{(1)} b_x^{(2)}}{2\pi(1-\nu)} \frac{x(x^2 - y^2)}{(x^2 + y^2)^2} = \frac{\mu b_x^{(1)} b_x^{(2)}}{2\pi(1-\nu)} \frac{x(x^2 - h^2)}{(x^2 + h^2)^2}. \tag{10.14}$$

Again, here we assume $b_x^{(1)} > 0$ and $b_x^{(2)} < 0$. And the force due to the applied stress $\sigma_{xy}^{\text{app}} = \tau$ is

$$f_x^{\text{app}} = \tau b_x^{(2)}. \tag{10.15}$$

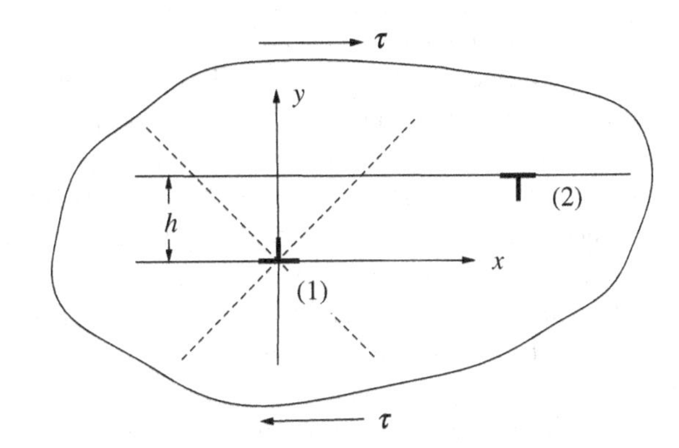

Figure 10.5. Bypassing of two edge dislocations under applied shear stress τ.

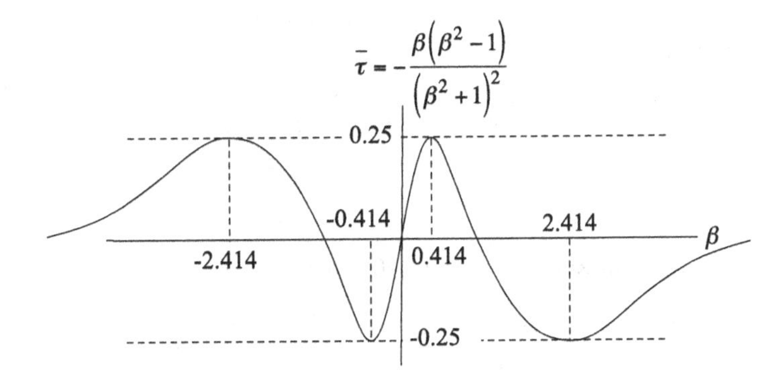

Figure 10.6. Interaction stresses for passing edge dislocations.

We find the applied stress needed to hold dislocation (2) in place by setting

$$f_x^{\text{int}} + f_x^{\text{app}} = 0,$$

$$\frac{\mu b_x^{(1)} b_x^{(2)}}{2\pi(1-\nu)} \frac{x(x^2-h^2)}{(x^2+h^2)^2} + \tau b_x^{(2)} = 0,$$

$$\tau = -\frac{\mu b_x^{(1)}}{2\pi(1-\nu)} \frac{x(x^2-h^2)}{(x^2+h^2)^2} = -\frac{\mu b_x^{(1)}}{2\pi(1-\nu)h} \frac{\beta(\beta^2-1)}{(\beta^2+1)^2}, \tag{10.16}$$

where $\beta \equiv x/h$. We can express this result in non-dimensional form. Define a non-dimensional stress

$$\bar{\tau} \equiv \frac{2\pi(1-\nu)h}{\mu b_x^{(1)}} \tau, \tag{10.17}$$

then

$$\bar{\tau} = -\frac{\beta(\beta^2-1)}{(\beta^2+1)^2}, \tag{10.18}$$

which is plotted in Fig. 10.6.

The interaction stresses in Fig. 10.6 can be interpreted as follows. Consider first that dislocation (2) starts off far to the right of the fixed dislocation (1), $\beta \gg 1$, and assume $h > 0$. Here the applied stress would have to be negative to prevent dislocation (2) from moving toward the oppositely signed dislocation (1). As dislocation (2) moves to the left and closer to dislocation (1), the attractive force grows larger and reaches a maximum at $\beta = 2.41$ before declining to zero at $\beta = 1$. This is a stable equilibrium position for the dipole. In order to push dislocation (2) still closer to dislocation (1) a positive shear stress needs to be applied. That required shear stress rises quickly and reaches $\overline{\tau} = 0.25$ at $\beta = 0.414$ before declining to zero at $\beta = 0$. The maximum in the shear stress represents the critical shear stress that would be needed to drive the two dislocations past each other. This critical passing stress, as discussed in Section 10.3.1, plays an important role in strain hardening.

As dislocation (2) moves to the left of dislocation (1) a negative shear stress is again needed to prevent the two dislocations from pushing away from each other to reach a new stable equilibrium position at $\beta = -1$. A positive shear stress is required to move the oppositely signed dislocations away from each other, to $\beta \ll -1$. Notice that the critical shear stress needed to pull the two dislocations apart is the same as the shear stress needed to make them pass in the first place.

It is easy to find the critical passing stress and the positions where the maximum interaction stress is found. These points (extrema) are characterized by the condition that

$$\frac{\partial \overline{\tau}}{\partial \beta} = 0. \tag{10.19}$$

Given the expression of $\overline{\tau}$ in Eq. (10.18), we have

$$\frac{\partial \overline{\tau}}{\partial \beta} = -\frac{(\beta^2+1)(3\beta^2-1) - 4\beta^2(\beta^2-1)}{(\beta^2+1)^3} = 0, \tag{10.20}$$

which simplifies to

$$\beta^4 - 6\beta^2 + 1 = 0. \tag{10.21}$$

Letting $\beta^2 = w$ we have

$$w^2 - 6w + 1 = 0, \tag{10.22}$$

which has the solutions

$$w = 3 \pm \sqrt{8} \tag{10.23}$$

so the roots are

$$\beta = \pm\sqrt{3+\sqrt{8}} = \pm(\sqrt{2}+1) = \pm 2.414,$$
$$\beta = \pm\sqrt{3-\sqrt{8}} = \pm(\sqrt{2}-1) = \pm 0.414.$$

All four roots give rise to $|\overline{\tau}| = 1/4$. Therefore, the dimensionless critical passing stress is $\overline{\tau}_c = 1/4$ and the critical passing stress is

$$\tau_c = \frac{\mu b}{8\pi(1-\nu)h}, \tag{10.24}$$

where we have replaced $b_x^{(1)}$ by b for brevity.

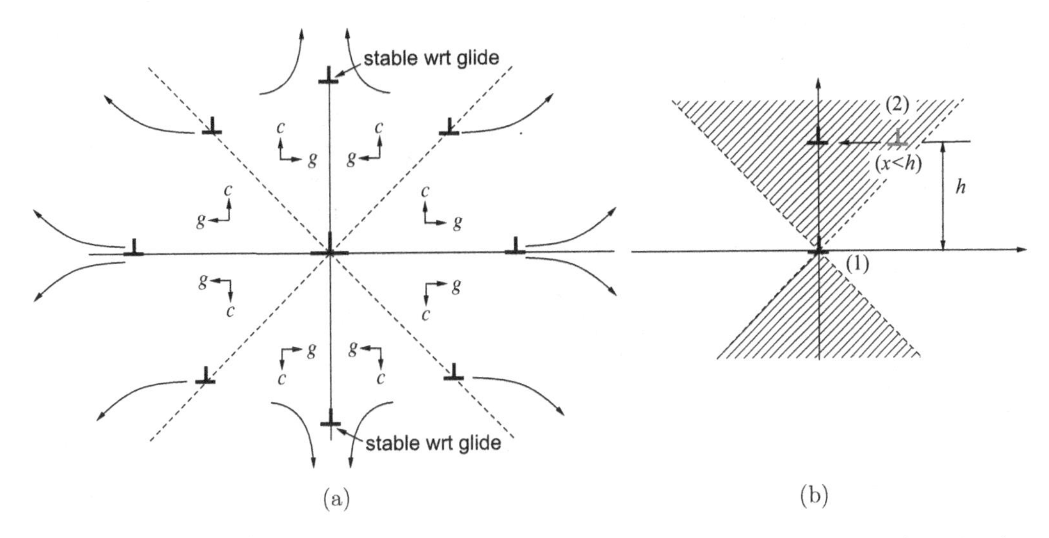

Figure 10.7. (a) Interaction between two same-signed edge dislocations. (b) If dislocation (2) is located in the diagonally hatched region, it will quickly glide to the $x = 0$ line, forming a configuration stable with respect to glide.

10.1.5 Same-signed edge dislocations

The interaction forces between same-signed edge dislocations have exactly the opposite direction as those between a dislocation dipole. We will see below that two same-signed edge dislocations also have stable equilibrium structures with respect to glide (similar to the dislocation dipole), but there is no stable equilibrium with respect to climb. Let $b_x^{(1)} = b_x^{(2)} = b$, and by using Eq. (10.12), the interaction force on dislocation (2) is

$$\begin{aligned} f_x^{(g)} &= \sigma_{xy}^{(1)} b = \frac{\mu b^2}{2\pi(1-\nu)} \frac{x(x^2-y^2)}{(x^2+y^2)^2}, \\ f_y^{(c)} &= -\sigma_{xx}^{(1)} b = \frac{\mu b^2}{2\pi(1-\nu)} \frac{y(3x^2+y^2)}{(x^2+y^2)^2}. \end{aligned} \tag{10.25}$$

The glide force (along x) has the same sign as $\sigma_{xy}^{(1)}$, and the climb force (along y) has the opposite sign as $\sigma_{xx}^{(1)}$. The domains where $\sigma_{xx}^{(1)}$ and $\sigma_{xy}^{(1)}$ are positive or negative are shown in Fig. 9.7. Therefore, the climb force points to the positive y direction if dislocation (2) is above the x axis, and points to the negative y direction if dislocation (2) is below the x axis. In other words, the climb force always tends to push dislocation (2) away from the x axis. The direction of the glide force, on the other hand, is more complex.

The directions of the glide and climb forces on dislocation (2) are indicated in Fig. 10.7a by arrows denoted as g and c, respectively, in each of the octants. The curved arrows indicate the path that the second dislocation might take relative to the first if it were able to both glide and climb in response to the interaction forces. As the curved arrows indicate, the ultimate fate of the second dislocation would be to move very far away from the same signed dislocation at the origin, if both glide and climb is allowed to occur. If the second dislocation lies anywhere along the dotted lines representing $|x| = |y|$, then the two dislocations are in unstable equilibrium with respect to glide. The glide force is zero there, but any perturbation of the dislocation

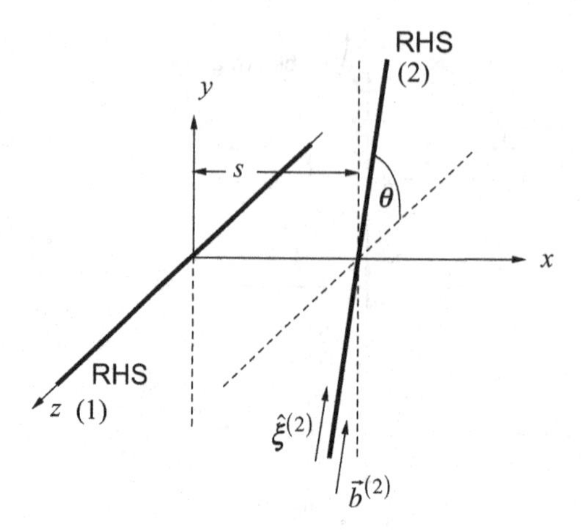

Figure 10.8. Pure screw dislocations offset by a distance s and skewed by an angle θ.

position will lead to a glide force that pushes the dislocation further away from the $|x| = |y|$ line. Similarly, when the second dislocation lies along $y = 0$ the dislocations are in unstable equilibrium with respect to climb for similar reasons.

When the second dislocation lies along the $x = 0$ line, the dislocations are in stable equilibrium with respect to glide. The glide force is not only zero there, but also increases on either side of these positions to restore the dislocation to the $x = 0$ line. Recall that the glide mobility is almost always much higher than the climb mobility. Therefore, when dislocation (2) is in the diagonally hatched regions shown in Fig. 10.7b, it will quickly glide to the $x = 0$ line, so that the two dislocations are aligned one above the other, and remain there in stable equilibrium (with respect to glide). Only at high temperatures, where dislocations can climb by diffusional processes, could the two dislocations move apart in the y direction.

10.1.6 Non-parallel screw dislocations

In the examples described above the interacting dislocations were straight and parallel to each other. Of course, these are the simplest kinds of interactions. Here we consider straight dislocations that are not parallel to each other and later we consider a case in which one of the dislocations is not straight.

Consider two pure RHS dislocations in Fig. 10.8. They lie in planes parallel to the yz plane but are separated at their closest points by a distance s. One dislocation lies along the z axis while the other is tilted from a line parallel to z by the angle θ. Define the sense vector for dislocation (1) as $\hat{\xi}^{(1)} = [0\,0\,\bar{1}]$, so that its Burgers vector is $\vec{b}^{(1)} = b^{(1)}[0\,0\,\bar{1}]$, where $b^{(1)}$ is the magnitude of the Burgers vector. Similarly, define the sense vector for dislocation (2) as $\hat{\xi}^{(2)} = [0\ \sin\theta\ -\cos\theta]$ so that its Burgers vector is $\vec{b}^{(2)} = b^{(2)}[0\ \sin\theta\ -\cos\theta]$. It is easy to calculate the forces on dislocation (2) using the Peach–Koehler formula, since we know the elastic field of dislocation (1), which is

$$\overleftrightarrow{\sigma}^{(1)} = \begin{bmatrix} 0 & 0 & \sigma_{xz}^{(1)} \\ 0 & 0 & \sigma_{yz}^{(1)} \\ \sigma_{zx}^{(1)} & \sigma_{zy}^{(1)} & 0 \end{bmatrix}, \tag{10.26}$$

where, from Eq. (9.8), we have

$$\sigma_{xz}^{(1)} = \sigma_{zx}^{(1)} = -\frac{\mu b^{(1)}}{2\pi}\frac{y}{x^2+y^2},$$
$$\sigma_{yz}^{(1)} = \sigma_{zy}^{(1)} = \frac{\mu b^{(1)}}{2\pi}\frac{x}{x^2+y^2}.$$

Using the Peach–Koehler formula, we have

$$\begin{aligned}
\vec{f}^{(2)} &= (\vec{b}^{(2)} \cdot \overset{\leftrightarrow}{\sigma}^{(1)}) \times \hat{\xi}^{(2)} \\
&= \left\{ b^{(2)}[0 \quad \sin\theta \quad -\cos\theta] \cdot \begin{bmatrix} 0 & 0 & \sigma_{xz}^{(1)} \\ 0 & 0 & \sigma_{yz}^{(1)} \\ \sigma_{zx}^{(1)} & \sigma_{zy}^{(1)} & 0 \end{bmatrix} \right\} \times \begin{bmatrix} 0 \\ \sin\theta \\ -\cos\theta \end{bmatrix} \\
&= b^{(2)} \begin{bmatrix} -\cos\theta\sigma_{zx}^{(1)} \\ -\cos\theta\sigma_{zy}^{(1)} \\ \sin\theta\sigma_{yz}^{(1)} \end{bmatrix} \times \begin{bmatrix} 0 \\ \sin\theta \\ -\cos\theta \end{bmatrix} \\
&= -b^{(2)} \left[(\sin^2\theta - \cos^2\theta)\sigma_{zy}^{(1)} \quad \cos^2\theta\sigma_{zx}^{(1)} \quad \sin\theta\cos\theta\sigma_{zx}^{(1)} \right].
\end{aligned} \tag{10.27}$$

It is interesting to focus on the x component of the interaction force on dislocation (2), $f_x^{(2)}$:

$$f_x^{(2)} = -b^{(2)}(\sin^2\theta - \cos^2\theta)\sigma_{zy}^{(1)} = b^{(2)}(\cos 2\theta)\sigma_{zy}^{(1)}. \tag{10.28}$$

Because dislocation (2) is located on the plane with $x > 0$, where $\sigma_{zy}^{(1)} > 0$, this produces the following results.

$\theta = 0$	$f_x^{(2)} = b^{(2)}\sigma_{zy}^{(1)} > 0$	repulsive
$\theta = \pi/4$	$f_x = 0$	no interaction
$\theta = \pi/2$	$f_x^{(2)} = -b^{(2)}\sigma_{zy}^{(1)} < 0$	attractive
$\theta = 3\pi/4$	$f_x = 0$	no interaction
$\theta = \pi$	$f_x^{(2)} = b^{(2)}\sigma_{zy}^{(1)} > 0$	repulsive

Obviously, as we have seen before, like-signed screws (e.g. two RHS or two LHS) repel each other when they are parallel to each other ($\theta = 0$ or $\theta = \pi$). But we now find that like-signed screws attract each other when they are orthogonal ($\theta = \pi/2$). This attractive interaction is responsible for the formation of low angle twist boundaries consisting of two sets of screw dislocation arrays that are orthogonal to each other (see Section 14.2.2).

As long as the screws are not parallel, the distribution of interaction forces is not uniform. Take the orthogonal case, $\theta = \pi/2$, as an example. Dislocation (2) is now oriented parallel to the y axis, with $x = s$. The stress field of dislocation (1) is

$$\sigma_{zy}^{(1)} = \frac{\mu b^{(1)}}{2\pi}\frac{x}{x^2+y^2}$$

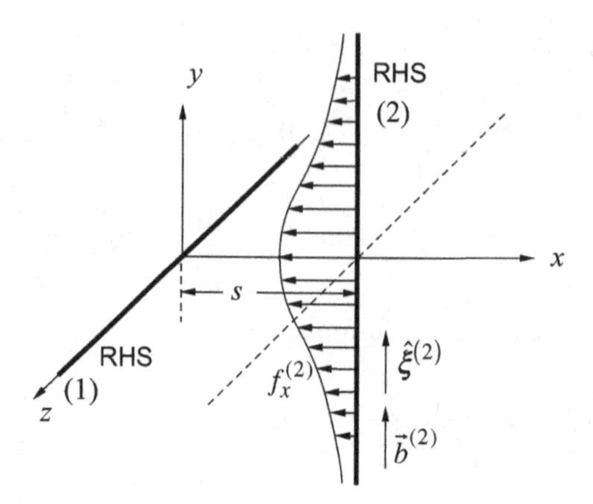

Figure 10.9. Interaction forces between orthogonal screw dislocations.

so the x component of the interaction forces is expressed as

$$f_x^{(2)} = -\sigma_{zy}^{(1)} b^{(2)} = -\frac{\mu b^{(1)} b^{(2)}}{2\pi} \frac{x}{x^2 + y^2} = -\frac{\mu b^{(1)} b^{(2)}}{2\pi} \frac{s}{s^2 + y^2}, \tag{10.29}$$

as shown in the Fig. 10.9.

We note that varying the angle θ between the two RHS dislocations discussed here does not correspond to rotating the dislocation line(s) in space. When a dislocation line in a real crystal rotates, its Burgers vector remain unchanged, so that the angle between the line direction and Burgers vector changes. Hence a RHS dislocation will cease to be a RHS after a small rotation. For example, consider the case where the two RHS dislocations are parallel to each other, i.e. $\theta = 0$. Obviously they repel each other in this configuration. If we rotate dislocation (2) by 180° (without changing its Burgers vector), it then becomes a LHS dislocation and the interaction with dislocation (1) is now attractive. This does not contradict the result in the table above, which shows two RHS dislocations repel each other when $\theta = 180^\circ$. Actually, two RHS dislocations at an angle $\theta = 180^\circ$ is the same configuration as two RHS dislocations at an angle $\theta = 0^\circ$ (see Section 8.3.2).

10.1.7 Straight dislocation with dislocation loop

The Peach–Koehler formula may also be used to compute the forces on curved dislocations of the kind shown in Fig. 10.10. Here an edge dislocation loop is near a positive edge dislocation.

The sense vector and Burgers vector of dislocation (1) are

$$\hat{\xi}^{(1)} = [0 \quad 0 \quad 1]$$
$$\vec{b}^{(1)} = b^{(1)}[1 \quad 0 \quad 0].$$

The sense vector and Burgers vector of the loop can be expressed as

$$\hat{\xi}^{(2)} = [0 \quad \sin\theta \quad -\cos\theta]$$
$$\vec{b}^{(2)} = b^{(2)}[1 \quad 0 \quad 0].$$

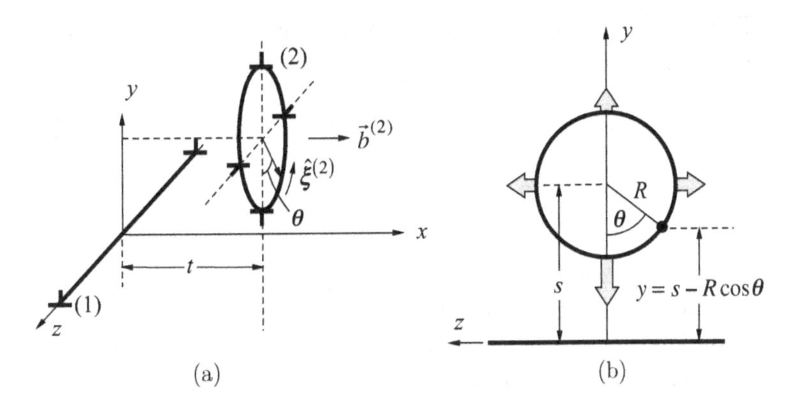

Figure 10.10. (a) Dislocation loop near an edge dislocation. The loop lies on the plane $x = t$. (b) Dislocation loop immediately above a positive edge dislocation. The block arrows indicate the Peach–Koehler force on the dislocation loop exerted by the straight dislocation line.

Because the extra half-plane of the loop points away from the center of the loop, this dislocation loop can be formed by condensing vacancies. Hence this loop is called a *vacancy loop*. The stress field of dislocation (1) is

$$\overleftrightarrow{\sigma}^{(1)} = \begin{bmatrix} \sigma_{xx}^{(1)} & \sigma_{xy}^{(1)} & 0 \\ \sigma_{yx}^{(1)} & \sigma_{yy}^{(1)} & 0 \\ 0 & 0 & \sigma_{zz}^{(1)} \end{bmatrix}. \tag{10.30}$$

So the interaction forces can be found using the Peach–Koehler formula

$$\begin{aligned} \vec{f}^{(2)} &= \left(\vec{b}^{(2)} \cdot \overleftrightarrow{\sigma}^{(1)}\right) \times \hat{\xi}^{(2)} \\ &= \left\{ b^{(2)} [1 \quad 0 \quad 0] \cdot \begin{bmatrix} \sigma_{xx}^{(1)} & \sigma_{xy}^{(1)} & 0 \\ \sigma_{yx}^{(1)} & \sigma_{yy}^{(1)} & 0 \\ 0 & 0 & \sigma_{zz}^{(1)} \end{bmatrix} \right\} \times \begin{bmatrix} 0 \\ \sin\theta \\ -\cos\theta \end{bmatrix} \\ &= b^{(2)} \begin{bmatrix} \sigma_{xx}^{(1)} \\ \sigma_{xy}^{(1)} \\ 0 \end{bmatrix} \times \begin{bmatrix} 0 \\ \sin\theta \\ -\cos\theta \end{bmatrix} \\ &= \left[-\sigma_{xy}^{(1)} b^{(2)} \cos\theta \quad \sigma_{xx}^{(1)} b^{(2)} \cos\theta \quad \sigma_{xx}^{(1)} b^{(2)} \sin\theta \right]. \end{aligned} \tag{10.31}$$

We can use this result to describe the attractive interaction when the dislocation loop is just above a positive edge dislocation, i.e. when the loop lies on the plane $x = 0$, as shown in Fig. 10.10b.

For a loop in this location the relevant stresses from dislocation (1) are

$$\begin{aligned} \sigma_{xy}^{(1)} &= 0, \\ \sigma_{xx}^{(1)} &= -\frac{\mu b^{(1)}}{2\pi(1-\nu)y} = -\frac{\mu b^{(1)}}{2\pi(1-\nu)(s - R\cos\theta)}. \end{aligned} \tag{10.32}$$

From the result above, the x component of the interaction force is zero, and the y and z components of the interaction force (per unit length) are

$$\begin{aligned} f_y^{(2)} &= \sigma_{xx}^{(1)} b^{(2)} \cos\theta = -\frac{\mu b^{(1)} b^{(2)} \cos\theta}{2\pi(1-\nu)(s - R\cos\theta)}, \\ f_z^{(2)} &= \sigma_{xx}^{(1)} b^{(2)} \sin\theta = -\frac{\mu b^{(1)} b^{(2)} \sin\theta}{2\pi(1-\nu)(s - R\cos\theta)}. \end{aligned} \tag{10.33}$$

These forces tend to cause the loop to expand, non-uniformly, as shown by Fig. 10.10b. It is obvious from these results that, since F_z is an odd function of θ, the net force on the entire loop in the z direction is zero. This result is expected because dislocation (1) is infinitely long along the z axis; translating the loop along z does not change the relative positions between the two dislocations and hence leaves the energy constant. On the other hand, the net force on the loop in the y direction is not zero; it is attractive because the attraction forces for the part of the loop nearest the edge dislocation are greater than the repulsive forces on the part of the loop far from the edge dislocation. The total force on the entire circular loop can be computed as follows:

$$\begin{aligned} F_y^{\text{total}} &= 2\int_0^{\pi} f_y^{(2)} R\, d\theta \\ &= -\frac{\mu b^{(1)} b^{(2)}}{\pi(1-\nu)} \int_0^{\pi} \frac{\cos\theta\, d\theta}{(s/R - \cos\theta)} \\ &= -\frac{\mu b^{(1)} b^{(2)}}{\pi(1-\nu)} \left(\frac{s/R}{\sqrt{(s/R)^2 - 1}} - 1 \right). \end{aligned} \tag{10.34}$$

The above expression is only valid for $s/R > 1$, i.e. when the vacancy loop has not yet touched dislocation (1). When this is the case, the term in the bracket is positive, so that $F_y^{\text{total}} < 0$, i.e. the vacancy loop is pulled toward the edge dislocation.

If the dislocations were able to climb, then the loop would extend toward the straight edge and the edge would bulge up toward the loop so that eventually the loop would annihilate in the dislocation. After a very long period of time we would have again a straight edge dislocation at a higher elevation (i.e. above the $y = 0$ plane), due to the absorption of the vacancy loop.

10.2 Dislocation arrays

The elastic interactions of dislocations cause them to organize into arrays of certain kinds that are useful in understanding the mechanical properties of crystals. Broadly speaking, dislocations can be found in attractive configurations, wherein they tend to collect together to form stable arrays, or in repulsive configurations, wherein they tend to push away from each other. Edge dislocations of the same sign can form both types of configurations, depending on their relative positions.

10.2.1 Dislocation array in an attractive configuration

In Section 10.1.5, we have discussed the interaction between two same-signed edge dislocations. Figure 10.7b shows that, if dislocation (2) is put in the diagonally hatched area ($|x| < |y|$), it will

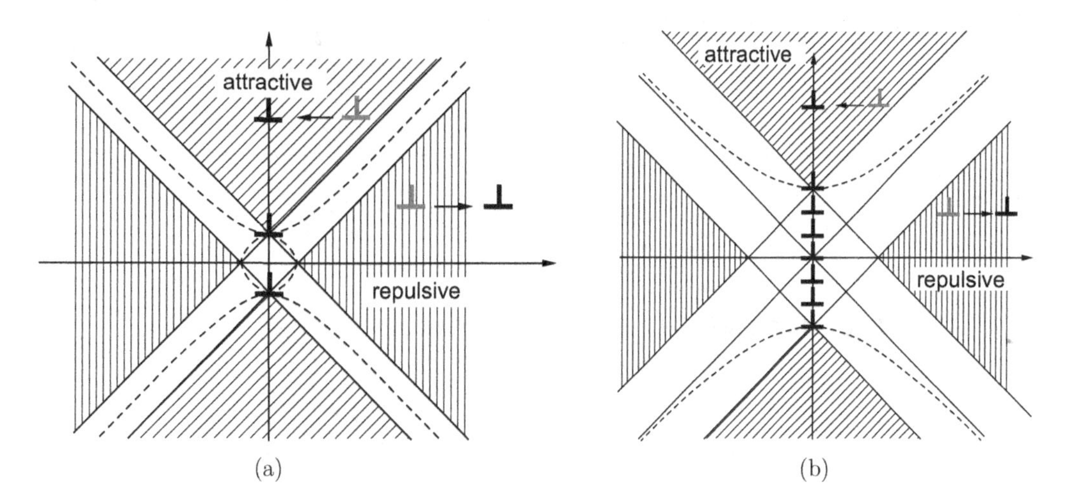

Figure 10.11. (a) Stability of three same-signed edge dislocations. (b) Stability of a stack of same-signed edge dislocations.

quickly glide to the location just above (or just below) dislocation (1), forming a configuration that is stable with respect to glide.

Now consider adding a third dislocation to the configuration, as shown in Fig. 10.11a. For similar reasons, if the third dislocation is placed in the diagonally hatched region it will be attracted by both of the existing dislocations and will find a glide equilibrium position just above (or just below) the first two dislocations. If the third dislocation is positioned to the sides of the first two dislocations, in the vertically hatched regions, then it will be repelled from the first two. In between the third dislocation could be either attracted or repelled depending on its exact position. The dotted lines show approximately the lines of demarcation between attraction and repulsion.

As additional edge dislocations are added to the array, we find that the added dislocations are in stable positions with respect to glide, as shown in Fig. 10.11b. The diagram shows the regions for which attractive or repulsive forces would act on an added dislocation. The arrays of same-signed edge dislocations shown in these figures are equivalent to low angle tilt grain boundaries, which will be discussed in Chapter 14. These configurations shown here are stable with respect to glide but they are unstable with respect to climb. It is easy to show that if the dislocations were able to climb, they would push away from each other causing the spacing between them to gradually increase with time.

10.2.2 Dislocation pile-ups

While an array of same-signed edge dislocations on parallel slip planes can attract (if climb is not allowed) and stack on top of each other, they repel each other when they are all on the same plane. A group of dislocations of the same Burgers vector can be found on the same glide plane when they are emitted by the same dislocation source, such as the Frank–Read source (see Section 8.5.3). The applied stress drives these dislocations in the same direction. When an obstacle is strong enough to stop the lead dislocation in the group, the entire group of dislocations can pile-up against the obstacle.

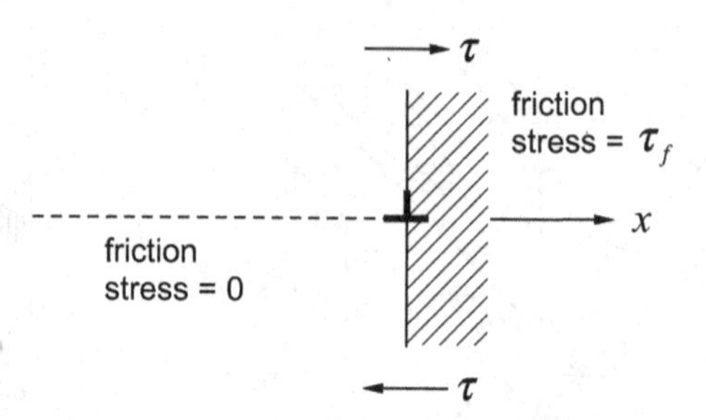

Figure 10.12. Single edge dislocation stopped at a phase boundary.

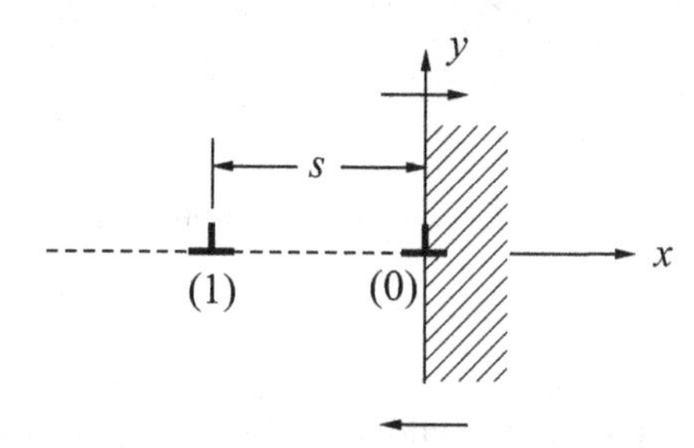

Figure 10.13. Two same-signed edge dislocations driven against a phase boundary.

Obstacle to dislocation motion

Let us start by considering a single edge dislocation held up at a phase boundary, as shown in Fig. 10.12. We may think of the second phase in front of the dislocation as a region with a high shear resistance, or friction stress, τ_f, so that the dislocation is not able to cut into that phase if the applied stress is below the critical value τ_f. For convenience we will assume that the elastic properties of the second phase are the same as the parent phase containing the dislocation so that the elastic field of the dislocation is the same as if it were in an infinite, homogeneous solid. We will assume that the friction stress in the parent phase is zero so that under a small applied shear stress, $\tau < \tau_f$, the dislocation will glide easily in the primary phase and stop at the interface.

For one dislocation, the applied stress needed to push the dislocation into the second phase and to cause slip there would be $\tau \geq \tau_f$. We can define the net force (per until length) on the dislocation as

$$f_x = (\tau - \tau_f)b \tag{10.35}$$

so that the dislocation begins to enter the second phase when $f_x = 0$. Therefore, $\tau_f b$ may be regarded as the force (per unit length) opposing glide of the dislocation in the second phase.

Reduction of penetration stress

Now consider that another dislocation with the same Burgers vector b follows the lead dislocation considered above (see Fig. 10.13). Let us label the lead dislocation as (0), and the trailing dislocation as (1). In this case the lead dislocation is pushed forward by the applied stress, as well as by the stress field of dislocation (1), and its motion forward is resisted by the friction stress, τ_f.

If the applied stress is small, the net force on dislocation (0) is

$$f_x^{(0)} = \tau b + \sigma_{xy}^{(1)} b - \tau_f b, \tag{10.36}$$

where $\sigma_{xy}^{(1)}$ at dislocation (0) is

$$\sigma_{xy}^{(1)} = \frac{\mu b}{2\pi(1-\nu)s}, \tag{10.37}$$

so that

$$f_x^{(0)} = \left[\tau + \frac{\mu b}{2\pi(1-\nu)}\frac{1}{s} - \tau_f\right] b. \tag{10.38}$$

But what is the stand-off distance, s? How close does the dislocation (1) get to dislocation (0) under an applied shear stress, τ? Dislocation (1) moves until the repulsive force from (0) exactly balances the force from the applied shear stress, τ. So the equilibrium position of dislocation (1) can be obtained by the force balance on (1),

$$f_x^{(1)} = \tau b - \frac{\mu b}{2\pi(1-\nu)}\frac{1}{s} b = 0. \tag{10.39}$$

Notice that dislocation (1) is in the parent phase and hence does not experience the resistance force $\tau_f b$. Therefore, the stand-off distance is

$$s = \frac{\mu b}{2\pi(1-\nu)\tau}. \tag{10.40}$$

Now the net force on the lead dislocation is

$$\begin{aligned} f_x^{(0)} &= \left[\tau + \frac{\mu b}{2\pi(1-\nu)}\frac{2\pi(1-\nu)\tau}{\mu b} - \tau_f\right] b \\ &= (2\tau - \tau_f) b. \end{aligned} \tag{10.41}$$

The lead dislocation begins to cuts into the second phase when

$$f_x^{(0)} = (2\tau - \tau_f) b = 0 \tag{10.42}$$

or

$$\tau = \frac{\tau_f}{2}. \tag{10.43}$$

With just one trailing dislocation, the critical applied stress for cutting into the second phase is reduced by a factor of 2.

Now consider N dislocations stacked behind the lead dislocation, as shown in Fig. 10.14. The net force on the lead dislocation is

$$f_x^{(0)} = \tau b + N\tau b - \tau_f b, \tag{10.44}$$

where the first term is the force arising from the applied stress on the lead dislocation, the second term is the force arising from the N dislocations in the pile-up, and the last term is the resisting force arising from the shear strength of the second phase.

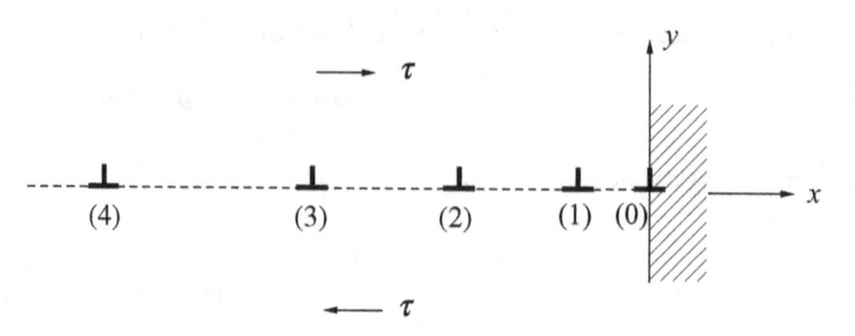

Figure 10.14. Pile-up of edge dislocations.

The first two terms on the right hand side of Eq. (10.44) can be derived by the following thought experiment, in which the second-phase obstacle is replaced by an auxiliary force f_{aux} (pointing to the left) applied solely to dislocation (0), to keep the entire train of dislocations at equilibrium under an applied stress τ. Now imagine that the entire train of dislocations, including the lead dislocation, advances to the right by a small virtual amount, δx. The total work done by all internal and external stresses and forces must be zero because the structure is assumed to be at equilibrium. The work done on the $N+1$ dislocations by the applied stress is $(N+1)\tau b\delta x$. The work done by the internal stress of the dislocations is zero because the relative distances between dislocations have not changed. The work done by the auxiliary force is $-f_{aux}\delta x$. Therefore,

$$(N+1)\tau b\delta x - f_{aux}\delta x = 0, \\ f_{aux} = (N+1)\tau b. \tag{10.45}$$

This means that the total driving force on the lead dislocation (excluding the resistance of the second phase) is $(N+1)\tau b$.

The lead dislocation cuts into the second phase when $f_x^{(0)} = 0$. From Eq. (10.44), this occurs when

$$\tau = \frac{\tau_f}{N+1}. \tag{10.46}$$

So even a small number of (say 20) dislocations in the pile-up greatly reduces the applied stress at which cutting into the second phase occurs.

Distribution within pile-up

It is sometimes of interest to determine how the dislocations in the pile-up are distributed. They are positioned so that the net force on each dislocation in the pile-up is zero. If we consider the ith dislocation in the pile-up, the net force acting on it, which is zero, can be expressed as

$$f_x^{(i)} = \tau b - \frac{\mu b^2}{2\pi(1-\nu)} \sum_{\substack{n=0 \\ n\neq i}}^{N} \frac{1}{x_n - x_i} = 0, \tag{10.47}$$

where the first term is the force arising from the applied shear stress and the second term is the interaction force arising from all of the other dislocations in the array. Assuming the lead dislocation is pinned at the obstacle, i.e. x_0 is known, the positions of all trailing dislocations

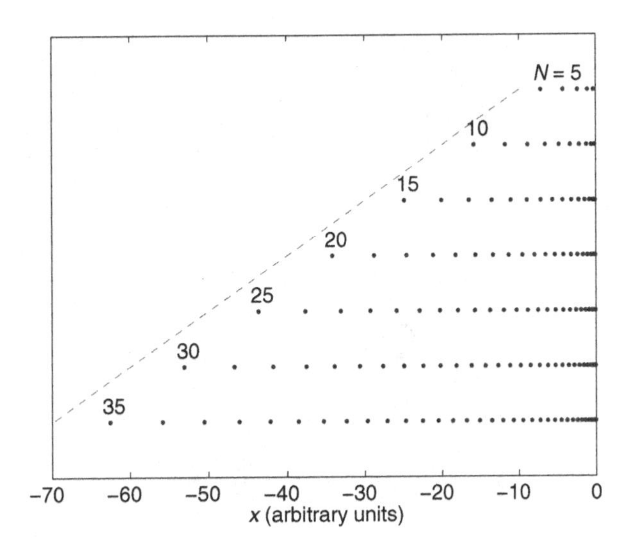

Figure 10.15. Numerical solution of the equilibrium positions of dislocation pile-ups with $\tau/A = 1$ (in arbitrary units) for $N = 5, 10, \ldots, 35$. The dashed line indicates the analytic result [71] of $|x_N - x_0| = 2NA/\tau + \mathcal{O}(N^{1/3})$.

are found by solving N coupled (non-linear) equations simultaneously

$$\sum_{\substack{n=0 \\ n\neq i}}^{N} \frac{1}{x_n - x_i} = \frac{\tau}{A} \quad \text{(for every } i = 1, \ldots, N\text{)}, \tag{10.48}$$

where $A \equiv \mu b/[2\pi(1-\nu)]$, with N unknowns $(x_1, x_2, \ldots, x_N)$ and $x_0 = 0$.

An analytic solution for x_i is difficult to obtain[1] for an arbitrary size of the pile-up, N, and the applied stress, τ. However, a numerical solution for Eq. (10.48) can be easily obtained, e.g. using Matlab (see Exercise problem 10.17). Figure 10.15 plots the numerical solution of equilibrium dislocation positions with $\tau/A = 1$ (in arbitrary units) for $N = 5, 10, \ldots, 35$. We can see that, under the same applied stress, the total length span of the dislocation pile-up is roughly proportional to the number of dislocations in the pile-up.

Given the solution for the equilibrium dislocation positions under one stress τ, the solution at another stress τ' can be easily obtained (as long as $\tau' < \tau_f/(N+1)$ so that the lead dislocation stays pinned). Given the form of Eq. (10.48), all we need to do is to scale the distance between every dislocation and the lead dislocation by a factor of τ/τ'. This will scale the distance between every pair of dislocations by the same factor, and Eq. (10.48) will be satisfied under the new stress τ'. This means that increasing the applied stress will compress all features of the pile-up uniformly, as long as the number of dislocations in the pile-up stays the same. The same scaling procedure can be applied to obtain the equilibrium positions of screw dislocations in a pile-up configuration, as long as all dislocations stay in one plane (i.e. no cross-slip). For screw dislocations, the equilibrium equation, (10.48), stays the same, except that $A = \mu b/(2\pi)$.

1 Eshelby, Frank, and Nabarro [71] showed that the x_is are the roots of the first derivative of Laguerre polynomials.

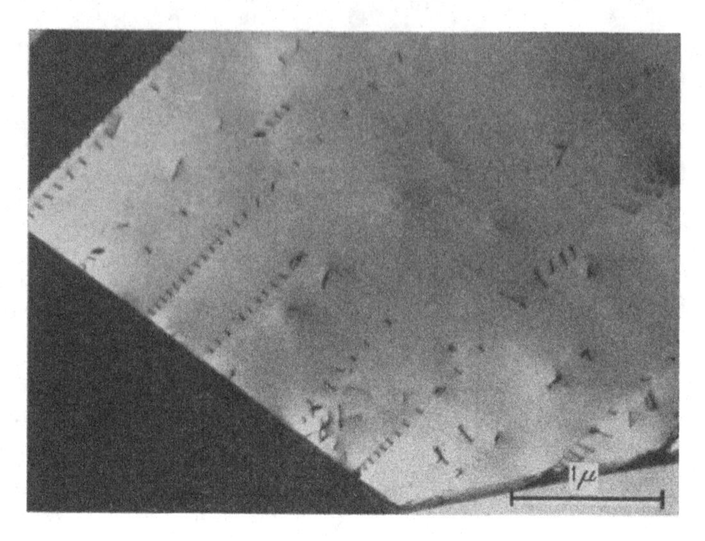

Figure 10.16. Dislocation pile-ups in a Cu–4.5%Al alloy [72, 73]. Permission granted by John Wiley & Sons.

Figure 10.16 is one of the classic TEM pictures of dislocation pile-ups in a copper-aluminum alloy. The alloy has a low stacking fault energy so that the dislocations are confined to their slip planes (see Section 11.1.2). The short line segments represent the individual dislocations extending from one surface of the thin TEM foil to the other. The distance between neighboring dislocations becomes progressively smaller as the obstacle is approached.

10.3 Strengthening mechanisms

We now have sufficient machinery to be able to derive useful relations for describing some mechanical properties of crystalline materials. In Section 8.1, we have seen that the existence of crystal dislocations successfully explains why the yield strengths of metals are typically orders of magnitude lower than their ideal strengths. Nonetheless, the yield strengths of engineering metals and alloys are still much higher than the critical stress required to move a dislocation in a pure metal. The strength of nominally the same metal or alloy can also be varied significantly through deformation and heat treatments, without changing its composition. This means that there are many mechanisms by which metals can be strengthened beyond the minimum stress required to move a single dislocation. Two such mechanisms are discussed below. For further details on strengthening mechanisms see [74].

10.3.1 Taylor hardening

The strength of a pure metal can be increased substantially by plastic deformation itself. This phenomenon is called *strain hardening* and is due to the interaction between dislocations. During plastic deformation, the dislocation density increases due to various multiplication mechanisms, such as those discussed in Section 8.5.3. The link between higher strength and higher dislocation density can be understood in terms of the critical stress needed to move a dislocation in the presence of all other dislocations in the crystal.

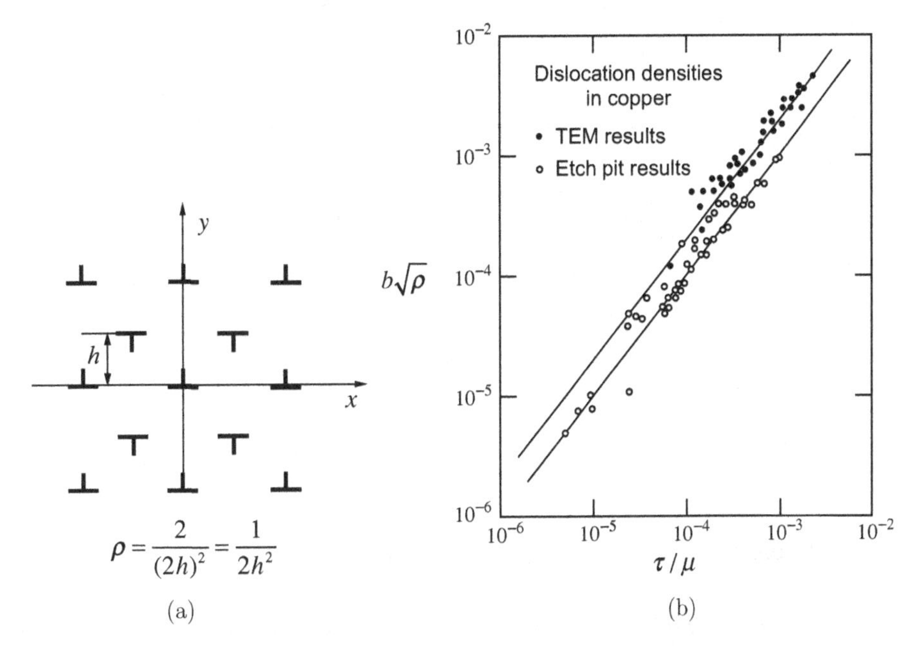

Figure 10.17. (a) Oppositely signed edge dislocations in an array (a Taylor lattice). (b) Relationship between dislocation density and flow stress in copper showing the validity of the Taylor relation [75].

The result we have derived in Section 10.1.4 for the passing stress for two oppositely signed edge dislocations can be used to derive a basic formula for the strengthening effect of dislocations – called the Taylor hardening formula. Let us assume that the passing stress we have derived applies to the passing of oppositely signed edge dislocations in an array of the kind shown in Fig. 10.17a. In reality, the passing stress when many dislocations are present differs from the result we have for just two dislocations, but only by a multiplicative constant – the form is the same.

For the simple array shown in Fig. 10.17a the density of dislocations, ρ (the number of dislocations per unit area), is of course related to the spacing between the slip planes, h, as $\rho = 1/(2h^2)$ or $h = 1/\sqrt{2\rho}$. Inserting this into Eq. (10.24) for the passing stress gives

$$\tau_c = \frac{\mu b}{8\pi(1-\nu)h} = \frac{\sqrt{2}\mu b\sqrt{\rho}}{8\pi(1-\nu)} = \alpha\mu b\sqrt{\rho}, \tag{10.49}$$

where $\alpha = \sqrt{2}/(8\pi(1-\nu)) \approx 0.1$. Figure 10.17b shows that the shear strength of pure copper scales almost perfectly with the square root of the dislocation density, as described by the Taylor relation (with $\alpha \sim 0.5$–1). This simple relation is ultimately linked to the fact that the stress field of dislocations is of the form $\sigma_{ij} \propto r^{-1}$.

10.3.2 Dispersion hardening

Another way to enhance the strength of a metal is to introduce non-shearable particles to impede the motion of dislocations. This phenomenon is called *dispersion hardening* because of the fine hard particles that are dispersed into the soft metal matrix. An example is a Ni-based superalloy

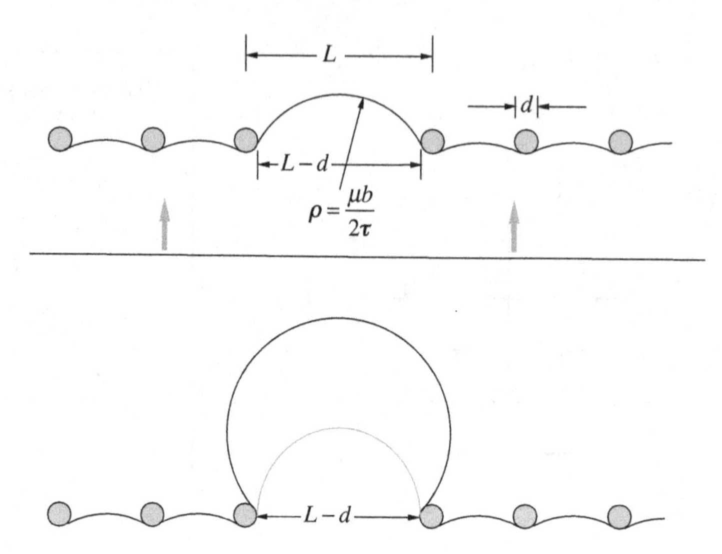

Figure 10.18. Orowan bowing between non-shearable particles.

containing 3 vol. % of spherical yttria (Y_2O_3) particles with mean diameter about 0.15 μm [76]. The strengthening effect can be understood in terms of the critical stress needed to move a dislocation over long distances on its slip plane in the presence of these particles.

As a simple model of the strengthening effect, we consider a linear array of non-shearable particles intersecting a slip plane in a pure metal. A glide dislocation (say an edge dislocation) glides toward and eventually comes into contact with the particles, as shown in Fig. 10.18. If the particles are simply non-shearable, and have no long-range stress fields, then the dislocation will move easily until it makes contact with the particles. As shown in Fig. 10.18, the dislocation line will bow in between the particles under a given applied shear stress.

To determine the approximate shape of the bowed dislocation under stress we make a very simple analysis of dislocation bowing. As shown in Section 9.3, if we ignore the orientation dependence of the line energy, i.e.

$$E_L \approx E_e \approx E_s \approx \frac{\mu b^2}{2}, \tag{10.50}$$

then the dislocation line tension is also

$$T \approx \frac{\mu b^2}{2}. \tag{10.51}$$

Hence we are now using the constant line tension model of Section 9.3.1. Under this approximation, the shape of the dislocation line bowed out between the obstacles is a circular arc with radius given by Eq. (9.70),

$$\rho = \frac{T}{f} = \frac{\mu b^2/2}{\tau b} = \frac{\mu b}{2\tau}. \tag{10.52}$$

For sufficiently small τ, a stable equilibrium state can be achieved, and Eq. (10.52) shows that the radius of the circular arc decreases with increasing τ. However, as the dislocation line continues to bow out, the minimum radius of the circular arc is $\rho_c = (L - d)/2$, as shown in

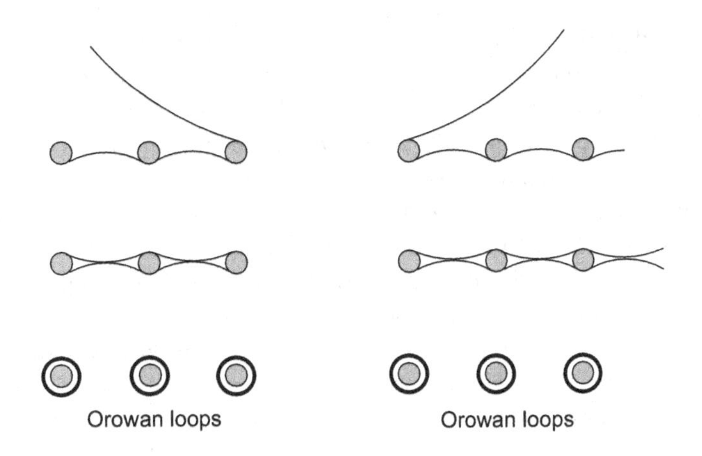

Figure 10.19. Orowan looping process leading to Orowan loops around each particle.

Fig. 10.18. Further bowing of the dislocation line actually results in a larger radius, so that a stable equilibrium state cannot be achieved. This defines a critical stress, as follows.

$$\boxed{\tau_c = \frac{\mu b}{L - d}} \tag{10.53}$$

Beyond this critical stress the pair of obstacles will not be able to hold back the dislocation line.

Equation (10.53) is called the Orowan bowing formula, after one of the discoverers of dislocations, Egon Orowan. According to this relation, very closely spaced particles can lead to very high strengths. This is the basis for precipitation or dispersion strengthening of crystalline solids.

In a real crystal, the spacing between particles is not constant. This is represented in Fig. 10.18 by letting the spacing of one particular pair be larger than the others. Naturally, the dislocation bows in between those more widely spaced particles more easily. Note that the curvatures of all of the segments are the same since they are all driven by the same shear stress. The segment extending between the more widely spaced particles is the critical segment as it will move past the particles before the others can.

If the line energy difference between edge and screw dislocations is taken into account, then the line shape can be described by de Wit and Koehler's solution, Eq. (9.82). Graphically, it can be obtained by scaling Fig. 9.13 by the factor $E_e/(\tau b)$ and cutting out a section of the loop to match the obstacle spacing. The resulting critical stress will depend on the initial orientation of the dislocation when it just touches the pair of obstacles. But the result is qualitatively the same as Eq. (10.53), which is also qualitatively the same as the critical stress to activate a Frank–Read source.

If the applied stress exceeds the critical stress to pass through the most widely spaced particles,

$$\tau > \tau_c = \frac{\mu b}{L - d},$$

then the curvature of the segment will decrease and the dislocation will continue its motion, as shown in Fig. 10.19. Eventually, as the loop expands, oppositely signed segments will move back toward the linear array of particles and approach the segments held up at the more closely spaced particles. If the stress is sufficiently high, these backward moving segments can annihilate with the held up segments and leave loops around each of the particles, as shown in Fig. 10.19. The Orowan loops left around each particle are shown in bold for clarity. Once one dislocation has moved past the particles by the Orowan looping process, it will be more difficult for another dislocation to follow the same path. A second dislocation of the same Burgers vector is repelled by Orowan loops, and would have to squeeze in between the Orowan loops left by the first dislocation in order to pass the particles in the same way. This effectively increases the particle diameter, d, and increases the critical stress according to Eq. (10.53). In this way it becomes ever more difficult for subsequent dislocations to bypass the particles. This causes very significant strain hardening. Dispersion strengthened metals are known to strain harden very rapidly by this mechanism. Through this mechanism plastic flow by dislocation glide leads to the production of dislocation debris in the crystal in the form of concentric Orowan loops around the particles.

10.4 Dislocation kinetics and plastic flow

Our study of dislocations started with the recognition that these imperfections are fundamental carriers of plasticity in crystalline materials. In Section 10.3, we have seen that the strength of the materials is enhanced if the motion of dislocations is impeded by interaction with other imperfections. However, the discussions have been limited to quasi-static scenarios and the focus has not yet been placed on how moving dislocations produce dynamic plastic flow. Here we describe some of the basic relations in this subject, which we call *dislocation kinetics*.[2] The discussion of dislocation kinetics and crystal plasticity given here is necessarily limited to avoid treating the much larger subject of mechanical properties of crystalline materials.

10.4.1 The Orowan equation

Plasticity in single-slip

We start the discussion on dislocation kinetics by establishing a relation between dislocation velocity and the plastic strain rate. Consider a very simple model in which a fixed number of mobile dislocations gliding on a single-slip system are responsible for plastic flow. Figure 10.20 shows N mobile dislocations with Burgers vector b intersecting an area defined by length l and height h. The mobile dislocation density is then $\rho_{\rm m} = N/(hl)$.

If the dislocations are gliding with an average velocity $\overline{v}$ then in a small increment of time Δt they will each glide a distance $\Delta x = \overline{v}\Delta t$ in their glide plane. The probability that such dislocations would intersect an arbitrarily placed line marker perpendicular to the glide plane is $\Delta x/l$. When the trajectory of a dislocation cuts across the line marker, the top of the marker is displaced by b relative to the bottom of the line marker. Given that, out of the N dislocations, on average $N\Delta x/l$ of them would cut across the line marker, the top of the line marker is expected to be displaced, on average, by $Nb\Delta x/l$ relative to the bottom during time increment Δt. In other words, the top of the element is displaced by $Nb\Delta x/l$ relative to the bottom.

2 While historically this subject has been called *dislocation dynamics*, we avoid this term here because it is now commonly used to refer to a computer simulation method for dislocations [2].

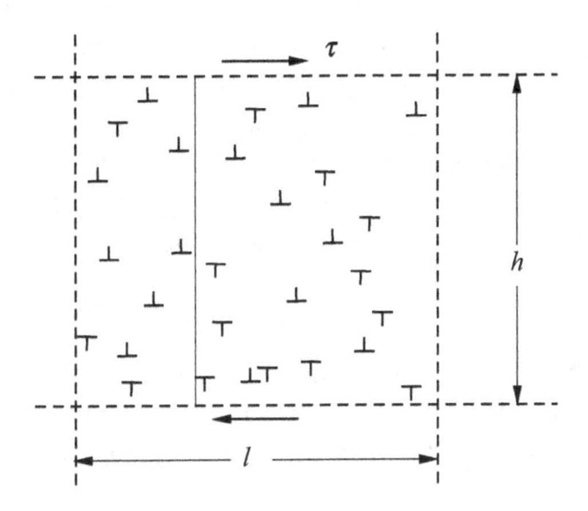

Figure 10.20. A group of N mobile dislocations intersecting a rectangular area of length l and height h.

The relative displacement between the top and bottom of the element amounts to a plastic shear strain increment of

$$\Delta\gamma_{\text{p}} = \frac{Nb(\Delta x/l)}{h}, \tag{10.54}$$

giving an average shear strain rate of

$$\dot{\gamma}_{\text{p}} = \frac{\Delta\gamma_{\text{p}}}{\Delta t} = \left(\frac{N}{hl}\right) b \left(\frac{\Delta x}{\Delta t}\right), \tag{10.55}$$

or it may be expressed as follows.

$$\boxed{\dot{\gamma}_{\text{p}} = \rho_{\text{m}} b \overline{v}} \tag{10.56}$$

Here ρ_{m} is the mobile dislocation density and $\overline{v}$ is the average glide velocity. This has been called the Orowan equation in honor of Egon Orowan who first described crystal plasticity in terms of dislocation kinetics in 1940.

While the Orowan relation is valid for all crystalline solids that plastically deform by dislocation glide, this particular form of the Orowan equation is especially convenient for hard crystals (such as semiconductors) with low densities of dislocations that move in a continuous, sluggish manner. For such crystals most of the dislocations can be regarded as being mobile most of the time and continuously in motion during plastic flow. When a constant strain rate is applied to such crystals (such as silicon at high temperature), a pronounced drop in flow stress is commonly observed right after plastic yielding starts. This can be understood in terms of Eq. (10.56) and the increase of dislocation density with strain, to be discussed in Section 10.4.2.

Alternative form of the Orowan equation

For soft metals with high dislocation densities, plastic flow usually occurs by jerky dislocation motion wherein dislocations move at high velocities after they are created or mobilized and come to an abrupt halt when they encounter a sufficiently big obstacle. In this case the mobile dislocation density in Eq. (10.56) refers to the density of dislocations in motion at any moment.

Let $\dot{\rho}_{\rm m}^{+}$ be the rate at which mobile dislocations are created (whose unit is $\rm m^{-2}s^{-1}$), and let $\dot{\rho}_{\rm m}^{-}$ be the rate at which dislocations become immobilized. Except for extremely high strain rates, the magnitudes of $\dot{\rho}_{\rm m}^{+}$ and $\dot{\rho}_{\rm m}^{-}$ are close to each other, and much greater than their difference, the net rate of mobile dislocation multiplication. If the mean distance traveled from the point of creation or mobilization to the point of stopping is called Λ (mean free path), then the Orowan relation can be expressed in a different form,

$$\dot{\gamma}_{\rm p} = \dot{\rho}_{\rm m}^{+}\Lambda b = \dot{\rho}_{\rm m}^{-}\Lambda b. \tag{10.57}$$

It should be emphasized that Eqs. (10.56) and (10.57) are alternate forms of the Orowan relation. They should not be considered independent or additive expressions for the plastic strain rate as they represent different ways to account for the same plastic deformation.

Plasticity in multi-slip

For plastic flow involving moving dislocations on multiple slip systems the axial strain rate in any particular direction, such as the loading axis in tension, can be found by resolving the strain rate for each slip system onto the loading axis and summing the contributions of each. From the transformation rules of the strain tensor described in Section 2.2, the plastic axial strain rate is

$$\dot{\varepsilon}_{\rm p} = \sum_i \cos\lambda_i \cos\varphi_i \dot{\gamma}_{\rm p}^i = \sum_i \cos\lambda_i \cos\varphi_i \rho_{\rm m}^i b\overline{v}^i, \tag{10.58}$$

where λ_i is the angle between the slip direction in slip system i and the loading axis, and φ_i is the angle between the slip plane normal and the loading axis. Note that $\rho_{\rm m}^i$ and $\overline{v}^i$ are the density and mean velocity of mobile dislocations on slip system i, respectively.

10.4.2 Application to yield point drop

The Orowan equation allows us to understand certain constitutive (i.e. stress–strain) behaviors of crystals in terms of dislocation kinetics. An example is the phenomenon of yield point drop when hard crystals are deformed at a constant rate. Here we consider the case of the ionic crystal LiF, which can be plastically deformed at room temperature and is transparent. The average dislocation velocity $\overline{v}$ in LiF has been measured as a function of applied shear stress, and the result can be described by a power law [77]

$$\overline{v} = v_0(\tau/\tau_0)^m, \tag{10.59}$$

where m ranges from 15 to 25, τ_0 is on the order of 10 MPa, and v_0 is on the order of 10^{-2} m · s^{-1}. Equation (10.59) is a good approximation as long as the dislocation velocity is several orders of magnitude lower than the sound velocity in the crystal, which is on the order of 10^3 m · s^{-1}. The dislocation velocity starts to level off as the sound velocity is approached.

In order to use the Orowan equation (10.56) to predict the stress–strain behavior of the crystal, we need to estimate how the dislocation density evolves with time (or with plastic strain). Consider N mobile dislocations intersecting a rectangular area, as shown in Fig. 10.20. A common model for dislocation multiplication is that every dislocation produces a new dislocation after it travels a distance λ. Assuming that in a small increment of time dt, all dislocations move by a distance $\overline{v}dt$, the average increase of the number of dislocations is

$$dN = N\frac{\overline{v}\,dt}{\lambda} = \delta N\overline{v}\,dt, \tag{10.60}$$

where $\delta \equiv 1/\lambda$ is called the "breeding" parameter. In the literature, δ is often assumed to be proportional to stress [78]. But here for simplicity we assume δ is a constant, following [77]. Recall that the mobile dislocation density in Fig. 10.20 is $\rho_{\rm m} = N/(hl)$. The rate of dislocation density increase is

$$\dot{\rho}_{\rm m} = \frac{1}{hl}\frac{dN}{dt} = \frac{\delta N\overline{v}}{hl} = \delta\rho_{\rm m}\overline{v}. \tag{10.61}$$

Comparing this with Orowan's equation (10.56), we have

$$\dot{\rho}_{\rm m} = \frac{\delta}{b}\dot{\gamma}_{\rm p}. \tag{10.62}$$

Integrating both sides of the equation with time, we obtain

$$\rho_{\rm m} = \rho_{\rm m0} + \frac{\delta}{b}\gamma_{\rm p}, \tag{10.63}$$

where $\rho_{\rm m0}$ is the initial mobile dislocation density. Equation (10.63) predicts that the dislocation density increases linearly with plastic strain. Of course, this relation is only valid if the breeding parameter δ is a constant, which is not generally true. Nonetheless, the linear relation between dislocation density and plastic strain has indeed been observed in LiF [77]. In ionic crystals (such as LiF) and semiconductors, the value of δ/b is on the order of 10^{13} m^{-2}. In metals, the value of δ/b is in the range of 10^{15}–10^{16} m^{-2} [79].

Let us first consider the case where the crystal is subjected to a constant shear strain rate $\dot{\gamma}$. Recognizing that the total strain rate is the sum of the elastic and plastic strain rates, we have

$$\dot{\gamma} = \dot{\gamma}_{\rm el} + \dot{\gamma}_{\rm p} = \frac{\dot{\tau}}{\mu} + \rho_{\rm m}b\overline{v} = \frac{\dot{\tau}}{\mu} + [\rho_{\rm m0} + (\delta/b)\gamma_{\rm p}]bv_0(\tau/\tau_0)^m. \tag{10.64}$$

Integration of this differential equation can give a prediction of the stress–strain relation for hard crystals.

In practice, the specimen is most conveniently tested under uniaxial loading. Owing to the compliance of the loading mechanism, the strain rate imposed on the specimen is not strictly a constant. This condition can be modeled by considering a crosshead moving at a constant velocity to deform the specimen. The compliance of the loading mechanism can be considered as an imaginary spring that lies between the crosshead and the specimen. A nominal total strain rate, $\dot{\varepsilon}$, can be defined as the velocity of the crosshead divided by the original length of the specimen. The motion of the crosshead is accommodated by the deformation of the spring, as well as the elastic and plastic strain rates of the specimen, i.e.

$$\dot{\varepsilon} = \dot{\varepsilon}_{\rm LM} + \dot{\varepsilon}_{\rm el} + \dot{\varepsilon}_{\rm p}, \tag{10.65}$$

where $\dot{\varepsilon}_{\rm LM}$ is the rate of length change of the spring divided by the original length of the specimen. Given that both $\varepsilon_{\rm LM}$ and $\varepsilon_{\rm el}$ are proportional to the axial stress σ, we can define an effective

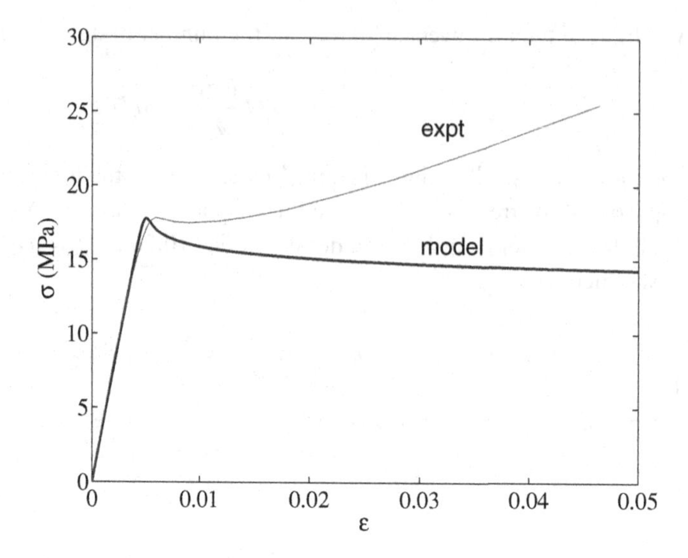

Figure 10.21. Axial stress–strain curves of LiF. Thin line: experimental data [77]. Thick line: prediction from dislocation kinetics model.

Young's modulus

$$E_{\text{eff}} = \frac{\sigma}{\varepsilon_{\text{LM}} + \varepsilon_{\text{el}}} \tag{10.66}$$

so that

$$\dot{\varepsilon} = \frac{\dot{\sigma}}{E_{\text{eff}}} + \dot{\varepsilon}_{\text{p}}. \tag{10.67}$$

Assuming that both the slip plane and dislocation Burgers vectors are oriented at 45° with respect to the loading axis, we have $\tau = \sigma/2$ and $\dot{\varepsilon}_{\text{p}} = \dot{\gamma}_{\text{p}}/2$, so that

$$\dot{\varepsilon} = \frac{\dot{\sigma}}{E_{\text{eff}}} + \frac{\rho_{\text{m}} b}{2} v_0 \left(\frac{\sigma}{2\tau_0}\right)^m = \frac{\dot{\sigma}}{E_{\text{eff}}} + \frac{b}{2}\left[\rho_{\text{m0}} + (\delta/b)\frac{\varepsilon_{\text{p}}}{2}\right] v_0 \left(\frac{\sigma}{2\tau_0}\right)^m. \tag{10.68}$$

Integration of these differential equations can give a prediction of the axial stress–strain relation for hard crystals.

Figure 10.21 plots the predicted (thick line) and measured (thin line) stress–strain curves for LiF. The parameters used in the dislocation kinetics model are: $\dot{\varepsilon} = 6 \times 10^{-5}\ \text{s}^{-1}$, $E_{\text{eff}} = 3.76\,\text{GPa}$, $b = 2.85\ \text{Å}$, $\rho_{\text{m0}} = 8 \times 10^9\ \text{m}^{-2}$, $\delta = 4000\ \text{m}^{-1}$, $m = 20$, $\tau_0 = 12\ \text{MPa}$, and $v_0 = 0.01\ \text{m} \cdot \text{s}^{-1}$. The dislocation kinetics model successfully captures the drop in flow stress immediately after the upper yield point. The upper yield point exists because the initial mobile dislocation density in the crystal is very low, so that at the onset of yield a high stress is needed to move the dislocations fast enough to maintain an appreciable plastic strain rate. As the mobile dislocations multiplies, a lower average dislocation velocity is needed to maintain the same plastic strain rate, according to Orowan's equation. This leads to a lower stress. The dislocation kinetics model described above predicts that the flow stress would decrease monotonically beyond the upper yield point. However, the experimental data show an increase of flow stress at large strains. This is due to the strain hardening effect that has been neglected in the simple dislocation kinetics model described above.

The yield point drop becomes even more pronounced in harder crystals where dislocation motion is very sluggish, such as in semiconductors (Si and Ge), which can be deformed plastically at high temperatures (see Exercise problem 10.18). The effect can also become more dramatic at higher strain rates and in crystals with a lower initial dislocation density. The dislocation kinetics model can also be generalized to account for strain hardening by replacing the stress τ in the dislocation mobility equation (10.59) with an effective stress that depends on plastic strain.

10.4.3 Application to sigmoidal creep

A creep test is a common technique to probe the time-dependent deformation behavior of materials. In a creep test, the specimen is subjected to a constant applied stress, and its strain is recorded as a function of time. For crystals that exhibit an upper yield point under a constant strain rate, the creep test is often conducted at a stress slightly below the upper yield stress. Under this condition, the initial plastic strain rate is very low. However, the strain rate can increase with time and reach a maximum that is much greater than the initial value. At later times the strain rate decreases with time. This behavior is called *sigmoidal creep* because the plastic strain as a function of time has a characteristic shape of the letter S (a sigmoid).

Sigmoidal creep can be explained by using the dislocation kinetics model. The initial increase of plastic strain rate is caused by the multiplication of mobile dislocations, while the later decrease of strain rate is due to strain hardening. In this section we develop such a model and apply it to the sigmoidal creep of Si.

At high enough temperature at which Si can be deformed plastically without fracture, the dislocation velocity v can be expressed as the following function of the local shear stress τ and temperature T [80]

$$v(\tau, T) = v_0 \left(\frac{\tau}{\tau_0}\right)^m \exp\left(-\frac{Q}{k_B T}\right), \tag{10.69}$$

where $m = 1.4$, $Q = 2.2$ eV, $v_0 = 3.8 \times 10^4$ m $\cdot$ s^{-1}, and $\tau_0 = 9.8$ MPa. To account for the interaction between the dislocations and the Taylor hardening effect (see Section 10.3.1), we define an effective stress

$$\tau_{\text{eff}} = \tau - \alpha\mu b\sqrt{\rho_{\text{m}}}, \tag{10.70}$$

where τ is the applied stress and the second term represents an averaged internal stress. For simplicity, we have also assumed that all dislocations are mobile, i.e. $\rho = \rho_{\text{m}}$. We assume that the average dislocation velocity $\overline{v}$ can be expressed as a function of τ_{eff} using Eq. (10.69), i.e.

$$\overline{v} = v(\tau_{\text{eff}}, T) = v_0 \left(\frac{\tau - \alpha\mu b\sqrt{\rho_{\text{m}}}}{\tau_0}\right)^m \exp\left(-\frac{Q}{k_B T}\right). \tag{10.71}$$

To describe the multiplication of mobile dislocations, we use a similar model as in the previous section,

$$\dot{\rho}_{\text{m}} = \delta\rho_{\text{m}}\overline{v}, \tag{10.72}$$

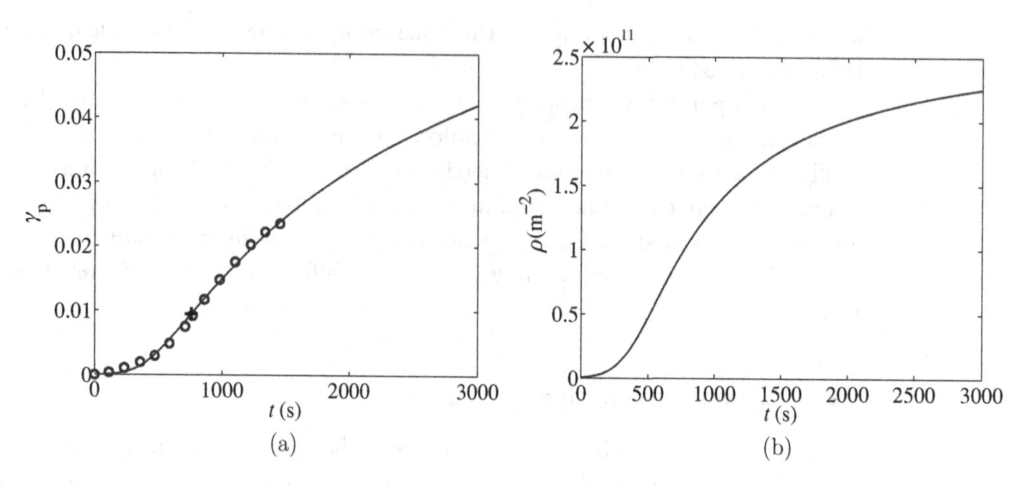

Figure 10.22. Sigmoidal creep of Si. (a) Plastic strain as a function of time. Circles: experimental data [80]. Line: prediction from dislocation kinetics model. The + symbol indicates the time at which the plastic strain rate reaches the maximum. (b) Dislocation density as a function of time predicted by the dislocation kinetic model.

except that the breeding parameter is no longer a constant, but is proportional to the effective stress, i.e.

$$\delta = K\tau_{\text{eff}}, \tag{10.73}$$

where K is a constant. Therefore,

$$\dot{\rho}_{\text{m}} = K(\tau - \alpha\mu b\sqrt{\rho_{\text{m}}})\rho_{\text{m}}\overline{v}. \tag{10.74}$$

Integrating this differential equation can predict the evolution of the dislocation density as a function of time during the creep test. Given the dislocation density, the plastic strain as a function of time can be obtained by integrating the Orowan equation, $\dot{\gamma}_{\text{p}} = \rho_{\text{m}} b\overline{v}$.

Figure 10.22 plots the predicted plastic strain and dislocation density evolution with time for Si under creep conditions with $\tau = 4.9$ MPa and $T = 1129$ K. The parameters used in this dislocation kinetics model are: $K = 1.4 \times 10^{-3}\ \text{m} \cdot \text{N}^{-1}, \alpha = 0.36, \rho_{\text{m0}} = 5 \times 10^{8}\ \text{m}^{-2}$, $\mu = 66.6$ GPa, and $b = 3.84$ Å. The plastic strain rate is the slope of the curve in Fig. 10.22a. We can see that the plastic strain rate starts out very small, but gradually increases to reach a maximum (at $t = 760$ s). At later times, the strain hardening effect significantly reduces the effective stress on the dislocations so that the plastic strain rate is gradually reduced and the dislocation density starts to level off.

10.5 Formation of dislocations at interfaces

10.5.1 Epitaxial film on substrate

The examples in the previous section illustrate the role of dislocation motion and multiplication in the plastic deformation of bulk crystals. We now discuss the role of dislocations on interfaces in relieving the misfit strain between two crystals.

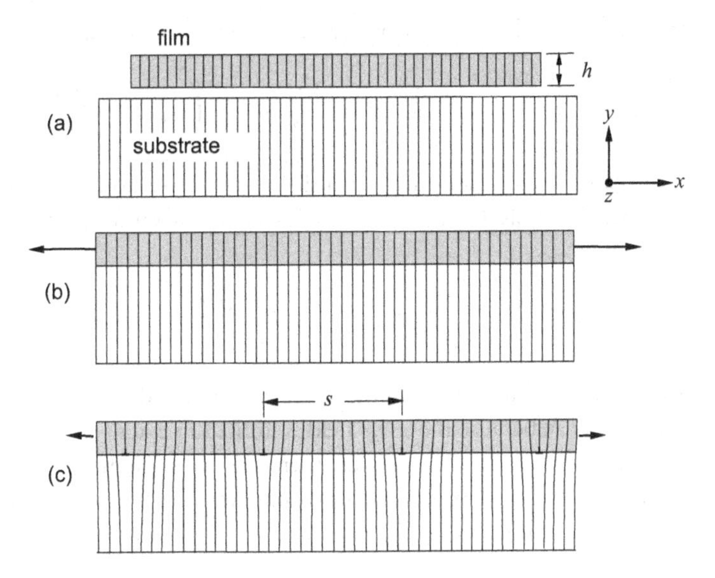

Figure 10.23. Epitaxial film and misfit dislocations. (a) The free-standing thin film has a smaller lattice constant than the substrate. (b) The film is stretched biaxially (along both x and z) so that it can be attached epitaxially to the substrate. (c) Two sets of edge misfit dislocations form on the interface to relieve the misfit strain. One set (shown here) is parallel to the z axis and the other set (not shown) is parallel to the x axis.

Consider a film of thickness h, which has a biaxial misfit strain $\varepsilon_{\text{misfit}} < 0$ relative to a much thicker substrate to which it is bonded epitaxially. If the film is sufficiently thin, below the critical thickness, h_c, for the formation of misfit dislocations, the equilibrium state of the film will be a perfectly coherent, uniformly strained film, as shown in Fig. 10.23b. In this state the biaxial elastic strain in the film is

$$\varepsilon_{\text{el}} = -\varepsilon_{\text{misfit}} = \left(\frac{1-\nu}{E}\right)\sigma = \frac{1-\nu}{2\mu(1+\nu)}\sigma, \tag{10.75}$$

where σ is the biaxial stress in the film ($\sigma_{xx} = \sigma_{zz} = \sigma$). The elastic energy of the fully coherent film, per unit area, is then

$$\begin{aligned} E_{\text{coherent}} &= \left(\frac{1}{2}\sigma_{xx}\varepsilon_{xx} + \frac{1}{2}\sigma_{zz}\varepsilon_{zz}\right)h \\ &= \sigma\varepsilon_{\text{el}}h = \frac{2\mu(1+\nu)}{1-\nu}\varepsilon_{\text{el}}^2 h = \frac{2\mu(1+\nu)}{1-\nu}\varepsilon_{\text{misfit}}^2 h. \end{aligned} \tag{10.76}$$

Above a critical film thickness, h_c, misfit dislocations will form and partially relax the biaxial strain in the film, which now becomes semi-coherent. Here we consider two sets of parallel edge dislocation arrays of spacing s. One set is parallel to the z axis, as shown in Fig. 10.23(c), and the other set is parallel to the x axis. These misfit dislocations reduce the misfit strain in the film by b/s in both the x and z directions, i.e.

$$|\varepsilon_{\text{el}}| = |\varepsilon_{\text{misfit}}| - \frac{b}{s}. \tag{10.77}$$

This reduces the elastic strain energy associated with the homogeneous strain in the film to

$$E_{\text{strain}} = \frac{2\mu(1+\nu)}{1-\nu}\left(|\varepsilon_{\text{misfit}}| - \frac{b}{s}\right)^2 h. \tag{10.78}$$

But the elastic energy associated with the misfit dislocations must also be taken into account in an energetic analysis. Using Eq. (9.63), the energy per unit length of each misfit dislocation can be written as

$$E_{\text{edge–misfit}} = \frac{\mu b^2}{4\pi(1-\nu)} \ln\left(\frac{\alpha_e h}{b}\right), \tag{10.79}$$

where the outer cutoff radius in the line energy of the dislocation is taken to be the thickness of the film, $R = h$, because the stress field is limited in extent by the thickness of the film. In each square area of size s on the interface, there are two edge misfit dislocations each of length s, so that the length of misfit dislocation per unit area is $2s/s^2 = 2/s$. Therefore, the dislocation line energy, per unit film area, is

$$E_{\text{dislocation}} = \left(\frac{2}{s}\right)\frac{\mu b^2}{4\pi(1-\nu)} \ln\left(\frac{\alpha_e h}{b}\right). \tag{10.80}$$

Then the total energy of the semi-coherent film is

$$\begin{aligned} E_{\text{semi–coh}} &= E_{\text{strain}} + E_{\text{dislocation}} \\ &= \frac{2\mu(1+\nu)}{1-\nu}\left(|\varepsilon_{\text{misfit}}| - \frac{b}{s}\right)^2 h + \left(\frac{2}{s}\right)\frac{\mu b^2}{4\pi(1-\nu)} \ln\left(\frac{\alpha_e h}{b}\right). \end{aligned} \tag{10.81}$$

The inverse of the distance between neighboring parallel misfit dislocations, $1/s$, can be taken as a measure of the density of misfit dislocations. The equilibrium density of misfit dislocations can be obtained by minimizing the total energy of the semi-coherent film with respect to $1/s$. For this we need,

$$\frac{\partial E_{\text{semi–coh}}}{\partial(1/s)} = -2b\frac{2\mu(1+\nu)}{1-\nu}\left(|\varepsilon_{\text{misfit}}| - \frac{b}{s}\right)h + \frac{\mu b^2}{2\pi(1-\nu)} \ln\left(\frac{\alpha_e h}{b}\right). \tag{10.82}$$

Setting this partial derivative to zero, we obtain the equilibrium density

$$\frac{1}{s_{\text{eq}}} = \frac{|\varepsilon_{\text{misfit}}|}{b} - \frac{1}{8\pi(1+\nu)h} \ln(\alpha_e h/b). \tag{10.83}$$

The critical film thickness h_c can be obtained by setting the equilibrium density of misfit dislocations, Eq. (10.83), to zero. This leads to the following condition for the critical thickness.

$$\boxed{\frac{h_c}{\ln(\alpha_e h_c/b)} = \frac{b}{8\pi(1+\nu)|\varepsilon_{\text{misfit}}|}} \tag{10.84}$$

Mathematically, for each value of $\varepsilon_{\text{misfit}}$, Eq. (10.84) has two solutions for h_c. It is the larger solution (for which $\ln(\alpha_e h_c/b) > 1$) that should be taken as the physical prediction of the critical film thickness.

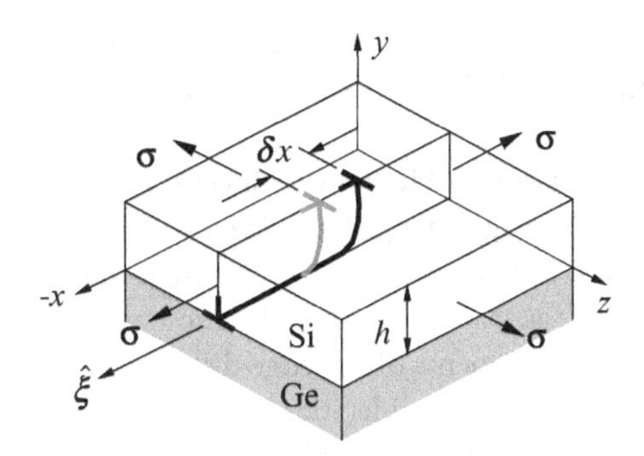

Figure 10.24. Incremental dislocation motion in the thin film on a substrate.

For $h > h_c$, the equilibrium density of misfit dislocations, as given by Eq. (10.83), is positive and increases with increasing film thickness. The reduced (average) strain in the film is

$$|\varepsilon_{\text{el}}| = |\varepsilon_{\text{misfit}}| - \frac{b}{s_{\text{eq}}} = \frac{b}{8\pi(1+\nu)h}\ln(\alpha_e h/b). \tag{10.85}$$

It can be shown that the corresponding energy of the semi-coherent film is lower than that of the fully coherent film. So when $h > h_c$ the film is partially relaxed and there are misfit dislocations at the interface between the film and substrate. An example is the misfit dislocations at the interface between CdTe and GaAs shown in Fig. 8.26. For $h < h_c$, the energy of the semi-coherent film is higher than that of the coherent film, so that the misfit dislocations do not form.

10.5.2 Dislocation motion in a thin film

Our treatment of the critical thickness of an epitaxial film and the equilibrium density of misfit dislocations in the previous section has been based entirely on thermodynamic considerations. No consideration has been given so far on the kinetics of how the misfit dislocations form and multiply to reach the equilibrium density. The kinetic processes are important because, if they are impeded in some way, the crystal may not be able to reach the full equilibrium state predicted from thermodynamic considerations. For example, the interface can remain fully coherent even when the film thickness has exceeded the critical thickness. Even for semi-coherent films, the relaxation of the misfit strain is often found to be lower than the value predicted from equilibrium considerations alone.

Here we consider an example of dislocation motion in a strained epitaxial film to relax the misfit strain, as shown in Fig. 10.24. To be specific, we can think of a thin film of Si grown epitaxially onto a Ge substrate. Because Si has a smaller lattice parameter than Ge, the Si film must be stretched to achieve a lattice match. The Ge substrate, however, is assumed to be infinitely thick and wide, so that it maintains the equilibrium lattice parameter of a pure Ge crystal.

For simplicity we will assume that an edge dislocation is present in the film as shown and that it might climb in response to the biaxial tension stress σ in the film. This will serve as a model of the more common glide motion of dislocations in thin films. The edge dislocation shown in Fig. 10.24 is curved. The straight section lying at the film/substrate is the misfit dislocation considered in Section 10.5.1. The segment that curves up from the film/substrate interface to the free surface is called the *threading segment.*

Energy changes due to virtual displacement

We consider the energy changes associated with an incremental movement of the edge dislocation in the film. As shown in Fig. 10.24, when the threading segment moves a small distance, δx, the net result is that the misfit dislocation extends in length by δx. The interaction energy between different segments of the edge dislocation remains unchanged. This greatly simplifies the analysis. The above consideration also shows that the motion of the threading segment is a possible mechanism for the multiplication of misfit dislocations.

The energy required to lengthen the misfit dislocation must be supplied by the stress in the film and that determines the critical stress needed to drive the threading segment forward. The increased energy for the incremental length of misfit dislocation can be expressed as

$$\delta E_{\text{misfit}} = E_e \delta x = \left[\frac{\mu b^2}{4\pi (1-\nu)} \ln \frac{\alpha_e h}{b} \right] \delta x. \tag{10.86}$$

Also, the work done by the internal stress in the film on the moving threading segment is

$$\Delta W = \sigma b h \delta x. \tag{10.87}$$

When the internal stress does work, the elastic energy of the strained film changes accordingly,

$$\delta E_{\text{int}} = -\Delta W = -\sigma b h \delta x. \tag{10.88}$$

The first two terms in this expression, σb, represent the climb force (per unit length) which when multiplied by the length of the threading segment, h, gives the total force pushing the threading segment forward. This times the incremental movement is the incremental change in elastic energy. For a system in equilibrium, $\delta E = \delta E_{\text{int}} + \delta E_{\text{misfit}} = 0$, so that the threading segment can advance spontaneously if

$$\begin{gathered} \delta E = \delta E_{\text{int}} + \delta E_{\text{misfit}} < 0, \\ -\sigma b h \delta x + \left[\frac{\mu b^2}{4\pi (1-\nu)} \ln \frac{\alpha_e h}{b} \right] \delta x < 0. \end{gathered} \tag{10.89}$$

Critical stress for generating misfit dislocation

From Eq. (10.89), the critical stress needed to drive the threading segment forward is

$$\sigma_c = \frac{\mu b}{4\pi (1-\nu) h} \ln \frac{\alpha_e h}{b}. \tag{10.90}$$

This result means that for a given film thickness h, the threading segment will advance spontaneously if the internal stress $\sigma > \sigma_c$. At a lower stress, $\sigma < \sigma_c$, the threading segment will move in the opposite direction and the film will become spontaneously more perfect as the misfit dislocation is erased.

This result can be interpreted from a force balance point of view, where the interaction between the misfit dislocation and the threading segment produces a force on the threading segment. This is often called the self-force, because it is due to the interaction between different segments of the same dislocation. The Peach–Koehler force from the stress σ acts in the opposite direction as the self-force. If σ is not large enough, the threading segment moves in the direction that shortens the misfit dislocation. Therefore, Eq. (10.90) can be interpreted as the maximum back-stress (or self-stress) that the curved dislocation can generate when it is confined to the film thickness h. Equation (10.90) has a similar form as the critical stress to activate a Frank–Read source of length L, if h is replaced by $L/2$.

Critical thickness for generating misfit dislocation

It is often useful to rewrite the critical stress condition, Eq. (10.90), into a condition of a critical thickness, h_c, for a given stress in the film, which is

$$\frac{h_c}{\ln(\alpha_e h_c/b)} = \frac{\mu b}{4\pi(1-\nu)\sigma}. \tag{10.91}$$

For an epitaxial film, the stress σ in the film is related to the biaxial misfit $\varepsilon_{\text{misfit}}$ strain through Eq. (10.75), so that

$$\sigma = -\frac{2\mu(1+\nu)}{1-\nu}\varepsilon_{\text{misfit}}. \tag{10.92}$$

Substituting Eq. (10.92) into Eq. (10.91) gives us a condition for the critical thickness h_c that is identical to Eq. (10.84).

That the force balance analysis here and the energy analysis in the previous section lead to the same prediction for the critical thickness indicates that the two approaches are consistent with each other. In fact, the process shown in Fig. 10.24 can be considered as a kinetic pathway to generate the first, infinitely long, misfit dislocation shown in Fig. 10.23c. However, the kinetic process considered here can be more useful than the energy analysis when discussing strain relaxation in real films. For example, if an obstacle exists on the plane of motion for the threading segment, the elongation of the misfit dislocation can be stopped until the film thickness increases to a higher value than h_c. The obstacle can be an inclusion on the interface, or simply an existing misfit dislocation perpendicular to the growing misfit dislocation. The interaction between the threading segment with existing misfit dislocations impedes the strain relaxation process, so that the residual (average) strain in the film is usually higher than that given by Eq. (10.85), which is predicted from thermodynamic considerations alone.

For the case of a Si film on a Ge substrate the misfit strain is negative so that the resulting stress is positive (tension). Equation (10.84) can be used to determine the critical thickness, below which a strained film without dislocations is thermodynamically stable. Even if a dislocation is artificially created in a film below the critical thickness, it would not be stable and would tend to spontaneously move out of the film.

If we consider the case of a Ge film on a Si substrate instead, the misfit strain is positive so that the resulting stress is negative (compression). Then the above analysis needs to be applied to a dislocation of the opposite sign as the one shown in Fig. 10.24. The result will be the same as Eq. (10.84).

As noted above, the model calculation described in this section is for a climbing edge dislocation. In a real thin film, the misfit dislocation and the threading segments are usually mixed dislocations; and the threading dislocation propagates by gliding on its slip plane, which is inclined with respect to the top surface. The analysis for the glide scenario is a little more complex but the rationale is the same, and the final result is of the same form as Eq. (10.84). In

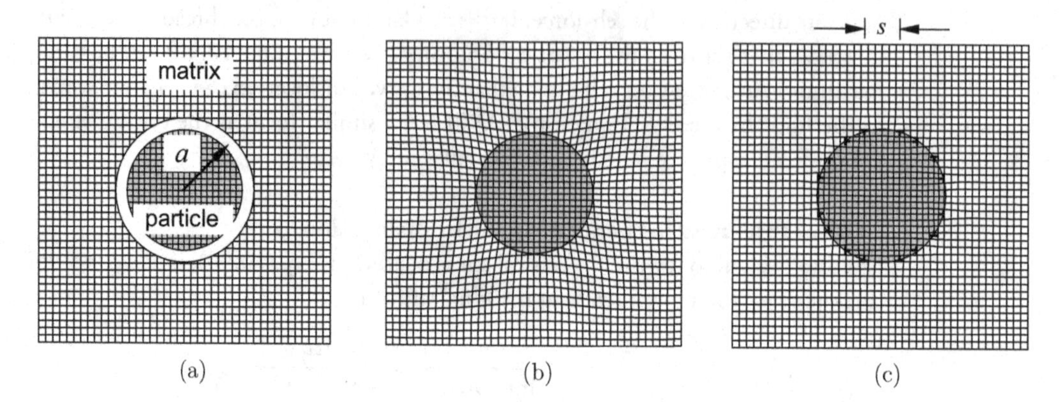

Figure 10.25. Particle in a matrix.

fact, this model gives quite a good result for mixed misfit dislocations if the b in Eq. (10.87) is replaced by the edge component of the Burgers vector in the plane of the interface, $b_{\mathrm{e}}^{\mathrm{in}}$, while the b in Eq. (10.86) remains as the magnitude of the total Burgers vector. In other words, the critical film thickness can be determined by

$$\frac{h_c}{\ln(\alpha_{\mathrm{e}} h_c/b)} = \frac{b_{\mathrm{e}}^{\mathrm{in}}}{8\pi(1+\nu)|\varepsilon_{\mathrm{misfit}}|}. \tag{10.93}$$

10.5.3 Particle in a matrix

Semi-coherent interfaces are also created when initially coherent precipitates are grown to sizes that exceed a critical size for misfit dislocation formation. The energy approach in Section 10.5.1 can be extended to predict the critical size of the precipitates.

Consider a cylindrical precipitate of radius a with an in-plane misfit strain $\varepsilon_{\mathrm{misfit}} < 0$ relative to the matrix in which it resides, caused by the smaller lattice parameter in the precipitate relative to the matrix, as illustrated in Fig. 10.25a.

For simplicity we will assume that the precipitate and matrix have the same isotropic elastic properties. If the radius of the precipitate is sufficiently small, the equilibrium state of the precipitate will be a perfectly coherent, uniformly strained cylinder, as shown in Fig. 10.25b.

Energy of a coherent particle

The configuration corresponds to a plane strain problem defined in Section 2.5.3. It can be shown that the stress field inside the particle is uniform,

$$\sigma_{rr}^{\mathrm{p}} = \sigma_{\theta\theta}^{\mathrm{p}} = -\frac{\mu}{1-\nu}\varepsilon_{\mathrm{misfit}}, \quad \sigma_{zz}^{\mathrm{p}} = \nu\left(\sigma_{rr}^{\mathrm{p}} + \sigma_{\theta\theta}^{\mathrm{p}}\right) = -\frac{2\mu\nu}{1-\nu}\varepsilon_{\mathrm{misfit}}, \tag{10.94}$$

while the stress state in the matrix is

$$\sigma_{rr}^{\mathrm{m}} = -\sigma_{\theta\theta}^{\mathrm{m}} = -\frac{\mu}{1-\nu}\varepsilon_{\mathrm{misfit}}\left(\frac{a}{r}\right)^2, \quad \sigma_{zz}^{\mathrm{m}} = \nu\left(\sigma_{rr}^{\mathrm{m}} + \sigma_{\theta\theta}^{\mathrm{m}}\right) = 0, \tag{10.95}$$

where μ is the shear modulus and ν is Poisson's ratio. The elastic energy per unit length in this fully coherent state can be expressed as

$$E_{\text{coherent}} = \frac{\mu}{1-\nu}\varepsilon_{\text{misfit}}^2 \pi a^2. \tag{10.96}$$

Energy of a semi-coherent particle

If the radius of the precipitate exceeds a critical value then it becomes energetically favorable for misfit dislocations to form at the particle–matrix interface. An orthogonal set of edge dipoles with spacing s can be expected to form, as shown in Fig. 10.25c. These dislocations lessen the misfit between the precipitate and the matrix by b/s just as in the case of an epitaxial film on a substrate, and the elastic energy per unit length then becomes

$$E_{\text{strain}} = \frac{\mu}{1-\nu}(|\varepsilon_{\text{misfit}}| - b/s)^2 \pi a^2. \tag{10.97}$$

But the energies of the misfit dislocation dipoles must also be taken into account. To obtain the energy of an edge dislocation dipole, we imagine the process of separating the two dislocations in the dipole (by climb) from a smaller distance of b to a larger distance of h. The work done in this process against the interaction force between the two dislocations is the energy of the dislocation dipole, because we assume that when the two dislocations are at distance b, they have effectively annihilated each other and the energy is zero. Hence the energy of the dislocation dipole is

$$E_{\text{dipole}} = \int_b^h \frac{\mu b^2}{2\pi(1-\nu)}\frac{1}{s}ds = \frac{\mu b^2}{2\pi(1-\nu)}\ln\frac{h}{b}. \tag{10.98}$$

As a rough estimate we may take the separation distance to be $h = a$ for all of the dipoles and ignore the interaction between the dipoles. Then the energies of all of the dipoles shown in Fig. 10.25c would be

$$E_{\text{all-dipoles}} = 2\left(\frac{2a}{s}\right)\frac{\mu b^2}{2\pi(1-\nu)}\ln\frac{a}{b}. \tag{10.99}$$

The total energy per unit length of the semi-coherent precipitate is then

$$E_{\text{semi-coh}} = \frac{\mu}{1-\nu}(|\varepsilon_{\text{misfit}}| - b/s)^2 \pi a^2 + 2\left(\frac{2a}{s}\right)\frac{\mu b^2}{2\pi(1-\nu)}\ln\frac{a}{b}. \tag{10.100}$$

The equilibrium density of misfit dislocations can be obtained by minimizing the total energy of the semi-coherent particle with respect to $1/s$. This leads to

$$\frac{1}{s_{\text{eq}}} = \frac{|\varepsilon_{\text{misfit}}|}{b} - \frac{1}{\pi^2 a}\ln\left(\frac{a}{b}\right). \tag{10.101}$$

The critical precipitate radius a_c can be obtained by setting the equilibrium density of misfit dislocations to zero, which leads to the following condition for the critical radius.

$$\boxed{\frac{a_c}{\ln(a_c/b)} = \frac{b}{\pi^2|\varepsilon_{\text{misfit}}|}} \tag{10.102}$$

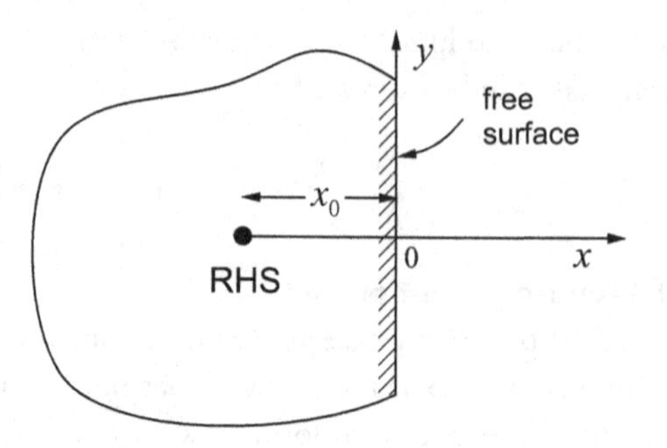

Figure 10.26. Right-handed screw dislocation near a free surface.

While the derivation given here is for a two-dimensional, cylindrical particle, it may be of interest to compare the predictions of this analysis with experiments involving initially coherent spherical precipitates of Co in Cu [81]. Very small spherical Co precipitates in Cu take the FCC structure and are fully coherent with the Cu matrix. Based on the atomic radii for Co and Cu given in Appendix A, the misfit strain for Co precipitates is expected to be $\varepsilon_{\rm misfit} = -0.0202$ and the Burgers vector of Cu is $b = 0.2826$ nm. Using Eq. (10.102) this gives a critical radius of $a_{\rm c} = 3.61$ nm (or a critical diameter of $d_{\rm c} = 7.2$ nm), largely consistent with the observation of fully coherent Co particles of the size (diameter) 3–9 nm in the Cu matrix [81].

10.6 Elastic fields of dislocations near interfaces

Our discussion of dislocation forces in Sections 10.1–10.3 has assumed that the solid in which interacting dislocations are found has uniform elastic properties and is infinite in extent. This is a good approximation for dislocations deep inside the bulk of a homogeneous crystal. In Section 10.5, we have seen a few examples in which it is important to understand the forces on dislocations near the crystal surface or an interface between two different crystals. Here we consider the case of straight screw dislocations lying parallel to planar boundaries. We will show that image solutions can be constructed to describe the elastic field satisfying the appropriate boundary conditions, and to obtain the corresponding forces that act on the dislocations.

10.6.1 Image dislocation for free surface

Consider a right-handed screw dislocation in a semi-infinite elastic solid with shear modulus μ, positioned a distance x_0 from a free surface, as shown in Fig. 10.26. The surface is traction free, meaning that the following components of stress must be zero along the interface:

$$\sigma_{xx}(x{=}0, y) = 0, \quad \sigma_{yx}(x{=}0, y) = 0, \quad \sigma_{zx}(x{=}0, y) = 0. \tag{10.103}$$

Since $\sigma_{xx} = \sigma_{yx} = 0$ for a RHS, we need to be concerned only with the boundary condition $\sigma_{zx}(x{=}0, y) = 0$. In addition, we only need to seek the solution in the domain occupied by the elastic solid, $x \le 0$.

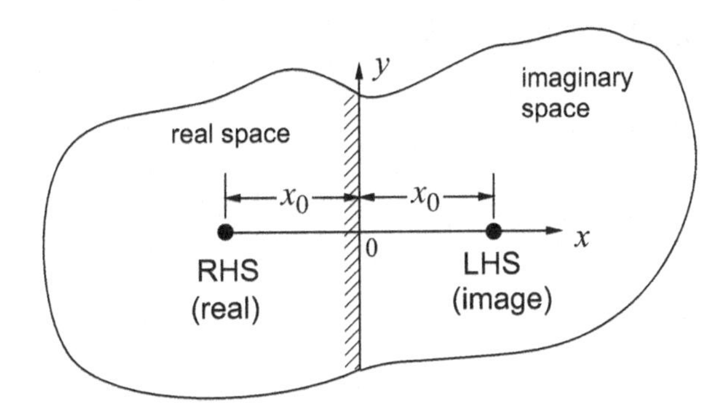

Figure 10.27. Image dislocation for a RHS dislocation near a free surface.

Recall that what we have here is an anti-plane strain problem, in which we seek the displacement field $u_z(x, y)$ that satisfies Laplace's equation $\nabla^2 u_z = 0$ (see Section 9.1.1). As in other similar problems involving surfaces, we start with the solution for a RHS in an infinite solid and then make corrections for the traction-free surface. For a RHS located at $(-x_0, 0)$ in an infinite solid, the stresses are

$$\begin{aligned} \sigma_{xz}^{\infty}(x, y) &= -\frac{\mu b}{2\pi} \frac{y}{(x + x_0)^2 + y^2}, \\ \sigma_{yz}^{\infty}(x, y) &= \frac{\mu b}{2\pi} \frac{x + x_0}{(x + x_0)^2 + y^2}, \end{aligned} \tag{10.104}$$

which obviously satisfies $\nabla^2 u_z^{\infty} = 0$, as we saw before. But this solution produces tractions on the plane located at $x = 0$,

$$\sigma_{xz}^{\infty}(x=0, y) = -\frac{\mu b}{2\pi} \frac{y}{x_0^2 + y^2}, \tag{10.105}$$

so the plane is not traction free. These tractions must be removed by a corrective field. We construct this correction by using an "image" or imaginary dislocation placed in imaginary space. We imagine that the elastic solid is extended into an imaginary domain, $x > 0$, to the right of the free surface, and that an image dislocation, a LHS dislocation in this case, is located at $(+x_0, 0)$, as shown in Fig. 10.27.

What we have in Fig. 10.27 is actually an infinite solid containing two dislocations. This configuration is *equivalent* to our original configuration shown in Fig. 10.26 as long as the two configurations have the same stress, strain, and displacement fields in the domain $x \leq 0$. This is guaranteed as long as the stress field satisfies the condition, Eq. (10.103), prescribed on the surface $x = 0$, because the solution is unique.

In the equivalent problem of an infinite medium, the image LHS dislocation produces a stress field, which will be called the image field,

$$\begin{aligned} \sigma_{xz}^{\mathrm{img}}(x, y) &= -\frac{\mu(-b)}{2\pi} \frac{y}{(x - x_0)^2 + y^2}, \\ \sigma_{yz}^{\mathrm{img}}(x, y) &= \frac{\mu(-b)}{2\pi} \frac{x - x_0}{(x - x_0)^2 + y^2}. \end{aligned} \tag{10.106}$$

Therefore, the total stress field is

$$\sigma_{xz}(x,y) = \sigma_{xz}^{\infty} + \sigma_{xz}^{\text{img}} = -\frac{\mu b}{2\pi}\frac{y}{(x+x_0)^2+y^2} + \frac{\mu b}{2\pi}\frac{y}{(x-x_0)^2+y^2},$$
$$\sigma_{yz}(x,y) = \sigma_{yz}^{\infty} + \sigma_{yz}^{\text{img}} = \frac{\mu b}{2\pi}\frac{x+x_0}{(x+x_0)^2+y^2} - \frac{\mu b}{2\pi}\frac{x-x_0}{(x-x_0)^2+y^2}. \qquad (10.107)$$

It can be easily shown that

$$\sigma_{xz}(x=0,y) = 0, \qquad (10.108)$$

so that the traction-free boundary condition is satisfied. Therefore, Eq. (10.107) gives the correct stress field of a screw dislocation in a semi-infinite medium with a traction-free surface. Although Eq. (10.107) gives the stress field in the entire space, only the part inside the solid domain $x < 0$ has physical meaning for the problem shown in Fig. 10.26.

With the image construction we can now compute the force on the real dislocation that arises from the presence of the free surface. Because a straight dislocation in an infinite space cannot exert a force on itself, the force on the dislocation is entirely the Peach–Koehler force from the image stress, so that it is often called the image force. The image force can be obtained from the Peach–Koehler formula and evaluating the image stress at $x = -x_0$ and $y = 0$,

$$f_x = \sigma_{zy}^{\text{img}} b = \frac{\mu(-b)}{2\pi}\frac{(-x_0-x_0)}{(-x_0-x_0)^2+0^2} b = \frac{\mu b^2}{4\pi x_0}. \qquad (10.109)$$

Therefore, the force on the real dislocation pulls it towards the free surface. This force can also be considered as the attractive force due to the image dislocation at location $x = x_0$ in the equivalent problem of an infinite body.

Unfortunately, the image dislocation construction cannot be directly extended to edge (or mixed) dislocations. If the real dislocation is a positive edge dislocation with Burgers vector along x and located at $x = -x_0$, simply placing a negative edge dislocation at $x = +x_0$ will not cancel all the tractions at the surface $x = 0$ (see Section 10.6.7). This is another example showing that the plane strain problem of an edge dislocation is more difficult to solve than the anti-plane strain problem of a screw dislocation. Interestingly, the image force on the positive edge dislocation in this case is still the same as the Peach–Koehler force produced by the image (i.e. negative edge) dislocation, i.e. $f_x = \mu b^2/[4\pi x_0(1-\nu)]$.

10.6.2 Image dislocation for a rigid surface

Now consider a right-handed screw dislocation parallel to a perfectly rigid surface. We may think of a coating that is perfectly rigid or infinitely stiff, with an infinite shear modulus. This means that the displacements must be zero at the interface between the solid and the rigid coating. As the real dislocation approaches that interface, there can be no displacements at the interface. In other words, the boundary condition is

$$u_z(x=0,y) = 0. \qquad (10.110)$$

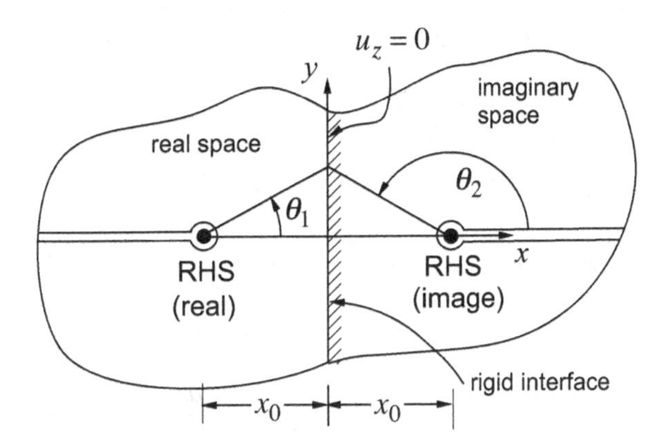

Figure 10.28. Image dislocation for a rigid interface.

Again, we start with the solution of a single dislocation located at $(-x_0, 0)$ in an infinite medium (see Section 9.1.1),

$$u_z^\infty = \frac{b}{2\pi}\theta_1 = \frac{b}{2\pi}\arctan\frac{y}{x+x_0}. \tag{10.111}$$

Note that θ_1 is a multi-valued function of (x, y). It is made into a single-valued function by introducing a branch-cut, which corresponds to the cut-plane when the dislocation is introduced. In Fig. 10.28, the branch-cut is introduced along the $-x$ direction, i.e. the range of θ_1 is $[-\pi, \pi)$. Obviously, the solution u_z^∞ does not satisfy the boundary condition of zero displacement on the plane $x = 0$.

The image solution needed to satisfy this boundary condition is a same-signed dislocation positioned at $x = x_0$, as shown in Fig. 10.28. The displacement field of the image dislocation is

$$u_z^{\text{img}} = \frac{b}{2\pi}\theta_2 - \frac{b}{2} = \frac{b}{2\pi}\arctan\frac{y}{x-x_0} - \frac{b}{2}. \tag{10.112}$$

Here we have introduced a constant $-\frac{b}{2}$ to satisfy the boundary condition given by Eq. (10.110). This constant term corresponds to a rigid-body translation along z that does not affect the equilibrium or the compatibility conditions. The branch-cut for θ_2 is introduced along the $+x$ direction, i.e. $0 \le \theta_2 < 2\pi$. This choice of the branch-cut guarantees that the displacement jump of the image dislocation does not enter the real domain defined by $x \le 0$.

The total displacement field is

$$u_z = u_z^\infty + u_z^{\text{img}} = \frac{b}{2\pi}(\theta_1 + \theta_2 - \pi). \tag{10.113}$$

On the plane $x = 0$, it can be easily proved that $\theta_1 + \theta_2 = \pi$, hence the boundary condition, Eq. (10.110), is satisfied. The displacements associated with the real dislocation are canceled by the image dislocation, so that the interface behaves as if it is perfectly rigid.

Since the image dislocation is now a RHS, the image stress for the rigid interface is simply the opposite of that for the traction free surface. Therefore, for the rigid interface problem, the

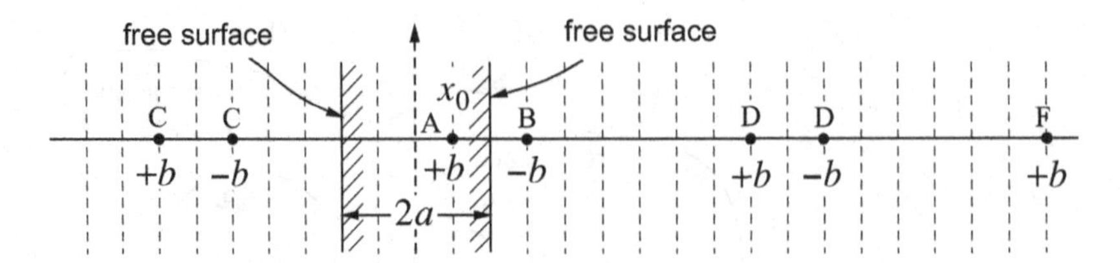

Figure 10.29. Multiple images for a dislocation in a thin foil.

image stress is

$$\sigma_{xz}^{\text{img}}(x,y) = -\frac{\mu b}{2\pi}\frac{y}{(x-x_0)^2+y^2},$$
$$\sigma_{yz}^{\text{img}}(x,y) = \frac{\mu b}{2\pi}\frac{x-x_0}{(x-x_0)^2+y^2}, \tag{10.114}$$

and the total stress field is

$$\sigma_{xz}(x,y) = \sigma_{xz}^{\infty} + \sigma_{xz}^{\text{img}} = -\frac{\mu b}{2\pi}\frac{y}{(x+x_0)^2+y^2} - \frac{\mu b}{2\pi}\frac{y}{(x-x_0)^2+y^2},$$
$$\sigma_{yz}(x,y) = \sigma_{yz}^{\infty} + \sigma_{yz}^{\text{img}} = \frac{\mu b}{2\pi}\frac{x+x_0}{(x+x_0)^2+y^2} + \frac{\mu b}{2\pi}\frac{x-x_0}{(x-x_0)^2+y^2}. \tag{10.115}$$

Since the real dislocation is repelled by the same-signed image, we may conclude that the real dislocation is repelled by the rigid interface,

$$f_x = \sigma_{zy}^{\text{img}} b = \frac{\mu b}{2\pi}\frac{(-x_0-x_0)}{(-x_0-x_0)^2+y^2} b = -\frac{\mu b^2}{4\pi x_0}. \tag{10.116}$$

10.6.3 Multiple images

For a dislocation in a thin foil with two free surfaces we need multiple images to account for the traction-free conditions on both surfaces. Figure 10.29 shows a thin foil of thickness $2a$ with a real dislocation at $x = x_0$. To construct the solution that satisfies traction-free boundary conditions at both surfaces, $x_s = \pm a$, we will need an infinite array of images.

The real dislocation is shown by a dot at $x = x_0$, marked as A, with Burgers vector $+b$. We know that to account for the traction-free condition on the nearest surface to the right (at $x_s = a$), an image dislocation of opposite sign needs to be placed at the position $2a - x_0$, marked as B, with Burgers vector $-b$. With the dislocations at A and B, the right surface ($x_s = a$) is traction free, but the left surface ($x_s = -a$) is not traction free.

To make the left surface ($x_s = -a$) traction free, we need to create two more images for the two dislocations at A and B. Both new images are marked as C in Fig. 10.29. The image for dislocation A is located at $x = -2a - x_0$ with Burgers vector $-b$; the image for dislocation B is located at $x = -2a - (2a - x_0) = -4a + x_0$ with Burgers vector $+b$. The combined stress field of the four dislocations considered so far makes the left surface traction free, but the right surface is no longer traction free.

To restore the traction-free condition for the right surface ($x_s = a$), we need to create two more images for the dislocations at C. These new images are marked as D in Fig. 10.29. They

Table 10.1. The Burgers vector (in parantheses) and location of multiple images for a dislocation in a thin foil. Dislocation A (corresponding to $n = 0$ and $i = 0$) is the real dislocation.

	$i = 0$		$i = 1$	
$n = 0$	A $(+b)$	x_0	B $(-b)$	$2a - x_0$
$n = 1$	C $(-b)$	$-2a - x_0$	C $(+b)$	$-2a - (2a - x_0) = -4a + x_0$
$n = 2$	D $(+b)$	$2a - (-2a - x_0) = 4a + x_0$	D $(-b)$	$2a - (4a + x_0) = 6a - x_0$
$n = 3$	E $(-b)$	$-2a - (4a + x_0) = -6a - x_0$	E $(+b)$	$-2a - (6a - x_0) = -8a + x_0$
$n = 4$	F $(+b)$	$2a - (-6a - x_0) = 8a + x_0$	F $(-b)$	$2a - (-8a + x_0) = 10a - x_0$
$n = 5$	G $(-b)$	$-2a - (8a + x_0) = -10a - x_0$	G $(+b)$	$-2a - (10a - x_0) = -12a + x_0$
...		...		...

in turn create tractions on the left that have to be corrected by more image dislocations (E, not shown). In this way an infinite array of images is needed to make both surfaces traction free. The Burgers vectors and the location of these dislocations are given in Table 10.1. All dislocations except A are image dislocations and are located outside the physical domain of the thin film.

The dislocations in this table can be generated using the following algorithm. Let us define the two dislocation locations on row n as $x_n^{(i)}$, with $i = 0, 1$. The top row is assigned as $n = 0$, with (A) $x_0^{(0)} = x_0$ and (B) $x_0^{(1)} = 2a - x_0$. Given $x_{n-1}^{(i)}$, we compute $x_n^{(i)}$ as follows.

(1) For each row $n > 0$, we consider the surface at $x_s = (-1)^n a$, and introduce two image dislocations to make this surface traction free.
(2) The locations of the two images are: $x_n^{(i)} = 2x_s - x_{n-1}^{(i)}$, for $i = 0, 1$.

The Burgers vectors in the table show an alternating pattern, which can be described by $b_n^{(i)} = (-1)^{n+i} b$.

It can be shown that the result of the above algorithm is two periodic arrays of dislocations: one periodic array with Burgers vector $+b$ at $x_0 + 4ma_0$, and another periodic array with Burgers vector $-b$ at $2a - x_0 + 4ma_0$, with $m = 0, \pm1, \pm2 \ldots$. Therefore, the total stress field can be constructed from the analytic expressions for the stress fields of periodic dislocation arrays given in Section 14.2.1.

For the stress field in the physical domain of the thin film, $-a \le x \le a$, the contributions from more and more distant dislocations become smaller and smaller. When the real dislocation is at the center of the foil the arrangements of the images would be perfectly symmetric and there would be no net force on the real dislocation. If the real dislocation is closer to one of the surfaces, then naturally the attraction of that surface will be greater than the attraction of the more distant surface.

10.6.4 General image problem for a phase boundary

We now consider a more general problem in which a right-handed screw dislocation lies parallel to a planar phase boundary that separates two phases having shear moduli μ_1 and μ_2, as shown in Fig. 10.30.[3] In this case, phase (1) lies to the left of the interface while phase (2) is on the

3 Note that a mismatch in ν does not affect the solution for the screw dislocation.

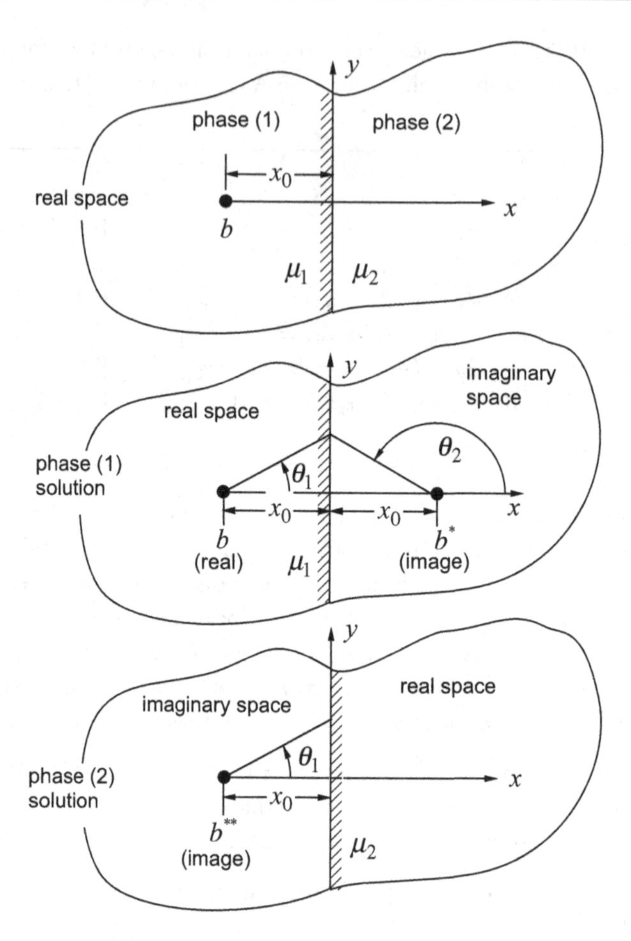

Figure 10.30. Image solution for a screw dislocation near a phase boundary.

right. The real dislocation exists only in phase (1); there are no real dislocations in phase (2). The presence of the dislocation in phase (1) creates displacements at and tractions on the interface.

We seek a solution for the full elastic fields in both phases that satisfies the boundary conditions that both the displacements and the tractions match at the interface,

$$\begin{aligned} u_z^{(1)}(x=0,y) &= u_z^{(2)}(x=0,y), \\ \sigma_{zx}^{(1)}(x=0,y) &= \sigma_{zx}^{(2)}(x=0,y). \end{aligned} \tag{10.117}$$

Construction of two image dislocations

We can construct the solution to this problem by using the method of images. The image construction is shown in Fig. 10.30. A different image dislocation is needed when seeking the solution in each of the two phases. As shown in Fig. 10.30, when considering the elastic field in phase (1), we imagine a problem where phase (1) occupies the entire space, and we are allowed to introduce an image dislocation (b^*) in the imaginary space $x > 0$ (previously occupied by phase (2)). Doing so will not introduce unwanted singularities in the physical space ($x < 0$) of phase (1). However, when considering the elastic field in phase (2), we imagine a different problem where phase (2) occupies the entire space, and we are allowed to introduce an image dislocation (b^{**}) in the imaginary space $x < 0$ (previously occupied by phase (1)). Doing so will not introduce unwanted singularities in the physical space ($x > 0$) of phase (2). The Burgers vectors of the two image dislocations will be chosen to satisfy the boundary conditions in Eq. (10.117).

Given the image constructions described above, the total stress field in phase (1) is

$$\sigma_{xz}^{(1)} = -\frac{\mu_1 b}{2\pi}\frac{y}{(x+x_0)^2+y^2} - \frac{\mu_1 b^*}{2\pi}\frac{y}{(x-x_0)^2+y^2}. \tag{10.118}$$

Notice that μ_1 is used in both terms (corresponding to the infinite and image solutions), because phase (1) occupies the entire space in the equivalent problem we are imagining. The stress field in phase (2) is

$$\sigma_{xz}^{(2)} = -\frac{\mu_2 b^{**}}{2\pi}\frac{y}{(x+x_0)^2+y^2}. \tag{10.119}$$

The traction boundary condition in Eq. (10.117) is satisfied if

$$-\frac{\mu_1(b+b^*)}{2\pi}\frac{y}{x_0^2+y^2} = -\frac{\mu_1 b^{**}}{2\pi}\frac{y}{x_0^2+y^2}. \tag{10.120}$$

or as follows.

$$\boxed{\mu_1(b+b^*) = \mu_2 b^{**}} \tag{10.121}$$

The total displacement field in phase (1) is

$$u_z^{(1)} = \frac{b}{2\pi}\theta_1 + \frac{b^*}{2\pi}(\theta_2 - \pi), \tag{10.122}$$

where the range of θ_1 is $[-\pi, \pi)$ and the range of θ_2 is $[0, 2\pi)$, similar to the choices made in Section 10.6.2. The displacement field in phase (2) is

$$u_z^{(2)} = \frac{b^{**}}{2\pi}\theta_1. \tag{10.123}$$

Notice that on the plane $x = 0$, $\theta_2 - \pi = -\theta_1$, so that the displacement continuity condition in Eq. (10.117) can be satisfied if

$$\frac{b-b^*}{2\pi}\theta_1 = \frac{b^{**}}{2\pi}\theta_1$$

or as follows.

$$\boxed{b - b^* = b^{**}} \tag{10.124}$$

Solution for the image Burgers vectors

Solving Eqs. (10.121) and (10.124) simultaneously gives

$$\begin{aligned} b^* &= b\frac{\mu_2-\mu_1}{\mu_2+\mu_1}, \\ b^{**} &= b\frac{2\mu_1}{\mu_2+\mu_1}. \end{aligned} \tag{10.125}$$

So the Burgers vectors of the image dislocations are now known, and the total stress and displacement fields are known explicitly. For convenience of notation we define

$$\frac{\mu_2 - \mu_1}{\mu_2 + \mu_1} \equiv \kappa, \tag{10.126}$$

so that

$$\frac{2\mu_1}{\mu_2 + \mu_1} = 1 - \kappa. \tag{10.127}$$

Therefore

$$\begin{aligned} b^* &= \kappa b, \\ b^{**} &= (1 - \kappa) b. \end{aligned} \tag{10.128}$$

With this notation if the real dislocation has unit Burgers vector, $b = +1$, then we have the following.

$$\boxed{\begin{aligned} b^* &= \kappa \\ b^{**} &= 1 - \kappa \end{aligned}} \tag{10.129}$$

Here b^* is the image Burgers vector for phase (1) (which is the phase that contains the real dislocation) and b^{**} is the image Burgers vector for phase (2), as shown in Fig. 10.31.

The solution we have constructed gives the full stress field in both phases (1) and (2) for a screw dislocation with a unit Burgers vector at a distance x_0 from the phase boundary. To compute the force exerted on the real dislocation by the interaction with the second phase, only the image dislocation b^* for phase (1) is required. The image dislocation b^{**} for phase (2) is only needed if we want to find out the elastic fields in phase (2).

It is easy to show that the general solution given here recovers the two special cases considered in Sections 10.6.1 and 10.6.2. The case of a free surface corresponds to $\mu_2 = 0$, so that the image dislocation for phase (1) is $b^* = \kappa = -1$, which attracts the real unit dislocation. The case of rigid surface corresponds to $\mu_2 = \infty$, so that the image dislocation for phase (1) is $b^* = \kappa = 1$, which repels the real dislocation.

10.6.5 Multiple phase boundaries

The methods described in Sections 10.6.3 and 10.6.4 can be combined to construct elastic fields in layered materials with multiple phase boundaries. Figure 10.32 shows a screw dislocation in a metal near an oxide layer at the surface. One may think of the oxide layer at the surface as being stiffer than the metal containing the dislocation. Intuitively, we would expect that when the dislocation is very far from the oxide layer it is attracted toward the surface, much like the case of a perfectly free surface. But when the dislocation gets very close to the stiffer oxide layer it is repelled, much like it would be if placed near a rigid phase. Such a problem is called Head's problem, after A. K. Head [82] who first solved it in the 1950s.

To construct the solution in the metal region and oxide region, two equivalent problems are imagined. In one problem, the entire space is occupied by the metal (middle row of Fig. 10.32, which contains not only the real dislocation (A), but also a set of image dislocations (B, D, F,

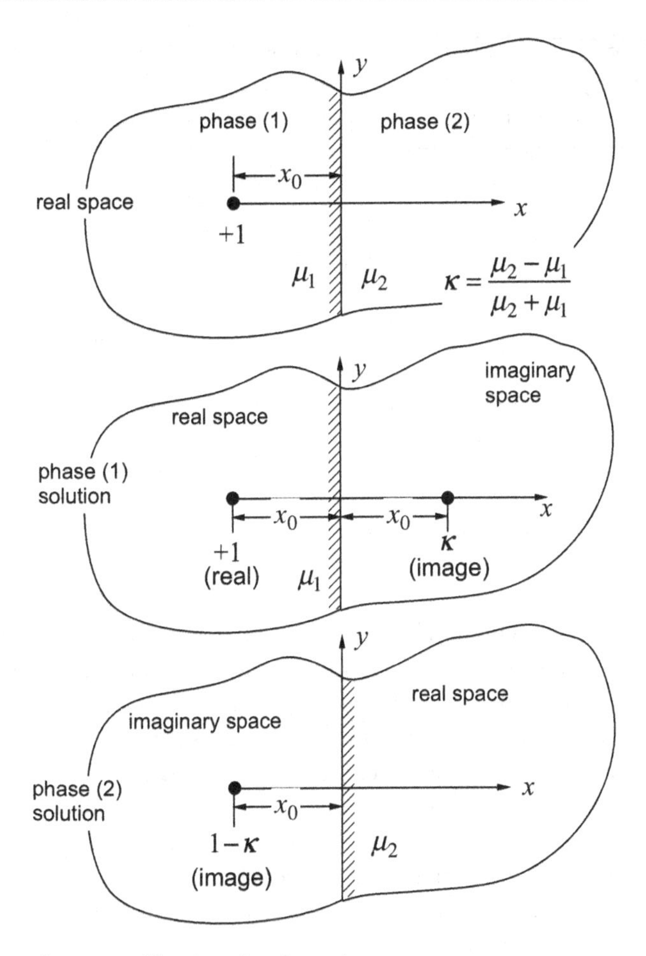

Figure 10.31. General image problem in reduced notation.

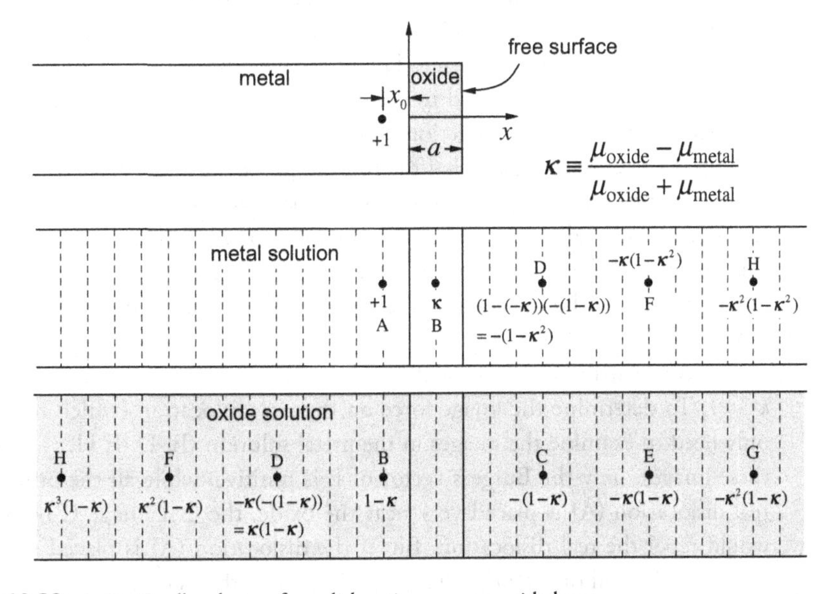

Figure 10.32. A. K. Head's solution for a dislocation near an oxide layer.

Table 10.2. The Burgers vector and location of multiple images for a dislocation near an oxide layer. Dislocation A (corresponding to $n = 0$ and $i = 0$) is the real dislocation; x_s is the location of the interface corresponding to the image dislocations.

		metal ($i = 0$)		oxide ($i = 1$)	
$n = 0$		A $(+1)$	$-x_0$		
$n = 1$	$x_s = 0$	B (κ)	x_0	B $(1-\kappa)$	$-x_0$
$n = 2$	$x_s = a$			C $-(1-\kappa)$	$2a + x_0$
$n = 3$	$x_s = 0$	D $-(1-\kappa^2)$	$2a + x_0$	D $\kappa(1-\kappa)$	$-2a - x_0$
$n = 4$	$x_s = a$			E $-\kappa(1-\kappa)$	$4a + x_0$
$n = 5$	$x_s = 0$	F $-\kappa(1-\kappa^2)$	$4a + x_0$	F $\kappa^2(1-\kappa)$	$-4a - x_0$
$n = 6$	$x_s = a$			G $-\kappa^2(1-\kappa)$	$6a + x_0$
$n = 7$	$x_s = 0$	H $-\kappa^2(1-\kappa^2)$	$6a + x_0$	H $\kappa^3(1-\kappa)$	$-6a - x_0$
...	...	...		...	

H, ...). The solution of this problem matches the solution of the original problem only in the physical domain of the metal. In the other problem, the entire space is occupied by the oxide (bottom row of Fig. 10.32), which contains another set of image dislocations (B, C, D, E, F, G, H, ...). The solution of this problem matches the solution of the original problem only in the physical domain of the oxide. Owing to the presence of multiple interfaces, we need to apply the image construction repeatedly in the same way as in Section 10.6.3. The Burgers vectors and locations of the image dislocations are given in Table 10.2.

The image dislocations B are introduced to satisfy the boundary condition (disrupted by dislocation A) at the metal–oxide interface using the results from Section 10.6.4. If the oxide layer were infinitely thick, the image dislocations B would be sufficient in satisfying all stress and displacement compatibility conditions for this problem. The image dislocation C is introduced (in the oxide solution only) to satisfy the traction-free boundary condition (disrupted by dislocation B) at the oxide surface. The image dislocations D are then introduced to satisfy the boundary condition (disrupted by dislocation C) at the metal–oxide interface. Again, we can use a similar approach as in Section 10.6.4, except that the starting dislocation is now in phase (2). Since the only change is that the roles of phase (1) and phase (2) have switched, the same formula can be used except that κ needs to be replaced by $-\kappa$. Subsequently, the image dislocation E is introduced (in the oxide solution only) to satisfy the traction-free boundary condition (disrupted by dislocation D) at the oxide surface. This is followed by the images labeled F, followed in turn by the image at G and then the images labeled H, as shown in Fig. 10.32 and Table 10.2.

Suppose that the oxide is much stiffer than the metal, $\mu_{\text{oxide}} \gg \mu_{\text{metal}}$, so that $\kappa \approx 1$ but $\kappa < 1$. To determine the image force on the real dislocation (which resides in the metal), we only need to examine the images in the metal solution (B, D, F, H, ...). We note that among these images, only the Burgers vector of B is positive, while all the others are negative. If the real dislocation (A) is placed very near the oxide, the first image (B) would dominate and it would repel the real dislocation. But if the dislocation (A) is placed very far from the oxide interface, then all of the other images would cause the net force to be attractive. Naturally, an equilibrium position can be found at an intermediate distance from the oxide layer (see Exercise problem 10.24).

The method we have described can be used to find the elastic field of a dislocation in other multilayered configurations. An example is a dislocation in a thin film on a substrate [83]. The situation is similar to Fig. 10.32 except that the dislocation lies between the internal interface and the free surface.

10.6.6 Screw dislocation near curved interfaces

The problem of a screw dislocation near a planar phase boundary, as shown in Fig. 10.30, can be further generalized so that the phase boundary has a cylindrical shape. Professor. D. M. Barnett gave the solutions to such problems in his dissertation [84].

Screw dislocation near a cylindrical inhomogeneity

Consider an infinite elastic medium (the matrix) with shear modulus μ_1 containing a cylindrical inhomogeneity of radius a with shear modulus μ_2. We will also refer to the matrix as phase (1) and the inhomogeneity as phase (2). We first consider the scenario in which the screw dislocation is located in the matrix at $(x_0, 0)$, with $x_0 > a$, as shown in Fig. 10.33. For simplicity, we assume the dislocation is a right-handed screw with Burgers vector $b = +1$.

The stress field in this configuration can be constructed from the stress field of real and image dislocations. Define $\kappa \equiv (\mu_2 - \mu_1)/(\mu_2 + \mu_1)$. The stress field in phase (1), i.e. the matrix, can be written as the sum of the stress field of three dislocations in a homogeneous, infinite medium with shear modulus μ_1: $b = +1$ at $(x_0, 0)$ (the real dislocation), $b = \kappa$ at $(a^2/x_0, 0)$ and $b = -\kappa$ at $(0, 0)$ (the image dislocations). This leads to the following stress state

$$\begin{aligned}\sigma_{xz} &= -\frac{\mu_1}{2\pi}\left\{\frac{y}{(x-x_0)^2+y^2} + \kappa\left[\frac{y}{(x-a^2/x_0)^2+y^2} - \frac{y}{x^2+y^2}\right]\right\}\\ \sigma_{yz} &= \frac{\mu_1}{2\pi}\left\{\frac{x-x_0}{(x-x_0)^2+y^2} + \kappa\left[\frac{x-a^2/x_0}{(x-a^2/x_0)^2+y^2} - \frac{x}{x^2+y^2}\right]\right\}\end{aligned} \tag{10.130}$$

in the region $x^2 + y^2 > a^2$.

The stress field in phase (2), i.e. the cylindrical inhomogeneity, is equal to the stress field of an image dislocation in a homogeneous, infinite medium with shear modulus μ_2. The image dislocation has Burgers vector $b = 1 - \kappa$ and is located at $(x_0, 0)$. The stress field is then

$$\begin{aligned}\sigma_{xz} &= -\frac{\mu_2}{2\pi}(1-\kappa)\frac{y}{(x-x_0)^2+y^2}\\ \sigma_{yz} &= \frac{\mu_2}{2\pi}(1-\kappa)\frac{x-x_0}{(x-x_0)^2+y^2}\end{aligned} \tag{10.131}$$

in the region $x^2 + y^2 < a^2$.

The stress field of a screw dislocation near a cylindrical void can be obtained from the above solution by letting $\mu_2 = 0$. In this case, the stress field outside the void can be obtained from Eq. (10.130) by letting $\kappa = -1$, and the stress field inside the void is zero.

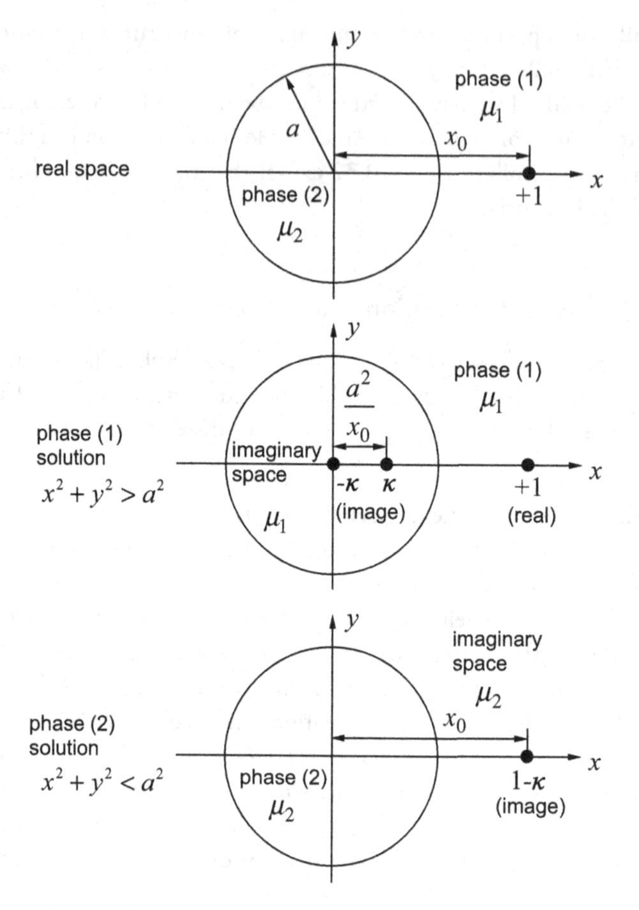

Figure 10.33. Image solution for a screw dislocation near a cylindrical inhomogeneity.

Screw dislocation inside a cylindrical inhomogeneity

We now consider the scenario in which the screw dislocation with Burgers vector $b = +1$ is located inside the inhomogeneity at $(x_0, 0)$, with $x_0 < a$, as shown in Fig. 10.34. To be consistent with the case considered above, let us still define $\kappa \equiv (\mu_2 - \mu_1)/(\mu_2 + \mu_1)$. The stress field in phase (1), i.e. the matrix, can be written as the sum of the stress field of two dislocations in a homogeneous, infinite medium with shear modulus μ_1: $b = 1 + \kappa$ at $(x_0, 0)$ and $b = -\kappa$ at $(0, 0)$. The stress field is then

$$\begin{aligned} \sigma_{xz} &= -\frac{\mu_1}{2\pi}\left\{(1+\kappa)\frac{y}{(x-x_0)^2+y^2} - \kappa\frac{y}{x^2+y^2}\right\} \\ \sigma_{yz} &= \frac{\mu_1}{2\pi}\left\{(1+\kappa)\frac{x-x_0}{(x-x_0)^2+y^2} - \kappa\frac{x}{x^2+y^2}\right\} \end{aligned} \tag{10.132}$$

in the region $x^2 + y^2 > a^2$.

The stress field in phase (2), i.e. the cylindrical inhomogeneity, can be written as the sum of the stress field of two dislocations in a homogeneous, infinite medium with shear modulus μ_2:

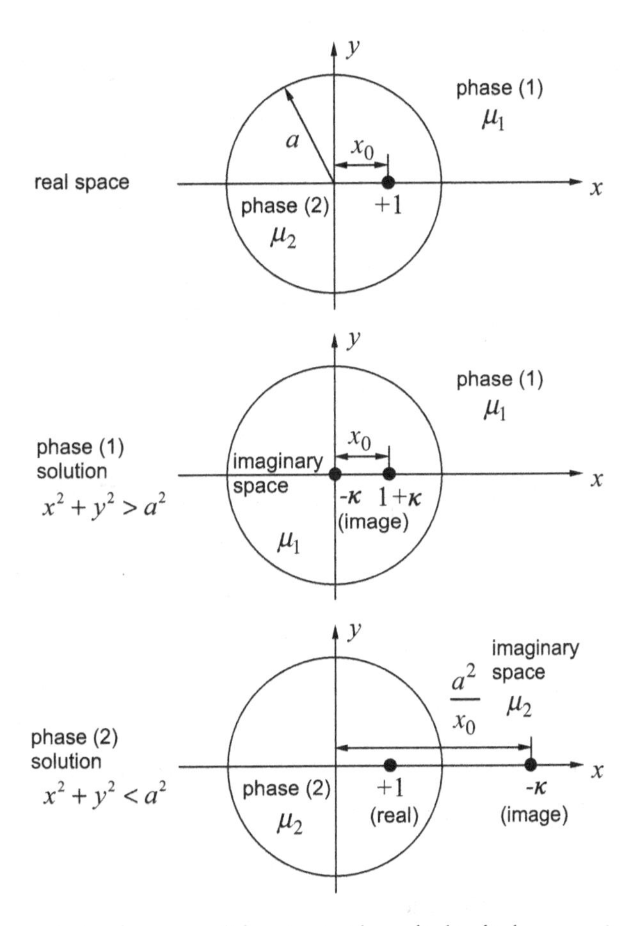

Figure 10.34. Image solution for a screw dislocation inside a cylindrical inhomogeneity.

$b = +1$ at $(x_0, 0)$ (the real dislocation) and $b = -\kappa$ at $(a^2/x_0, 0)$. The stress field is then

$$\begin{aligned} \sigma_{xz} &= -\frac{\mu_2}{2\pi}\left\{\frac{y}{(x-x_0)^2+y^2} - \kappa\frac{y}{(x-a^2/x_0)^2+y^2}\right\} \\ \sigma_{yz} &= \frac{\mu_2}{2\pi}\left\{\frac{x-x_0}{(x-x_0)^2+y^2} - \kappa\frac{x-a^2/x_0}{(x-a^2/x_0)^2+y^2}\right\} \end{aligned} \qquad (10.133)$$

in the region $x^2 + y^2 < a^2$.

The stress field of a screw dislocation inside an infinitely long elastic cylinder with traction-free surfaces can be obtained from the above solution by letting $\mu_1 = 0$. In this case, the stress field inside the cylinder can be obtained from Eq. (10.133) by letting $\kappa = 1$, and the stress field outside the cylinder is zero.

The stress field of a screw dislocation inside a cylindrical void of an infinite elastic solid can be obtained from the above solution by letting $\mu_2 = 0$. Such a situation occurs when a screw dislocation in the solid moves towards a void (attracted by the image stress) and eventually enters the void through its surface. In this case, the stress field outside the void can be obtained from Eq. (10.132) by letting $\kappa = -1$, and the stress field inside the void is zero. It is interesting to

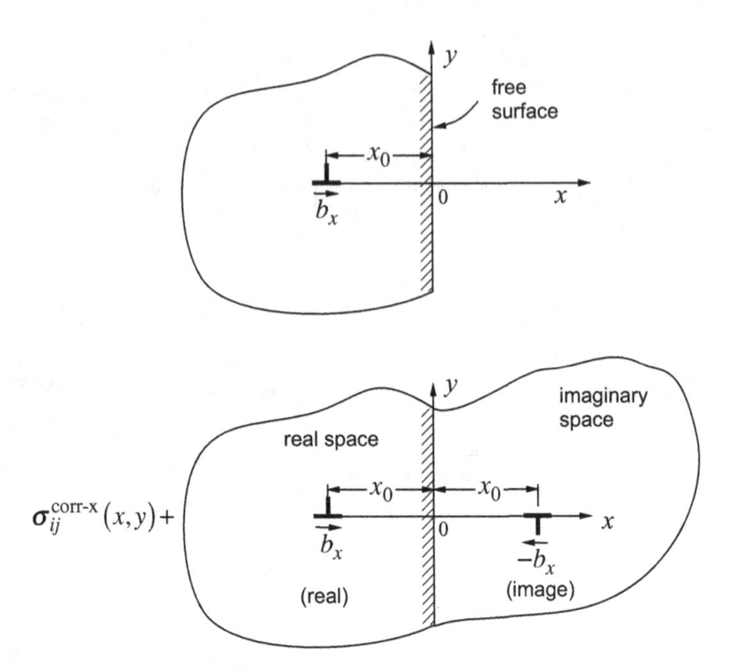

Figure 10.35. An edge dislocation with Burgers vector perpendicular to a free surface.

observe that, given $-\kappa = 1$ and $1 + \kappa = 0$, the stress field in the matrix is the same as the stress field of a screw dislocation with $b = +1$ at the origin.

10.6.7 Edge dislocation near free surface

The near-interface dislocations that we have considered so far are all screw dislocations. The stress field of an edge dislocation near a planar phase boundary has been obtained by A. K. Head [85]. As an example, here we give Head's solution for an edge dislocation near a flat free surface.

Burgers vector perpendicular to surface

Consider a positive edge dislocation with Burgers vector b_x in a semi-infinite elastic solid with shear modulus μ, positioned a distance x_0 from a free surface ($x = 0$), as shown in Fig. 10.35. The traction-free boundary condition of the free surface requires that $\sigma_{xx} = \sigma_{xy} = 0$ on the $x = 0$ plane. Similar to the approach discussed above, we will try to construct the stress field in the elastic half space by superimposing the stress fields of dislocations in an infinite elastic solid. For example, the infinite solid can contain the real dislocation with Burgers vector b_x at $(-x_0, 0)$, as well as an image dislocation with Burgers vector $-b_x$ at $(x_0, 0)$. The superposition of the stress field of these two dislocations satisfies the condition that $\sigma_{xx}(0, y) = 0$, meaning that the x component of the traction force on the surface is zero. However, the y component of the surface traction force is still non-zero. In order to make both components of the traction force vanish on the free surface, it is necessary to superimpose another corrective stress field. The analytic expressions of the resulting stress field are given below.

From Eq. (9.31), the stress field in an infinite elastic medium of an edge dislocation with Burgers vector b_x (line sense along the positive z axis) located at the origin is

$$\begin{aligned}\sigma_{xx}^{\text{edge-x}}(x,y) &= -\frac{\mu b_x}{2\pi(1-\nu)}\frac{y(3x^2+y^2)}{(x^2+y^2)^2},\\ \sigma_{yy}^{\text{edge-x}}(x,y) &= \frac{\mu b_x}{2\pi(1-\nu)}\frac{y(x^2-y^2)}{(x^2+y^2)^2},\\ \sigma_{xy}^{\text{edge-x}}(x,y) &= \frac{\mu b_x}{2\pi(1-\nu)}\frac{x(x^2-y^2)}{(x^2+y^2)^2}.\end{aligned} \tag{10.134}$$

The stress field of the edge dislocation in the elastic half space shown in Fig. 10.35 is

$$\sigma_{ij}(x,y) = \sigma_{ij}^{\text{edge-x}}(x+x_0,y) - \sigma_{ij}^{\text{edge-x}}(x-x_0,y) + \sigma_{ij}^{\text{corr-x}}(x,y), \tag{10.135}$$

where $\sigma_{ij}^{\text{edge-x}}(x+x_0,y)$ is the stress field of the real dislocation in an infinite solid, $\sigma_{ij}^{\text{edge-x}}(x-x_0,y)$ is the stress field of the image dislocation, and $\sigma_{ij}^{\text{corr-x}}(x,y)$ is the corrective stress field [85, 86]:

$$\begin{aligned}\sigma_{xx}^{\text{corr-x}}(x,y) &= -\frac{\mu b_x}{2\pi(1-\nu)}(4x_0xy)\frac{3(x-x_0)^2-y^2}{\left((x-x_0)^2+y^2\right)^3},\\ \sigma_{yy}^{\text{corr-x}}(x,y) &= \frac{\mu b_x}{2\pi(1-\nu)}(4x_0y)\frac{(x-x_0)^2(x+2x_0)-(3x-2x_0)y^2}{\left((x-x_0)^2+y^2\right)^3},\\ \sigma_{xy}^{\text{corr-x}}(x,y) &= \frac{\mu b_x}{2\pi(1-\nu)}(2x_0)\frac{(x-x_0)^3(x+x_0)-6x(x-x_0)y^2+y^4}{\left((x-x_0)^2+y^2\right)^3}.\end{aligned} \tag{10.136}$$

Burgers vector parallel to surface

We now consider a semi-infinite elastic medium containing an edge dislocation with Burgers vector b_y located at $(-x_0, 0)$, as shown in Fig. 10.36. To construct the stress field, we again consider an infinite elastic solid containing the real dislocation with Burgers vector b_y at $(-x_0, 0)$ and an image dislocation with Burgers vector $-b_y$ at $(x_0, 0)$. The superposition of the stress field of these two dislocations satisfies the condition that $\sigma_{xy}(0, y) = 0$, meaning that the y component of the traction force on the surface is zero. However, the x component of the surface traction force is still non-zero, so that another corrective stress field, $\sigma_{ij}^{\text{corr-y}}$ is needed.

The stress field in an infinite elastic medium of an edge dislocation with Burgers vector b_y (line sense along the positive z axis) located at the origin can be obtained from Eq. (10.134) by a 90° rotation,

$$\begin{aligned}\sigma_{xx}^{\text{edge-y}}(x,y) &= \frac{\mu b_y}{2\pi(1-\nu)}\frac{x(x^2-y^2)}{(x^2+y^2)^2}\\ \sigma_{yy}^{\text{edge-y}}(x,y) &= \frac{\mu b_y}{2\pi(1-\nu)}\frac{x(x^2+3y^2)}{(x^2+y^2)^2}\\ \sigma_{xy}^{\text{edge-y}}(x,y) &= \frac{\mu b_y}{2\pi(1-\nu)}\frac{y(x^2-y^2)}{(x^2+y^2)^2}.\end{aligned} \tag{10.137}$$

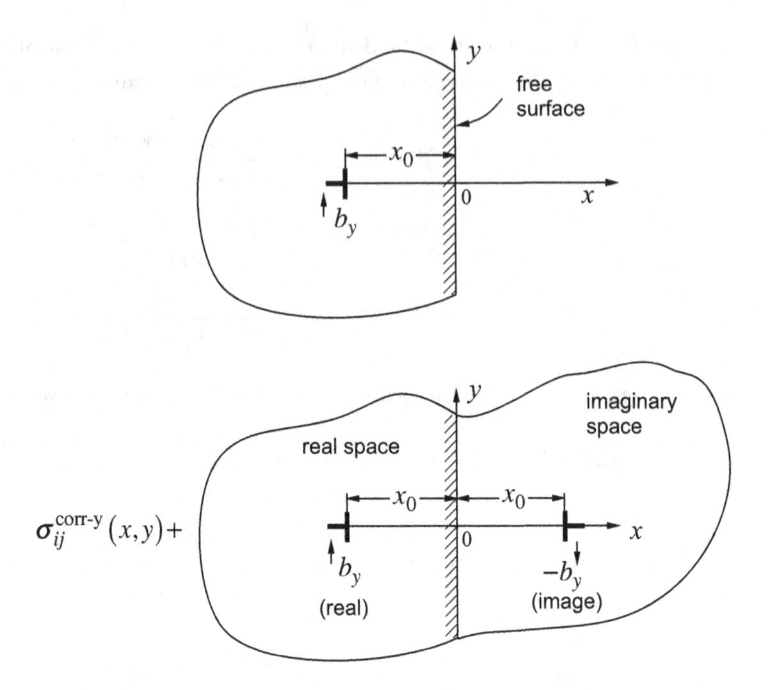

Figure 10.36. An edge dislocation with Burgers vector parallel to a free surface.

The stress field of the edge dislocation in the elastic half space shown in Fig. 10.36 is

$$\sigma_{ij}(x, y) = \sigma_{ij}^{\text{edge-y}}(x + x_0, y) - \sigma_{ij}^{\text{edge-y}}(x - x_0, y) + \sigma_{ij}^{\text{corr-y}}(x, y), \tag{10.138}$$

where $\sigma_{ij}^{\text{edge-y}}(x + x_0, y)$ is the stress field of the real dislocation in an infinite solid, $\sigma_{ij}^{\text{edge-y}}(x - x_0, y)$ is the stress field of the image dislocation, and $\sigma_{ij}^{\text{corr-y}}(x, y)$ is the corrective stress field [85]:

$$\begin{aligned}
\sigma_{xx}^{\text{corr-y}}(x, y) &= -\frac{\mu b_y}{2\pi(1-\nu)}(2x_0)\frac{(x-x_0)^3(3x-x_0) - 6x(x-x_0)y^2 - y^4}{\left((x-x_0)^2+y^2\right)^3},\\
\sigma_{yy}^{\text{corr-y}}(x, y) &= \frac{\mu b_y}{2\pi(1-\nu)}(2x_0)\frac{(x-x_0)^3(x+x_0) - 6x(x-x_0)y^2 + y^4}{\left((x-x_0)^2+y^2\right)^3},\\
\sigma_{xy}^{\text{corr-y}}(x, y) &= -\frac{\mu b_y}{2\pi(1-\nu)}(4x_0xy)\frac{3(x-x_0)^2 - y^2}{\left((x-x_0)^2+y^2\right)^3}.
\end{aligned} \tag{10.139}$$

10.7 Summary

The stress field of a dislocation produces Peach–Koehler forces on other dislocations. A myriad of dislocation behaviors may be described using the Peach–Koehler formula, including the interactions of straight dislocations in different orientation relationships with each other, and the interactions of straight dislocations with other arbitrarily shaped dislocation lines.

These applications of the Peach–Koehler formula show that oppositely signed, parallel screw dislocations attract while same-signed screws repel. Oppositely signed, parallel edge dislocations on different glide planes attract to form dipoles that are stable with respect to glide and unstable with respect to climb. A sufficiently high applied shear stress can cause oppositely signed edge dislocations to move past each other; that interaction leads to a critical passing stress which depends inversely on the separation distance between the glide planes.

Same-signed edge dislocations on parallel glide planes attract in the glide direction when a line perpendicular to the two dislocations and connecting them makes an angle less than 45° with the normal of the two glide planes. This attractive interaction among a group of same-signed edge dislocations on parallel slip planes promotes the formation of a dislocation array, perpendicular to the slip planes, which is stable with respect to glide but unstable with respect to climb. Same-signed edge dislocations on the same glide plane repel. This repulsive interaction leads to a pile-up configuration, when a train of same-signed dislocations emitted from a single source is blocked at some point. The pile-up configuration serves as a good model for a blocked slip band in a crystal and provides an account of the huge stress concentrations that can develop at the head of the blocked slip band.

The interaction of dislocations with themselves and other defects can be used to explain several strengthening mechanisms in crystals. The critical passing stress for two oppositely signed edge dislocations on parallel glide planes leads to the famous Taylor hardening equation, $\tau_c = \alpha\mu b\sqrt{\rho}$, describing the dependence of the flow stress on the dislocation density. The interaction of gliding dislocations with non-shearable particles on the glide plane leads to dispersion hardening, one of the important strengthening mechanisms for structural alloys. An analysis based on the line tension of the dislocation leads to the Orowan relation, $\tau_c = \mu b/(L - d)$, for the critical stress of bowing between non-shearable particles.

The plastic strain rate produced by the motion of dislocations is described by the famous Orowan equation, $\dot{\gamma}_{\mathrm{p}} = \rho_{\mathrm{m}} b\bar{v}$. Based on the Orowan equation, dislocation kinetics models can be constructed to explain a number of deformation behaviors of crystals. When the initial mobile dislocation density is low, and/or the dislocation mobility is low, the crystal can exhibit a yield point drop when loaded under constant strain rate. The reduction of flow stress beyond the upper yield point is caused by the multiplication of the mobile dislocation density ρ_{m}. When the crystal is subjected to a creep test under a stress below its upper yield point, it can exhibit sigmoidal creep. The creep rate initially increases with time due to multiplication of mobile dislocations, but eventually decreases with time due to strain hardening.

The dislocation mechanics described here can also be used to compute the critical thickness for a strained epitaxial film, below which the film is stable with respect to dislocation formation and above which the formation of misfit dislocations is energetically favorable. The analysis provides a quantitative account of how the strain in the film relaxes by the formation of more and more misfit dislocations. It can also be extended to strain relaxation in semi-coherent precipitate particles growing in a matrix.

Dislocations interact with free surfaces or any phase boundary where the elastic properties abruptly change. For screw dislocations lying parallel to such interfaces it is possible to use the image methods found in electrostatics to construct solutions for the elastic fields of dislocations near the interfaces, and, in that way, compute the "image" forces acting on real dislocations near such interfaces. Multiple sets of images are needed, one set describing the elastic field in each phase, when multiple phases are present.

10.8 Exercise problems

In all the exercise problems below, we model the solid as an elastically isotropic medium with shear modulus μ and Poisson's ratio ν, unless otherwise mentioned.

10.1 Two infinitely long straight edge dislocations of opposite sign on the same slip plane are initially separated along the x axis by a distance of $2s$. Assuming that the glide velocity of each dislocation can be described by $v_x = M f_x$, where M is the glide mobility and f_x is the glide force per unit length acting on the dislocation, derive an expression for the time, t_c, required for the dislocations to annihilate.

10.2 Two positive edge dislocations are positioned one above the other and separated by $2s$ as shown in Fig. 10.37a. The crystal is under no external stress and is held at sufficiently high temperature that both glide and climb are possible.

(a) Indicate how the dislocations would move in response to the interaction forces acting between them.
(b) Taking the equilibrium vacancy concentration in an unstressed crystal to be c_v^0, write expressions for the vacancy concentration near the cores of the two dislocations. The vacancy volume can be considered to be equal to the atomic volume for this expression.
(c) Using the solution to Exercise problem 7.2 as a guide, and assuming steady state diffusion of vacancies between the climbing dislocations, derive an expression for the climb velocity of each dislocation for a given separation distance, $2s$. The core radius of the dislocations can be taken to be b, the magnitude of the Burgers vector.

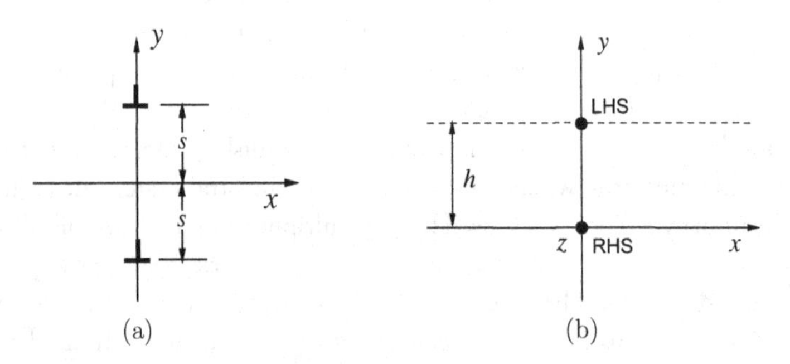

Figure 10.37. (a) Two edge dislocations. (b) A screw dislocation dipole.

10.3 Two infinitely long, oppositely signed screw dislocations, one a right-handed screw (RHS) and the other a left-handed screw (LHS), are shown in Fig. 10.37b. The RHS dislocation is fixed and the LHS dislocation is allowed to glide only on the plane parallel to the x–z plane.

(a) Indicate how the LHS dislocation would move if a shear stress $\sigma_{yz}^{\text{applied}} = \tau$ were applied, where τ is a positive quantity.
(b) Write an expression for the critical shear stress τ_c that would be required to cause the LHS dislocation to glide indefinitely.
(c) Give the coordinates of the LHS dislocation at the point of the critical passing stress.

10.4 A mixed dislocation dipole is shown in Fig. 10.38a. Dislocation (1) has a positive edge component and a RHS component. Dislocation (2) has the opposite sign. Dislocation (1) is locked at the origin and the dislocation (2) is free to move by glide on a plane parallel to the x–z plane but it is not free to climb.

(a) Write an expression for the x component of force on dislocation (2) due to dislocation (1) as a function of the x coordinate of dislocation (2).
(b) Starting dislocation (2) in the approximate position shown, indicate the glide equilibrium positions for that dislocation for the following two cases: $\varphi = 0$ and $\varphi = \pi/2$.
(c) Make a sketch of the x component of the edge–edge and screw–screw interaction forces in the domain $0 < x < h$ and show how these curves can be used to find glide equilibrium positions.
(d) Write an expression for the critical value of φ, φ_c, below which the equilibrium position of dislocation (2) is just above dislocation (1), at $x = 0$.

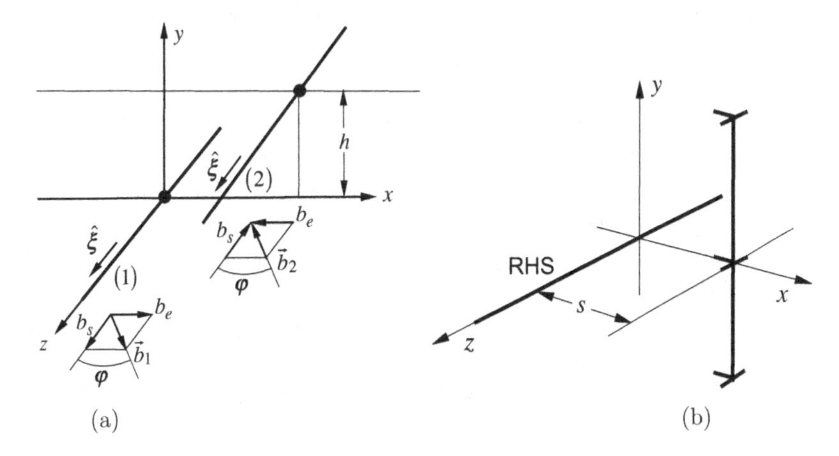

Figure 10.38. (a) A mixed dislocation dipole. (b) A screw dislocation and an edge dislocation perpendicular to each other.

10.5 Two long straight dislocations with the same Burgers vector magnitude are shown in Fig. 10.38b. Derive an expression for the local forces on the edge dislocation due to the stress field of the screw. Make a sketch of the forces on each of the dislocations due to the mutual interactions. Assuming that the RHS dislocation is able to glide only on the x–z plane and that the edge can only glide and not climb, what would be the ultimate fate of these dislocations after their interactions. The stress field of the edge dislocation can be obtained by introducing a new coordinate system $X' - Y' - Z'$ whose Z'-axis is along the dislocation line.

10.6 Two straight edge dislocations are shown in Fig. 10.39a. The Burgers vectors of the dislocations have the same magnitude b.

(a) Write expressions for the components of the distributed forces acting on dislocation (2) due to the stress field of dislocation (1).
(b) Make a sketch of the interaction forces acting on each dislocation.
(c) Using the results of (a), show how the dislocations would move if they were free to climb and glide. Show what would eventually happen to the dislocations if they were allowed to climb and glide.

10.7 Two infinitely long straight edge dislocations are shown in Fig. 10.39b. Dislocation (1) lies parallel to the z axis while dislocation (2) is in a skew orientation.

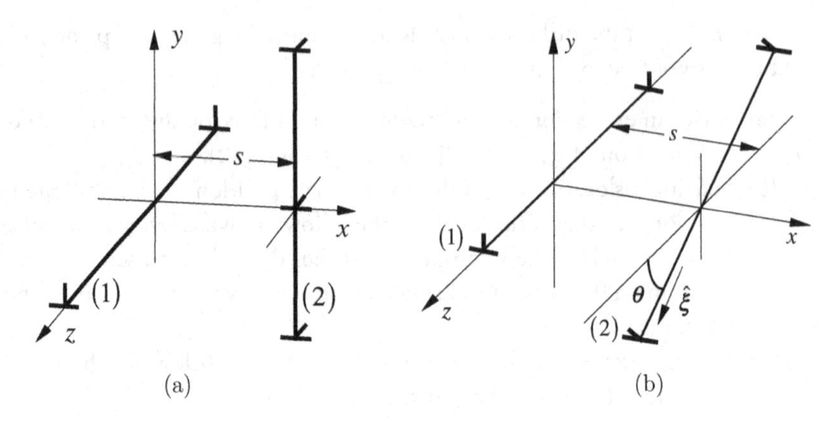

Figure 10.39. (a) Two edge dislocations perpendicular to each other. (b) Two edge dislocations at an angle θ with respect to each other.

(a) Write expressions for the sense vector (shown) and Burgers vector of the skewed dislocation, using the RH/SF convention.
(b) Use the Peach–Koehler formula to derive an expression for the vector force (per unit length) exerted on dislocation (2) by the elastic field of dislocation (1), for any angle θ. You may leave this expression in terms of the components of stress found in the field of dislocation (1).
(c) Show that the result obtained in (b) reduces to physically expected results for the special cases of $\theta = 0$ and $\theta = \pi/2$.
(d) Write an expression for the x component of force on dislocation (2) as a function of y position for any θ and make a sketch of that force distribution for the case of $\theta = \pi/4 = 45°$.

10.8 A pure edge dislocation lies along the z axis as shown in Fig. 10.40a. A circular vacancy loop in the x–y plane is positioned such that the straight edge dislocation goes through the center of the loop. The dislocations have Burgers vectors of length b.

(a) Write an expression for the components of the force per unit length acting on one point on the loop (the one point indicated). Express your result in terms of the components of force in the polar coordinate system.
(b) If the loop were constrained to remain circular with radius R, indicate how it would move in response to the interaction forces exerted by the straight edge dislocation.
(c) Indicate the equilibrium position of the loop under the constraints described in (b).
(d) Suppose the loop is not constrained to remain circular; indicate the final equilibrium configuration of the dislocation.

10.9 A jogged dislocation is shown in Fig. 10.40b. The long straight pieces of the dislocation are assumed to be much longer than the jog heights ($\lambda \gg h$). We wish to consider how the dislocation responds when a positive shear stress of the type σ_{zy} is applied.

(a) Indicate how the dislocation would move in response to the applied stress if the temperature is low and glide, but not climb, is allowed to occur.

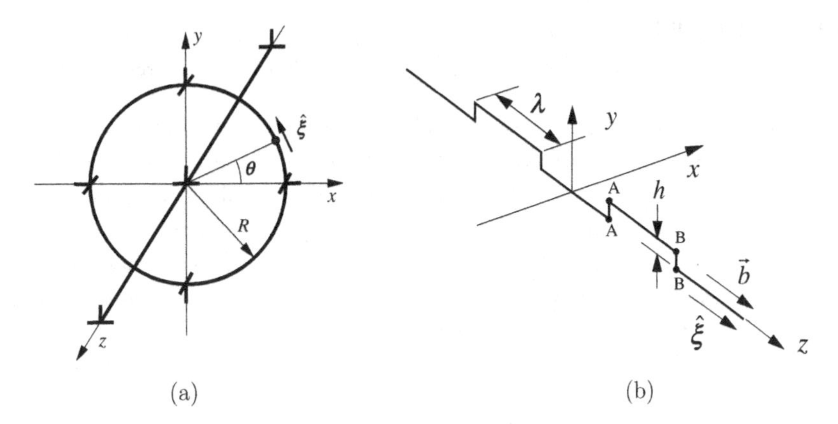

Figure 10.40. (a) A straight edge dislocation and a circular vacancy loop. (b) A jogged dislocation.

(b) Write an expression for the critical shear stress at which the dislocation would begin to move steadily away from the initial position. For simplicity, the line energy of the dislocation can be approximated as $\mu b^2/2$. Show the configuration of the dislocation when this critical stress is exceeded.
(c) Explain what happens to the dislocation when the stress exceeds $\mu b/[8\pi(1-\nu)h]$.
(d) Explain how the dislocation can move at very high temperatures even if the applied stress is very small.

10.10 A positive edge dislocation under an applied shear stress approaches a linear array of circular particles as shown Fig. 10.41. As shown in the figure, if the applied shear stress is small, the dislocation pushes part way in between the particles and then comes to equilibrium. The friction stress for dislocation glide in the matrix is zero while the friction stress for glide in the particles is τ_f. The particle and matrix may be assumed to have the same elastic properties.

(a) Write an expression for the applied shear stress that would be needed to cause the dislocation to shear into the particles and pass the particles by cutting through them.
(b) Assuming that the volume fraction of particles is constant and given by $\chi = \pi(R/L)^2$ and that it is much less than 1, write an expression for the critical particle radius, R^*, at which the stress needed for cutting is equal to that for bowing between the particles, as given by Eq. (10.53).
(c) Write an expression for the shear strength of the solid when the particles have the critical radius, again assuming that the volume fraction of particles is fixed at χ.
(d) Make a sketch of the shear strength of the solid as a function of particle radius, again assuming that the volume fraction of particles is fixed at χ.
(e) Give a physical explanation of the surprising behavior at $R = 0$.

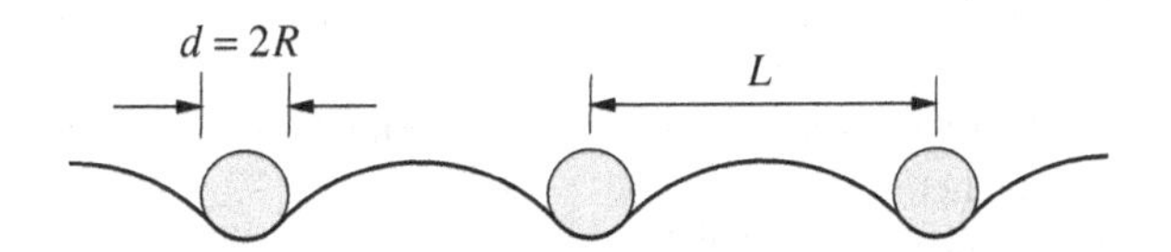

Figure 10.41. Dislocation interaction with shearable particles.

10.11 An edge dislocation tripole is shown in Fig. 10.42a. The edge dislocation at the origin is assumed to be locked and is not free to glide or climb. No external stresses are applied.

(a) Show how dislocations (1) and (2) move if they are free to glide, but not free to climb.
(b) Describe the stable equilibrium positions for dislocations (1) and (2). It is not possible to give an exact answer to this question without making an extensive analysis. Instead of making this analysis, describe the results you would expect from such an analysis.
(c) Show with a sketch how the dislocations would move if they were free to glide and climb. Assume that the glide mobility is much greater than the climb mobility. What happens to the dislocations after a very long period of time?

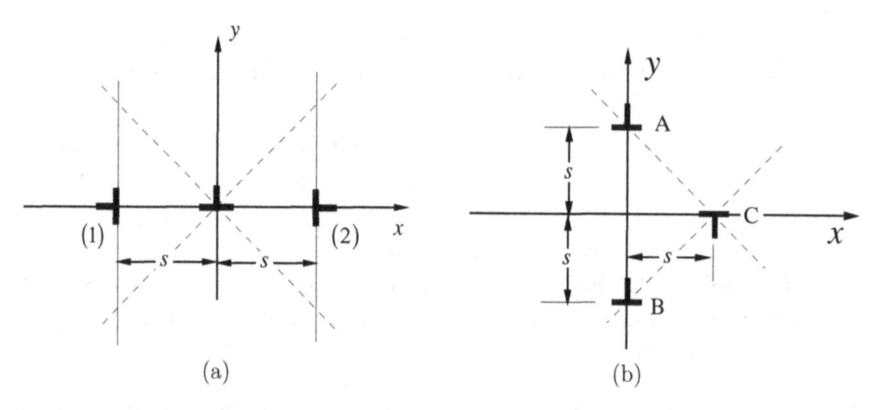

Figure 10.42. An edge dislocation tripole.

10.12 Three edge dislocations in an infinite crystal are shown in Fig. 10.42b. Dislocations A and B have the same sign, while dislocation C is of opposite sign. All of the dislocations have Burgers vectors of the same magnitude b. The dislocations are free to glide easily and quickly in their slip planes (no friction stress) and can also climb, but only very slowly. The climb velocity of any dislocation can be expressed as $v_c = M_c F_c$, where F_c is the climb force per unit length and M_c is the climb mobility.

(a) Write expressions for the x and y components of the velocities of all of the dislocations when they are in the positions shown.
(b) Indicate the directions of motion and what happens to each of the dislocations in the course of time.

10.13 Four edge dislocations, all parallel to the z axis, are shown in Fig. 10.43a. We wish to consider what happens to this particular configuration when the dislocations are allowed to glide but not climb. For convenience, we shall assume that the two positive edge dislocations lying in the y–z plane are fixed and not free to move at all. Only the dislocations lying in the x–z plane are allowed to glide.

(a) Describe qualitatively how the dislocations glide in response to the interaction forces between them.
(b) Indicate the paths the dislocations take during glide, paying particular attention to any asymmetry in the positions of the dislocations.

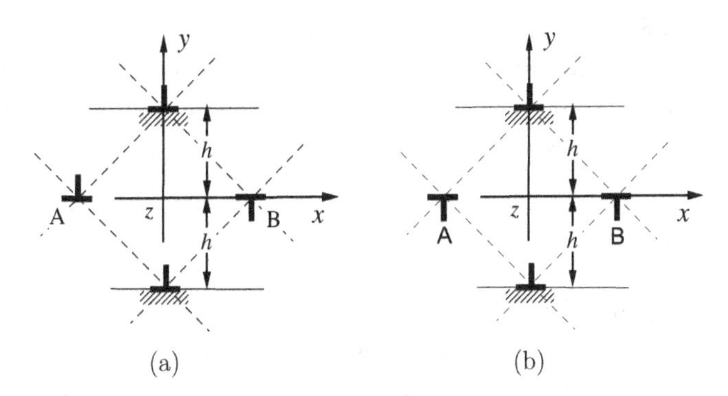

Figure 10.43. Four edge dislocations.

(c) Describe the final equilibrium state of the system and justify your description with estimates of the glide components of the interaction forces.

10.14 Consider four edge dislocations shown in Fig. 10.43b. The dislocations are all pure edge and they are all parallel to the z axis. Assume that the positive edge dislocations lying in the y–z plane are fixed and not free to move at all and we concern ourselves with the glide motion of the dislocations lying in the x–z plane.

(a) Describe qualitatively how the dislocations glide in response to the interaction forces between them.

(b) Determine the final equilibrium state of the system, indicating (quantitatively) the glide equilibrium positions of all of the dislocations.

10.15 Three screw dislocations are shown in Fig. 10.44. The two RHS dislocations are assumed to be held fixed at their positions while the LHS dislocation will be free to glide on its slip plane, which is parallel to the x axis. The dislocation is not allowed to cross-slip (see Section 8.4.2).

(a) Write an expression for the σ_{zy} component of stress at the location of dislocation (3) due to dislocations (1) and (2).

(b) Develop an expression for the equilibrium position of dislocation (3), x_{eq}, assuming that it is free to glide easily along the x axis.

(c) Assuming the position of dislocation (2) in the slip plane, x_0, could be changed, determine the minimum value of x_0 for which an equilibrium position can be found for dislocation (3).

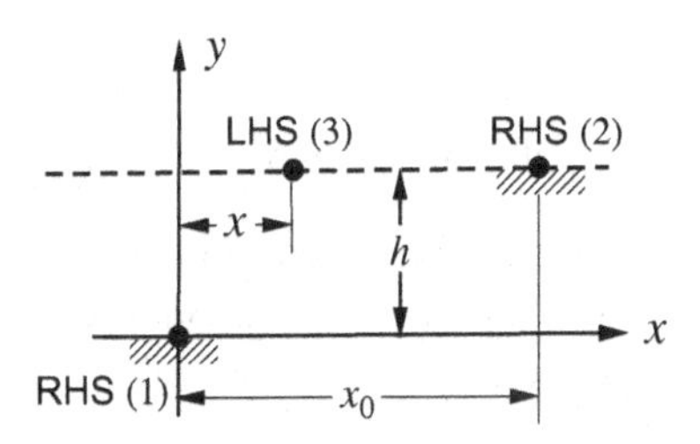

Figure 10.44. Three screw dislocations.

10.16 Three edge dislocations under a shear stress τ are shown in Fig. 10.45. The leading dislocation has been blocked by a high-angle grain boundary and the trailing dislocations have taken glide equilibrium positions, as shown. The two grains are assumed to have identical elastic properties characterized by μ and ν.

(a) Show how the equilibrium positions of the two trailing dislocations can be found for a given applied stress (it is not necessary to solve for these positions; just write the governing equations).
(b) Develop an expression for the equilibrium position of the second trailing dislocation, s_2, in terms of the equilibrium position of the first trailing dislocation, s_1.
(c) Write an expression for the tensile stress, σ_{xx}, acting on the grain boundary at small distance, h, below the leading dislocation in terms of the applied stress and the two stand-off distances, s_1 and s_2. This stress might cause grain boundary cracking.

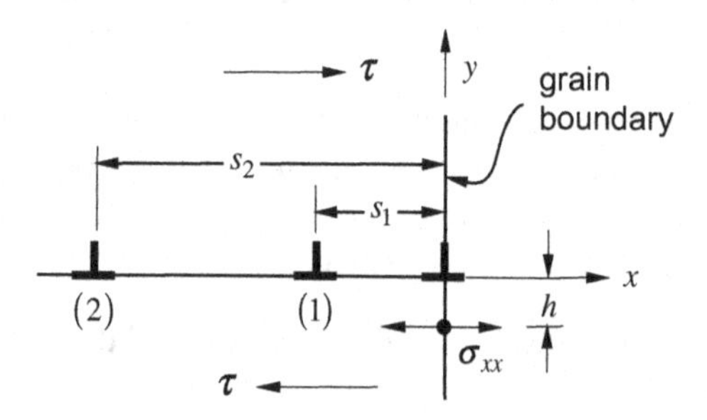

Figure 10.45. Three edge dislocations piling up against a grain boundary.

10.17 Modify the Matlab program (which creates Fig. 10.15) on the book website to obtain the equilibrium positions of $N = 100$ edge dislocations piling up against the lead dislocation $x_0 = 0$. What is the physical span of this dislocation array in Cu under the applied stress of $\tau = 10$ MPa? The elastic constants of Cu are provided in Appendix B. Assume that each dislocation x_i introduces a displacement jump of magnitude b in the $x < x_i$ region of the x axis, where b is the Burgers vector. Plot the displacement jump caused by the dislocation array as a function of x.

10.18 Modify the Matlab program (which creates Fig. 10.21) on the book website to predict the axial stress–strain curve in Si using the dislocation kinetics model. Use the average dislocation velocity function given in Eq. (10.71) and the mobile dislocation multiplication rate given in Eq. (10.74). Consider the experimental condition in which the total strain rate is $\dot{\varepsilon} = 6.6 \times 10^{-5}\ \mathrm{s}^{-1}$ and temperature is 815 °C. The effective Young's modulus that subsumes the compliance of the loading mechanism is $E_{\mathrm{eff}} = 12$ GPa. Assume the dislocation breeding constant is $K = 5 \times 10^{-4}\ \mathrm{m \cdot N^{-1}}$, the initial dislocation density is $\rho_0 = 10^7\ \mathrm{m}^{-2}$, and the Taylor hardening coefficient is $\alpha = 0.7$. Compare your prediction with the experimental stress–strain curve available on the book website.

10.19 Modify the Matlab program (which creates Fig. 10.22) on the book website to predict the strain as a function of time in LiF under creep conditions at room temperature using the dislocation kinetics model. Use the average dislocation velocity function given in Eq. (10.59) with $m = 20$, $\tau_0 = 12$ MPa, and $v_0 = 0.01\ \mathrm{m \cdot s^{-1}}$, but with τ replaced by τ_{eff} defined in Eq. (10.70) to account for strain hardening. Choose the Taylor hardening coefficient to be $\alpha = 0.36$. Use the mobile dislocation multiplication rate given in Eq. (10.61) with $\delta = 4000\ \text{m}^{-1}$. Assume the initial dislocation density is $\rho_0 = 10^9\ \text{m}^{-2}$. Compute the plastic strain as a function of time under constant shear stress $\tau = 7.35$ MPa for a period of 30 000 s.

10.20 Figure 10.46a shows a model representation of a dislocation moving in a strained film. For modeling purposes we assume that the film is subjected to a fixed misfit shear stress, τ, as shown, and that a pure screw dislocation is deposited at the film/substrate interface as the threading edge segment glides in the film. The dislocation is not allowed to glide into the substrate. Both the film and substrate are assumed to have the same isotropic shear modulus, μ.

(a) Derive an equation for the energy per unit length of the screw dislocation at the film/substrate interface by computing the work (per unit length) needed to move a screw dislocation from the surface to the interface. You may assume that the dislocation starts a small distance, b, from the free surface, in this process, as shown in Fig. 10.46b.

(b) Derive an equation for the critical film thickness, h_c, above which the edge dislocation will move spontaneously and deposit the screw dislocation in its wake.

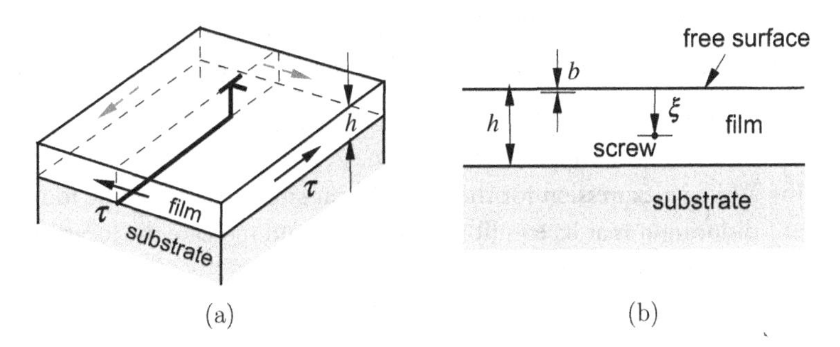

Figure 10.46. A misfit dislocation in strained film.

10.21 An elastically isotropic thin film of thickness h on a perfectly rigid ($\mu = \infty$) semi-infinite substrate contains a single right-handed screw dislocation (with $b = 1$) as shown in Fig. 10.47a. The dislocation is parallel to the plane of the film.

(a) Make a sketch showing the positions and strengths of the image dislocations that are needed to describe the displacements, strains, and stresses in the film.

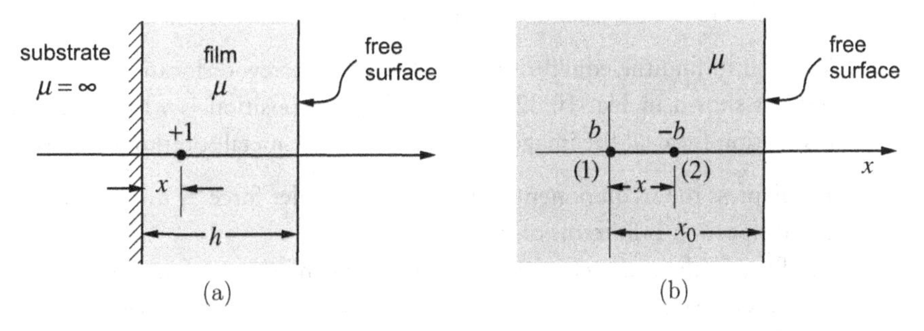

Figure 10.47. (a) A screw dislocation in a thin film on a rigid substrate. (b) Oppositely signed screw dislocations near a free surface.

(b) Write an expression for the image force (per unit length) acting on the dislocation. Include all of the image dislocations in the analysis and write the result in a compact form. It is not necessary to find analytical expressions for any infinite series solutions you may find.
(c) Derive an expression for the dislocation position x_{min} for which the image force has the smallest value.

10.22 Two parallel, oppositely signed screw dislocations are located near the free surface of an elastically isotropic solid as shown in Fig. 10.47b. Dislocation (1) is positioned at a fixed distance, x_0, from the free surface and is not allowed to move. Dislocation (2) is placed between the first dislocation and the free surface, a distance x from the first dislocation. Dislocation (2) is allowed to move by gliding along the x-axis.

(a) Write an expression for the critical value of x, x_c, at which the second dislocation would experience no glide force and would thus not move if placed in that position.
(b) Describe what happens if the second dislocation is placed at $x > x_c$. What if $x < x_c$?

10.23 A right-handed screw dislocation is positioned in phase (1) with shear modulus, μ_1, a distance s away from a planar interface with a semi-infinite phase (2) having a higher shear modulus $\mu_2 = 3\mu_1$, as shown in Fig. 10.48. The two semi-infinite half spaces are bonded together and are subjected to an applied shear strain $\gamma_{yz} = \gamma$, where γ is a positive quantity.

(a) Write an expression for the equilibrium stand-off distance, s_{eq}, for the dislocation as a function of the applied shear strain, γ.
(b) Write an expression for the largest shear stress that would be found in phase (2) when the dislocation is at its equilibrium position and indicate the location of that point of highest shear stress in phase (2).

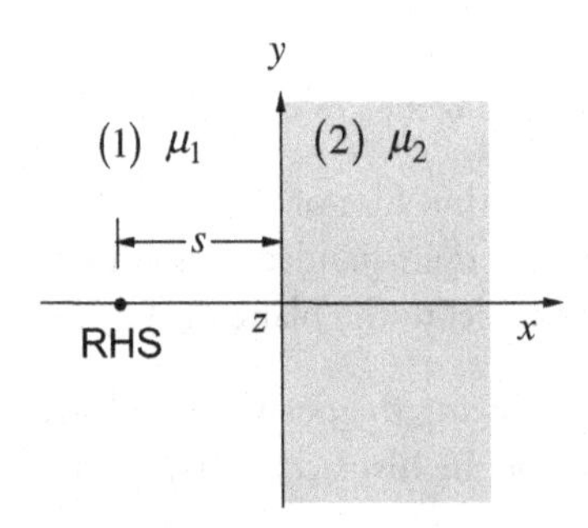

Figure 10.48. A right-handed screw dislocation near a phase boundary.

10.24 We wish to find the equilibrium position of the screw dislocation in a metal beneath an oxide film, as shown in Fig. 10.32. The equilibrium position is where the total force on the real dislocation from all the image dislocations (in the metal solution only) is zero.

(a) Express the x component of the Peach–Koehler force f_1 on dislocation A due to image dislocation B in terms of x_0 and κ.
(b) Express the x component of the Peach–Koehler force f_2 on dislocation A due to all the other image dislocations, D, F, H, etc. Leave the expression as an infinite sum.
(c) Write a Matlab program to plot $-f_1$ and f_2 as functions of x_0/a in the range of $0.5 \le x_0/a \le 2$ for $\kappa = 0.5$. Truncate the infinite summation in the expression for f_2 to

the first 10 000 terms. Determine the equilibrium dislocation position x_0^{eq}/a for which $f_1 + f_2 = 0$.

(d) Numerically compute x_0^{eq}/a for $\kappa = 0.1, 0.2, 0.3, \ldots, 0.9$. Discuss the asymptotic behavior of x_0^{eq}/a in the limits of $\kappa \to 0$ and $\kappa \to 1$.

10.25 A single RHS dislocation resides in a thin film of thickness h, which has one free surface and is perfectly bonded to an elastic substrate as shown in Fig. 10.49. The dislocation is positioned a distance $a = 2h/3$ from the free surface as shown. The shear modulus of the film is μ_{f} and that of the substrate is μ_{s}.

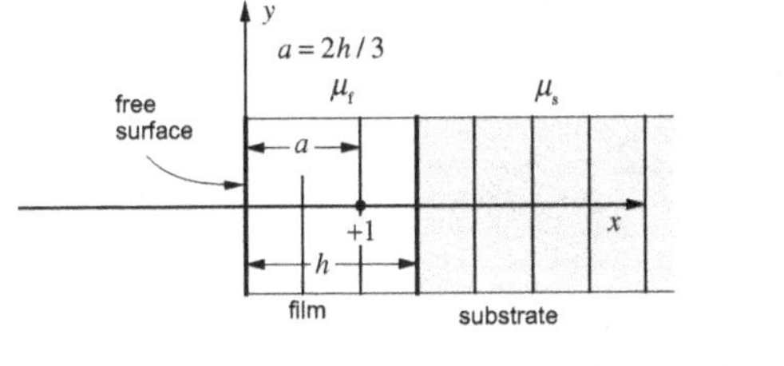

Figure 10.49. A screw dislocation in a thin film.

(a) Considering that the Burgers vector of the real dislocation is of unit strength, +1, determine the positions and the Burgers vectors of all of the image dislocations that would be needed to describe the stress field in the film. Use the template given in the figure to show where the dislocations are located. You need to include only those images that fit onto the template.

(b) Write an expression for the force acting on the real dislocation using the elastic fields of the two most important image dislocations.

11 Partial and extended dislocations

Our treatment of dislocations in previous chapters has focused on perfect dislocations, line defects that are surrounded by perfect crystal and have Burgers vectors equal to the shortest complete lattice translation vector. When perfect dislocations glide in a crystal they cause the atoms on either side of the slip plane to be displaced relative to each other by exactly a lattice translation vector, so that the crystal is perfect both ahead of and behind the gliding dislocation. Here we study the atomic motions associated with slip in more detail and observe that the sliding of atomic planes relative to each other often does not go immediately from one perfect state to the next. Instead, the slipping of atomic planes from one perfect state to the next may be broken up into two or more steps by the movement of partial dislocations separated by faults in the atomic stacking. We will see that partial dislocations and stacking faults in different crystal structures can be anticipated from the atomic packing arrangements in the crystal by treating the atoms as hard spheres and taking account of the bonding between them.

In Section 11.1, we consider the dissociation of perfect dislocations in FCC metals into Shockley partials, and obtain the equilibrium separation between the partials from a force-balance analysis. We introduce Thompson's notation to conveniently label the Burgers vectors of various perfect and partial dislocations in FCC metals. We also examine non-planar dislocation structures such as the transient core structure during cross-slip of a screw dislocation and the stable core structure of a Lomer–Cottrell dislocation. Finally, we consider the Frank partial dislocation loop formed from the condensation of vacancies, and the transformation of the Frank partial loop into a perfect dislocation loop or a stacking fault tetrahedron.

In Section 11.2, we discuss the types of dislocations in HCP metals. Because the atomic arrangements on the basal planes of HCP metals are very similar to those on the {1 1 1} planes in FCC metals, perfect dislocations on the basal planes are also dissociated into Shockley partials in HCP metals. In Sections 11.3 and 11.4, we discuss dislocations in $CrCl_3$ and Ni_3Al, respectively, which are crystals formed by more than one chemical species. The chemical complexity of these crystals leads to much longer perfect lattice vectors than those in elemental crystals. The layered structure of these crystals also make the dislocations dissociate into partials separated by various types of stacking faults and antiphase boundaries. In the case of Ni_3Al, the perfect dislocations are called superdislocations, which dissociate into superpartials. The lowest energy configuration

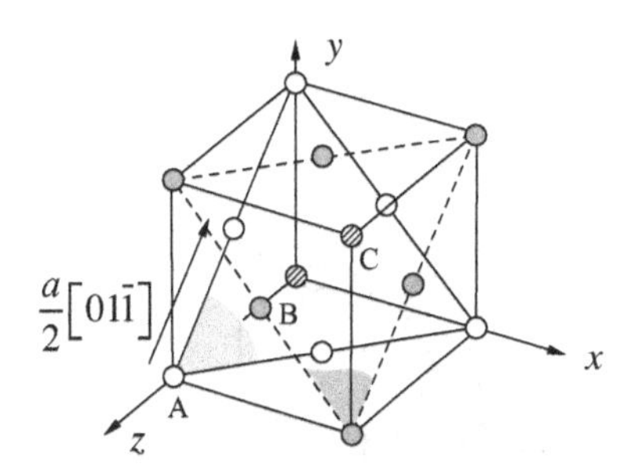

Figure 11.1. An FCC slip system: the (111) slip plane and the $\frac{a}{2}[01\bar{1}]$ slip direction. The atoms are colored according to which atomic layer (A, B, or C) they belong.

of a screw superdislocation has a non-planar structure, which results in the anomalous behavior of increasing yield strength with increasing temperature.

11.1 Partial dislocations in FCC metals

There is a general approach that we can follow to gain a basic understanding of partial dislocations and stacking faults in different crystal structures.

(1) Know the crystal structure, i.e. the positions of atoms of all types.
(2) Know the active slip planes, which are usually the most widely spaced and smoothest atomic planes.
(3) Know the slip directions, which are always the shortest lattice translation vectors on the active slip planes.
(4) Treat the atoms as hard spheres and determine how atoms are likely to slide past each other.
(5) Pay attention to the bonding between unlike near neighbor atoms. We will see that this approach can be quite helpful in understanding partial dislocations in many different crystal structures.

In this section, we apply this general approach to FCC metals and show why perfect dislocations prefer to dissociate into two partial dislocations. We also introduce Thompson's notation, which allows us to quickly determine the Burgers vectors of the partial dislocations for a given perfect dislocation. Finally, we use Thompson's notation to analyze the various stable and transient non-planar dislocation core structures found in FCC metals.

11.1.1 Slip planes and directions

We first apply steps 1, 2, 3 of the general approach outlined above to determine the slip planes and directions in FCC crystals. The FCC crystal structure was described in Chapter 1, and an FCC unit cell is shown again in Fig. 11.1. The most widely spaced atomic planes in an FCC structure are the {1 1 1} planes; they are indeed the most active slip planes for plastic deformation. The atomic layers (A, B, C) on parallel (1 1 1) planes are shown in different patterns in Fig. 11.1. The relative sliding between adjacent atomic layers of this type is the dominant mechanism for plastic deformation.

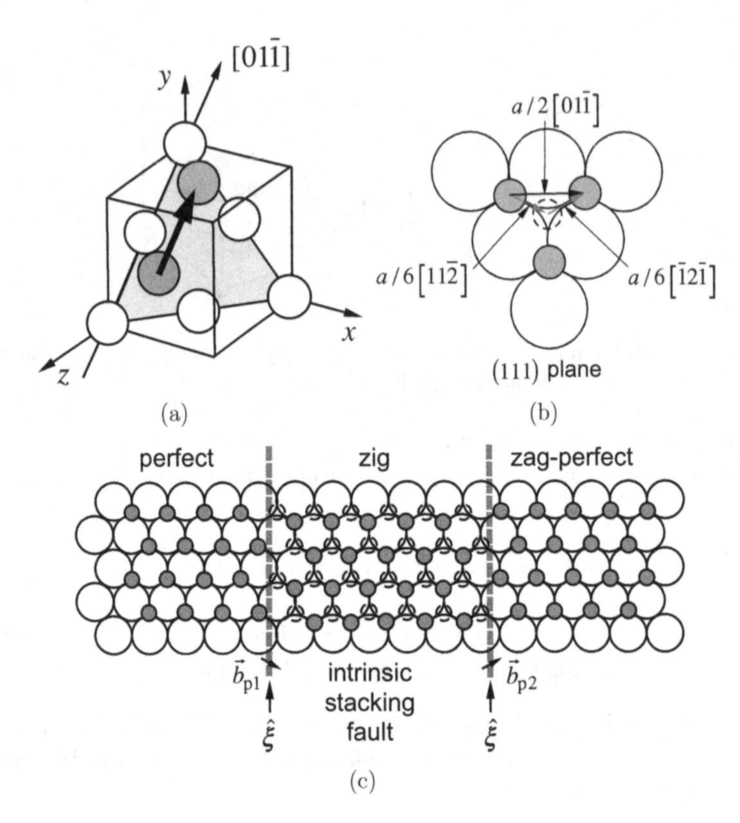

Figure 11.2. (a) Atoms on top of $(1\,1\,1)$ plane sliding by $\frac{a}{2}[0\,1\,\bar{1}]$ in three-dimensional view. (b) Splitting of the perfect slip vector $\frac{a}{2}[0\,1\,\bar{1}]$ into two partial slip vectors: $\vec{b}_{p1} = \frac{a}{6}[1\,1\,\bar{2}]$ and $\vec{b}_{p2} = \frac{a}{6}[\bar{1}\,2\,\bar{1}]$. (c) A group of atoms on top of $(1\,1\,1)$ plane sliding by $\frac{a}{2}[0\,1\,\bar{1}]$ in plan view. Left: before sliding. Middle: after partial sliding by $\vec{b}_{p1}$. Right: after another partial sliding by $\vec{b}_{p2}$. The original locations of the gray spheres are indicated by small circles in dashed line.

The shortest lattice translation vectors on the $\{1\,1\,1\}$ planes are the $\frac{a}{2}\langle 1\,1\,0\rangle$ vectors; they are the Burgers vectors of the most abundant dislocations in FCC metals. As an example, the $\frac{a}{2}[0\,1\,\bar{1}]$ direction is shown in Fig. 11.1. It is clear that the atoms on every $\{1\,1\,1\}$ plane form a triangular lattice, and that there are three different $\frac{a}{2}\langle 1\,1\,0\rangle$ slip directions (i.e. Burgers vectors) available on each $\{1\,1\,1\}$ plane. Each combination of slip plane and slip direction is called a *slip system*. Because there are four different $\{1\,1\,1\}$ planes: $(1\,1\,1)$, $(1\,1\,\bar{1})$, $(1\,\bar{1}\,1)$, and $(\bar{1}\,1\,1)$, there are $4 \times 3 = 12$ slip systems in every FCC crystal.

11.1.2 Shockley partials

We now apply step (4) of the general approach outlined at the beginning of Section 11.1, and determine how atoms, when treated as hard spheres, are likely to slide past each other on the slip plane.

Figure 11.2a plots an atom (gray spheres) on layer B sliding over the underlying $(1\,1\,1)$ layer A (white spheres) by $\frac{a}{2}[0\,1\,\bar{1}]$ (indicated by an arrow). Figure 11.2b shows the same structure in plan view, where the $\frac{a}{2}[0\,1\,\bar{1}]$ slip direction now points horizontally to the right. Figure 11.2c shows more atoms on the two atomic planes that slide against each other. The structure before slip occurs is shown on the left, and the structure after slip occurs is shown on the right. As the group of gray spheres slides half-way to the right, they are likely to fall into the pocket at the center of three white spheres, as shown by the middle structure (indicated as "zig") in Fig. 11.2c. This means that the trajectory of the atoms on top of the slip plane is unlikely to follow a straight line along $\frac{a}{2}[0\,1\,\bar{1}]$. Instead, it is more likely to first take a step of $\frac{a}{6}[1\,1\,\bar{2}]$, and then take another step of $\frac{a}{6}[\bar{1}\,2\,\bar{1}]$, as indicated in Fig. 11.2b.

In the middle ("zig") configuration shown in Fig. 11.2c, the atoms above the slip plane are translated by a vector that is not the repeat vector of the FCC lattice. The result is a planar defect over the area where this type of (partial) slip occurs. This defect is called a *stacking fault* and leads to an excess energy per unit area, called the stacking fault energy $\gamma_{\rm sf}$. Stacking faults in FCC crystals will be discussed in more detail in Section 11.1.3.

Recall that the dislocation line is the boundary between areas over which different amounts of slip have occurred. Now imagine a dislocation with Burgers vector $\frac{a}{2}[0\,1\,\bar{1}]$, which separates two areas on the $(1\,1\,1)$ plane. To the left of the dislocation, assume that slip has not occurred, corresponding to the left configuration in Fig. 11.2c. To the right of the dislocation, assume that atoms on top of the slip plane have slipped by $\frac{a}{2}[0\,1\,\bar{1}]$ relative to the atoms below, corresponding to the right configuration in Fig. 11.2c. In other words, the dislocation moves to the left as the slipped area expands. According to the analysis above, it is likely that there is an intermediate zone between these two areas for which the atoms on top have slipped by $\frac{a}{6}[1\,1\,\bar{2}]$ relative to the atoms below. Therefore, we can think of two boundary lines separating these three areas, as shown in Fig. 11.2c. Each line is identified as a dislocation; they are called Shockley *partial dislocations*, whose Burgers vectors are $\vec{b}_{p1} = \frac{a}{6}[1\,1\,\bar{2}]$ and $\vec{b}_{p2} = \frac{a}{6}[\bar{1}\,2\,\bar{1}]$, respectively, when the sense vector $\hat{\xi}$ points to the north. The term partial dislocation refers to the fact that its Burgers vector is not a full translation vector of the lattice (see Section 1.4.1). Therefore, in FCC metals, a *perfect dislocation* with $\frac{a}{2}\langle 1\,1\,0\rangle$-type Burgers vectors dissociates into two partial dislocations with $\frac{a}{6}\langle 1\,1\,2\rangle$-type Burgers vectors, bounding an area of stacking fault.

Figure 11.3 plots the atomic structure corresponding to a dissociated dislocation loop. In the centermost region, the atoms above the glide plane have slipped by a perfect Burgers vector $\vec{b} = \frac{a}{2}[0\,1\,\bar{1}]$. However, within an annulus region, the atoms have only slipped by a partial Burgers vector $\vec{b}_{p1} = \frac{a}{6}[1\,1\,\bar{2}]$, creating an area of stacking fault. The boundaries between the three regions are the two partial dislocations.

In the analysis above, we have made the simplifying assumption that the transition from one area with a certain amount of slip to another area with a different amount of slip is atomically sharp. This would correspond to very narrow core regions for the partial dislocations, similar to Volterra's dislocations. However, we expect that the partial dislocations would also have a finite core width, over which the transition to (or from) the stacking fault area occurs gradually, based on the Peierls–Nabarro model to be discussed in Section 12.1. Therefore, the glide dislocations in FCC metals are expected to have an extended core consisting of two finite-width partial dislocations bounding a stacking fault area.

The extended dislocation core and the metallic bonding make dislocations in FCC metals very mobile. As a result, FCC metals are generally very soft. The dislocation velocity in pure FCC metals can be well described by a linear mobility function, $v = M \cdot f$, where M is the mobility parameter and f is the Peach–Koehler force, as long as v is much smaller than the sound velocity. The mobility is mainly limited by the interaction between moving dislocations

Table 11.1. Mobility M of $\vec{b} = \frac{a}{2}\langle 1\,1\,0 \rangle$ dislocations on $\{1\,1\,1\}$ planes in several FCC metals.

Metal	M (Pa^{-1}s^{-1})	T (K)	Reference
Cu	1.25×10^6	4.2	[88]
Cu	5.9×10^4	296	[89]
Al	6.7×10^4	123	[90]
Al	4.0×10^4	296	[90]
Pb	6.6×10^4	4.2	[91]
Pb	2.9×10^4	296	[91]

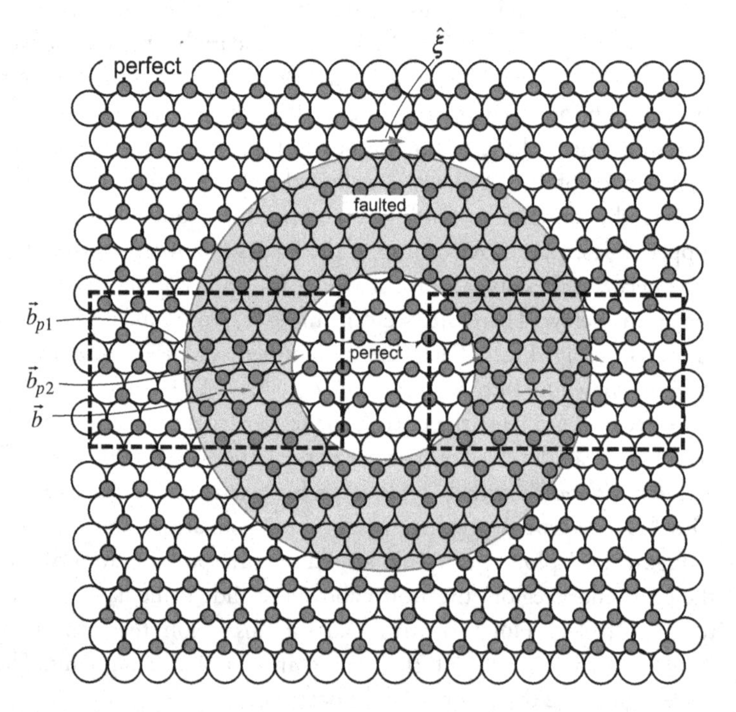

Figure 11.3. The atoms immediately above (gray) and below (white) the slip plane containing a perfect dislocation loop dissociated into two partial dislocation loops. The arrows indicate the direction of the Burgers vectors of the partial dislocations and the perfect dislocation. The region inside the dashed rectangle on the left is equivalent to that in Fig. 11.2c. The region inside the dashed rectangle on the right is equivalent to the dislocation shown in Fig. 11.7.

with lattice vibrations (i.e. phonon scattering) and decreases with increasing temperature [87]. At very low temperatures, the mobility is limited by the interaction between moving dislocations and electrons. The mobility parameters of several FCC metals are listed in Table 11.1.

11.1.3 Stacking faults

We now explain why the area between two Shockley partials of a perfect dislocation is called a stacking fault. To do so, we first need to define the perfect stacking sequence of an FCC crystal.

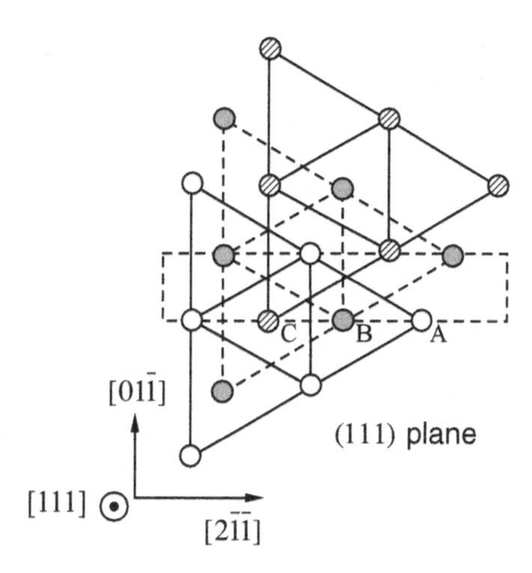

Figure 11.4. Perfect FCC crystal viewed along the [1 1 1] direction, showing the layers A, B, C stacked on top of one another.

Consider all the atoms on a (1 1 1) plane in an FCC crystal, such as the white circles in Fig. 11.1 and Fig. 11.4. Let us refer to this layer of atoms as layer A. They form a triangular lattice on the (1 1 1) plane. The next layer of atoms on the (1 1 1) plane also forms a triangular lattice. They are shown as gray circles in Fig. 11.1 and Fig. 11.4. Let us refer to these atoms as layer B. When viewed along the [1 1 1] direction, the atoms in layer B appear to occupy the center of the triangles left by the atoms in layer A. These are low energy positions because each atom is bonded to three atoms in the layer below. The next layer of atoms beyond layer B also forms a triangular lattice. When viewed along the [1 1 1] direction, the atoms in the third layer of an FCC crystal appear to occupy the center of the triangles left by atoms in both layer B and layer A, and this layer of atoms shall be called layer C. The atoms in the fourth layer occupy the same sites as the first layer when viewed along [1 1 1], so that they will be called layer A again. Therefore, the perfect FCC structure can be considered as layers of triangular lattices on top of each other in the stacking sequence of ABCABC..., as shown in Fig. 11.4.

Another way to illustrate the ABCABC... stacking sequence in the FCC lattice is to view the crystal along the $[0\,1\,\bar{1}]$ direction (the perfect Burgers vector) with the (1 1 1) plane lying horizontal, as shown in Fig. 11.5a. Here the unshaded circles are meant to represent the atoms in the plane of the page while the shaded circles that are partly covered by neighboring circles are the atoms either just below or just above the plane of the page. The Burgers vector of a dislocation gliding in the $[0\,1\,\bar{1}](111)$ slip system would point into the page in this view.

Consider, for example, a perfect edge dislocation gliding on the (1 1 1) plane, between atomic layers A (below) and B (above). The dislocation line is along the $[2\,\bar{1}\,\bar{1}]$ direction, pointing east in Fig. 11.5. The gliding of the dislocation would shift all of the atoms in the layer above the plane (relative to the layer below) into the page by a unit lattice translation, $\frac{a}{2}[0\,1\,\bar{1}]$ (pointing into the page of Fig. 11.5) so that the crystal would remain perfect after the dislocation has gone by. Figure 11.5(b) shows what happens when only the first Shockley partial has cut through the plane of the page. The upper half of the crystal is moving into the plane of the page and a little to the right so that the atoms that were above the plane of the page (shaded circles) are now brought into the plane of the page (unshaded circles). The displacement of these atoms perpendicular to

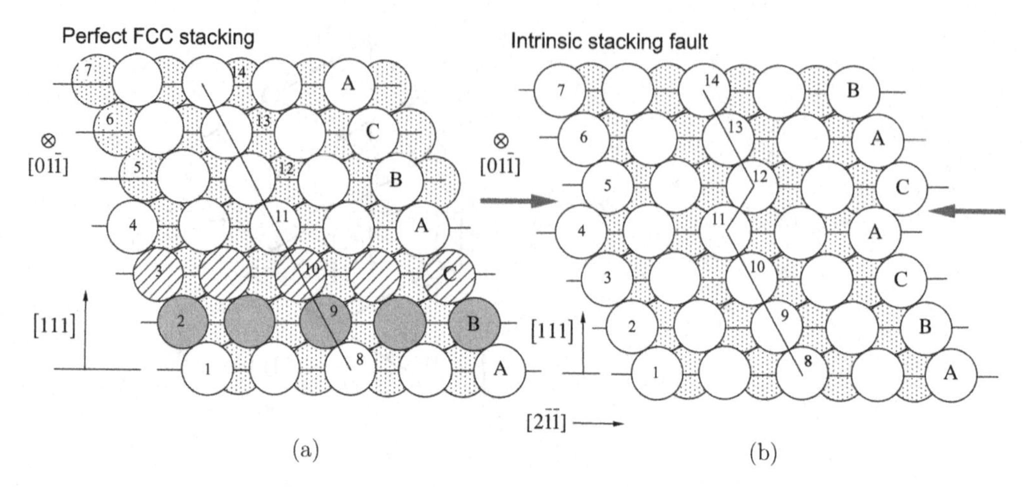

Figure 11.5. (a) Perfect FCC crystal corresponding to the stacking sequence of ABCABCA. . . . The bottom three rows of atoms are colored according to the layers, consistent with Fig. 11.1. (b) An intrinsic stacking fault created by sliding a layer B into the C position, and all the layers above it by the same amount, creating the stacking sequence of ABCA|CAB. . . . Small numbers 1 to 14 are labels for atoms to show which atoms have moved going from (a) to (b).

the plane of the page represents the edge component of the partial dislocation while the small displacement to the right represents the screw component of that partial dislocation. We see that cutting the crystal by this partial dislocation has left a mistake in the stacking sequence. The new stacking sequence after the sliding is ABCA|CABCA. . . , where | indicates the plane of sliding, as indicated by the two arrows in Fig. 11.5b. Because the perfect stacking sequence is interrupted, the resulting structure is called a *stacking fault*, which is a planar defect with an excess energy that is proportional to its area.

Because there are multiple ways the perfect stacking sequence can be interrupted, there are multiple types of stacking faults. The stacking fault created by sliding over a partial Burgers vector is called the *intrinsic stacking fault*. It is usually of the lowest energy among all types of stacking faults. It has the distinctive feature that the stacking sequence is as if one layer is missing. We shall see in Section 11.1.12 that the intrinsic stacking fault can indeed be created by removing a layer of atoms through vacancy condensation.

The *extrinsic stacking fault* is obtained when a new layer is inserted between two existing layers, leading to a stacking sequence of the type: ABCA|C|BCAB. . . , where the two | signs indicate the location where the new layer is inserted. The extrinsic stacking fault usually has a higher energy per unit area than the intrinsic stacking fault, owing to the higher degree of disruption to the stacking sequence. As shown in Fig. 11.6a, the extrinsic stacking fault can also be obtained by sliding over two adjacent (1 1 1) planes. The sliding on the lower plane moves atoms from layer B positions to layer C positions. The sliding on the higher plane then moves the next layer from A to B. In other words, the extrinsic stacking fault can be considered as two intrinsic stacking faults on top of each other.

If sliding occurs between all neighboring atomic planes in the upper half crystal, stacking sequence would be changed to ABCA|CBACBA. . . , as shown in Fig. 11.6b. This can be thought of as an infinite layer of stacking faults stacked on top of each other. However, instead of creating a highly defective structure, the top half crystal becomes a perfect crystal again, only a

Table 11.2. The energy per unit area of intrinsic stacking fault (γ_{ISF}) and twin boundary (γ_{TB}) of several FCC metals, in units of $mJ \cdot m^{-2}$ [92].

Metal	γ_{ISF}	γ_{TB}
Ag	22	8
Al	166	75
Au	32	15
Cu	78	24
Ni	128	43
Pb	25	10
Pt	322	161

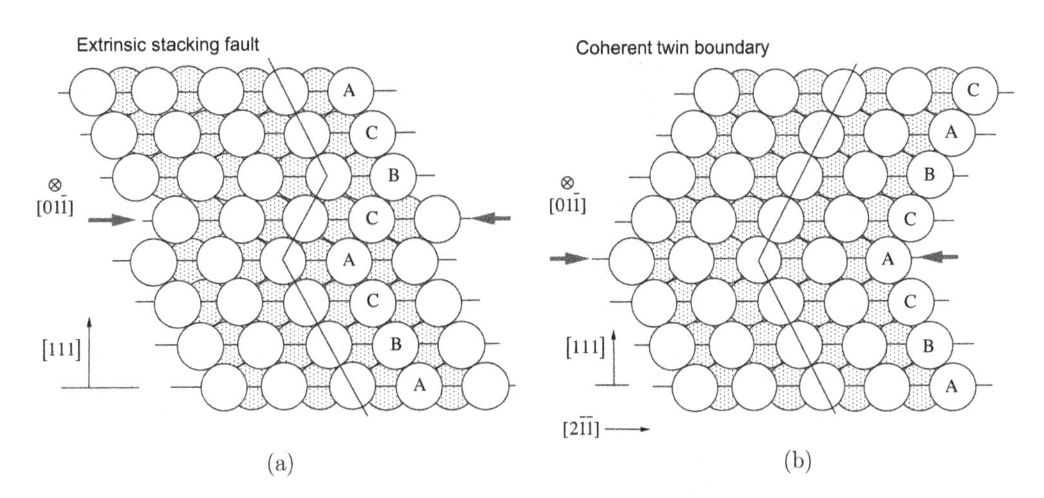

Figure 11.6. (a) An extrinsic stacking fault created by sliding over two consecutive (1 1 1) planes, creating the stacking sequence of ABCA|C|BCAB.... (b) A twin boundary created by sliding over all (1 1 1) planes above a given layer, creating the stacking sequence of ABCA|CBACBA....

mirror image of the lower half crystal. There is only one defective plane in the entire crystal, i.e. between the two crystal halves; and it is called the ***twin boundary***. The energy per unit area of the twin boundary is approximately half of that of the intrinsic stacking fault. This is because an intrinsic stacking fault can be considered as two twin boundaries on adjacent planes (e.g. compare Fig. 11.5b and Fig. 11.6b). Table 11.2 lists the energies of the intrinsic stacking fault and twin boundary of several FCC metals.

11.1.4 Equilibrium separation between partials

It is of interest to determine the equilibrium separation between the partials, i.e. the width of the stacking fault ribbon. We will first determine the separation under zero stress, and then discuss how the separation changes with stress.

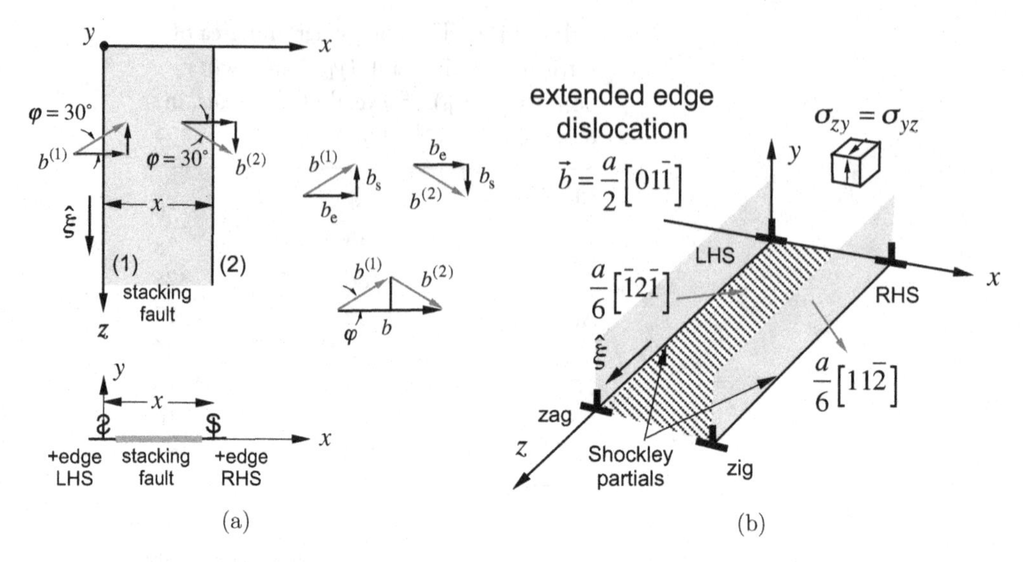

Figure 11.7. Extended edge dislocation. Each partial dislocation has both edge and screw components. (a) Plan view. (b) Three-dimensional view showing the extra half-planes of the edge components and type of screw components of the partials.

Perfect edge dislocation

Consider a perfect edge dislocation with Burgers vector $\vec{b} = \frac{a}{2}[0\,1\,\bar{1}]$ on the $(1\,1\,1)$ plane, as shown in Fig. 11.7a. Owing to the choice of the line sense vector, this dislocation is of the opposite sign compared to the one shown in Fig. 11.2c. The dislocation in Fig. 11.7a actually corresponds to the region inside the dashed rectangular region on the right side of the dislocation loop shown in Fig. 11.3.

The hard sphere model indicates that partial Burgers vectors make angles of $\varphi = 30^\circ$ with the perfect Burgers vector; hence they make angles of 60° with the dislocation lines, as shown in Fig. 11.7a. Thus the partials are 60° mixed dislocations. Figure 11.7b is a three-dimensional view of the dissociated dislocation. The extra half-planes point to the same direction, because the edge components of the two partials have the same sign. The screw components of the two partials have opposite signs, so that one partial contains a LHS and the other contains a RHS.

To find the equilibrium separation between the two partials, we assume that partial dislocation (1) is fixed at the origin, and the partial dislocation (2) is free to move in the x direction. The goal is to find the equilibrium position x_{eq} of partial dislocation (2). In the absence of external stress, there are three force contributions (per unit length) to partial dislocation (2). The elastic repulsion between the (like-signed) edge components of the two partials gives rise to

$$f_{\text{edge}} = +\frac{\mu b_e b_e}{2\pi(1-\nu)}\frac{1}{x}, \tag{11.1}$$

where $b_e = b/2$. The elastic attraction between the (opposite-signed) screw components of the two partials gives rise to

$$f_{\text{screw}} = -\frac{\mu b_s b_s}{2\pi}\frac{1}{x}, \tag{11.2}$$

where $b_s = (b/2)\tan\varphi$, and $\varphi = 30^\circ$. The stacking fault area gives rise to a constant attractive force (per unit length) between the partials. To see how this force arises, we note that the energy of the stacking fault area per unit length along the dislocation is $E_{SF} = \gamma_{sf}\, x$. Thus the force exerted by the stacking fault on partial dislocation (2) is

$$f_{SF} = -\frac{dE_{SF}}{dx} = -\gamma_{SF}. \tag{11.3}$$

This force is analogous to the surface tension force exerted by a thin film of liquid.

At equilibrium, the total force on partial dislocation (2) must vanish, i.e.

$$\begin{aligned} f^{(2)} &= f_{edge} + f_{screw} + f_{SF} \\ &= \frac{\mu b_e^2}{2\pi(1-\nu)}\frac{1}{x_{eq}} - \frac{\mu b_s^2}{2\pi}\frac{1}{x_{eq}} - \gamma_{SF} = 0. \end{aligned} \tag{11.4}$$

Therefore,

$$x_{eq} = \frac{1}{\gamma_{SF}}\left[\frac{\mu b_e^2}{2\pi(1-\nu)} - \frac{\mu b_s^2}{2\pi}\right] = \frac{1}{\gamma_{SF}}\frac{\mu b^2}{8\pi}\left[\frac{1}{1-\nu} - \frac{1}{3}\right]. \tag{11.5}$$

It is instructive to make an estimate of x_{eq} in a real FCC metal, such as gold, where (from Tables 11.2 and B.1)

$$\begin{aligned} \mu &\approx 28.2\,\text{GPa}, \\ \nu &\approx 0.423, \\ b &= 0.287\,\text{nm}, \\ \gamma_{SF} &= 32\,\text{mJ/m}^2, \\ x_{eq} &\approx 4.0\,\text{nm}. \end{aligned} \tag{11.6}$$

If an external stress σ_{yz} is present, this stress component exerts forces on two partials with equal magnitudes but in opposite directions. Therefore, this stress component does not exert a net force on the perfect dislocation as a whole, but modifies the equilibrium separation between the partials. This stress component is sometimes called the *Escaig* stress, after B. Escaig. The force on partial dislocation (2) from the Escaig stress is

$$f_{Escaig} = \sigma_{yz}\, b_s. \tag{11.7}$$

The equilibrium condition now becomes

$$\begin{aligned} f^{(2)} &= f_{edge} + f_{screw} + f_{SF} + f_{Escaig} \\ &= \frac{\mu b_e^2}{2\pi(1-\nu)}\frac{1}{x_{eq}} - \frac{\mu b_s^2}{2\pi}\frac{1}{x_{eq}} - \gamma_{SF} + \sigma_{yz}\, b_s = 0. \end{aligned} \tag{11.8}$$

Therefore,

$$\begin{aligned} x_{eq} &= \frac{1}{\gamma_{SF} - \sigma_{yz}\, b_s}\left[\frac{\mu b_e^2}{2\pi(1-\nu)} - \frac{\mu b_s^2}{2\pi}\right] \\ &= \frac{1}{\gamma_{SF} - \sigma_{yz}\, b/(2\sqrt{3})}\frac{\mu b^2}{8\pi}\left[\frac{1}{1-\nu} - \frac{1}{3}\right]. \end{aligned} \tag{11.9}$$

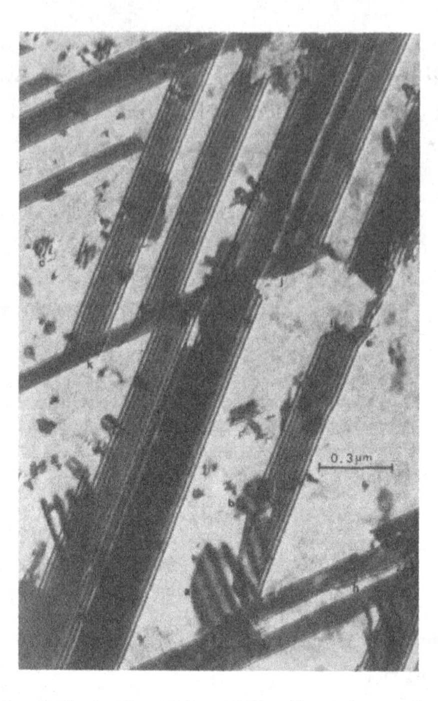

Figure 11.8. Extended stacking faults in Cu–12.2wt.%Ge alloy solution [93]. Used with permission from Elsevier.

Hence the Escaig stress has the same effect as modifying the stacking fault energy γ_{SF}. At the critical stress of $\sigma_{yz} = 2\sqrt{3}\,\gamma_{SF}/b$, $x_{eq} \to \infty$. Using the above values for gold, the critical stress is estimated to be 0.38 GPa. At this stress and beyond, the two partials are no longer bound together. The partials can move about independent of each other, creating very large stacking fault areas. Figure 11.8 is a transmission electron microscopy image of a Cu–Ge alloy in which the stacking fault areas are very wide because the stacking fault energy is very low.

Perfect screw dislocation

We now consider a perfect screw dislocation with Burgers vector $\vec{b} = \frac{a}{2}[0\,1\,\bar{1}]$ on the $(1\,1\,1)$ plane. The partial Burgers vectors now make angles of $\varphi = 30°$ with the partial dislocations, as shown in Fig. 11.9a. Thus the partials are 30° mixed dislocations. Figure 11.9b is a three-dimensional view of the dissociated dislocation. The extra half-planes point to opposite directions, because the edge components of the two partials have opposite signs. The screw components of the two partials have the same sign, so that both partials have RHS components.

The equilibrium separation between the two partials can be obtained following the same analysis as above for the edge dislocation. The elastic repulsion between the partials now comes from the (like-signed) screw components of the two partials,

$$f_{\text{screw}} = +\frac{\mu b_s b_s}{2\pi}\frac{1}{x}, \tag{11.10}$$

where $b_s = b/2$. The elastic attraction between the (opposite-signed) edge components of the two partials gives rise to

$$f_{\text{edge}} = -\frac{\mu b_e b_e}{2\pi(1-\nu)}\frac{1}{x}, \tag{11.11}$$

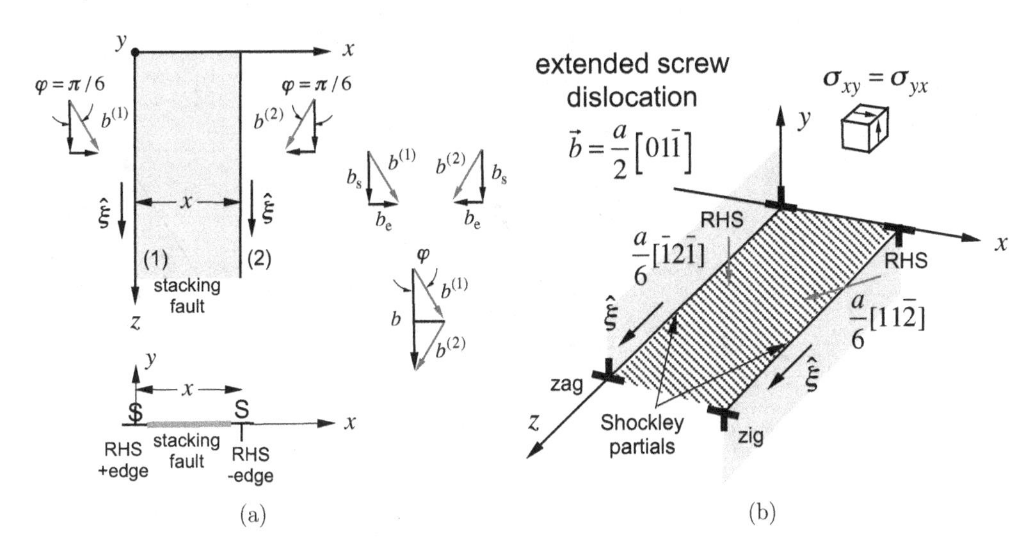

Figure 11.9. Extended screw dislocation. Each partial dislocation has both edge and screw components. (a) Plan view. (b) Three-dimensional view showing the extra half-plane of the edge components and type of screw components of the partials.

where $b_e = (b/2)\tan\varphi$, and $\varphi = 30°$. The stacking fault area gives rise to the same constant attractive force (per unit length) between the partials,

$$f_{SF} = -\frac{dE_{SF}}{dx} = -\gamma_{SF}. \tag{11.12}$$

The Escaig stress in the present geometry is σ_{xy}, which exerts forces on the two partials with equal magnitudes but in opposite directions. The force on partial dislocation (2) from the Escaig stress is

$$f_{\text{Escaig}} = -\sigma_{xy}\, b_e. \tag{11.13}$$

The equilibrium condition now becomes

$$\begin{aligned} f^{(2)} &= f_{\text{screw}} + f_{\text{edge}} + f_{SF} + F_{\text{Escaig}} \\ &= \frac{\mu b_s^2}{2\pi}\frac{1}{x_{eq}} - \frac{\mu b_e^2}{2\pi(1-\nu)}\frac{1}{x_{eq}} - \gamma_{SF} - \sigma_{xy}\, b_e = 0. \end{aligned} \tag{11.14}$$

Therefore,

$$\begin{aligned} x_{eq} &= \frac{1}{\gamma_{SF} + \sigma_{xy}\, b_e}\left[\frac{\mu b_s^2}{2\pi} - \frac{\mu b_e^2}{2\pi(1-\nu)}\right] \\ &= \frac{1}{\gamma_{SF} + \sigma_{xy}\, b/(2\sqrt{3})}\frac{\mu b^2}{8\pi}\left[1 - \frac{1}{3(1-\nu)}\right]. \end{aligned} \tag{11.15}$$

Using the above values for gold, we can estimate that, under zero stress, the equilibrium separation between the two partials in a screw dislocation is $x_{eq} \approx 1.2$ nm. Hence the equilibrium width of a screw dislocation (in the absence of stress) is narrower than of an edge dislocation.

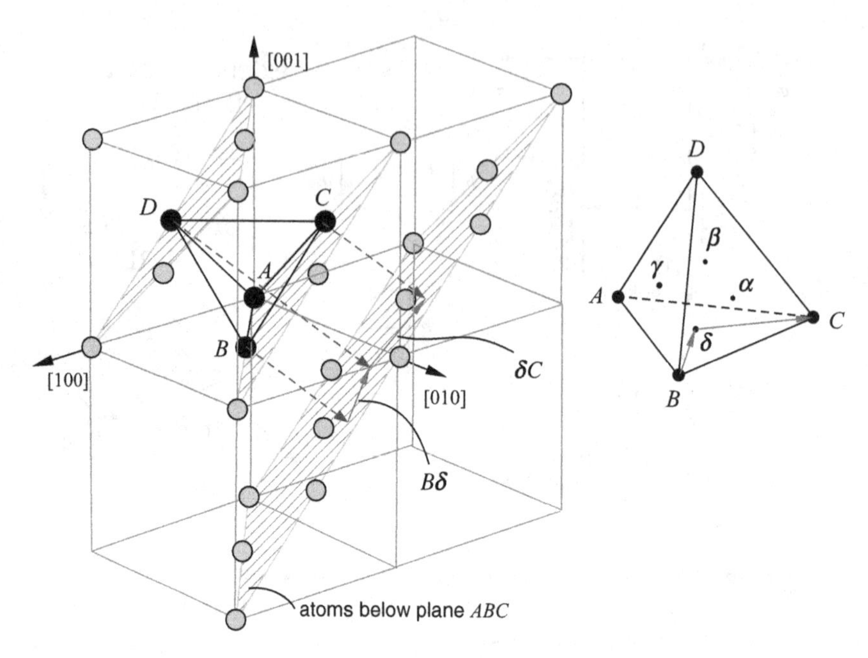

Figure 11.10. Construction of Thompson's tetrahedron in FCC crystals.

The underlying cause for this difference in the core width is that the elastic interaction between edge dislocations is stronger than that between screw dislocations by a factor of $1/(1-\nu)$.

In the above analysis we have treated the partials as Volterra dislocations with zero width. Such an assumption is most valid when the separation between partials are much larger than the width of their core (see Exercise problem 12.2).

11.1.5 Thompson's notation

The Thompson notation was invented to concisely represent all the perfect and partial Burgers vectors and the slip planes in FCC crystals. It is designed in such a way that the lengths and the geometric relationships between different vectors can be easily gleaned from the two letters used to represent each vector.

We first introduce Thompson's tetrahedron *ABCD*, shown in dark lines in Fig. 11.10. Point *A* is located at the origin, *B* at $\frac{a}{2}[110]$, *C* at $\frac{a}{2}[011]$, and *D* at $\frac{a}{2}[101]$. Here we use the italic font (*A*, *B*, *C*, *D*) for the vertices of the Thompson tetrahedron, to distinguish from the atomic layers (A, B, C) introduced in Section 11.1.1. Note that point *A* is at the corner and points *B*, *C*, *D* are at the face centers of the unit cell. An important aspect of these choices is that, when the fingers of the right hand curl around in the direction of $A \to B \to C$, the right thumb points towards point *D*.

The center of Thompson's tetrahedron is a tetrahedral site for interstitial impurities. In Section 4.2.2 we have mentioned that all tetrahedra and octahedra formed by nearest neighbor FCC lattice points fill up the entire space. Here we add that all such octahedra have the same orientation; they can all be obtained from an arbitrarily chosen octahedron by translation. On the other hand, there are two types of tetrahedra. All tetrahedra of the same type are related to each other by translation. Tetrahedra of different types are related to each other by mirror reflection, e.g. with respect to the [0 0 1] plane. Therefore, Thompson's tetrahedron is related to half of the tetrahedral sites in an FCC crystal by translation.

The center of every triangular face of Thompson's tetrahedron is represented by a Greek letter (α, β, γ, or δ) corresponding to the vertex (A, B, C, or D) facing the triangle. For example, the center point of triangle CBD (facing vertex A) is called α. Now every {111} plane in the FCC structure can be represented by three Roman letters, as listed below.

$(\bar{1}\,\bar{1}\,\bar{1})$	CBD
$(1\,1\,\bar{1})$	ACD
$(1\,\bar{1}\,1)$	ABC
$(\bar{1}\,1\,1)$	ADB

This representation also allows us to uniquely specify the direction of the surface normal, by the order of the three Roman letters. Following the right hand rule, when the fingers of the right hand curl in the direction of the three letters, the right thumb points in the direction of the surface normal. In the above table, all surface normals are the inward normal vectors of the Thompson tetrahedron. As we will see below, there is an advantage in adopting this convention when using Thompson's notation.

In Thompson's notation, each vector is specified by two letters corresponding to the beginning and end points of the vector. Vectors specified by two Roman letters (AB, AD, BC, ...) are the Burgers vectors of perfect dislocations. For example, BC is the same as vector $\frac{a}{2}[\bar{1}\,0\,1]$. Thus whenever we see a vector designated by two Roman letters, we know it is of the type $\frac{a}{2}\langle 1\,1\,0\rangle$ and has length $a/\sqrt{2}$.

Vectors specified by two letters that contain at least one Greek letter ($A\beta$, $A\alpha$, $\alpha\beta$, ...) are the Burgers vectors of partial dislocations. Among these, if one letter is Roman and the other is Greek, and if they are in different locations in their respective alphabet, such as $A\beta$, $B\gamma$, $C\alpha$, ..., then they are Burgers vectors of Shockley partial dislocations. For example, $B\delta$ is the same as vector $\frac{a}{6}[\bar{2}\,\bar{1}\,1]$ and δC is the same as vector $\frac{a}{6}[\bar{1}\,1\,2]$. These vectors are of the type $\frac{a}{6}\langle 1\,1\,2\rangle$ and have length $a/\sqrt{6}$. The dissociation of the perfect Burgers vector $\frac{a}{2}[\bar{1}\,0\,1]$ into two Shockley partial Burgers vectors, i.e.

$$\frac{a}{2}[\bar{1}\,0\,1] = \frac{a}{6}[\bar{2}\,\bar{1}\,1] + \frac{a}{6}[\bar{1}\,1\,2] \tag{11.16}$$

can be conveniently written in Thompson's notation as

$$BC = B\delta + \delta C. \tag{11.17}$$

Vectors specified by $A\alpha$, $B\beta$, $C\gamma$, and $D\delta$ are the Burgers vectors of *Frank partials* to be discussed in Section 11.1.12. For example, $A\alpha = \frac{a}{3}[1\,1\,1]$. These vectors are of the type $\frac{a}{3}\langle 111\rangle$ and have length $a/\sqrt{3}$.

Vectors specified by two Greek letters ($\alpha\beta$, $\beta\gamma$, $\gamma\delta$, ...) are the Burgers vectors of the *stair-rod* partial dislocations to be discussed in Sections 11.1.10 and 11.1.11. For example, $\alpha\beta = \frac{1}{6}[0\,\bar{1}\,\bar{1}]$. These vectors are of the type $\frac{1}{6}\langle 1\,1\,0\rangle$ and have length $a\sqrt{2}/6$. The name stair-rod refers to the fact that these dislocations exist at a bend in a stacking fault.

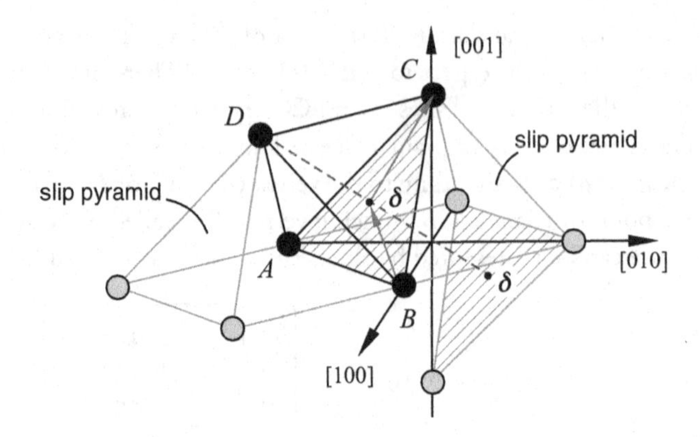

Figure 11.11. Slip pyramids adjacent to Thompson's tetrahedron.

From the hard sphere analysis, we have seen that when two $\{1\,1\,1\}$ atomic planes slip against each other by one perfect Burgers vector, the slip is likely to happen in two steps, corresponding to the two Burgers vectors of the partial dislocations. Although it is relatively easy to determine the set of the two partial Burgers vectors (e.g. $BC = B\delta + \delta C$), more analysis is needed to determine which partial has which Burgers vector. As described below, a rule can be established to determine the partial Burgers vectors unambiguously using Thompson's notation.

As an example, let us consider the plane ABC in Fig. 11.10. According to our convention, the surface normal is along $[1\,\bar{1}\,1]$, i.e. along δD. The atoms lying on the positive side of the surface normal vector are said to be "above" the atomic plane ABC; e.g. atom D is above the plane ABC. The atoms lying on the negative side of the surface normal vector are said to be "below" the atomic plane ABC.

We now consider the sliding of the atomic plane ABC by vector BC relative to the atomic plane "below" it. All the atoms "above" the atomic plane ABC travel together with the atomic plane ABC. In other words, the tetrahedron $ABCD$ translates as a rigid body. Similar to the analysis in Section 11.1.2, we can easily show that the translation will take place in two steps: first along $B\delta$ and then along δC.

Although this may seem obvious, it is important to specify that the slip is relative to the atomic plane "below" ABC. As a counter example, let us suppose that the slip along BC is actually against the atomic plane "above" plane ABC, i.e. relative to atom D. Then moving the atom initially at B into position δ would require it to ride directly against the atom at D, a very energetic process.

Therefore, the atom at B only slides into δ when it slides against atoms outside the tetrahedron, while the tetrahedron translates as a rigid body during the sliding process. In other words, all of the atoms *inside the tetrahedron* slide together *relative to the atoms outside* of the tetrahedron. To help us remember the correct sign of the sliding, we can think of atoms sliding on the inside surface of the tetrahedron relative to the atoms outside the tetrahedron.

11.1.6 Slip pyramids

Figure 11.11 shows another illustration of "sliding on the inside surface of the tetrahedron." Here Thompson's tetrahedron is positioned between two "slip pyramids." Each slip pyramid

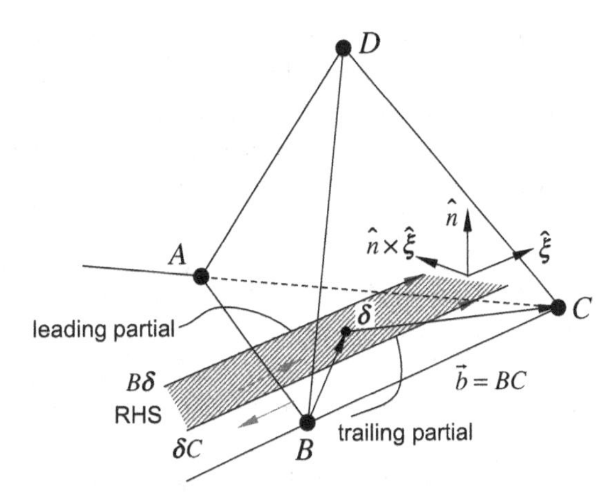

Figure 11.12. Dissociation of a right-handed screw with Burgers vector BC on plane ABC.

is simply half of the octahedron enclosing the octahedral site defined in Section 4.2.2. Each Thompson tetrahedron shares a face with four octahedra, and hence four slip pyramids, two of which are shown in Figure 11.11.

Again we consider the slip along the direction BC, which happens in two steps: $B\delta$ and then δC, as indicated by the arrows. It is important to specify that the positive surface normal of the slip plane is the *outward surface normal of the slip pyramid*, which is the *inward surface normal of the tetrahedron*. Hence we say that slip occurs on the outside surface of the slip pyramid (equivalent to the inside surface of the tetrahedron). Figure 11.11 shows that when atoms slide from B to δ they do so by moving directly on top of the midpoint of the three gray atoms on the shaded plane below.

11.1.7 Leading and trailing partials

The findings above allow us to determine the Burgers vectors of each of the two partials in a dissociated dislocation. Knowing the Burgers vectors of the partials is important, because, for example, it allows us to determine whether a given applied stress widens or narrows the stacking fault width, which influences the rate of cross-slip (see Section 11.1.10). As an example, consider a right-handed screw (RHS) dislocation on plane ABC, being driven toward the vertex A by the applied stress, as shown in Fig. 11.12. Here Thompson's tetrahedron is drawn to be much larger than the atomic dimension. For clarity, it appears even larger than the dissociation width of the dislocation. For the chosen sense vector $\hat{\xi}$, the Burgers vector of this RHS dislocation is BC. Note that we have again chosen the positive normal vector $\hat{n}$ of the slip plane to be the inward normal of the tetrahedron, i.e. along δD. Define the shear stress to be "positive" when the force acting on the positive surface of the slip plane ABC is along the Burgers vector BC. When such a "positive" shear stress is applied (as indicated by a pair of arrows above and below the slip plane), the RHS dislocation moves toward the corner A. This can be seen from the way the standard loop expands in Fig. 8.21. From the direction of dislocation motion under the "positive" shear stress, the partial closer to the corner A (when the dislocation is drawn within the triangle ABC) is called the *leading partial* and the partial closer to the edge BC is called the *trailing partial*. Here the words leading and trailing refer to the fact that the leading partial is ahead of the trailing partial in the direction of dislocation motion when a "positive" shear stress is applied. Notice that if we reverse the sense vector, then the Burgers vector, the direction of the "positive" shear stress, as well as the designation of leading versus trailing partials, would all reverse.

The designation of leading versus trailing partials can be determined quickly by an Axiom proposed by Hirth and Lothe [5]. For the slip plane normal $\hat{n}$ defined here (as the inward normal of the tetrahedron), the Axiom states that *the vector* $\hat{n} \times \hat{\xi}$ *points to the leading partial*, when a "positive" shear stress is applied. In this example, $\hat{n} \times \hat{\xi}$ points in the direction of δA. Hence the partial closer to the corner A is the leading partial, consistent with the designation above.

As the dislocation passes through a material element, the atoms above the slip plane slide by the Burgers vector BC relative to the atoms below the slip plane. We know that the slip will occur in two steps: first along $B\delta$ and then along δC. These two slip events must correspond to the moments when the leading partial and then the trailing partial passes through the material element. Therefore the leading partial must have Burgers vector $B\delta$ and the trailing partial must have Burgers vector δC. From the designation of leading and trailing partials above, we can now conclude that the partial near the corner A (when the dislocation is drawn within the triangle ABC) has the Burgers vector $B\delta$ and the partial near the edge BC has the Burgers vector δC, when the sense vector $\hat{\xi}$ is along BC, as shown in Fig. 11.12.

The above example can be easily generalized to arbitrary mixed dislocations with $\frac{a}{2}\langle 110 \rangle$ Burgers vectors on (111) planes. The conclusions can be summarized as follows. For a "positive" shear stress, the leading partial always has the Burgers vector starting from a Roman letter and ending with a Greek letter (e.g. $B\delta$, $A\gamma$, etc.), while the trailing partial always has the Burgers vector starting from a Greek letter and ending with a Roman letter (e.g. δC, γD, etc.). The leading partial is pointed to by the $\hat{n} \times \hat{\xi}$ vector where $\hat{n}$ is the inward normal of Thompson's tetrahedron (or the outward normal of a slip pyramid).

11.1.8 Extended and constricted nodes

We now apply the general principles established above to analyze the structure of a planar dislocation network shown in Fig. 11.13. This structure can form when a small twist exists between two crystalline regions (see Section 14.2.2). Two layers of atoms are shown. The atoms above the slip plane are shown as small gray circles. The atoms below the slip plane are shown as large white circles. With this example, we demonstrate how the rules described above can be used to analyze arbitrary partial dislocation structures.

We form Thompson's tetrahedron by three atoms on the lower plane (A, B, C) and one atom on the upper plane (D). Notice that as the fingers of the right hand curls around $A \to B \to C$, the right thumb points in the δD direction. The positive normal vector $\hat{n}$ of the slip plane is the inward surface normal of the tetrahedron, and points out of the page (parallel to δD), as shown in Fig. 11.13.

The dislocation network consists entirely of RHS perfect dislocations, at angles of 120° from each other. It can be easily verified that such an arrangement guarantees the conservation of Burgers vector at every node where three dislocations meet. We now analyze how the partial dislocations of the network are connected to each other.

Let us start with dislocation (1) at the south end. For the chosen sense vector ($\hat{\xi}$ along CA), the perfect Burgers vector is CA. Given the direction of $\hat{n} \times \hat{\xi}$, we know that the partial to the west must have Burgers vector $C\delta$. The partial to the east has Burgers vector δA.

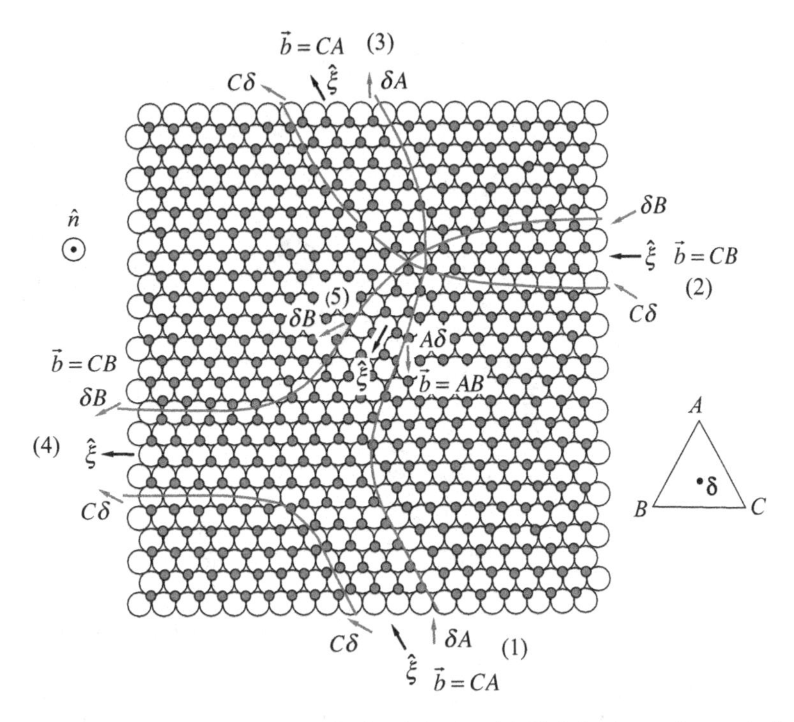

Figure 11.13. A planar network formed by right-handed screw (RHS) dislocations, showing both constricted and extended nodes.

For the dislocation (2) at the east end, we choose its sense vector to point to the left ($\hat{\xi}$ along CB), so that its Burgers vector is CB. Given the direction of $\hat{n} \times \hat{\xi}$, we know that the partial to the south must have Burgers vector $C\delta$. The partial to the north has Burgers vector δB.

Dislocation (3) is identical to dislocation (1), and dislocation (4) is identical to dislocation (2). The partial Burgers vectors of dislocation (5) can be obtained similarly. Notice that, in Fig. 11.13, the sense vector of partial $A\delta$ of dislocation (5) is opposite to the sense vector of partial δA of dislocation (1). Hence these two partials would have the same Burgers vector if the same sense vectors had been chosen for them.

The partial $C\delta$ of dislocation (1) smoothly joins the partial $C\delta$ of dislocation (4). Similarly, the partial δA of dislocation (1) joins the partial $A\delta$ of dislocation (5); and the partial δB of dislocation (5) joins the partial δB of dislocation (4). In all three cases, the partials remain on the same side of the stacking fault area as it goes from one dislocation to the next. Consequently, the node at which the three perfect dislocations meet is an extended area of stacking fault. This node is called the *extended node*.

The partial δB of dislocation (2) has the same Burgers vector as the partial δB of dislocation (5). Owing to the conservation of Burgers vectors, we can imagine this partial continues as a smooth line from dislocation (2) to dislocation (5), as shown in Fig. 11.13. However, in doing so, this partial dislocation must cut across the stacking fault area of dislocation (3), creating a high energy region. The length of such crossing is very short (on the order of an atomic spacing) in order to minimize the energy. Similarly, the partial $A\delta$ of dislocation (5) needs to cross dislocation (2) before joining dislocation (3); and the partial $C\delta$ of dislocation (2) needs to cross dislocation (5) before joining dislocation (3). To minimize the length of these three

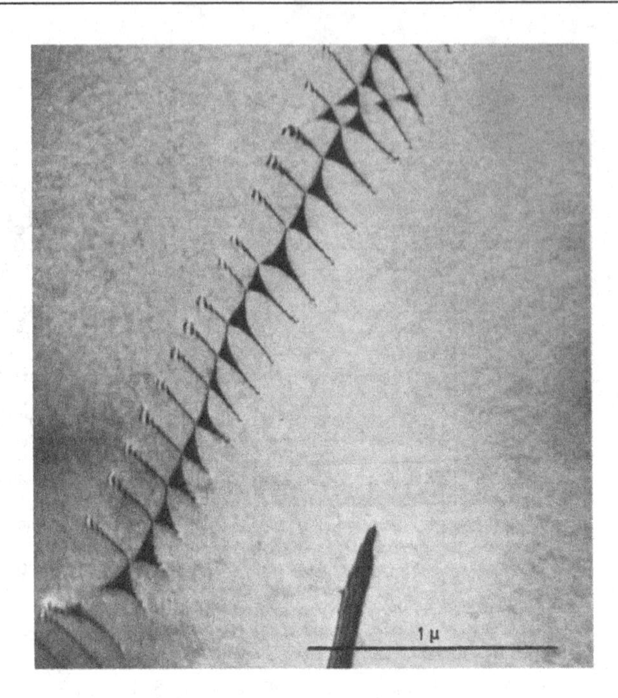

Figure 11.14. Extended and constricted nodes in dislocation network observed by transmission electron microscopy (TEM) in a Cu–8%Al alloy [40]. Used with permission from Elsevier.

crossings, the node where dislocations (2), (3), (5) meet has a narrow extension, and is called the *constricted node.*

The extended and constricted nodes in planar dislocation networks have been observed in FCC metals by transmission electron microscopy (TEM), as shown in Fig. 11.14.

11.1.9 Climb of extended non-screw dislocations

We now examine some non-planar dislocation structures. In this section, we consider a dissociated edge dislocation with a jog, which enables the edge dislocation to climb.

Consider a dissociated edge dislocation subjected to stress causing climb to occur, as shown in Fig. 11.15. The climb of partial dislocations P to P′ and Q to Q′ leaves faults (i.e. planar defects) along PP′ and QQ′. These faults are not on {111} planes and consequently have very high energy. Consequently, the climb mechanism shown in Fig. 11.15 is not likely to occur.

A more likely mechanism for climb to occur is for the partials to constrict first, i.e. to combine into a perfect dislocation. This may occur at jogs in dislocations where the partials are constricted, as shown in Fig. 11.16. As the jog moves to the left, the net effect is that a short segment of the extended dislocation climbs down by the height of the jog. As the jog moves over the length of the dislocation, the entire dislocation climbs down by the height of the jog. No additional fault is created during the climb process, because it occurs by the motion of the jog, which is a short segment of perfect dislocation. The motion of the jog absorbs atoms in the vicinity of the dislocation to the extra half-plane, extending the half-plane in the downward direction. This process leaves vacancies in the lattice, which diffuse away from the dislocation. Hence the jog motion shown in Fig. 11.16 emits vacancies.

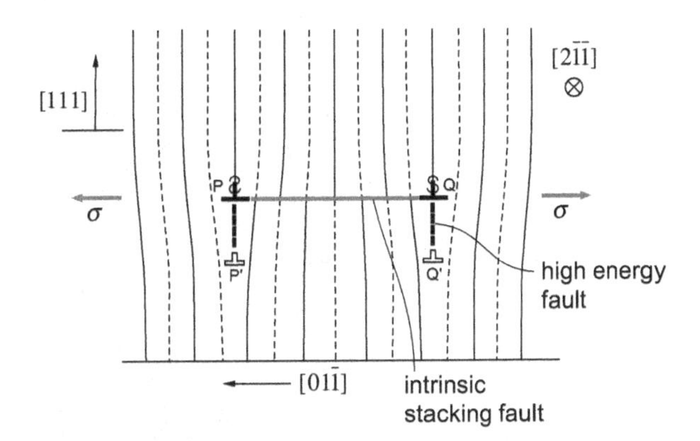

Figure 11.15. Climb of dissociated edge dislocation under tensile stress σ. Climb of partial dislocations P to P′ and Q to Q′ leaves very energetic faults along PP′ and QQ′.

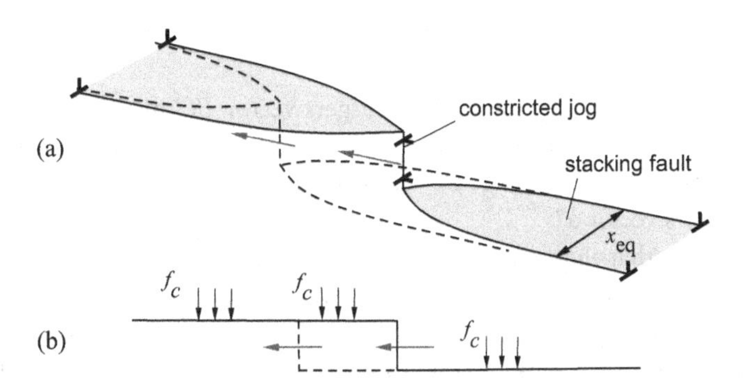

Figure 11.16. Dislocation climb by jog motion. (a) Three-dimensional view. (b) View along the glide plane. Large arrows indicate the direction of jog motion. Small arrows indicate the direction of climb force on the edge dislocation.

11.1.10 Cross-slip of extended screw dislocations

Dislocations can only move conservatively (i.e. without diffusion) on glide planes simultaneously containing the Burgers vector and line direction. In principle, a screw dislocation can glide conservatively on any plane that contains its Burgers vector. However, in FCC metals, each perfect screw dislocation dissociates into two mixed partial dislocations on a $\{1\,1\,1\}$ plane, so that the extended screw dislocation can only glide conservatively on the plane of dissociation. Nonetheless, the glide motion of the screw dislocation is not entirely confined on one $\{1\,1\,1\}$ plane, because there is another $\{1\,1\,1\}$ plane on which the screw dislocation can also dissociate and glide.

Figure 11.17 shows the two possible dissociated structures of the right-handed screw dislocation, with line direction $\hat{\xi}$ parallel to the Burgers vector BC, onto the ABC and DCB planes, respectively. On plane ABC, the perfect Burgers vector BC dissociates into partial Burgers vectors $B\delta$ and δC. The inward normal $\hat{n}$ of the tetrahedron is along δD. From $\hat{n} \times \hat{\xi}$, we know that in Fig. 11.17 the partial closer to the corner A has Burgers vector $B\delta$. The other partial on the ABC plane has Burgers vector δC.

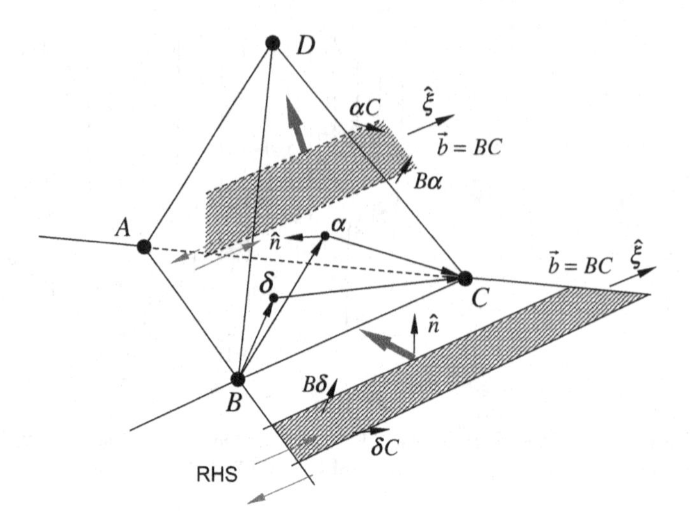

Figure 11.17. Cross-slip of a right-handed screw with Burgers vector BC from plane ABC to plane CBD.

On plane DCB, the perfect Burgers vector BC dissociates into partial Burgers vectors $B\alpha$ and αC. The inward normal $\hat{n}$ of the tetrahedron is now along αA. From $\hat{n} \times \hat{\xi}$, we know that the partial closer to edge BC has Burgers vector $B\alpha$. The other partial on the DCB plane has Burgers vector αC.

Note that the two partial Burgers vectors point away from each other on plane ABC, but they point toward each other on plane DCB. This is a generic feature of all extended screw dislocations in FCC metals on the two available $\{1\,1\,1\}$ planes. Of course, if we reverse the direction of the sense vector (to point along CB instead), then all the partial Burgers vectors would reverse their directions too. As a result, the two partial Burgers would point toward each other on plane ABC and away from each other on plane DCB.

Imagine that the screw dislocation BC is originally dissociated on plane ABC. The plane DCB is then called the *cross-slip plane*. Note that the partials $B\delta$ and δC are mixed dislocations and hence cannot cross-slip to plane DCB. One way for cross-slip to occur is through the Friedel–Escaig mechanism. The two partials $B\delta$ and δC first constrict to a point on the original plane ABC. The dislocation then re-dissociates onto plane DCB, creating a short extended dislocation segment with partials $B\alpha$ and αC, as shown in Fig. 11.18. Assisted by appropriate shear stresses, the length of the segment dissociated on the DCB plane increases, until the entire dislocation extends and moves on the cross-slip plane. The rate of cross-slip is higher if the separation between the partials on the original slip plane is reduced. Because the equilibrium separation between the partials depends on the component of stress that exerts opposite forces on the two partials (see Section 11.1.4), this stress component (namely the Escaig stress on the original slip plane) has a strong effect on the rate of cross-slip by the Friedel–Escaig mechanism.

An alternative cross-slip mechanism in FCC metals is the Fleischer mechanism. As shown in Fig. 11.19, one of the partials (δC) is emitted on the cross-slip plane before the two partials on the original glide plane constrict to a point. This leads to a short segment along the screw dislocation with a non-planar core, because stacking fault areas exist on both planes. The intersection line between the two stacking fault areas is a stair-rod dislocation with Burgers vector $\delta\alpha$, so that the total Burgers vector is still $B\delta + \delta\alpha + \alpha C = BC$. As the partial dislocation

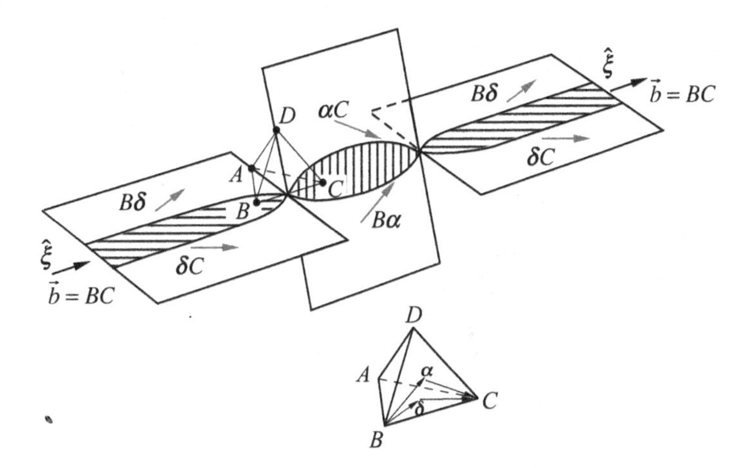

Figure 11.18. The Friedel–Escaig mechanism of cross-slip.

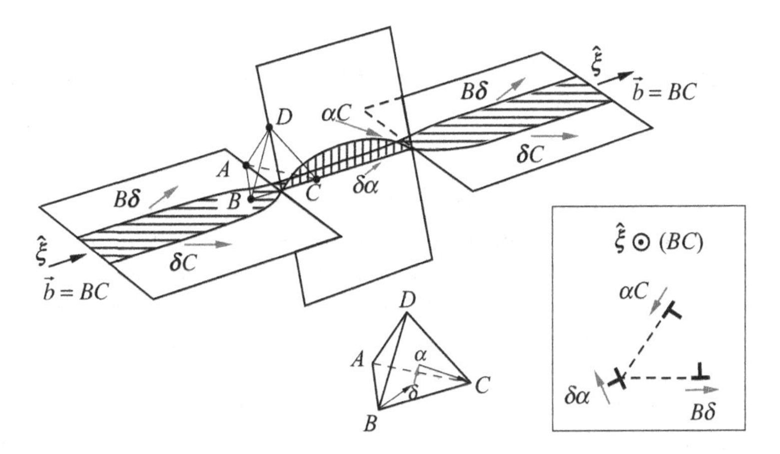

Figure 11.19. The Fleischer mechanism of cross-slip. The inset shows the direction of Burgers vectors and extra half-planes of the non-planar segment of the screw dislocation during cross-slip when viewed along BC.

αC becomes longer and bows out on the cross-slip plane, the partial $B\delta$ moves closer to the stair-rod dislocation $\delta\alpha$, and eventually combines with it to form the other partial dislocation on the cross-slip plane: $B\delta + \delta\alpha = B\alpha$. Because the Fleischer mechanism requires the formation of a high energy configuration (two Shockley partial and one stair-rod dislocations), it is expected to occur only under very high stresses.

11.1.11 Lomer–Cottrell dislocation

The Lomer–Cottrell dislocation is an edge dislocation with $\frac{a}{2}\langle 1\,1\,0\rangle$ Burgers vector on a $\{1\,0\,0\}$ glide plane, and a non-planar core structure. The Lomer–Cottrell dislocation can be formed in an FCC metal in several ways.

Lomer–Cottrell reaction

One way to form the Lomer–Cottrell dislocation is through the reaction of two 60° dislocations, as shown in Fig. 11.20a. The sense vectors of both dislocations are chosen as AB. The Burgers

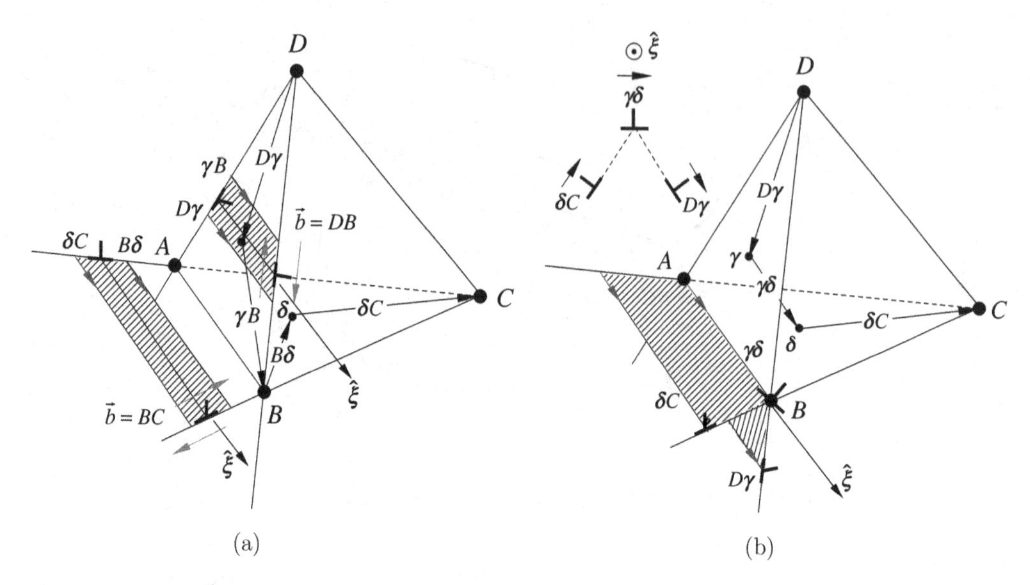

Figure 11.20. Lomer–Cottrell reaction. (a) Two 60° dislocations before the reaction. (b) The non-planar core structure of the Lomer–Cottrell dislocation after the reaction.

vector of the perfect dislocation on plane ABC is BC. From the $\hat{n} \times \hat{\xi}$ rule, the partial on the right has Burgers vector $B\delta$, and the partial on the left has Burgers vector δC. The Burgers vector of the perfect dislocation on plane ADB is DB. From the $\hat{n} \times \hat{\xi}$ rule, the partial below has Burgers vector $D\gamma$, and the partial above has Burgers vector γB.

When the two perfect dislocations react, they combine to form a new perfect dislocation with Burgers vector $DB + BC = DC$. Notice that the new Burgers vector (DC) has the same length as the two Burgers vectors before the reaction. Recall that the line energy of the dislocation is proportional to the square of the Burgers vector. Hence the reaction of two dislocations reduces the dislocation energy (by a factor of 2), so that the reaction is energetically favorable. The resulting perfect dislocation from this reaction is called a Lomer–Cottrell dislocation if it is dissociated in a non-planar manner, as shown in Fig. 11.20b. If the dislocation is not dissociated, it is called a *Lomer* dislocation instead.

The glide plane of the Lomer–Cottrell dislocation is not a $\{1\,1\,1\}$ plane because it must simultaneously contain the line direction AB and the Burgers vector DC. In the Miller index notation, $AB = \frac{a}{2}[1\,1\,0]$ and $DC = \frac{a}{2}[\bar{1}\,1\,0]$, so that the glide plane normal is $\hat{n} = [0\,0\,1]$, see Fig. 11.10. Dislocations are less mobile on $\{0\,0\,1\}$ planes than on $\{1\,1\,1\}$ planes.

The non-planar core structure of the Lomer–Cottrell dislocation, as shown in Fig. 11.20b, further reduces its mobility. As a result, the Lomer–Cottrell dislocation is often called a *Lomer–Cottrell lock*. When the two 60° perfect dislocations react, the leading partial $B\delta$ of one 60° dislocation and the trailing partial γB of the other react to form the stair-rod $\gamma\delta$ at the center of the Lomer–Cottrell dislocation. The inset of Fig. 11.20b shows the direction of the Burgers vector and extra half-plane of the two partials and the stair-rod when viewed along the sense vector $\hat{\xi} = AB$. The structure has some similarity with the transient three-dimensional core structure of the screw dislocation undergoing the Fleischer mechanism of cross-slip, shown in Fig. 11.19, but there are several substantial differences. Here in Fig. 11.20b the two Shockley

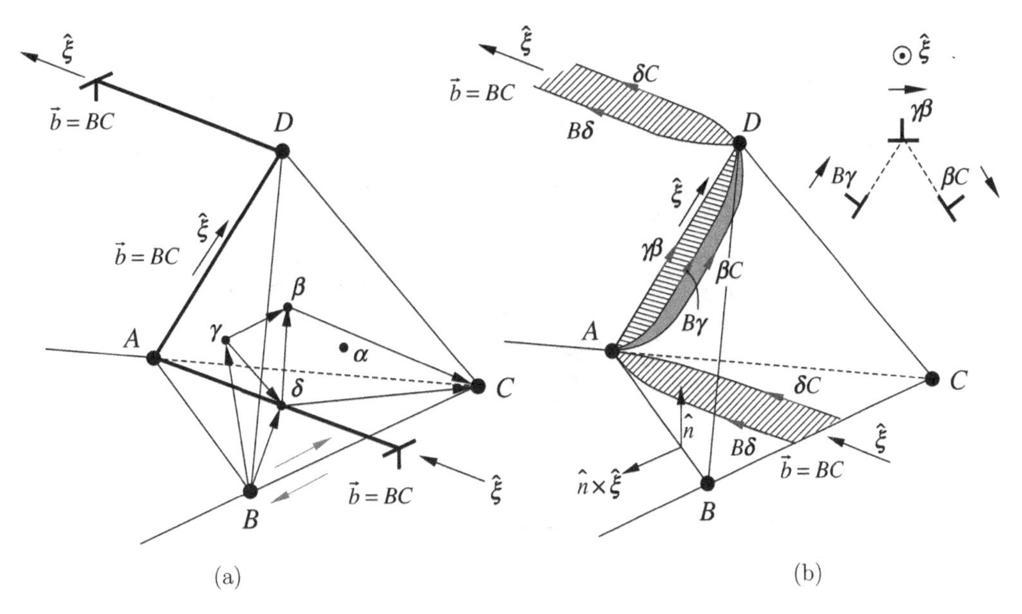

Figure 11.21. Lomer–Cottrell jog. (a) Non-dissociated jog as a segment of Lomer dislocation. (b) Dissociated jog as a segment of Lomer–Cottrell dislocation. The node A is not fully relaxed (see Fig. 11.22).

partials and the stair-rod of the Lomer–Cottrell dislocation are all pure edge, and the total Burgers vector is also pure edge. However, in Fig. 11.19, the two Shockley partials are of the mixed type, only the stair-rod is pure edge, and the total Burgers vector is pure screw. Furthermore, the Lomer–Cottrell dislocation shown in Fig. 11.20b is a stable structure, while the non-planar core of the screw dislocation in Fig. 11.19 is an unstable structure (i.e. an activated state) that the dislocation goes through during cross-slip.

Lomer–Cottrell jog

Lomer–Cottrell dislocations have the same $\frac{a}{2}\langle 110\rangle$-type Burgers vectors as ordinary glide dislocations in FCC metals. Therefore, a Lomer–Cottrell dislocation can be formed from a single glide dislocation when its line is re-oriented along the appropriate $\langle 1\,1\,0\rangle$ orientation (out of its original glide plane).

Figure 11.21a shows a perfect edge dislocation with a jog. The sense vector for the segment on the ABC slip plane is chosen to be along δA, and the Burgers vector is BC. The jog segment has line sense vector AD and Burgers vector BC. In the Miller index notation, $AD = \frac{a}{2}[1\,0\,1]$ and $BC = \frac{a}{2}[\bar{1}\,0\,1]$, so that the glide plane is $\hat{n} = [0\,1\,0]$.

Figure 11.21b shows how the jogged edge dislocation will dissociate. On the ABC plane, the direction of $\hat{n} \times \hat{\xi}$ indicates that the partial closer to the corner B has Burgers vector $B\delta$ and the partial closer to the corner C has Burgers vector δC. The Lomer–Cottrell dislocation dissociates into two Shockley partials and a stair-rod: $BC = B\gamma + \gamma\beta + \beta C$. The direction of these partial Burgers vectors are shown in Fig. 11.21a.

The Lomer–Cottrell dislocation shown in Fig. 11.21b is of the opposite sign than that shown in Fig. 11.20b, in the sense that in Fig. 11.21b the extra half-plane of the stair-rod points away from Thomposon's tetrahedron while in Fig. 11.20b it points towards Thompson's tetrahedron.

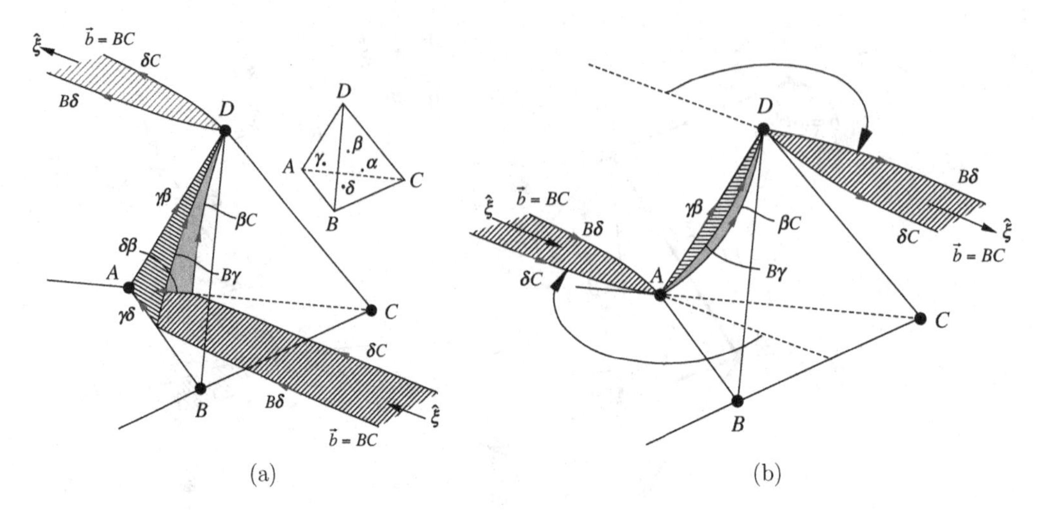

Figure 11.22. (a) Reaction between the dissociated Lomer–Cottrell jog with partials on the glide plane to form an extended node. (b) Rotation of the partial on the bottom plane makes the node constricted.

In fact, the core of the Lomer–Cottrell dislocation shown in Fig. 11.20b spreads on the faces of a different type of tetrahedron than Thompson's tetrahedron. (See Section 11.1.5 for a discussion of the two types of tetrahedral sites in an FCC crystal.)

When the dislocation segment on the glide plane ABC is suitably oriented (such as in the edge orientation), the partial $B\delta$ prefers to react with one of the partials γB (with line sense vector reversed) of the Lomer–Cottrell dislocation: $B\delta + \gamma B = \gamma\delta$, as shown in Fig. 11.22a. So the reaction product is another stair-rod dislocation. This reaction is energetically favorable because the square of $\gamma\delta$ is smaller than the sum of the squares of $B\delta$ and γB. Similarly, another stair-rod is formed by the reaction: $\delta C + C\beta = \delta\beta$. Hence the turning point from the glide dislocation to the Lomer–Cottrell dislocation develops into an extended node, with three stair-rod dislocations converging at point A.[1]

Figure 11.22a shows that the other end of the Lomer–Cottrell jog at point D is constricted. This is analogous to the nodal structure of the planar dislocation network in Fig. 11.13, where one end of the dislocation is extended and the other end is constricted.

Owing to the non-planar core structure of the Lomer–Cottrell jog, it is often assumed to be immobile and hence provide a pinning point to the mobile dislocation lines on the glide plane ABC. This can lead to spiral sources shown in Fig. 8.29. However, as the glide segment around node A rotates by 180°, the reaction between partials on the Lomer–Cottrell dislocation and the partials on the glide dislocation is no longer favorable, and the node at A becomes constricted [94]. In this case, both ends of the Lomer–Cottrell dislocation become constricted.

Because of the non-planar structure, an infinitely long Lomer–Cottrell dislocation (i.e. without ends) requires very high stress to move (on the order of several GPa [95]). However, every Lomer–Cottrell dislocation in a real crystal must have two end nodes, and one or both of these nodes is constricted. The Lomer–Cottrell dislocation can start moving from a constricted node,

1 This structure is equivalent to a corner of the so-called *stacking fault tetrahedron* structure, in which all six edges of the tetrahedron are occupied by stair-rod dislocations, and all four faces of the tetrahedron are covered by intrinsic stacking faults (see Section 11.1.13).

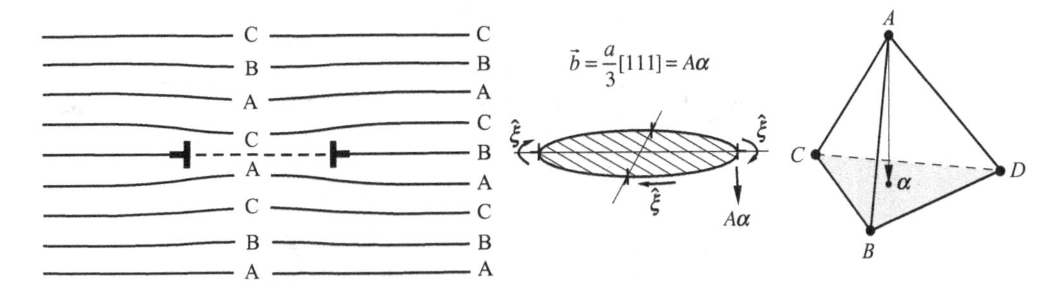

Figure 11.23. Frank partials – edge loop containing intrinsic stacking fault.

and gradually rotate itself away from the special ⟨1 1 0⟩ orientation. Once this happens, the dislocation is no longer a Lomer–Cottrell dislocation, but a dislocation with a compact core on the {1 0 0} glide plane. The stress required to move such a dislocation is still higher than that required to move dislocations on {1 1 1} planes, but not nearly as high as that required to move a Lomer–Cottrell without ends [95]. Therefore, the Lomer–Cottrell jog does not act as a permanent (or indestructable) source of dislocations, even if no other dislocations are allowed to come and react with the Lomer–Cottrell segment.

11.1.12 Frank partial dislocation

In Chapter 7, we have seen that, when metals are quenched from high temperatures, the crystal is supersaturated with excess vacancies. There is not enough time for the vacancies to diffuse out of the crystal or annihilate by going to vacancy sinks. Instead, they form multiple-vacancy clusters. When a vacancy disc on a {1 1 1} plane reaches a critical size, the two opposing surfaces of the vacant volume collapse toward each other, creating an edge dislocation loop.

Figure 11.23 shows that, within the area enclosed by the dislocation loop, the stacking sequence is changed from ABCABCAB to ABCA|CAB, where | indicates the plane of the dislocation loop. Hence the area enclosed by the dislocation loop is an intrinsic stacking fault area.

Since Thompson's tetrahedron is made by four nearest neighboring lattice positions in an FCC lattice, the Burgers vector of the dislocation loop formed by a vacancy disc is simply the height of the tetrahedron. For the sense vector chosen in Fig. 11.23, the Burgers vector is $A\alpha = \frac{a}{3}[1\,1\,1]$. This Burgers vector is shorter than the perfect Burgers vector, and the corresponding dislocation loop is called the *Frank partial* dislocation loop. In summary, the Frank partial loop is a pure edge dislocation loop with an intrinsic stacking fault inside. Unlike the Shockley partial, the Burgers vector of the Frank partial is perpendicular to the stacking fault plane. The critical size at which the vacancy disc collapses into a Frank partial loop depends on the intrinsic stacking fault energy.

The Frank partial dislocation loop expands by climb as more vacancies join in. The energy of the structure contains not only the strain energy of a dislocation loop, but also the stacking fault energy times the area of the loop. In materials with a high stacking fault energy, it is energetically favorable to convert the Frank partial loop to a perfect dislocation loop, thus removing the stacking fault area and the associated energy term that is proportional to the square of the loop radius. This process is sometimes called "unfaulting."

As shown in Fig. 11.24, the unfaulting process occurs by nucleating a Shockley partial dislocation loop within the stacking fault area. The Burgers vector of the nucleated Shockley

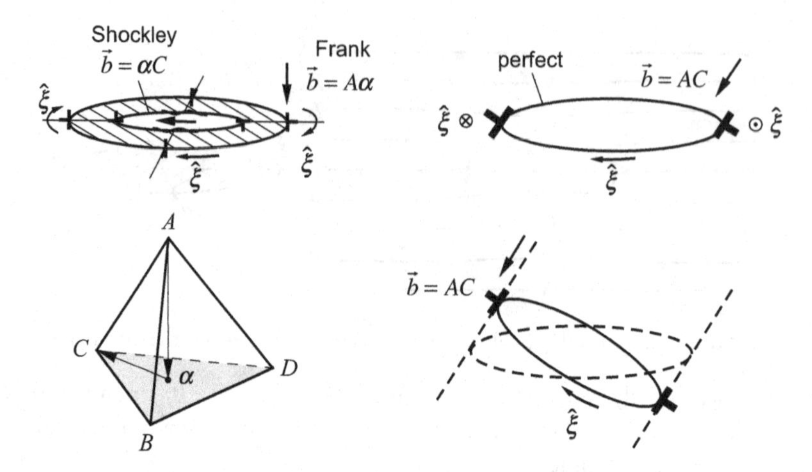

Figure 11.24. Formation of a Shockley partial loop inside the Frank partial loop eventually converts the loop into a perfect dislocation loop and removes the stacking fault area.

partial dislocation can be either αB, αC, or αD. The case of αC is discussed below and shown in Fig. 11.24. The shear corresponding to the Shockley partial loop changes the stacking type of all atomic planes above the loop and restores the stacking sequence to ABCA|BCA, i.e. the stacking sequence of a perfect FCC lattice. When the Shockley partial loop combines with the Frank partial loop, the result is a dislocation with Burgers vector $\alpha C + A\alpha = AC$. This is a perfect dislocation loop, and the stacking fault area within the loop is completely removed. The loop is now free to glide on the cylindrical glide surface that contains the Burgers vector AC and the local dislocation line. When no external stress is present, the loop prefers to rotate itself to be perpendicular to the Burgers vector, thus minimizing its total length. The result is a perfect edge dislocation loop, as shown in Fig. 11.24. If a different Shockley partial (e.g. αB or αD) were nucleated within the stacking fault, the "unfaulted" loop would have a different Burgers vector (e.g. *AB* or *AD*) and final orientation.

11.1.13 Stacking fault tetrahedron

Instead of unfaulting, a Frank partial loop can also evolve into a *stacking fault tetrahedron* (SFT). An SFT is a tetrahedron with all four faces occupied by stacking faults and all six edges occupied by stair-rod partial dislocations. SFTs have been observed in metals of low stacking fault energy (such as gold and silver) after vacancy condensation by quenching [96, 97]. Figures 11.25a–f show how a Frank partial loop can evolve into an SFT.

First, a segment of the Frank partial locally parallel to the *BC* orientation dissociates into a Shockley partial and a stair-rod dislocation, according to reaction $A\alpha = A\delta + \delta\alpha$, as shown in Fig. 11.25b. The Shockley partial $A\delta$ moves on the *ABC* plane, and elongates the stair-rod dislocation $\delta\alpha$ along the *BC* direction. Similarly, a segment of the Frank partial locally parallel to *DB* dissociates according to reaction $A\alpha = A\gamma + \gamma\alpha$, and a segment locally parallel to *CD* dissociates according to reaction $A\alpha = A\beta + \beta\alpha$.

As the Shockley partials move on their {1 1 1} glide planes, the stair-rod dislocations eventually become sufficiently long and meet at points *B*, *C*, and *D*, thus forming the triangular base

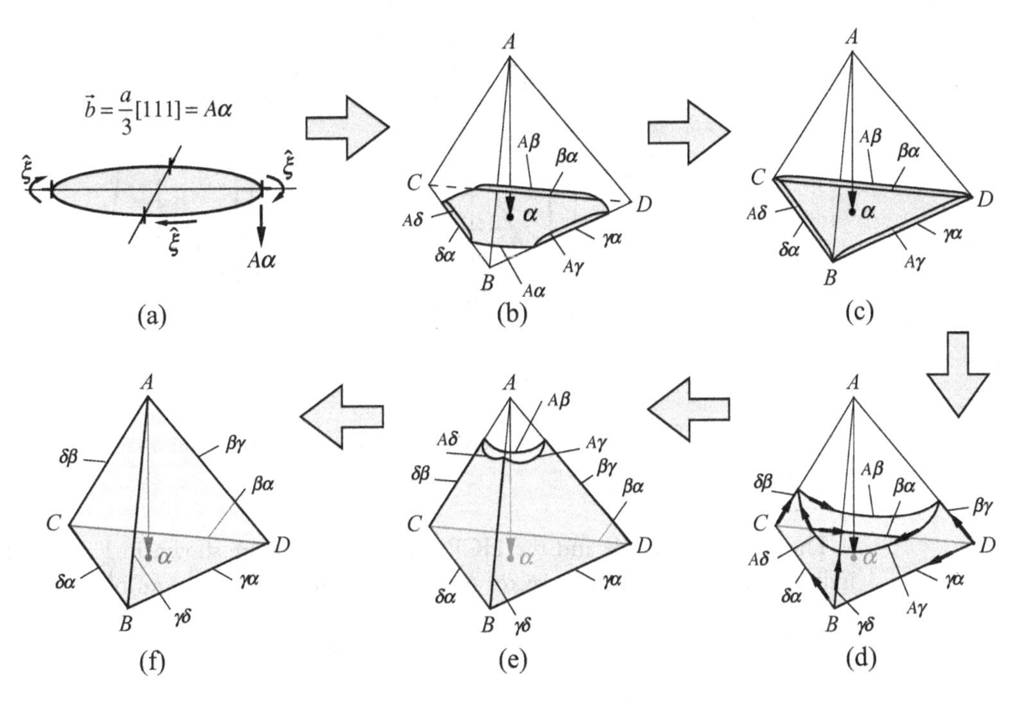

Figure 11.25. Formation of a stacking fault tetrahedron from a Frank partial loop. The line sense vectors for the partial dislocations are indicated by arrows in (d).

of the SFT, as shown in Fig. 11.25c. In Fig. 11.25d, the Shockley partials $A\delta$ and $A\gamma$ meet along the intersection line BA to produce another stair-rod dislocation, according to reaction $\gamma A + A\delta = \gamma\delta$. Similarly, a stair-rod dislocation forms along CA by reaction $\delta A + A\beta = \delta\beta$; another stair-rod dislocation forms along DA by reaction $\beta A + A\gamma = \beta\gamma$. These reactions are energetically favorable so that the stair-rod dislocations become progressively longer at the expense of the Shockley partials. Eventually, the three stair-rod dislocations meet at the apex A and the Shockley partials completely disappear, resulting in an SFT structure.

11.2 Dislocations in HCP metals

The HCP structure is closely related to the FCC structure. In particular, the dislocations on the basal plane of an HCP crystal are very similar to those on the $\{1\,1\,1\}$ planes in FCC crystal. However, there is only one basal plane in an HCP crystal, in contrast to four different $\{1\,1\,1\}$ planes in FCC. Consequently, slip on planes other than the basal plane is needed to accommodate an arbitrary state of plastic strain.

11.2.1 Miller–Bravais indices in hexagonal lattice

Before we discuss the preferred slip systems in HCP crystals, we need to introduce the Miller–Bravais indices notation in the hexagonal lattice. It is different from Miller indices in that four (instead of three) numbers are used to specify a crystallographic direction or plane.

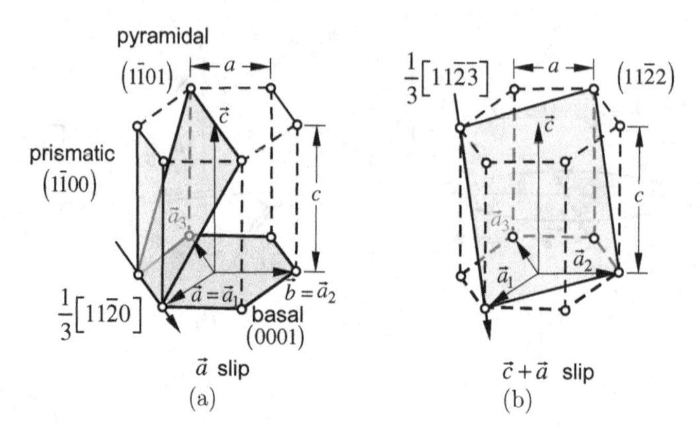

Figure 11.26. Slip systems in an HCP crystal. (a) Three slip planes for the $\vec{a}$ slip direction. (b) A slip plane for the $\vec{c}+\vec{a}$ slip direction.

The hexagonal lattice and the HCP crystal structure are shown in Fig. 1.6. In the Miller indices notation, an arbitrary vector $\vec{R}$ can be represented by three numbers, i, j, k, where

$$\vec{R} = i\,\vec{a} + j\,\vec{b} + k\,\vec{c} \tag{11.18}$$

and $\vec{a}$, $\vec{b}$, and $\vec{c}$ are lattice repeat vectors. In particular, the plane spanned by $\vec{a}$ and $\vec{b}$ is called the *basal plane*. The angle between $\vec{a}$ and $\vec{b}$ is 120° and $\vec{c}$ is perpendicular to both $\vec{a}$ and $\vec{b}$.

A drawback for using the Miller indices notation for hexagonal lattices is that vectors related by symmetry do not have the same type of indices. For example, $\vec{a} = [1\,0\,0]$, $\vec{b} = [0\,1\,0]$ and $-\vec{a} - \vec{b} = [\bar{1}\,\bar{1}\,0]$. However, the three vectors $\vec{a}$, $\vec{b}$ and $-\vec{a} - \vec{b}$ all have the same length and are related to each other by a 120° rotation within the basal plane. This symmetry is not apparent in the Miller indices notation.

To overcome this deficiency, we introduce three lattice repeat vectors in the basal plane, $\vec{a}_1 = \vec{a}$, $\vec{a}_2 = \vec{b}$, and $\vec{a}_3 = -\vec{a}_1 - \vec{a}_2$, as shown in Fig. 11.26. An arbitrary vector $\vec{R}$ can now be represented as a linear combination of four vectors,

$$\vec{R} = h\,\vec{a}_1 + k\,\vec{a}_2 + i\,\vec{a}_3 + l\,\vec{c} \tag{11.19}$$

or, $\vec{R} = [h\,k\,i\,l]$, which is called the Miller–Bravais indices notation.

Because $\vec{a}_1 + \vec{a}_2 + \vec{a}_3 = 0$, there are multiple indices that correspond to the same vector $\vec{R}$. In particular, $[h\,k\,i\,l]$ and $[h+m\;k+m\;i+m\;l]$ correspond to the same vector for arbitrary m. To remove this redundancy, the convention is to choose the indices such that $h+k+i=0$. Using this convention, the three shortest lattice vectors on the basal plane can be represented as $\vec{a}_1 = \frac{1}{3}[2\,\bar{1}\,\bar{1}\,0]$, $\vec{a}_2 = \frac{1}{3}[\bar{1}\,2\,\bar{1}\,0]$, and $\vec{a}_3 = \frac{1}{3}[\bar{1}\,\bar{1}\,2\,0]$. The symmetries between these three vectors are now apparent in their Miller–Bravais indices.

Similar to the Miller indices, a group of directions related to each other by (rotation or reflection) symmetry are represented using angular brackets, i.e. $\langle h\,k\,i\,l\rangle$. A plane with normal vector along $[h\,k\,i\,l]$ is represented as $(h\,k\,i\,l)$. A group of planes related to each other by symmetry are represented using braces $\{h\,k\,i\,l\}$.

Figure 11.26 shows some important slip directions and slip planes in HCP crystals. The $\vec{a}$ slip corresponds to the slip directions of the type $\frac{1}{3}[1\,1\,\bar{2}\,0]$ and can occur on the basal, *pyramidal*,

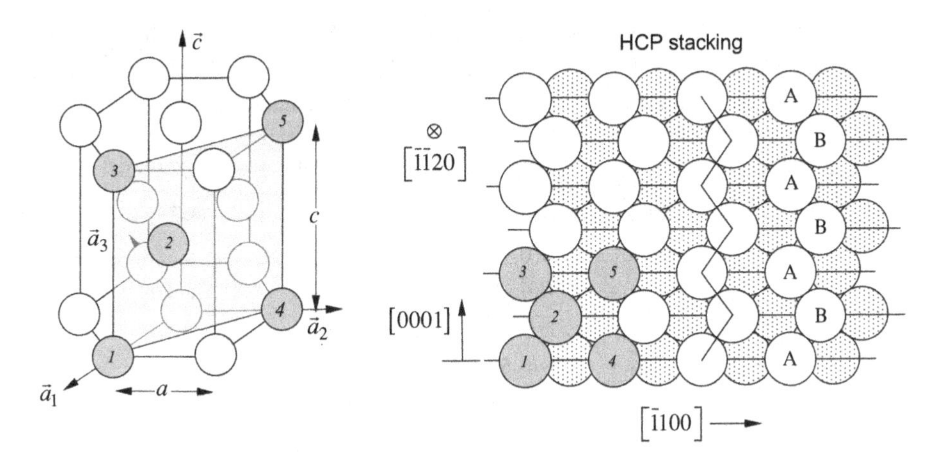

Figure 11.27. (a) The HCP crystal structure. (b) The stacking of atomic layers parallel to the basal plane.

or *prismatic* planes. The $\vec{c} + \vec{a}$ slip corresponds to the slip directions of the type $\frac{1}{3}[1\ 1\ \bar{2}\ 3]$ and can also occur on several planes, such as $(1\ 1\ \bar{2}\ 2)$.

11.2.2 HCP stacking

The ideal HCP structure is closely related to the FCC structure. In Section 11.1.3, we have seen that the FCC structure can be considered as atomic layers, each forming a triangular lattice parallel to the $\{1\ 1\ 1\}$ plane, stacked on top of each other in the sequence of ABCABC.... The ideal HCP structure is obtained if the atomic layers are stacked on top of each other in the sequence of ABABAB..., as shown in Fig. 11.27. Consequently, the c/a ratio of the ideal HCP structure is $(2/\sqrt{3})/(1/\sqrt{2}) = \sqrt{8/3} = 1.633$, where $c = |\vec{c}|$ and $a = |\vec{a}|$. The c/a ratio of real HCP metals can be either larger or smaller than the ideal value.

Because the $\{1\ 1\ 1\}$ planes are the most widely spaced atomic planes and hence the most active slip planes in an FCC structure, we expect the basal plane to be the most active slip plane in an HCP structure, if its c/a ratio is close to or greater than the ideal value. Table 11.3 shows that the critical resolved shear stress (CRSS) values on the basal plane is indeed very small for Cd, Zn, and Mg, whose c/a ratio is either close to or greater than the ideal value of 1.633. The CRSS values on the basal planes become much higher in Zr, Ti, and Be, whose c/a ratio is much smaller. Table 11.3 shows that the CRSS values are generally the smallest on the basal plane for most HCP metals. This suggests that basal slip should be the most active in HCP metals. However, as noted earlier, non-basal slip must also be present in order to accommodate a general plastic strain.

11.2.3 Stacking faults in HCP metals

An HCP crystal can contain stacking faults when its perfect stacking sequence, ABABAB..., is altered. Similar to an FCC crystal, both intrinsic and extrinsic stacking faults can exist in an HCP crystal. However, there are two types of intrinsic stacking faults in HCP, I_1 and I_2.

Figure 11.28a shows an intrinsic stacking fault I_2. To create this fault from a perfect HCP crystal, consider a plane parallel to the basal plane between two adjacent atomic layers B and A, dividing the HCP crystal into two halves. Define the positive side of $[0\ 0\ 0\ 1]$ as the upward

Table 11.3. Critical resolved shear stress (CRSS) on different slip systems for various HCP metals (in MPa). Basal = $\langle 1\,1\,\bar{2}\,0\rangle\{0\,0\,0\,1\}$. Prismatic = $\langle 1\,1\,\bar{2}\,0\rangle\{1\,\bar{1}\,0\,0\}$. Pyramidal = $\langle 1\,1\,\bar{2}\,0\rangle\{1\,\bar{1}\,0\,1\}$. $c+a = \langle 1\,1\,\bar{2}\,\bar{3}\rangle\{1\,1\,\bar{2}\,2\}$. Calculated data are indicated with *.

Metal	c/a	Basal	Prismatic	Pyramidal	$c+a$	Ref.
Cd	1.886	<0.1	—	≈7	≈7	[98]
Zn	1.856	<0.1	—	—	≈5	[98]
		0.4	6–15	4–10	—	[99]
		—	—	—	≈2.8	[100]
Mg	1.624	≈0.5	≈40	—	≈40	[98]
		0.52–0.81	39–50	—	—	[101]
Zr	1.589	3.5*	21*	14*	—	[102]
Ti	1.587	≈80	≈20	—	—	[98]
		3.1*	12*	4.2*	—	[102]
Be	1.568	≈5	—	—	—	[98]
		6	50	—	—	[103]

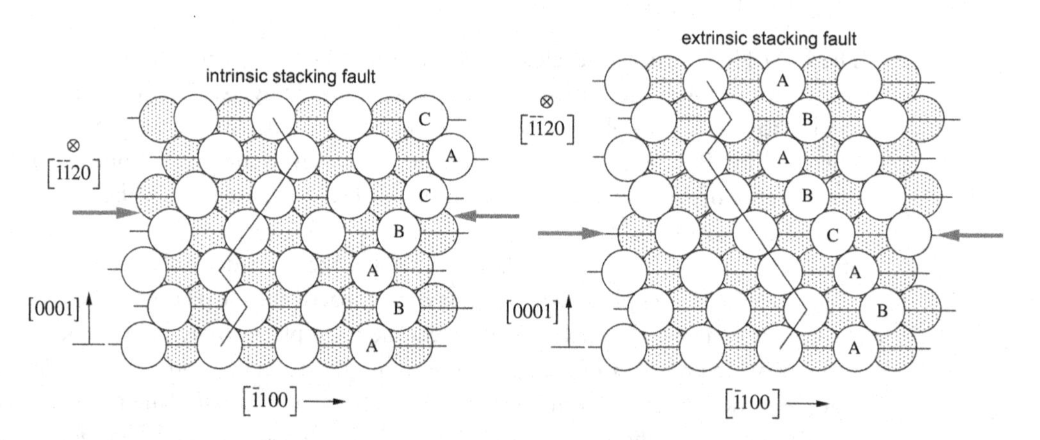

Figure 11.28. Stacking faults in HCP crystal. (a) Intrinsic stacking fault I_2. (b) Extrinsic stacking fault.

direction. Here the layer A is on top of the layer B. Let the upper half of the crystal slide relative to the lower half, such that the atomic layer A moves to the position C. The new stacking sequence after the sliding is ABAB|CACA, where | indicates the plane of sliding, as indicated by the two arrows in Fig. 11.28a. The result is an intrinsic stacking fault I_2. Notice that near an intrinsic stacking fault the local stacking sequence corresponds to that of an FCC crystal, i.e. AB|CA. In fact, if an intrinsic stacking fault I_2 is introduced between every other available sliding plane between adjacent atomic layers, the perfect FCC stacking is created. In other words, the HCP and FCC crystal structures can be converted to each other by sliding along the basal (or $\{1\,1\,1\}$) planes.

If we remove the layer B immediately below an intrinsic stacking fault I_2, the result is an intrinsic stacking fault I_1, with a stacking sequence of ABA|CACA. An extrinsic stacking fault is created if an extra layer of atoms is inserted between layers B and A, leading to a stacking sequence of ABACBABA, as shown in Fig. 11.28b. The extrinsic stacking fault can also be

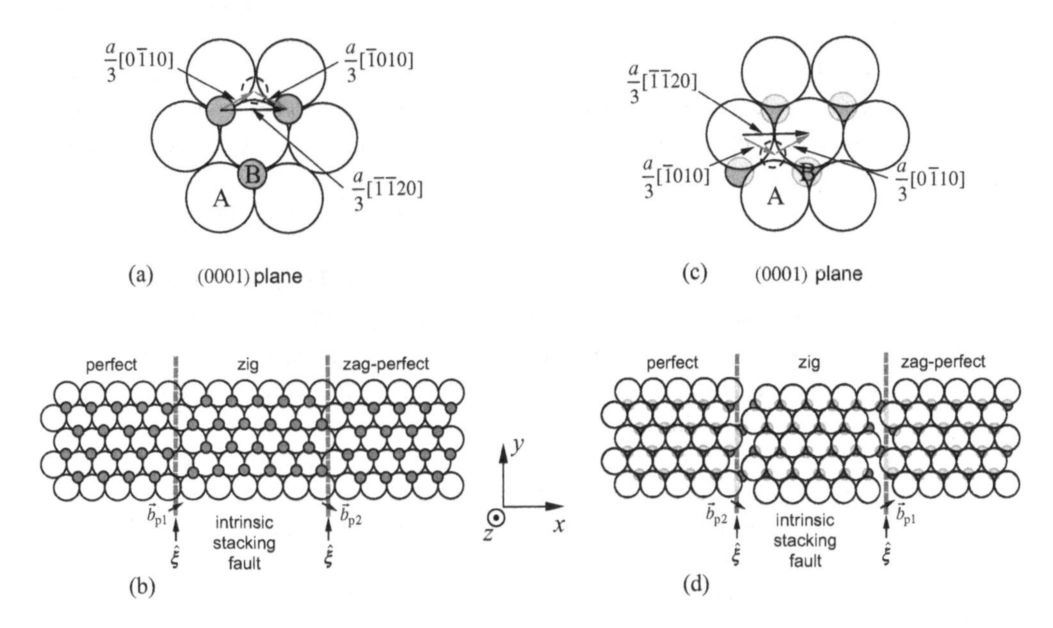

Figure 11.29. Dissociation of perfect dislocations into Shockley partials in an HCP crystal. Two layers of atoms immediately above and below the glide plane are shown. (a) Sliding of layer B above layer A. (b) Dissociation of a perfect dislocation whose glide plane is between the two atomic layers shown in (a). (c) Sliding of layer A above layer B. (d) Dissociation of a perfect dislocation whose glide plane is between the two atomic layers shown in (c).

created by shearing on the plane immediately above an intrinsic stacking fault I_1. Starting from the stacking sequence for I_1, ABA|CACA, if we slide the layer A above the lowest layer C to position B, and move all the layers above it by the same amount, the resulting stacking sequence is ABA|C|BAB, which is identical to that of the extrinsic stacking fault. Hence an extrinsic stacking fault in HCP can be regarded as two intrinsic stacking faults, I_1 and I_2, on top of each other.

11.2.4 Partial dislocations in HCP metals

We are now ready to consider the dissociation of perfect dislocations on the basal plane in HCP crystals. Similar to FCC crystals, a perfect dislocation on the basal plane also dissociates into two Shockley partials bounding an intrinsic stacking fault area. However, the ordering of the two partials depends on the location of the dislocation glide plane in the stacking sequence.

Define the positive $[0\,0\,0\,1]$ direction as the upward direction. Figure 11.29 shows two layers of atoms immediately above and below the glide plane. In Fig. 11.29a, layer B is above layer A. Imagine the process in which layer B needs to slide relative to the underlying layer A by a lattice vector $\frac{a}{3}[\bar{1}\,\bar{1}\,2\,0]$. Figure 11.29b shows more atoms on the two atomic planes that slide against each other. Consider atoms as hard spheres; the trajectory of the atoms on top of the slip plane is unlikely to follow a straight line along $\frac{a}{3}[\bar{1}\,\bar{1}\,2\,0]$. Instead it is more likely to follow a straight line along $\frac{a}{3}[0\,\bar{1}\,1\,0]$ and then take another step of $\frac{a}{3}[\bar{1}\,0\,1\,0]$, as indicated by Fig. 11.29a.

Now imagine an edge dislocation with Burgers vector $\frac{a}{3}[\bar{1}\,\bar{1}\,2\,0]$. To the left of the dislocation, assume that slip has not occured. To the right of the dislocation, assume that atoms on top of the slip plane have slipped by $\frac{a}{3}[\bar{1}\,\bar{1}\,2\,0]$ relative to the atoms below, corresponding to the right configuration in Fig. 11.29b. Similar to Section 11.1.2, we expect that there is an intermediate zone between these two areas. This is the middle ("zig") configuration shown in Fig. 11.29b corresponding to the intrinsic stacking fault I_2. The two boundary lines separating these three areas correspond to two Shockley partial dislocations, whose Burgers vectors are $\vec{b}_{p1} = \frac{a}{3}[0\,\bar{1}\,1\,0]$ and $\vec{b}_{p2} = \frac{a}{3}[\bar{1}\,0\,1\,0]$.

It is possible that a perfect dislocation exists on a glide plane next to which layer A is above layer B, as shown in Fig. 11.29c. In this case, we need to consider sliding layer A relative to the underlying layer B by a lattice vector $\frac{a}{3}[\bar{1}\,\bar{1}\,2\,0]$. The atoms are likely to first take a step of $\frac{a}{3}[\bar{1}\,0\,1\,0]$ and then take another step of $\frac{a}{3}[0\,\bar{1}\,1\,0]$. Figure 11.29d shows that an edge dislocation of Burgers vector $\frac{a}{3}[\bar{1}\,\bar{1}\,2\,0]$ also dissociates into two Shockley partials bounding an area of intrinsic stacking fault. However, the ordering of the two partials is opposite to that shown in Fig. 11.29b, for which layer B is above layer A.

Because of the difference in the ordering of the partials, the dissociation width responds to shear stresses differently for the two dislocations shown in Figs. 11.29b, d. For example, a positive shear stress σ_{yz} would expand the equilibrium dissociation width for the dislocation in Fig. 11.29b (layer B above layer A), but would reduce the equilibrium dissociation width for the dislocation in Fig. 11.29d (layer A above layer B).

In HCP metals no stacking faults exist on planes other than the basal plane. Therefore dislocations on non-basal planes of HCP metals do not dissociate into partial dislocations, but have a compact core instead. This is consistent with the observation that the critical resolved shear stress on non-basal planes is generally much higher than that on the basal plane, as shown in Table 11.3.

11.3 Partial dislocations in $CrCl_3$

We now discuss dislocations in compound crystals containing multiple elements, but with structures still similar to the close-packed structure of FCC and HCP crystals. In this section, we consider the case of the layered crystal $CrCl_3$. The $CrCl_3$ crystal has a monoclinic Bravais lattice and a large basis. Even though the lattice and basis together define a unique crystal structure, they are not very useful here to develop an intuitive understanding of the crystal structure of $CrCl_3$. Instead, we will consider $CrCl_3$ as a crystal formed by Cl^- ions, which contains Cr^{3+} ions in interstitial sites. The Cl^- ions form an FCC sub-lattice, and the Cr^{3+} ions occupy 1/3 of its octahedral sites. The pattern that the octahedral sites occupy gives the $CrCl_3$ crystal a layered structure. First, the octahedral sites in an FCC crystal are arranged as (1 1 1) layers stacked on top of each other, and only 1/2 of these layers contain Cr^{3+} ions. Within the layers that do contain Cr^{3+} ions, only 2/3 of the sites are occupied. This is how 1/3 of all the octahedral sites are occupied by Cr^{3+} ions.

Figure 11.30(a) shows the layered structure of $CrCl_3$. The Cl^- ions form (1 1 1) layers that are stacked on top of each other in the sequence of ABCABC.... The Cr^{3+} ions are intercalated in the octahedral sites between the (1 1 1) layers of Cl^- ions. The Cr^{3+} layer between layers B and C is called layer α; that between layers A and B is called layer γ; and that between layers C and A is called layer β. Because only one of every two inter-layer gaps are filled by Cr^{3+} ions, the Cr^{3+} layers are stacked in the sequence of $\alpha\gamma\beta\alpha\gamma\beta\ldots$.

Figure 11.30(b) shows the Cl^- layers B and C and the Cr^{3+} layer α in between. Three octahedral sites between Cl^- ions are shown as shaded hexagons. It can be seen that only 2/3 of

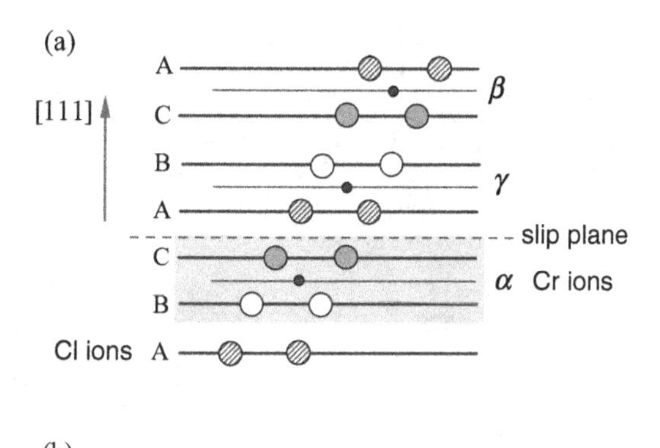

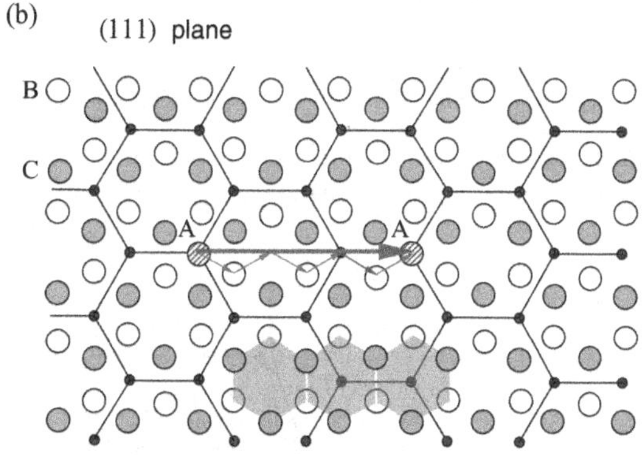

Figure 11.30. Layered crystal structure of $CrCl_3$. (a) Viewed along a direction perpendicular to [1 1 1]. The Cl^- ions are arranged into layers ABCABC.... The Cr^{3+} ions are arranged into layers $\alpha\gamma\beta\alpha\gamma\beta\ldots$. (b) Viewing along [1 1 1]. Only layers B, α, C, together with two atoms on layer A, are shown. The large arrow shows the perfect Burgers vector $\vec{b}$, which is a shortest lattice repeat vector in the (1 1 1) plane. The short arrows show the six steps a slip by $\vec{b}$ on the (1 1 1) plane is likely broken into.

the octahedral sites in this layer are occupied by Cr^{3+} ions, which form a hexagonal pattern. Because the octahedral sites are only partially filled in this layer, the shortest repeat vector $\vec{b}$ of the $CrCl_3$ crystal structure in the (1 1 1) plane is three times that of the FCC sub-lattice of the Cl^- ions. Hence the Burgers vector $\vec{b}$ of dislocations in the $CrCl_3$ crystal is very large. An example of vector $\vec{b}$ is shown in Fig. 11.30(b).

The slip planes in the $CrCl_3$ crystal are the {1 1 1} planes between Cl^- layers not occupied by Cr^{3+} ions, such as the plane between the layers A and C in Fig. 11.30a. Figure 11.30b plots two atoms in layer A that are separated by $\vec{b}$ apart. Now imagine that layer A (together with all layers above it) slides relative to layer C by $\vec{b}$. Considering atoms as hard spheres, we can see that trajectory of layer A will not follow a straight line. Instead, it will break into six steps, as shown by the six small arrows in Fig. 11.30b. The first step takes it to the position directly on top of layer B. The second step brings it back to layer A. The third step takes it to the position on top of layer B again, etc. Because of the arrangement of Cr^{3+} ions, the crystal is not restored to perfect structure until the sixth step is completed, at which the layer A has slipped relative to layer C by $\vec{b}$. Therefore, the perfect dislocation on the {1 1 1} planes of the $CrCl_3$ crystal with Burgers vector $\vec{b}$ is expected to dissociate into six partial dislocations, each connected to its neighbors by stacking faults. This is similar to the dissociation of perfect dislocations into two Shockley partial dislocations in FCC crystals, except that there are multiple types of stacking

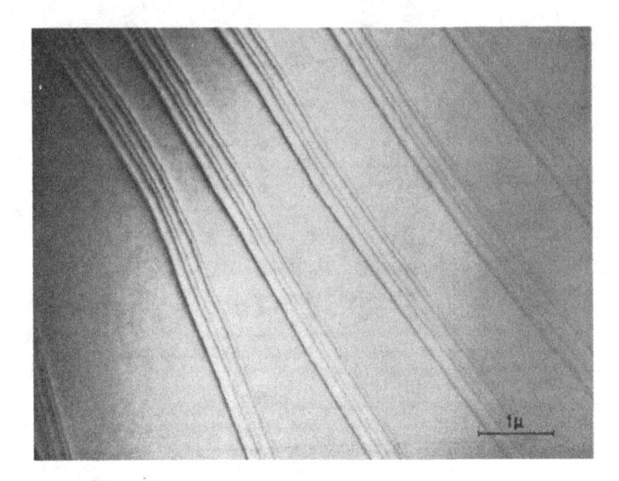

Figure 11.31. Transmission electron microscopy (TEM) image of dissociated dislocations in $CrCl_3$ [40]. Reproduced with permission from [104]. Copyright 1962, AIP Publishing LLC.

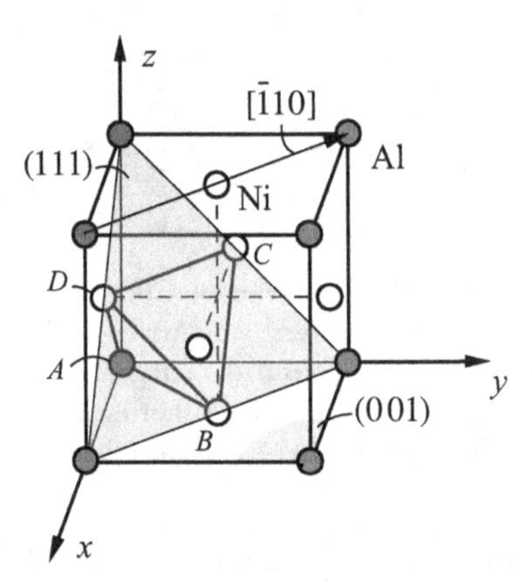

Figure 11.32. The unit cell of the $L1_2$ crystal structure of Ni_3Al. The Ni atoms (white spheres) occupy the face centers and the Al atoms (gray spheres) occupy the cube corners. Thompson's tetrahedron is also shown.

faults between the six partial dislocations in $CrCl_3$ with different stacking fault energies per unit area. A transmission electron micrograph of dissociated dislocations in $CrCl_3$ is shown Fig. 11.31. Five out of the six partials can be clearly observed.

11.4 Superdislocations in ordered Ni_3Al

Ni_3Al is an important high temperature material. The $L1_2$ crystal structure of Ni_3Al has been introduced in Section 1.1.3. The unit cell of the $L1_2$ structure is re-plotted in Fig. 11.32, where the Al atoms are represented by gray spheres at the corners and the Ni atom are represented by white spheres at the face centers. The $L1_2$ crystal structure has the simple cubic Bravais lattice and a four-atom basis (three Ni and one Al). However, to understand the slip systems in Ni_3Al, it is useful to first ignore the difference between Ni and Al atoms. If all Al atoms in Ni_3Al were replaced by Ni atoms, the result would be an FCC Ni crystal, with $\{1\ 1\ 1\}$ slip planes and

$\frac{1}{2}\langle 1\ 1\ 0\rangle$ Burgers vectors. Hence we may think of Ni_3Al as an FCC Ni crystal containing Al atoms as substitutional impurities. Of course, the Al "impurities" are not randomly distributed; instead their distribution is ordered and they form a simple cubic sub-lattice. The presence of Al does not change the fact that the $\{1\ 1\ 1\}$ planes are still the most widely spaced and smoothest atomic planes in the crystal. So the active slip planes in Ni_3Al are the $\{1\ 1\ 1\}$ planes.

From Fig. 11.32, we observe that the nearest neighbor of every Al atom is a Ni atom, while the nearest neighbor of every Ni atom is either Al or Ni. While the $\frac{1}{2}\langle 1\ 1\ 0\rangle$ vectors are the shortest translation vectors in FCC Ni, the presence of Al in Ni_3Al makes $\frac{1}{2}\langle 1\ 1\ 0\rangle$ no longer a lattice repeat vector. Instead, the shortest repeat vector on the $\{1\ 1\ 1\}$ planes are $\langle 1\ 1\ 0\rangle$ vectors, as shown in Fig. 11.32. Translating half of the crystal by $\frac{1}{2}\langle 1\ 1\ 0\rangle$ relative to the other half would place Al atoms on Ni sites, which leads to a planar defect, called an *antiphase boundary* (APB), on the slip plane. Because the perfect dislocations in Ni_3Al have Burgers vectors that have twice the length (in units of lattice constants) of those in an FCC crystal, the perfect dislocations in Ni_3Al are called superdislocations. Similar to what we have seen in Section 11.3, we expect the superdislocations in Ni_3Al to be dissociated into multiple partial dislocations connected by faults of various kinds.

11.4.1 Stacking of $\{1\ 1\ 1\}$ planes

To understand how superdislocations dissociate into partials, we need to examine the stacking pattern of the Ni_3Al crystal. In this section, we examine the stacking pattern of $\{1\ 1\ 1\}$ planes.

Figure 11.33a plots two $(1\ 1\ 1)$ atomic layers of the Ni_3Al crystal. On each $(1\ 1\ 1)$ layer, the atoms form a triangular lattice (if we ignore the difference between Ni and Al atoms). Furthermore, the Al atoms form a triangular sub-lattice that has twice the lattice spacing of the triangular lattice formed by Ni and Al atoms together. Again, it is clear from Fig. 11.33a that every Al atom is surrounded by Ni atoms as nearest neighbors, and the nearest neighbors of every Ni atom can be either Al or Ni.

The perfect Burgers vector, $[\bar{1}\ 1\ 0]$, of a superdislocation is drawn as a long arrow between two Al atoms on the top layer. Imagine that the top layer slides relative to the bottom layer by $[\bar{1}\ 1\ 0]$. Considering the atoms as hard spheres, we can see that the trajectory of the top layer is unlikely to follow a straight path. Instead, it would follow a zig-zag path indicated by the four shorter arrows in Fig. 11.33a. The first step would take the atoms in the top layer to the adjacent center of the triangle formed by the bottom layer. The result is a planar defect, called a *complex stacking fault* (CSF), shown in Fig. 11.33b. It is a stacking fault in the sense that, had we ignored the difference between Ni and Al atoms, the structure would be the same as the intrinsic stacking fault in FCC Ni. However, Fig. 11.33b shows that every Al atom in these two layers now has an Al atom as a nearest neighbor, which is a high energy configuration. Hence the complex stacking fault is a defect with both stacking disorder and chemical disorder.

After the top $(1\ 1\ 1)$ layer slides by two steps following the short arrows in Fig. 11.33a, the result is another planar defect, called an *antiphase boundary* (APB), as shown in Fig. 11.33c. Had we ignored the difference between Ni and Al atoms, the ABC stacking sequence of FCC would have been restored in this configuration. However, in Fig. 11.33c every Al atom in these two layers still has one Al atom as a nearest neighbor. Hence the APB has chemical disorder but no stacking disorder. We expect the APB energy on the $\{1\ 1\ 1\}$ plane to have the same order of magnitude as, but smaller than, the CSF energy, because CSFs have both chemical

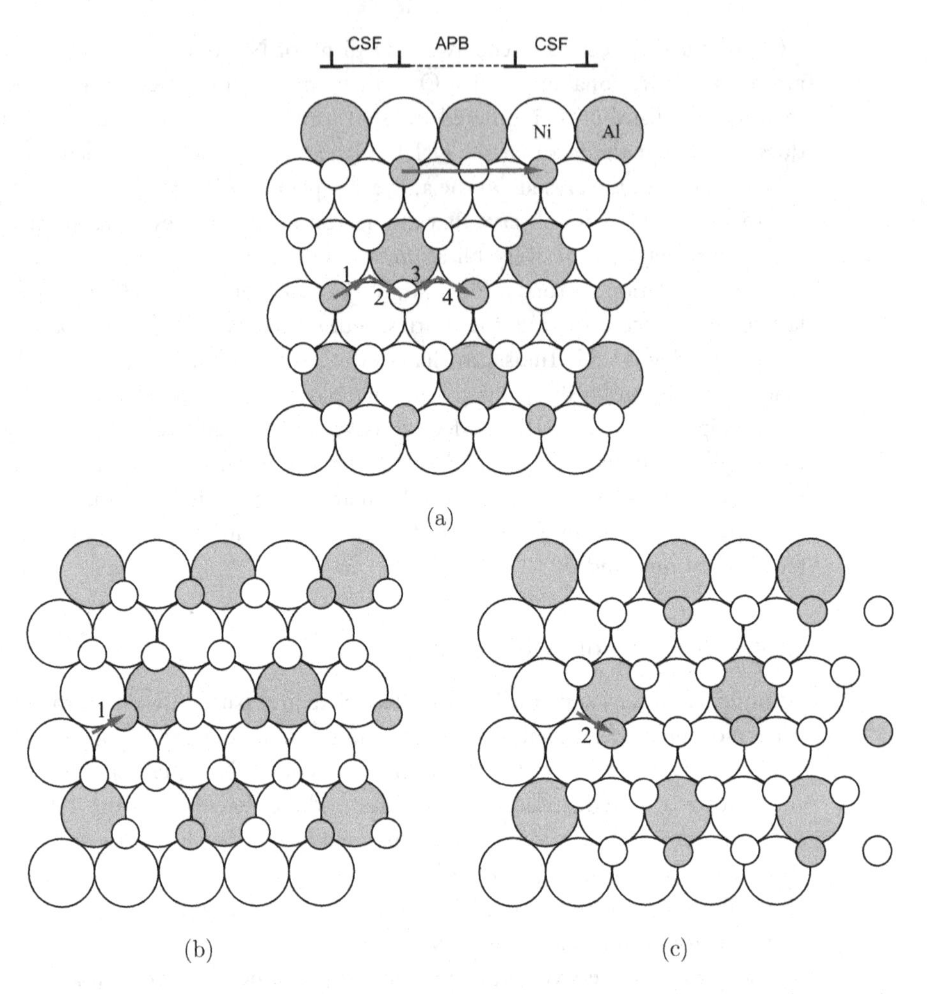

Figure 11.33. Two (1 1 1) layers of atoms of the Ni_3Al crystal. The atoms at the bottom layer are shown as larger spheres; the atoms at the top layer are shown as smaller spheres. The Ni atoms are shown as white spheres and the Al atoms are shown as gray spheres. The sliding of the top layer by $[\bar{1}\,1\,0]$ (long arrow) is broken into four steps, as indicated by the four short arrows. (b) The complex stacking fault (CSF) formed after the top layer slides by the first short vector shown in (a). (c) The antiphase boundary (APB) formed after the top layer slides by the second short vector shown in (a).

and stacking disorder. Experimental estimates for these energies in Ni_3Al, obtained from the observed separation distances between the superpartials, are: $\gamma_{APB}(1\,1\,1) = 175 \pm 15\,\mathrm{mJ}\cdot\mathrm{m}^{-2}$ and $\gamma_{CSF} = 235 \pm 40\,\mathrm{mJ}\cdot\mathrm{m}^{-2}$ [105].

After the top (1 1 1) layer slides by three steps, the result is a CSF again. The perfect crystal structure is restored only after the fourth sliding step is completed. Therefore, a superdislocation on the (1 1 1) plane with Burgers vector $[\bar{1}\,1\,0]$ dissociates into four partial dislocations, which are connected to each other by a CSF, an APB, and another CSF, as shown in Fig. 11.33a. Because both CSFs and APBs on the $\{1\,1\,1\}$ planes have relatively high energies, the separation between partial dislocations on the $\{1\,1\,1\}$ plane is not very large. The separation between partials is expected to be smaller than the APB width, consistent with $\gamma_{APB} < \gamma_{CSF}$ (see the discussion of partial separation in FCC metals in Section 11.1.4).

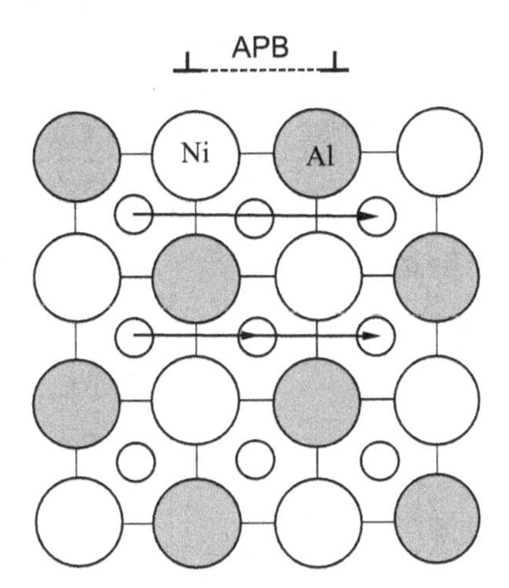

Figure 11.34. Two (0 0 1) layers of atoms of the Ni_3Al crystal. The atoms at the bottom layer are shown as larger spheres; the atoms at the top layer are shown as smaller spheres. The Ni atoms are shown as white spheres and the Al atoms are shown as gray spheres.

11.4.2 Stacking of {0 0 1} planes

The {1 1 1} plane is not the only plane on which the superdislocation can dissociate. A competing dissociation plane is the {0 0 1} plane. As we will see below, the APB on the {0 0 1} plane has a much lower energy than that on the {1 1 1} plane.

Figure 11.34 shows two (0 0 1) layers of atoms of the Ni_3Al crystal. The bottom layer (larger spheres) contains both Ni and Al atoms, while the top layer contains only Ni atoms. Imagine that the top layer slides relative to the bottom layer by $[\bar{1}\,1\,0]$, shown as the long arrow in Fig. 11.34. Considering the atoms as hard spheres, we can see that the trajectory of the top layer is likely to follow two steps along a straight path. The slip vector of each step is $[\bar{1}\,1\,0]/2$, shown as the short arrows in Fig. 11.34. After the first slip of $[\bar{1}\,1\,0]/2$, an APB is again created, but now on the (0 0 1) plane. Because the top layer consists entirely of Ni atoms, the nearest environment of every atom in the APB remains unchanged from that in the perfect crystal. The second nearest neighbor interactions make a small contribution to the APB energy. Hence the APB on the {0 0 1} plane is expected to be smaller than that on the {1 1 1} plane. Experimental estimates for the APB energies in Ni_3Al are: $\gamma_{APB}(0\,0\,1) = 104 \pm 15\,\mathrm{mJ} \cdot \mathrm{m}^{-2}$ versus $\gamma_{APB}(1\,1\,1) = 175 \pm 15\,\mathrm{mJ} \cdot \mathrm{m}^{-2}$ [105].

Note that there are no stacking faults on the {0 0 1} plane. Therefore, a superdislocation on the (0 0 1) plane with Burgers vector $[\bar{1}\,1\,0]$ dissociates into only two partial dislocations (called superpartials), which are connected to each other by an APB, as shown in Fig. 11.34. The lower energy of APB on the {0 0 1} plane makes the separation between the partial dislocations larger than that on the {1 1 1} plane.

11.4.3 Non-planar structure of Kear–Wilsdorf lock

We have now discussed two competing modes of dissociation, i.e. two competing core structures, for a superdislocation in Ni_3Al. These two competing models require the superdislocation core to spread onto two different planes. For non-screw superdislocations, the plane that

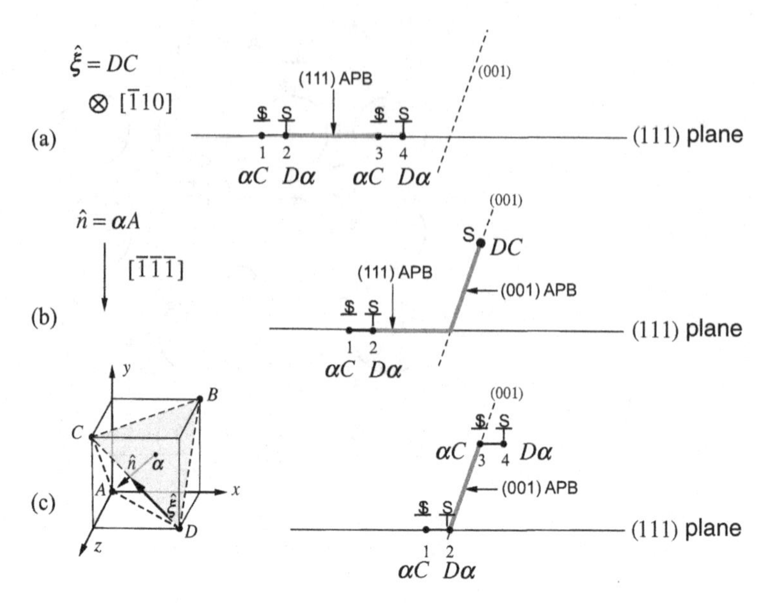

Figure 11.35. Cross-slip pinning in Ni_3Al. (a) A screw superdislocation on the (1 1 1) plane. (b) The Shockley partials 3 and 4 combine into a screw superpartial DC, which then cross-slips on the (0 0 1) plane. (c) The superpartial DC dissociates into Shockley partials on a parallel (1 1 1) plane, resulting in the non-planar structure of the Kear–Wilsdorf lock. Symbols: ⊥ indicates positive edge dislocations, T indicates negative edge dislocations, and S indicates right-handed screw dislocations. The Burgers vectors of the partial dislocations are specified in Thompson's notation: $DC = \frac{a}{2}[\bar{1}\,1\,0]$, $D\alpha = \frac{a}{6}[\bar{1}\,2\,\bar{1}]$, $\alpha C = \frac{a}{6}[\bar{2}\,1\,1]$.

simultaneously contains the line direction $\hat{\xi}$ and Burgers vector $\vec{b}$ is the glide plane, which is usually the plane of dissociation. If the superdislocation has the screw orientation, then both modes of dissociation are available for the same superdislocation. Because the superdislocation prefers the core structure with the lowest energy, it is of interest to compare the energies of these two dissociation modes. The energy of a dissociated superdislocation has contributions from the line energies of the partial dislocations, which are proportional to the square of the partial Burgers vectors, as well as the contributions from the faults between the partials.

Comparing the dissociation modes in the two different planes, the {1 1 1} dissociation has the advantage that the partial dislocations have shorter Burgers vectors. However, the {0 0 1} dissociation has the advantage that it has a much lower APB energy. It turns out that the lowest energy core structure of a superdislocation in the screw orientation is a combination of both dissociation modes. The result is a non-planar structure, called the *Kear–Wilsdorf lock* [106], shown in Fig. 11.35. The superdislocation is dissociated into four partials. The Burgers vectors of the partials are identical to those when dissociation is entirely on the (1 1 1) plane. The partials 1 and 2 are still connected by a CSF on the (1 1 1) plane; so are partials 3 and 4. However, the partials 2 and 3 are connected by an APB on the (0 0 1) plane, which has lower energy than the APB on the (1 1 1) plane. Figure 11.35 also sketches the process by which a superdislocation originally on the (1 1 1) plane can evolve into this non-planar structure by cross-slip.

The non-planar core structure is called the Kear–Wilsdorf (KW) lock because it is impossible for this dislocation to move on either the {1 1 1} or {0 0 1} plane. Because this structure has lower energy than the planar configuration, it is unlikely for the dislocation to revert to the planar (and hence glissile) configurations once the non-planar structure is formed. The dislocation

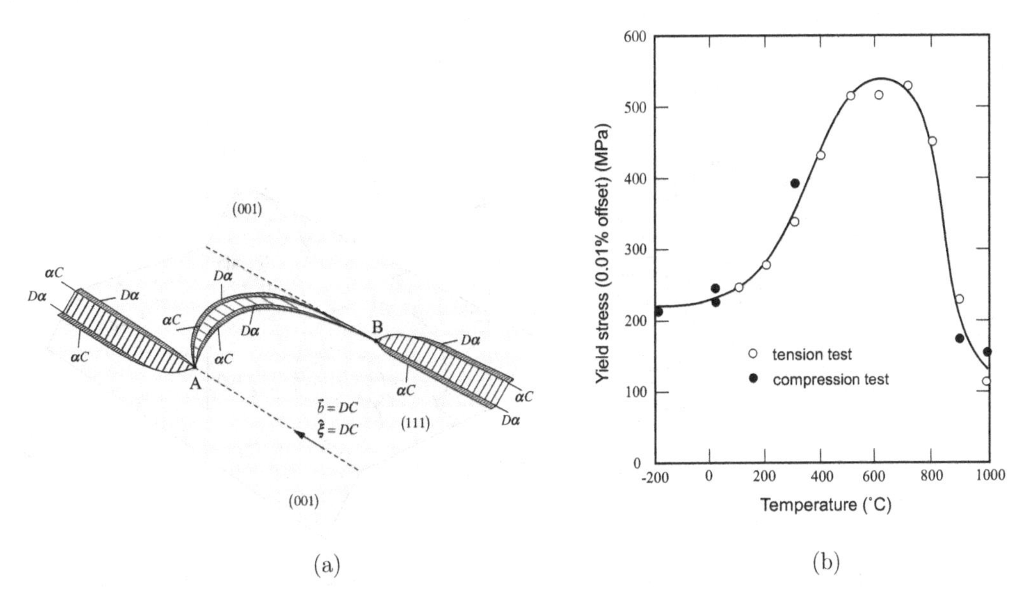

Figure 11.36. (a) Superkink in Ni_3Al connecting two non-planar KW locks. (b) Temperature dependence of the proportional limit of Ni_3Al [107, 108]. Open circles correspond to compression test, and filled circles correspond to tension test, at strain rate of $10^{-3}\ s^{-1}$.

thus stops contributing to the plastic deformation and acts as an obstacle to block the motion of other dislocations.

The transformation from the planar (glissile) core structure to the non-planar KW lock structure occurs gradually along the dislocation line through the *superkink* mechanism, in which the node between the planar and non-planar core moves along the dislocation line. As an illustration, Fig. 11.36a shows a structure in which two KW locks are connected by a planar superdislocation between nodes A and B. Here the glissile superdislocation AB is dissociated on a $\{1\ 1\ 1\}$ plane, which coincides with the upper $\{1\ 1\ 1\}$ plane of the KW lock on the left and with the lower $\{1\ 1\ 1\}$ plane of the KW lock on the right. The planar segment AB is called a superkink.

Even though the KW locks cannot move perpendicular to their own line directions, the nodes A and B can move conservatively along the Burgers vector direction (dashed lines). Suppose node B moves along the dashed line to the left, while node A remains fixed. As a result, the planar segment AB becomes shorter in length and the KW lock on the right becomes longer. In other words, the planar core structure is gradually converted to the non-planar KW lock structure.

Now suppose both node A and node B move along the dashed line to the left at the same speed. The result is that the KW lock on the left becomes shorter and the KW lock on the right becomes longer. Eventually, the entire KW lock on the left disappears and is replaced by the KW lock on the right. Effectively, the KW lock moves from one location to another. Therefore, KW locks are not permanently frozen in space, but can still migrate from one location to another after they are formed.

Because the dislocation multiplication process usually involves dislocation motion, the superdislocations emitted by sources (such as Frank–Read sources, see Section 8.5.3) probably have planar core structures, which have higher energies than the KW Lock. Given sufficient time and with the assistance of thermal fluctuation, the superdislocations with screw orientation can evolve into the KW lock configuration and impede plastic flow, thus acting as a

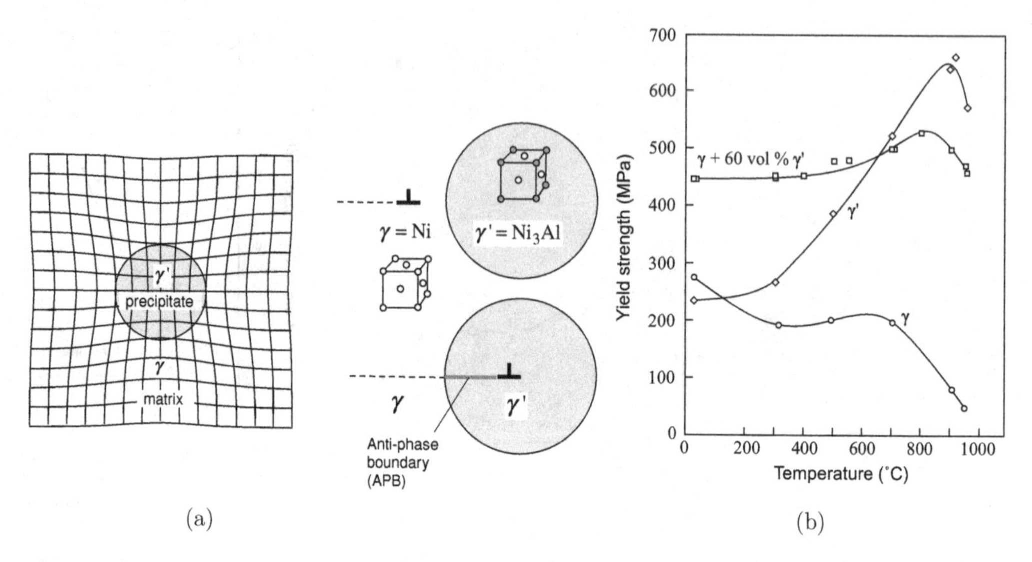

Figure 11.37. (a) Shearing of ordered, coherent γ' precipitates in superalloys. (b) Temperature dependence of the yield strength of γ, γ', and γ + 60 vol.% γ' [109, 110]. Here the yield strength is defined as the 0.2% offset flow stress. The maximum yield strength for the γ' phase occurs at a higher temperature than that shown in Fig. 11.36b due to the presence of Cr solutes here.

strengthening mechanism. Because higher temperatures make it easier for screw superdislocations to "discover" the non-planar configuration, this mechanism actually causes the flow stress of Ni_3Al alloys to increase with increasing temperature up to 600 °C, as shown in Fig. 11.36b. Such a temperature dependence is anomalous because the strength of most structural materials decreases with increasing temperature. The anomalous temperature dependence of the strength of Ni_3Al makes it an attractive material for high temperature applications.

11.4.4 Ni–Ni_3Al superalloy

Ni_3Al is a component of the Ni–Ni_3Al superalloy that is used in the blades of gas turbine engines. Here the word "super" refers to the excellent mechanical strength of the alloy, especially at high temperature. The Ni–Ni_3Al superalloy has a two-phase microstructure usually referred to as γ–γ'. The γ-phase is FCC Ni and the γ' phase is $L1_2$ Ni_3Al in the form of precipitates. Figure 11.37a shows a coherent γ' precipitate particle embedded in a γ-matrix. Because Ni_3Al has a smaller lattice constant than FCC Ni, the precipitate particle induces an elastic strain field around it that interacts with dislocations.

Now consider a perfect dislocation with a $\frac{1}{2}\langle 1\,1\,0\rangle$ Burgers vector moving on a $\{1\,1\,1\}$ in the γ phase (FCC Ni), as shown in Fig. 11.37a. As it enters the γ' phase, it must leave an APB behind, because a perfect Burgers vector in Ni_3Al is $\langle 1\,1\,0\rangle$. The dislocation with Burgers vector $\frac{1}{2}\langle 1\,1\,0\rangle$ is only a superpartial in the γ' phase. The APB connects the superpartial and the γ–γ' interface. Part of the APB can be removed when a second perfect dislocation in the γ phase, with the same Burgers vector and on the same slip plane as the first one, enters the γ' phase. Therefore, two perfect dislocations in the γ phase together form a single superdislocation after they enter the γ' phase.

The elastic strain near the precipitate particle and the APB that is created when a dislocation enters the precipitate both provide barriers to dislocation motion and act as strengthening mechanisms. The yield strengths of γ, γ', and γ + 60 vol.% γ' as functions of temperature are plotted in Fig. 11.37b. At low temperature, the two-phase γ–γ' alloy is stronger than both the homogeneous γ and γ' phases. At temperatures above 650 °C, the yield strength of the γ–γ' alloy is below that of the γ' alloy, but it still retains a significant fraction of the latter. In fact, the γ–γ' alloy has a high strength in the entire temperature range of 0° to 1000 °C. This is why the γ–γ' alloy is the material of choice for turbine blades which must maintain a high strength over a wide range of temperatures.

11.5 Summary

In close-packed metals, perfect dislocations tend to dissociate into partial dislocations separated by stacking faults in order to reduce their energy. The nature of partial dislocations and stacking faults can be rationalized by considering the atoms as hard spheres that slide on top of each other when slip occurs. Close-packed layers of spheres invariably slide on each other in a zig-zag manner, where the zig motion may be associated with the passage of a leading partial dislocation and the zag motion with the passage of a trailing partial. The equilibrium separation between the partials can be obtained by balancing the elastic interaction forces between the partials and the attraction force exerted by the stacking fault. The separation distance is affected by the component of stress that causes the partials to move in opposite directions, an effect that plays an important role in cross-slip of screw dislocations.

The hard sphere picture of FCC metals leads to a variety of different kinds of partial dislocations, extending from Shockley partials involved in the slip process and having Burgers vectors of the type $\frac{a}{6}\langle 1\,1\,2\rangle$, to Frank partials, with Burgers vectors of the type $\frac{a}{3}\langle 1\,1\,1\rangle$, that are created when vacancies condense on a close-packed plane to form a faulted dislocation loop. Reactions between gliding dislocations are also affected by the dissociation into partials and this leads to stair-rod dislocations, with Burgers vectors of the type $\frac{a}{6}\langle 1\,1\,0\rangle$, at the junctions between stacking faults on different slip planes.

The description of partial dislocations in FCC metals is greatly simplified using Thompson's tetrahedron and the associated Thompson notation. The apexes of the tetrahedron, denoted as A, B, C, and D, represent neighboring lattice points in the FCC lattice, while the centers of the triangular faces of the tetrahedron, denoted by α, β, γ, and δ, represent fault positions in the lattice. With this notation the Burgers vectors of perfect dislocations, $\frac{a}{2}\langle 1\,1\,0\rangle$, are represented by vectors of the type AB while the partial dislocations are represented by, e.g. $A\delta$ – Shockley partials, $D\delta$ – Frank partials and $\alpha\delta$ – stair-rod partials. A careful consideration of the atomic packing in FCC crystals leads to the rule that slip occurs "on the inside surface of the tetrahedron." This can be codified with a $\hat{n} \times \hat{\xi}$ rule that determines the Burgers vectors of the leading and trailing Shockley partial dislocations, where $\hat{n}$ is inward pointing normal of the tetrahedron surface and $\hat{\xi}$ is the dislocation sense vector. This useful rule permits the determination of the relative positions of partial dislocations in complex reactions, such as at nodes where different extended dislocations meet.

The HCP structure is closely related to the FCC structure. If an intrinsic stacking fault is introduced on every other $\{1\,1\,1\}$ plane in an FCC structure, the result is an HCP structure whose basal plane corresponds to the $\{1\,1\,1\}$ plane of FCC. Therefore, the critical resolved shear stress in HCP metals is usually the lowest on the basal plane. Perfect dislocations on the basal plane dissociates into two Shockley partials separated by stacking faults in HCP metals.

The hard sphere picture can also be applied to the compound crystal of $CrCl_3$, which has a layered structure. The $CrCl_3$ structure can be considered as an FCC crystal formed by Cl^- ions containing Cr^{3+} ions as interstitials. The ordered distribution of Cr^{3+} ions between $\{1\,1\,1\}$ planes make the shortest lattice repeat vector of $CrCl_3$ three times larger than that of the FCC crystal formed by Cl^- ions. Hence the perfect Burgers vectors in $CrCl_3$ are of the type $\frac{3a}{2}\langle 1\,1\,0\rangle$. As a result, a perfect dislocation on a $\{1\,1\,1\}$ plane of $CrCl_3$ dissociates into six Shockley partials connected by different types of stacking faults.

For important intermetallic compounds like Ni_3Al, which constitute the hardening phase in superalloys, the Burgers vectors are of the type $a\langle 1\,1\,0\rangle$ and the dislocations are called superdislocations. They are composed of superpartials, each with Burgers vectors of the type $\frac{a}{2}\langle 1\,1\,0\rangle$, which are separated by antiphase boundaries. The superpartials, in turn, are separated into Shockley partials with complex faults between them, much like they would be in disordered FCC solid solutions. Simple considerations of near neighbor intermetallic bonds show that antiphase boundaries lying on $\{0\,0\,1\}$ planes have smaller energies that those on $\{1\,1\,1\}$ planes. This can cause screw-oriented superdislocations gliding on $\{1\,1\,1\}$ planes to cross-slip onto $\{0\,0\,1\}$ planes where they become immobilized as Kear–Wilsdorf locks. This represents a principal hardening mechanism for both Ni_3Al and the superalloys containing that phase.

11.6 Exercise problems

11.1 A positive edge dislocation and a right-handed screw dislocation, both of which are extended into Shockley partials, are shown in the two separate Thompson tetrahedrons in Fig. 11.38. The sense vectors shown define the Burgers vectors of the dislocations. Using the Thompson notation, determine the Burgers vectors of the two Shockley partials in each dislocation.

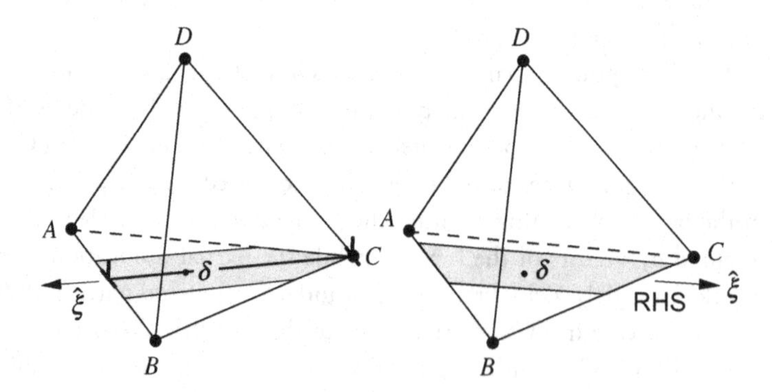

Figure 11.38. A dissociated edge dislocation and a dissociated screw dislocation shown relative to Thompson's tetrahedrons.

11.2 Two perfect dislocations with Burgers vectors BC and DB form a junction as shown in Fig. 11.39. Redraw the dislocation structure with all dislocations dissociated into partials. Note that the junction dissociates into a non-planar Lomer–Cottrell dislocation. Mark the Burgers vectors on all partial dislocations. Which end of the Lomer–Cottrell dislocation has an extended node and which end has a constricted node?

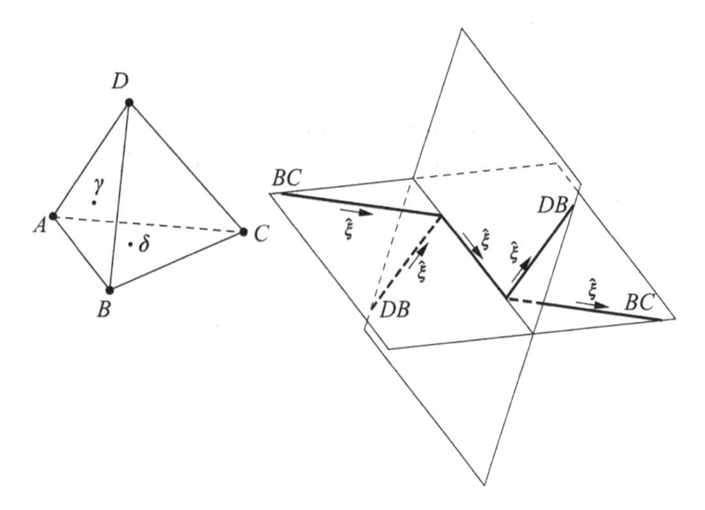

Figure 11.39. A dislocation junction in an FCC metal.

11.3 Consider a 60° dislocation dissociated into two Shockley partials in FCC Cu as shown in Fig. 11.40a. Write an expression for the equilibrium spacing, s, between the Shockley partial dislocations in terms of the lattice parameter, a, the elastic constants and the stacking fault energy. Calculate the equilibrium spacing using the shear modulus and Poisson's ratio from Table B.1, stacking fault energy from Table 11.2, and lattice parameter $a = 0.361$ nm.

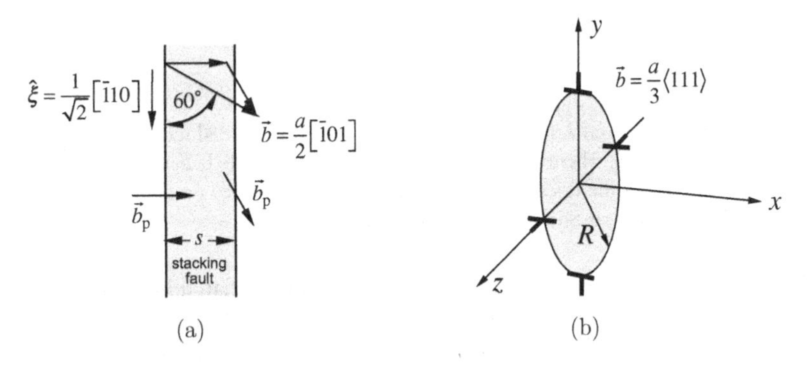

Figure 11.40. (a) A dissociated 60° dislocation. (b) A Frank partial dislocation loop.

11.4 An FCC metal with a low stacking fault energy is rapidly quenched from a high temperature. This causes excess vacancies to condense as Frank partial dislocation loops on $\{1\,1\,1\}$ planes. One such Frank partial dislocation loop is shown in Fig. 11.40b. We assume that vacancies are destroyed only at Frank partial loops and that all of the loops in the crystal have the same radius, R. Eventually a state of equilibrium will be reached in which the total free energy of the system reaches a minimum. We wish to determine the vacancy chemical potential in the crystal when this equilibrium state is reached. From this chemical potential the supersaturation of vacancies caused by the loops in the crystal could be calculated.

To simplify the algebra you may assume that the line tension of the dislocations is $\mu b^2/2$. Also, the intrinsic stacking fault energy can be represented by $\gamma_{\rm ISF}$. It is useful to know that the atomic density on the {1 1 1} plane is $4/(\sqrt{3}a^2)$ atoms per unit area, where a is the lattice parameter. Derive an expression for the vacancy chemical potential in the crystal at equilibrium.

11.5 In Fig. 11.13, we considered a network of RHS dislocations on a {1 1 1} plane of an FCC metal. If the same dislocation network occurs on the basal plane of an HCP crystal, the dislocations will dissociate in the same way, provided that the layer A is immediately below the slip plane and layer B is immediately above the slip plane, as shown in Fig.11.41a. Draw a new diagram to show the dissociation pattern when layer B is immediately below the slip plane and layer A is immediately above the slip plane. Mark the Burgers vectors of the partial dislocations on your diagram and show how the partial dislocation lines are connected to form the extended and constricted nodes.

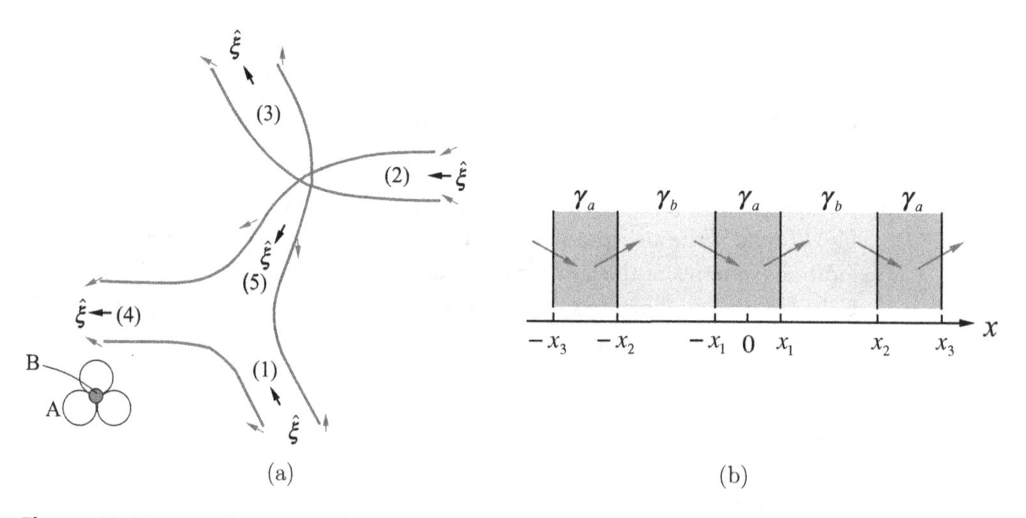

Figure 11.41. (a) A planar network formed by right-handed screw (RHS) dislocations on the basal plane of an HCP crystal, where layer A and layer B are immediately below and above the slip plane, respectively. The arrows next to the lines are Burgers vectors of the Shockley partial dislocations. (b) Dissociation of an edge dislocation into six Shockley partial dislocations in $CrCl_3$.

11.6 Consider a perfect edge dislocation in $CrCl_3$ dissociated into six partial dislocations as shown in Fig.11.41b. For simplicity, assume that the five different stacking faults only take two possible values, γ_a and γ_b, as shown. Then the equilibrium positions of the partials must be symmetric and can be specified by x_1, x_2, x_3 and $-x_1$, $-x_2$, $-x_3$, respectively. Let b be the magnitude of the Burgers vector for each partial, and let μ and ν be the shear modulus and Poisson's ratio of the crystal.

(a) Write down the systems of equations for x_1, x_2, and x_3 at equilibrium.
(b) Assuming $\gamma_a = 2\gamma$ and $\gamma_b = \gamma$, and $\nu = 0.3$, write a Matlab program (using the `fsolve` function) to find the equilibrium values of x_1, x_2, and x_3 in units of $\mu b^2/\gamma$.

11.7 Figure 11.42 shows a screw superdislocation in Ni_3Al near a free surface, with a slip plane parallel to that surface. The dislocation is composed of two superpartial dislocations, each with Burgers vector $b/2$, separated by an antiphase boundary (APB) with energy (per unit area) γ_{APB}. We will assume that the Shockley partials in each of these superpartials are so close together that each superpartial can be considered a single dislocation of Burgers vector $b/2$.

(a) Write the equations that would be needed to determine the equilibrium spacing between the superpartials, s_{eq}, as a function of the distance, h, between the free surface and the slip plane of the dislocation.
(b) Determine the equilibrium spacing when the superdislocation is very far from the surface, $h \gg s_{eq}$.
(c) Determine the equilibrium spacing when the superdislocation is very close to the surface, $h \ll s_{eq}$.

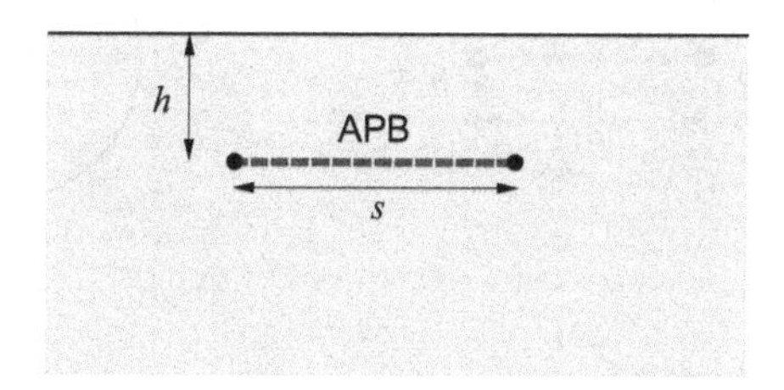

Figure 11.42. A screw superdislocation in Ni_3Al dissociated into two superpartials with APB parallel to the surface.

11.8 Figure 11.43 shows a pair of edge dislocations near the γ–γ' interface of the Ni–Ni_3Al superalloy. The applied shear stress exerts Peach–Koehler forces on both dislocations in the $+x$ direction. When the leading dislocation (on the right) enters the γ' phase (Ni_3Al), it becomes a superpartial with an antiphase boundary (APB) trailing behind, whose energy per unit area is γ_{APB}.

Derive an expression for the critical shear stress τ_c required to initiate shear of the γ' phase, i.e. when the leading dislocation enters the γ' phase. Also calculate the separation distance between the partials, s_c, when the critical condition is satisfied. Assume that the superpartials are not dissociated into Shockley partials (i.e. the complex stacking fault energy is extremely high).

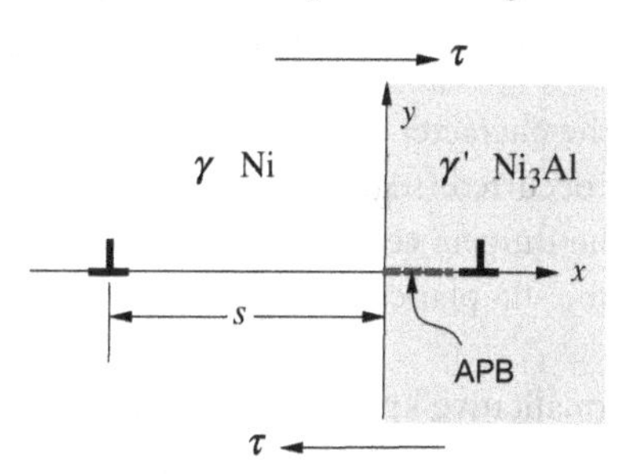

Figure 11.43. A pair of edge dislocations in a Ni–Ni_3Al superalloy.

11.9 A pure right-handed screw dislocation in Ni_3Al is shown in Fig. 11.44a. The dislocation consists of two superpartial dislocations separated by an antiphase boundary. Assume that the Shockley partial dislocations within each superpartial are so close together that their separate identities can be ignored.

(a) Assuming the dislocation to be located in an otherwise perfect crystal, develop an expression for the equilibrium separation distance, s^0_{eq}, between the superpartials. For this calculation you may assume isotropic elasticity, that the Burgers vector of each superpartial is b, and that the energy per unit area of the antiphase boundary is γ_{APB}.
(b) Now consider that the dislocation lies near a tiny cylindrical void as shown in Fig. 11.44b. The diameter d of the void is less than twice the equilibrium spacing between

superpartials, as calculated above. Indicate what would happen to the dislocation if it were free to glide easily on its slip plane. Describe the final configuration/position of the dislocation as a result of its interaction with the hole. The stress field of a screw dislocation near or inside a cylindrical void is given in Section 10.6.6.

(c) Describe the final equilibrium state if $d > 2s^0_{eq}$.

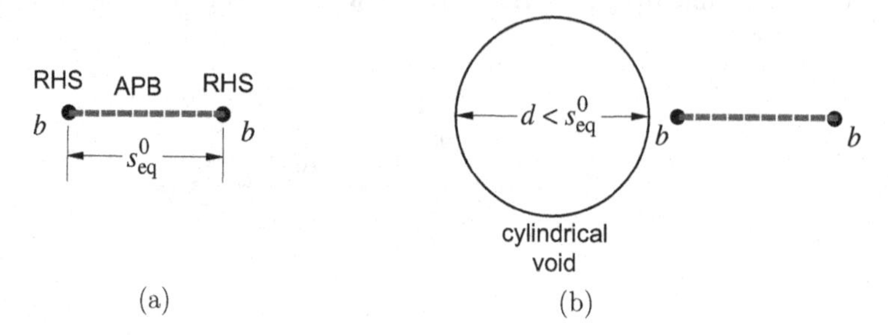

Figure 11.44. (a) A screw superdislocation in an otherwise perfect Ni_3Al crystal. (b) A screw superdislocation in Ni_3Al near a cylindrical void.

11.10 Figure 11.45 shows misfit dislocations in an epitaxial (0 0 1) SiGe film on a Si substrate. The SiGe film itself is not shown; only misfit dislocations and diagrams indicating the crystallography are shown. Because the misfit dislocations form by slip on {1 1 1}-type planes, they lie along ⟨1 1 0⟩-type directions in the interface between the film and the substrate. The two mutually perpendicular misfit dislocations (along the lines AI and GD) shown in the diagram serve to partially relieve the biaxial compressive stress in the SiGe film. In some cases the crossing misfit dislocations are observed to "react" and break apart, as shown in the diagram. As the points C and F move away from the substrate they can separate from each other when they reach the free surface of the film (not shown). This is known as the Hagen–Strunk reaction. The pyramids represent the various slip planes and directions in this FCC lattice, as discussed in Section 11.1.6.

(a) Determine the character of all of the dislocation segments after the Hagen–Strunk reaction has occurred (segments ABC, DEF, and GHI). Choose sense vectors and determine the Burgers vectors for each of these dislocation segments.

(b) Determine the slip planes that were active when the two crossing misfit dislocations were created.

(c) Using your qualitative knowledge of dislocation line energy (or line tension), explain why the dislocation AHD can break apart while the dislocation GHI does not.

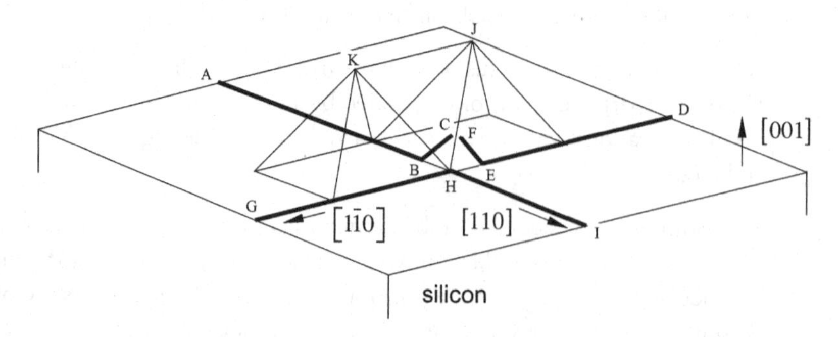

Figure 11.45. Misfit dislocations at the interface between a Si–Ge film on a Si substrate.

12 Dislocation core structure

In our treatment of dislocations thus far, we have avoided the dislocation core. For example, in Volterra's dislocation model, the stress–strain fields diverge on the dislocation line, so that a cylindrical region of material is usually removed around the dislocation line to avoid the singularity. In the line tension model, the dislocation is modeled as a string that carries a line energy per unit length, but is otherwise featureless. In Chapter 11, we have seen that perfect dislocations in close-packed metals tend to dissociate into partial dislocations, but the partial dislocations were still treated as Volterra's dislocation lines. In reality, every (perfect or partial) crystal dislocation has a core region, which possesses a specific atomistic structure, called the *core structure*. The core structure is determined by non-linear interatomic interactions and the crystal structure, and, in turn, strongly influences the energetics and mobility of the dislocations. In this chapter, we discuss typical dislocation core structures and their effects on dislocation properties in several crystal structures.

In Section 12.1, we start our discussion with the classical Peierls–Nabarro (PN) model, which was the first physical model for the dislocation core and naturally predicts that the dislocation core should have a finite width. In Section 12.2, we generalize the original PN model to account for the presence of stacking faults in FCC metals. Consistent with the hard sphere model in Chapter 11, the generalized PN model also predicts dissociation of perfect dislocations into partials, except that each partial now has a finite width.

For crystals whose structures are sufficiently different from close-packed, hard spheres are no longer a good model for the atoms. Nonetheless, the geometry of the stacking of atomic layers is still useful for understanding the dislocation core structures in these crystals, as discussed in Section 12.3 (diamond cubic crystals) and Section 12.4 (BCC crystals). Finally, in Section 12.5 we discuss the interaction between dislocations and point defects, which usually leads to segregation of point defects around the dislocation core.

12.1 Peierls–Nabarro model

The classical model by Peierls and Nabarro [111, 112] considers the spreading of the dislocation over the glide plane. It provides a good description of dislocations with planar cores, such

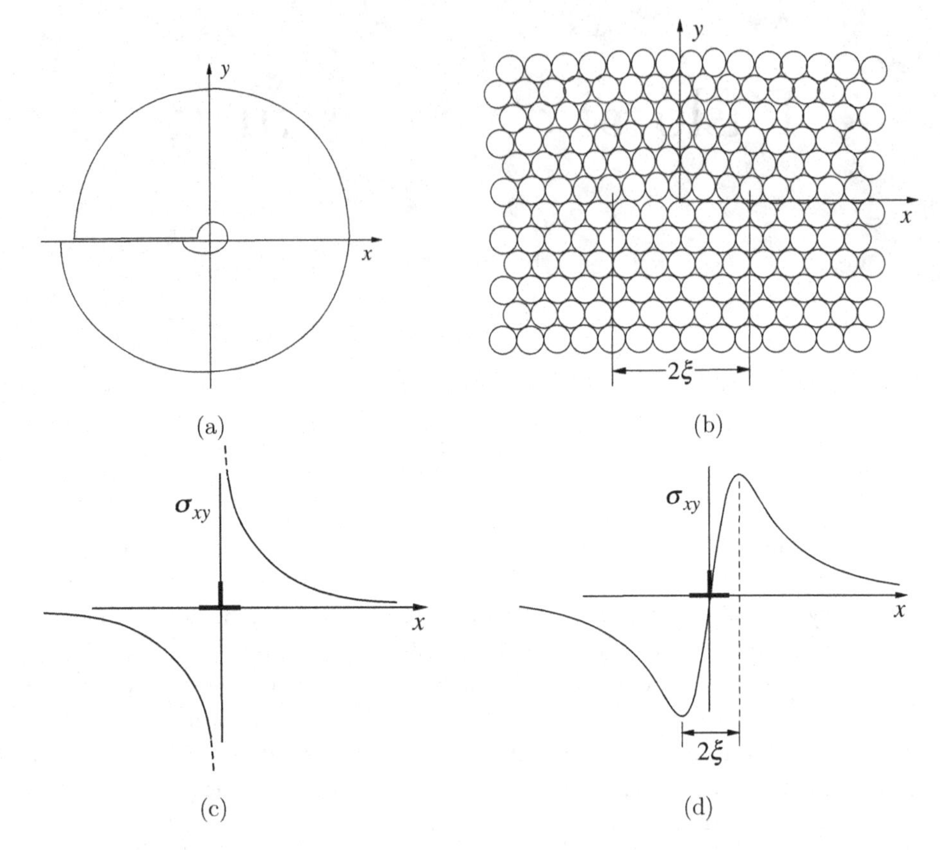

Figure 12.1. (a) Volterra's edge dislocation model. (b) An edge dislocation in a triangular crystal lattice. (c) Singular stress field of a Volterra's dislocation along the x axis. (d) Non-singular stress field of the Peierls–Nabarro dislocation along the x axis.

as ordinary dislocations on $\{1\,1\,1\}$ planes in FCC metals.[1] The Peierls–Nabarro model also provides a starting point to understand non-planar dislocation core structures.

12.1.1 Motivation

In Volterra's edge dislocation model, as shown in Fig. 12.1a, the relative displacement between material above and below the slip plane experiences a discontinuity at the dislocation line. This discontinuity is responsible for the stress singularity. For example, along the x axis,

$$\sigma_{xy}(x,0) = \frac{\mu b}{2\pi(1-\nu)}\frac{1}{x}. \tag{12.1}$$

In the limit of $x \to 0$, $\sigma_{xy} \to \infty$, as shown in Fig. 12.1c.

However, a more realistic model of an edge dislocation in a crystal is shown in Fig. 12.1b. It can be seen that the atoms gradually slide past each other over a finite region near the dislocation

1 The original Peierls–Nabarro model does not predict dissociation of perfect dislocations into partials. But the dissociation can be captured by a simple extension of the Peierls–Nabarro model, as shown in Section 12.2.

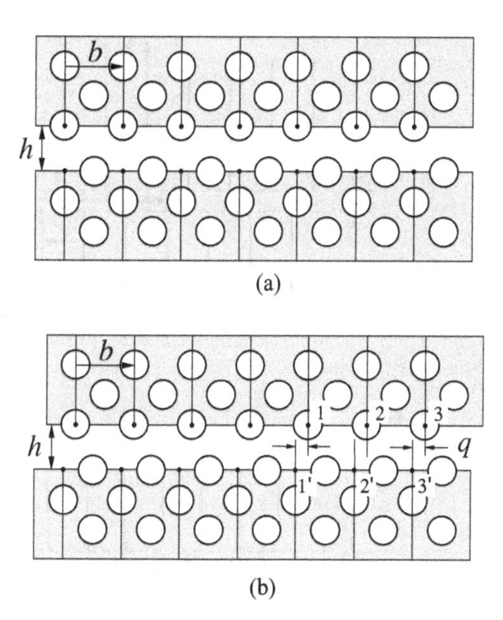

Figure 12.2. (a) A perfect crystal cut into two halves. The vertical lines drawn in the two halves are aligned, indicating zero disregistry. (b) Shearing the upper half crystal leads to misalignment between the vertical lines. The disregistry q is the misalignment between marker points on the vertical lines immediately above and below the glide plane.

center. As a result, the stress field is no longer singular. For example, it may be approximated by the following function, shown in Fig. 12.1d,

$$\sigma_{xy}(x, 0) = \frac{\mu b}{2\pi(1-\nu)} \frac{x}{x^2 + \xi^2}, \tag{12.2}$$

where ξ is called the half width of the dislocation core. At distances far away from the dislocation, the stress value given by the non-singular expression, Eq. (12.2), is nearly the same as that given by the singular expression, Eq. (12.1). However, as $x \to 0$, the non-singular expression gives a more physical description of the dislocation core. The Peierls–Nabarro model not only leads to a non-singular stress expression of the form of Eq. (12.2), but also provides a physics-based prediction of the core width 2ξ.

12.1.2 Disregistry

The Peierls–Nabarro model uses the concept of *disregistry*. Consider a perfect crystal cut into two halves, as shown in Fig. 12.2a. Draw a set of vertical lines that pass through the atoms. The spacing between neighboring vertical lines is b, the nearest neighbor distance between atoms in the horizontal direction. The vertical lines in the upper and lower halves are perfectly aligned in Fig. 12.2a, indicating there is no offset between the two halves.

We now move the upper crystal to the right by a distance q, as shown in Fig. 12.2b. Then the vertical lines in the upper and lower crystal halves are offset by q from each other. To give a more precise definition of disregistry, we put marker points at the intersection between the vertical lines and the surface of the two crystal halves, as shown in Fig. 12.2b. For each marker point in

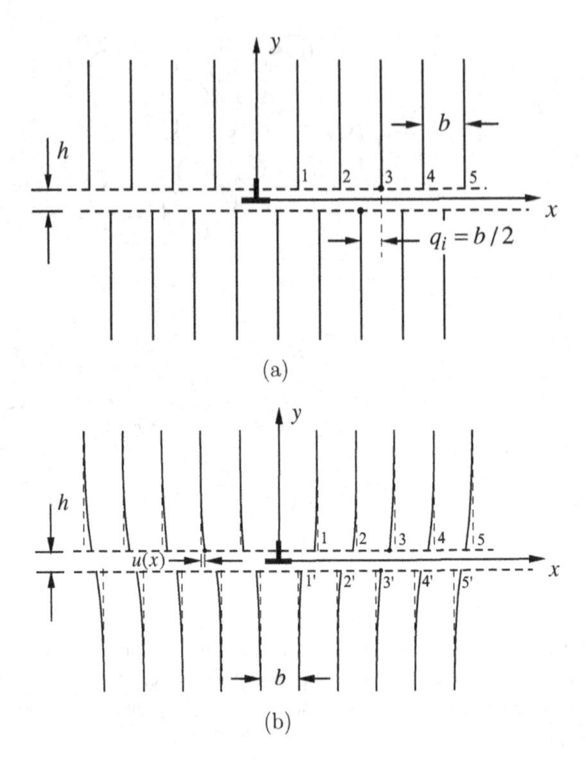

Figure 12.3. Joining two crystal halves. (a) Before relaxation. There is maximum disregistry at the interface, but there is no strain energy. (b) After relaxation. The disregistry is reduced and there is finite strain energy.

the lower half, say 1′, we can find the closest marker point in the upper half, say 1. The relative position between the two marker points in the nearest neighbor pair is the local disregistry, e.g.

$$q_i = x_i - x_i'. \tag{12.3}$$

Owing to the periodicity of the crystal lattice, $|q_i| \leq b/2$. In Fig. 12.2b, the disregistry is a constant (q) everywhere along the x axis, because the upper crystal is only rigidly translated.

As in Frenkel's model (see Section 8.1.1), when two crystal halves slide relative to each other by a uniform amount q, the shear stress σ_{xy} that develops at the interface must be a periodic function of q. It is natural to assume σ_{xy} is a sinusoidal function of q. Similar to Eqs. (8.1) and (8.4), we have

$$\sigma_{xy} = \frac{\mu}{2\pi}\frac{b}{h}\sin\left(\frac{2\pi q}{b}\right), \tag{12.4}$$

where μ is the shear modulus of the crystal and h is the distance between the surfaces of the two half-crystals in the y direction (the interplanar spacing).

12.1.3 Formulation of the Peierls–Nabarro model

We now consider the formation of an edge dislocation by joining two crystal halves together, where the upper half crystal contains one more vertical line (and the associated atomic planes) than the lower half. Figure 12.3a shows the configuration where the atoms in the two crystal halves are at their perfect crystal positions and not allowed to relax. For each marker point in the lower half, there are two marker points in the upper half at equal distance, so that there is some ambiguity as to whether the disregistry in this configuration is $0.5b$ or $-0.5b$. Anticipating how the marker points will pair up after the relaxation, we assign the disregistry in this configuration as follows:

$$\begin{aligned} q_i &= +\frac{b}{2}, \quad \text{for} \quad x_i > 0, \\ q_i &= -\frac{b}{2}, \quad \text{for} \quad x_i < 0. \end{aligned} \tag{12.5}$$

Since the local shear stress σ_{xy} is a periodic function of the disregistry q, the local stress stays the same if one adds to q an integer multiple of b. The above choices of q_i only ensure that, after the relaxation, its magnitude will be less than $0.5b$ everywhere, which may appear more agreeable to our intuition.

In the configuration before the relaxation, as shown in Fig. 12.3a, there is no strain energy in either crystal half, but the disregistry between the two halves has the maximum possible magnitude. This leads to a high misfit energy (per unit area) at the interface between the two crystal halves. When atoms are allowed to relax, the magnitude of disregistry and the misfit energy will be reduced, but a non-zero strain energy will develop. The total energy is still reduced by the relaxation. The minimum energy structure is the equilibrium core structure of the dislocation.

Let $u(x)$ represent the displacement of marker points on the upper half of the slip plane after the relaxation. Even though the marker points are spaced at discrete locations, we shall consider $u(x)$ as a continuous function. By symmetry, the displacement of marker points on the lower half of the slip plane is $-u(x)$. Hence the local relative displacement between marker points on the upper and lower halves near location x is $2u(x)$. Therefore, the remaining disregistry after relaxation is

$$\begin{aligned} q(x) &= +\frac{b}{2} + 2u(x), \quad \text{for} \quad x > 0, \\ q(x) &= -\frac{b}{2} + 2u(x), \quad \text{for} \quad x < 0, \end{aligned} \tag{12.6}$$

where we have also approximated q as a continuous function of x. From Fig. 12.3b, we can see that both $u(x)$ and $q(x)$ are odd functions of x.

From Eq. (12.4), the local shear stress on the interface between the two crystal halves is

$$\begin{aligned} \sigma_{xy}(x) &= \frac{\mu}{2\pi}\frac{b}{h}\sin\left(\frac{2\pi\, q(x)}{b}\right) \\ &= \frac{\mu}{2\pi}\frac{b}{h}\sin\left(\frac{4\pi\, u(x)}{b} \pm \pi\right) \\ &= -\frac{\mu}{2\pi}\frac{b}{h}\sin\left(\frac{4\pi\, u(x)}{b}\right). \end{aligned} \tag{12.7}$$

This is true for both $x > 0$ and $x < 0$. The shear stress given in Eq. (12.7) is due to the cohesive forces acting through the disregistry across the plane of the dislocation.

The displacements and shear stresses must be consistent with the elastic fields in the two crystal halves, both subjected to a non-uniform displacement on their surfaces ($u(x)$ for the upper half and $-u(x)$ for the lower half). The stress field due to such distortions can be conveniently expressed in terms of the stress fields of dislocations.

Recall that the relative displacement at location x between the upper and lower crystal halves during the relaxation is $2u(x)$. If $2u(x)$ is not uniform, then there is a Burgers vector content of $b'(x)\,dx = 2u(x) - 2u(x+dx)$ in the domain $[x, x+dx]$, which can be shown by drawing a Burgers circuit around point x (choosing the sense vector as pointing out of plane). Hence we can interpret

$$b'(x) = -2\,du(x)/dx \tag{12.8}$$

as the Burgers vector density at x. Recall that the stress field at point $(x, 0)$ of an edge dislocation with Burgers vector b located at the origin is

$$\sigma_{xy}(x,0) = \frac{\mu b}{2\pi(1-\nu)}\frac{1}{x}.$$

Hence the stress field at point $(x, 0)$ of an infinitesimal edge dislocation with Burgers vector $b'(x')dx'$ located at point $(x', 0)$ is

$$\sigma_{xy}(x,0) = \frac{\mu b'(x')\,dx'}{2\pi(1-\nu)}\frac{1}{x-x'}.$$

Therefore, the stress field set up by the relaxation can be written in terms of the stress field of all the infinitesimal dislocations on the glide plane,

$$\sigma_{xy}(x) = \frac{\mu}{2\pi(1-\nu)}\int_{-\infty}^{+\infty}\frac{b'(x')}{x-x'}\,dx'. \tag{12.9}$$

With this integral we sum up the stress field contribution from all infinitesimal pieces of Burgers vectors. From Eq. (12.8), we have

$$\begin{aligned}\sigma_{xy}(x) &= \frac{\mu}{2\pi(1-\nu)}\int_{-\infty}^{+\infty}\frac{-2\,du(x')/dx'}{x-x'}\,dx'\\ &= \frac{\mu}{\pi(1-\nu)}\int_{u(-\infty)}^{u(+\infty)}\frac{-du(x')}{x-x'}.\end{aligned} \tag{12.10}$$

Combining Eqs. (12.7) and (12.10), we arrive at an integral equation for $u(x)$,

$$\begin{aligned}\frac{\mu}{\pi(1-\nu)}\int_{u(-\infty)}^{u(+\infty)}\frac{-du(x')}{x-x'} &= -\frac{\mu}{2\pi}\frac{b}{h}\sin\left(\frac{4\pi u(x)}{b}\right),\\ \int_{u(-\infty)}^{u(+\infty)}\frac{du(x')}{x-x'} &= \frac{(1-\nu)b}{2h}\sin\left(\frac{4\pi u(x)}{b}\right).\end{aligned} \tag{12.11}$$

Peierls found the solution of this equation,

$$\begin{aligned}u(x) &= -\frac{b}{2\pi}\arctan\left(\frac{2(1-\nu)x}{h}\right),\\ u(x) &= -\frac{b}{2\pi}\arctan\left(\frac{x}{\xi}\right),\end{aligned} \tag{12.12}$$

where $\xi = \frac{h}{2(1-\nu)}$ is defined as the dislocation half width.

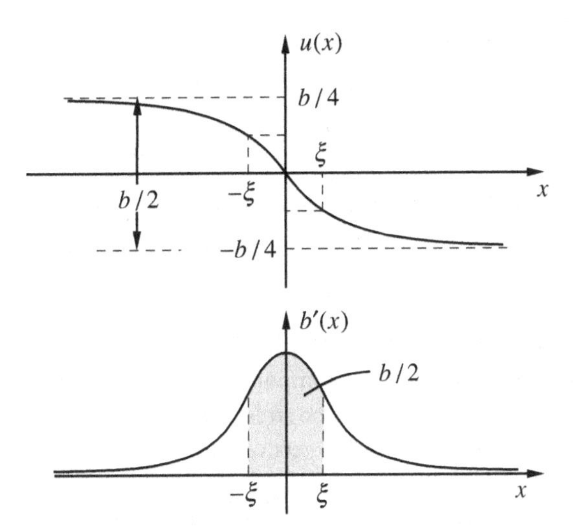

Figure 12.4. (a) Peierls solution for the displacement during relaxation. (b) Distribution of the Burgers vector density.

12.1.4 Properties of the Peierls–Nabarro dislocation

Figure 12.4a plots the displacement $u(x)$ of the surface of the upper half crystal. It changes monotonically from $b/4$ at $x = -\infty$ to 0 at $x = 0$ to $-b/4$ at $x = +\infty$. The derivative of $u(x)$ times -2 is the local Burgers vector density $b'(x) = (b/\pi)\xi/(x^2 + \xi^2)$, which is plotted in Fig. 12.4b. It shows that the Burgers vector of a Peierls–Nabarro dislocation is not concentrated at the dislocation line ($x = 0$), but is spread out on the glide plane.

Since $u(x)$ changes by $-b/2$ as x goes from $-\infty$ to $+\infty$, the integral $\int_{-\infty}^{+\infty} b'(x)\,\mathrm{d}x$ equals b, which means that the total Burgers vector content of the distributed dislocation core is b, as it should be. Even though the Burgers vector density $b'(x)$ is non-zero everywhere, we can see that $b'(x)$ becomes very small for $|x|$ much larger than ξ. Furthermore, half of the area under the $b'(x)$ curve is contained within the domain $-\xi \leq x \leq \xi$. This means that half of the Burgers vector content of the entire dislocation is contained within the domain $-\xi \leq x \leq \xi$. This is why 2ξ is an appropriate measure of the width of the dislocation core.

The stress field of the distributed dislocation core can be obtained either from Eq. (12.10) or from Eq. (12.7). Using the latter, we have

$$
\begin{aligned}
\sigma_{xy}(x, 0) &= -\frac{\mu}{2\pi}\frac{b}{h}\sin\left[\frac{4\pi}{b}\left(-\frac{b}{2\pi}\arctan\left(\frac{x}{\xi}\right)\right)\right] \\
&= \frac{\mu}{2\pi}\frac{b}{h}\sin\left[2\arctan\left(\frac{x}{\xi}\right)\right].
\end{aligned}
\tag{12.13}
$$

Define $\theta \equiv \arctan(x/\xi)$, we have $\sin\theta = x/\sqrt{x^2 + \xi^2}$ and $\cos\theta = \xi/\sqrt{x^2 + \xi^2}$, so that

$$
\begin{aligned}
\sigma_{xy}(x, 0) &= \frac{\mu}{2\pi}\frac{b}{h}\sin 2\theta = \frac{\mu}{2\pi}\frac{b}{h}2\sin\theta\cos\theta \\
&= \frac{\mu}{\pi}\frac{b}{h}\frac{x\xi}{x^2 + \xi^2}.
\end{aligned}
\tag{12.14}
$$

By using $\xi = \frac{h}{2(1-\nu)}$, the stress field becomes the following.

$$\boxed{\sigma_{xy}(x, y=0) = \frac{\mu b}{2\pi(1-\nu)} \frac{x}{x^2+\xi^2}} \tag{12.15}$$

This is the non-singular stress expression, Eq. (12.2), promised earlier. Note that the stress of the Peierls–Nabarro dislocation reduces to the stress field of a Volterra dislocation, Eq. (12.1), for $|x| \gg \xi$.

Equation (12.15) gives one component of the stress field of a Peierls–Nabarro dislocation on its glide plane. By integrating the contributions from the entire Burgers vector distribution along the x axis, we can obtain all components of the stress field of the Peierls–Nabarro dislocation in the entire space. Based on the stress field of a single dislocation given by Eq. (9.31), we can write the stress field due to Burgers vector distribution $b'(x')$ as

$$\begin{aligned}
\sigma_{xx}(x, y) &= -\frac{\mu}{2\pi(1-\nu)} \int_{-\infty}^{\infty} b'(x') \frac{y[3(x-x')^2+y^2]}{[(x-x')^2+y^2]^2}\, \mathrm{d}x', \\
\sigma_{yy}(x, y) &= \frac{\mu}{2\pi(1-\nu)} \int_{-\infty}^{\infty} b'(x') \frac{y[(x-x')^2-y^2]}{[(x-x')^2+y^2]^2}\, \mathrm{d}x', \\
\sigma_{xy}(x, y) &= \frac{\mu}{2\pi(1-\nu)} \int_{-\infty}^{\infty} b'(x') \frac{(x-x')[(x-x')^2-y^2]}{[(x-x')^2+y^2]^2}\, \mathrm{d}x'.
\end{aligned} \tag{12.16}$$

The results are [5],

$$\begin{aligned}
\sigma_{xy}(x, y) &= \frac{\mu b}{2\pi(1-\nu)} \left\{ \frac{x}{x^2+(y\pm\xi)^2} - \frac{2xy(y\pm\xi)}{[x^2+(y\pm\xi)^2]^2} \right\}, \\
\sigma_{xx}(x, y) &= -\frac{\mu b}{2\pi(1-\nu)} \left\{ \frac{3y\pm 2\xi}{x^2+(y\pm\xi)^2} - \frac{2y(y\pm\xi)^2}{[x^2+(y\pm\xi)^2]^2} \right\}, \\
\sigma_{yy}(x, y) &= -\frac{\mu b}{2\pi(1-\nu)} \left\{ \frac{y}{x^2+(y\pm\xi)^2} - \frac{2x^2 y}{[x^2+(y\pm\xi)^2]^2} \right\}, \\
\sigma_{zz}(x, y) &= \nu(\sigma_{xx}+\sigma_{yy}) = -\frac{\mu b \nu}{\pi(1-\nu)} \frac{y\pm\xi}{x^2+(y\pm\xi)^2},
\end{aligned} \tag{12.17}$$

where the $+$ sign is used for $y > 0$ and the $-$ sign is used for $y < 0$. The stress components of the Peierls–Nabarro model are non-singular. However, σ_{xx} and σ_{zz} are discontinous across the $y = 0$ plane.

The analysis above for edge dislocations can also be applied to screw dislocations. The only difference is that the disregistry now measures displacement discontinuity in the z direction. The displacement incurred during the relaxation is

$$u_z(x) = -\frac{b}{2\pi} \arctan\left(\frac{2x}{h}\right) = -\frac{b}{2\pi} \arctan\left(\frac{x}{\xi_s}\right), \tag{12.18}$$

where $\xi_s = \frac{h}{2}$ is the half width of the screw dislocation. Notice that ξ_s is smaller than the half width ξ of the edge dislocation (for the usual case of $\nu > 0$). This is because the interaction between edge dislocations is stronger than that between screw dislocations, by a factor of $1/(1-\nu)$. The stronger repulsion between like-signed (distributed) edge components leads to a wider core width for the edge dislocation.

The non-singular stress field of the Peierls–Nabarro screw dislocation is

$$\begin{aligned}\sigma_{xz}(x,y) &= -\frac{\mu b}{2\pi}\frac{y \pm \xi_s}{x^2 + (y \pm \xi_s)^2},\\ \sigma_{yz}(x,y) &= \frac{\mu b}{2\pi}\frac{x}{x^2 + (y \pm \xi_s)^2},\end{aligned} \tag{12.19}$$

where the + sign is used for $y > 0$ and the − sign is used for $y < 0$.

12.2 Dislocations in FCC metals

The original Peierls–Nabarro (PN) model predicts that the dislocation core spreads out on the slip plane. But the predicted dislocation core does not split into partial dislocations, as observed in FCC metals (see Chapter 11). This is because the original PN model does not account for the possibility of stacking faults on the slip plane. In this section, we show that once the existence of a stacking fault is taken into account, the PN model predicts that a perfect dislocation would dissociate into partial dislocations with finite core widths.

In Eq. (12.4), we assumed a simple sinusoidal relation for the stress σ_{xy} between two crystal halves that have slid relative to each other by a uniform amount q. We now define $\gamma(q)$ as the work done (per unit area of the interface), i.e. the free energy cost, in order to induce the slide q. The energy function $\gamma(q)$ is called the *generalized stacking fault* (GSF) energy. The shear stress σ_{xy} is simply the derivative of the GSF function,

$$\sigma_{xy} = \frac{\partial \gamma(q)}{\partial q}. \tag{12.20}$$

The GSF function that corresponds to the shear stress given in Eq. (12.4) is then

$$\gamma(q) = \frac{\mu b^2}{2\pi^2 h}\sin^2\left(\frac{\pi q}{b}\right). \tag{12.21}$$

Figure 12.5a plots the GSF and shear stress as functions of q. We can see that $\gamma(q) = 0$ if q is an integer multiple of b, as expected. Furthermore, the GSF curve $\gamma(q)$ in Fig. 12.5a contains no intermediate minimum in the domain $0 < q < b$. As a result, there is no stacking fault in this simple model.

In FCC crystals stacking faults can exist on $\{1\,1\,1\}$ planes, which means that a metastable state can be created when two crystal halves slide against each other by a partial Burgers vector, which is shorter than the perfect Burgers vector. In reality, the partial Burgers vector is not parallel to the perfect Burgers vector. However, here we ignore this complexity and consider the hypothetical situation in which the partial Burgers vector is exactly half the perfect Burgers vector.

A GSF curve that allows a stacking fault to exist at $q = b/2$ is sketched in Fig. 12.5b. It has a double-humped shape, with a local minimum at $q = b/2$. The value of the GSF curve at $q = b/2$ is the stacking fault energy, i.e. $\gamma(b/2) = \gamma_{SF}$. For simplicity, we consider the following GSF function, which is shown in Fig. 12.5b,

$$\gamma(q) = \gamma_{SF}\sin^2\left(\frac{\pi q}{b}\right) + \left(\frac{\mu b^2}{8\pi^2 h} - \frac{\gamma_{SF}}{4}\right)\sin^2\left(\frac{2\pi q}{b}\right), \tag{12.22}$$

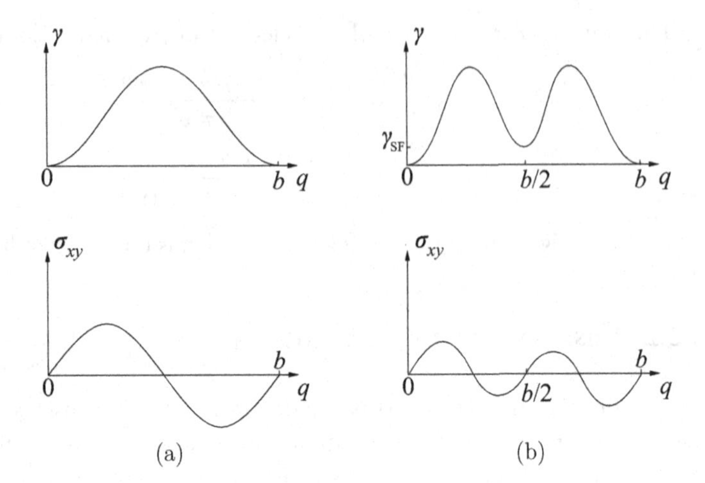

Figure 12.5. Generalized stacking fault energy γ and its derivative (the resulting shear stress) as a function of the relative slip q. (a) The original PN model, in which $\gamma(q)$ has no intermediate minimum in the domain $0 < q < b$. (b) A PN model representation of an extended dislocation with partials, in which $\gamma(q)$ has an intermediate minimum at $q = b/2$.

whose corresponding shear stress function is

$$\sigma_{xy}(q) = \frac{\partial \gamma(q)}{\partial q} = \frac{\pi \gamma_{\mathrm{SF}}}{b} \sin \frac{2\pi q}{b} + \left(\frac{\mu b}{4\pi h} - \frac{\pi \gamma_{\mathrm{SF}}}{2b} \right) \sin \left(\frac{4\pi q}{b} \right). \tag{12.23}$$

It can be verified that this chosen functional form not only satisfies the condition of $\gamma(b/2) = \gamma_{\mathrm{SF}}$, but also $\partial\sigma_{xy}(q)/\partial q|_{q=0} = \mu/h$, which is the case in Frenkel's model (see Section 8.1.1).

We are now ready to discuss how an edge dislocation between two crystal halves, as shown in Fig. 12.3, would relax given the new interface stress relation, Eq (12.23). Again, we assume the disregistry after relaxation is given by Eq. (12.6). In terms of the displacement $u(x)$ during relaxation, the local shear stress can be written as (using $q = 2u(x) \pm b/2$)

$$\sigma_{xy}(x) = -\frac{\pi \gamma_{\mathrm{SF}}}{b} \sin \frac{4\pi u(x)}{b} + \left(\frac{\mu b}{4\pi h} - \frac{\pi \gamma_{\mathrm{SF}}}{2b} \right) \sin \left(\frac{8\pi u(x)}{b} \right). \tag{12.24}$$

This stress field must equal the stress field due to the continuous distribution of dislocations on the glide plane, as given by Eq. (12.10), i.e.

$$\sigma_{xy}(x) = \frac{\mu}{\pi(1-\nu)} \int_{u(-\infty)}^{u(+\infty)} \frac{-du(x')}{x - x'}.$$

Combining Eqs. (12.24) and (12.10), we again arrive at an integral equation for $u(x)$:

$$\int_{u(-\infty)}^{u(+\infty)} \frac{du(x')}{x - x'} = \frac{\pi^2(1-\nu)\gamma_{\mathrm{SF}}}{\mu b} \sin \left(\frac{4\pi u(x)}{b} \right) - (1-\nu) \left(\frac{b}{4h} - \frac{\pi^2 \gamma_{\mathrm{SF}}}{2\mu b} \right) \sin \left(\frac{8\pi u(x)}{b} \right). \tag{12.25}$$

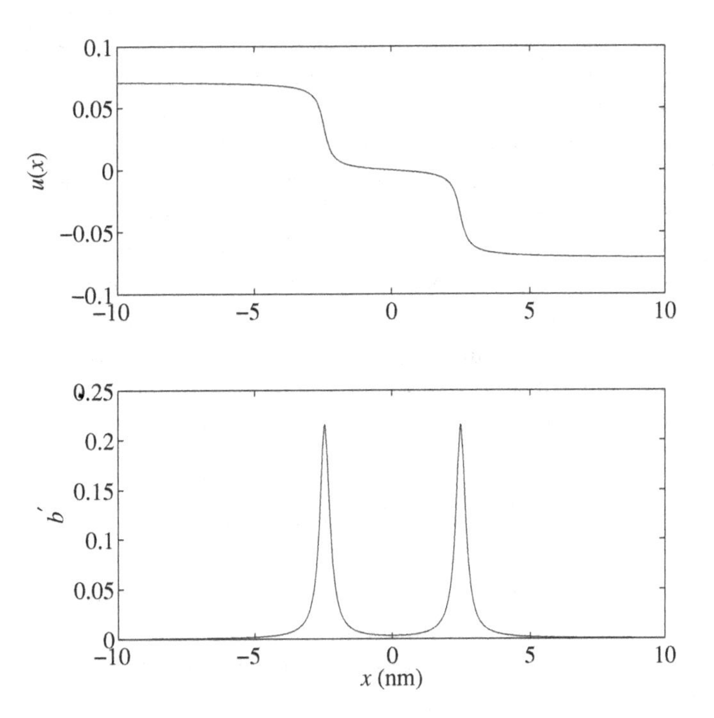

Figure 12.6. Numerical solution for the displacement $u(x)$ during the relaxation, and the distribution of the Burgers vector density $b'(x) = -2\,du(x)/dx$.

The solution of this integral equation can no longer be found analytically. However, approximate solutions can be obtained by numerical methods. For example, anticipating that the dislocation would dissociate into partials, we can try solutions of the following form,[2]

$$u(x) = -\frac{b}{4\pi}\arctan\left(\frac{x + w/2}{\xi_{\rm p}}\right) - \frac{b}{4\pi}\arctan\left(\frac{x - w/2}{\xi_{\rm p}}\right), \tag{12.26}$$

and search for w and $\xi_{\rm p}$ that minimizes the error in the integral equation (see Exercise problem 12.3). In this trial solution, w represents the dissociation width (i.e. width of the stacking fault), and $\xi_{\rm p}$ represents the core width of the Shockley partials. Using the parameters for FCC gold listed in Eq. (11.6) and $h = \sqrt{2/3}b = 0.234$ nm, the numerical solution is

$$\begin{aligned} w &= 4.9 \text{ nm}, \\ \xi_{\rm p} &= 0.2 \text{ nm}. \end{aligned} \tag{12.27}$$

The dissociation width w is larger than the previous estimate ($x_{\rm eq} \approx 4.0$ nm) in Eq. (11.6), mainly because here we have ignored the screw component of the partial dislocations. The displacement field $u(x)$ and the Burgers vector density $b'(x) = -2\,du(x)/dx$ are plotted in Fig. 12.6.

The core half width $\xi_{\rm p}$ of the partial dislocation agrees very well with the analytic prediction $\xi = h/(2(1-\nu)) = 0.2$ nm for perfect edge dislocations by the original PN model. This

2 The quality of the approximate solution depends on the functional form of the trial solution. A more general approach is to discretize the $u(x)$ function.

suggests that as long as the partial dislocations are well separated, so that their core regions do not overlap, it is acceptable to apply the original PN model for perfect dislocations to the core structure of partial dislocations.

For a more accurate description of FCC metals, the model considered in this section can be further extended to allow the partial Burgers vectors to be non-parallel to the perfect Burgers vector. To do so, the free energy of the interface needs to be a function of two slip components, and this two-dimensional function is often called the γ-surface. The equilibrium profile of the dislocation core can be obtained by matching the two shear stress components on the slip plane obtained from the partial derivatives of the γ-surface with those produced by the distribution of edge and screw Burgers vectors [2]. As long as the partials are well separated, the result can be well approximated by assuming that the partial dislocation cores spread out on the slip plane in the same way as predicted by the original PN model (see Exercise problem 12.2).

12.3 Dislocations in diamond cubic structures

We now discuss dislocations in crystals that do not have close-packed structures. We begin with dislocations in diamond cubic crystals. The diamond cubic crystal structure is called an open structure, because it contains lots of open spaces between atoms (see Exercise problem 4.2). The density of diamond cubic crystals (such as Si and Ge) is lower than that of the liquid. As a result, these crystals float on their own liquid upon melting.[3]

12.3.1 Glide-set and shuffle-set planes

As shown in Chapter 1, the diamond cubic crystal structure has the FCC lattice and a two-atom basis. One atom in the basis is located at the origin, and the other atom is located at $\frac{a}{4}[1\,1\,1]$. These two atoms are shown as white and gray spheres, respectively, in Fig. 12.7. Each atom forms four covalent bonds, shown as rods, with its neighbors. The angle between every two such bonds is 109°.

In Fig. 12.7a, the crystal is oriented in such a way that the atomic layers on $(1\,1\,1)$ planes can be clearly identified (see shaded triangles). Since the white spheres form an FCC lattice, they can be regarded as $(1\,1\,1)$ atomic layers stacked on top of each other in the sequence of ABCABC.... The gray spheres form another FCC lattice, and can also be regarded as $(1\,1\,1)$ atomic layers stacked on top of each other. We will label the atomic layer that is offset by $\frac{a}{4}[1\,1\,1]$ from the layer A (B, and C) as the layer a (b, and c). Hence the entire diamond cubic crystal structure can be regarded as $(1\,1\,1)$ atomic layers stacked on top of each other in the sequence of AaBbCcAaBbCc..., as shown in Fig. 12.7b.

We note that the atoms in layer a are directly on top of layer A (when viewed along $[1\,1\,1]$), and that the distance between layers a and A is greater than the distance between layers a and B. Because the definition of dislocations often involves slip between adjacent atomic layers, it is important to distinguish the *shuffle-set* planes between the widely spaced atomic layers, and *glide-set* planes between the closely spaced atomic layers. The shuffle-set planes can be

3 Ice floats on water for the same reason. Even though ice crystals do not have the diamond cubic structure, they also have an open structure.

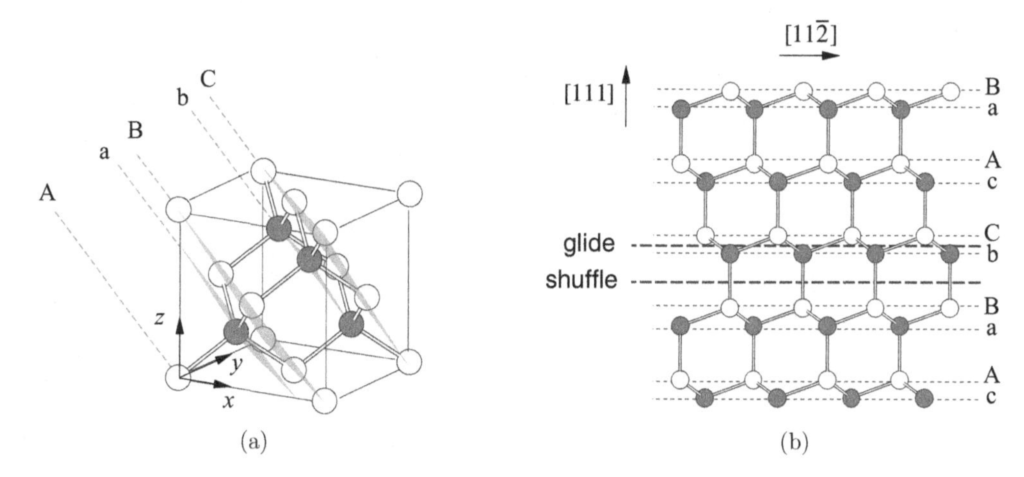

Figure 12.7. (a) Unit cell of the diamond cubic crystal structure. One atom in the basis forms the atomic layers ABC. . . and the other atom in the basis forms the atomic layers abc. . . . (b) The shuffle-set plane is between more widely spaced atomic layers and the glide-set plane is between more closely spaced atomic layers.

represented as A|a, B|b, or C|c, where | indicates the location of the slip plane. Similarly, the glide-set planes can be represented as a|B, b|C, or c|A. Of the four covalent bonds connected to the same atom, one is cut by a shuffle-set plane and three are cut by a glide-set plane. Dislocations on the shuffle-set planes and those on the glide-set planes have very different core structures.

12.3.2 Partial dislocations on glide-set planes

Because the diamond cubic crystal has the same FCC lattice as the FCC crystal, it also has the same slip systems. The active slip planes in diamond cubic crystals are $\{1\,1\,1\}$ planes, and the slip directions are $\frac{a}{2}\langle 1\,1\,0\rangle$.

We now adapt the approach introduced in Section 11.1 to analyze the dislocation core structure in diamond cubic crystals, starting with dislocations on glide-set slip planes. In particular, we will determine how atoms are likely to slide past each other on the slip plane, using a picture similar to Fig. 11.2. However, we can no longer treat the atoms as hard spheres. Instead, we need to account for the fact that each atom prefers to form four covalent bonds that are at 109° from each other.

Figure 12.8 plots three layers of atoms adjacent to a glide-set $(1\,1\,1)$ plane. The white spheres immediately above the slip plane represent atoms in layer A. The black spheres immediately below the slip plane represent atoms in layer c. The gray spheres represent atoms in layer C. Layer A slips relative to layer c in the $\frac{a}{2}[1\,0\,\bar{1}]$ direction. In Fig. 12.8c, the structure before slip occurs is shown on the left, and the structure after slip occurs is shown on the right. As the group of white spheres slides half-way to the right, they are likely to fall into the pocket at the center of three black spheres, as shown by the middle structure (indicated as "zig") in Fig. 12.8c. In doing so, each white sphere can still form four covalent bonds with its neighbors: three bonds with the black spheres in layer c below and one bond with the atoms in layer a above (not shown). This means that the trajectory of the atoms on top of the slip plane is unlikely to follow a straight line along $\frac{a}{2}[1\,0\,\bar{1}]$. Instead, it is more likely to first take a step of $\frac{a}{6}[1\,1\,\bar{2}]$, and then take another step of $\frac{a}{6}[2\,\bar{1}\,\bar{1}]$, as indicated in Fig. 12.8b.

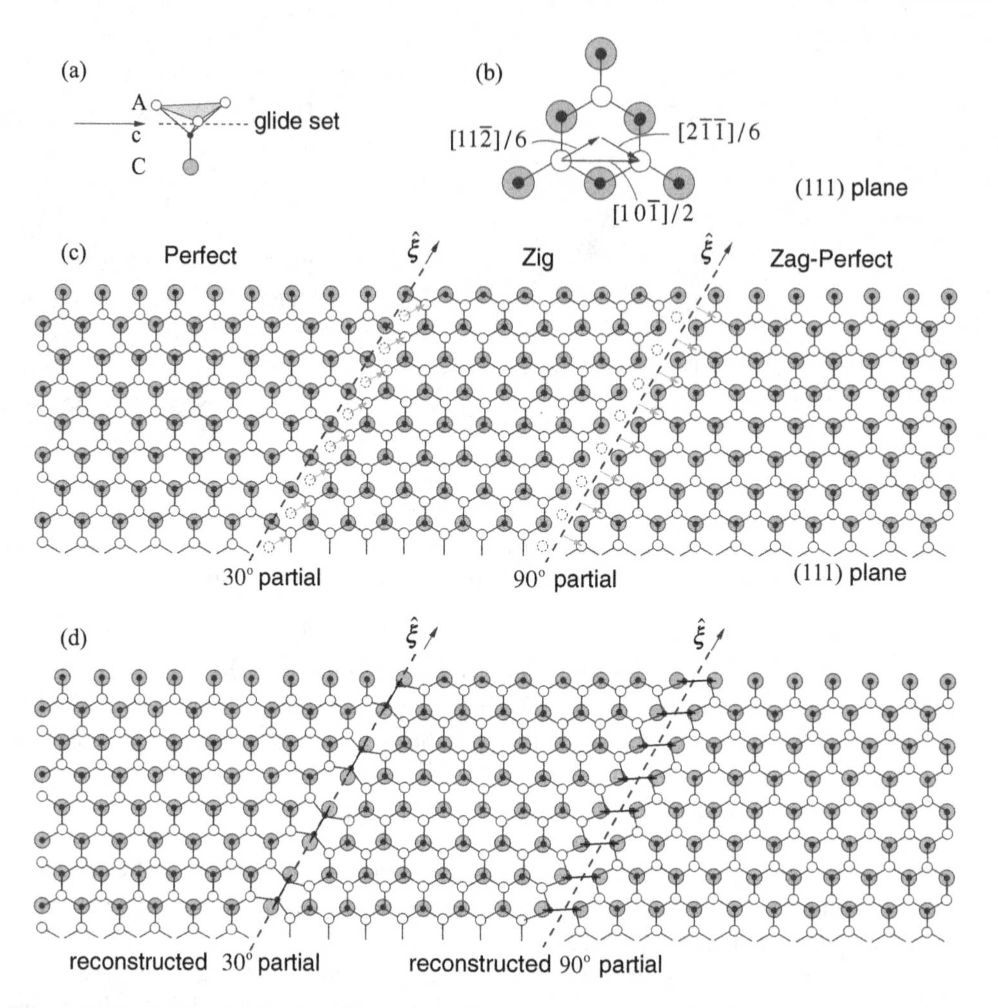

Figure 12.8. (a) Three layers of atoms adjacent to a glide-set plane. (b) Plan view of these three atomic layers. (c) A group of atoms on top of (1 1 1) glide-set plane sliding by $\frac{a}{2}[1\,0\,\bar{1}]$ in plan view. Left: before sliding. Middle: after partial sliding by $\frac{a}{6}[11\bar{2}]$. Right: after another partial sliding by $\frac{a}{6}[2\bar{1}\bar{1}]$. The original locations of the white spheres are indicated by dashed circles in the cores. (d) Atomic positions after core reconstruction.

In the middle ("zig") configuration shown in Fig. 12.8c, the atoms in layer A above the slip plane are translated by a vector that is not the repeat vector of the FCC lattice. After the translation, these atoms occupy the location of layer B. Notice that all atoms above this layer are translated by the same amount. Hence the stacking sequence in this region becomes AaBbCc|BbCcAa . . . , corresponding to the intrinsic stacking fault.

The boundary between the perfect (unslipped) area and the zig (stacking fault) area is a partial dislocation with Burgers vector $\vec{b}_{p1} = \frac{a}{6}[11\bar{2}]$ for the sense vector $\hat{\xi}$ shown. The boundary between the zig (stacking fault) area and the area where slip by a lattice repeat vector has occurred is another partial dislocation with Burgers vector $\vec{b}_{p2} = \frac{a}{6}[2\,\bar{1}\,\bar{1}]$. For the dislocation line direction shown in Fig. 12.8c, the partial dislocation ($\vec{b}_{p1}$) on the left is a 30° mixed dislocation, commonly called a 30° partial. The partial dislocation ($\vec{b}_{p2}$) on the right is a 90° partial.

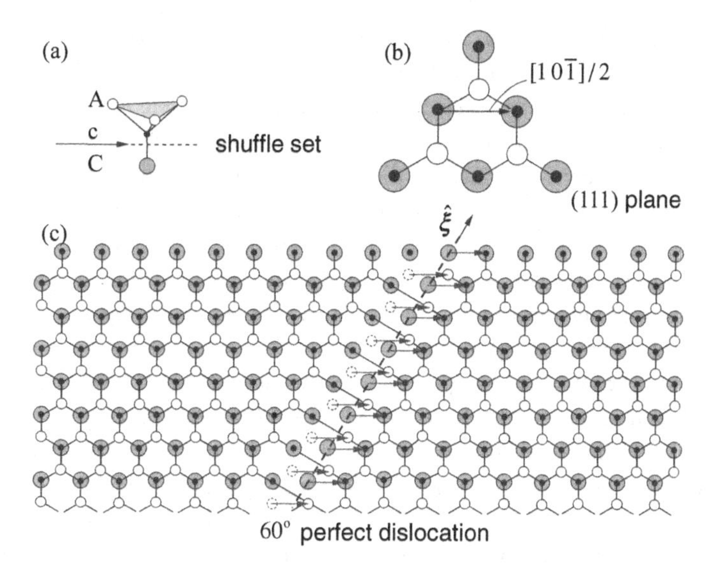

Figure 12.9. (a) Three layers of atoms adjacent to the shuffle-set plane. (b) Plan view of these atomic planes. (c) A group of atoms on top of (1 1 1) shuffle-set plane sliding by $\frac{a}{2}[1\,0\,\bar{1}]$ in plan view. Left: before sliding. Right: after sliding. The original locations of the white spheres are indicated by dashed circles.

In Fig. 12.8c, each black atom at the core of the 30° partial forms only three covalent bonds with their neighbors (two with white atoms above and one with a gray atom below). This is a high energy configuration. The dislocation can lower its energy by slightly adjusting the positions of the atoms near the core. In this case, neighboring black atoms in the core can pair up and form covalent bonds with each other, as shown in Fig. 12.8d. This process is called *core reconstruction* and leads to lower energy configurations that are believed to be the actual atomistic configurations in the dislocation core [113]. Similarly, each black atom at the core of the 90° partial shown in Fig. 12.8c also forms only three covalent bonds with their neighbors. They can bond with each other by core reconstruction to form a lower energy configuration. A possible core reconstruction of the 90° partial is shown in Fig. 12.8d.

The directionality of covalent bonds makes the core contribution to the dislocation line energy strongly dependent on the line orientation. The perfect glide-set dislocations prefer the screw (0°) and 60°-mixed orientations. The line direction is of the $\langle 1\,1\,0\rangle$ type in both cases. Each perfect screw dislocation dissociates into two 30° partials, and each 60°-mixed perfect dislocation (as shown in Fig. 12.8) dissociates into one 30° partial and one 90° partial.

Motion of dislocations in diamond cubic crystals requires breaking strong covalent bonds, which requires a higher energy than that for breaking metallic bonds. Hence dislocations in diamond cubic crystals usually have very low mobility at room temperature (see Eq. (10.69)). Consequently diamond cubic semiconductors are typically brittle at room temperature.

12.3.3 Perfect dislocations on shuffle-set planes

We now examine what would happen if slip occurs on the shuffle-set plane, between two widely spaced atomic layers. Figure 12.9 plots the scenario when shuffle-set (1 1 1) plane is between layers C and c. In this case, the atoms in layer A and those in layer c translate together as a rigid block; the covalent bonds between them are not disrupted.

Consider an atom in layer c, shown as a black sphere in Fig. 12.9a. The three covalent bonds it forms with atoms in layer A translate together with the atom in layer c. To keep the angle between every two bonds on the same atom at 109°, the direction of the fourth covalent bond on the atom in layer c must remain in the [1 1 1] direction. Therefore, for the covalent bond between the layers c and C to form again after the slip, the slip must occur by a lattice repeat vector, such as the $\frac{a}{2}[1\,0\,\bar{1}]$ vector shown in Fig. 12.9b. In other words, there are no stable stacking faults on the shuffle-set plane. Consequently, dislocations on the shuffle-set plane do not dissociate into partials.

On the left side of Fig. 12.9c, slip has not occurred; on the right side, atoms in layer c (and all atoms above them) have moved by $\frac{a}{2}[1\,0\,\bar{1}]$ relative to the atoms in layer C. The boundary between these two regions is a perfect dislocation with Burgers vector $\frac{a}{2}[1\,0\,\bar{1}]$ for the sense vector shown. Note that each black atom in the dislocation core shown in Fig. 12.9c forms only three covalent bonds, two with white atoms above and one with a gray atom below. To the right of these black atoms, a row of gray atoms also forms only three covalent bonds (with atoms further below, not shown). More covalent bonds may be formed during core reconstruction to lower the dislocation core energy. It has been observed that the perfect shuffle-set dislocations prefer the screw ⟨1 1 0⟩ orientation and the 41°-mixed ⟨1 2 3⟩ orientation [114], possibly due to favorable core reconstructions for dislocations along these orientations.

12.4 Dislocations in BCC metals

The body-centered cubic (BCC) crystals also have non-close-packed structures. The core structure and behavior of dislocations in BCC metals are more complex than those in FCC metals. In fact, certain aspects of dislocation behavior in BCC metals still remain controversial even to date. This means that the dislocation core structure in BCC metals are not dictated by geometry alone, but are influenced by the details of interatomic interactions. Hence the hard sphere model we have employed in FCC metals will only be partially successful for gaining a basic understanding of dislocations in BCC metals. In the following, we discuss those aspects of dislocation behavior in BCC metals that can be understood through the geometric concepts we have introduced so far.

The shortest lattice vectors in BCC metals are of the $\frac{a}{2}\langle 1\,1\,1\rangle$ type, and slip is indeed observed to occur in the ⟨1 1 1⟩ direction. Hence we will focus on dislocations with $\frac{a}{2}\langle 1\,1\,1\rangle$ Burgers vectors here. The most widely spaced atomic planes in BCC metals are the {1 1 0} planes. Hence we expect them to be the active slip planes on which dislocations move, and if so, we expect the perfect dislocation cores to spread on {1 1 0} planes according to the Peierls–Nabarro model. It has been generally observed that the most active slip planes in BCC metals are indeed of the {1 1 0} type at low temperatures [115]. However, at higher temperatures, slip on {1 1 2} and {1 2 3} planes has also been observed.[4] This is because the screw dislocations in BCC metals do not have a planar core and they are able to change their slip planes easily. The non-planar core

4 Note that {1 1 2} planes have the second largest spacing and {1 2 3} planes rank the third among the planes that contain the ⟨111⟩ direction.

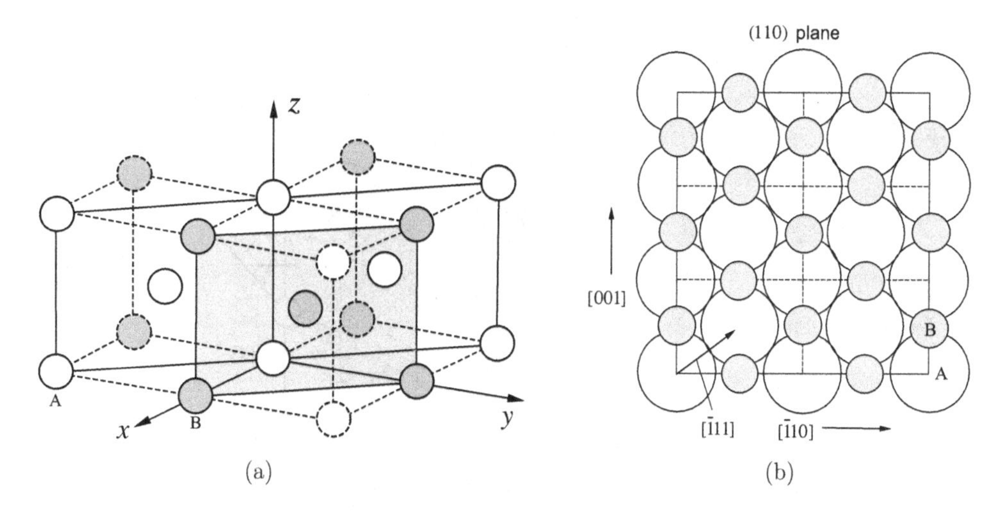

Figure 12.10. Stacking of BCC crystal on the (1 1 0) plane. (a) Three-dimensional view. (b) Plan view.

structure of screw dislocations in BCC metals is a major difference from FCC metals, and the behavior of screw dislocations is primarily responsible for a variety of mechanical properties of BCC metals.

Screw dislocations in BCC metals do not dissociate onto a plane (as they do in FCC metals) because there are no stable stacking faults in BCC metals. Under low temperature and high stress conditions, BCC metals can deform by twinning on {1 1 2} planes. This is because based on geometrical considerations alone stacking faults can exist in BCC crystals on {1 1 2} planes. The nature of interatomic interactions in BCC metals is such that a stacking fault on a single {1 1 2} plane is unstable. However, when several stacking fault planes form on top of each other, the resulting structure becomes metastable and is geometrically equivalent to a micro-twin, similar to the twinning mechanism in FCC metals (see Section 11.1.3). In the following, we discuss the stacking sequence of BCC metals on {1 1 0} and {1 1 2} planes.

12.4.1 Stacking sequence on {1 1 0} planes

Figure 12.10a shows atoms in three adjacent unit cells of the BCC crystal structure. We can see that the atoms on each (1 1 0) plane form a centered-rectangle lattice. The atomic layer passing through the origin is shown as white spheres with solid borders in Fig. 12.10a. We will label this layer as layer A. The next atomic layer in the [1 1 0] direction is shown as gray spheres with solid borders, and will be labeled as layer B. The atoms in the next layer are exactly on top of those in layer A when viewed along [1 1 0]. Therefore, the entire BCC crystal structure can be considered as (1 1 0) atomic layers stacked on top of each other in the sequence of ABAB.... Figure 12.10b shows the BCC crystal structure viewed in the [1 1 0] direction. The atoms in layer A are shown as white spheres and the atoms in layer B are shown as gray spheres. Layer B is displaced from layer A in the out-of-plane direction by $a/\sqrt{2}$, where a is the lattice constant.

Because there are only two possible types of {1 1 0} layers, based on a hard sphere cohesion picture any mistake in the stacking sequence would lead to very high energy configurations (either A|A or B|B) where atoms sit exactly on top of each other. Therefore, no stacking faults

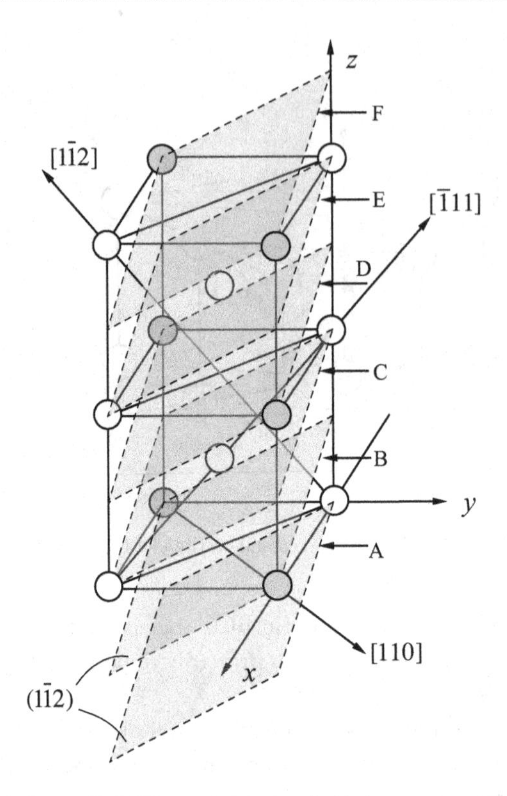

Figure 12.11. The $(1\,\bar{1}\,2)$ planes of a BCC crystal showing the six layers comprising the ABCDEF... stacking sequence.

can exist on the $\{1\,1\,0\}$ plane in BCC metals. Consequently, dislocations do not dissociate into partials on $\{1\,1\,0\}$ planes.

12.4.2 Stacking sequence on $\{1\,1\,2\}$ planes

Figure 12.11 shows a set of $(1\,\bar{1}\,2)$ atomic planes in a BCC crystal. There are six types of $(1\,\bar{1}\,2)$ layers, stacked on top of each other in the periodic pattern of ABCDEFABCDEF.... Figure 12.12a shows the BCC crystal structure viewed along the $[1\,1\,0]$ direction. This is the same viewing direction as that used in Fig. 12.10b. White atoms are in the plane of the paper, and the coordinates of the gray atoms have an out-of-plane component of $a/\sqrt{2}$. The crystal is reoriented in Fig. 12.12a so that the horizontal axis is along the $[\bar{1}\,1\,1]$ direction and horizontal dashed lines now represent the $(1\,\bar{1}\,2)$ planes. In this orientation, it is easier to identify the six different types of $(1\,\bar{1}\,2)$ layers, stacked on top of each other in the sequence of ABCDEF... to form a perfect BCC structure.

Suppose that layer B (together with all layers above it) slips relative to layer A by $\frac{a}{6}[\bar{1}\,1\,1]$, then it becomes layer F, as shown in Fig. 12.12b. The stacking sequence becomes ABCDEFA|FABCDE..., where | indicates the plane on which slip occurs. The result is an intrinsic stacking fault, and the same stacking sequence can be obtained if four layers (BCDE) were removed. This result is analogous to the intrinsic stacking fault in FCC metals, where the stacking sequence ABCA|CAB... can also be obtained by either slip or removing an atomic layer.

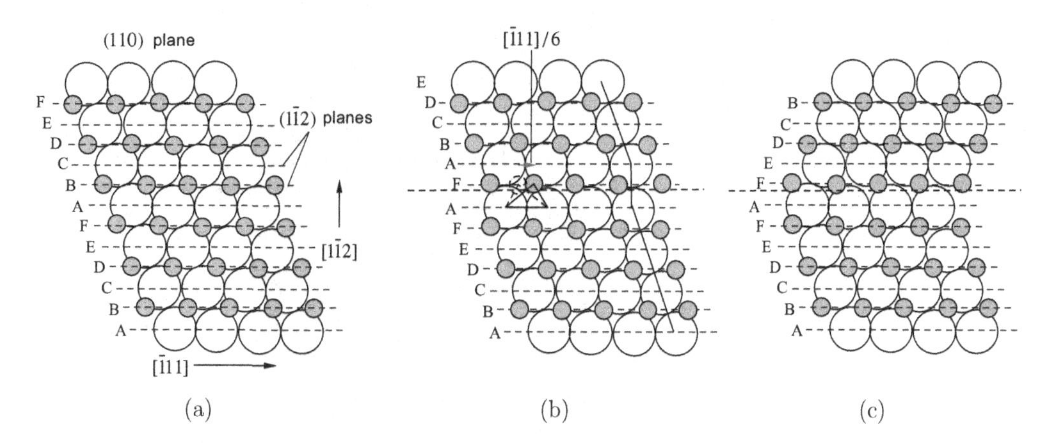

Figure 12.12. BCC crystal viewed along the [1 1 0] direction, showing the stacking sequence on the $(1\,\bar{1}\,2)$ plane. (a) Perfect stacking sequence ABCDEF. . . . (b) If the layer B slips relative to layer A by $\frac{a}{6}[\bar{1}\,1\,1]$, it becomes layer F, creating an intrinsic stacking fault. (c) If every layer above A slips by $\frac{a}{6}[\bar{1}\,1\,1]$ relative to the layer below it, then a twin boundary is formed.

The intrinsic stacking fault in BCC crystals, as shown in Fig. 12.12b, is unstable. This means that if left alone, the layer F would spontaneously slip backward (by $\frac{a}{6}[1\,\bar{1}\,\bar{1}]$) to return to the original position of layer B. Because the stacking fault is unstable, dislocations on $\{1\,1\,2\}$ planes do not dissociate into partials either.

However, if every layer above layer A slips relative to the layer below it by $\frac{a}{6}[\bar{1}\,1\,1]$, then the crystal above layer A becomes a mirror image of the crystal below, as shown in Fig. 12.12c. The result is a twin boundary, with stacking sequence ABCDEFA|FEDCBA. . . . The twin boundary is meta-stable. Figure 12.12c also shows that a bi-crystal with a twin boundary is equivalent to a single crystal containing an infinite number of intrinsic stacking faults on top of each other. If a small number (N) of intrinsic stacking faults form on adjacent planes, the structure is equivalent to a thin twinned crystal bounded by two twin boundaries. The structure becomes meta-stable if the twinned crystal is sufficiently thick. Computer models [116] have indicated that the structure becomes meta-stable for N as small as 3. This means that the intrinsic stacking fault becomes meta-stable when a small number of them appear on adjacent planes to form a twin plate. The twin plate thickens by forming new stacking fault planes in the neighboring crystal. This can be accomplished by nucleating loops of partial dislocations with Burgers vector $\frac{a}{6}\langle 1\,1\,1\rangle$, which are called *twinning partials*. Therefore, even though perfect dislocations in BCC metals do not dissociate into well-defined partial dislocations, the partial dislocations play an important role in deformation twinning.

12.4.3 Edge dislocation core

Non-screw dislocations in BCC metals have well-defined slip planes. Atomistic simulations predict that non-screw dislocations with orientations sufficiently different from screw (e.g. more than 10°) indeed spread on their slip planes [117], consistent with the Peierls–Nabarro model.[5]

5 Non-screw dislocations very close to the screw orientation have a zig-zag structure with longer segments aligned along the screw orientation connected by shorter non-screw segments, called "kinks."

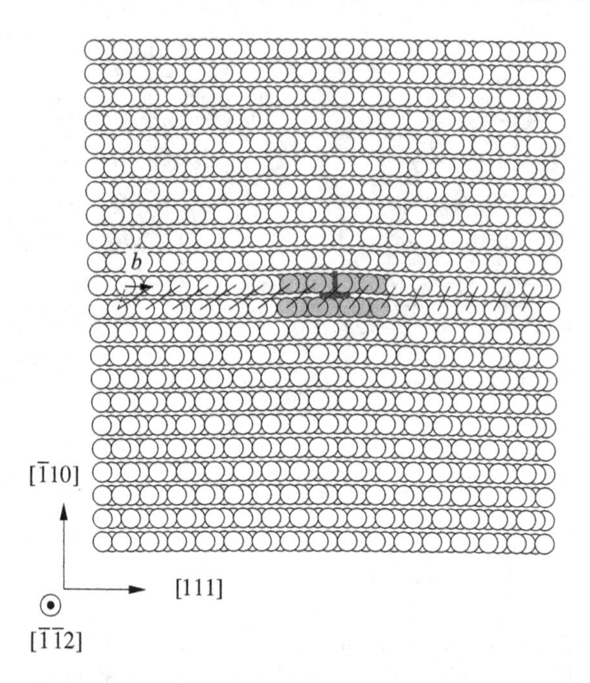

Figure 12.13. Core structure of an edge dislocation predicted by the Finnis–Sinclair potential of tantalum. Atoms with local central symmetry parameter [118, 2] greater than 0.6 Å^2 are colored gray to show the dislocation core region. Short line segments connect atoms immediately above and below the slip plane that would have been nearest neighbors in a perfect crystal.

Figure 12.13 shows the atomic positions around an edge dislocation with Burgers vector $\vec{b} = \frac{a}{2}[1\,1\,1]$ and slip plane $\hat{n} = (\bar{1}\,1\,0)$ in BCC tantalum. The structure is viewed along the $\hat{\xi} = [\bar{1}\,\bar{1}\,2]$ line direction of the dislocation. When the two atomic layers immediately above and below the slip plane are examined from right to left, the atoms on the top layer gradually move over the bottom layer by $\vec{b}$. This is similar to the schematic shown in Fig. 12.1b. Hence we expect the Peierls–Nabarro model to provide a good description of the core structure of non-screw dislocations in BCC metals.

12.4.4 Screw dislocation core

Given that both the line direction and the Burgers vector of screw dislocations in BCC crystals are along $\langle 1\,1\,1 \rangle$ directions, it is useful to consider BCC crystals as a collection of atomic rows along $\langle 1\,1\,1 \rangle$ when visualizing screw dislocations.

Figure 12.14a shows that the BCC crystal can be considered as a collection of [1 1 1] atomic rows. The spacing between neighboring atoms in each row is $\frac{a}{2}[1\,1\,1]$. There are three types of atomic rows, designated as A, B, and C. Atomic rows of different types have relative offset from each other along the [1 1 1] direction. Atomic rows of the same type form a triangular lattice when viewed along the [1 1 1] direction, as shown in Fig. 12.14b.

Figure 12.15a shows three atomic rows in a perfect BCC crystal, A, B, and C, that are closest to each other. These three atomic rows form a triangle when viewed along [1 1 1], as shown in Fig. 12.14b. Figure 12.15a shows that atoms in row B are offset by $\frac{a}{6}[1\,1\,1]$ relative to atoms in

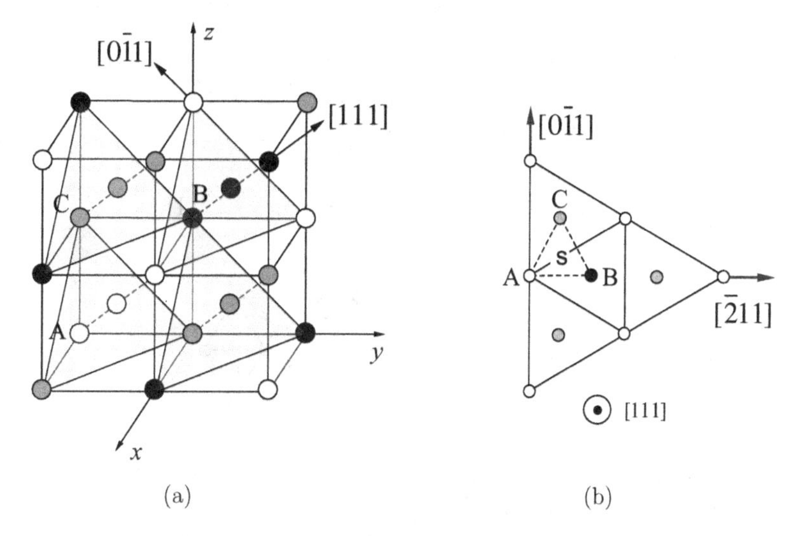

Figure 12.14. (a) BCC crystal structure with (1 1 1) planes highlighted as shaded triangles. (b) BCC crystal structure viewed along [1 1 1], showing three types of atomic columns A, B, and C. In this view, the screw dislocation s (if introduced) resides at the center of the triangle ABC.

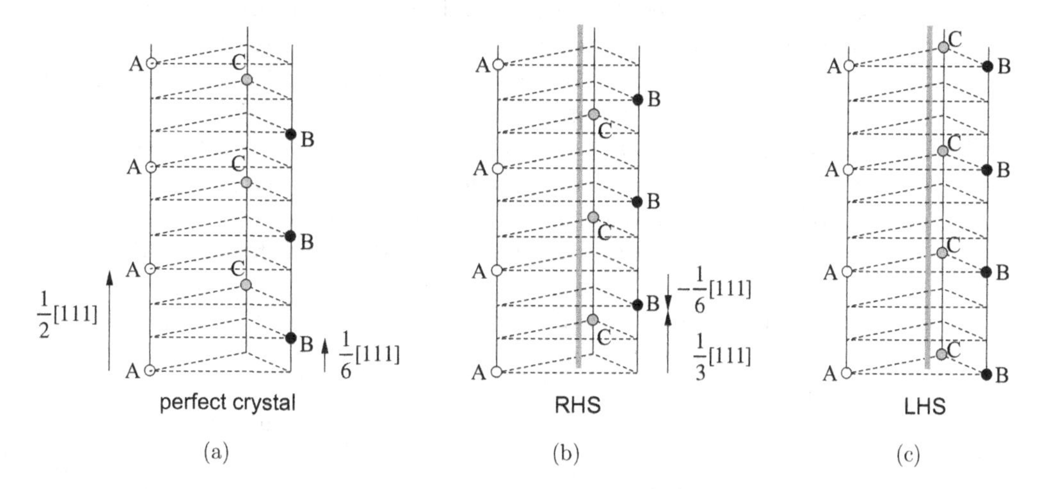

Figure 12.15. (a) Three-dimensional view of the [1 1 1] atomic rows in a perfect BCC structure. (b) Three-dimensional view of the [1 1 1] atomic rows containing a right-handed screw dislocation in the "easy core" structure. (c) Three-dimensional view of the [1 1 1] atomic rows containing a left-handed screw dislocation in the "hard core" structure.

row A. Atoms in row C are offset by the same amount relative to atoms in row B. The same is true for atoms in row A relative to atoms in row C.

We now consider what would happen if we insert a right-handed screw (RHS) dislocation with Burgers vector $\vec{b} = \frac{a}{2}[1\ 1\ 1]$ at the center of the three atomic rows. The RHS will introduce additional relative displacements between neighboring atomic rows. The displacement field of the RHS dislocation is such that as one goes around the dislocation line by 360°, e.g. from row A to B, C and back to A, one accumulates a total displacement of $\vec{b} = \frac{a}{2}[1\ 1\ 1]$. By symmetry, row B is further shifted by $\vec{b}/3 = \frac{a}{6}[1\ 1\ 1]$ relative to row A, and row C is also further shifted by

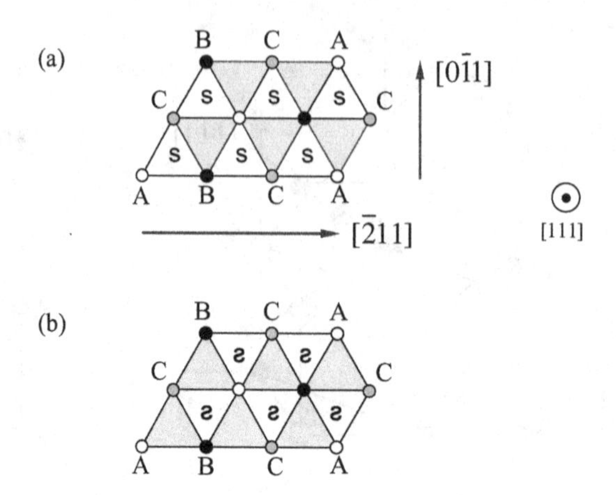

Figure 12.16. Atomic rows of the BCC crystal viewed along [1 1 1]. (a) The triangular regions are shaded according to whether a RHS takes the "easy core" (white) or "hard core" (gray) structure. (b) The triangular regions are shaded according to whether a LHS takes the "easy core" (white) or "hard core" (gray) structure.

the same amount relative to row B. As a result, atoms in row B are now offset by $2\vec{b}/3 = \frac{a}{3}[1\ 1\ 1]$ relative to atoms in row A. Given that each atomic row has periodicity of $\vec{b}$, we can also say that atoms in row B are now offset by $-\vec{b}/3 = -\frac{a}{6}[1\ 1\ 1]$ relative to atoms in row A, as shown in Fig. 12.15b. The end result is that the nearest neighbor distance between atoms in row A and atoms in row B has not changed by the introduction of the RHS. For the same reason, atoms in row C are now also offset by $-\vec{b}/3$ relative to atoms in row B, and the nearest neighbor distance between atoms in row C and those in row B has not changed either. The same is true between atoms in row A and atoms in row C. Hence the dislocation core structure shown in Fig. 12.15b is expected to be a low energy configuration, and is often referred to as the "easy core" structure.

If, instead, we insert a left-handed screw (LHS) dislocation at the center of the triangle ABC, then row B is further shifted by $-\frac{a}{6}[1\ 1\ 1]$ relative to row A, and row C is also further shifted by the same amount relative to row B. The result is that all three atomic rows at the dislocation core are perfectly aligned in the [1 1 1] direction, as shown in Fig. 12.15c. The nearest neighbor distance between the atomic rows is now smaller than that in the perfect BCC crystal. As a result, the dislocation core structure shown in Fig. 12.15c is expected to be a high energy configuration, and is often referred to as the "hard core" structure.

Figure 12.16a shows the locations in a BCC crystal where a RHS can assume the "easy core" structure (white triangles). It can be easily shown that, if the same RHS is put at the neighboring triangles, then it must take the "hard core" structure (gray triangles). The situation is reversed for a LHS, as shown in Fig. 12.16b. Atomistic simulations often predict that the "hard core" structure is unstable. In this case, the "hard core" structure can be considered as an energy barrier between two adjacent "easy core" structures.

First principles models (based on quantum mechanics) [119] predict that the screw dislocation in BCC metals has the "easy core" structure shown in Fig. 12.15b. There is no visible spreading of the dislocation core on any ({1 1 0} or {1 1 2}) planes. The compact core structure of the screw dislocation in BCC metals makes it difficult to move. For example, its Peierls stress, which is the critical stress required to move the dislocation in the zero temperature limit (without the aid of thermal flucutations), is very high (on the order of 1 GPa), except for metals

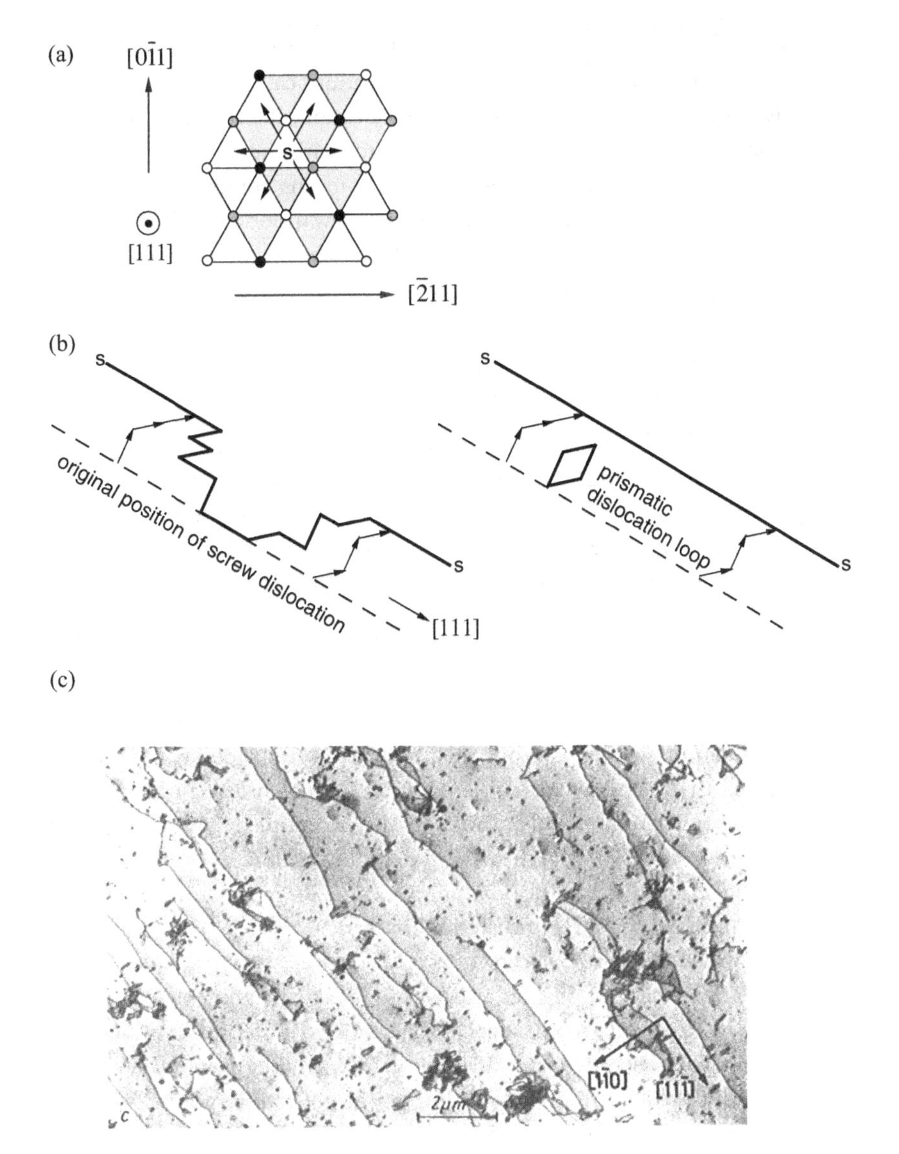

Figure 12.17. (a) Multiple possible directions in which a screw dislocation in a BCC crystal can move. (b) Formation of a dislocation loop if different sections of a long screw dislocation follow different trajectories. (c) TEM image of dislocation loop debris formed in the wake of moving screw dislocations in BCC molybdenum [120]. Copyright Wiley-VCH Verlag GmbH & Co. kGaA. Reproduced with permission.

with low melting points (such as sodium). On the other hand, the non-screw dislocations have a planar core structure and are much more mobile than the screw dislocations. The motion of the more mobile, non-screw dislocations creates long segments of screw dislocations, which dominate dislocation microstructures in BCC metals at medium to low temperatures. As a result, the plastic deformation of BCC metals is largely controlled by the motion of screw dislocations.

As shown in Fig. 12.17a, the shortest distance that a screw dislocation can move from one easy core configuration to the next is $\frac{a}{3}\langle 2\ 1\ 1\rangle$, which is on a $\{1\ 1\ 0\}$ plane. There are six possible $\frac{a}{3}\langle 2\ 1\ 1\rangle$ steps that a screw dislocation can take. Because the screw dislocation in a BCC crystal is not dissociated on any particular plane, its motion is not confined to a single plane. The variation of local stress and thermal fluctuations can easily change the plane on which the screw dislocation moves. For a long screw dislocation, the sequence of planes on which one section chooses to move does not necessarily agree with the choices made by a different section. The conflicts between these choices creates self-pinning points on a moving dislocation and prismatic dislocation loops in the wake of a moving screw dislocation [116], as illustrated in Fig. 12.17b. A transmission electron microscopy (TEM) image of screw dislocations and the dislocation loops they created is shown in Fig. 12.17c.

12.5 Dislocation–point defect interactions

We close this chapter with a discussion on the interaction between point defects and dislocations. We shall see that, due to the elastic interactions, there is a tendency for solutes to segregate near the dislocation core. This phenomenon was first studied by Cottrell and the solute cloud near the dislocation core is now called the *Cottrell atmosphere*. The Cottrell atmosphere can have a strong influence on the mechanical properties of crystals, by modifying the energy and the mobility of dislocations. It provides a mechanism for solid solution hardening mentioned in Section 4.2.4.

In Sections 11.3 and 11.4 we have discussed dislocations in crystals containing two chemical species forming an ordered compound. The crystal to be discussed in this section also contains two chemical species, with one species appearing as solutes (i.e. point defects) in the host crystal formed by the other species. The solutes are not ordered in the sense that whether or not a specific (interstitial or substitutional) site is occupied by a solute atom is a random variable. However, at a larger length scale, the solute distribution can be described by a concentration function, which is usually a continuous function in space. If the solutes are mobile, their local concentration can adapt to the inhomogeneous stress fields of dislocations, causing them to segregate around the dislocation core.

For simplicity, we shall consider only interstitial point defects. We shall also limit our discussions to solutes that produce a spherically symmetric strain field, such as that discussed in Section 4.4.1. We shall also assume that the solutes are sufficiently mobile that they are able to reach the equilibrium distribution for the given dislocation configuration. Therefore, a good example to keep in mind is hydrogen (H) interstitials that occupy the octahedral sites of FCC palladium (Pd).

12.5.1 Finite solid under zero traction

Figure 12.18 illustrates a simple elasticity-based model of solutes in a crystalline solid. The solid is considered as an isotropic elastic medium with shear modulus μ and Poisson's ratio ν. The solid contains a three-dimensional periodic array of available solute sites (e.g. octahedral sites in an FCC crystal). When a site is occupied by a solute atom, it is modeled as first creating a spherical hole of volume v_0 and then inserting a sphere made of the same material but with volume $v_0 + \Delta \tilde{v}_f$ into the hole.[6] Such an arrangement leads to two significant effects: (1) no two

6 In this model, the exact value of v_0 is left unspecified. It needs to be small enough so that neighboring inclusions would not overlap. Hence v_0 usually needs to be smaller than the atomic volume Ω.

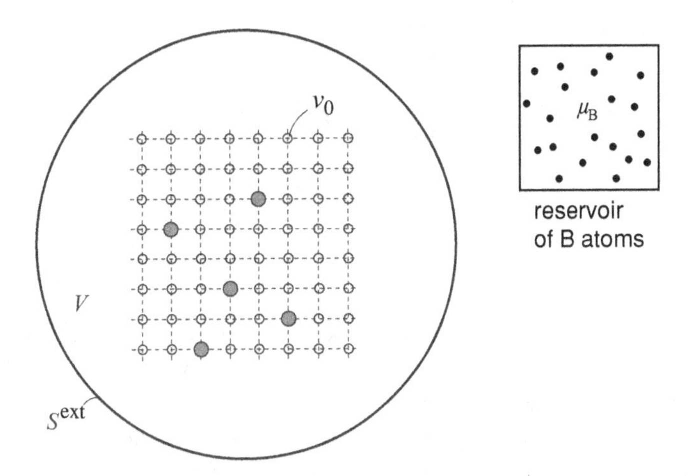

Figure 12.18. A schematic showing a solid of volume V and surface S^{ext} containing a periodic array of sites where inclusions can be inserted. These sites are a model of the interstitial sites in a crystal. To insert an inclusion into a site, a spherical hole of volume v_0 is first created and then a sphere of volume $v_0 + \Delta\tilde{v}_f$ is inserted into the hole. The sites containing inclusions are shown in gray.

solutes can occupy the same solute site, and (2) there is a maximum solute concentration limit (c_{max}). The quantity $\Delta\tilde{v}_f$ has been discussed in Section 5.4. For interstitial solutes, $\Delta\tilde{v}_f = \Omega_B^\dagger$.

Let V be the total volume of the solid and N_s be the total number of sites for solutes (inclusions), so that $c_{\text{max}} = N_s/V$. The purpose of this section is to find the equilibrium concentration c_0 of solutes when the solid is under zero stress (other than the stress generated by the solutes themselves). The solid under consideration is an open system that is allowed to exchange solutes with its environment. An example is a palladium crystal immersed in hydrogen gas. Let $N_{i,0}$ be the equilibrium number of solutes in the solid, so that $c_0 = N_{i,0}/V$. Define $\chi_0 \equiv N_{i,0}/N_s$ as the equilibrium fraction of sites occupied by the solutes, so that $c_0 = \chi_0 c_{\text{max}}$.

The equilibrium fraction of interstitials is given by the Fermi–Dirac distribution, Eq. (5.77). We assume that the interstitials are atoms of type B and are in thermodynamic equilibrium with a reservoir of B atoms. As a result, the chemical potential of B atoms as interstitials in the solid, $\Delta\mu_i$, equals the chemical potential of B atoms in the reservoir, μ_B. Therefore,

$$\chi_0 = \frac{1}{1 + \exp\left[\frac{1}{k_B T}(\Delta\tilde{e}_f + p\Delta\tilde{v}_f - T\Delta\tilde{s}_f - \mu_B)\right]}. \tag{12.28}$$

The 1 term in the denominator guarantees that $\chi_0 \leq 1$, so that $c_0 \leq c_{\text{max}}$. The quantities $\Delta\tilde{e}_f$, $\Delta\tilde{v}_f$, and $\Delta\tilde{s}_f$, have been defined in Section 5.4; $\Delta\tilde{e}_f$ is the energy cost of introducing one solute (inclusion) in an infinite solid. It can be shown that the elastic energy contribution to $\Delta\tilde{e}_f$ is

$$\Delta\tilde{e}_f^{\text{el}} = \frac{1}{2}p^{\text{I},\infty}\Delta\tilde{v}_f, \tag{12.29}$$

where $p^{\text{I},\infty}$ is the pressure inside the inclusion. Let $v_0 + \Delta v_A$ be the volume of the sphere after it has been inserted into the spherical hole. Then, Hooke's law gives

$$p^{\text{I},\infty} = B\frac{\Delta\tilde{v}_f - \Delta v_A}{v_0}. \tag{12.30}$$

Using Eqs. (4.59) and (2.44), we have

$$p^{\mathrm{I},\infty} = \frac{4\mu(1+\nu)}{9(1-\nu)} \frac{\Delta\tilde{v}_{\mathrm{f}}}{v_0} \tag{12.31}$$

and

$$\Delta\tilde{e}_{\mathrm{f}}^{\mathrm{el}} = \frac{2\mu(1+\nu)}{9(1-\nu)} \frac{(\Delta\tilde{v}_{\mathrm{f}})^2}{v_0}. \tag{12.32}$$

However, differences in chemical bond energies also contribute to $\Delta\tilde{e}_{\mathrm{f}}$ and can even dominate the formation energy. Therefore, we shall treat $\Delta\tilde{e}_{\mathrm{f}}$ as a constant in the following discussions. The formation entropy $\Delta\tilde{s}_{\mathrm{f}}$ will also be treated as a constant.

In Section 4.4, we have seen that in an infinite solid, the stress field outside an inclusion is purely shear. Therefore, there is no interaction between two inclusions in an infinite solid as long as they do not touch each other. However, Section 4.4 shows that in order to satisfy the zero traction boundary condition of a finite solid, an image stress must be superimposed for each inclusion introduced to the solid. For a single inclusion at the center of a spherical solid, the image stress is a constant and equals

$$\sigma_{ij}^{\mathrm{img}} = \frac{4\mu(1+\nu)}{9(1-\nu)} \frac{\Delta\tilde{v}_{\mathrm{f}}}{V} \delta_{ij}, \tag{12.33}$$

where V is the volume of the entire solid. Because $\sigma_{ij}^{\mathrm{img}}$ is positive, the image stress is a tensile field, and corresponds to a negative image pressure, i.e.

$$p^{\mathrm{img}} = -\frac{4\mu(1+\nu)}{9(1-\nu)} \frac{\Delta\tilde{v}_{\mathrm{f}}}{V}. \tag{12.34}$$

Provided that solid contains a large number of solutes, and the variation of the solute concentration occurs over a length scale that is much smaller than the dimension of the solid, the image stress is still uniformly distributed inside the solid (except for a very thin layer near the surface). Given $N_{\mathrm{i},0} = N_{\mathrm{s}}\chi_0$ solutes at equilibrium, the image pressure is simply $N_{\mathrm{i},0}$ times the image pressure due to a single solute, i.e.

$$p = N_{\mathrm{i},0} p^{\mathrm{img}} = -N_{\mathrm{s}}\chi_0 \frac{4\mu(1+\nu)}{9(1-\nu)} \frac{\Delta\tilde{v}_{\mathrm{f}}}{V}. \tag{12.35}$$

Note that the image pressure only depends on the total number of solutes, but not on their locations; it does not induce any pairwise interaction between two inclusions. However, as we shall see below, the image stress does influence the equilibrium solute concentration in the solid. Define

$$\eta \equiv \frac{4\mu(1+\nu)}{9(1-\nu)} (\Delta\tilde{v}_{\mathrm{f}})^2 \frac{N_{\mathrm{s}}}{V} = \frac{4\mu(1+\nu)}{9(1-\nu)} (\Delta\tilde{v}_{\mathrm{f}})^2 c_{\mathrm{max}}; \tag{12.36}$$

we have

$$p\Delta\tilde{v}_{\mathrm{f}} = -\eta\chi_0. \tag{12.37}$$

Combining Eqs. (12.28)) and (12.37), we have

$$\chi_0 = \frac{1}{1 + \exp\left[\frac{1}{k_B T}(\Delta \tilde{e}_f - T \Delta \tilde{s}_f - \eta \chi_0 - \mu_B)\right]}. \tag{12.38}$$

Because χ_0 appears on both sides of Eq. (12.38), it is an implicit equation. Notice that $\eta > 0$, so that the $-\eta \chi_0$ term has the effect of lowering the energy cost to insert solutes and enhancing the solute concentration. In other words, the tensile image stress of existing solutes promotes the introduction of more solutes into the solid [121].

12.5.2 Solid with inhomogeneous stress field

We now consider a finite-sized solid under the influence of an inhomogeneous pre-existing stress field $\sigma_{ij}^{d}(\vec{x})$ before the inclusions are introduced. While the superscript ^{d} indicates our primary interest in the stress fields produced by dislocations, $\sigma_{ij}^{d}(\vec{x})$ can also include the stress fields generated by other defects as well as by external loads.

Because the pre-existing stress field is often inhomogeneous, we expect the equilibrium solute concentration to be inhomogeneous as well. Based on Eqs. (12.28) and (5.77), we have,

$$\chi(\vec{x}) = \frac{1}{1 + \exp\left[\frac{1}{k_B T}(\Delta \tilde{e}_f - T \Delta \tilde{s}_f + p(\vec{x}) \Delta \tilde{v}_f - \mu_B)\right]}. \tag{12.39}$$

The key is to find the pressure field $p(\vec{x})$ at sites unoccupied by solutes. In general, $p(\vec{x})$ has contributions from both the pre-existing stress field $\sigma_{ij}^{d}(\vec{x})$ and the solutes themselves. As noted above, the pressure field caused by an inclusion is zero outside the inclusion itself in an infinite medium. Hence the only contribution to $p(\vec{x})$ from the solutes is an image stress field.

We assume that the variation of the internal stress field $\sigma_{ij}^{d}(\vec{x})$ occurs over a length scale that is much smaller than the dimension of the solid. Under this condition, the image stress field is still uniform and simply proportional to the total number of solutes, i.e. the same as that in the previous section. Therefore,

$$p(\vec{x}) = -\frac{1}{3}\sigma_{ii}^{d}(\vec{x}) + N_i p^{\text{img}} = -\frac{1}{3}\sigma_{ii}^{d}(\vec{x}) + N_i \frac{4\mu(1+\nu)}{9(1-\nu)} \frac{\Delta \tilde{v}_f}{V}, \tag{12.40}$$

where N_i is the total number of solutes in the solid. Define

$$\overline{\chi} \equiv \frac{N_i}{N_s} = \frac{1}{V} \int \chi(\vec{x})\, d^3\vec{x} \tag{12.41}$$

as the volume average of the solute fraction over the entire solid. Hence,

$$p(\vec{x}) = -\frac{1}{3}\sigma_{ii}^{d}(\vec{x}) + N_s \overline{\chi} \frac{4\mu(1+\nu)}{9(1-\nu)} \frac{\Delta \tilde{v}_f}{V} \tag{12.42}$$

and

$$p(\vec{x}) \Delta \tilde{v}_f = -\frac{1}{3}\sigma_{ii}^{d}(\vec{x}) \Delta \tilde{v}_f - \eta \overline{\chi}. \tag{12.43}$$

Plugging this into Eq. (12.39), we obtain

$$\chi(\vec{x}) = \frac{1}{1 + \exp\left[\frac{1}{k_B T}\left(\Delta\tilde{e}_{\mathrm{f}} - T\Delta\tilde{s}_{\mathrm{f}} - \frac{1}{3}\sigma^{\mathrm{d}}_{ii}(\vec{x})\Delta\tilde{v}_{\mathrm{f}} - \eta\overline{\chi} - \mu_{\mathrm{B}}\right)\right]}. \quad (12.44)$$

From Eq. (12.38), the equilibrium solute fraction χ_0 under zero stress satisfies the following equation,

$$\frac{\chi_0}{1-\chi_0} = \exp\left[-\frac{1}{k_B T}(\Delta\tilde{e}_{\mathrm{f}} - T\Delta\tilde{s}_{\mathrm{f}} - \eta\chi_0 - \mu_{\mathrm{B}})\right], \quad (12.45)$$

allowing us to write the solute distribution field as [121]

$$\chi(\vec{x}) = \left\{1 + \frac{1-\chi_0}{\chi_0}\exp\left[\frac{1}{k_B T}\left(-\frac{1}{3}\sigma^{\mathrm{d}}_{ii}(\vec{x})\Delta\tilde{v}_{\mathrm{f}} - \eta(\overline{\chi} - \chi_0)\right)\right]\right\}^{-1}. \quad (12.46)$$

We note that the heterogeneous part of $\chi(\vec{x})$ only depends on the stress field $\sigma^{\mathrm{d}}_{ii}(\vec{x})$ which was present before the solutes were introduced. The existence of solutes only alters the equilibrium distribution uniformly through the image stress, as represented by the $-\eta(\overline{\chi} - \chi_0)$ term. This is consistent with the notion that there is no interaction between solutes that produce spherically symmetric strain fields in an infinite isotropic solid.

12.5.3 Solute distribution around an edge dislocation

As an example, we consider the equilibrium solute distribution around an infinitely long, straight edge dislocation in a very large solid under zero traction on the surface. Let χ_0 be the equilibrium fraction of solutes in the solid before the dislocation is introduced, when the solid is subjected to no stress other than those produced by the solutes.

The material parameters correspond to hydrogen interstitials occupying octahedral interstitial sites of the FCC metal palladium (Pd): $\mu = 48.3$ GPa, $\nu = 0.385$, $b = 2.75$ Å, and $\Delta\tilde{v}_{\mathrm{f}}/\Omega = 0.186$, where $\Omega = 14.72\,\text{Å}^3$ is the volume per Pd atom. We consider a positive edge dislocation along the z axis, with the extra-half-plane pointing in the $+y$ direction. The stress field of this dislocation is given in Eq. (9.31).

Let $\overline{\chi}$ be the average volume fraction of solutes after the dislocation is introduced. As the volume of the solid goes to infinity, the difference between $\overline{\chi}$ and χ_0 vanishes, so that the image term $-\eta(\overline{\chi} - \chi_0)$ in Eq. (12.46) can be ignored. In this sense, we reach the limit of the infinite solid.[7] In this limit, the equilibrium hydrogen fraction field is

$$\chi(x, y) = \left\{1 + \frac{1-\chi_0}{\chi_0}\exp\left[\frac{1}{k_B T}\left(-\frac{1}{3}\sigma^{\mathrm{d}}_{ii}(x, y)\Delta\tilde{v}_{\mathrm{f}}\right)\right]\right\}^{-1}. \quad (12.47)$$

Notice that this is an explicit expression, since the unknown $\chi(x, y)$ does not appear on the right hand side. Equation (12.47) is essentially the same as the expression given by Cottrell and Bilby [122, 123], except that here the constraint of at most one solute per site is taken into account explicitly.

Figure 12.19 shows the predicted $\chi(x, y)$ field with a zero-stress fraction of $\chi_0 = 0.01$. The contour lines of the H density are circles whose highest points all go through the center of the

7 However, the image term cannot be ignored if the solid contains a finite density of dislocations, so that $\overline{\chi} - \chi_0$ remains finite even as the volume of the solid goes to infinity.

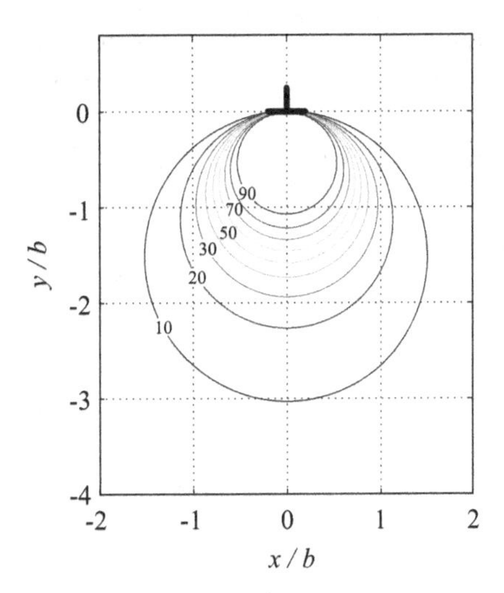

Figure 12.19. Contour lines of H fraction around an edge dislocation in Pd at 300 K at a background fraction of $\chi_0 = 0.01$. The numbers indicate the value of χ / χ_0, the multiplicative factor for the enhancement of solute fraction.

dislocation. These are essentially the contour lines of the pressure field of an edge dislocation. The solute concentration is enhanced beneath the glide plane where the pressure field is negative, and is reduced above the glide plane, where the pressure field is positive. Obviously the effect is stronger closer to the dislocation center, due to the $1/r$ stress singularity. But the Fermi–Dirac distribution guarantees that $0 \leq \chi \leq 1$. Therefore, our model predicts that χ becomes very close to 1 within a circular region beneath the glide plane. For example, $\chi \geq 0.9$ within the inner most contour line in Fig. 12.19.

The equilibrium solute distribution is highly non-uniform near the edge dislocation. There is not only a net rearrangement of solute atoms from above the glide plane to below the glide plane, but an overall net accumulation of solutes (due to the non-linear relationship between solute concentration and pressure). The solute distribution around the dislocation is usually called a *Cottrell atmosphere*. As the dislocation moves, it may drag the Cottrell atmosphere with it. Since solute motion requires diffusion, this slows the dislocation down [124], unless the dislocation moves fast and breaks away from the solute atmosphere. However, when the dislocation stops, the solutes can rearrange around the dislocation, so that a higher stress may be required to move the dislocation again. This effect is commonly referred to as *dynamic strain aging*.

The equilibrium point defect fraction field expression, Eq. (12.47), is applicable to not only interstitial solutes, but also to substitutional solutes and vacancies, provided that the appropriate values for $\Delta\tilde{v}_\mathrm{f}$ are used. In Section 5.4, we have seen that for interstitial solutes $\Delta\tilde{v}_\mathrm{f} = \Omega_\mathrm{B}^\dagger$, for substitutional solutes $\Delta\tilde{v}_\mathrm{f} = \Omega_\mathrm{B}^* - \Omega_\mathrm{A}$, and for vacancies $\Delta\tilde{v}_\mathrm{f} = \Delta v_\mathrm{v} - \Omega = \Delta V_\mathrm{v}^\mathrm{rlx}$.

12.5.4 Solute stress around an edge dislocation

The inhomogeneous concentration field of the solutes around the dislocation produces a stress field of its own. Recall that the stress field of each solute (inclusion) is pure shear outside the inclusion and hence produces no force on other solutes. However, the shear components of the solute stress field can produce Peach–Koehler forces on dislocations. In this section, we discuss the stress field σ^{c}_{ij} produced by the equilibrium solute distribution obtained in the previous section.

The stress field σ^{c}_{ij} can be obtained by integrating the stress field caused by excess solute concentration (relative to c_0) over the entire space. In two dimensions, the excess solute in the differential domain $\mathrm{d}x'\mathrm{d}y'$ around point (x', y') is equivalent to a line source of dilatation with excess volume

$$\Delta\tilde{v}_{\mathrm{f}}[c(x', y') - c_0]\,\mathrm{d}x'\mathrm{d}y'$$

per unit length. Therefore the stress field σ^{c}_{ij} can be written as

$$\begin{aligned}\sigma^{c}_{ij}(x, y) &= \iint_{-\infty}^{\infty} \Delta\tilde{v}_{\mathrm{f}}[c(x', y') - c_0]\sigma^{\mathrm{dila}}_{ij}(x - x', y - y')\,\mathrm{d}x'\mathrm{d}y' \\ &= \iint_{-\infty}^{\infty} c_{\max}\Delta\tilde{v}_{\mathrm{f}}[\chi(x', y') - \chi_0]\sigma^{\mathrm{dila}}_{ij}(x - x', y - y')\,\mathrm{d}x'\mathrm{d}y',\end{aligned} \tag{12.48}$$

where $\sigma^{\mathrm{dila}}_{ij}(x, y)$ is the stress field of a line dilatation center with unit excess volume per unit length along the z axis. The expression of $\sigma^{\mathrm{dila}}_{ij}(x, y)$ is known analytically. Only the deviatoric components are needed for our purpose, and are given below:

$$\begin{aligned}\sigma^{\mathrm{dila}}_{xy} &= -\frac{\mu(1+\nu)}{3\pi(1-\nu)}\frac{2xy}{(x^2+y^2)^2}, \\ \sigma^{\mathrm{dila}}_{xx} - \sigma^{\mathrm{dila}}_{yy} &= -\frac{\mu(1+\nu)}{3\pi(1-\nu)}\frac{2(x^2-y^2)}{(x^2+y^2)^2}.\end{aligned} \tag{12.49}$$

Given Eqs. (12.47), (12.48), and (12.49), the deviatoric part of the Cottrell atmosphere stress field can be obtained by numerical integration [121]. Figure 12.20 presents the shear stress σ^{c}_{xy} along the positive x axis for the case of $\chi_0 = 0.01$. The stress field of the edge dislocation itself, σ^{d}_{xy}, is also plotted as a solid line, which is a straight line here due to its $1/r$ dependence. We can see that at distances greater than $10b$ from the dislocation center, the atmosphere shear stress become proportional to the dislocation stress, i.e. it also develops a $1/r$ dependence.

It can be shown [121] that far away from the dislocation line, the deviatoric stress field $\tilde{\sigma}^{c}_{ij}$ of the atmosphere is always proportional to the deviatoric stress field $\tilde{\sigma}^{d}_{ij}$ of the dislocation itself. The proportionality constant, which is the same for all deviatoric stress components, is

$$\lim_{x^2+y^2\to\infty}\frac{\tilde{\sigma}^{c}_{ij}(x, y)}{\tilde{\sigma}^{d}_{ij}(x, y)} = -\frac{1+\nu}{2}\frac{\eta}{k_{\mathrm{B}}T}\chi_0(1-\chi_0). \tag{12.50}$$

This is true even for arbitrarily curved dislocation lines in three dimensions.[8]

Recall that $\eta > 0$ and $0 < \chi_0 < 1$ (for $T > 0$). Hence the far-field stress ratio given in Eq. (12.50) is always negative. This means that the long-range stress field of the edge dislocation

8 The linear relationship between $\tilde{\sigma}^{c}_{ij}(x, y)$ and $\tilde{\sigma}^{d}_{ij}(x, y)$ is ultimately caused by the linear relationship between $\chi(x, y) - \chi_0$ and $\sigma^{d}_{ii}(x, y)$, when $\sigma^{d}_{ii}(x, y)$ is sufficiently small, which can be obtained through a Taylor expansion of Eq. (12.47).

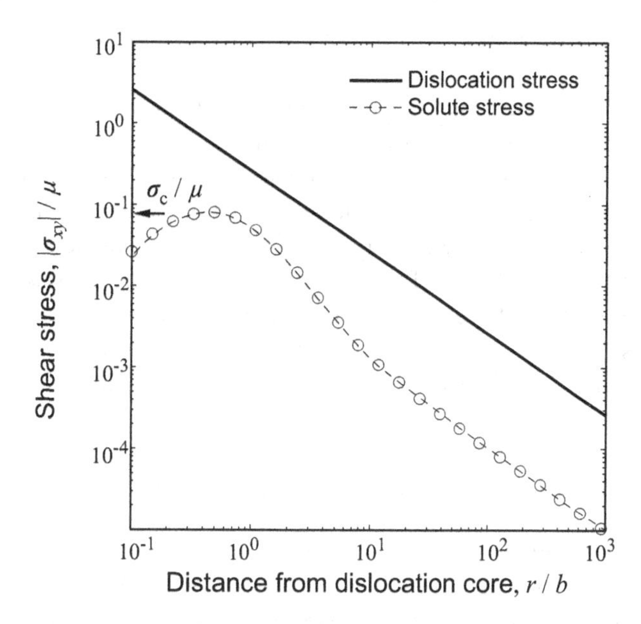

Figure 12.20. Shear stress field σ_{xy} around an edge dislocation at the origin in an infinite medium along the positive x axis with $\chi_0 = 0.01$. The stress due to the dislocation is plotted as a solid line. The stress generated by the solute is plotted in circles. Note that the solute stresses are negative and only their absolute values are plotted.

is reduced (or screened) by the equilibrium solute distribution, as predicted by Larché and Cahn [125]. Hence the long-range elastic interaction between edge dislocations is weakened by the solutes. This can cause increase of the equilibrium separation between partials of screw dislocations,[9] and suppress cross-slip. We note that because η is proportional to $(\Delta\tilde{v}_f)^2$, the effect mentioned here is fairly small, being of the second order in $(\Delta\tilde{v}_f/\Omega)$.

Figure 12.20 also shows that, near the dislocation center ($r < 10b$), the shear stresses from the solute distribution deviate from the $1/r$ shape. In fact, the solute-produced shear stress is zero at $r = 0$, meaning that the solute atmosphere produces no Peach–Koehler force on the dislocation when the solute distribution is at thermodynamic equilibrium with the dislocation's pressure field. In Fig. 12.20 the magnitude of the solute-produced shear stress reaches a maximum at around $r \approx b$, where the magnitude is $\sigma_c \approx 0.1\mu$ for this set of parameters (χ_0, $\Delta\tilde{v}_f$, b, etc.); σ_c is the critical stress required to break the dislocation from the solute atmosphere, if the solute distribution is fixed and not allowed to move with the dislocation. This situation can occur if the diffusivity of the solutes is very low (e.g. for large solute atoms or at very low temperature), and/or if the applied strain rate is very high so that the dislocation must move very fast. In these cases, the "frozen-in" solute atmosphere generates a shear stress that pulls the dislocation back toward its original position. The applied stress must exceed σ_c to break the dislocation free from the solute atmosphere. If the solutes are mobile, the dislocation can move (slowly) at stresses lower than σ_c by dragging the solute atmosphere along.

9 The solutes can also increase the separation between partials by reducing the stacking fault energy, which is an effect not considered here.

12.6 Summary

The singular elastic fields of dislocations break down in the core of the dislocation where non-linear interatomic interactions determine the stresses, strains and energies. Understanding the structure and energy of dislocation cores and the variations in energy associated with dislocation motion is essential for understanding dislocation mobility. The nature of the interatomic bonding and the associated crystal structure determine the properties of dislocation cores.

The Peierls–Nabarro (PN) model of the dislocation core was the first to show that accounting for interatomic forces active across the glide plane, together with the associated elastic fields, leads naturally to a non-singular expression for the stress field, involving the half width of the dislocation core, ξ. For a positive edge dislocation the stress field in its slip plane is then expressed as

$$\sigma_{xy}(x, y = 0) = \frac{\mu b}{2\pi(1-\nu)} \frac{x}{x^2 + \xi^2} \tag{12.51}$$

where the long-range tail is retained in the far field, $\sigma_{xy} \propto 1/x$, but where the singularity is removed at the core. The lattice resistance to dislocation motion depends strongly on the width of the core, with little resistance for wide dislocation cores where the atoms in the slip plane slide very gradually over each other in the course of dislocation motion.

The original PN model can be easily extended to dislocations on $\{1\,1\,1\}$ planes of FCC metals, by introducing a generalized stacking fault (GSF) energy function that contains a local minima corresponding to the stacking fault energy. The PN model then naturally predicts dissociation of a perfect dislocation into two partial dislocations enclosing an area of stacking fault, with both partials spreading out their cores on the slip plane. When the two partials are well separated so that their core regions do not overlap, their core widths agree well with the prediction of the original PN model.

Dislocation cores in non-close-packed diamond cubic (DC) solids, such as Si and Ge, are dominated by considerations of covalent bonding. Like other crystal structures based on the FCC lattice, slip in DC crystals occurs on $\{1\,1\,1\}$ planes and in $\langle 1\,1\,0 \rangle$ directions. If the stacking of the DC crystal in the $\langle 1\,1\,1 \rangle$ direction is described as ... CcAaBbCcAa ... then dislocations may be one of two types: *glide-set* dislocations involving shear between c|A, a|B, b|C layers or *shuffle-set* dislocations involving shear between A|a, B|b, C|c layers. The glide-set dislocations dissociate into partials with stacking faults in between while the shuffle-set dislocations do not dissociate.

Dislocation core structures in BCC metals are much more complex than those in close-packed crystals, in part because there are no stable stacking faults in BCC structures. As a result, perfect dislocations do not dissociate into partials. While non-screw dislocations spread out their cores on the slip plane, as predicted by the PN model, screw dislocations have compact, non-planar core structures. Screw dislocations take the lower energy "easy" core structure, or the higher energy "hard" core structure, depending on its position within the triangular lattice formed by the $\langle 1\,1\,1 \rangle$ atomic rows. As the screw dislocation moves from one "easy" core site to the adjacent site, the "hard" core configuration in between acts as a barrier. This causes screw dislocations to be much less mobile than non-screw dislocations, with the consequence that plastic deformation usually leaves long, straight screw dislocations in the crystal. In addition, because the cores of screw dislocations are inherently non-planar, different parts of the same dislocation can move on different slip planes and this leads to self-pinning of the dislocation and the production of dislocation loops in the wake of the dislocation motion.

The interactions between the elastic field of solutes and the elastic field of dislocations cause the solutes to segregate to the dislocations. Solutes modeled as elastic inclusions with a dilatational misfit interact with the hydrostatic component of the stress field of edge dislocations and lead to a cloud of solutes near the core of the dislocation called a Cottrell atmosphere. This produces a pinning effect if the solutes are not able to move with the moving dislocation or a drag effect on the dislocations if they are. In addition to pinning, the solute cloud alters the shear stress field of the dislocation, though, for the case of elastic homogeneity, the hydrostatic component of the dislocation field that attracted the solutes in the first place is unaffected.

12.7 Exercise problems

12.1 From Eq. (10.12), it is easy to show that the glide force acting between two like-signed Volterra edge dislocations on the same slip plane and separated by a distance s can be expressed as $f_x = \mu b^2/[2\pi(1-\nu)s]$. This interaction force is singular and is not defined at $s = 0$. Here we consider the repulsive interaction between two non-singular edge dislocations, each with core width 2ξ described by the Peierls–Nabarro (PN) model, also on the same slip plane and separated by a distance s. We shall assume that the interaction between the two dislocations does not alter the widths of the cores. Assume one dislocation is centered at the origin. Its stress field on the glide plane is given by Eq. (12.15).

(a) Write an expression for the glide force acting between two PN edge dislocations. To obtain an explicit expression, you need the following formula:

$$\int_{-\infty}^{+\infty} \frac{t}{t^2+a^2}\frac{c}{\pi}\frac{1}{(x-t)^2+c^2}\,dt = \frac{x}{x^2+(a+c)^2}. \tag{12.52}$$

(b) Show that the result in (a) reduces to the result for singular dislocations when $s \gg \xi$.
(c) Write an expression for the maximum repulsive force that can act between two like-signed PN edge dislocations on the same slip plane.
(d) Determine the separation distance for which the repulsive force in (c) is found.

12.2 Equation (11.5) gives an expression for the equilibrium separation between two Shockley partials of a perfect edge dislocation in an FCC metal, when the partials are modeled as (singular) Volterra dislocations. Here we consider the partials as (non-singular) PN dislocations, with the half widths of the edge and screw components of the partials given by $\xi_e = h/[2(1-\nu)]$ and $\xi_s = h/2$, respectively, where h is the spacing between adjacent $\{1\,1\,1\}$ planes.

(a) Write an equation that the equilibrium separation between the two partials, x_{eq}, must satisfy.
(b) Using the properties of FCC gold, as given in Eq. (11.6), solve the equation in (a) numerically (e.g. using the `fsolve` function in Matlab) and obtain the value of x_{eq}. Compare the result with the prediction given in Eq. (11.6).

(c) Obtain the predicted values of x_{eq} using the properties of FCC aluminum (Tables 11.2 and B.1), with the partials considered as Volterra dislocations and as PN dislocations, respectively.

12.3 The Matlab code available on the book website solves Eq. (12.25) numerically and obtains the separation w between and core half width ξ_p of the partials, using the parameters for FCC gold. Modify the code to find the numerical solution using the parameters for FCC aluminum (see Tables 11.2 and B.1). Plot the displacement field during relaxation $u(x)$ and Burgers vector distribution $b'(x)$ and compare them with Fig. 12.6.

12.4 An epitaxial thin film of a SiGe solid solution alloy on a Si substrate is shown in Fig. 12.21. The lattice mismatch produces a biaxial compressive stress in the SiGe film. This stress causes dislocations to be created at the top surface of the film at high temperatures. If the stacking fault energy were sufficiently low, the first dislocation to form would be a Shockley partial. Figure 12.21 shows a Shockley partial half loop with an associated stacking fault extending from the top surface. The accompanying diagram shows the crystallographic orientation of the SiGe film, and the various possible slip directions and slip planes in this crystal. The slip plane of the half loop in the film is shaded in the crystallographic diagram.

(a) Determine the possible Burgers vectors of the Shockley partial half loop.
(b) Determine the Burgers vector of the half loop if the film were subjected to tension instead of compression.
(c) On the basis of your analysis, indicate whether you would expect the film to sustain more compression or more tension at high temperatures before dislocations start to form.
(d) Repeat the analyses in (a)–(c) now assuming that the dislocation nucleated from the surface is a perfect dislocation on the shuffle-set plane.

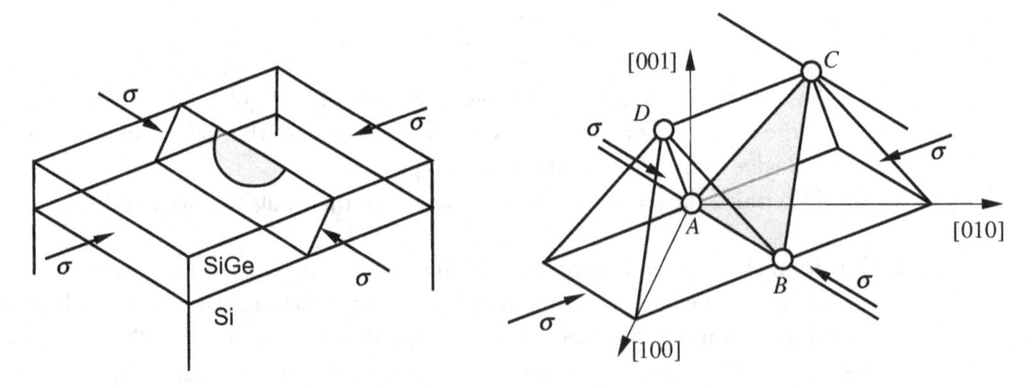

Figure 12.21. Leading partial dislocation in SiGe film under biaxial compressive stress.

12.5 The generalized stacking fault (GSF) energy for BCC Mo on the $(1\,\bar{1}\,0)$ plane along the $[1\,1\,1]$ direction has been calculated using atomistic models [126] and the result can be approximately fitted to the following equation

$$\gamma(q) = A[1 - \exp(-B\sin^2(\pi q/b))], \tag{12.53}$$

where $A = 1.12\ \text{J/m}^2$ and $B = 2.5$, see Fig. 12.22. To find the core shape of a perfect edge dislocation on the $(1\,\bar{1}\,0)$ plane, we need to solve the stress balance equation

$$\frac{\mu}{\pi(1-\nu)} \int_{u(-\infty)}^{u(+\infty)} \frac{-du(x')}{x-x'} = \frac{\partial \gamma(q)}{\partial q}, \tag{12.54}$$

where $q(x) = \pm b/2 + 2u(x)$. Here we seek an approximate solution of the following form

$$u(x) = -\frac{b}{4\pi} \arctan\left(\frac{x}{\xi_1}\right) - \frac{b}{4\pi} \arctan\left(\frac{x}{\xi_2}\right). \tag{12.55}$$

Modify the Matlab code on the book website to find the numerical solution for ξ_1 and ξ_2. Plot the displacement field $u(x)$ during relaxation and Burgers vector distribution $b'(x)$.

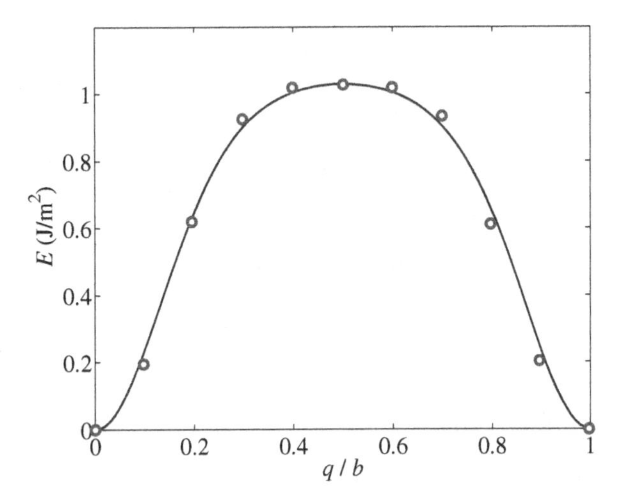

Figure 12.22. GSF energy of BCC Mo on the $(1\,\bar{1}\,0)$ plane along the $[1\,1\,1]$ direction. Circles: atomistic calculations [126]. Solid line: Eq. (12.53).

12.6 The energy difference between "hard" and "easy" core structures of screw dislocations in BCC Mo has been estimated to be $0.08\ \text{eV}/b$ by density functional theory (DFT) calculations [119]. Assuming that the screw dislocation must go through the "hard" core structure as it moves from one "easy" core structure to the next, estimate the critical stress required to move the screw dislocation at zero temperature (i.e. the Peierls stress). Assume the core energy varies with dislocation position as a simple sinusoidal function.

12.7 Figure 12.19 shows the equilibrium distribution of H solute atoms around an edge dislocation in FCC Pd. Figure 12.20 plots the shear stress produced by solute distribution on the slip plane of the dislocation. Here we would like to see to what extent this shear stress field can be approximated by assuming all the excess solute atoms are localized within a region beneath the dislocation.

(a) Figure 12.19 shows that the contour lines of equilibrium solute fraction are circles whose highest point overlaps with the dislocation. Find the radius r_c of the circle that corresponds to the contour line for $\chi = 50\%$.

(b) As a simple approximation, consider a solute distribution in which $\chi = 100\%$ within the circle of radius r_c, and $\chi = 0\%$ outside the circle. The highest point of the circle overlaps with the dislocation. Write an expression for the shear stress σ_{xy} for this solute distribution, with the help of Eq. (10.95).

(c) Plot the stress field in (b) on the dislocation glide plane. Compare your plot with Fig. 12.20 and discuss its behavior at both short distances and long distances from the dislocation.

12.8 Figure 12.23 shows an infinitely long single crystal cylinder containing a positive edge dislocation at its center. In Fig. 12.23a, the cylinder is held at a very high temperature, T, and vacancy equilibrium is established everywhere in the cylinder. We then apply a radial pressure, p, to the cylinder, as shown in Fig. 12.23b, and again allow a new vacancy equilibrium to be established. The free surface of the crystal is assumed to be a good source/sink for vacancies.

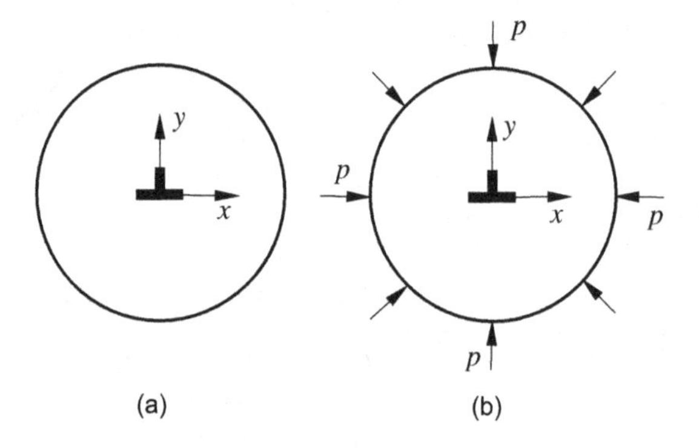

Figure 12.23. An edge dislocation in a cylinder under (a) zero pressure, and (b) pressure p.

(a) Write an expression for the equilibrium concentration of vacancies in the cylinder when no pressure is applied to the cylinder, assuming the vacancies are rigid, i.e. their relaxation volume $\Delta V_v^{\mathrm{rlx}} = 0$. Assume that the vacancy formation energy Δe_v, vacancy formation entropy Δs_v, and atomic volume Ω are known constants.

(b) Write an expression for the equilibrium concentration of vacancies after the pressure is applied and a new equilibrium is established.

(c) Show how the position of the dislocation will be changed when the pressure is applied and indicate whether the dislocation comes to rest or not, long after the pressure is applied.

(d) Repeat the analyses in (a)–(c) now assuming that the vacancy relaxation volume $\Delta V_v^{\mathrm{rlx}} = 0$ has a negative value.

PART IV

GRAIN BOUNDARIES

A grain boundary (GB) is the interface between two single crystals (i.e. grains) of the same material with different orientations. We mentioned in Chapter 1 that most engineering metals and alloys are polycrystals, which are aggregates of a large number of single crystal grains separated by grain boundaries. Grain boundaries play an important role in a wide range of material behaviors and properties. For example, grain boundaries can increase the strength of metals by blocking the motion of lattice dislocations, leading to the Hall–Petch behavior, in which the strength of crystalline materials increases with decreasing grain size. We have mentioned in Chapter 7 that grain boundaries act as sources and sinks of vacancies, and that vacancy diffusion along GBs is the mechanism of Coble creep. Because impurities tend to segregate at grain boundaries, the chemical environment is often different at grain boundaries, which can be preferential sites for crack initiation or chemical attack. Grain boundary engineering [127], i.e. the keeping of "good" GBs and removal of "bad" GBs through processing, has led to significant improvements in material strength and corrosion resistance.

In Chapter 13 (Grain boundary geometry), we first define the five orientation variables that specify the misorientation between the two grains and the direction of the boundary plane normal. We then introduce the coincidence site lattice (CSL) and the associated Σ number that are very useful to characterize special (low energy) grain boundaries. We will see that special grain boundaries usually have low Σ numbers but low Σ numbers do not necessarily mean the boundaries are special. The displacement shift complete (DSC) lattice, of which the CSL is a sublattice, and prescribes the allowable Burgers vectors of GB dislocations, which can be much shorter than those of lattice dislocations.

In Chapter 14 (Grain boundary mechanics), we explore the relationship between grain boundaries and dislocations. First, low angle grain boundaries are equivalent to arrays of lattice dislocations, as described by the famous Read–Shockley model. Second, a grain boundary with misorientation vicinal to a low-Σ GB can be considered as a superposition of the low-Σ GB and an array of GB dislocations. Finally, we discuss disconnections, which are steps on the GBs with a non-zero Burgers vector content, i.e. they are simultaneously GB steps and GB dislocations.

13 Grain boundary geometry

At the length scale where a crystallite (grain) may be considered as a continuum, the geometry of a grain boundary can be specified by five rotational degrees of freedom: three for the relative misorientation between the two grains, and two for the direction of the grain boundary plane normal. This is analogous to the Burgers vector $\vec{b}$ and the line direction $\hat{\xi}$ specifying the geometry of a dislocation. In Section 13.1, we introduce the notation for describing the geometry of the grain boundary at the continuum level.

The periodicity of the crystal lattice makes certain misorientations between a pair of crystallites more special than others. Grain boundaries between such crystallite pairs are likely to have special properties. For example, some of them are much more resistant to corrosion than others. These special misorientations are described by the theory of the coincidence site lattice (CSL). In Section 13.2, we introduce the CSL theory, which leads to the designation of GB misorientation by the Σ number.

Within the CSL theory, two interpenetrating lattices with special misorientation form a periodic pattern in space. The CSL theory also predicts the vectors by which one lattice can be translated relative to the other while keeping the periodic pattern unchanged. These displacement vectors also form a lattice, called the *displacement shift complete* (DSC) lattice, to be introduced in Section 13.3. The smallest repeat vectors of the DSC lattice are the Burgers vectors of GB dislocations. They are usually much smaller than the Burgers vectors of crystal dislocations within the grains. The physical consequence is that a crystal dislocation (or part of it) with appropriate Burgers vectors can spread into GBs by dissociating into many GB dislocations with much shorter Burgers vectors.

13.1 Grain boundary orientation variables

13.1.1 Misorientation between two grains

Let us label the crystallites on the two sides of the grain boundary as crystal A and crystal B. There are multiple ways to specify the misorientation between these two crystals. One way is to imagine that crystal B starts with the same orientation as crystal A, and specify the rotation needed to transform crystal B into its designated orientation.

Axis–angle representation

An arbitrarily oriented crystal B can be obtained from crystal A by a rotation around some axis $\hat{t}_R$ by some angle θ. Hence the misorientation between the two crystals can be described by specifying the axis $\hat{t}_R$ and the angle θ of rotation, where $\hat{t}_R$ is a unit vector. Because $\hat{t}_R$ exists on a unit sphere, it can be represented by two angular variables, α_R and ϕ_R, i.e.

$$\hat{t}_R = \begin{bmatrix} \cos\alpha_R \sin\phi_R \\ \sin\alpha_R \sin\phi_R \\ \cos\phi_R \end{bmatrix}. \tag{13.1}$$

Therefore, the misorientation between the two crystals can be represented by three angular variables: α_R, ϕ_R, and θ.

A more concise notation can be introduced by defining a *rotation vector* $\vec{T}_R$ as the product of the unit vector $\hat{t}_R$ and the rotation angle θ (in radians)

$$\vec{T}_R = \hat{t}_R \cdot \theta. \tag{13.2}$$

Now $\vec{T}_R$ is an arbitrary vector in three-dimensional space and completely and uniquely specifies the misorientation of the grain boundary.

For symmetric grain boundaries, it is often convenient to specify the misorientation by assuming that both crystals A and B rotate from a reference orientation around a common axis $\hat{t}_R$. Crystal A is rotated from the reference orientation by $-\theta/2$, and crystal B by $\theta/2$. Since the rotation around the axis $\hat{t}_R$ does not change the vector $\hat{t}_R$ itself, $\hat{t}_R$ can be specified by a coordinate system fixed on crystal A, or crystal B, or the reference crystal. In other words, $\hat{t}_R$ is a common vector in both crystals on the two sides of the grain boundary.

Orientation matrix

We have seen that three variables (such as α_R, ϕ_R, and θ) can uniquely specify the rotation from crystal A to crystal B. However, for computational purposes, it is often more convenient to represent the rotation by a 3×3 orientation matrix containing nine numbers.

The orientation matrix is defined as the dot products between the axes of two coordinate systems. One coordinate system is fixed on crystal A. Let $\hat{e}_x$, $\hat{e}_y$, and $\hat{e}_z$ be the unit vectors along the [1 0 0], [0 1 0], and [0 0 1] directions of crystal A, respectively. The other coordinate system is fixed on crystal B. Let $\hat{e}_{x'}$, $\hat{e}_{y'}$, and $\hat{e}_{z'}$ be the unit vectors along the [1 0 0], [0 1 0], and [0 0 1] directions of crystal B after the rotation, respectively. The misorientation between crystals A and B can be uniquely specified by the 3×3 orientation matrix of dot products,

$$g_{AB} = \begin{bmatrix} (\hat{e}_x \cdot \hat{e}_{x'}) & (\hat{e}_x \cdot \hat{e}_{y'}) & (\hat{e}_x \cdot \hat{e}_{z'}) \\ (\hat{e}_y \cdot \hat{e}_{x'}) & (\hat{e}_y \cdot \hat{e}_{y'}) & (\hat{e}_y \cdot \hat{e}_{z'}) \\ (\hat{e}_z \cdot \hat{e}_{x'}) & (\hat{e}_z \cdot \hat{e}_{y'}) & (\hat{e}_z \cdot \hat{e}_{z'}) \end{bmatrix}. \tag{13.3}$$

Note that the first column of the matrix g_{AB} is just the coordinates of unit vector $\hat{e}_{x'}$ in the x–y–z coordinate system (fixed on crystal A). Similarly, the second and third column of matrix g_{AB} contain the coordinates of $\hat{e}_{y'}$ and $\hat{e}_{z'}$, respectively.

The matrix g_{AB} is also called a rotation matrix because it relates the coordinates of every point in crystal B before and after the rotation. Consider an arbitrary point before the rotation, specified by vector $\vec{r}_i = (x_i, y_i, z_i)$. After it follows the rotation of crystal B, its location is specified by a different vector $\vec{r}_f = (x_f, y_f, z_f)$, where

$$\begin{bmatrix} x_f \\ y_f \\ z_f \end{bmatrix} = g_{AB} \cdot \begin{bmatrix} x_i \\ y_i \\ z_i \end{bmatrix}. \tag{13.4}$$

For example, if $x_i = 1$, $y_i = 0$, $z_i = 0$, then $x_f = (\hat{e}_x \cdot \hat{e}_{x'})$, $y_f = (\hat{e}_y \cdot \hat{e}_{x'})$, $z_f = (\hat{e}_z \cdot \hat{e}_{x'})$. This means that if $\vec{r}_i = \hat{e}_x$, then $\vec{r}_f = \hat{e}_{x'}$, as it should.

We now give some specific examples of the orientation matrix g_{AB}. If crystal B can be obtained from crystal A by rotating around the z axis by angle θ_1, then

$$g_{AB}^{(1)} = \begin{bmatrix} \cos\theta_1 & -\sin\theta_1 & 0 \\ \sin\theta_1 & \cos\theta_1 & 0 \\ 0 & 0 & 1 \end{bmatrix}. \tag{13.5}$$

On the other hand, if crystal B can be obtained from crystal A by rotating around the x axis by angle θ_2, then

$$g_{AB}^{(2)} = \begin{bmatrix} 1 & 0 & 0 \\ 0 & \cos\theta_2 & -\sin\theta_2 \\ 0 & \sin\theta_2 & \cos\theta_2 \end{bmatrix}. \tag{13.6}$$

The advantage of the matrix notation can be illustrated by the following example. If crystal B can be obtained from crystal A by first rotating around the z axis by angle θ_1, and then rotating around the x axis by angle θ_2 (the x–y–z coordinate system remains fixed on crystal A), then

$$g_{AB}^{(3)} = g_{AB}^{(2)} \cdot g_{AB}^{(1)}. \tag{13.7}$$

Notice that the matrix corresponding to the first rotation (by θ_1) is on the right, so that when $g_{AB}^{(3)}$ multiplies a column vector, as in Eq. (13.4), the $g_{AB}^{(1)}$ term multiplies the column vector first.

Even though there are nine matrix elements in g_{AB}, the rotation matrix has only three degrees of freedom due to the six constraints that it must satisfy. The constraints are: every column of the matrix is a unit vector, and every two columns are orthogonal to each other (i.e. their dot product is zero). Consequently, g_{AB} is called an orthogonal matrix, whose inverse is simply its transpose, $\left(g_{AB}\right)^T$, i.e.

$$g_{AB} \cdot (g_{AB})^T = I, \tag{13.8}$$

where I is the 3×3 identity matrix. This can be easily verified for the examples given in Eqs. (13.5)–(13.7).

The two representations of misorientation introduced above are obviously related to each other. For convenience, let us define an orientation matrix

$$g_{AB}^{(4)} = \begin{bmatrix} \cos\alpha_R \cos\phi_R & \sin\alpha_R \cos\phi_R & -\sin\phi_R \\ -\sin\alpha_R & \cos\alpha_R & 0 \\ \cos\alpha_R \sin\phi_R & \sin\alpha_R \sin\phi_R & \cos\phi_R \end{bmatrix}. \tag{13.9}$$

It can be easily verified that

$$g_{AB}^{(4)} \cdot \hat{t}_R = \begin{bmatrix} 0 \\ 0 \\ 1 \end{bmatrix}, \tag{13.10}$$

which means that $g_{AB}^{(4)}$ turns the $\hat{t}_R$ direction into the z axis. Then the orientation matrix g_{AB} corresponding to the axis–angle representation $\hat{t}_R$, θ_1 can be written as the following matrix product:

$$g_{AB} = \left(g_{AB}^{(4)}\right)^T \cdot g_{AB}^{(1)} \cdot g_{AB}^{(4)}. \tag{13.11}$$

From right to left, the three matrices above mean the following series of rotations of crystal B: (1) rotate crystal B until the axis originally along $\hat{t}_R$ is aligned with the z axis, (2) rotate crystal B around z axis by angle θ_1, and (3) rotate crystal B in the reverse direction of (1). The net effect of these three rotations is the rotation of crystal B around $\hat{t}_R$ by angle θ_1.

Conversely, the rotation axis $\hat{t}_R$ and the rotation angle θ can be quickly obtained from the rotation matrix g_{AB}. The rotation angle can be found from the relation

$$\cos\theta = \frac{1}{2}\left[(g_{AB})_{11} + (g_{AB})_{22} + (g_{AB})_{33} - 1\right] \tag{13.12}$$

and the rotation axis is given by

$$\begin{aligned} \left(\hat{t}_R\right)_x &= \frac{(g_{AB})_{32} - (g_{AB})_{23}}{2\sin\theta}, \\ \left(\hat{t}_R\right)_y &= \frac{(g_{AB})_{13} - (g_{AB})_{31}}{2\sin\theta}, \\ \left(\hat{t}_R\right)_z &= \frac{(g_{AB})_{21} - (g_{AB})_{12}}{2\sin\theta}. \end{aligned} \tag{13.13}$$

13.1.2 Orientation of grain boundary normal

We have seen that there are three degrees of freedom (such as α_R, ϕ_R, and θ) in the relative misorientation between the grains on the two sides of the grain boundary. The grain boundary has two more degrees of freedom corresponding to the local direction of the grain boundary plane normal. The GB plane normal can be represented by a unit vector $\hat{n}_B$, which can be specified by two angular variables, α_B and ϕ_B, i.e.

$$\hat{n}_B = \begin{bmatrix} \cos\alpha_B \sin\phi_B \\ \sin\alpha_B \sin\phi_B \\ \cos\phi_B \end{bmatrix}. \tag{13.14}$$

The normal vector $\hat{n}_B$ specifies how the regions occupied by crystal A and crystal B are divided up in space. We can take the convention that the region on the positive side of the $\hat{n}_B$ vector is occupied by crystal B. The relationship between the boundary plane normal $\hat{n}_B$ and the rotation axis $\hat{t}_R$ determines the type of the grain boundary. If $\hat{n}_B$ is parallel to $\hat{t}_R$, then the GB is of the *pure twist* type, as shown in Fig. 13.1a. If $\hat{n}_B$ is perpendicular to $\hat{t}_R$, i.e. the rotation axis is in the GB plane, then the GB is of the *pure tilt* type, as shown in Fig. 13.1b. In general, a grain boundary has both twist and tilt components and will be called *mixed*.

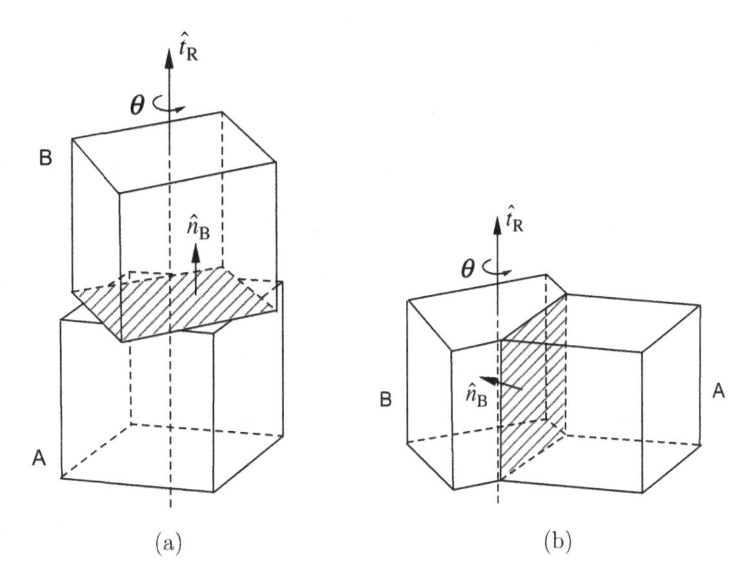

Figure 13.1. (a) Pure twist grain boundary with $\hat{n}_B$ parallel to $\hat{t}_R$. (b) Pure tilt grain boundary with $\hat{n}_B$ perpendicular to $\hat{t}_R$.

13.1.3 Comparing grain boundary and dislocation formalisms

The role of $\hat{n}_B$ in a grain boundary may be likened to the role of the local line direction $\hat{\xi}$ in a dislocation, both being unit vectors with two degrees of freedom. The role of the rotation vector $\vec{T}_R$ in the grain boundary may be likened to the role of the Burgers vector $\vec{b}$ in a dislocation, both having three degrees of freedom. Therefore, both a dislocation and a grain boundary have five degrees of freedom in the continuum description.

However, grain boundaries are often perceived as much more complicated than dislocations for several reasons. First, the dislocation Burgers vector $\vec{b}$ describes a translational relationship between two lattices of the same orientation, while the rotation vector $\vec{T}_R$ of a grain boundary describes a rotational relationship between two lattices. The mathematics of rotation is more complex (involving matrix multiplication, see Section 13.1.1) than the mathematics of translation (describable by addition and subtraction). Second, the dislocation's line energy is proportional to the magnitude of the Burgers vector squared, so that in a given crystal, only a few Burgers vectors of the shortest lengths need to be considered. In comparison, many grain boundaries with a wide range of misorientations have comparable energies. It is nearly impossible to enumerate all relevant misorientations that one can encounter in a polycrystal. The functional relationship between GB energy and GB misorientation is very complex and will be discussed further in this and the next chapter.

13.2 Coincidence site lattice

The grain boundary energy (per unit area) is a complex function of the five degrees of freedom of the GB, and is often plotted against only one degree of freedom, while keeping the other four fixed. Figure 13.2a shows the grain boundary energy γ as a function of misorientation angle θ

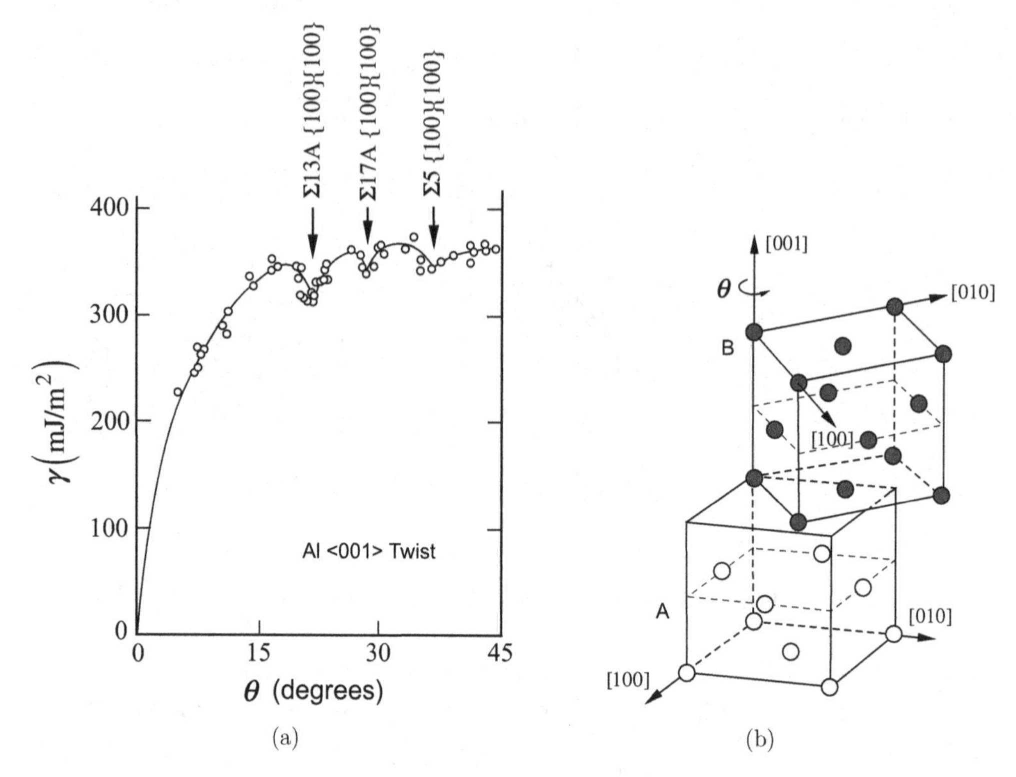

Figure 13.2. (a) Energy per unit area of Al (0 0 1) pure twist grain boundary as a function of misorientation angle θ [128]. (b) Pure twist GB between two FCC crystals with [0 0 1] rotation axis.

for (0 0 1) pure twist boundaries in Al. The designation of (0 0 1) pure twist means that all grain boundaries shown in Fig. 13.2a have the rotation axis $\hat{r}_R$ and the GB plane normal $\hat{n}_B$ along the [0 0 1] axis, as shown in Fig. 13.2b. The grain boundary energy is a non-smooth function of θ, with several cusps visible at around 22.62°, 28.07°, and 36.87°. These are pointed out by arrows and designated by Σ numbers in Fig. 13.2a. The Miller indices following the Σ numbers indicate the orientation of the GB plane in terms of each crystal's coordinate system.[1] In this case, the GB plane is the {1 0 0} plane in both crystals.

Figure 13.2a shows that there are special GBs whose energy is lower than other GBs with similar misorientations. These special boundaries show up as cusps in the energy plots like Fig. 13.2a. They are called *singular* GBs [129] due to the discontinuity in the slope of the energy function at such misorentations. Furthermore, these singular boundaries have well-defined Σ numbers.[2] In this section, we introduce the coincidence site lattice (CSL) theory [130] that describes the geometrical properties of the singular boundaries, and defines the Σ number.

1 The letter A following the Σ numbers is used to indicate the structure (by convention) when there are multiple GB structures with the same Σ.

2 The converse is not necessarily true, i.e. boundaries with well-defined Σ numbers are not necessarily singular.

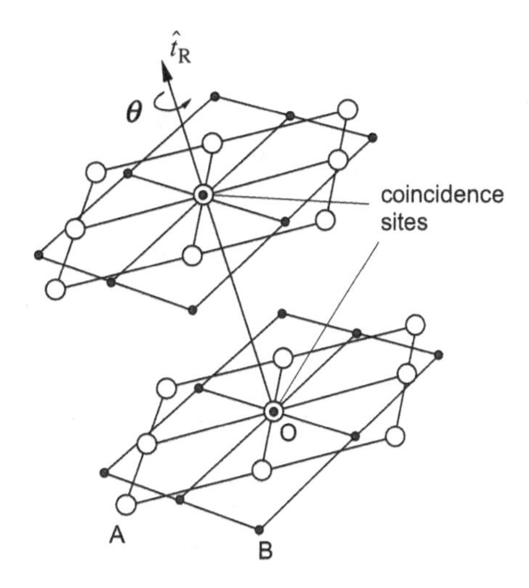

Figure 13.3. Two lattices A and B related to each other by rotation around axis $\hat{t}_R$ by angle θ. All lattice points along axis $\hat{t}_R$, including the origin O, are coincidence sites.

13.2.1 Definition of CSL and Σ number

When discussing the misorientation between two crystals in Section 13.1, we have considered them as continuous media and ignored their discrete crystal structures. In the coincidence site lattice (CSL) theory, we consider two three-dimensional Bravais lattices (lattice A and lattice B) that are misoriented from each other (note that the atomic basis is still ignored). Both Bravais lattices occupy the entire space, so that the two lattices interpenetrate each other. Only at a later stage, do we introduce a grain boundary plane (with normal vector $\hat{n}_B$) to truncate half of each lattice.

We assume that both lattice A and lattice B contain the origin as a lattice point. Hence the two lattices *coincide* at the origin, which is called a *coincidence site*. Because each lattice is an infinite periodic array of points, the two lattices would coincide either at a single point (i.e. the origin) only, or at an infinite number of points, depending on the misorientation between them.

Let us further assume that the rotation axis $\hat{t}_R$ passes through the origin and another lattice point in lattice A (and lattice B), as shown in Fig. 13.3. In other words, the line along $\hat{t}_R$ can be represented by integer Miller indices. Since rotation around $\hat{t}_R$ does not displace any points on the line along $\hat{t}_R$, the one-dimensional periodic array of lattice points along the line of $\hat{t}_R$ comprises coincidence sites. Given the above two assumptions, there are an infinite number of coincidence sites, containing (at least) a one-dimensional periodic lattice along $\hat{t}_R$.

For certain choices of misorientation angle θ, the one-dimensional array along $\hat{t}_R$ may be the only coincidence sites. However, for special values of misorientation angle, other coincidence sites can be found, and they form a three-dimensional periodic lattice, called the *coincidence site lattice* (CSL). The volume of the smallest repeat cell (i.e. the *primitive cell*) of the CSL divided by that of the lattice A (or lattice B) is the Σ number for the CSL. In other words, $1/\Sigma$ is the ratio of the volume density of CSL points over the volume density of the points in lattice A (or lattice B). A smaller Σ value means that a larger fraction of the points in the two lattices coincide. For example, a $\Sigma 5$ CSL means that one out of every five points in lattice A (or lattice

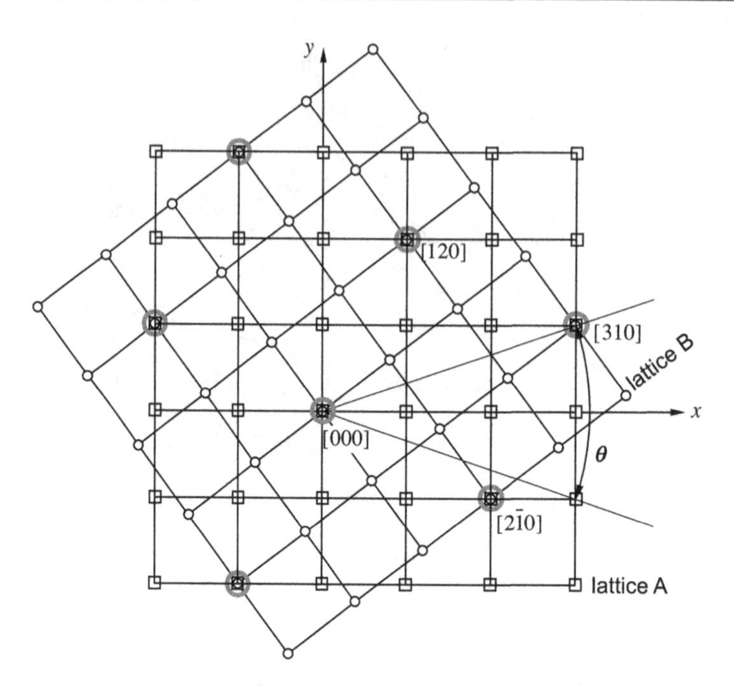

Figure 13.4. The (0 0 1) plane of two simple cubic lattices with misorientation $\theta = 36.87°$ around the [0 0 1] rotation axis. Points of lattice A are shown in white squares. Points of lattice B are shown in white circles. The coincidence sites are highlighted in large gray circles, and some of them are marked by Miller indices of lattice A.

B) are in coincidence. It can be shown that the Σ value for a CSL between two cubic lattices must be an odd number [131, 132].

13.2.2 Σ5 CSL of simple cubic lattice

To start with a simple example, let us consider the CSL between two simple cubic lattices, A and B, with misorientation angle θ around the rotation axis $\hat{r}_R = [0\,0\,1]$. Since all (0 0 1) planes of a simple cubic lattice look exactly the same, it suffices to plot only one plane of lattice points for each lattice, as shown in Fig. 13.4. When $\theta = 2\arctan(1/3) = 36.87°$, point [3 1 0] of lattice A coincides with a point of lattice B.[3] Other points, such as [1 2 0] and [2 $\bar{1}$ 0] also coincide with certain points of lattice B. Obviously, the points [0 0 0] and [0 0 1] (not shown) along the rotation axis are also coincidence sites. Due to the periodicity of lattices A and B, the coincidence sites form a 3D periodic lattice, i.e. the CSL, with shortest repeat vectors being [2 $\bar{1}$ 0], [1 2 0], and [0 0 1].

The CSL considered here is a primitive tetragonal Bravais lattice. The unit cell of the CSL is a rectangular box, with a square base on the (0 0 1) plane. The length of the square base is $\sqrt{5}\,a$ and the height of the rectangular box is a, where a is the lattice constant of the lattice A. Hence the volume of the CSL unit cell is $5a^3$. Obviously the volume of the unit cell of lattice A is a^3. Here, both lattice A and the CSL have only one lattice point in the unit cell, i.e. the unit cell is

3 This point would be labelled [3 $\bar{1}$ 0] if we use the x'–y'–z' coordinate system that rotates with lattice B.

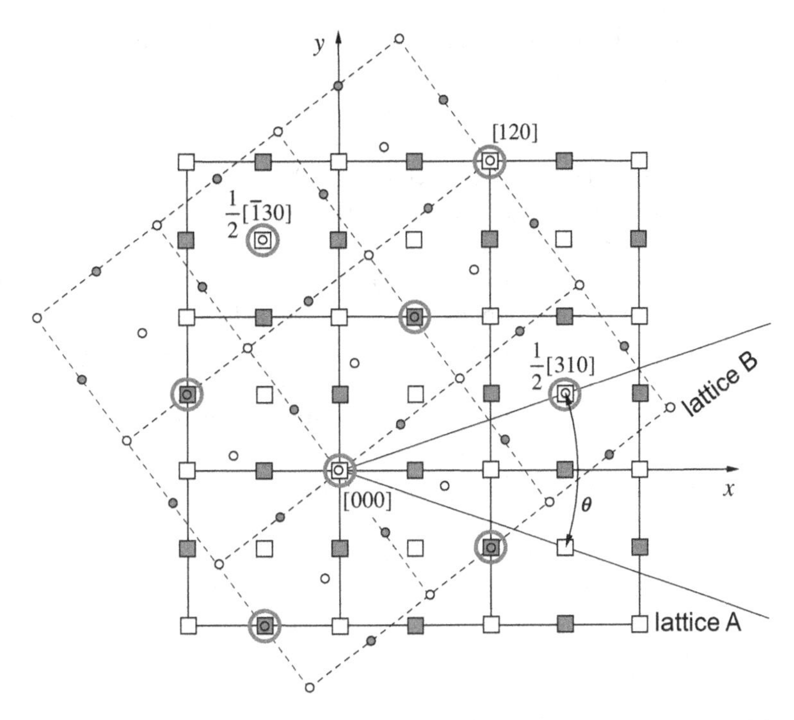

Figure 13.5. The (0 0 1) planes of two FCC lattices with misorientation $\theta = 36.87^\circ$ around the [0 0 1] rotation axis. Points of lattice A are shown in white squares (bottom plane) and gray squares (top plane). Points of lattice B are shown in white circles (bottom plane) and gray circles (top plane). The coincidence sites are highlighted in large gray circles, and some of them are marked by Miller indices of lattice A.

the same as the primitive cell. Therefore, the volume density of CSL points is 1/5 of the volume density of lattice A points, i.e. one out of every five points in lattice A are in coincidence with lattice B. This CSL is thus labeled Σ5.

Given the rotation angle $\theta = 2\arctan(1/3)$, we have $\tan(\theta/2) = 1/3$, $\cos\theta = (1 - \tan^2(\theta/2))/(1 + \tan^2(\theta/2)) = 4/5$, $\sin\theta = 2\tan(\theta/2)/(1 + \tan^2(\theta/2)) = 3/5$. Therefore, using Eq. (13.5), the rotation matrix of the CSL is

$$g_{\mathrm{AB}} = \frac{1}{5}\begin{bmatrix} 4 & -3 & 0 \\ 3 & 4 & 0 \\ 0 & 0 & 5 \end{bmatrix}. \tag{13.15}$$

Notice that all components of the rotation matrix g_{AB} are rational numbers, and that after multiplying these numbers by Σ, they all become integers. This is a general feature for CSLs between cubic lattices [133–135].

13.2.3 Σ5 CSL of FCC lattice

We now consider the CSL between two FCC lattices with the same [0 0 1] rotation axis and the 36.87° misorientation angle as in the previous section. There are two different (0 0 1) planes in the FCC lattice. Figure 13.5 plots both planes in the two FCC lattices to show the CSL pattern. The points in both lattices on the bottom (0 0 1) plane, with $z = 0$, are plotted in white symbols (squares for A and circles for B). The points in both lattices on the top (0 0 1) plane, with $z = \frac{a}{2}$, are plotted in gray symbols.

On the bottom plane, the coincidence points identified earlier in the simple cubic lattices remain coincidence points in the FCC lattices. Furthermore, we observe additional coincidence points on this plane at points such as $\frac{a}{2}[3\,1\,0]$, which are in the center of the squares formed by the CSL between two simple cubic lattices. On the top plane, the coincidence sites also form a square lattice with the same density as that on the bottom plane. Hence the CSL considered here is a body-centered tetragonal Bravais lattice. The repeat vectors of the unit cell are: $\frac{a}{2}[3\,1\,0]$, $\frac{a}{2}[\bar{1}\,3\,0]$, $[0\,0\,1]$ (in terms of the coordinate system fixed on lattice A). There are two lattice points per unit cell: one on the corner and the other on the body center.

The volume of the CSL unit cell is $\frac{5}{2}a^3$. Since there are two CSL points per unit cell, the density of CSL points is $(\frac{5}{4}a^3)^{-1}$. Recall that the FCC lattice has four points per unit cell, whose volume is a^3, so that the density of lattice A points is $(\frac{1}{4}a^3)^{-1}$. Hence the ratio of lattice point density between the CSL and lattice A remain at the same value (1/5) as in the previous section. In other words, the CSL between the two FCC lattices considered here should also be labeled as $\Sigma 5$. The misorientation (36.87°) represented by this CSL corresponds to the $\Sigma 5$ cusp in the GB energy curve shown in Fig. 13.2a.

13.2.4 $\Sigma 5$ twist boundary

The Σ value only describes the misorientation between the two grains. It does not completely determine the grain boundary type because the GB plane normal is still unspecified. In this section, we construct an atomistic model of the $\Sigma 5$ twist boundary, in which the grain boundary normal $\hat{n}_B$ is parallel to the rotation axis $\hat{t}_R$, as shown in Fig. 13.1a and Fig. 13.2b. This grain boundary corresponds to the cusp indicated as the $\Sigma 5\{100\}\{100\}$ twist boundary in Fig. 13.2a. An atomistic model of a $\Sigma 5$ tilt boundary will be discussed in the next section.

For simplicity, we shall first assume that the atoms on either sides of the grain boundary occupy positions in the perfect crystal structure of the grain on that side of the boundary. Because the FCC crystal structure has a one-atom basis, the perfect crystal structure is obtained by placing one atom (e.g. Al) at every lattice site. To construct an atomsitic model for the twist GB, we assume that crystal B is above crystal A when viewed along [0 0 1], and that the GB plane is between two (0 0 1) atomic layers. Therefore, the bi-crystal structure can be obtained by stacking (0 0 1) atomic layers according to lattice A positions below the GB plane, and then according to lattice B positions above the GB plane. Figure 13.6 plots the top layer of crystal A (below the GB plane) in gray squares and the bottom layer of crystal B (above the GB plane) in white circles.

Figure 13.6 shows that one fifth of the atoms in these two layers occupy coincidence sites, highlighted by large gray circles. Owing to the coincidence, the nearest neighbor atoms (indicated by arrows) around each atom on the coincidence sites are located at exactly at the same distance away as those in the perfect crystal. Hence we expect the atoms on the coincidence sites to have low energy.

We note that the atoms in these two layers that are not on coincidence sites have some of their nearest neighbors at distances different from those in a perfect crystal. Therefore, in a real GB structure, we expect atoms to adjust their positions away from the idealized structure shown in

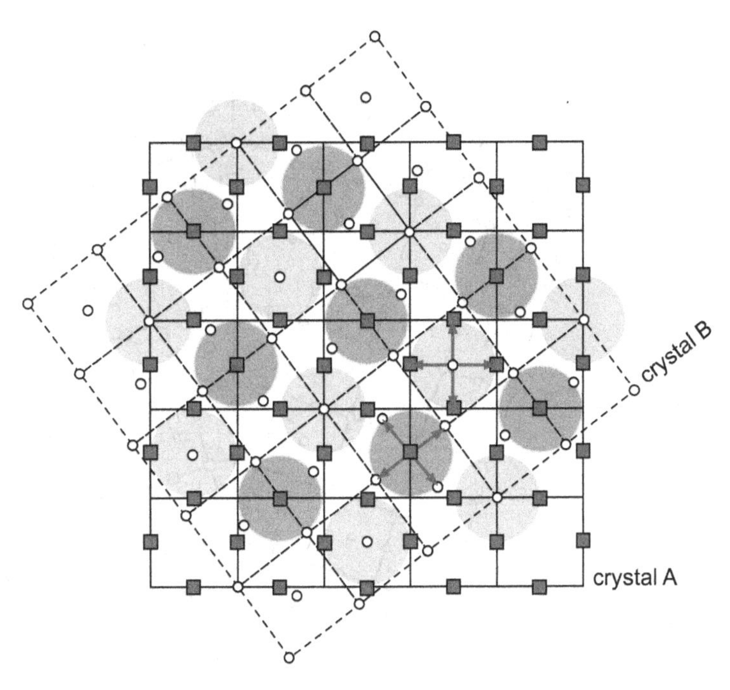

Figure 13.6. The top atomic layer of FCC crystal A and the bottom layer of FCC crystal B of a $\Sigma 5$ pure twist GB with [0 0 1] rotation axis. Arrows indicate the nearest neighbor atoms around atoms at coincidence sites. Dark gray circles denote coincidence sites occupied by atoms in crystal A and light gray circles denote coincidence sites occupied by atoms in crystal B. Because each crystal only occupies half of the space either above or below the GB plane, coincidence between points in the two lattices cannot be seen in this illustration, but they can be found in Fig. 13.5.

Fig. 13.6, in order to lower the GB energy. This process is called reconstruction, and is similar to the core reconstruction of dislocations discussed in Section 12.3.2. During the reconstruction, we expect the atoms on the coincidence sites shown in Fig. 13.6 to move very little, because they are already in low energy positions. The four atoms surrounding each atom on the CSL site are expected to experience more displacement during the reconstruction. Furthermore, the reconstructed GB structure is expected to have a periodic pattern with the same periodicity as that shown in Fig. 13.6. In other words, the relaxed atomistic structure of the $\Sigma 5$ pure twist is expected to contain periodic units whose area is $5a^2/2$. These periodic units are called the *structural units* of singular GBs [129].

13.2.5 $\Sigma 5$ tilt boundary

We construct an atomistic model of a $\Sigma 5$ tilt boundary, in which the grain boundary normal $\hat{n}_B$ is perpendicular to the rotation axis $\hat{t}_R$, as shown in Fig. 13.1b. Figure 13.7 shows an atomistic model of a symmetric tilt $\Sigma 5$ GB between FCC crystals. The atoms occupy the points of lattice A to the right of the GB and occupy the points of lattice B to the left of the GB. The GB plane itself is assumed to pass through a plane of CSL points, so that the atoms on the GB plane can be said to belong to either crystal A or crystal B. These atoms are likely in energetically favorable positions. However, the two atoms (enclosed by the dashed oval) next to every coincidence site

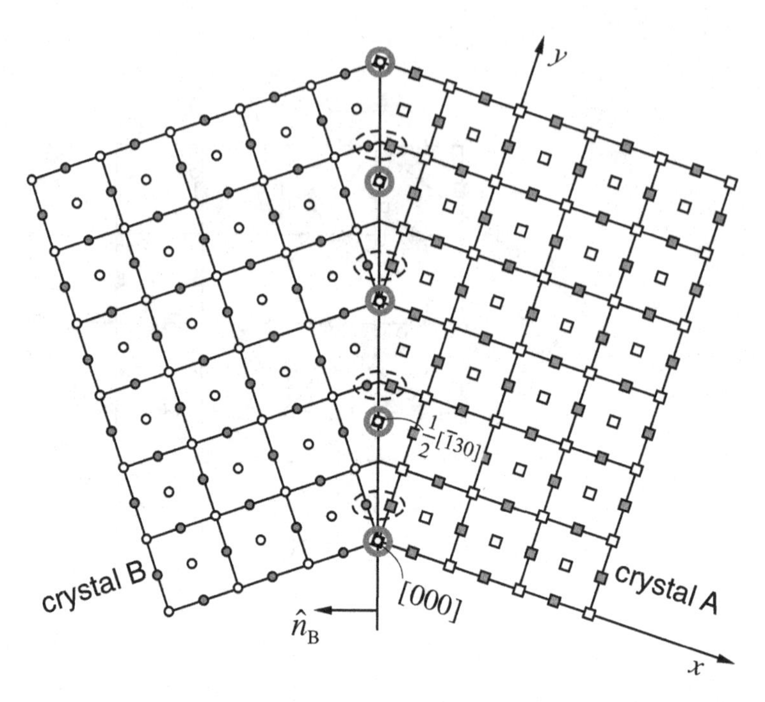

Figure 13.7. The atomic structure of a $\Sigma5$ symmetric tilt boundary with [0 0 1] rotation axis. The normal vector of the GB plane is $\hat{n}_B$.

on the GB are very close to each other, and may lead to a very high energy. In a real GB structure, one of these two atoms may be absent, by absorbing a lattice vacancy, in order to reduce the GB energy. In this case, the remaining atom would move to the GB plane, and the resulting structure would have a lower atomic density than that in the bulk. This is consistent with the notion that in the GB region there is more "free volume" available to gather impurities. The resulting (reconstructed) GB structure still has a periodicity dictated by the pattern of coincidence sites on the GB plane.

The energy of symmetric tilt GBs in FCC Al is shown in Fig. 13.8a as a function of misorientation angle θ around the $\langle 0\,0\,1\rangle$ rotation axis. No visible cusp is observed at $\theta = 36.86^\circ$ (corresponding to the $\Sigma5$ CSL), contrary to the pure twist GB energy curve shown in Fig. 13.2a. This shows that a GB with a low Σ number does not necessarily have a significantly lower energy than other GBs with similar misorientation.

13.2.6 $\Sigma11$ tilt boundary

As an example of special (i.e. singular) tilt boundaries, we now consider a $\Sigma11$ symmetric tilt GB with rotation axis along the [0 1 1] direction. Figure 13.8b plots the energy of symmetric tilt boundaries in Al as a function of the misorientation angle θ with rotation axis along [0 1 1]. Two cusps can be clearly observed at non-zero θ values. The cusp at $\theta = 70.53^\circ$ corresponds to the CSL of $\Sigma3$, with the GB plane being the {1 1 1} planes in both grains. This is the twin boundary discussed in Section 11.1.3. The other cusp at $\theta = 129.52^\circ$ (or equivalently 50.48°) corresponds to the CSL of $\Sigma11$, with the GB plane being the {1 1 3} planes in both grains.

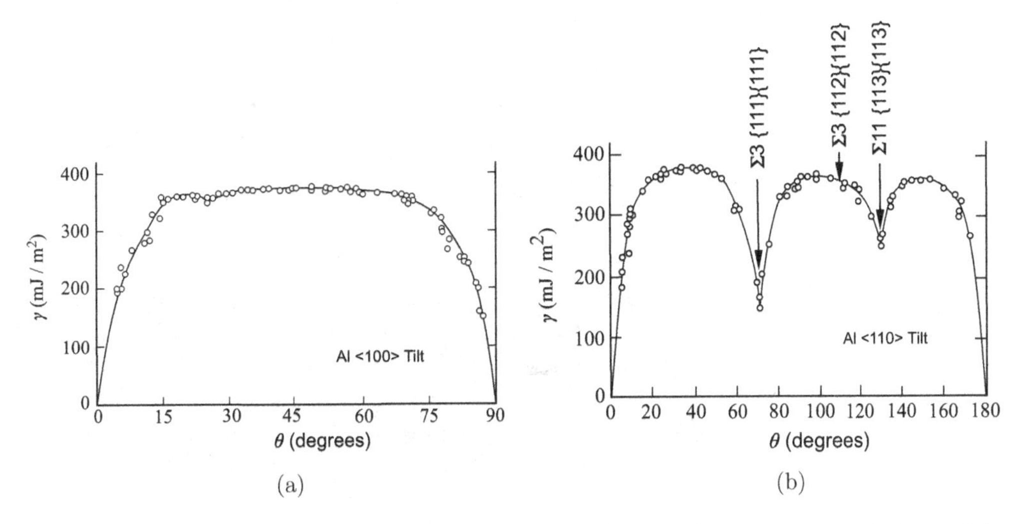

Figure 13.8. Energy per unit area of Al symmetric tilt GB as a function of misorientation angle θ around the (a) $\langle 0\,0\,1\rangle$ rotation axis and (b) $\langle 0\,1\,1\rangle$ rotation axis [128].

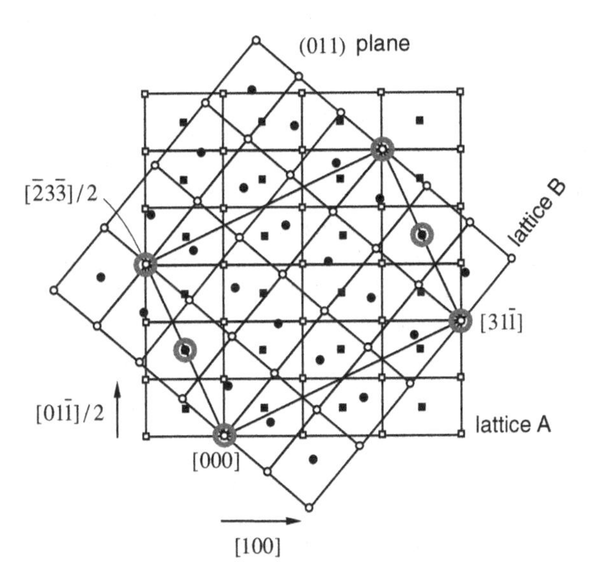

Figure 13.9. Two FCC lattices A (squares) and B (circles) forming a $\Sigma 11$ CSL viewed along $[0\,1\,1]$. The FCC lattices have two different layers. The bottom layers are shown in white symbols; the top layers are shown in black symbols. The CSL points are highlighted by large gray circles.

Figure 13.9 shows two FCC lattices viewed along $[0\,1\,1]$. Lattice B is rotated by $\theta = 50.48^\circ$ relative to lattice A around the $[0\,1\,1]$ axis. The repeat vectors of the CSL unit cell are: $[3\,1\,\overline{1}]$, $[\overline{2}\,3\,\overline{3}]/2$, and $[0\,1\,1]/2$ (out of plane). The CSL is a base-centered orthorhombic Bravais lattice, with two coincidence sites per CSL unit cell. The volume of the CSL unit cell is $\frac{11}{2}a^3$. Hence the density of CSL points is $(\frac{11}{4}a^3)^{-1}$. Recall that the density of FCC lattice points is $(\frac{1}{4}a^3)^{-1}$. Hence the ratio of the density of CSL and lattice A points is $1 : 11$, i.e. one out of every 11 points in lattice A is in coincidence.

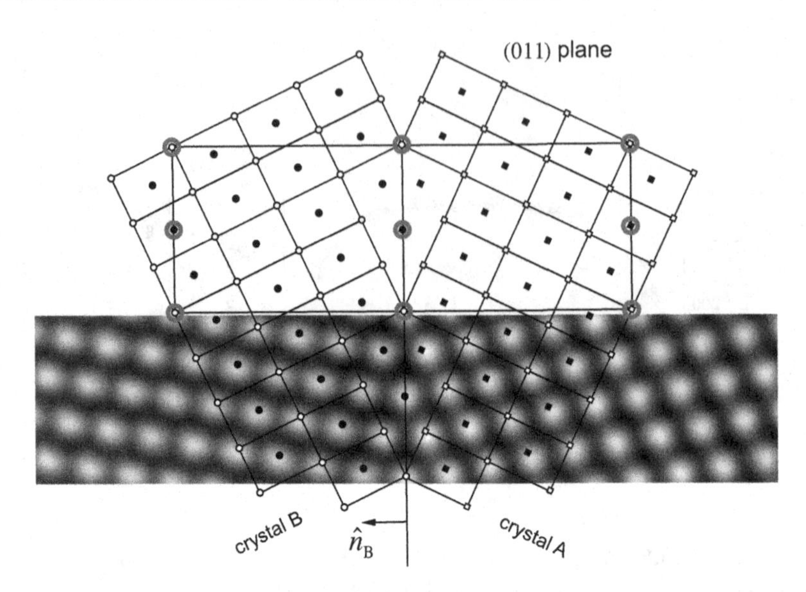

Figure 13.10. A $\Sigma 11$ symmetric tilt boundary in an FCC crystal. A unit cell of CSL is shown on either side of the GB with coincidence points highlighted in gray circles. A high resolution transmission electron microscopy (HRTEM) image of this GB in FCC Al [136] is shown in the background. Permission granted by W. E. King.

The atomistic structure of the singular $\Sigma 11$ symmetric tilt GB is shown in Fig. 13.10. The GB plane is the $(3\,1\,\bar{1})$ plane in lattice A, and is the $(3\,\bar{1}\,1)$ plane when expressed in the x'–y'–z' coordinate system fixed on lattice B. The atoms occupy the sites of lattice A to the right of the GB, and the sites of lattice B to the left of the GB. The atoms on the GB plane occupy coincidence sites between the two lattices. There is a high density of coincidence sites on the GB plane, and no atoms are too close to their neighbors (unlike those shown in Fig. 13.7). All these factors are consistent with the singular nature of the $\Sigma 11$ symmetric tilt GB. The GB structure consists of periodic units with repeat vectors $[\bar{2}\,3\,\bar{3}]/2$ and $[0\,1\,1]/2$ in the GB plane. A high-resolution transmission electron microscopy (HRTEM) image of this GB in FCC Al is also shown in Fig. 13.10.

13.3 Displacement shift complete lattice

We now introduce the displacement shift complete (DSC) lattice, which predicts the allowed Burgers vectors for GB dislocations in singular GBs. The DSC lattice contains all vectors by which one of the lattices (A or B) can be shifted relative to the other while regenerating the coincidence pattern of the CSL.

13.3.1 DSC lattice for $\Sigma 5$ CSL of simple cubic lattices

To start the discussion, we plot the two interpenetrating simple cubic lattices of Fig. 13.4 again in Fig. 13.11a, highlighting the periodic pattern formed by the two lattices. The square connecting four coincidence sites shows the unit cell of the CSL, inside of which we observe the fundamental

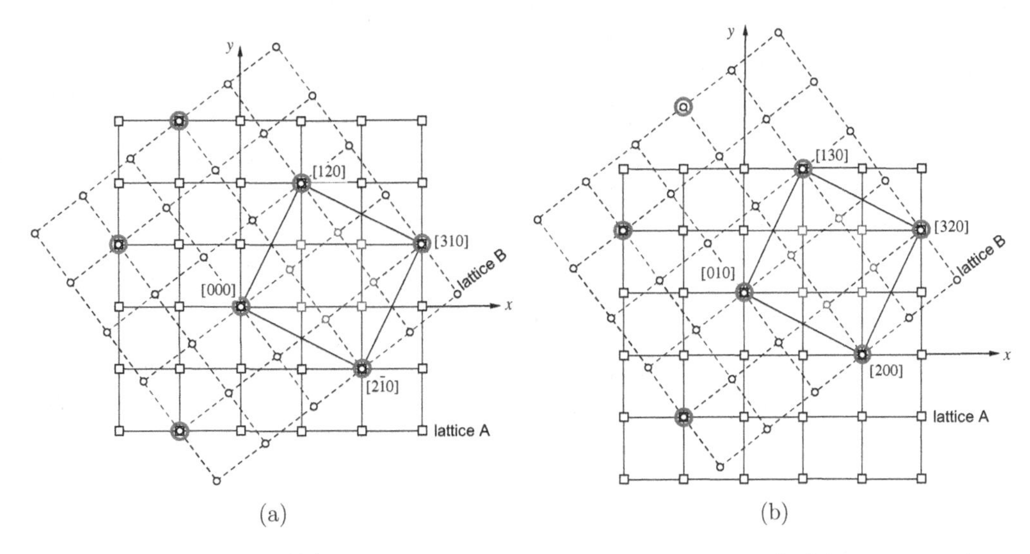

Figure 13.11. (a) Two interpenetrating simple cubic lattices A and B. A unit cell of the Σ5 CSL is shown as a square connecting four coincidence sites (large gray spheres). (b) After lattice B is shifted by $a[0\,1\,0]$, the periodic pattern in (a) is preserved, and only shifted by $a[0\,1\,0]$.

unit of the periodic pattern formed by the two lattices. Each unit cell of the CSL contains five points of lattice A; they are: $[0\,0\,0]$, $a[1\,0\,0]$, $a[2\,0\,0]$, $a[1\,1\,0]$, $a[2\,1\,0]$. Each unit cell of the CSL also contains five points of lattice B; expressed in the coordinate system fixed on lattice A,[4] they are: $[0\,0\,0]$, $\frac{a}{5}[4\,3\,0]$, $\frac{a}{5}[8\,6\,0]$, $\frac{a}{5}[7\,\bar{1}\,0]$, $\frac{a}{5}[11\,2\,0]$.

The periodic pattern may be destroyed when lattice B is shifted by some arbitrary vector. However, if lattice B is shifted by a repeat vector $\vec{r}_A^{\,i}$ of lattice A, such as $a[0\,1\,0]$, the periodic pattern is preserved, only shifted by $\vec{r}_A^{\,i}$, as shown in Fig. 13.11b. Obviously, if lattice B is shifted by a repeat vector $\vec{r}_B^{\,j}$ of lattice B, then there would be no change to the periodic pattern at all (not even a shift can be noticed).

We now look for the set of all vectors by which lattice B can be shifted relative to lattice A while preserving the periodic pattern formed by the two lattices. We note that the entire periodic pattern is preserved whenever one point in lattice B is brought into coincidence with some point in lattice A. Suppose that $\vec{r}_1$ and $\vec{r}_2$ are two vectors by which lattice B can be shifted while preserving the periodic pattern, then the periodic pattern must also be preserved when lattice B is shifted by vector $\vec{r}_1 + \vec{r}_2$. Therefore, the collection of all vectors by which lattice B can be shifted while preserving the periodic pattern form a lattice, which is called the *displacement shift complete* (DSC) lattice.

Obviously, all points of lattice A and all points of lattice B belong to the DSC lattice. Hence lattice A and lattice B are both *sub-lattices* of the DSC lattice.[5] However, the DSC lattice contains more points than all the points in lattice A and lattice B combined. For example, if point $\vec{r}_A^{\,i}$

4 When expressed in the x'–y'–z' coordinate system that rotates together with lattice B, these points are: $[0\,0\,0]$, $a[1\,0\,0]$, $a[2\,0\,0]$, $a[1\,\bar{1}\,0]$, $a[2\,\bar{1}\,0]$. The coordinates in the x–y–z coordinate system fixed on lattice A can be obtained from Eqs. (13.4) and (13.15).

5 The CSL is a sub-lattice of lattice A, lattice B and the DSC lattice.

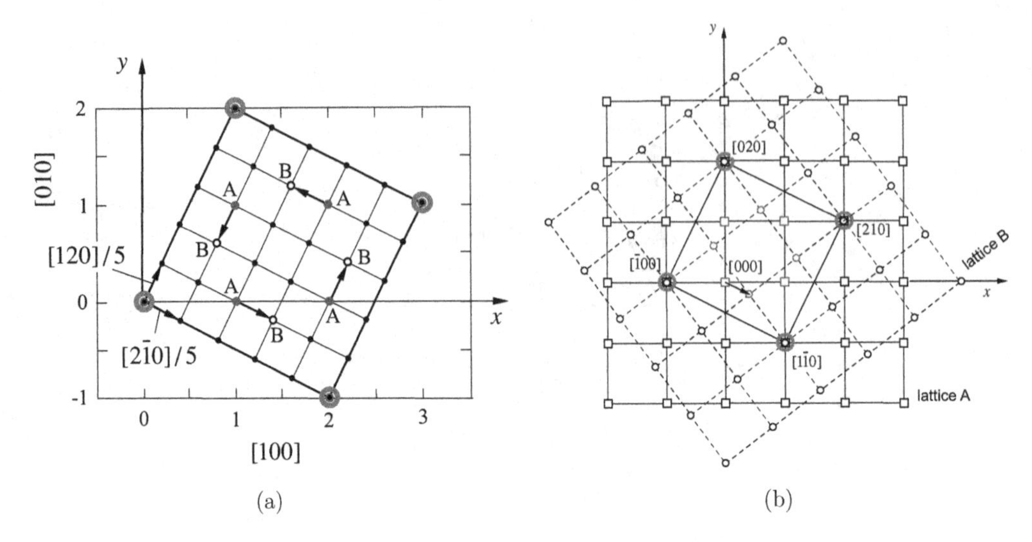

Figure 13.12. (a) The DSC lattice inside the unit cell of a $\Sigma 5$ CSL between two simple cubic lattices. The coincidence sites are highlighted in large gray circles. Points belonging to lattice A and lattice B are marked accordingly. The DSC lattice points are the corners of all the small squares. (b) After lattice B is shifted by a DSC vector $\frac{a}{5}[2\,\bar{1}\,0]$, the periodic pattern in Fig. 13.11a is preserved, and only shifted by $a[\bar{1}\,0\,0]$. For clarity see color figure on the book website.

belongs to lattice A, and point $\vec{r}_B^{\,j}$ belongs to lattice B, then point $\vec{r}_A^{\,i} + \vec{r}_B^{\,j}$ must belong to the DSC lattice, even thought it may not belong to either lattice A or lattice B.

We can generate all DSC lattice points inside a CSL primitive cell using the following procedure.

(1) Choose a primitive cell of the CSL. Since the CSL considered here contains only one CSL lattice point per unit cell, the unit cell is also the primitive cell. Here we choose the three repeat vectors of the primitive cell to be: $a[2\,\bar{1}\,0]$, $a[1\,2\,0]$, $a[0\,0\,1]$. The bottom (square) face of the primitive cell is shown in Fig. 13.11(a).
(2) Find all points $\vec{r}_A^{\,i}$, $i = 1, \ldots, \Sigma$, of lattice A that fall within the CSL primitive cell. In this example, there are five points: $[0\,0\,0]$, $a[1\,0\,0]$, $a[2\,0\,0]$, $a[1\,1\,0]$, $a[2\,1\,0]$.
(3) Find all points $\vec{r}_B^{\,j}$, $j = 1, \ldots, \Sigma$ of lattice B that fall within the CSL primitive cell. In this example, there are also five points: $[0\,0\,0]$, $\frac{a}{5}[4\,3\,0]$, $\frac{a}{5}[8\,6\,0]$, $\frac{a}{5}[7\,\bar{1}\,0]$, $\frac{a}{5}[11\,2\,0]$.
(4) Find all points $\vec{r}_A^{\,i} + \vec{r}_B^{\,j}$, for $i, j = 1, \ldots, \Sigma$, and map them back into the CSL primitive cell. They are the DSC lattice points inside the CSL primitive cell.

By mapping back we mean adding or subtracting an integer multiple of any repeat vectors of the CSL primitive cell until the vector falls inside the CSL primitive cell. It can be shown that all points generated by step (4) are different, and that they contain all possible points of the DSC lattice within the chosen CSL primitive cell. Therefore, there are Σ^2 DSC lattice points in each CSL primitive cell. Consequently, the ratio of volume density of lattice points of the DSC lattice, lattice A, and the CSL is $\Sigma : 1 : \Sigma^{-1}$.

Figure 13.12a plots the DSC lattice points inside the CSL primitive cell in the example considered here. It is seen that the $(0\,0\,1)$ plane of the DSC lattice is a square lattice with repeat length 1/5 of the repeat length of CSL on this plane. While many DSC lattice points

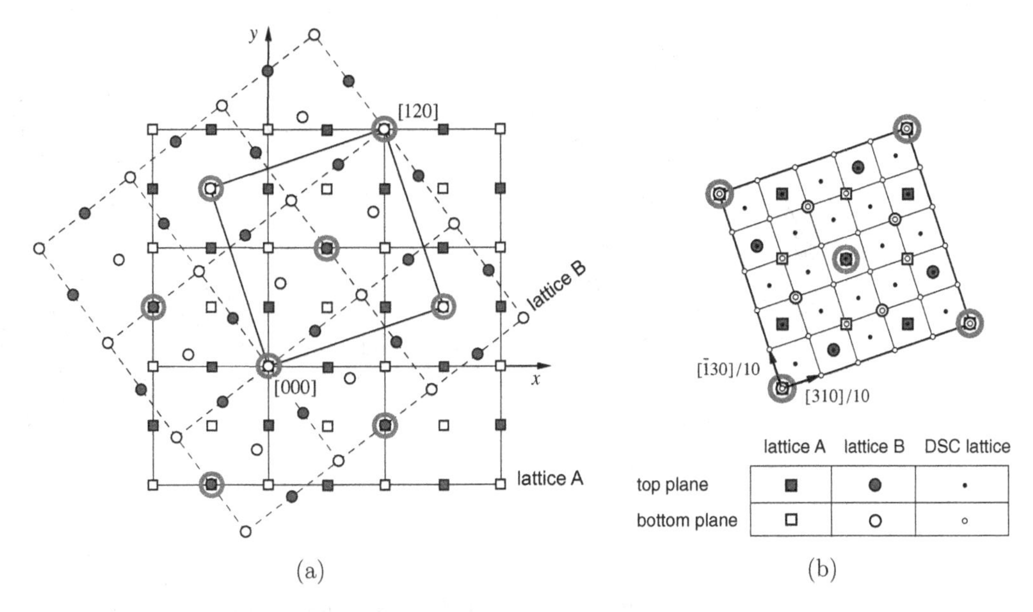

Figure 13.13. (a) Two interpenetrating FCC lattices A and B. A unit cell of the Σ5 CSL is shown as a square connecting four coincidence sites (large gray spheres). (b) The DSC lattice inside the unit cell of the Σ5 CSL.

do not belong to either lattice A or lattice B, every DSC lattice vector is a linear combination of vectors in lattice A and B. For example, one of the shortest repeat vector of the DSC lattice, $\frac{a}{5}[2\,\bar{1}\,0]$, can be obtained by $\frac{a}{5}[7\,\bar{1}\,0] - a[1\,0\,0]$. Another shortest repeat vector of the DSC lattice, $\frac{a}{5}[1\,2\,0]$ can be obtained by $\frac{a}{5}[11\,2\,0] - a[2\,0\,0]$. The three repeat vectors of the DSC lattice are: $\frac{a}{5}[2\,\bar{1}\,0]$, $\frac{a}{5}[1\,2\,0]$, and $a[0\,0\,1]$. Hence the DSC lattice is also a primitive tetragonal Bravais lattice, i.e. of the same type as the CSL.

As an illustration, Fig. 13.12b shows what would happen if lattice B is shifted by a DSC vector, $\frac{a}{5}[2\,\bar{1}\,0]$, which does not belong to either lattice A or lattice B. After the shift, coincidence is re-established at points $a[\bar{1}\,0\,0]$, $a[1\,\bar{1}\,0]$, $a[2\,1\,0]$, and $a[0\,2\,0]$. Therefore, the periodic pattern is preserved, and only shifted by $a[\bar{1}\,0\,0]$.

13.3.2 DSC lattice for Σ5 CSL of FCC lattices

The DSC lattice between two interpenetrating FCC lattices can be similarly defined as that between two simple cubic lattices. We plot the two FCC lattices of Fig. 13.5 again in Fig. 13.13a, highlighting the periodic pattern formed by the two lattices. The coincidence sites are shown as large gray spheres. The square connecting four coincidence sites show the unit cell of the CSL. Recall that the CSL is a body-centered tetragonal lattice. Hence there is also a coincidence site in the body center of the CSL unit cell. Each unit cell of the CSL now contains 10 points of lattice A and 10 points of lattice B.

Figure 13.13b shows that the DSC lattice is a body-centered tetragonal Bravais lattice, which is the same type as the CSL. The shortest repeat vectors of the DSC lattice on the (0 0 1) plane is 1/5 of those of the CSL. The three repeat vectors of the unit cell of the DSC lattice are:

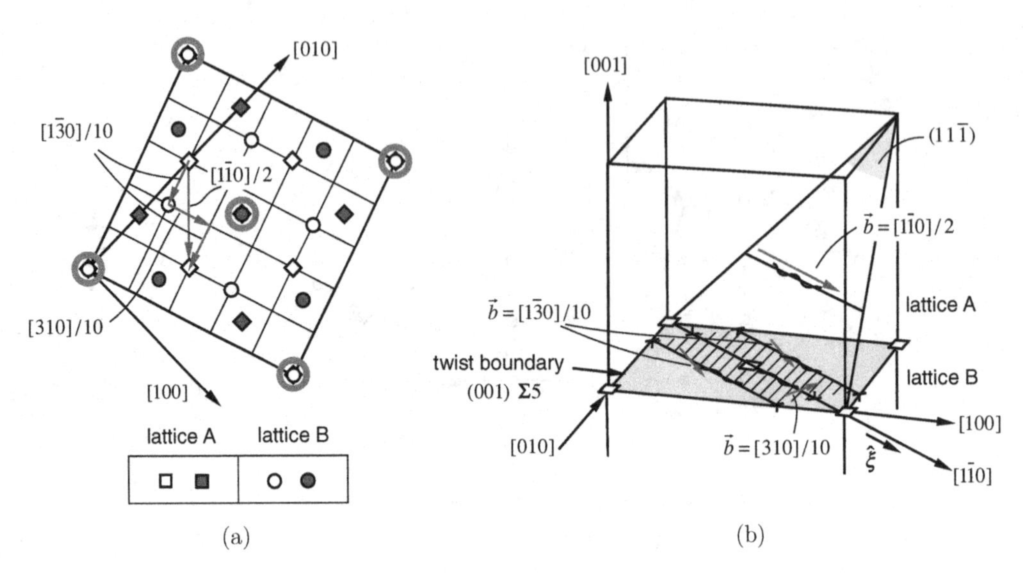

Figure 13.14. (a) The Burgers vector of a screw dislocation in the lattice equals the sum of three Burgers vectors of GB dislocations. (b) Dissociation of a lattice screw dislocation into three mixed GB dislocations as it enters the Σ5 twist boundary. For clarity see color figure on the book website.

$\frac{a}{10}[3\,1\,0]$, $\frac{a}{10}[\bar{1}\,3\,0]$, $a[0\,0\,1]$. There are two lattice points per unit cell of the DSC lattice: one at the corner and one at the body center.

For Σ5 grain boundaries with rotation axis $[0\,0\,1]$ in FCC crystals, the periodic pattern of the grain boundary structure is unchanged if grain B slides over grain A by vector $\frac{a}{10}[3\,1\,0]$, or by vector $\frac{a}{10}[\bar{1}\,3\,0]$. The pattern merely shifts by a repeat vector of lattice A in the $(0\,0\,1)$ plane. Therefore, the grain boundary can support dislocations with Burgers vectors $\frac{a}{10}[3\,1\,0]$ and $\frac{a}{10}[\bar{1}\,3\,0]$. These are called *grain boundary dislocations*, to be discussed in Sections 14.4 and 14.5.

When the GB plane is parallel to $(0\,0\,1)$, the sliding of grain B relative to grain A by vectors $\frac{a}{10}[3\,1\,0]$ and $\frac{a}{3}[\bar{1}\,3\,0]$ shifts the periodic pattern formed by the two lattices within the GB plane. Since the coincidence sites are preferred locations of the GB plane, there is no need for the GB to move out of its plane. This is true regardless of whether the boundary is pure twist, pure tilt, or mixed. However, if the grain boundary plane is not parallel to $(0\,0\,1)$, then the sliding of grain B by these vectors will require migration of the GB normal to its own plane. For such boundaries, GB dislocations with Burgers vectors $\frac{a}{10}[3\,1\,0]$ and $\frac{a}{10}[\bar{1}\,3\,0]$ are accompanied by a step in the grain boundary, and would be called disconnections (see Section 14.6).

Compared with the $\frac{1}{2}\langle 1\,1\,0\rangle$ Burgers vectors of lattice dislocations, the Burgers vectors of GB dislocations are much shorter, and hence contain much lower line energy. The GB dislocations cannot exist outside the grain boundary. On the other hand, when a lattice dislocation enters the grain boundary, the two components of the Burgers vector in the $(0\,0\,1)$ plane can dissociate into multiple GB dislocations with $\frac{a}{10}[3\,1\,0]$ and $\frac{a}{10}[\bar{1}\,3\,0]$ Burgers vectors, in order to reduce the line energy.

Figure 13.14 sketches the scenario when a right-handed screw lattice dislocation with Burgers vector $\vec{b} = \frac{a}{2}[1\,\bar{1}\,0]$ gliding on the $(1\,1\,\bar{1})$ plane intersects the Σ5 $(0\,0\,1)$ twist boundary. The

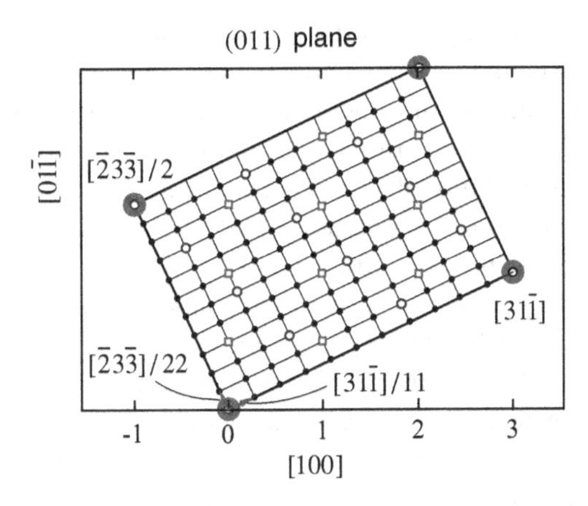

Figure 13.15. The bottom layer of the $\Sigma 11$ CSL unit cell containing lattice A points (open squares), lattice B points (open circles), and DSC lattice points (dots). For clarity see color figure on the book website.

Burgers vector of the screw dislocation can be decomposed into the Burgers vectors of three GB dislocations, according to the reaction

$$\frac{a}{2}[1\,\bar{1}\,0] = \frac{a}{10}[1\,\bar{3}\,0] + \frac{a}{10}[3\,1\,0] + \frac{a}{10}[1\,\bar{3}\,0]. \tag{13.16}$$

Since the dislocation line energy is proportional to the length of the Burgers vector squared, the decomposition given above leads to a reduction of the line energy by approximately 40%.

13.3.3 DSC lattice for $\Sigma 11$ CSL of FCC lattices

The DSC lattice for the $\Sigma 11$ CSL defined in Section 13.2.6 can be obtained by using the same procedure as described above. Similar to the CSL, the DSC lattice is also a base-centered orthorhombic Bravais lattice, and has two different (0 1 1) planes. For clarity, only the bottom plane is plotted in Fig. 13.15. The points in lattice A (open squares) and lattice B (open circles) are also plotted. The shortest repeat vectors in the DSC lattice are the shortest vector that connects a lattice A point and a lattice B point. Figure 13.15 shows that the shortest repeat vectors of the DSC lattices on the (0 1 1) plane are $[3\,1\,\bar{1}]/11$, $[\bar{2}\,3\,\bar{3}]/22$. They are 1/11 of the corresponding repeat vectors of the CSL. The shortest repeat vector of the DSC lattice in the out-of-plane direction is [0 1 1]/2, the same as that in the CSL. There are two lattice points per DSC unit cell. The density of DSC lattice points is Σ^2 times that of the CSL, the same as in Section 13.3.1. Notice that the length of the DSC lattice vector along the $[3\,1\,\bar{1}]$ direction is $a/\sqrt{\Sigma}$. The repeat vectors of this DSC lattice are the allowable Burgers vectors of the GB dislocation on $\Sigma 11$ grain boundaries corresponding to this CSL. A GB dislocation on a $\Sigma 11$ symmetric tilt GB is discussed in Section 14.4.

13.4 Summary

Grain boundaries (GBs) are the interfaces between crystals having the same crystal structure and composition but different relative orientations. The geometry of GBs can be specified with five degrees of freedom: three for the relative misorientation of the crystals and two for the direction of the grain boundary plane normal. Twist boundaries are created when the axis of rotation of one crystal relative to the other is normal to the plane of the boundary while tilt boundaries are formed when the axis of rotation lies in the plane of the boundary. A boundary with both twist and tilt components is called a mixed boundary. But these specifications provide no information about the structure or properties of the interface itself. For this the structure of the adjoining crystals needs to be taken into account.

The periodicity of the crystal lattice allows certain grain boundaries to have special structures and properties. These special misorientations are described in terms of the coincidence site lattice (CSL). In this picture, the lattices of the differently oriented crystals are imagined to interpenetrate and to form two superimposed lattices in space. For special misorientations, a fraction, $1/\Sigma$, of the lattice points coincide, which leads to the designation of grain boundary misorientation by the Σ number. For grain boundaries with low Σ numbers, a large fraction of the atoms in the plane of the grain boundary are positioned exactly as they would be in both of the lattices, and this leads to low energy configurations because fewer atoms in the grain boundary are displaced from their equilibrium lattice positions. Cusps in the plot of energy versus angle of misorientation are found for some of these special misorientations.

The CSL theory also predicts the vectors by which one lattice can be translated relative to the other while keeping the periodic pattern unchanged. These displacement vectors also form a lattice, called the displacement shift complete (DSC) lattice. The smallest repeat vectors of the DSC lattice are the Burgers vectors of grain boundary dislocations. They are usually much smaller than the Burgers vectors of crystal dislocations within the grains. The physical consequence is that a crystal dislocation with an appropriate Burgers vector can spread out in the GB by dissociating into many GB dislocations with much shorter Burgers vectors and lower energies.

13.5 Exercise problems

13.1 In Fig. 13.4 we showed that a $\Sigma 5$ CSL is created when a simple cubic lattice B is rotated by $\theta = 2\arctan(1/3)$ relative to another simple cubic lattice A about the [0 0 1] axis. Show that a CSL is also formed if the rotation angle is $\theta = 2\arctan(1/2)$. Determine the Σ number for this CSL. Obtain the rotation matrix g_{AB}.

13.2 Show that a CSL is formed when a simple cubic lattice B is rotated by $\theta = 2\arctan(1/n)$ relative to another simple cubic lattice A about the [0 0 1] axis, where $n = 4, 5, 6, \ldots$. Determine the Σ number and the rotation matrix g_{AB} as a function of n. Recall that Σ must be an odd number for a CSL between two cubic lattices.

13.3 Figure 13.16 shows the basal plane of a hexagonal lattice. Show that a CSL is formed when a hexagonal lattice B is rotated about the [0 0 0 1] axis so that the lattice point $\frac{a}{3}[5\,\bar{4}\,\bar{1}\,0]$ of lattice B coincides with the lattice point $\frac{a}{3}[5\,\bar{1}\,\bar{4}\,0]$ of lattice A. Determine the Σ number of this CSL. See Section 11.2 for the definition of the Miller–Bravais indices used here.

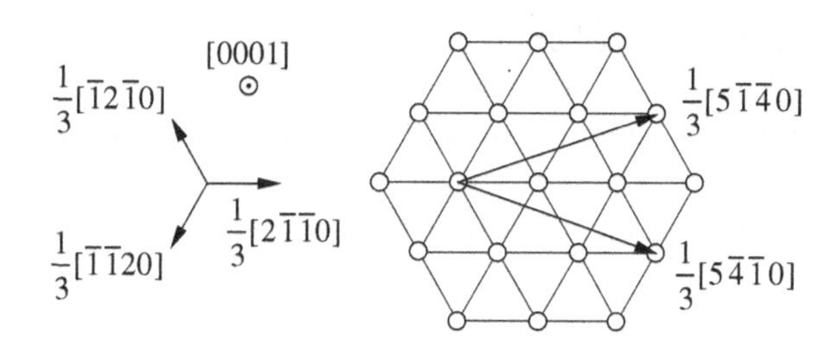

Figure 13.16. A hexagonal lattice viewed along the [0 0 0 1] axis.

13.4 In Fig. 13.5 we showed that a $\Sigma 5$ CSL is created when an FCC lattice B is rotated about the [0 0 1] axis relative to another FCC lattice A so that the lattice point $\frac{a}{2}[3\ \bar{1}\ 0]$ of B is coincident with the lattice point $\frac{a}{2}[3\ 1\ 0]$ of A. Determine the Σ number for a boundary created by rotation of an FCC lattice B about the [0 0 1] axis so that the lattice point at $\frac{a}{2}[5\ \bar{1}\ 0]$ of B is coincident with the lattice point $\frac{a}{2}[5\ 1\ 0]$ of A. At what misorientation angle would this CSL be expected?

13.5 A twin boundary in an FCC lattice can be created by rotating a crystal by 180° about an axis parallel to the [1 1 1] direction. Using your knowledge of the arrangements of lattice points (ABCABC...) in an FCC crystal, show that the boundary created must be a $\Sigma 3$ boundary. For your convenience, Fig. 13.17a shows an FCC lattice viewed along the [1 1 1] direction.

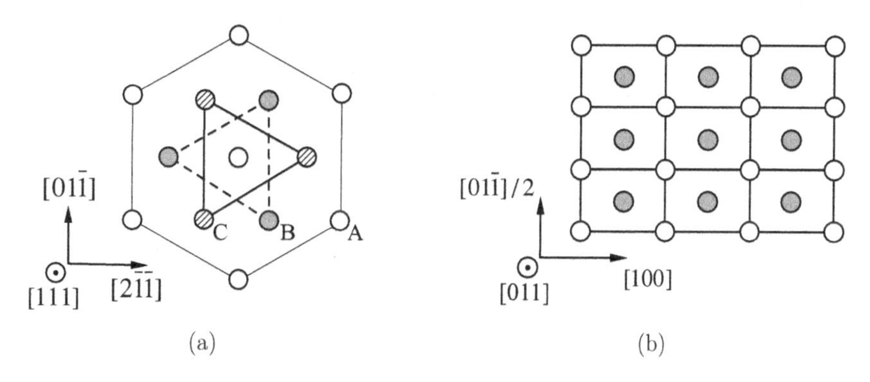

Figure 13.17. An FCC lattice viewed along (a) the [1 1 1] axis and (b) the [0 1 1] axis. Different symbols represent atoms on different layers.

13.6 Figure 13.17b shows an FCC lattice viewed along the [0 1 1] direction. Show that a CSL is formed when an FCC lattice is rotated by $\theta = 2\arctan(1/\sqrt{2})$ relative to another FCC lattice about the [0 1 1] axis.

(a) Sketch the unit cell of the CSL.
(b) Determine the Σ number for a boundary created by this misorientation.
(c) Find the three repeat vectors of the unit cell of the DSC lattice.
(d) Determine the volume density of the DSC lattice points.

13.7 Figure 12.10 shows a BCC lattice viewed along the [1 1 0] direction. Show that a CSL is formed when a BCC lattice is rotated by $\theta = 2\arctan(1/\sqrt{2})$ relative to another BCC lattice about the [1 1 0] axis.

(a) Sketch the unit cell of the CSL and describe its symmetry type.
(b) Determine the Σ number for this CSL.
(c) Construct an atomistic model for a symmetric tilt boundary corresponding to this CSL and the $(1\,\bar{1}\,2)$ boundary plane.
(d) Find the three repeat vectors of the unit cell of the DSC lattice.

14 Grain boundary mechanics

Our study of grain boundaries to this point has focused on their geometry and special misorientations that lead to periodic patterns in the GB structure. We now turn to another important aspect of grain boundaries: their energies and possible elastic fields. A planar grain boundary usually does not have a long range stress field by itself. However, certain grain boundaries contain periodic dislocation arrays as part of their structures. In such cases, there is an appreciable stress field around the GB at distances comparable to the inter-dislocation spacing in the GB. The GB model based on dislocation arrays, combined with the theory of coincidence site and DSC lattices, provides a way to understand the GB energy as a function of its misorientation angle.

We have seen that the GB energy as a function of the misorientation angle has a complex structure, as shown in Fig. 13.2 and Fig. 13.8. Nonetheless, such plots suggest a classification of grain boundaries broadly into three types: *singular*, *vicinal*, and *general* [129]. The *singular* GBs correspond to the sharp minima on the energy plots. Their misorientations usually correspond to low-Σ CSLs. The singular GBs are usually special in other properties as well, such as mobility and point defect segregation. The *vicinal* GBs have both misorientation and GB plane direction sufficiently close to the singular GBs, and they can be considered as singular GBs superimposed with one or more GB dislocation arrays. The spacing between the nearest dislocations in the array reduces as the misorientation deviates further away from that of the singular GB. The *general* GBs are those boundaries that are sufficiently different from the singular GBs that the dislocation array is no longer a useful model as the necessary dislocation density would be so large that the dislocation cores would overlap.

In this classification of GBs, the case of zero misorientation and its vicinal range deserves extra attention. On the one hand, when the misorientation angle θ equals zero, the grain boundary disappears and the GB energy is zero, because the two crystals are perfectly aligned with each other. On the other hand, the GB energy plots always exhibit a strong cusp at $\theta = 0$, similar to (and in fact more prominent than) the cusps at other singular misorientations with low Σ values. Therefore, it is customary to designate the case of $\theta = 0$ as a "singular GB of $\Sigma 1$," even though there is no boundary at all.

The vicinal GBs around $\theta = 0$ (usually within $\pm 15^\circ$) are mathematically and physically equivalent to arrays of lattice dislocations. They are also called *low angle grain boundaries*

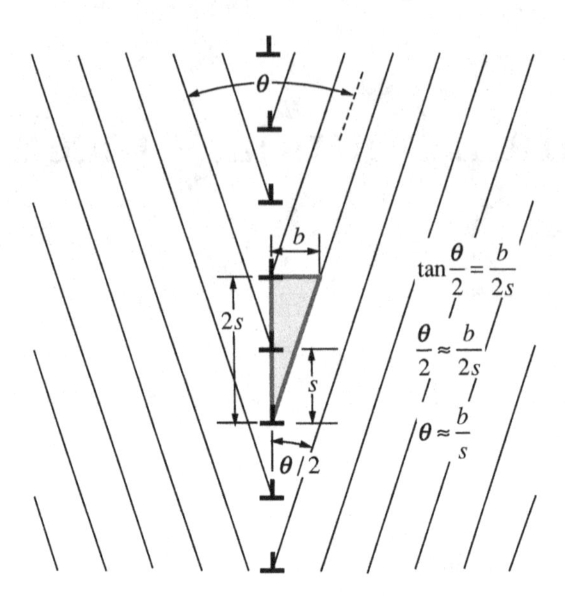

Figure 14.1. Low angle tilt boundary. Thin lines indicate extra-half-planes of the edge dislocation.

(LAGB) and their energy can be predicted from dislocation mechanics as done in the classic Read–Shockley model [137]. GBs with larger misorientation angles are called *high-angle grain boundaries* (HAGB), which, according to the above, can be classified into singular, vicinal, and general types. In this chapter, we will first discuss edge and screw lattice dislocation arrays in LAGBs and then discuss GB dislocation arrays in vicinal HAGBs.

14.1 Low angle tilt boundaries

14.1.1 Read–Shockley model

In Section 10.2.1, we have shown that an array of edge dislocations with the same sign stacked on top of each other is a stable low energy configuration. This configuration is equivalent to a low angle symmetric tilt grain boundary. As shown in Fig. 14.1, the extra-half-planes that extend from each of the edge dislocations must diverge and this causes one side of the crystal to be misoriented relative to the other side by an angle θ. This is a tilt grain boundary because the rotation axis lies in the plane of the boundary.

As shown in Fig. 14.1, the angle of misorientation is related to the spacing between the edge dislocations that comprise the boundary and the magnitude of the Burgers vector. By considering the shaded triangle we can write

$$\tan\frac{\theta}{2} = \frac{b}{2s} \tag{14.1}$$

or, for small θ, $\theta/2 \approx b/(2s)$, or, finally, $\theta \approx b/s$.

The angle of misorientation increases as the spacing of dislocation decreases. Typically this model for tilt grain boundaries holds up to a misorientation of about $\theta = 15°$, above which the dislocations are so close together that the discrete dislocation model breaks down.

One of the early successes of the theory of dislocations was the quantitative description of low angle grain boundaries, and, in particular, the energies of low angle boundaries. A model for the energies of low angle boundaries was published in 1950 by W. T. Read and W. Shockley [137]. This is now called the Read–Shockley model. Using Fig. 14.1 we may state that the energy (per unit area) of a low angle tilt boundary is simply the energies of the dislocations that make up that boundary. Recalling that the energy (per unit length) of an edge dislocation is

$$E_{\mathrm{e}} = \frac{\mu b^2}{4\pi(1-\nu)} \ln\left(\frac{\alpha_{\mathrm{e}} R}{b}\right) \tag{14.2}$$

then, if such edge dislocations are spaced at a distance s apart, the corresponding grain boundary energy (per unit area) is

$$E_{\mathrm{LAGB}} = \frac{1}{s} E_{\mathrm{e}}. \tag{14.3}$$

Since the stress fields of the dislocations tend to cancel in between the dislocations, the outer cutoff radius in the expression for the energy of the dislocation should be taken as $R \approx s/2$. We note that the compressive σ_{xx} stresses above one dislocation, at $\Delta y = s/2$, are canceled by the tensile σ_{xx} stresses from the dislocation above, at $\Delta y = -s/2$. In this way we see that the stress fields cancel in between like-signed edge dislocations and this justifies using $R \approx s/2$ for the cutoff radius. The stress field of dislocation arrays will be further discussed in Section 14.1.3.

With these considerations the energy of the boundary is

$$E_{\mathrm{LAGB}} = \frac{1}{s}\frac{\mu b^2}{4\pi(1-\nu)} \ln\left(\frac{\alpha_{\mathrm{e}} s}{2\, b}\right) = \frac{1}{s}\frac{\mu b^2}{4\pi(1-\nu)}\left[\ln\left(\frac{s}{b}\right) + \ln\left(\frac{\alpha_{\mathrm{e}}}{2}\right)\right]. \tag{14.4}$$

Since we have shown that $\theta \approx b/s$ or $s = b/\theta$, it follows that

$$E_{\mathrm{LAGB}}(\theta) = \frac{\theta}{b}\frac{\mu b^2}{4\pi(1-\nu)}\left[\ln\left(\frac{1}{\theta}\right) + \ln\left(\frac{\alpha_{\mathrm{e}}}{2}\right)\right], \tag{14.5}$$

which can be written in the form

$$E_{\mathrm{LAGB}}(\theta) = A\theta\,[B - \ln\theta], \tag{14.6}$$

where

$$A = \frac{\mu b}{4\pi(1-\nu)}, \tag{14.7}$$

$$B = \ln\left(\frac{\alpha_{\mathrm{e}}}{2}\right). \tag{14.8}$$

Equation (14.6) shows that the energy of the tilt boundary varies nonlinearly with the angle of misorientation, and reaches a (shallow) maximum at a critical angle of misorientation, $\theta_{\max}$, as shown in Fig. 14.2a. Looking back at Fig. 13.8a, we can see that in Al the energy of the symmetric tilt GB around the [0 0 1] rotation axis indeed shows a maximum at around $\theta_{\max} \approx 18^\circ$.

The maximum in the energy can be determined by finding the extremum in E_{LAGB} vs θ. By setting $dE_{\mathrm{LAGB}}/d\theta = 0$, we get

$$B = 1 + \ln\theta_{\max}. \tag{14.9}$$

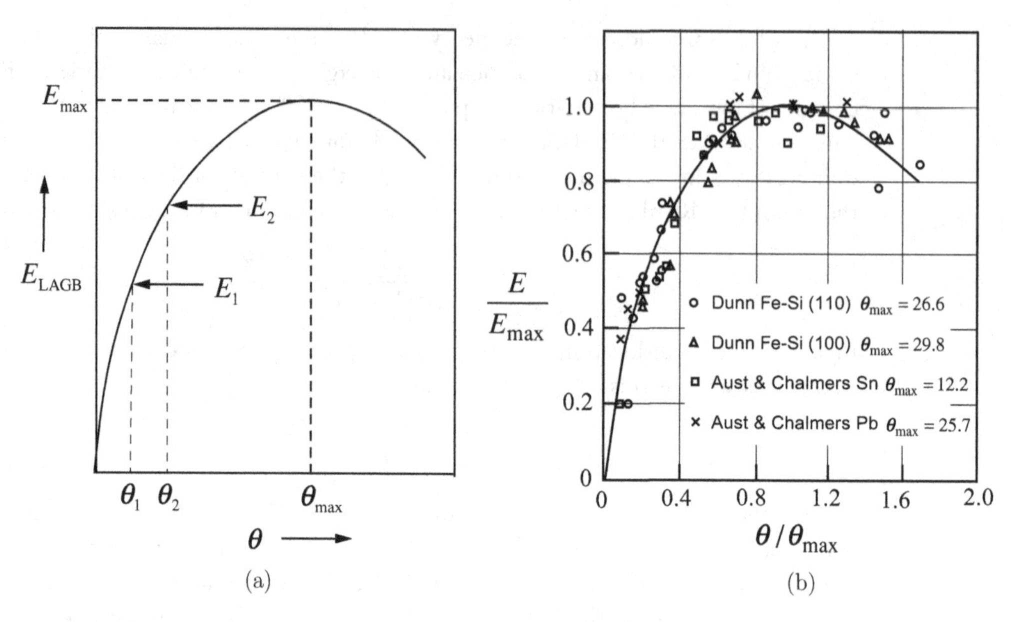

Figure 14.2. (a) Boundary energy vs angle for the Read–Shockley model. (b) Comparison of measured grain boundary energies with the predictions of the Read–Shockley model [6].

With this we can write

$$E_{max} = A\theta_{max}(B - \ln\theta_{max}) = A\theta_{max}. \tag{14.10}$$

Finally the energy of the boundary can be written in compact form as

$$\frac{E}{E_{max}} = \frac{\theta}{\theta_{max}}\left[1 - \ln\left(\frac{\theta}{\theta_{max}}\right)\right]. \tag{14.11}$$

As shown in Fig. 14.2b, early measurements of the energies of grain boundaries showed that this relationship is very well obeyed.

14.1.2 Polygonization

Low angle grain boundaries are also called sub-grain boundaries, to distinguish them from the high-angle grain boundaries. While most of the grain boundaries in as-grown crystals are high-angle grain boundaries, the sub-grain boundaries start to appear within the original grains after plastic deformation, through the accumulation and self-organization of dislocations. The sub-grain boundaries become more clearly defined when the crystal is annealed following plastic deformation, by a process called *polygonization.*

The Read–Shockley model provides a good understanding of the phenomenon of polygonization, the process by which deformed crystals naturally subdivide into differently oriented domains. Early observations of deformed metal crystals with X-ray diffraction revealed that while the Laue diffraction spots are smeared or spread-out by plastic deformation, on annealing these smeared Laue spots subdivide into multiple discrete spots, representing differently

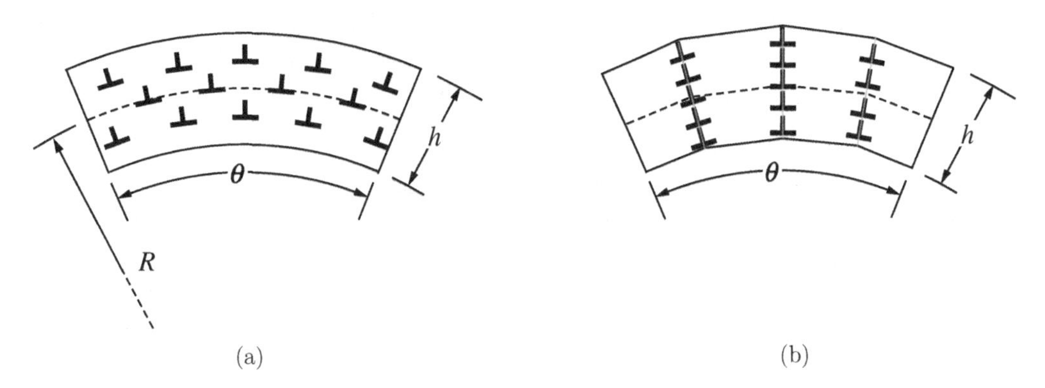

Figure 14.3. (a) Geometrically necessary dislocations in a uniformly bent crystal. (b) Rearrangement of dislocations after polygonization.

oriented sub-domains in the crystal. To describe this process we first consider a crystal uniformly bent by plastic deformation, as shown in Fig. 14.3a.

For the simple case of slip on planes parallel to the long dimension of the crystal, edge dislocations of the type shown would be needed to accommodate the plastic bending. These are called *geometrically necessary dislocations* because they are necessary to account for the permanent bending. The density of such geometrically necessary dislocations can be determined as follows. Suppose there are n dislocations in the element ($n = 14$ in the example here). The extra-half-planes terminate at the top of the element but not at the bottom and that accounts for the length of the top, $(R + (h/2))\,\theta$, being greater than the length of the element at the bottom, $(R - (h/2))\,\theta$. This length difference is simply the number of dislocations times the size of the Burgers vector (the thickness of the half-planes). With this we can write

$$\begin{aligned} nb &= [(R + (h/2))\theta] - [(R - (h/2))\theta] \\ &= h\theta, \end{aligned} \tag{14.12}$$

so that

$$n = h\theta / b. \tag{14.13}$$

The geometrically necessary dislocation density is simply the number of geometrically necessary dislocations per unit area, so that

$$\rho_G = \frac{n}{R\theta h} = \frac{h\theta}{b}\frac{1}{R\theta h}, \tag{14.14}$$

which is

$$\rho_G = \frac{1}{Rb} = \frac{\kappa}{b}, \tag{14.15}$$

where $\kappa = 1/R$ is the curvature of the lattice. This called the Cahn–Nye relation after Robert Cahn [138] and J. F. Nye [139]. This indicates that as the curvature of the crystal increases with increased plastic bending, the corresponding density of geometrically necessary dislocations also increases.

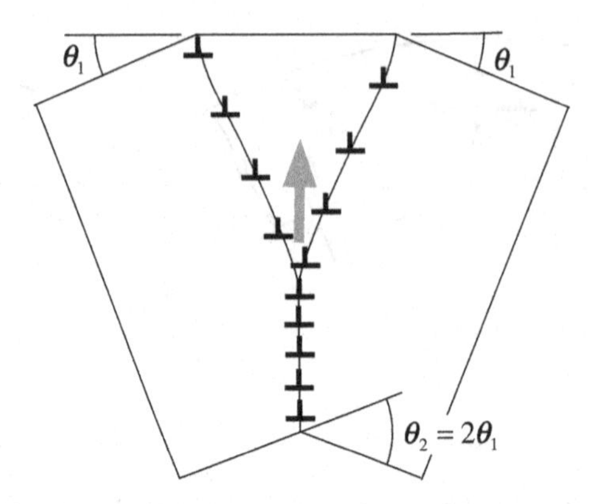

Figure 14.4. Polygonization by the coalescence of low angle tilt boundaries.

The model we have discussed would give rise to streaked Laue spots when plastic deformation occurs. The bent crystal no longer has a single Laue spot because the crystal orientation changes gradually with position and that leads to streaking of Laue spots [140].

As we have already discussed, like-signed dislocations are more energetically favorable (with respect to glide) when they are stacked one above the other in the form of low angle tilt boundaries. Thus one can expect the geometrically necessary dislocations in the bent crystal to rearrange themselves so that they line up, one above another. For this to happen, some of the dislocations may need to climb (hence annealing at an elevated temperature is typically required) so that the dislocations can get onto different glide planes. When this happens, the gradually bent crystal becomes discretely bent where the low angle boundaries are located, as shown in Fig. 14.3b.

Figure 14.4 shows a bent crystal containing three low angle tilt grain boundaries. We call this configuration a triple junction because three grain boundaries are joined at one point (i.e. a line in three dimensions). We see one boundary with a misorientation θ_2 in the lower half of the element and two boundaries, each with misorientation $\theta_1 = \theta_2/2$ in the upper half of the element. The energies associated with these boundaries, E_2 and E_1, are depicted in Fig. 14.2. Because of the non-linear (concave) relationship between the energy and the angle of misorientation, it is clear that the total energy of the two boundaries in the upper half of the element, $2E_1$, is greater than the energy of the boundary in the lower half of the element, E_2. It follows that the triple junction will move spontaneously upward as the two boundaries coalesce to make a single boundary.

This analysis shows that when deformed (bent) crystals are annealed, low angle tilt boundaries tend to coalesce so that the number of differently oriented domains decrease with annealing while the misorientation between those remaining domains increases. The phenomenon is called polygonization because each domain (i.e. sub-grain) is separated from neighboring domains by boundaries that becomes better defined and straighter with time, and the crystal looks more and more like a polygon. In the long time limit, a bent crystal might be expected to subdivide into two differently oriented domains, with a single low angle boundary separating the two regions.

14.1.3 Stress field of an edge dislocation array

Given that a low angle tilt boundary is equivalent to an array of edge dislocations, the stress field of the boundary can be obtained by superimposing the stress field of individual edge dislocations.

The stress field of a positive edge dislocation along the z axis with Burgers vector along the x axis is given in Eq. (9.31). To facilitate the discussion of LAGBs with different boundary orientations, we generalize this equation by allowing the Burgers vector to have both x and y components. The stress field of such an edge dislocation is given below:

$$
\begin{aligned}
\sigma_{xy}(x,y) &= \frac{\mu}{2\pi(1-\nu)\,r^2}\left[b_x x\left(1-\frac{2y^2}{r^2}\right) - b_y y\left(1-\frac{2x^2}{r^2}\right)\right],\\
\sigma_{yy}(x,y) &= \frac{\mu}{2\pi(1-\nu)\,r^2}\left[b_x y\left(1-\frac{2y^2}{r^2}\right) + b_y x\left(1+\frac{2y^2}{r^2}\right)\right],\\
\sigma_{xx}(x,y) &= \frac{-\mu}{2\pi(1-\nu)\,r^2}\left[b_x y\left(1+\frac{2x^2}{r^2}\right) + b_y x\left(1-\frac{2x^2}{r^2}\right)\right],
\end{aligned}
\tag{14.16}
$$

where $r = \sqrt{x^2+y^2}$, and μ and ν are the shear modulus and Poisson's ratio, respectively.

For a dislocation array along the y direction with periodicity L_y, the stress fields can be obtained from an infinite summation,

$$
\sigma_{ij}^{\text{yPBC}}(x,y) = \sum_{n=-\infty}^{+\infty} \sigma_{ij}(x, y-nL_y),
\tag{14.17}
$$

where $\sigma_{ij}(x,y)$ is the stress field of a single dislocation, as given in Eq.(14.16). The superscript yPBC indicates that the resulting stress field satisfies periodic boundary conditions (PBC) in the y direction. The summation in Eq. (14.17) can be obtained analytically [5, 141], and the result is

$$
\begin{aligned}
\sigma_{xx}^{\text{yPBC}} &= \frac{\mu}{2(1-\nu)L_y}\left\{-b_x s_Y\left[\frac{2\pi X S_X + C_X - c_Y}{(C_X-c_Y)^2}\right] + b_y 2\pi X\left[\frac{C_X c_Y - 1}{(C_X-c_Y)^2}\right]\right\},\\
\sigma_{yy}^{\text{yPBC}} &= \frac{\mu}{2(1-\nu)L_y}\left\{b_x s_Y\left[\frac{2\pi X S_X - C_X + c_Y}{(C_X-c_Y)^2}\right]\right.\\
&\qquad \left. -b_y\left[\frac{2\pi X(C_X c_Y - 1) - 2S_X(C_X - c_Y)}{(C_X-c_Y)^2}\right]\right\},\\
\sigma_{xy}^{\text{yPBC}} &= \frac{\mu}{2(1-\nu)L_y}\left\{b_x 2\pi X\left[\frac{C_X c_Y - 1}{(C_X-c_Y)^2}\right] + b_y s_Y\left[\frac{2\pi X S_X - C_X + c_Y}{(C_X-c_Y)^2}\right]\right\},
\end{aligned}
\tag{14.18}
$$

where $X \equiv x/L_y$, $Y \equiv y/L_y$, $s_Y \equiv \sin 2\pi Y$, $c_Y \equiv \cos 2\pi Y$, $S_X \equiv \sinh 2\pi X$, $C_X \equiv \cosh 2\pi X$.

Even though the expressions in Eq. (14.18) seem complicated, it is easy to see that all the stress components depend on y through $\sin(2\pi y/L_y)$ and $\cos(2\pi y/L_y)$, so that they are all periodic functions of y with periodicity L_y, as they should be. It is of interest to find out how the stress field depends on x for large $|x|$. Recall that the stress field of a single dislocation has a $1/r$ dependence, i.e. a long-range tail. As a result, the line energy of a single dislocation at the center of a cylinder of radius R is proportional to $\ln R$, which diverges as R goes to infinity. If grain boundaries are to have a well-defined energy per unit area (independent of the size of the solid), then it must not have a long-range stress field, as indicated in Section 14.1.1. We shall see that the stress field of a periodic array of edge dislocations is often short ranged, due to cancellation of the stress field from different dislocations. However, the dislocation array must satisfy certain geometric conditions for the long-range field to cancel completely.

As an example, Fig. 14.5a plots the stress field $\sigma_{yy}^{\text{yPBC}}$ for an edge dislocation array with periodicity $L_y = 10$ and Burgers vector $b_x = 1$, $b_y = 0$. The stress field is well localized within the

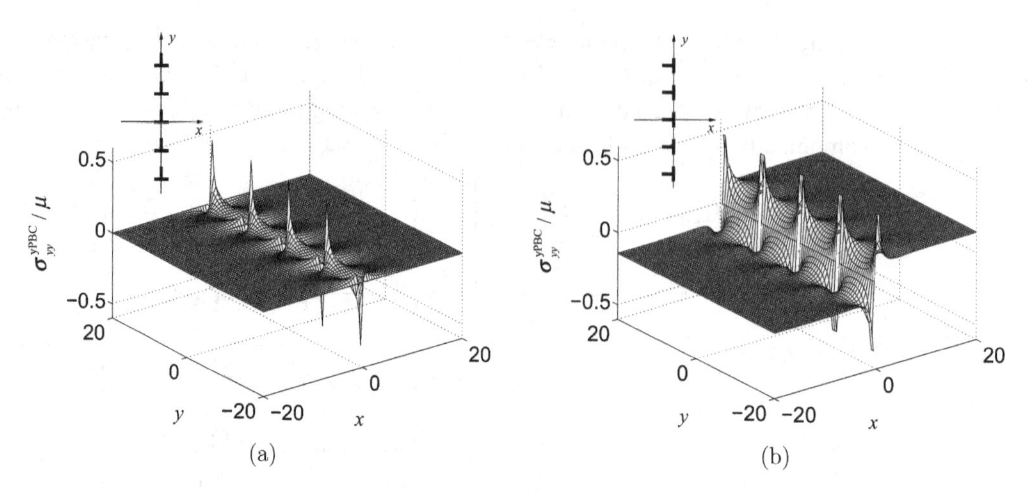

Figure 14.5. The σ_{yy} component of the stress field of an edge dislocation array along y axis with periodicity $L_y = 10$. (a) For Burgers vector $b_x = 1$, $b_y = 0$. (b) For Burgers vector $b_x = 0$, $b_y = 1$. The Poisson's ratio ν is taken to be 0.3 here.

domain of $|x| \leq L_y$. At larger $|x|$, the stress field quickly decays to zero. It can be shown, based on Eq. (14.18), that the stress field decays with $|x|$ exponentially fast. The other two stress components, $\sigma_{xx}^{\rm yPBC}$ and $\sigma_{xy}^{\rm yPBC}$, have the same behavior.

On the other hand, Fig. 14.5b plots the stress field $\sigma_{yy}^{\rm yPBC}$ for an edge dislocation array with periodicity $L_y = 10$ and Burgers vector $b_x = 0$, $b_y = 1$. At large $|x|$, the stress field goes to a non-zero constant. It can be shown, based on Eq. (14.18), that the limiting stress field is [129, 141]

$$\lim_{|x|\to\infty} \sigma_{yy}^{\rm yPBC}(x, y) = \frac{\mathrm{sgn}(x)\, \mu\, b_y}{(1-\nu)L_y}, \tag{14.19}$$

where sgn(x) equals $+1$ for $x > 0$ and equals -1 for $x < 0$. The other two stress components, $\sigma_{xx}^{\rm yPBC}$ and $\sigma_{xy}^{\rm yPBC}$, decay to zero similar to Fig. 14.5a.

In summary, as $|x|$ goes to infinity, $\sigma_{yy}^{\rm yPBC}$ is the only component of $\sigma_{ij}^{\rm yPBC}$ that may stay non-zero; and if $b_y = 0$, then all components of $\sigma_{ij}^{\rm yPBC}$ go to zero. Hence the condition for a low angle tilt GB consisting of a single periodic array of edge dislocations to have no long-range stress is that the Burgers vectors should be orthogonal to the GB plane.

The non-zero far-field stress of $\sigma_{yy}^{\rm yPBC}$ when $b_y \neq 0$ is due to the fact that the edge dislocations in Fig. 14.5b are, after all, on the same glide plane and hence in a repulsive configuration. It may appear that we should not have considered such a dislocation array as a model for a grain boundary in the first place. However, this type of array is formed by misfit dislocations at the interfaces of heteroepitaxial films on substrates (see Fig. 8.26). The non-zero far-field stress exists to cancel the stress caused by the lattice mismatch between the two materials. A mismatch strain can also exist between two grains of the same material for non-coincident grain boundaries. For example, if the grain boundary plane is the (1 0 0) plane for one crystal and is the (1 1 0) plane for the other crystal, then no three-dimensional coincidence site lattice can be found, unless the grains are stretched or compressed to match the lattice spacing of each other. Misfit

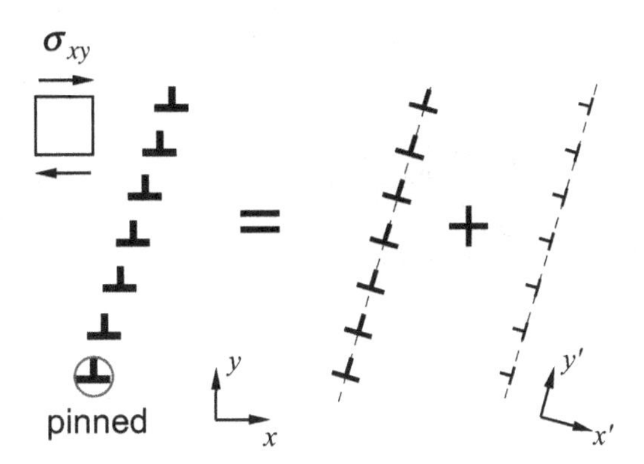

Figure 14.6. A low angle tilt boundary deformed by an external stress can be considered as a superposition of two dislocation arrays, one with Burgers vector normal to the boundary plane, and the other with Burgers vector in the boundary plane.

dislocation arrays can form at the grain boundaries to reduce the elastic energy caused by the lattice mismatch.

Another scenario where dislocation arrays of the kind in Fig. 14.5b can appear in GBs is when a symmetric tilt LAGB is deformed by stress. Because a low angle tilt boundary is equivalent to an array of edge dislocations, it experiences a Peach–Koehler force when an appropriate stress is applied. Figure 14.6 shows that when σ_{xy} is applied, the positive edge dislocations experience forces to the right. If the bottom dislocation is pinned, the boundary may deviate from its original position, so that the Burgers vectors have components both out of the boundary plane and in the boundary plane. The latter component has a long-range uniform stress field that can be obtained more easily in the rotated coordinate system x'–y' (see Exercise problem 14.1).

14.2 Low angle twist boundaries

While low angle tilt boundaries are equivalent to arrays of edge dislocations, low angle twist boundaries are equivalent to arrays of screw dislocations. However, a low angle twist boundary must consist of two or more periodic arrays of screw dislocations with different orientations. This is because, unlike edge dislocations, a single periodic array of screw dislocations always has a long-range stress field, which must be canceled by another array (or more arrays) to form a LAGB with no long-range stress.

14.2.1 Stress field of a screw dislocation array

The stress field of a single right-handed screw dislocation lying along the z axis is given in Eq. (9.8). For a screw dislocation array along the y direction with periodicity L_y, the stress fields can be obtained from the same infinite summation given in Eq. (14.17), with the stress field of a single dislocation given by Eq. (9.8). Again, the summation in Eq. (14.17) can be obtained

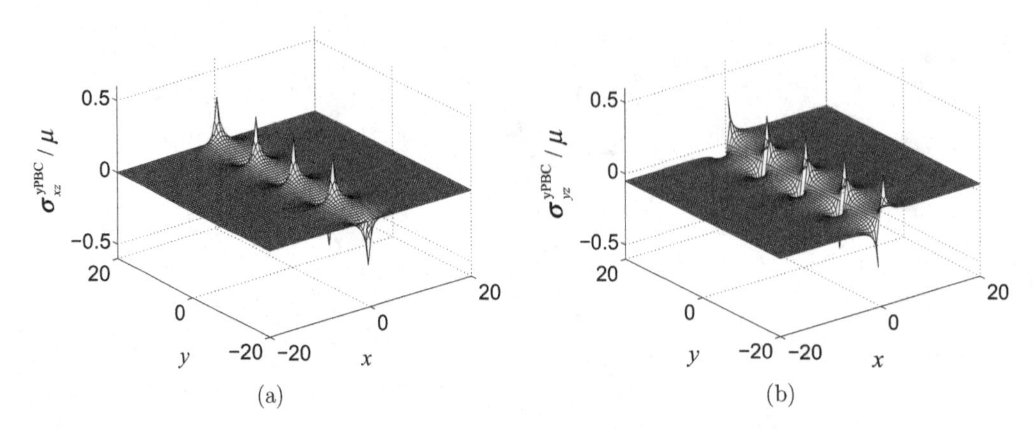

Figure 14.7. The (a) σ_{xz} and (b) σ_{yz} components of the stress field of a right-handed screw dislocation array along y axis with periodicity $L_y = 10$.

analytically [5, 141], and the result is

$$\begin{aligned}\sigma_{xz}^{\text{yPBC}} &= -\frac{\mu b_z}{2L_y}\left[\frac{\sin(2\pi y/L_y)}{\cosh(2\pi x/L_y) - \cos(2\pi y/L_y)}\right],\\ \sigma_{yz}^{\text{yPBC}} &= \frac{\mu b_z}{2L_y}\left[\frac{\sinh(2\pi x/L_y)}{\cosh(2\pi x/L_y) - \cos(2\pi y/L_y)}\right].\end{aligned} \tag{14.20}$$

It can be shown that

$$\lim_{|x|\to\infty} \sigma_{xz}^{\text{yPBC}}(x, y) = 0, \tag{14.21}$$

$$\lim_{|x|\to\infty} \sigma_{yz}^{\text{yPBC}}(x, y) = \frac{\text{sgn}(x)\,\mu\, b_z}{2\,L_y}. \tag{14.22}$$

Figure 14.7 plots the stress field $\sigma_{xz}^{\text{yPBC}}$ and $\sigma_{yz}^{\text{yPBC}}$ for an array of right-handed screws with periodicity $L_y = 10$ and Burgers vector $b_z = 1$. We can see that at large $|x|$, $\sigma_{xz}^{\text{yPBC}}$ goes to zero, while $\sigma_{yz}^{\text{yPBC}}$ goes to a non-zero constant that depends on the sign of x, consistent with Eqs. (14.21) and (14.22).

Therefore, a single periodic array of screw dislocations always has a long-range stress field. This long-range stress field can be canceled by introducing another periodic array of right-handed screw dislocations with Burgers vector b_y and periodicity L_z. It can be shown that the far-field stress of the second dislocation array is

$$\lim_{|x|\to\infty} \sigma_{xz}^{\text{zPBC}}(x, y) = 0, \tag{14.23}$$

$$\lim_{|x|\to\infty} \sigma_{yz}^{\text{zPBC}}(x, y) = -\frac{\text{sgn}(x)\,\mu\, b_y}{2\,L_z}. \tag{14.24}$$

Therefore, the long-range stress field σ_{yz} will be exactly cancelled if the "Burgers vector density" b_z/L_y of the first array equals the "Burgers vector density" b_y/L_z of the second array.

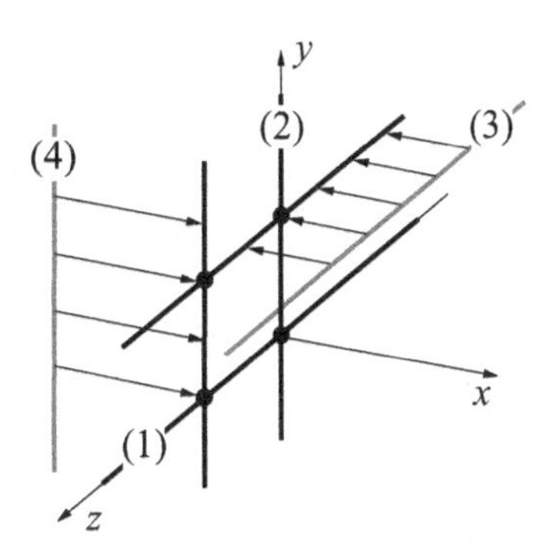

Figure 14.8. Two sets of like-signed screw dislocations forming a fragment of a low angle twist boundary.

14.2.2 Screw dislocation network

In Section 10.2.1, we have seen that two right-handed screw dislocations attract each other if their orientation angle θ is between $\pi/4$ and $3\pi/4$, with the strongest attraction occurring at $\theta = \pi/2$. This is consistent with the conclusion in the previous section that two periodic arrays of screw dislocations at 90° cancel the long-range stress fields of each other.

Figure 14.8 shows how the crisscross network of screw dislocations can form by dislocation interactions. Suppose that dislocations (1) and (2) have already reacted to form a stable configuration. Then, if dislocation (3) arrives as shown, it will be attracted to dislocation (2) and will form a stable configuration with it. If dislocation (4) subsequently arrives, it will be attracted to both (1) and (3) and will also join the growing planar network. In this way a low angle twist boundary can be formed.

To see how such a network of screw dislocations constitutes a low angle twist boundary, we examine the displacement field of these screw dislocations assuming that their glide planes are identical to the boundary plane. Figure 14.9a shows the deformed shape of a solid containing a periodic array of right-handed screw (RHS) dislocations with lines along the y axis. The solid immediately above the slip plane experiences an overall shear strain $\gamma_{yz} < 0$ relative to the solid immediately below the slip plane. This is consistent with the fact that a single periodic array of screw dislocation has a non-zero long-range stress field.

Figure 14.9b shows the displacement field of another array of right-handed screw (RHS) dislocations, whose lines are along the z axis. The solid immediately above the slip plane now experiences an overall shear strain $\gamma_{yz} > 0$ relative to the solid immediately below the slip plane.

Figure 14.9c shows the displacement field of the two dislocation arrays superimposed. The solid above the slip plane now experiences an overall rotation around the x axis by angle θ. The spacing between the dislocations is again inversely proportional to the angle of misorientation and the same relation discussed above applies, $\tan(\theta/2) = b/(2s)$, or

$$\theta \approx \frac{b}{s}. \tag{14.25}$$

Similar to the previous section, the energy of a low angle twist boundary as a function of misorientation angle also has the following form

$$E_{\mathrm{LAGB}}(\theta) = A_s\,\theta\,[B_s - \ln\theta], \tag{14.26}$$

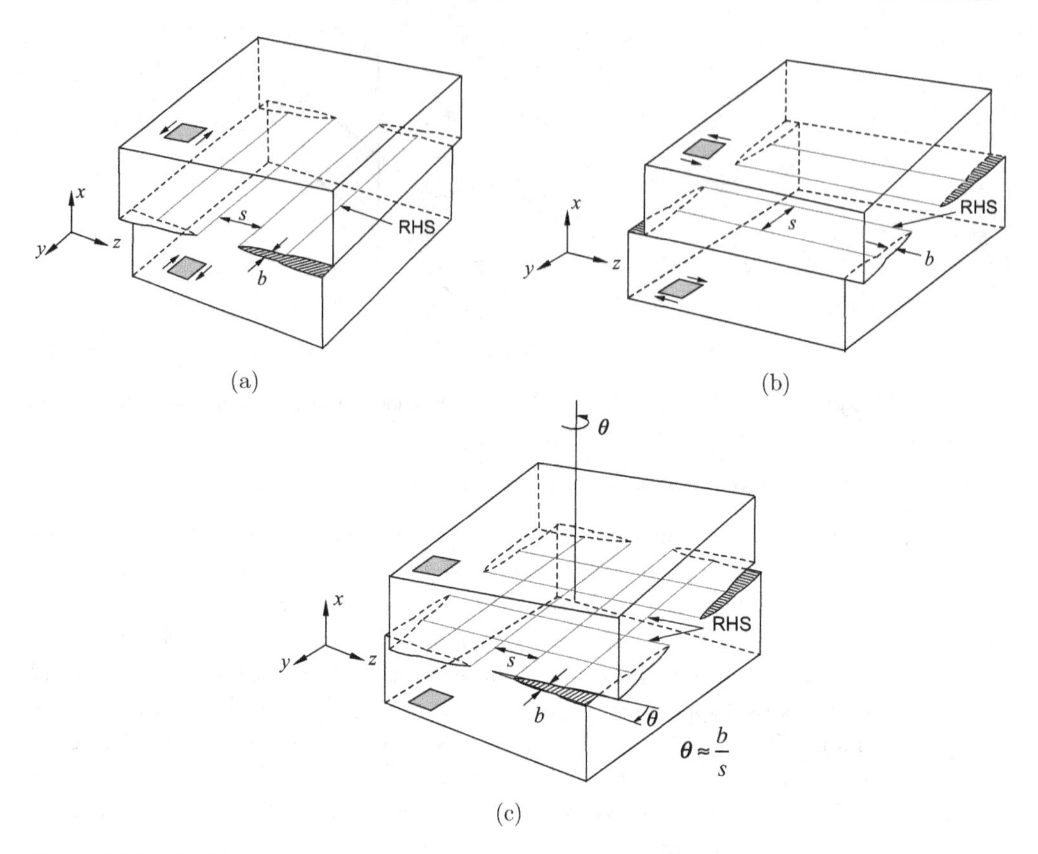

Figure 14.9. The distortion of a solid containing a network of screw dislocations. (a) An array of right-handed screw (RHS) dislocations with line direction along the y axis. (b) An array of RHS dislocations with line direction along the z axis. (c) The superposition of two sets of screw dislocations in (a) and (b), leading to a rotation of angle θ.

where

$$A_s = \frac{\mu b}{2\pi}, \tag{14.27}$$

$$B_s = \ln\left(\frac{\alpha_s}{2}\right). \tag{14.28}$$

The changes from Eqs. (14.7) and (14.8) reflect the difference between edge and screw dislocations, as well as a factor of 2 due to the presence of two sets of screw dislocations in a twist boundary. Similar to the low angle tilt boundaries, we also expect the GB energy to reach a maximum for low angle twist boundaries. An energy maximum is indeed observed at around $\theta \approx 18^\circ$ in Fig. 13.2, although this is arguably at the border where the dislocation network model applies.

Since the screw dislocation network corresponds to a low angle twist boundary, we expect such a dislocation structure to form in response to an applied torque. Figure 14.10 shows a number of screw dislocations in a circular cylinder subjected to a torque around its axis. We note that the torque applied causes shear stresses of the type $\sigma_{\theta z}$ to be created which

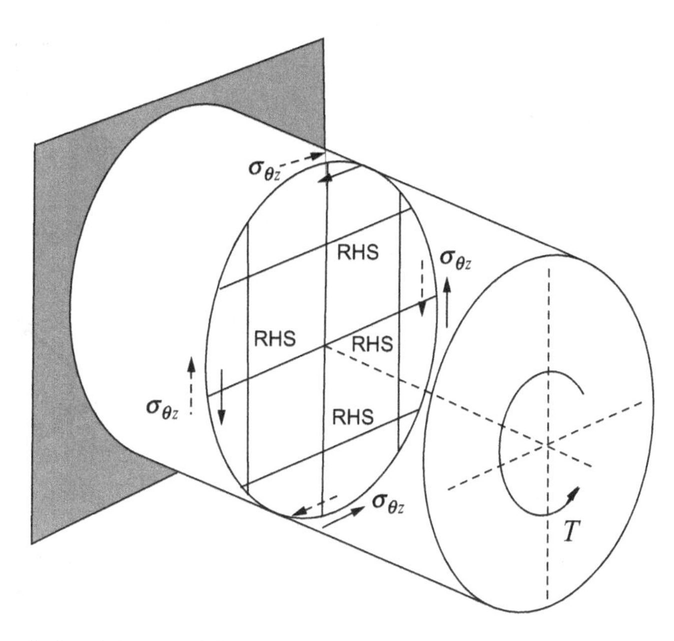

Figure 14.10. Right-handed screw dislocation array caused by twisting a cylinder, leading to a low angle twist boundary. All RHS dislocations are driven toward the center of the cylinder when the torque is applied.

drives the RHS dislocations toward the center of the cylinder. With more and more twisting, more and more dislocations are formed at the surface and move toward the center of the cylinder.

In the hypothetical case considered in Fig. 14.10, we have assumed that there exist two sets of screw dislocations whose Burgers vectors are perpendicular to each other and to the cylinder axis, and that the slip planes of these dislocations coincide with the plane of the twist boundary. In a real crystal, the type of screw dislocation network which forms depends on the availability of Burgers vectors on the boundary plane, and the way the network forms also depends on the orientation of the slip plane with respect to the boundary plane. For example, if a cylinder of an FCC crystal is twisted along the [1 0 0] axis, then two sets of screw dislocations with Burgers vectors $\frac{1}{2}[0\,1\,1]$ and $\frac{1}{2}[0\,1\,\bar{1}]$ can form a twist boundary. However, the slip planes of these dislocations (at ambient temperature) are of the {1 1 1} type, which are inclined with respect to the (1 0 0) plane of the twist boundary. As a result, the screw dislocations nucleate on the cylinder surface away from the twist boundary and glide on {1 1 1} planes to meet on the (1 0 0) twist boundary [142] (see Exercise problem 14.3).

However, when the twist is applied along the [1 1 1] axis of an FCC crystal, the available Burgers vectors on the (1 1 1) boundary plane, $\frac{1}{2}[0\,1\,\bar{1}]$, $\frac{1}{2}[\bar{1}\,0\,1]$, and $\frac{1}{2}[1\,\bar{1}\,0]$, are at 120° to each other. As a result, the twist boundary consists of a hexagonal network of three sets of screw dislocations, as shown in Fig. 14.11a. In this case, the plane of the twist boundary is also the slip plane of the dislocations, so that the dislocations directly nucleate on the twist boundary from the cylinder surface. Figure 14.11(b) shows that the node at which three screw dislocations meet has an atomic structure that is either extended or constricted, as we have seen in Fig. 11.13.

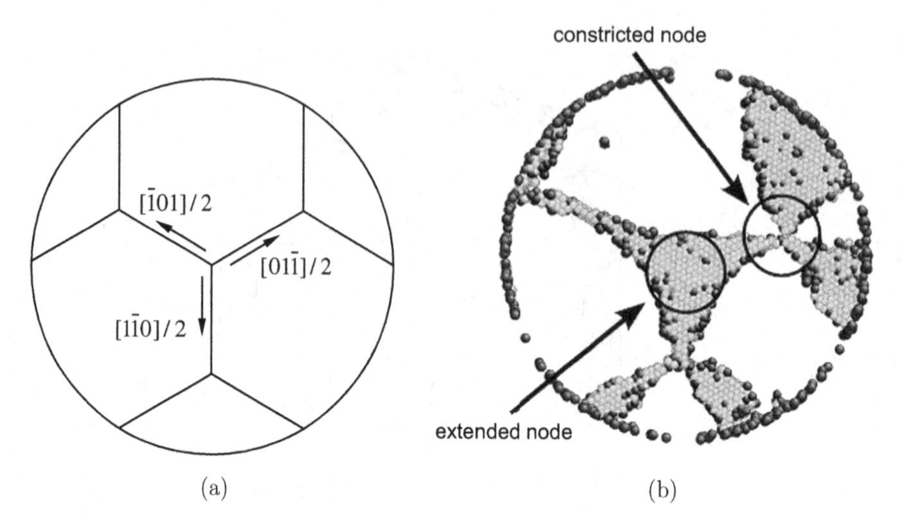

Figure 14.11. (a) A hexagaonal network of screw dislocations forming a (1 1 1) low angle twist boundary when a [1 1 1] Au nanowire is twisted around its axis. (b) Snapshot from an atomistic simulation of this process in a nanowire of diameter $D = 17.5$ nm. Only atoms in the core of the screw dislocations are plotted, showing extended and constricted nodes where three screw dislocations meet [143].

14.3 Dislocation content of arbitrary low angle grain boundaries

By now we have considered dislocation models for the symmetric tilt and pure twist LAGBs. These only account for two degrees of freedom of a grain boundary, while a general grain boundary has five degrees of freedom. To allow extra degrees of freedom, we start by considering an asymmetric tilt LAGB in a simple cubic lattice, as shown in Fig. 14.12.

The grain boundary now makes angles $\theta/2 + \varphi$ and $\theta/2 - \varphi$, respectively, with the [0 1 0] directions of the two grains. It can be shown that the spacing between dislocations with the $\vec{b}_{\perp} = [1\,0\,0]$ Burgers vector is [135]

$$L_{\perp} = \frac{b}{\theta \, \cos\varphi}, \tag{14.29}$$

and that the spacing between dislocations with the $\vec{b}_{\vdash} = [0\,\bar{1}\,0]$ Burgers vector is

$$L_{\vdash} = \frac{b}{\theta \, \sin\varphi}. \tag{14.30}$$

Note that the average Burgers vector density on the GB is perpendicular to the boundary plane, guaranteeing that the long range stress field vanishes.

For the asymmetric tilt boundary considered above, the axis of rotation must be perpendicular to both sets of Burgers vectors of the edge dislocations. For tilt LAGBs whose axis of rotation is along an arbitrary orientation, the above procedure would not work. The construction of such general tilt LAGBs requires mixed dislocations, whose edge components give rise to the required misorientation, and screw components are arranged to cancel each other. Similarly, twist boundaries can be constructed from pure screw dislocations only if the boundary plane contains two (or more) required Burgers vectors. The construction of twist LAGBs whose axis of

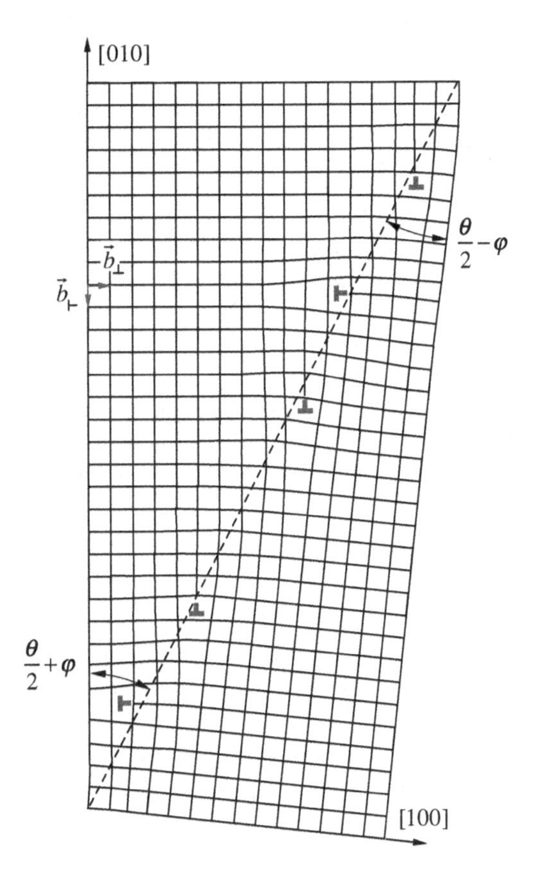

Figure 14.12. Dislocation model of an asymmetric tilt LAGB [6, 135].

rotation is along an arbitrary orientation requires mixed dislocations, whose screw components give rise to the required misorientation, and edge components are arranged to cancel each other.

It has been shown [135, 144] that it is possible to specify the net dislocation content of a general grain boundary. Figure 14.13a shows an arbitrary boundary between two lattices. The origin O is chosen to be one of the coincidence sites. Suppose lattice B can be obtained from lattice A by applying rotation matrix g_{AB}. Let $\vec{p}$ = OP be an arbitrarily large vector in the boundary. Frank's procedure allows us to determine the total Burgers vector of all dislocations in the boundary that cross vector $\vec{p}$.

To find the total Burgers vector, we draw a Burgers circuit PLOMP. According to the right-handed rule, this means that the line sense vectors of the dislocations to be considered point out of the page. We now imagine rotating lattice B back to the same orientation of lattice A, so that points M and P become points N and Q, respectively, as shown in Fig. 14.13b. This means that the Burgers circuit, when drawn entirely in lattice A, is PLONQ, where vector OQ $= g_{AB}^{-1}\, p$. The closure failure is QP $=$ OP $-$ OQ $= (I - g_{AB}^{-1})\, \vec{p}$. The net Burgers vector is

$$\vec{b}_{A} = \left(I - g_{AB}^{-1}\right) \vec{p}. \tag{14.31}$$

This is the Burgers vector expressed in terms of lattice A. It means that if we start from lattice A, as shown in Fig. 14.13b, then collect a set of dislocations whose total Burgers vector is

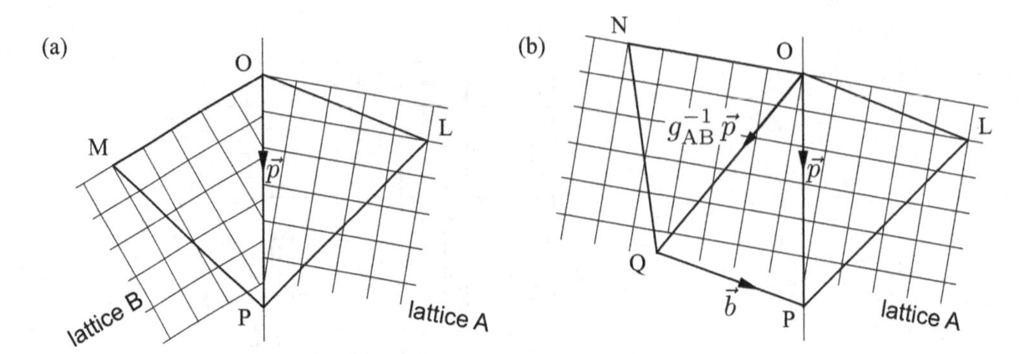

Figure 14.13. Frank's procedure for defining the net Burgers vector content of a general grain boundary [135]. (a) Burgers circuit drawn in the final configuration. (b) The corresponding Burgers circuit when lattice B is rotated back to the orientation of lattice A. The closure failure QP is the Burgers vector expressed in lattice A.

$\vec{b}_A$ and distribute them uniformly on OP, the result is the grain boundary structure shown in Fig. 14.13a.

In the Frank–Bilby equation [144, 145], the net Burgers vector is specified in a reference lattice C, from which lattice A is obtained by applying rotation matrix g_{CA}, and lattice B is obtained by g_{CB}. The net Burgers vector in the reference lattice C is

$$\vec{b}_C = \left(g_{CA}^{-1} - g_{CB}^{-1}\right)\vec{p}. \tag{14.32}$$

Consider a rotation from lattice C to lattice A (g_{CA}), followed by a rotation from lattice A to lattice B (g_{AB}), the result is simply a rotation from lattice C to lattice B (g_{CB}). Therefore

$$\begin{aligned} g_{CB} &= g_{AB} \cdot g_{CA} \\ g_{CB}^{-1} &= g_{CA}^{-1} \cdot g_{AB}^{-1} \\ \vec{b}_C &= \left(g_{CA}^{-1} - g_{CA}^{-1} \cdot g_{AB}^{-1}\right)\vec{p} = g_{CA}^{-1}\,\vec{b}_A, \end{aligned} \tag{14.33}$$

so that

$$\vec{b}_A = g_{CA}\,\vec{b}_C. \tag{14.34}$$

The net Burgers vector predicted by the Frank–Bilby equation may not be an integer multiple of a single Burgers vector of the corresponding lattice. In this case, multiple types of dislocations are required so that the total sum of their Burgers vectors equals $\vec{b}$. There are multiple ways in which a group of dislocations can be assembled to produce the net Burgers vector predicted by the Frank–Bilby equation. The exact type of dislocations contained in the grain boundary depends not only on the energy of the dislocation structure, but also the history of how the GB is formed.

14.4 Grain boundary edge dislocations

The LAGBs discussed above can be regarded as vicinal GBs near the singular $\Sigma 1$ misorientation. We have seen that the GB energy function exhibits other cusps at non-zero misorientation angles. The HAGBs vicinal to these cusps can be regarded as singular HAGBs plus GB dislocations.

As an example, let us consider the symmetric tilt boundaries vicinal to the $\Sigma 11$ special boundary with the $\langle 0\,1\,1 \rangle$ rotation axis. The corresponding GB energy cusp is shown in Fig. 13.8b, at $\theta = 129.52°$ (or equivalently 50.48°). The atomic structure of the $\Sigma 11$ special boundary is shown in Fig. 13.10. The atomic structure is reproduced in Fig. 14.14, together with a Burgers circuit around the boundary plane.

The misorientation of the GB can deviate from the special value 50.48° by incorporating a periodic edge dislocation array into the GB plane. The Burgers vector of the edge dislocations needs to be normal to the GB plane. While for the low angle tilt boundary the Burgers vectors of the dislocation array must be the FCC lattice repeat vectors, here the Burgers vectors of the GB dislocation array can be the DSC lattice repeat vectors. The smallest DSC lattice repeat vector normal to the GB plane is $\frac{1}{11}[3\,1\,\bar{1}]$. Figure 14.14b shows a negative GB edge dislocation with Burgers vector $\vec{b}_{\rm GB} = \frac{2}{11}[\bar{3}\,\bar{1}\,1]$ for line sense vector along $[0\,1\,1]$. Note that $b_{\rm GB} = |\vec{b}_{\rm GB}| = 2a/\sqrt{11}$. This Burgers vector is twice the length of the shortest DSC vector in the $[3\,1\,\bar{1}]$ direction.[1] If the GB contains a periodic array of such GB dislocations stacked on top of each other with separation s between nearest neighbors, then the misorientation angle between the two grains will be modified by

$$\Delta\theta \approx -\frac{b_{\rm GB}}{s}. \tag{14.35}$$

The dislocation array introduces additional energy to the grain boundary. The elasticity theory of lattice dislocations applies equally well to the elastic fields of GB dislocations. Following Eq. (14.5), the energy of the vicinal grain boundary at misorientation angle θ can be written in terms of the energy of the singular grain boundary at misorientation angle $\theta_{\rm S}$ as

$$\begin{aligned} E_{\rm HAGB}(\theta) &= E_{\rm HAGB}(\theta_{\rm S}) + |\Delta\theta|\,\frac{\mu\, b_{\rm GB}}{4\pi(1-\nu)}\left[\ln\left(\frac{1}{|\Delta\theta|}\right) + \ln\left(\frac{\alpha_{\rm e}}{2}\right)\right] \\ &= E_{\rm HAGB}(\theta_{\rm S}) + A'\,|\Delta\theta|\,[B - \ln|\Delta\theta|], \end{aligned} \tag{14.36}$$

where $\Delta\theta = \theta - \theta_{\rm S}$ and

$$A' = \frac{\mu\, b_{\rm GB}}{4\pi(1-\nu)}. \tag{14.37}$$

Because, in general, $b_{\rm GB} < b$, A' is less than the A defined in Eq. (14.7) for LAGBs. This means that on the energy versus misorientation angle plot, the cusps at singular GBs are expected to be shallower than the cusp at $\theta = 0$, as can be seen from Fig. 13.8b.

As a rule of thumb [146], the maximum misorientation deviation $\theta_{\rm m}$ around an arbitrary singular GB may be estimated from $\theta_{\rm m} \approx \theta_0/\Sigma^n$, with n between 1/2 and 1, where $\theta_0 \approx 15°$ delineates the vicinal range around $\Sigma 1$ (i.e. $\theta = 0$). The reduction in $\theta_{\rm m}$ with increasing Σ is due to the decreasing magnitude of the Burgers vector $b_{\rm GB}$. Assuming that the cores of GB dislocations will overlap if their separation is smaller than a constant value (e.g. lattice spacing),

1 We consider this Burgers vector because an edge dislocation with Burgers vector $\frac{1}{11}[3\,1\,\bar{1}]$ would create too large of a step in the grain boundary and it would be difficult to create a vertical array of such dislocations. A GB dislocation associated with a step is called a disconnection, which will be discussed in Section 14.6.

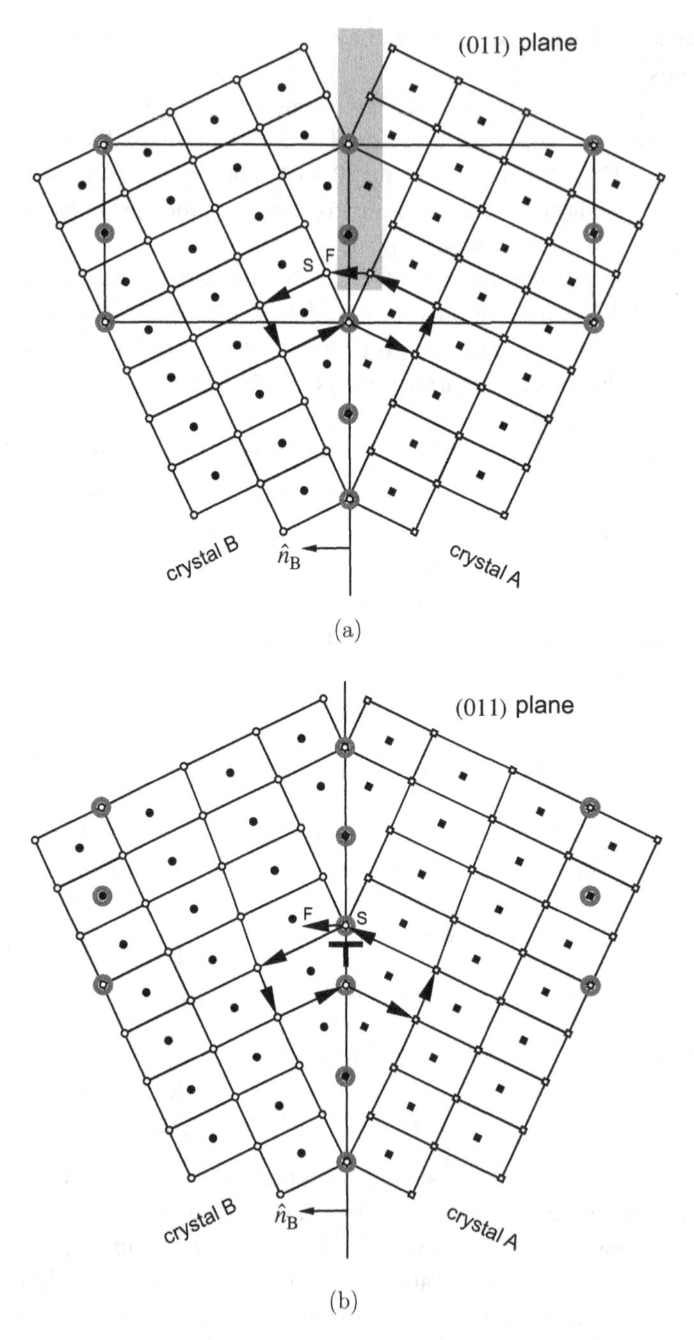

Figure 14.14. (a) The $\Sigma 11$ tilt boundary in an FCC crystal. A Burgers circuit is drawn (with arrows) where the starting point S overlaps with the final point F. A large rectangle is drawn to show the repeat unit in the neighborhood of the GB. (b) A negative edge dislocation created by removing the atoms in the shaded area in (a). The Burgers vector of the GB edge dislocation is the vector pointing from S to F, which is $\frac{2}{11}[\bar{3}\,\bar{1}\,1]$, and corresponds to the thickness of the shaded area in (a).

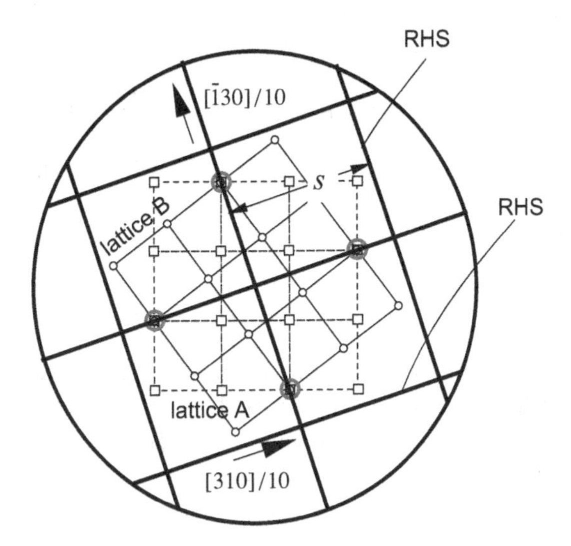

Figure 14.15. Screw dislocation network in vicinal GB near Σ5 pure twist GB viewed along the [0 0 1] rotation axis. For brevity only lattice points on the corner of the unit cells of each lattice are shown.

a smaller b_{GB} would lead to a smaller change in misorientation when the GB dislocation cores start to overlap.

Note the edge dislocation shown in Fig. 14.14b cannot leave the GB, because its Burgers vector is not a repeat vector in the FCC lattice. In fact, this dislocation cannot move at all without climbing, because its glide plane is perpendicular to the GB plane. Alternatively, a GB dislocation may glide along the GB if its Burgers vector is in the GB plane.

14.5 Grain boundary screw dislocations

In this section, we discuss the vicinal pure twist GBs near the $\theta = 36.87^\circ$ cusp in Fig. 13.2a, which is the singular pure twist GB of Σ5 in FCC Al. Figure 13.13 shows that the shortest DSC lattice vectors normal to the [0 0 1] twist axis are: [3 1 0]/10 and [$\bar{1}$ 3 0]/10, which are orthogonal to each other. This means that the Σ5 pure twist GB can support two perpendicular arrays of GB screw dislocations. The Burgers vector magnitude of the GB dislocations are $b_{GB} = |[3\,1\,0]/10| = a/\sqrt{10} = b/\sqrt{\Sigma}$, where $b = a/\sqrt{2}$ is the Burgers vector magnitude of lattice dislocations.

Figure 14.15 shows the rectangular grid structure of the GB screw dislocation network on a vicinal GB near the Σ5 pure twist GB. Without the dislocations, crystal B is already rotated with respect to crystal A by $\theta = 36.87^\circ$. When the network of right-handed screw dislocations form on the twist boundary, a change is introduced to the misorientation angle by

$$\Delta\theta \approx \frac{b_{GB}}{s}. \tag{14.38}$$

Similar to Eq. (14.36), the energy of the vicinal twist boundary at misorientation angle θ can be written in terms of the energy of the singular twist boundary at misorientation angle

θ_S as

$$
\begin{aligned}
E_{\mathrm{HAGB}}(\theta) &= E_{\mathrm{HAGB}}(\theta_S) + |\Delta\theta| \frac{\mu\, b_{\mathrm{GB}}}{2\pi} \left[\ln\left(\frac{1}{|\Delta\theta|}\right) + \ln\left(\frac{\alpha_s}{2}\right) \right] \\
&= E_{\mathrm{HAGB}}(\theta_S) + A'_s\, |\Delta\theta| \, \left(B_s - \ln|\Delta\theta|\right), \qquad (14.39)
\end{aligned}
$$

where $\Delta\theta = \theta - \theta_S$ and

$$
A'_s = \frac{\mu\, b_{\mathrm{GB}}}{2\,\pi}. \qquad (14.40)
$$

When lattice B is shifted by the Burgers vectors of these GB screw dislocations, the shift of the coincidence sites is also in the GB plane. Hence the GB screw dislocations considered here do not create GB steps. Furthermore, the GB screw dislocations can glide in the GB plane. Therefore, if a sufficient torque is applied on crystal B around the rotation axis, new GB screw dislocations can nucleate from the surface and move toward the center of the crystal, in the same way as lattice dislocations in the low angle twist boundary shown in Fig. 14.10.

As a rough estimate, the vicinal range around a $\Sigma 5$ twist GB is $\theta_m \approx \theta_0 \cdot (b_{\mathrm{GB}}/b) \approx 15^\circ/\sqrt{5} \approx 7^\circ$. This means that the core of the parallel GB screw dislocations would start to overlap at $\Delta\theta \approx \pm 7^\circ$. This is consistent with the angular range around the $\Sigma 5$ cusp in Fig. 13.2a.

14.6 Disconnections and disclinations

We have seen that a grain boundary, such as a LAGB formed by a periodic array of dislocations, does not have the long-range $(1/r)$ stress field that is a characteristic of a dislocation. However, the grain boundaries we have considered are perfectly planar. In this section, we shall see that if a grain boundary contains a step, it is possible (although not necessary) to have a long-range $(1/r)$ stress field. In this case, we say that the GB step is also a GB dislocation, and contains a non-zero Burgers vector. A defect on the GB that has both the character of a step and the character of a dislocation is called a *disconnection* [147, 148].

As an example, consider a low angle tilt boundary equivalent to an infinite array of edge dislocations stacked on top of each other. After a gliding edge dislocation cuts through the grain boundary, a step is created on the grain boundary, as shown in Fig. 14.16. This step has a long-range stress field that is equivalent to the stress field of an edge dislocation (with a Burgers vector perpendicular to that of the edge dislocations forming the LAGB). There are several ways to show the existence of the long-range stress field, and to determine the corresponding Burgers vector of the disconnection. It is instructive to show that the same result can be obtained using several different approaches.

In Section 14.6.1, we consider the LAGB with the disconnection as two semi-infinite arrays of edge dislocations. We first show that the long-range stress field of a semi-infinite array of edge dislocations is equivalent to that of a wedge *disclination*. Hence the long-range stress field of the LAGB with the disconnection is equivalent to that of a disclination dipole. The latter is then shown to be equivalent to the stress field of an edge dislocation. This equivalence allows us to obtain the Burgers vector of the disconnection.

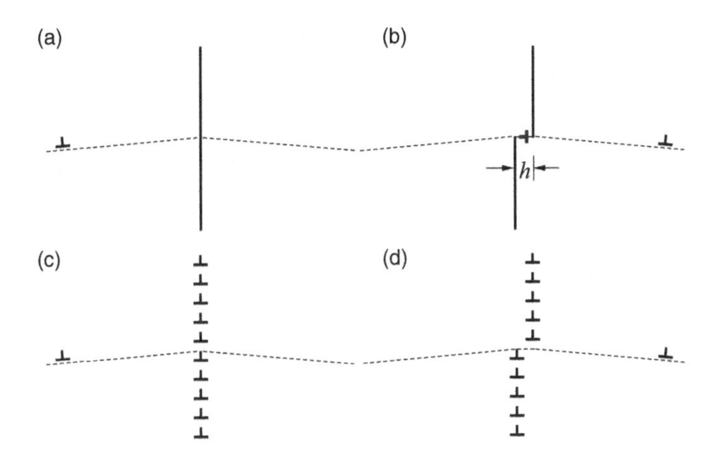

Figure 14.16. (a) A gliding edge dislocation approaching a low angle tilt boundary. (b) A step (disconnection) is created on the tilt boundary after the gliding dislocation has cut through it. The ⊣ symbol indicates the Burgers vector content of the disconnection. (c) The tilt boundary in (a) is represented by an infinite edge dislocation array. (d) The tilt boundary with a disconnection in (b) is represented by two semi-infinite dislocation arrays.

In Section 14.6.2, we consider the LAGB as a singular GB with a CSL and a DSC lattice. The repeat vectors of the DSC lattice define the allowed Burgers vector of GB dislocations. Some of the DSC lattice vectors cause a translation of the CSL. The corresponding translation vectors of the CSL are equivalent to jump vectors of GB steps. The Burgers vector of the disconnection can be found by matching the corresponding translation vector of the CSL lattice with the step vector of the disconnection.

In Section 14.6.3, we follow the original approach by Hirth [147], in which the disconnection is formed by joining two stepped surfaces together. The Burgers vector content of the disconnection is simply the difference between the two vectors characterizing the two surface steps. All of these approaches give the same prediction on the Burgers vector of the disconnection.

14.6.1 Disclination

The long-range stress field of a semi-infinite array of edge dislocations stacked on top of each other is equivalent to that of a wedge disclination. Disclinations (along with dislocations) in an elastic medium were first studied by Volterra. The wedge disclination can be introduced to an elastic cylinder by making a cut from the surface to the cylinder axis, and inserting a solid wedge of angle α made of the same material.[2] We shall call this disclination a *positive* wedge disclination, as shown in Fig. 14.17a. The strength of this wedge disclination is characterized by the angle α of the wedge before it is inserted into the cylinder. A *negative* wedge disclination can be created by removing a wedge of angle α from the cylinder and gluing the two surfaces together, as shown in Fig. 14.17b.

Intuitively, we expect the stress field of the disclination to be rotationally symmetric. In other words, the same stress field should be produced regardless of the orientation of the wedge being

2 If a solid slab of uniform thickness is inserted instead, the result is an edge dislocation.

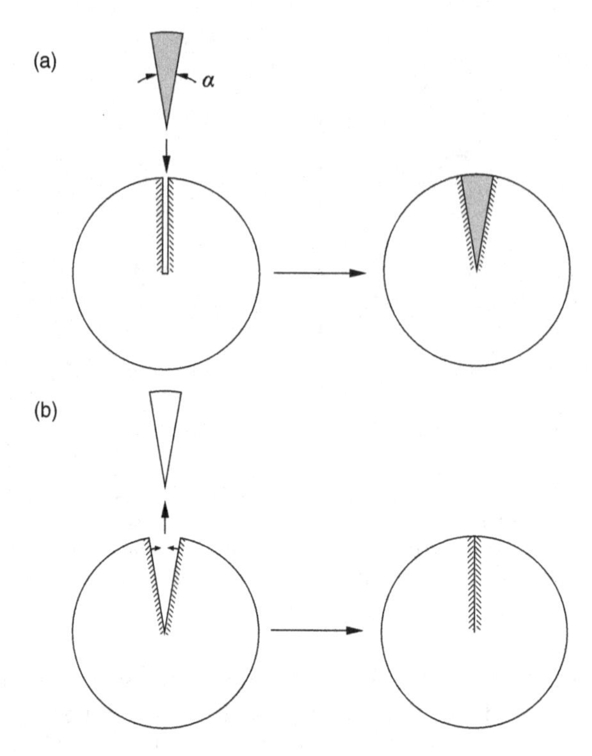

Figure 14.17. (a) Making a positive wedge disclination by inserting a wedge into the cylinder. (b) Making a negative wedge disclination by removing a wedge from the cylinder and gluing the exposed surfaces together.

inserted. This means that there are multiple ways to create the same wedge disclination, as shown in Fig. 14.18a.

Because an edge dislocation can be created by inserting a slab of material (i.e. a half-plane) into the cylinder, a wedge disclination is equivalent to a continuous distribution of edge dislocations, as shown in Fig. 14.18b. When the edge dislocations are distributed along the positive y axis, their Burgers vector is along the x direction. The Burgers vector content inside a region of $[y', y' + dy']$ is (for $y' > 0$),

$$db_x = \alpha\, dy'. \tag{14.41}$$

Equivalently, the same disclination can also be represented by a continuous distribution of edge dislocations along the positive x axis, with their Burgers vector along the $-y$ direction. The Burgers vector content inside a region of $[x', x' + dx']$ is (for $x' > 0$),

$$db_y = -\alpha\, dx'. \tag{14.42}$$

To find the stress field of the wedge disclination requires the solution of a plane strain problem, and a good strategy is to use the Airy stress function (see Section 2.5.3). Instead of trying to find the stress function $\phi_{\text{wedge}}(x, y)$ of the wedge disclination directly, we shall construct it from the known solution of the stress function ϕ_{edge} of an edge dislocation by superposition.

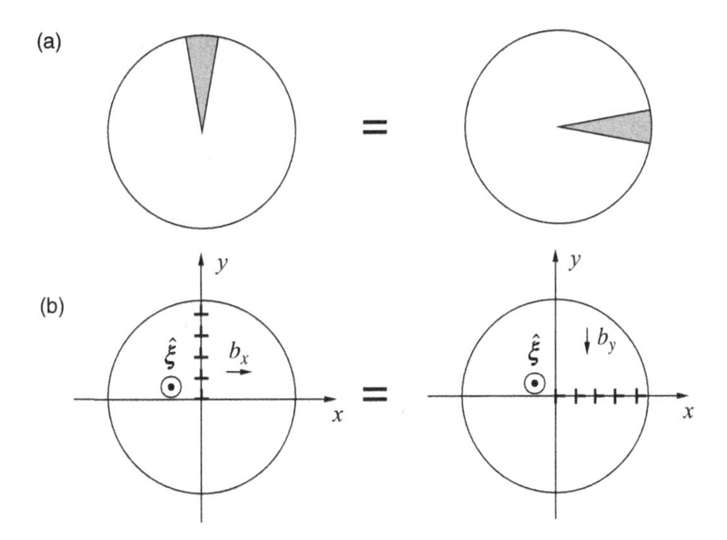

Figure 14.18. (a) Two ways to create the same disclination by inserting a wedge into the cylinder. (b) Two equivalent edge dislocation arrays that produce the same long-range stress field as the disclination in (a).

The stress function of a positive edge dislocation at the origin with Burgers vector b_x is given in Eq. (9.26), which is reproduced below:

$$\phi^x_{\text{edge}}(x, y) = -\frac{\mu\, b_x}{2\pi(1-\nu)}\, y \ln\sqrt{x^2 + y^2}.$$

Hence the stress function corresponding to the Burgers vectors contained inside the region of $[y', y' + dy']$ is (for $y' > 0$),

$$d\phi_{\text{wedge}}(x, y) = -\frac{\mu\,\alpha\, dy'}{2\pi(1-\nu)}\,(y - y') \ln\sqrt{x^2 + (y - y')^2}. \tag{14.43}$$

The stress function of the wedge disclination can now be written as an integral of the stress function of the edge dislocation over y',

$$\phi_{\text{wedge}}(x, y) = \int_0^{\infty} -\frac{\mu\,\alpha\, dy'}{2\pi(1-\nu)}\,(y - y') \ln\sqrt{x^2 + (y - y')^2}. \tag{14.44}$$

This integral can be performed analytically to give[3]

$$\phi_{\text{wedge}}(x, y) = -\frac{\mu\alpha}{2\pi(1-\nu)}\, r^2 \left(\ln r - \frac{1}{2}\right), \tag{14.45}$$

3 This result can be verified by showing $\phi^x_{\text{edge}}(x, y) = (b_x/\alpha)\, \partial\phi_{\text{wedge}}(x, y)/\partial y$.

where $r = \sqrt{x^2 + y^2}$. The stress field of the wedge disclination can be easily obtained by differentiation:

$$\begin{aligned}
\sigma_{xx} &= \frac{\partial^2}{\partial y^2}\phi_{\text{wedge}}(x, y) = -\frac{\mu\alpha}{\pi(1-\nu)}\left(\ln r + \frac{y^2}{r^2}\right), \\
\sigma_{yy} &= \frac{\partial^2}{\partial x^2}\phi_{\text{wedge}}(x, y) = -\frac{\mu\alpha}{\pi(1-\nu)}\left(\ln r + \frac{x^2}{r^2}\right), \\
\sigma_{xy} &= -\frac{\partial^2}{\partial x \partial y}\phi_{\text{wedge}}(x, y) = \frac{\mu\alpha}{\pi(1-\nu)}\frac{xy}{r^2}.
\end{aligned} \tag{14.46}$$

The rotational symmetry of the stress field of the wedge disclination can be clearly seen from the stress components in cylindrical coordinates:

$$\begin{aligned}
\sigma_{rr} &= \left(\frac{1}{r}\frac{\partial}{\partial r} + \frac{1}{r^2}\frac{\partial^2}{\partial \theta^2}\right)\phi_{\text{wedge}}(r, \theta) = -\frac{\mu\alpha}{\pi(1-\nu)}\ln r, \\
\sigma_{\theta\theta} &= \frac{\partial^2}{\partial r^2}\phi_{\text{wedge}}(r, \theta) = -\frac{\mu\alpha}{\pi(1-\nu)}(\ln r + 1), \\
\sigma_{r\theta} &= -\frac{\partial}{\partial r}\left(\frac{1}{r}\frac{\partial}{\partial \theta}\phi_{\text{wedge}}(r, \theta)\right) = 0.
\end{aligned} \tag{14.47}$$

Notice that all these stress components are independent of θ.

When the tilt GB is cut by a glide dislocation, the resulting disconnection is equivalent to a disclination dipole arranged along the x direction. The stress function of the disconnection is the superposition of the stress function of two disclinations,

$$\phi_{\text{dscn}}(x, y) = \phi_{\text{wedge}}(x - h, y) - \phi_{\text{wedge}}(x, y). \tag{14.48}$$

At field points far away from the origin (much greater than h), the disconnection stress field can be approximated as,

$$\phi_{\text{dscn}}(x, y) \approx -h\frac{\partial}{\partial x}\phi_{\text{wedge}}(x, y) = h\frac{\mu\alpha}{2\pi(1-\nu)}x \ln r. \tag{14.49}$$

This is precisely the stress function of an edge dislocation with Burgers vector $b_y = h\alpha$. Therefore, we conclude that the disconnection shown in Fig. 14.16b has an equivalent Burgers vector of $b_y = h\alpha$.

14.6.2 Disconnection as GB dislocation

While in the above we have considered the solid as an elastic continuum, we now account for the discrete lattices on both sides of the grain boundary. For simplicity, consider a symmetric tilt LAGB in a simple cubic crystal with lattice constant a, as shown in Fig. 14.19a. The tilt boundary is formed by a periodic array of positive edge dislocations on one out of every three planes. The misorientation angle θ can be calculated from the following relation,

$$\tan(\theta/2) = 1/6, \tag{14.50}$$

so that $\theta = 18.92°$.

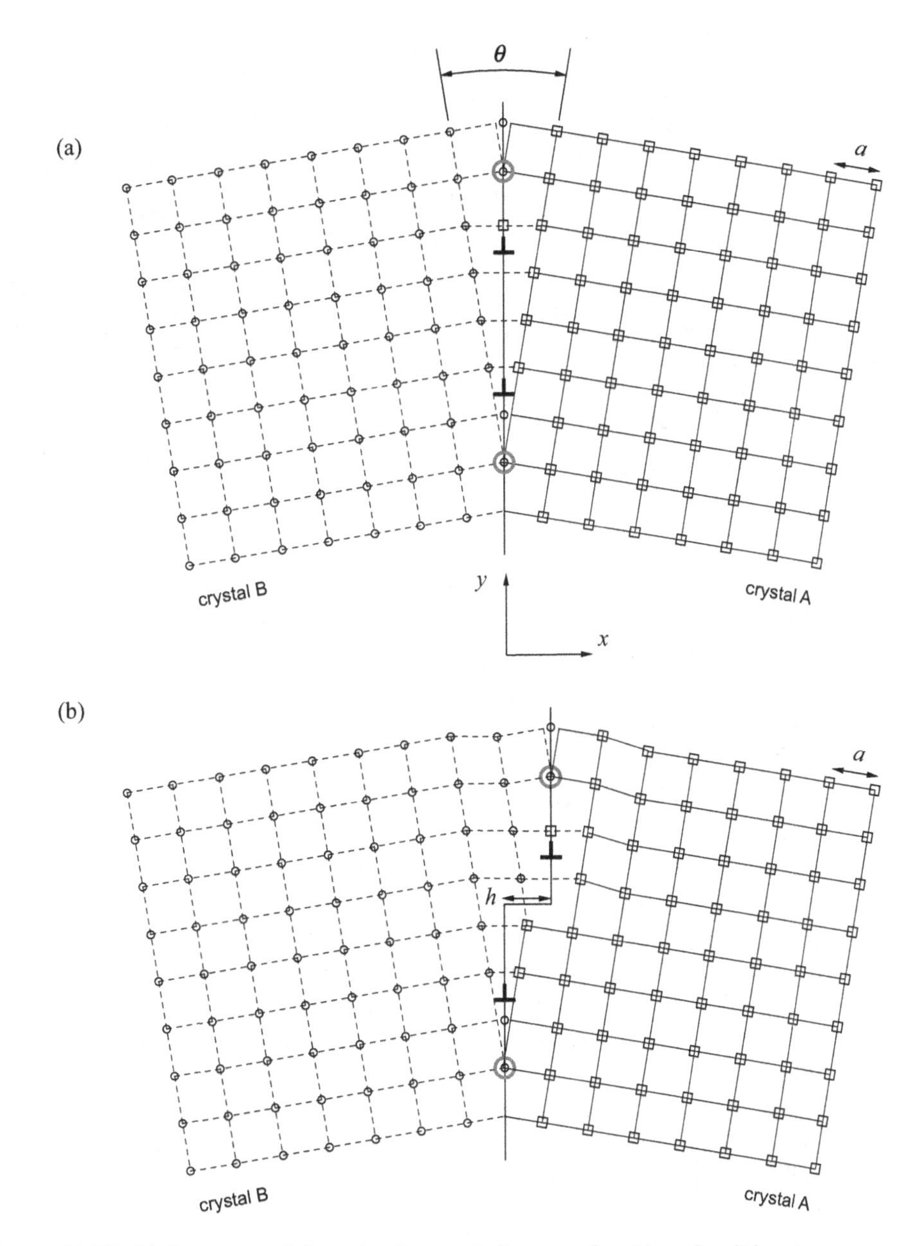

Figure 14.19. (a) Symmetric tilt boundary by a periodic array of positive edge dislocations on one out of every three planes. It is equivalent to a $\Sigma 37$ GB. (b) $\Sigma 37$ GB with a step of height h. Atoms on both sides of the GB need to be slightly displaced due to the strain field of the disconnection.

Figure 14.19b shows the same GB now containing a step of height h. In the following, we will show that this step contains a Burgers vector, i.e. this step is a disconnection. We can notice that atoms near the GB need to be slightly adjusted due to the strain field of the disconnection.

To obtain the Burgers vector of the disconnection, we start with the CSL corresponding to the two grains in Fig. 14.19a. Figure 14.20a shows the periodic pattern formed by the two lattices A and B. The origin O is one of the coincidence sites. The unit cell of the CSL is

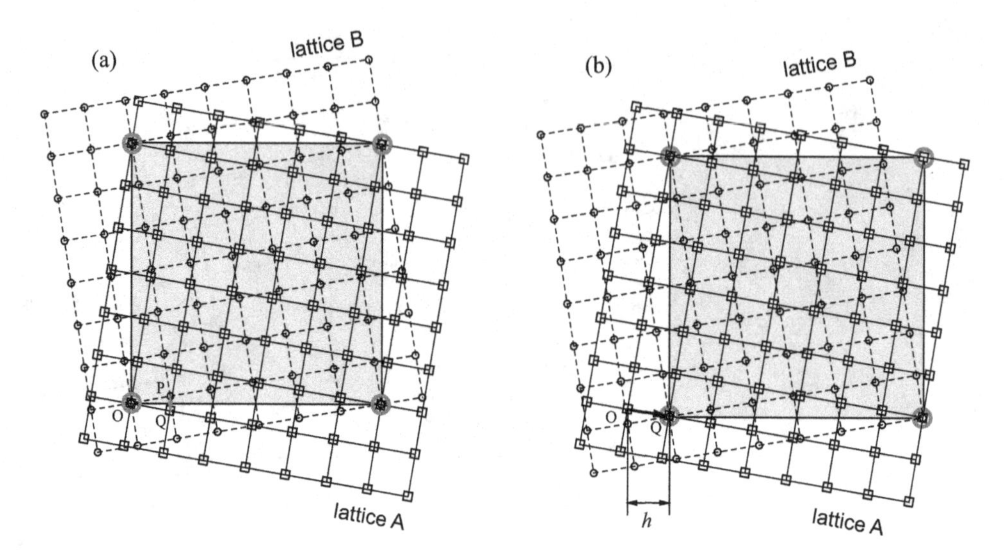

Figure 14.20. (a) The unit cell of the Σ37 CSL between the two lattices is shown as a shaded square. (b) If lattice B is shifted by vector PQ, then the coincidence site shifts from point O to point Q. This corresponds to an offset of the GB by h in the x direction. For clarity see color figure on the book website.

shown as a shaded square, whose dimension is $\sqrt{37}\,a$. The volume of the CSL unit cell is $\sqrt{37}\,a \times \sqrt{37}\,a \times a = 37\,a^3$. Hence the Σ value of the CSL is 37. In other words, the low angle tilt boundary formed by the dislocation array shown in Fig. 14.19a is equivalent to a Σ37 GB.

In Fig. 14.20a, points P and Q are two points next to point O, and they belong to lattice B and lattice A, respectively. Now suppose that lattice B is shifted downward by vector PQ. Then point Q becomes a coincidence site. As shown in Fig. 14.20b, the CSL and the periodic pattern formed by the two lattices now shifts to the right by vector OQ, which corresponds to an offset of the GB in the x direction by distance $h = a\cos(\theta/2)$.

If the grain boundary in Fig. 14.16a is cut by a gliding dislocation whose Burgers vector magnitude is $b = a$, then a GB step of height $h = a\cos(\theta/2)$ is formed on the GB, as shown in Fig. 14.16b. This step size is precisely the same h shown in Fig. 14.20b. Hence we conclude that along the GB region above the step, grain B needs to move relative to grain A by vector PQ, as illustrated in Fig. 14.20b. This corresponds to a grain boundary dislocation with a Burgers vector

$$b_y = |\mathrm{PQ}| = 2\,h\,\tan(\theta/2). \qquad (14.51)$$

In the limit of θ being very small, $b_y \approx h\theta$, which is consistent with the result in the previous section (letting $\alpha = \theta$).

14.6.3 Disconnection from joining stepped surfaces

We now follow the original approach of Hirth [147] to determine the Burgers vector content of the disconnection on the grain boundary. In Fig. 14.21a, we consider a process in which the Σ37 tilt boundary can be formed by joining together two semi-infinite crystals, A and B, with

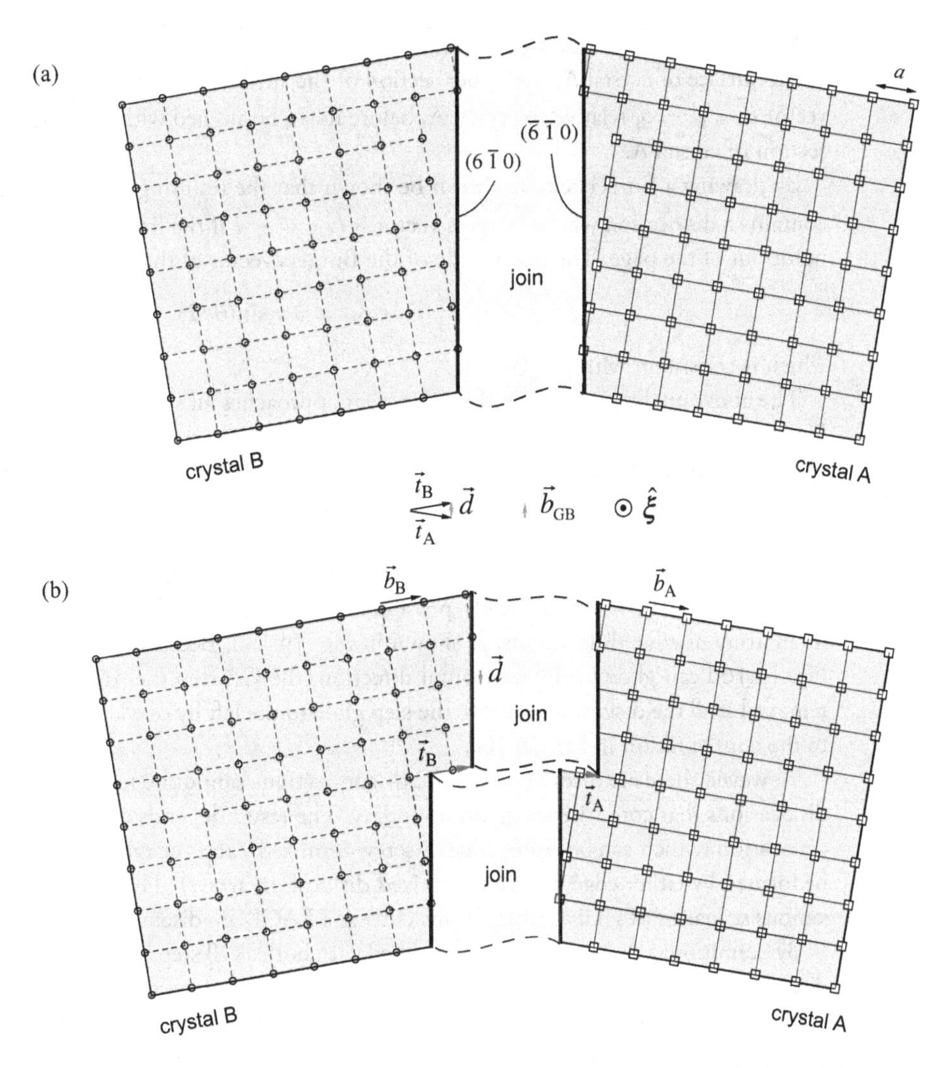

Figure 14.21. (a) The Σ37 symmetric tilt boundary shown in Fig. 14.20a can be formed by joining two crystals with {6 1 0} surfaces. (b) A GB with step can be formed by joining two crystals each having a surface step. The step is characterized by the lattice repeat vectors, $\vec{t}_A = \vec{b}_A$ and $\vec{t}_B = \vec{b}_B$, for crystal A and crystal B, respectively.

{6 1 0} surfaces. The orientations of the two surfaces are expressed in terms of the local coordinate systems fixed on the two crystals, respectively. Before joining the two crystals together, the exposed surface of crystal A is $(\bar{6}\,\bar{1}\,0)$, and the exposed surface of crystal B is $(6\,\bar{1}\,0)$. There are no long-range stresses when the two crystals are joined together to form a flat grain boundary.

Now suppose that a step is created on both surfaces before the two crystals are joined together, as shown in Fig. 14.21b. Owing to the periodicity of the crystal structure, each surface step is characterized by a repeat vector of the lattice, i.e. $\vec{t}_A = \vec{b}_A$ for crystal A and $\vec{t}_B = \vec{b}_B$ for crystal B. Each surface is now broken into two sections, and the corresponding sections of the two crystals need to join together to form the grain boundary.

When the lower section of the surface of crystal B is aligned to join with the lower section of the surface of crystal A, the upper section of the surface of crystal B must be displaced by a vector $\vec{d} = \vec{t}_A - \vec{t}_B$ relative to crystal A, before it can be aligned with the corresponding surface section of crystal A.

By drawing a Burgers circuit, it can be shown that the resulting step on the grain boundary contains a dislocation whose Burgers vector is $\vec{b}_{GB} = -\vec{d}$ if the line sense vector $\hat{\xi}$ is chosen to point out of the page. The magnitude of the Burgers vector of the disconnection is

$$b_{GB} = |\vec{t}_A - \vec{t}_B| = 2\,a\,\sin(\theta/2), \tag{14.52}$$

which is consistent with Eq. (14.51).

The above analyses using the three different approaches all show that the GB step shown in Fig. 14.16b contains a net Burgers vector, and hence is a disconnection. The Burgers vector content of the disconnection is caused by the mismatch of the Burgers vectors of the cutting edge dislocation in the two grains. The net Burgers vector of the disconnection produces a long-range ($1/r$) stress field and leads to an increase of elastic energy relative to the flat grain boundary. There is a tendency for the grain boundary to reduce its energy by removing the disconnection if it is kinematically possible. In our example, the tilt boundary is equivalent to an array of edge dislocations, as shown in Fig. 14.16d. Because all the edge dislocations in Fig. 14.16d can glide in the horizontal direction, the GB step (i.e. the disconnection) can be removed if all the dislocations above the step glide to the left by one lattice vector, i.e. returning to the configuration in Fig. 14.16c.

However, there are cases in which the disconnections cannot be removed by glide motion of dislocations that comprise the grain boundary. The result depends on both the type of cutting dislocation (which can be either edge or screw or mixed), and the type of the LAGB (which can be formed by either edge or screw or mixed dislocation arrays). The disconnections created by various scenarios of gliding dislocations cutting a LAGB are discussed by Hirth *et al.* [149].

By definition, a disconnection is associated with both a GB step and a GB dislocation. Certain disconnections have their Burgers vectors within the GB plane so that they can glide conservatively in response to the applied stress. Because of the step nature of the disconnection, its motion on the GB plane leads to the migration of the GB normal to its own plane. Hence these disconnections play an active role in GB migration, in the same way that twinning partials promote twinning (see Sections 11.1.3 and 12.4.2).

However, GB steps may not have any Burgers vector content, and certain GB dislocations do not lead to a step on the GB. Neither of these defects satisfy the definition of disconnections. An example of GB dislocations without a step is shown in Fig. 14.14b. When the GB is represented by an array of dislocations, the effect of this type of GB dislocation is to alter the distance between two neighboring dislocations in an otherwise periodic array. Consequently, such GB dislocations are also called *spacing defects* [149] (see Exercise problem 14.2).

14.7 Summary

Grain boundaries (GBs) may be classified according to how the energy and structure of the boundary vary with misorientation angle. Sharp minima in the plot of GB energy vs. misorientation angle indicate the presence of *singular* GBs, wherein both the relative orientations of the crystals and the orientation of the boundary plane combine to create a low energy interface. Singular GBs usually correspond to low-Σ coincident site lattices. *Vicinal* GBs, which have relative misorientation angles and GB plane orientations close to the singular GBs, may be considered as singular GBs superimposed with one or more GB dislocation arrays. Grain boundaries far removed from these special misorientations are considered *general* GBs.

In this classification scheme the case of zero misorientation and its vicinal range deserves special attention. When the misorientation angle is zero, the grain boundary disappears, so its energy is zero. This state could be called a "singular GB of $\Sigma 1$," even though there is no boundary at all. GBs with misorientations in the vicinal range of this "GB" are characterized by a very sharp and deep cusp in the plot of energy vs misorientation angle. Such boundaries are also called low angle GBs (LAGB); their energies can be well described by energies of the dislocation arrays that make up their structure.

The structure and energy of symmetric low angle tilt GBs can be well described by the famous Read–Shockley model wherein the boundary is composed of same-signed edge dislocations stacked on top of one another. An analysis of the energy of this dislocation array leads to an expression for the energy of the tilt boundary as a function of misorientation angle that is in excellent agreement with measurements. The dislocation model is valid up to a misorientation angle of about 15°, where the energy reaches a maximum. Uniformly dispersed edge dislocations of the same sign that produce a uniform lattice curvature can rearrange to form low angle tilt GBs, a process called *polygonization*. As the polygonization process continues, the non-linear dependence of the GB energy on misorientation angle causes multiple low angle GBs to combine to form fewer, higher angle boundaries.

Low angle pure twist boundaries are composed of two or more periodic arrays of screw dislocations with different orientations. For a more general low angle grain boundary (LAGB), having both tilt and twist components, the net Burgers vector content of the LAGB can be found from the geometry using the Frank–Bilby equation. For high angle GBs (HAGBs) vicinal to singular orientations, the structure and energy can be found by superimposing singular HAGBs with arrays of GB dislocations with Burgers vectors that are multiples of the DSC lattice repeat vectors. As with LAGBs, tilt misorientations away from the singular boundary involve edge GB dislocations, while twist misorientations are created by screw GB dislocations.

Steps are formed in GBs when crystal dislocations cut through the plane of the boundary. When the slip mismatch on the two sides of the boundary leaves residual dislocation content at the point of intersection, the step is called a *disconnection* and has the long-range stress field of a dislocation line. The step in the boundary can also be regarded as the site where two semi-infinite GBs terminate. The termination of a single grain boundary is called a *disclination*, which, for the case of a semi-infinite tilt boundary amounts to a wedge of excess material inserted into the crystal. Thus the disconnection may also be described as a disclination dipole with a strength depending on the misorientation angle of the boundary and the size of the step.

14.8 Exercise problems

14.1 Consider an infinite periodic array of edge dislocations with Burgers vector b. As shown in Fig. 14.22a, the array is oriented at an angle φ with respect to the y axis. The spacing between the adjacent glide planes of the dislocations is h. Obtain an expression of the stress field of the dislocation array in the limit of $x \to \pm\infty, y = 0$.

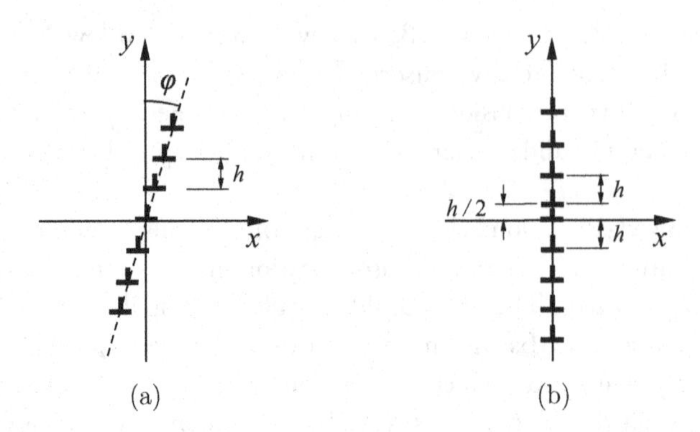

Figure 14.22. (a) A infinite periodic array of edge dislocations oriented at an angle φ relative to the y axis. (b) An infinite array of edge dislocations oriented along the y axis with a spacing defect.

14.2 Consider an infinite array of edge dislocations with Burgers vector b. As shown in Fig. 14.22b, the array is aligned along the y axis, with dislocations located at $y = 0, -h, -2h, \ldots$ and $y = h/2, 3h/2, 5h/2, \ldots$. In other words, there is a spacing defect in the array near the origin. Determine the long-range $(1/r)$ part of the stress field of this dislocation array. You can replace a Volterra dislocation located at y_0 by a continuously distributed dislocation with Burgers vector uniformly distributed in the region $[y_0 - h/2, y_0 + h/2]$ without changing its long-range stress field. Using this approach, you can replace the dislocation array by a disclination dipole with the same long-range stress field.

14.3 A low angle twist boundary consisting of a planar array of screw dislocations is formed on a $(0\,0\,1)$ plane in FCC aluminum by slip on octahedral planes of the type $\{1\,1\,1\}$. Make a sketch to show the Burgers vectors and the slip plane of the dislocations that are required to form the twist boundary, and to show the final dislocation arrangement in such a boundary. Make your sketch based on Fig. 14.23, which shows several slip pyramids on the $(0\,0\,1)$ plane.

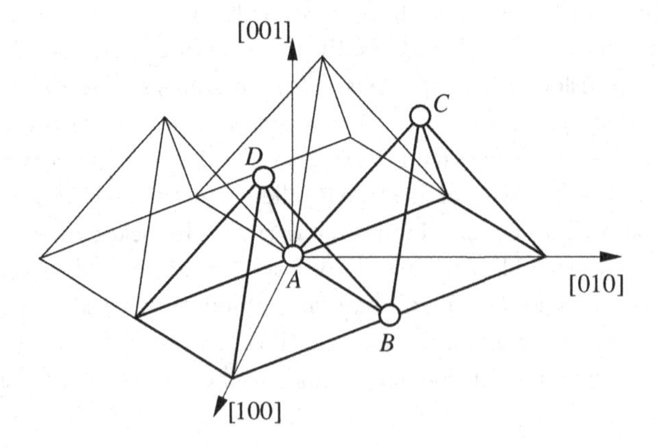

Figure 14.23. Slip pyramids on the $(0\,0\,1)$ plane of an FCC crystal.

14.4 A low angle twist grain boundary consisting of an infinite square array of infinitely long right-handed screw dislocations is illustrated in Fig. 14.24a. The dislocation spacing in the plane of the boundary is s, as shown. All of the Burgers vectors are of magnitude b. We wish to consider separating the vertical set of dislocations from the horizontal set in the manner shown in the diagram. With this operation in mind, all of the vertical dislocations in the y–z plane are shown as dotted lines to indicate where the vertical dislocations were located in the boundary. The solid vertical lines indicate the location of the vertical set of dislocations after they have been displaced from the horizontal set by a distance x_0 in the x direction. We wish to determine the stress that would be required to separate the dislocations in this way.

(a) First calculate the total interaction force that would be exerted on one of the vertical dislocations due to the stress field of one of the horizontal dislocations. For this calculation you will need to know that

$$\int_{-\infty}^{\infty} \frac{d\alpha}{1+\alpha^2} = \pi .$$

(b) Determine the type or component of applied stress that would be required to separate the dislocations in the manner shown in the diagram.

(c) Assuming that all of the dislocations remain absolutely straight, calculate the magnitude of the shear stress that would be needed to cause the two sets of dislocations to move apart in the manner shown in the diagram.

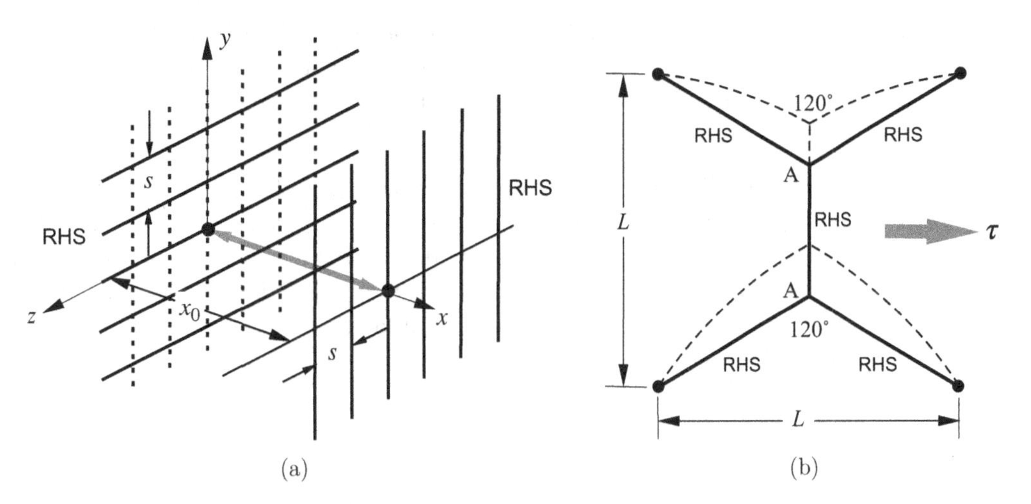

Figure 14.24. (a) Two arrays of right-handed screw dislocations. (b) A planar screw dislocation network. The dashed lines indicate the shape of the dislocations under a shear stress τ.

14.5 A planar screw dislocation network is shown in Fig. 14.24b. The corners of the network, as indicated by the solid points, are assumed to be fixed. We wish to consider what happens to the segment A–A when a shear stress is applied as shown. We assume that the dislocations are free to glide in the plane of the network and that all of the dislocations have a constant line tension $\mu b^2/2$. A possible equilibrium shape of the dislocations under shear stress τ is shown in dashed lines. Derive an expression for the shear stress at which the length of segment A–A would be reduced to zero.

14.6 A pure twist boundary which lies on the (001) plane of an Al single crystal is shown as a screw dislocation array in Fig. 14.25. The boundary was formed by twisting the crystal at high temperatures in the direction shown. The dislocations shown were created at the surface of the crystal and moved by glide on an (0 0 1) plane to the positions shown. The length of the Burgers vectors is b and the elastic properties of the crystal are described by E, μ, and ν. The separation distance between the dislocations is s and the angle of twist is θ.

(a) Determine the character of the dislocations in the grid.
(b) Briefly explain why the torque T must be maintained for the dislocation array to be stable.
(c) Derive an expression for the torque T_c needed to hold the boundary in equilibrium.

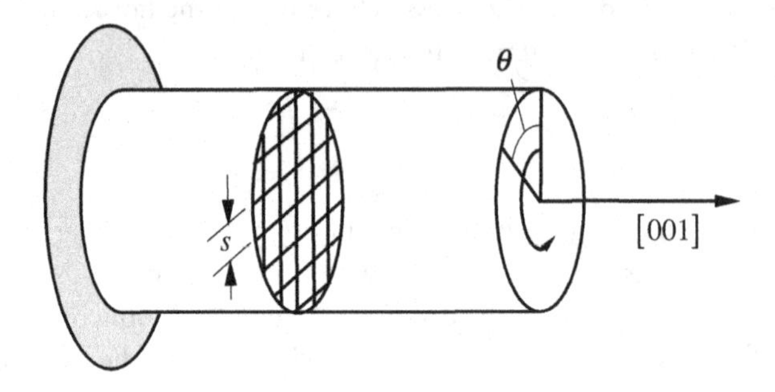

Figure 14.25. A pure twist boundary formed by twisting a cylindrical single crystal of Al at high temperatures.

14.7 Apply the Frank–Bilby equation (14.32) to obtain the dislocation content of a low angle symmetric tilt boundary and a low angle pure twist boundary of misorentation angle θ. Use a reference lattice C whose orientation is half-way in between the orientations of lattice A and lattice B.

14.8 A pure screw dislocation with a Burgers vector $(a/2)[0\,1\,\bar{1}]$ glides on a (1 1 1) plane and runs into a $\Sigma 11$ twist boundary lying parallel to the (0 1 1) plane of the FCC lattice A, as shown in Fig. 14.26. Where the dislocation lies in the plane of the boundary, it can dissociate into grain boundary dislocations that are free to glide in the plane of the twist boundary (the (0 1 1) plane). The Burgers vectors of the grain boundary dislocations must be lattice translation vectors of the DSC lattice.

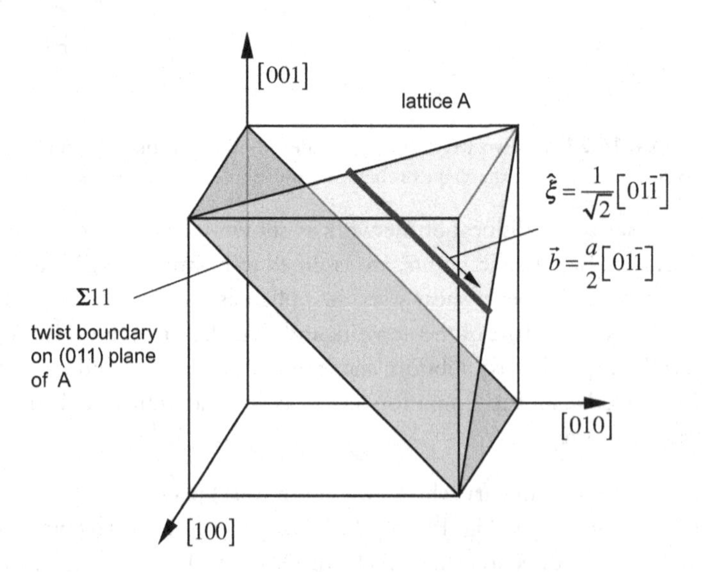

Figure 14.26. A screw lattice dislocation entering a $\Sigma 11$ twist boundary.

(a) Determine the number of grain boundary dislocations created when the lattice dislocation runs into the $\Sigma 11$ boundary.
(b) Determine the magnitude of the Burgers vector of each of the grain boundary dislocations (take the Burgers vector of the lattice dislocation to be b).
(c) Calculate the total energy (per unit length) of the grain boundary dislocations and compare that energy with the energy of the lattice dislocation: $\mu b^2/2$. Here we wish to find the energy change that occurs when the dislocation enters the boundary and dissociates into grain boundary dislocations.

14.9 For the boundary considered in Exercise problem 13.7 (which is a $(1\,\bar{1}\,2)$ twin boundary), find the step heights corresponding to the three repeat vectors of the unit cell of the DSC lattice, $\vec{b}_{\rm DSC}$. The results will tell us whether GB dislocations with $\vec{b}_{\rm DSC}$ as Burgers vectors are also GB steps, and hence disconnections. The step height is related to the shift $\vec{d}$ of the coincidence pattern when lattice B is translated relative to lattice A by $\vec{b}_{\rm DSC}$. The out-of-plane component of vector $\vec{d}$ (relative to the GB plane) corresponds to the height of GB step. If the shift of the coincidence pattern is along a direction that is within the GB plane, then the corresponding step height is zero and the GB dislocation does not qualify as a disconnection.

14.10 Figure 14.27 shows a tilt boundary containing a macroscopic step. Let the misorientation between the two crystals be θ. Construct a dislocation model for the three straight sections of the tilt boundary and determine their dislocation content. Explain why this boundary does not have a long-range $(1/r)$ stress field, contrary to the stepped boundary shown in Fig. 14.16d.

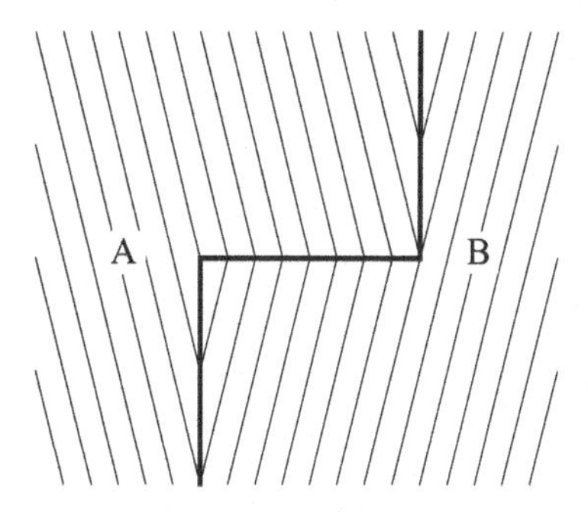

Figure 14.27. A non-straight tilt boundary (thick lines) separating crystal A and crystal B. The thin lines indicate the orientation of atomic planes in the two crystals.

APPENDIX A

King table for solid solutions

Table A.1. Atomic volumes and Seitz radii for elements at room temperature [20].

Element	Atomic number	Ω (Å^3)	r_0 (Å)	Element	Atomic number	Ω (Å^3)	r_0 (Å)
Ac	89	37.48	2.076	K	19	75.31	2.618
Ag	47	17.06	1.598	La	57	37.12	3.335
Al	13	16.60	1.582	Li	3	21.61	1.728
Am	95	33.77	2.005	Lu	71	29.50	1.917
As	33	21.54	1.726	Mg	12	23.23	2.853
Au	79	16.96	1.594	Mn	25	12.21	1.428
B	5	7.67	1.22	Mo	42	15.58	1.550
Ba	56	62.59	2.463	Na	11	39.50	2.113
Be	4	8.112	1.246	Nb	41	17.98	1.625
Bi	83	35.38	2.036	Nd	60	34.18	2.013
C	6	5.678	1.107	Ni	28	10.94	1.377
Ca	20	43.48	2.181	Np	93	19.22	1.662
Cd	48	21.58	1.726	Os	76	13.99	1.495
Ce	58	34.37	2.017	P	15	16.59	1.582
Co	27	11.13	1.385	Pa	91	24.94	1.812
Cr	24	12.00	1.423	Pb	82	30.33	1.949
Cs	55	115.17	4.865	Pd	46	14.72	1.521
Cu	29	11.81	1.413	Po	84	37.43	2.075
Dy	66	31.52	1.960	Pr	59	34.15	2.013
Er	68	30.64	1.941	Pt	78	15.10	1.534
Eu	63	48.86	2.268	Pu	94	23.4	1.77
Fe	26	11.77	1.411	Rb	37	92.67	2.80
Ga	31	19.59	1.672	Re	75	14.70	1.520
Gd	64	33.10	1.992	Rh	45	13.77	1.487
Ge	32	22.64	1.755	Ru	44	13.57	1.480
Hf	72	22.16	1.743	S	16	25.52	1.826
Hg	80	23.42	1.775	Sb	51	30.20	1.932
Ho	67	31.12	1.951	Sc	21	23.41	1.775
In	49	26.15	1.841	Se	34	27.27	1.863
Ir	77	14.14	1.500	Si	14	20.02	1.669

Table A.1. *(cont.)*

Element	Atomic number	Ω (Å^3)	r_0 (Å)	Element	Atomic number	Ω (Å^3)	r_0 (Å)
Sm	62	33.01	1.990	Tl	81	28.58	1.892
Sn	50	27.65	1.862	Tm	69	30.10	1.930
Sr	38	56.32	2.378	U	92	20.81	1.706
Ta	73	18.01	1.626	V	23	13.88	1.491
Tb	65	31.14	1.954	W	74	15.85	1.549
Tc	43	14.213	1.503	Y	39	33.01	1.990
Te	52	33.98	2.005	Yb	70	33.02	1.990
Th	90	32.86	1.987	Zn	30	15.24	1.538
Ti	22	17.65	1.614	Zr	40	23.27	1.771

Table A.2. King table for substitutional solid solutions [20]. χ_{max} is the maximum atomic fraction for which Eq. (4.21) is valid. Ω_{sf} is the volume size factor and l_{sf} is the linear size factor.

Solvent	Solute	χ_{max} (at.%)	Ω_{sf} (%)	l_{sf} (%)	Solvent	Solute	χ_{max} (at.%)	Ω_{sf} (%)	l_{sf} (%)
Ag	Al	18	−9.18	−3.16	Al	Cu	2.2	−37.77	−14.62
	As	8.5	+10.35	+3.33		Ga	0.5	+4.94	+1.62
	Au	34	−1.78	−0.60		Ge	2.0	+13.13	+4.19
	Bi	2.6	+70.92	+19.56		Li	12	−2.10	−0.70
	Cd	12	+14.84	+4.71		Mg	15	+40.82	+12.08
	Cu	12	−27.75	−10.27		Mn	3.5	−46.81	−18.98
	Ga	18	−5.09	−1.73		Pb	0.2	−53.63	−22.66
	Ge	7.5	+1.66	+0.53		Si	1.0	−15.78	−5.56
	Hg	23	+14.00	+4.46		Sn	0.1	+24.09	+7.45
	In	12	+23.50	+7.28		Th	0.4	+156.61	+36.91
	Mg	25	+7.13	+2.32		Ti	1.0	−15.06	−5.22
	Mn	15	+0.09	+0.02		V	0.6	−41.42	−16.37
	Pb	2.2	+54.52	+15.60		Zn	25	−5.74	−1.95
	Pd	21	−17.21	−6.10					
	Pt	10	−20.05	−7.19	As	Ge	16	−6.89	−2.35
	Sb	6	+44.93	+13.16		Sb	10	+6.49	+2.12
	Sn	10	+32.40	+9.81		Sn	22	+14.19	+4.51
	Tl	6.5	+39.42	+11.71					
	Zn	35	−13.74	−4.80	Au	Ag	32	−0.64	−0.21
						Al	14	−10.17	−3.51
Al	Ag	20	+0.12	+0.03		As	0.2	+17.69	+5.57
	Ca	1.1	+177.10	+40.46		Cd	24	+13.14	+4.20
	Cr	0.8	−57.23	−24.66		Co	8.5	−25.22	−9.23

(cont.)

Table A.2. *(cont.)*

Solvent	Solute	χ_{max} (at.%)	Ω_{sf} (%)	l_{sf} (%)	Solvent	Solute	χ_{max} (at.%)	Ω_{sf} (%)	l_{sf} (%)
Au	Cr	20	−16.45	−5.82	Ce	Pu	30	−28.49	−10.58
	Cu	46	−27.81	−10.29		Th	50	−7.29	−2.49
	Fe	16	−19.87	−7.12					
	Ga	11	−4.32	−1.46	Co	Al	17	+16.98	+5.36
	Ge	3.0	+5.54	+1.81		Au	30	+59.12	+16.74
	Hf	5.0	−3.30	−1.11		Cu	24	+6.99	+2.28
	Hg	18	+18.90	+5.94		Fe	20	+5.24	+1.71
	In	12	+20.57	+6.43		Ga	13	+17.42	+5.49
	Li	15	−19.24	−6.88		Ge	17	+15.08	+4.79
	Mn	7.0	−5.35	−1.82		Ir	50	+25.85	+7.96
	Mo	1.2	−14.86	−5.22		Mn	40	+7.33	+2.39
	Ni	10	−21.92	−7.92		Mo	18	+29.44	+8.98
	Pd	40	−14.20	−4.98		Ni	100	−1.73	−0.61
	Pt	30	−12.62	−4.40		Os	25	+21.93	+6.82
	Sb	1.0	+34.62	+10.41		Pd	20	+40.08	+11.89
	Sn	5.0	+28.78	+8.80		Pt	25	+40.01	+11.87
	Ta	6.5	+3.40	+1.11		Re	25	+30.50	+9.28
	Ti	12	−7.76	−2.66		Rh	40	+30.74	+9.34
	Tl	0.8	+23.82	−7.43		Ru	50	+21.67	+6.75
	V	8.0	−8.94	−3.08		Si	13	−6.39	−2.18
	Zn	18	−13.82	−4.84		Sn	3.0	+69.12	+19.13
	Zr	7.0	+13.19	+4.21		W	7.0	+22.52	+6.99
						Zn	10	+17.40	+5.48
Ba	Ca	11	−10.03	−3.46					
	Sr	20	−9.32	−3.21	Cr	Al	38	+25.95	+7.99
						Be	5.0	−18.17	−6.43
Be	Al	0.6	+74.81	+20.46		Co	25	−5.15	−1.75
	Ni	1.5	+45.65	+13.36		Fe	30	−2.07	−0.70
						Ir	2.5	+53.34	+15.31
Bi	Sb	100	−14.12	+4.95		Mn	10	+0.65	+0.20
						Mo	40	+33.66	+10.14
Ca	Ba	30	+49.84	+14.44		Ni	11	−4.80	−1.63
	Sr	100	+30.51	+9.28		Rh	15	+17.57	+5.53
						Si	7.5	−15.05	−5.29
Cd	Ag	6.0	−34.45	−13.13		Ta	2.5	+43.98	+12.91
	Hg	25	+4.43	+1.45		V	18	+8.79	+2.86
	Mg	45	−1.60	−0.54		W	20	+37.35	+11.15
Ce	Eu	5.0	+9.32	+3.01	Cs	K	25	−40.94	−16.10
	La	25	+10.05	+3.24		Rb	80	−20.80	−7.58

Table A.2. *(cont.)*

Solvent	Solute	χ_{max} (at.%)	Ω_{sf} (%)	l_{sf} (%)
Cu	Ag	4.2	+43.52	+12.79
	Al	20	+19.99	+6.26
	As	4.2	+38.77	+11.53
	Au	40	+47.59	+13.85
	Be	13	−26.45	−9.73
	Cd	2.5	+67.40	+18.74
	Co	15	−3.78	−1.28
	Cr	1.6	+19.72	+6.18
	Fe	2.6	+4.57	+1.50
	Ga	20	+24.11	+7.46
	Ge	11	+27.77	+8.51
	Hg	3.0	+5.44	+1.78
	In	10	+79.03	+21.42
	Mg	3.2	+50.80	+14.67
	Mn	15	+34.19	+10.30
	Ni	32	−8.45	−2.90
	P	2.0	+16.51	+5.22
	Pd	31	+27.96	+8.56
	Pt	40	+31.19	+9.47
	Sb	6.0	+91.87	+24.25
	Si	12	+5.08	+1.68
	Sn	9.0	+83.40	+22.41
	Th	1.5	+49.12	+14.25
	Ti	0.25	+25.74	+7.93
	Tl	2.5	+129.16	+31.84
	Zn	15	+17.10	+5.40
Er	Gd	25	+10.14	+3.27
Fe	Al	25	+12.79	+4.08
	Au	0.6	+44.16	+12.96
	Be	20	−26.23	−9.64
	Co	20	+1.54	+0.50
	Cr	9.0	+4.36	+1.43
	Cu	0.7	+17.53	+5.52
	Ge	10	+16.48	+5.21
	Mn	10	+4.89	+1.60
	Mo	2.0	+27.51	+8.43
	Nb	3.0	+17.58	+6.10
Fe	Ni	5.0	+4.65	+1.52
	P	3.5	−13.16	−4.59
	Pd	5.5	+62.19	+17.49
	Ru	1.8	+19.92	+6.24
	Sb	2.5	+36.40	+10.90
	Si	10	−7.88	−2.70
	Sn	10	+67.70	+18.81
	Te	0.6	+8.66	+2.80
	Ti	1.8	+14.44	+4.39
	V	30	+10.51	+3.38
	W	7.0	+32.99	+9.97
	Zn	24	+21.07	+6.58
Gd	Er	55	−6.94	−2.37
	Lu	20	−12.03	−4.18
	Y	10	+0.21	+0.06
Ge	Ga	1.0	+2.21	+0.72
	Si	25	−20.65	−7.42
	Sn	1.0	+49.53	+14.35
In	Bi	4.0	+31.09	+9.44
	Cd	4.0	−15.24	−5.36
	Hg	6.0	−11.55	−4.01
	Li	14	−13.61	−4.76
	Mg	35	−1.16	−0.39
	Pb	12	+21.52	+6.70
	Sn	7.5	+5.22	+1.70
	Tl	20	+10.97	+3.52
Ir	Cr	16	−16.12	−5.69
	Mn	40	+6.04	−2.06
	Mo	18	+6.48	+2.11
	Pd	26	+2.83	+0.93
	Pt	25	+8.09	+2.90
	Rh	100	−2.64	−0.88
	Ru	25	−4.43	−1.50
	Ta	10	−14.14	+4.50
	Ti	10	+8.62	+2.79
	W	20	+6.93	+2.26

(cont.)

Table A.2. (*cont.*)

Solvent	Solute	χ_{max} (at.%)	Ω_{sf} (%)	l_{sf} (%)
K	Cs	25	+41.53	+12.27
La	Ce	25	−7.04	−2.40
	Y	15	−8.97	−3.08
Li	Mg	50	−4.16	−1.41
Lu	Gd	20	+14.03	+4.46
	Tb	23	+10.09	+3.25
Mg	Ag	3.0	−63.42	−28.48
	Al	9.3	−35.80	−13.73
	Bi	1.0	+23.41	+7.25
	Cd	16	−21.08	−7.59
	Cu	0.4	−6.59	−2.25
	Ga	2.5	−35.58	−13.63
	Ge	0.1	+9.47	+3.06
	Hg	1.2	−41.78	−16.50
	In	10	−7.12	−2.43
	La	0.5	+80.80	+21.82
	Li	5.8	−12.53	−4.38
	Mn	2.5	−17.43	−6.18
	Pb	7.0	+14.11	+4.49
	Sn	2.5	−1.25	−0.42
	Tl	9.1	−4.82	−1.63
	Yb	1.2	+23.11	+7.17
	Zn	3.0	−48.79	−20.00
Mn	Al	10	+16.20	+5.13
	As	3.7	+24.70	+7.63
	Co	13	+3.06	+1.01
	Cr	8.0	−4.78	−1.62
	Cu	16	+21.16	+6.60
	Fe	10	−3.72	−1.25
	Ga	20	+27.12	+8.32
	Ge	7.5	+34.51	+10.39
	In	15	+88.11	+23.44
	Ni	14	+16.90	+5.34
	Pd	18	+79.74	+21.59
	Pt	11	+67.34	+18.72
Mn	Re	5.0	+20.27	+6.35
	Ru	10	+58.34	+16.55
	Si	14	−15.44	−5.44
	Zn	18	+36.79	+11.00
Mo	Cr	30	−17.87	−6.35
	Hf	20	+25.77	+7.94
	Ir	7.0	−6.98	−2.38
	Nb	20	+12.80	+4.09
	Os	10	−9.62	−3.32
	Pd	5.0	−4.49	−1.52
	Re	15	−5.56	−1.89
	Rh	20	−10.40	−3.59
	Ru	18	−5.92	−1.98
	Ta	40	+12.68	+4.05
	Ti	8.0	+0.56	+0.18
	V	50	−10.64	−3.69
	W	34	+1.74	+0.57
Nb	Fe	1.0	−38.37	−14.92
	Hf	50	+18.58	+5.84
	Mo	30	−16.45	−5.81
	Re	20	−25.37	−9.29
	Ru	25	−32.48	−12.27
	Si	2.0	+4.96	+1.62
	Ta	100	−0.26	−0.09
	Ti	45	−3.00	−1.01
	U	20	+5.03	+1.65
	V	30	−17.81	−6.33
	W	30	−98.22	−6.49
	Zr	50	+27.11	+8.32
Ni	Al	10	+14.70	+4.67
	Au	15	+63.60	+17.83
	Co	100	+1.76	+0.58
	Cr	40	+10.34	+3.33
	Cu	68	+7.18	+2.33
	Fe	58	+10.57	+3.40
	Ga	20	+17.16	+5.41
	Ge	10	+14.76	+4.68

Table A.2. (*cont.*)

Solvent	Solute	χ_{max} (at.%)	Ω_{sf} (%)	l_{sf} (%)
Ni	In	5.0	+36.66	+10.96
	Mn	25	+23.20	+7.20
	Mo	22	+22.27	+6.93
	Nb	8.0	+51.24	+14.79
	Os	12	+7.71	+2.51
	Pd	30	+41.33	+12.21
	Pt	25	+45.68	+13.37
	Ru	20	+28.76	+8.79
	Sb	5.5	+21.32	+6.65
	Si	12	−5.81	−1.98
	Sn	10	+74.08	+20.29
	Ti	9.0	+29.43	+8.97
	V	20	+13.34	+4.26
	W	15	+36.93	+11.04
	Zn	23	+19.90	+5.24
Np	Pu	40	+5.58	+1.82
	U	10	+11.14	+3.58
Os	Mn	30	+10.83	+3.48
	Mo	40	+10.23	+3.30
	Re	40	+4.27	+1.40
	Ru	50	−2.49	−0.84
	Ta	20	+29.57	+9.02
	Tc	40	+1.83	+0.60
	W	20	+11.04	+3.35
Pb	Bi	20	+7.04	+2.29
	Ca	1.0	−7.78	−2.66
	Cd	5.0	−43.11	−17.14
	Hg	10	−5.25	−1.78
	In	14	−11.16	−3.87
	Na	9.0	−2.49	−0.84
	Sb	0.2	−14.08	−4.93
	Sn	5.0	−8.25	−2.83
	Te	0.3	−28.89	−10.74
	Tl	40	−4.53	−1.53
Pd	Ag	20	+13.12	+4.19
	Au	20	+16.31	+5.10
Pd	Bi	0.2	+52.59	+15.12
	Cd	30	+21.63	+6.73
	Co	20	−15.37	−5.41
	Cr	10	−1.45	−0.49
	Cu	48	−18.60	−6.63
	Fe	20	−11.71	−4.07
	Hg	12	+32.09	−9.72
	Ir	20	−5.68	−1.93
	Mn	28	+3.68	+2.20
	Mo	10	−4.40	−1.49
	Nb	12	+6.00	+1.96
	Ni	12	−14.41	−5.06
	Pb	14	+38.72	+11.52
	Pt	40	+1.52	+0.49
	Rh	26	−4.97	−1.68
	Ru	10	−8.27	−2.84
	Sn	26	+27.08	+8.31
	Ta	14	+6.47	+2.11
	U	11	+9.43	+3.04
	V	20	−2.45	−0.82
	Zn	20	−8.40	−2.98
	Zr	21	+27.20	+8.35
Pt	Ag	30	+8.88	+2.88
	Al	10	−9.25	−3.18
	Au	25	+10.99	+3.52
	Cd	27	+16.99	+5.36
	Co	22	−18.65	−6.65
	Cr	44	−16.16	−5.70
	Cu	38	−20.11	−7.21
	Fe	40	−10.02	−3.46
	Hg	10	+19.89	+6.23
	Ir	45	−6.76	−2.31
	Mo	38	−0.55	−0.18
	Nb	22	+6.06	+1.97
	Ni	18	−23.39	−8.50
	Pd	20	−4.01	−1.36
	Re	42	+2.02	+0.66
	Rh	100	−8.86	−3.04
	Ru	60	−7.05	−2.41

(*cont.*)

Table A.2. (*cont.*)

Solvent	Solute	χ_{max} (at.%)	Ω_{sf} (%)	l_{sf} (%)	Solvent	Solute	χ_{max} (at.%)	Ω_{sf} (%)	l_{sf} (%)
Pt	Si	1.0	−14.56	−5.11	Ru	Ta	25	+26.23	+8.07
	Sn	8.0	−1.32	−0.44		Tc	25	+4.46	+1.46
	U	4.0	+77.88	+21.15		V	33	+1.48	+0.48
	Zn	25	−7.68	−2.63		W	36	+19.95	+6.25
						Zr	0.5	−4.13	−1.40
Pu	Al	11	−46.00	−18.57					
	Ce	20	+13.54	+4.32	Sb	As	80	−17.97	−6.39
	Hf	8.0	+0.93	+0.29		Bi	100	+16.45	+5.20
	In	2.0	+29.33	+8.95		Ge	2.5	−45.31	−18.22
	Np	60	−16.87	−5.97		Sn	4.0	−15.10	−5.31
	Sc	6.0	+12.45	+3.98					
	U	0.3	−8.41	−2.89	Sc	Y	22	+42.70	+12.58
	Zn	3.0	−39.47	−15.41		Zr	40	−8.67	−2.98
	Zr	10	+0.48	+0.15					
					Si	Ge	25	+4.68	−1.53
Rb	Cs	80	+24.43	+7.54					
					Sn	Bi	5.0	+22.40	+6.96
Re	Os	30	−7.08	−0.34		Cd	1.0	−14.79	−5.20
	Pt	20	+15.35	+4.32		Hg	7.2	−17.46	−6.20
	Ru	25	−9.02	−3.10		In	6.0	−3.14	−1.06
	Tc	40	−2.85	−0.86		Pb	1.0	+29.05	+8.87
						Sb	8.2	+5.47	+1.79
Rh	Au	2.4	+9.15	+2.95		Zn	0.7	−5.70	−1.94
	Cr	8.0	−7.27	−2.46					
	Ir	100	+2.72	+0.89	Sr	Ba	20	+21.65	+6.74
	Mn	30	+5.73	+1.88		Ca	100	−23.38	−8.49
	Mo	15	+8.03	+2.61					
	Pd	50	+7.56	+2.46	Ta	Mo	35	−16.56	−5.86
	Pt	100	+10.52	+4.84		Nb	100	−0.23	−0.08
	Sn	10	+35.94	+10.77		Os	10	−28.47	−10.57
	Ta	8.0	+23.85	+7.38		Ru	25	−28.44	−10.56
	W	14	+9.60	+3.20		Ti	28	−3.59	−1.21
						V	20	−10.02	−3.46
Ru	Ir	30	+5.36	+1.94		W	75	−13.13	−4.58
	Mo	41	+12.20	+3.91		Zr	22	+12.91	+4.12
	Nb	2.4	+13.19	+4.21					
	Ni	45	−16.30	−5.76	Tb	Lu	23	−6.12	−2.08
	Os	30	+4.64	+1.52					
	Pd	3.0	+2.28	+0.74	Tc	Co	40	−27.24	−10.06
	Re	42	+7.29	+2.36		Ir	15	−2.43	−2.19
	Rh	5.0	−0.37	−0.12		Ni	30	−24.32	−8.87

Table A.2. (*cont.*)

Solvent	Solute	χ_{max} (at.%)	Ω_{sf} (%)	l_{sf} (%)	Solvent	Solute	χ_{max} (at.%)	Ω_{sf} (%)	l_{sf} (%)
Tc	Os	50	−2.97	−1.00	V	Fe	10	−18.86	−6.73
	Pt	34	+0.38	+0.12		Mo	50	+9.83	+3.18
	Re	60	+2.24	+0.73		Nb	20	+27.93	+8.55
	Rh	50	−6.61	−2.28		Ru	20	−14.56	−5.11
	Ru	25	−6.39	−2.18		Ta	70	+36.06	+10.80
						W	50	+12.64	+4.04
Th	Ce	25	−1.21	−0.41					
	Pu	40	−19.71	−7.06	W	Cr	20	−21.73	−7.84
	Rh	1.0	−5.78	−1.89		Mo	25	−1.55	−0.52
	Ru	1.0	+3.65	+1.19		Nb	25	+7.33	+2.39
	U	5.0	−6.09	−2.07		Os	7.5	−7.82	−2.68
	Y	30	+0.20	−0.05		Pt	4.0	−2.71	−0.91
	Zr	10	−7.68	−10.25		Ru	12	−5.35	−1.82
						Si	20	+0.81	+0.27
Ti	Ag	5.0	−9.28	−3.20		Ta	45	+11.01	+3.54
	Al	27	−20.09	−7.20		U	2.0	−0.82	−0.74
	Cr	20	−37.71	−14.60		V	30	−10.60	−3.66
	Fe	20	−53.62	−22.59					
	Hf	25	+3.91	+1.28	Y	Gd	25	+0.65	+0.20
	Mn	13	−51.17	−21.25		La	25	+17.30	+5.46
	Mo	40	−20.96	−7.64		Sc	22	−20.95	−7.54
	Re	50	−30.16	−11.28		Th	27	−2.38	−0.80
	Rh	15	−32.15	−12.13					
	Ru	20	−33.75	−12.82	Zn	Ag	5.5	−10.19	−3.52
	Sn	11	+13.60	+4.26		Al	1.5	−6.25	−2.13
	Ta	5.5	+0.28	+0.08		Au	5.5	−10.98	−3.80
	V	8.0	−15.40	−5.42		Cd	1.2	+49.23	+14.27
	Zr	30	+30.08	+9.16		Cu	3.0	−54.57	−23.13
						Ga	2.0	+25.33	+7.79
Tl	In	6.5	−7.06	−2.41		Hg	4.0	+39.52	+11.74
U	Cr	1.4	+4.03	+1.33	Zr	Dy	7.5	+26.92	+8.27
	Nb	30	−24.49	−8.94		Nb	2.0	−6.40	−2.18
	Os	12	+15.99	+5.06		Sc	24	+4.51	+1.48
	Pu	20	+1.41	−0.46		Th	50	+51.67	+14.89
	Ti	3.0	−8.00	−2.74		Ti	10	−22.33	−8.08
V	Al	50	+8.84	+2.87					
	Cr	40	−15.91	−5.61					

Table A.3. King table for interstitial solid solutions [22]. The data for hydrogen (H) solute are taken from [150]. χ_{lmt} is the atomic fraction limit to which appropriate size factors may be applied for computing lattice strains. Ω_{sf} is the volume size factor and l_{sf} is the linear size factor.

Crystal structure	Solvent	Solute	χ_{lmt} (at.%)	Ω_{sf} (%)	l_{sf} (%)
BCC	Fe	C	0.1	78.1	21.2
		N	0.4	70.5	19.5
	Mo	C	0.1	62.9	17.7
	Nb	H	30	17.4	5.5
		C	0.3	24.1	7.5
		N	0.3	22.2	6.9
		O	4.8	25.5	7.9
	Ta	H	30	15.5	4.9
		C	0.1	12.8	4.1
		N	4.8	40.7	12.1
		O	3.6	43.0	12.6
	V	H	30	19	6.0
		O	3.2	79.5	21.5
FCC	Ce	C	11.5	−15.5	−5.46
	Pd	H	33	19	6.0
DC	Ge	O	56.9	0.2	0.1
	Si	B	3.6	−80.0	−41.5

APPENDIX B
Thermoelastic properties of common crystalline solids

In Table B.1 we present elastic constants, coefficient of thermal expansion, and melting temperature for a set of common crystalline solids of pure elements. Because for most materials each crystal grain is elastically anisotropic, isotropic elasticity is only an approximation. The values listed here correspond to the averaged elastic properties of a polycrystal consisting of a large number of randomly oriented single-crystalline grains.

There is a significant scatter in the averaged isotropic elastic constants in the literature. The scatter in the anisotropic elastic constants of single crystals is comparably less. Therefore, the values in the table are computed values from anisotropic elastic constants of single crystals. Specifically, the bulk modulus B and shear modulus μ are computed from the self-consistent averaging [151, 152] of single crystal values. Given B and μ, the Young's modulus is computed as $E = 9B\mu/(3B + \mu)$ and the Poisson's ratio as $\nu = (3B - 2\mu)/(6B + 2\mu)$.

The anisotropic elastic constants of single crystals, together with the thermal expansion coefficients and melting temperature are obtained from [153]. Both the elastic constants and coefficients of thermal expansion are room-temperature values. Given the experimental scatter and temperature dependence of these properties, an uncertainty of a few percent is to be expected for the values reported here. This should be sufficiently accurate for most calculations and estimates within the isotropic approximation. If very precise calculations are required, the original literature should be consulted and the temperature dependence of these properties should be taken into account. Three or more significant digits are often reported to provide self-consistency among B, E, μ, and ν, even though the data themselves are probably not accurate down to the last significant digit. The melting points are accurate to within 1 K.

Table B.1. Isotropic thermoelastic properties of common crystalline solids [153]. B: bulk modulus (in GPa). E: Young's modulus (in GPa). μ: shear modulus (in GPa). ν: Poisson's ratio. α: coefficient of linear thermal expansion (in 10^{-6} K^{-1}). T_m: melting temperature (in K).

Element	Structure	B	E	μ	ν	α	T_m
Ag	FCC	103.8	82.4	30.1	0.368	18.9	1235
Al	FCC	75.9	70.4	26.2	0.345	23.1	933
Au	FCC	172.8	80.3	28.2	0.423	14.2	1337
Be	HCP	114.4	311.1	148.6	0.047	11.3	1560
Cd	HCP	54.1	62.9	24.1	0.306	30.8	594
Co	HCP	190.3	215.3	82.1	0.311	13.0	1768
Cr	BCC	152.3	273.5	113.9	0.201	3.9	2180
Cu	FCC	137.5	129.0	48.0	0.344	16.5	1358
Fe	BCC	168.7	204.4	78.7	0.298	11.8	1811
Ge	DC	74.9	131.4	54.4	0.208	5.8	1211
Mg	HCP	35.5	44.4	17.2	0.292	24.8	923
Mo	BCC	259.8	323.2	125.0	0.293	4.8	2895
Nb	BCC	171.8	105.4	37.7	0.398	7.3	2750
Ni	FCC	186.0	220.6	84.7	0.302	13.4	1728
Pb	FCC	44.8	25.0	8.9	0.407	28.9	601
Pd	FCC	193.1	133.7	48.3	0.385	11.8	1828
Pt	FCC	282.7	177.7	63.7	0.395	8.8	2041
Si	DC	97.9	162.9	66.6	0.223	2.6	1687
Ta	BCC	189.7	185.1	69.2	0.337	6.3	3290
Ti	HCP	107.3	114.7	43.4	0.322	8.6	1943
V	BCC	155.6	129.5	47.6	0.361	8.4	2183
W	BCC	310.4	409.8	160.1	0.280	4.5	3687
Zn	HCP	70.9	102.2	40.6	0.260	30.2	693
Zr	HCP	95.3	96.0	36.0	0.332	5.7	1854

APPENDIX C

Thermodynamic and kinetic properties of vacancies

Table C.1. Properties of vacancies in pure metals [26]. $\chi_v(T_m)$: vacancy fraction at the melting temperature (in 10^{-4}). Δs_v: activation entropy of formation (in units of k_B). Δh_v: activation enthalpy of formation (in eV). ΔV_v^{rlx}: relaxation volume of the vacancy (in units of atomic volume Ω). Δh_v^m: activation enthalpy of migration (in eV). R stands for rhombohedral crystal structure. T stands for tetragonal crystal structure. BCT stands for body-centered tetragonal crystal structure. Para and ferro stand for the paramagnetic and ferromagnetic phases of iron (Fe), respectively.

Element	Structure	$\chi_v(T_m)$	Δs_v	Δh_v	ΔV_v^{rlx}	Δh_v^m
Ag	FCC	1.7	1.5	1.11±0.05		0.66±0.05
Al	FCC	9.4	0.7	0.67±0.03	−0.05	0.61±0.03
Au	FCC	7.2	0.72	0.93±0.04	−0.15	0.71±0.05
Be	HCP					0.8
Bi	R	6.17	0.3	0.35		0.5
Cd	HCP	4.5	0.5	0.46±0.05	−0.54	0.4±0.05
Co	HCP				−0.05	0.72
Cr	BCC			2.0±0.2		~0.95
Cu	FCC	2.1±0.1	1.6–2.8	1.28±0.05	−0.25	0.70±0.02
Fe	BCC				−0.05	
	(para)			1.8±0.1		
	(ferro)			1.59–1.73		0.55–1.3
In	T		5	0.52±0.04	−0.6	(0.31)
K	BCC			0.34		0.038
Li	BCC			0.48		0.038
Mg	HCP	7.2	0±0.3	0.79		0.5–0.6
Mo	BCC	0.13	1.6	3.0±0.2	−0.1	1.35
Na	BCC	7.8±0.3	3.9±0.3	0.354±0.035		0.03
Nb	BCC			3.07		0.55
Ni	FCC			1.79±0.05	−0.2	1.04±0.04
Pb	FCC	1.7	0.7–2.6	0.58±0.04		0.43±0.02
Pd	FCC	0.14		1.85		1.03±0.3
Pt	FCC	9.4	0.4	1.35±0.05	−0.28	1.43±0.05

(cont.)

Table C.1. *(cont.)*

Element	Structure	$\chi_v(T_m)$	Δs_v	Δh_v	ΔV_v^{rlx}	Δh_v^m
Re	HCP					0.5–2.2
Sb	R	10		1.4		~0.3
β-Sn	BCT	≤0.3	1.1	0.51		0.3 - 0.4
Ta	BCC			3.1		0.7
Th	FCC			1.28±0.2		2.04
V	BCC			2.2±0.4		0.5–0.7
W	BCC	3	3.2	3.6±0.2		1.7±0.1
Zn	HCP	5±0.3	1±1	0.54±0.03	−0.6	0.42±0.02
Zr	HCP			>1.5	−0.05	0.54–0.7

APPENDIX D

Diffusion coefficients in common crystals

Here we list the data concerning self-diffusion and impurity diffusion in a set of common crystals. For both kinds of diffusion, the diffusivity D can often be written in the form of

$$D = D_0 \exp\left(-\frac{Q}{k_B T}\right), \tag{D.1}$$

within a certain range of temperature, T, where D_0 and Q are constants. Sometimes it is necessary to express D using two exponential terms within the given temperature range,

$$D = D_{01} \exp\left(-\frac{Q_1}{k_B T}\right) + D_{02} \exp\left(-\frac{Q_2}{k_B T}\right), \tag{D.2}$$

where D_{01}, D_{02}, Q_1, and Q_2 are constants.

In Table D.1, we list the information regarding the coefficient of self-diffusion in a set of common crystals. The diffusion coefficient D is isotropic for crystals with cubic symmetry (e.g. FCC, BCC, DC). However, for HCP crystals, the diffusion coefficients are different in the directions perpendicular ($\perp$) to and parallel ($\|$) to the c axis, as indicated in the table. Arrhenius plots of the diffusion data in this appendix are available on the book website.

Table D.1. Self-diffusion data in common crystals. D_0, D_{01}, and D_{02} are in units of $m^2\ s^{-1}$. Q, Q_1, and Q_2 are in units of eV. The range of T is in units of K. Data for most metals, except for V [154], are taken from [155].

Element	Structure		D_0 (or D_{01})	Q (or Q_1)	D_{02}	Q_2	T range
Ag	FCC		5.5×10^{-6}	1.77	1.51×10^{-3}	2.35	550–1228
Al	FCC		2.25×10^{-4}	1.50			573–923
Au	FCC		2.25×10^{-6}	1.70	8.3×10^{-5}	2.20	603–1333
Be	HCP	$\parallel c$	6.2×10^{-5}	1.71			836–1343
Be	HCP	$\perp c$	5.2×10^{-5}	1.63			836–1343
Cd	HCP	$\parallel c$	1.18×10^{-5}	0.81			420–600
Cd	HCP	$\perp c$	1.83×10^{-5}	0.85			420–600
Co	FCC		2.54×10^{-4}	3.15			923–1743
Cr	BCC		1.28×10^{-1}	4.58			1073–2090
Cu	FCC		1.3×10^{-5}	2.06	4.5×10^{-4}	2.46	573–1334
Fe	BCC		1.21×10^{-2}	2.92			1067–1169
Fe	FCC		4.9×10^{-5}	2.94			1444–1634
Fe	BCC		2.01×10^{-4}	2.49			1701–1765
Ge	DC		2.48×10^{-3}	3.14			822–1164
Mg	HCP	$\parallel c$	1.78×10^{-4}	1.44			773–903
Mg	HCP	$\perp c$	1.75×10^{-4}	1.43			773–903
Mo	BCC		1.26×10^{-5}	4.53	1.39×10^{-2}	5.69	1363–2723
Nb	BCC		8×10^{-7}	3.62	3.7×10^{-4}	4.54	1353–2693
Ni	FCC		8.5×10^{-5}	2.87	1.35×10^{-1}	4.15	879–1673
Pb	FCC		9.95×10^{-5}	1.11			473–596
Pd	FCC		2.05×10^{-5}	2.76			1323–1723
Pt	FCC		3.4×10^{-6}	2.64	8.86×10^{-3}	4.05	773–1998
Si	DC		1.46×10^{-1}	5.02			1318–1663
Si	DC		1.54×10^{-2}	4.65			1128–1448
Ta	BCC		2.1×10^{-5}	4.39			1261–2893
V	BCC		2.88×10^{-5}	3.21			997–1915
V	BCC		1.73×10^{-2}	4.24			1915–2115
W	BCC		4×10^{-6}	5.45	4.6×10^{-3}	6.90	1703–3413
Zn	HCP	$\parallel c$	1.3×10^{-5}	0.95			513–691
Zn	HCP	$\perp c$	1.8×10^{-5}	1.00			513–691

Table D.2. Diffusion data for substitutional point defects in various crystals. D_0 is in units of $m^2\ s^{-1}$. Q is in units of eV. The range of T is in units of K. Diffusion data in metals correspond to tracer diffusivity taken from [155], i.e. the diffusivity is measured when the solute fraction is vanishingly small. Diffusion data in Si and Ge are from [156].

Solvent	Solute	D_0	Q	T range
Ag	Cu	1.23×10^{-4}	2.00	990–1218
	Cu	2.9×10^{-6}	1.70	699–897
	Au	8.5×10^{-5}	2.09	991–1198
	Zn	5.4×10^{-5}	1.81	916–1197
	Cd	4.4×10^{-5}	1.81	866–1210
	Al	1.3×10^{-5}	1.65	873–1223
	Ge	8.4×10^{-6}	1.58	943–1123
	Pb	2.2×10^{-5}	1.65	973–1098
	Pd	9.57×10^{-4}	2.46	1008–1212
	Pt	6.0×10^{-4}	2.47	923–1223
Al	Cu	6.54×10^{-5}	1.41	594–928
	Ag	1.18×10^{-5}	1.21	644–928
	Zn	3.25×10^{-5}	1.22	688–928
	Si	2.48×10^{-4}	1.42	753–893
	Ge	4.81×10^{-5}	1.26	674–926
	Pb	5×10^{-3}	1.51	777–876
	Mg	6.23×10^{-6}	1.19	598–923
Au	Cu	1.05×10^{-5}	1.76	973–1179
	Ag	7.2×10^{-6}	1.74	972–1281
	Zn	8.2×10^{-6}	1.64	969–1287
	Al	5.2×10^{-6}	1.49	773–1223
	Ge	7.3×10^{-6}	1.50	1010–1287
	Fe	8.2×10^{-6}	1.81	1027–1221
	Ni	3×10^{-5}	2.00	1153–1210
	Pd	7.6×10^{-6}	2.02	973–1273
	Pt	9.5×10^{-6}	2.09	973–1273
Cu	Ag	6.1×10^{-5}	2.02	873–1273
	Au	2.43×10^{-5}	2.05	633–1350
	Zn	3.4×10^{-5}	1.98	878–1322
	Cd	1.27×10^{-4}	2.02	1032–1346
	Si	7×10^{-6}	1.78	973–1323
	Ge	3.97×10^{-5}	1.94	975–1289
	Fe	1.01×10^{-4}	2.21	989–1329
	Ni	7.6×10^{-5}	2.33	600–1000
	Pd	1.71×10^{-4}	2.36	1080–1329
	Pt	5.6×10^{-5}	2.41	1149–1352

(cont.)

Table D.2. *(cont.)*

Solvent	Solute	D_0	Q	T range
Fe	Cu	1.9×10^{-5}	2.83	1198–1323
	Au	3.1×10^{-3}	2.71	1055–1174
	Zn	6×10^{-3}	2.72	1072–1169
	V	1.24×10^{-2}	2.84	1058–1172
	V	6.2×10^{-5}	2.83	1210–1607
	Nb	5.02×10^{-3}	2.61	1059–1162
	Nb	8.3×10^{-5}	2.76	1210–1604
	Ni	9.9×10^{-4}	2.69	1054–1173
	Ni	3.0×10^{-4}	3.25	1409–1673
	Pd	4.1×10^{-5}	2.91	1373–1573
Ge	P	3.3×10^{-4}	2.50	874–1174
	Ga	1.4×10^{-2}	3.35	828–1190
	As	2.1×10^{-4}	2.39	974–1174
	Au	2.25×10^{-2}	2.50	874–1174
	Si	2.4×10^{-5}	2.90	924–1174
Mo	V	2.9×10^{-4}	4.90	1803–1998
	Nb	1.4×10^{-3}	4.69	2123–2623
	Ta	3.5×10^{-8}	3.60	1193–1423
	Ta	1.9×10^{-4}	4.90	2098–2449
	W	1.7×10^{-4}	4.77	1973–2533
	Fe	1.5×10^{-5}	3.59	1273–1623
Nb	V	4.7×10^{-5}	3.91	1898–2348
	V	2.21×10^{-4}	3.69	1273–1673
	Ta	1.0×10^{-4}	4.31	1376–2346
	Mo	1.3×10^{-6}	3.63	1973–2298
	W	5×10^{-8}	3.98	2073–2473
	Fe	1.4×10^{-5}	3.05	1663–2168
	Ni	7.7×10^{-6}	2.74	1433–2168
Ni	Cu	2.7×10^{-5}	2.65	1048–1323
	Au	2.0×10^{-4}	2.82	1173–1373
	Al	1.0×10^{-4}	2.69	914–1212
	Ge	2.1×10^{-4}	2.74	939–1675
	V	8.7×10^{-5}	2.89	1073–1573
	W	2.87×10^{-4}	3.19	1346–1668
	Fe	1.0×10^{-4}	2.79	1478–1669
	Pt	2.5×10^{-4}	2.97	1354–1481
Pb	Zn	1.65×10^{-6}	0.50	453–773
	Cd	4.09×10^{-5}	0.92	423–593
	Cd	9.2×10^{-5}	0.96	523–823

Table D.2. (*cont.*)

Solvent	Solute	D_0	Q	T range
	In	3.3×10^{-3}	1.16	437–493
	Sn	4.1×10^{-5}	0.98	523–723
Pt	Ag	1.3×10^{-5}	2.68	1473–1873
	Au	1.3×10^{-5}	2.61	850–1265
	Al	1.3×10^{-7}	2.01	1373–1872
	Fe	2.5×10^{-6}	2.52	1273–1673
Si	P	1.1×10^{-4}	3.40	1174–1474
	Ga	6×10^{-3}	3.89	1174–1324
	As	6×10^{-3}	4.20	1224–1624
	Au	2.75×10^{-7}	2.05	974–1574
	Ge	3.5×10^{-5}	3.92	1129–1274
Ta	Nb	2.3×10^{-5}	4.28	1194–2757
	Mo	1.8×10^{-7}	3.51	2923–2493
	Fe	5.05×10^{-5}	3.10	1203–1513
	Fe	5.9×10^{-6}	3.42	2053–2330
V	Ta	2.44×10^{-5}	3.12	1373–2073
	Fe	3.73×10^{-5}	3.08	1233–1618
	Fe	2.74×10^{-2}	4.00	1688–2090
	Ni	1.8×10^{-5}	2.76	1175–1948
W	Nb	3.01×10^{-4}	5.97	1578–2640
	Ta	3.05×10^{-4}	6.07	1578–2648
	Mo	1.4×10^{-4}	5.88	1909–2658
	Fe	1.4×10^{-6}	2.86	1213–1513

Table D.3. Diffusion data for interstitial point defects in several crystals. D_0 is in units of $m^2\ s^{-1}$. Q is in units of eV. The range of T is in units of K.

Solvent	Solute	D_0	Q	T range	Reference
Co	C	8.72×10^{-6}	1.55	723–1073	[155]
Ni	H	6.9×10^{-7}	0.42	627–1774	[150]
	H	4.8×10^{-7}	0.41	285–627	[150]
	C	1.2×10^{-5}	1.42	873–1673	[155]
Nb	H	5×10^{-8}	0.11	273–873	[150]
	C	4.0×10^{-7}	1.43	413–534	[157]
	N	8.6×10^{-7}	1.51	423–568	[157]
	O	2.12×10^{-6}	1.17	313–423	[157]
Ta	H	4.4×10^{-8}	0.14	273–673	[150]
	C	6.1×10^{-7}	1.67	463–623	[157]
	N	5.6×10^{-7}	1.64	463–623	[157]
	O	4.4×10^{-7}	1.10	312–425	[157]
Si	H	6×10^{-5}	1.03	394–1481	[156]
	Li	2.5×10^{-7}	0.65	299–1624	[156]
	B	2.4×10^{-3}	3.87	1114–1524	[156]
	C	3.3×10^{-5}	2.92	1344–1674	[156]
	N	2.7×10^{-7}	2.80	1074–1474	[156]
	O	7×10^{-6}	2.44	974–1524	[156]
	Ni	8×10^{-5}	0.90	944–1174	[156]
Ge	Li	1.3×10^{-7}	0.46	624–1074	[156]
	B	1.8×10^{5}	4.55	874–1174	[156]
	O	4×10^{-5}	2.08	924–1124	[156]
	Ni	8×10^{-5}	0.90	944–1174	[156]

Bibliography

[1] N. W. Ashcroft and N. D. Mermin. *Solid State Physics*. Saunders College, Philadelphia, 1976.

[2] Vasily V. Bulatov and Wei Cai. *Computer Simulations of Dislocations*. Oxford University Press, 2006.

[3] C. Kittel. *Introduction to Solid State Physics*. Wiley, New York, 7th edition, 1996.

[4] W. Shockley, J. H. Hollomon, R. Maurer, and F. Seitz, editors. *Imperfections in Nearly Perfect Crystals*. Wiley, 1952.

[5] J. P. Hirth and J. Lothe. *Theory of Dislocations*. Wiley, New York, 1982.

[6] W. T. Read, Jr. *Dislocations in Crystals*. McGraw-Hill, 1953.

[7] D. B. Williams and C. B. Carter. *Transmission Electron Microscopy: A Textbook for Materials Science*. Springer, 1996.

[8] B. Modéer and R. Lagneborg. Creep deformation in a 20% Cr–35% Ni steel. *Jernkontorets Annaler*, **155**:363–367, 1971.

[9] P. B. Hirsch, A. Howie, R. B. Nicholson, D. W. Pashley, and M. J. Whelan. *Electron Microscopy of Thin Crystals*. Butterworths, Washington, 1965.

[10] M. H. Sadd. *Elasticity: Theory, Applications and Numerics*. Elsevier, 2005.

[11] J. R. Barber. *Elasticity*. Springer, 3rd edition, 2010.

[12] T. J. R. Hughes. *The Finite Element Method: Linear Static and Dynamic Finite Element Analysis*. Dover, Mieola, New York, 2000.

[13] F. Reif. *Fundamentals of Statistical and Thermal Physics*. McGraw-Hill, 1965.

[14] D. Chandler. *Introduction to Modern Statistical Mechanics*. Oxford University Press, Oxford, 1987.

[15] Y. Wang, J. Zhang, H. Xu, *et al.* Thermal equation of state of copper. *Applied Physics Letters*, **94**:071904, 2009.

[16] P. K. Sharma and N. Singh. Grüneisen parameters of cubic metals. *Physical Review B*, **1**:4635–4638, 1970.

[17] R. Shankar. *Principles of Quantum Mechanics*. Springer, 2nd edition, 1994.

[18] J. P. Hirth. Effects of hydrogen on the propertie of iron and steel. *Metallurgical and Materials Transactions A*, **11**A:861–890, 1980.

[19] R. L. Fleischer. Solid-solution hardening. In D. Peckner, editor, *The Strengthening of Metals*. Reinhold Publishing Corporation, New York, 1964.

[20] H. W. King. Quantitative size-factors for metallic solid solutions. *Journal of Materials Science*, **1**:79–90, 1966.

[21] W. B. Pearson. *Handbook of Lattice Spacings and Structures of Metals and Alloys*. Pergamon Press, New York, 1958.

[22] H. W. King. Quantitative size-factors for interstitial solid solutions. *Journal of Materials Science*, **6**:1157–1167, 1971.

[23] J. D. Eshelby. Distortion of a crystal by point imperfections. *Journal of Applied Physics*, **25**:255–261, 1954.

[24] J. D. Eshelby. Elastic inclusions and inhomogeneities. *Progress in Solid Mechanics*, **2**:89, 1961.

[25] V. L. Indenbom. Dislocations and internal stresses. In V. L. Indenbom and J. Lothe, editors, *Elastic Strain Fields and Dislocation Mobility*. Elsevier, 1992.

[26] H. J. Wollenberger. Point defects. In R. W. Cahn and P. Haasen, editors, *Physical Metallurgy*, volume II. North-Holland, 1996.

[27] W. M. Haynes, editor. *Handbook of Chemistry and Physics*. CRC, http://www.hbcpnetbase.com/, 94th edition, 2013–2014.

[28] W. D. Nix. Interface stresses. In K. H. Jürgen Buschow, R. W. Cahn, M. C. Flemings, *et al.*, editors, *Encyclopedia of Materials: Science and Technology*, pp. 4136–4141. Elsevier, 2001.

[29] A. C. Damask and G. J. Dienes. *Point Defects in Metals*. Gordon and Breach, 1963.

[30] J. D. Eshelby. The determination of the elastic field of an ellipsoidal inclusion and related problems. *Proceedings of the Royal Society of London A*, **241**:376–396, 1957.

[31] P. M. Fahey, P. B. Griffin, and J. D. Plummer. Point defects and dopant diffusion in silicon. *Reviews of Modern Physics*, **61**:289–384, 1989.

[32] D. V. Ragone. *Thermodynamics of Materials*, volume II. John Wiley & Sons, 1995.

[33] F. A. Kröger and H. J. Vink. Relations between the concentrations of imperfections in crystalline solids. *Solid State Physics*, **3**:307–435, 1956.

[34] J. Bloem, F. A. Kröger, and H. J. Vink. The physical chemistry of lead sulphide in relation to its semiconducting properties. In *Report of the Conference on Defects in Crystalline Solids* (The Physical Society, University of Bristol, 1955), pp. 273.

[35] J. Bloem. Controlled conductivity in lead sulphide single crystals. *Philips Research. Reports*, **11**:273–336, 1956.

[36] H. Okamoto. Al–Ni (aluminium–nickel). *Journal of Phase Equilibria*, **14**:257–259, 1993.

[37] R. D. Noebe, R. R. Bowman, and M. V. Nathal. Physical and mechanical properties of the B2 compound NiAl. *International Materials Reviews*, **8**:193–232, 1993.

[38] Y.-M. Chiang, III Birnie, D. P., and W. D. Kingery. *Physical Ceramics: Principles for Ceramic Science and Engineering*. John Wiley & Sons, 1997.

[39] A. Dwivedi and A. N. Cormack. A computer simulation study of the defect structure of calcia-stabilized zirconia. *Philosophical Magazine A*, **61**:1–22, 1990.

[40] S. Amelinckx. *The Direct Observation of Dislocations*. Academic Press, 1964.

[41] R. Becker and W. Döring. The kinetic treatment of nuclear formation in supersaturated vapors. *Annales de Physique*, **24**:719, 1935.

[42] H. Eyring. The activated complex in chemical reactions. *Journal of Chemical Physics*, **3**:107, 1935.

[43] P. Hanggi, P. Talkner, and M. Borkovec. Reaction-rate theory: fifty years after Kramers. *Reviews of Modern Physics*, **62**(2):251–42, 1990.

[44] N. L. Peterson. Diffusion in metals. *Solid State Physics*, **22**:409–512, 1969.

[45] D. Gupta and P. S. Ho, editors. *Diffusion Phenomena in Thin Films and Microelectronic Materials*. Noyes Publications, Park Ridge, New Jersey, 1989.

[46] J. L. Bocquet, G. Brebec, and Y. Limoge. Diffusion in metals and alloys. In R. W. Cahn and P. Haasen, editors, *Physical Metallurgy*, volume I. North-Holland, 1996.
[47] R. W. Balluffi, S. M. Allen, and W. C. Carter. *Kinetics of Materials*. Wiley, 2005.
[48] J. Crank. *The Mathematics of Diffusion*. Claredon Press, Oxford, 1975.
[49] W. D. Nix. The effects of grain shape on Nabarro–Herring and Coble creep processes. *Metals Forum*, **4**:38, 1981.
[50] R. Raj and M. F. Ashby. On grain boundary sliding and diffusional creep. *Metallurgical and Materials Transactions*, **2**:1113–1127, 1971.
[51] G. I. Taylor. Plastic deformation of crystals. *Proceedings of the Royal Society of London*, **145**:362–404, 1934.
[52] E. Orowan. Plasticity of crystals. *Zeitschrift fur Physik*, **89**(9–10):605–659, 1934.
[53] M. Polanyi. Lattice distortion which originates plastic flow. *Zeitschrift fur Physik*, **89**(9–10):660–662, 1934.
[54] P. B. Hirsch, R. W. Horne, and M. J. Whelan. Direct observations of the arrangement and motion of dislocations in aluminium. *Philosophical Magazine*, **1**(7):677–684, 1956.
[55] D. Cherns, W. T. Young, J. W. Steeds, F. A. Ponce, and S. Nakamura. Observation of coreless dislocations in α-GaN. *Journal of Crystal Growth*, **178**:201–206, 1997.
[56] P. H. Kitabjian. *High Temperature Deformation Behavior of NiAl(Ti) Single Crystals*. PhD thesis, Stanford University, 1998.
[57] G. E. Dieter. *Mechanical Metallurgy*. McGraw-Hill, 3rd edition, 1986.
[58] R. W. K. Honeycombe. *The Plastic Deformation of Metals*. Arnold, London, 1984.
[59] A. F. Schwartzman and R. Sinclair. Metastable and equilibrium defect structure of II–VI/GaAs interfaces. *Journal of Electronic Materials*, **20**:805–814, 1991.
[60] F. C. Frank and W. T. Read, Jr. Multiplication processes for slow moving dislocations. *Physical Review*, **79**:722–723, 1950.
[61] F. C. Frank. The Frank–Read source. *Proceedings of the Royal Society of London A*, **371**:136–138, 1980.
[62] M. J. Mills and D. B. Miracle. The structure of $a\langle 1\,0\,0\rangle$ and $a\langle 1\,1\,0\rangle$ dislocation cores in NiAl. *Acta Metallurgica et Materialia*, **41**:85–95, 1993.
[63] J. D. Eshelby. Screw dislocations in thin rods. *Journal of Applied Physics*, **24**:176–179, 1953.
[64] J. Zhu, H. Peng, A. F. Marshall, *et al.* Formation of chiral branched nanowires by the Eshelby twist. *Nature Nanotechnology*, **3**:477–481, 2008.
[65] M. J. Bierman, Y. K. A. Lau, A. V. Kvit, A. L. Schmitt, and S. Jin. Dislocation-driven nanowire growth and Eshelby twist. *Science*, **23**:1060–1063, 2008.
[66] W. Cai, V. V. Bulatov, J. Chang, J. Li, and S. Yip. Periodic image effects in dislocation modelling. *Philosophical Magazine*, **83**(5):539–567, 2003.
[67] Wei Cai, Athanasios Arsenlis, Christopher R. Weinberger, and Vasily V. Bulatov. A non-singular continuum theory of dislocations. *Journal of the Mechanics and Physics of Solids*, **54**(3):561–587, 2006.
[68] I. M. Gelfand and S. V. Fomin. *Calculus of Variation*. Dover, 2000.
[69] W. D. Cash and W. Cai. Dislocation contribution to acoustic nonlinearity: The effect of orientation-dependent line energy. *Journal of Applied Physics*, 109:014915, 2011.
[70] G. De Wit and J. S. Koehler Interaction of dislocations with an applied stress in anisotropic crystals. *Physics Review*, **116**:1113–1120, 1959.

[71] J. D. Eshelby, F. C. Frank, and F. R. N. Nabarro. The equilibrium of linear arrays of dislocations. *Philosophical Magazine*, **42**:351, 1951.

[72] P. R. Swann and J. Nutting. The influence of stacking-fault energy on the modes of deformation of polycrystalline copper alloys. *Journal of the Institute of Metals*, **90**:133–138, 1961.

[73] P. R. Swann. Dislocation arrangements in face-centered cubic metals and alloys. In G. Thomas and J. Washburn, editors, *Electron Microscopy and Strength of Crystals*. Interscience Publishers, 1963.

[74] A. S. Argon. *Strengthening Mechanisms in Crystal Plasticity*. Oxford University Press, 2008.

[75] H. Mecking and U. F. Kocks. Kinetics of flow and strain-hardening. *Acta Metallurgica*, **29**:1865–1875, 1981.

[76] R. J. Asaro. Elastic-plastic memory and kinematic-type hardening. *Acta Metallurgica*, **23**:1255–1265, 1975.

[77] W. G. Johnston and J. J. Gilman. Dislocation velocities, dislocation densities, and plastic flow in lithium fluoride crystals. *Journal of Applied Physics*, **30**:129, 1959.

[78] H. Alexander and P. Haasen. Dislocations and plastic flow in the diamond structure. *Solid State Physics*, **22**:27–158, 1969.

[79] J. J. Gilman. *Micromechanics of Flow in Solids*. McGraw-Hill, 1969.

[80] B. Reppich, P. Haasen, and B. Ilschner. Creep of silicon single crystals. *Acta Metallurgica*, **12**:1283–1288, 1964.

[81] G. Y. Yang, J. Zhu, W. D. Wang, Z. Zhang, and F. W. Zhu. Precipitation of nanoscale Co particles in a granular CuCo alloy with giant magnetoresistance. *Materials Research Bulletin*, **35**:875–885, 2000.

[82] A. K. Head. The interaction of dislocations and boundaries. *Philosophical Magazine*, **44**:92–94, 1953.

[83] M. L. Ovecoglu, M. F. Doerner, and W. D. Nix. Elastic interactions of screw dislocations in thin films on substrates. *Acta Metallurgica*, **35**:2947–2957, 1987.

[84] D. M. Barnett. *A Theoretical Investigation of Dislocation Distributions in Two-Phase Systems*. PhD thesis, Stanford University, 1967.

[85] A. K. Head. Edge dislocations in inhomogeneous media. *Proceedings of the Physical Society Section B*, **66**(9):793–801, 1953.

[86] J. Oswald, E. Wintersberger, G. Bauer, and T. Belytschko. A higher-order extended finite element method for dislocation energetics in strained layers and epitaxial islands. *International Journal for Numerical Methods in Engineering*, **85**:920–938, 2011.

[87] V. I. Alshits and V. L. Indenbom. Mechanisms of dislocation drag. In F. R. N. Nabarro, editor, *Dislocations in Solids*. Elsevier, 1986.

[88] K. M. Jassby and T. Vreeland, Jr. Dislocation mobility in pure copper at 4.2K. *Physical Review B*, **8**:3537–3541, 1973.

[89] K. M. Jassby and T. Vreeland, Jr. An experimental study of the mobility of edge dislocations in pure copper single crystals. *Philosophical Magazine*, **21**:1147–1168, 1970.

[90] J. A. Gorman, D. S. Wood, and T. Vreeland, Jr. Mobility of dislocations in aluminum. *Journal of Applied Physics*, **40**:833, 1969.

[91] V. R. Parameswaran and J. Weertman. Dislocation mobility in lead single crystals. *Scripta Metallurgica*, **3**:477–479, 1969.

[92] N. Bernstein and E. B. Tadmore. Tight-binding calculations of stacking energies and twinnability in FCC metals. *Physical Review B*, **69**:094116, 2004.

[93] P. J. Moroz, R. Taggart, and D. H. Polonis. Defect structures in low stacking fault energy Cu–Ge solid solutions. *Materials Science and Engineering*, **79**:201–210, 1986.

[94] S.-W. Lee and W. D. Nix. Geometrical analysis of 3D dislocation dynamics simulations of FCC micro-pillar plasticity. *Materials Science and Engineering A (Structural Materials: Properties, Microstructure and Processing)*, **527**:1903–1910, 2010.

[95] C. R. Weinberger and W. Cai. The stability of Lomer–Cottrell jogs in nanopillars. *Scripta Materialia*, **64**:529–532, 2011.

[96] J. Silcox and P. B. Hirsch. Direct observations of defects in quenched gold. *Philosophical Magazine*, **4**:72, 1959.

[97] R. E. Smallman, K. H. Westmacott, and J. A. Coiley. Clustered vacancy defects in some face-centered cubic metals and alloys. *Journal of the Institute of Metals*, **88**:127, 1959.

[98] H. Tonda and S. Ando. Effect of temperature and shear direction on yield stress by $\{11\bar{2}2\}\langle\bar{1}\bar{1}23\rangle$ slip in HCP metals. *Metallurgical and Materials Transactions A*, **33A**:831–836, 2002.

[99] D. H. Lassila, M. M. Leblanc, and J. N. Florando. Zinc single-crystal deformation experiments using a "6 degrees of freedom" apparatus. *Metallurgical and Materials Transactions A*, **38A**:2024–2032, 2007.

[100] K. H. Adams, R. C. Blish, and T. Vreeland, Jr. Second-order pyramidal slip in zinc single crystals. *Materials Science and Engineering*, **2**:201–207, 1967.

[101] W. B. Hutchinson and M. R. Barnett. Effective values of critical resolved shear stress for slip in polycrystalline magnesium and other hcp metals. *Scripta Materialia*, **63**:737–740, 2010.

[102] A. Poty, J.-M. Raulot, H. Xu, *et al.* Classification of the critical resolved shear stress in the hexagonal-close-packed materials by atomic simulation: Application to α-zirconium and α-titanium. *Journal of Applied Physics*, **110**:014905, 2011.

[103] A. Couret and D. Caillard. Prismatic slip in beryllium. *Philosophical Magazine A*, **59**:801–819, 1989.

[104] S. Amelinckx and P. Delavignette. Multiple ribbons of partial dislocations in chromium chloride. *Journal of Applied Physics*, **33**:1458–1460, 1962.

[105] H. P. Karnthaler, E. Th. Mühlbacher, and C. Rentenberger. The influence of the fault energies on the anomalous mechanical behaviour of Ni_3Al alloys. *Acta Materialia*, **44**:547–560, 1996.

[106] B. H. Kear and H. G. F. Wilsdorf. Dislocation configurations in plastically deformed polycrystalline Cu_3Au alloys. *Transactions of the Metallurgical Society of AIME*, **224**:382–386, 1962.

[107] P. A. Flinn. *Theory of deformation of superlattices. Transactions of the Metallurgical Society of AIME*, 218:145–154, 1960.

[108] D. P. Pope. Mechanical properties of intermetallic compounds. In R. W. Cahn and P. Haasen, editors, *Physical Metallurgy*, p. 2085. Elsevier, 4th edition, 1996.

[109] P. Beardmore, R. G. Davies, and T. L. Johnston. On the temperature dependence of the flow stress of nickel-base alloys. *Transactions of the Metallurgical Society of AIME*, **245**:1537–1545, 1969.

[110] R. C. Reed. *The Superalloys: Fundamentals and Applications*. Cambridge University Press, 2006.

[111] R. E. Peierls. The size of a dislocation. *Proceedings of the Physical Society*, **52**:34, 1940.

[112] F. R. N. Nabarro. Dislocations in a simple cubic lattice. *Proceedings of the Physical Society*, **59**:256, 1947.

[113] W. Cai, V. V. Bulatov, J. Chang, J. Li, and S. Yip. Dislocation core effects on mobility. In F. R. N. Nabarro and Hirth. J. P., editors, *Dislocations in Solids*, volume 12, pp. 1–80. Elsevier, Amsterdam, 2004.

[114] J. Rabier, P. Cordier, T. Tondellier, J. L. Demenet, and H. Garem. Dislocation microstructures in Si plastically deformed at RT. *Journal of Physics: Condened Matter*, **12**:10 059–10 064, 2000.

[115] C. R. Weinberger, B. L. Boyce, and C. C. Battaile. Slip planes in BCC transition metals. *International Materials Reviews*, **58**:296–314, 2013.

[116] J. Marian, W. Cai, and V. V. Bulatov. Dynamic transitions from smooth to rough to twinning in dislocation motion. *Nature Materials*, **3**(3):158–163, 2004.

[117] K. Kang, V. V. Bulatov, and W. Cai. Singular orientations and faceted motion of dislocations in body-centered cubic crystals. *Proceedings of the National Academy of Sciences USA*, **109**:15 174, 2012.

[118] C. L. Kelchner, S. J. Plimpton, and J. C. Hamilton. Dislocation nucleation and defect structure during surface indentation. *Physical Review B (Condensed Matter)*, **58**(17):11 085–11 088, 1998.

[119] S. Ismail-Beigi and T. A. Arias. Ab initio study of screw dislocations in Mo and Ta: a new picture of plasticity in BCC transition metals. *Physical Review Letters*, **84**:1499–1502, 2000.

[120] H.-J. Kaufmann, A. Luft, and D. Schulze. Deformation mechanism and dislocation structure of high-purity molybdenum single-crystals at low temperatures. *Crystal Research and Technology*, **19**:357–372, 1984.

[121] W. Cai, R. B. Sills, D. M. Barnett, and W. D. Nix. Modeling a distribution of point defects as misfitting inclusions in stressed solids. *Journal of the Mechanics and Physics of Solids*, **66**:154–171, 2014.

[122] A. H. Cottrell. Effect of solute atoms on the behaviour of dislocations. *Report of a Conference on Strength of Solids*. The Physical Society, London, 1948.

[123] A. H. Cottrell and B. A. Bilby. Dislocation theory of yielding and strain ageing of iron. *Proceedings of the Physical Society A*, **62**:49–62, 1949.

[124] R. Fuentes-Samaniego, R. Gasca-Neri, and J. P. Hirth. Solute drag on moving edge dislocation. *Philosophical Magazine A*, **49**:31–43, 1984.

[125] F. Larché and J. W. Cahn. A linear theory of thermochemical equilibrium of solids under stress. *Acta Metallurgica*, **21**:1051–1063, 1973.

[126] L. H. Yang, M. Tang, and J. A. Moriaty. Dislocations and plasticity in BCC transition metals at high pressure. In *Dislocations in Solids*, volume 16, pp. 1–46. Elsevier, 2010.

[127] T. Watanabe. Grain boundary engineering: historical perspective and future prospects. *Journal of Materials Science*, **46**:4095–4115, 2011.

[128] A. P. Sutton and R. W. Balluffi. On geometric criteria for low interfacial energy. *Acta Metallurgica*, **35**:2177–2201, 1987.

[129] A. P. Sutton and R. W. Balluffi. *Interfaces in Crystalline Materials*. Oxford University Press, 2006.

[130] W. Bollmann. *Crystal Defects and Crystalline Interfaces*. Springer-Verlag, 1970.

[131] G. Friedel. *Leçons de Cristallographie*. Paris, BergerLevrault, 1926.

[132] S. Ranganathan. On the geometry of coincidence-site lattices. *Acta Crystallographica*, **21**:197–199, 1966.

[133] D. H. Warrington and P. Bufalini. The coincidence site lattice and grain boundaries. *Scripta Metallurgica*, **5**:771–776, 1971.

[134] H. Grimmer. Coincidence-site lattices. *Acta Crystallographica*, **A32**:783–785, 1976.

[135] J. W. Christian. *The Theory of Transformations in Metals and Alloys*. Pergamon Press, Oxford, 1975.

[136] W. E. King, G. H. Campbell, S. M. Foiles, D. Cohen, and K. M. Hanson. Quantitative HREM observation of the $\Sigma 11(113)/[\bar{1}10]$ grain-boundary structure in aluminium and comparison with atomistic simulation. *Journal of Microscopy*, **190**:131–143, 1998.

[137] W. T. Read and W. Shockley. Dislocation models of crystal grain boundaries. *Physical Review*, **78**:275–289, 1950.

[138] R. W. Cahn. Recrystallization of single crystals after plastic bending. *Journal of the Institute of Metals*, **76**:121, 1949.

[139] J. F. Nye. Some geometrical relations in dislocated crystals. *Acta Metallurgica*, **1**:153–162, 1953.

[140] B. D. Cullity. *Elements of X-Ray Diffraction*. Addison-Wesley, 1956.

[141] W. P. Kuykendall and W. Cai. Conditional convergence in 2-dimensional dislocation dynamics. *Modelling and Simulation in Materials Science and Engineering*, **21**:055003, 2013.

[142] C. R. Weinberger and W. Cai. Orientation-dependent plasticity in metal nanowires under torsion: twist boundary formation and Eshelby twist. *Nano Letters*, **10**:139–142, 2010.

[143] C. R. Weinberger and W. Cai. Plasticity of metal wires in torsion: molecular dynamics and dislocation dynamics simulations. *Journal of the Mechanics and Physics of Solids*, **50**:1011–1025, 2010.

[144] F. C. Frank. In *Symposium on the Plastic deformtation of Crystalline Solids*, p. 159, Pittsburgh, Pennsylvania, 1950. The Physical Society, London.

[145] B. A. Bilby, R. Bullough, and E. Smith. Continuous distributions of dislocations: a new application of the methods of non-Riemannian geometry. *Proceedings of the Royal Society of London A*, **231**:263–273, 1955.

[146] D. Brandon. The structure of high-angle grain boundaries. *Acta Metallurgica*, **14**:1479–1484, 1966.

[147] J. P. Hirth. Dislocations, steps and disconnections at interfaces. *Journal of Physics and Chemistry of Solids*, **55**:985–989, 1994.

[148] J. P. Hirth and R. C. Pond. Steps, dislocations and disconnections as interface defects relating to structure and phase transformations. *Acta Materialia*, **44**:4749–4763, 1996.

[149] J. P. Hirth, R. C. Pond, and J. Lothe. Spacing defects and disconnections in grain boundaries. *Acta Materialia*, **55**:5428–5437, 2007.

[150] G. Alefeld and J. Völkl. *Hydrogen in Metals I. Basic Properties*. Springer-Verlag, 1978.

[151] T. R. Middya and A. N. Basu. Self-consistent T-matrix solution for the effective elastic properties of noncubic polycrystals. *Journal of Applied Physics*, **59**:2368, 1986.

[152] P. Sisodia and M. P. Verma. Polycrystalline elastic moduli of some hexagonal and tetragonal materials. *Physica Status Solidi (a)*, **122**:525, 1990.

[153] *CRC Handbook of Chemistry and Physics*. CRC, 95th edition, 2014–2015.

[154] J. Pelleg. Self diffusion in vanadium single crystals. *Philosophical Magazine*, **29**:383–393, 1974.

[155] Y. Sohn. Diffusion in metals. In W. F. Gale and T. C. Totemeier, editors, *Smithells Metals Reference Book*. Elsevier, 8th edition, 2004.

[156] B. L. Sharma. Diffusion data for semiconductors. In W. M. Haynes, editor, *CRC Handbook for Chemistry and Physics*. CRC, 2014–2015.

[157] N. L. Peterson. *Diffusion in Refractory Metals*. WADD Technical Report 60–793, 1960.

Index

activated state, 178
Airy stress function, 38, 251
  positive edge dislocation, 253, 477
  wedge disclination, 476
Al_2O_3, 10
Albenga's law, 92
amorphous, 2
anion
  interstitial, 143
  vacancy, 143, 149
antiphase boundary (APB), 14, 383
antiplane problem, 37, 245
anti-site defect, 161
anti-structure defect, 161
Arrhenius plot, 180, 187, 198, 501
atomic fraction, 80
atomic radius, *see* Seitz radius
atomic size factor, 79
atomic volume, 79
auxetic material, 29

band gap
  energy, 149, 150
  engineering, 9
basal plane, 376
biharmonic equation, 39, 251
biharmonic function, 39
binding energy, 165, 166
Boltzmann factor, 60
Boltzmann relation, 109, 127
Boltzmann's constant, 55, 57, 106
Boltzmann's distribution, 58, 60
Boltzmann's entropy, *see* entropy expression
Bravais lattice, *see* lattice
Brouwer approximation, 157
Brouwer diagram, 158, 160
bulk modulus, 29, 33, 42, 88, 93, 130, 138, 497
Burgers circuit, 219
Burgers vector, 12, 216, 219, 271
  density, 401

Cahn–Nye relation, 459
capillarity effect, 118
cation
  interstitial, 143
  vacancy, 142, 143, 149
CdTe, 232
charge neutrality, 140, 155
chemical equilibrium, 140
  constant, 147, 151
chemical potential, 51, 54, 55, 128
Coble creep, 198
coherent
  film, 313
  particle, 318
coincidence site lattice (CSL), 437, 439
  Σ number, 438
  $\Sigma 5$ CSL
    FCC lattice, 441
    simple cubic lattice, 440
compatibility condition, *see* strain
complex stacking fault, 384
conduction band, 150, 153
constitutional defect, 161
coordination number, 116, 164, 178
Cottrell atmosphere, 423
$CrCl_3$, 380
creep
  diffusional, 193, 198
  grain boundary sliding, 199, 201
  sigmoidal, 311
critical passing stress, 288, 290
critical stress, 316
critical thickness, 313, 314, 317
crystal structure
  B1 (NaCl), 17
  B2 (NiAl, CsCl), 7
  BCC/body-centered cubic, 5
  C1 (CaF_2), 17
  DC/diamond cubic, 6, 407
  FCC/face-centered cubic, 5
  HCP/hexagonal close-packed, 7, 377
  $L1_2$ (Ni_3Al, Cu_3Cl), 7, 382
  perfect, 1
  simple cubic, 2
  Zincblende (GaAs, CdTe), 6

diffusion, 9
  grain boundary, 198
  lattice, 198
diffusion coefficient, 185, 501
  interstitial solutes, 187, 506
  self-diffusivity, 187, 502
  substitutional solutes, 187, 503
  vacancies, 186, 190
diffusion equation, 183, 186, 188
dilatometry, 121
disclination, 12, 474, 475, 478
  wedge, 476
disconnection, 474, 480
dislocation
  array, 296, 460, 463
  BCC metals, 410
  bowing, 304
  breeding parameter, 309
  character angle, 214
  climb, 225, 229, 366, 367
  climb force, 268, 269
  climb plane, 271
  constricted node, 364, 366, 468

dislocation (*cont.*)
  cross-slip, 225, 228, 229, 367, 369, 386
  cross-slip pinning, 386
  cross-slip plane, 368
  density, 218
    mobile, 309
  dissociation, 356, 359, 450
  edge dipole, 215, 288
  edge dislocation, 12, 13, 214, 220, 251, 260
  equilibrium shape, 266
  extended node, 364, 365, 468
  geometrically necessary, 459
  glide, 225
  glide force, 268, 269
  glide plane, 220, 225, 226, 233, 236
  grown-in, 231
  infinitesimal dislocation, 400
  interaction, 284
    edge–edge, 286
    non-parallel screw, 292
    screw–screw, 284
    with point defect, 77, 418
  jog, 233, 367, 371
  jog motion, 367
  kinetics, 306
  kink, 413
  left-handed screw (LHS), 220
  line energy, 257, 260
  line tension, 262, 263, 265, 304
  line tension approximation, 265
  loop, 176, 295
  misfit dislocation, 232, 313, 315–317
  mixed dislocation, 214, 257, 262
  mobile, 308
  motion
    conservative, 225
    in thin film, 315
    non-conservative, 226
  multiplication, 235
  partial dislocation, 12, 13, 216, 349, 350, 361, 364, 371, 373
    in HCP crystal, 379
    leading, 363
    trailing, 363
  perfect dislocation, 12, 216
  pile-up, 297
  right-handed screw (RHS), 217, 220
  screw dislocation, 12, 13, 214, 220, 245
  sense vector, 219
  source, 231, 233
  spiral source, 234
  stair-rod dislocation, 361, 368, 371
  standard glide loop, 227, 228
  superdislocation, 382, 387
  superkink, 387
  superpartial, 388
  tangles, 13
  threading segment, 317
dislocation core
  distributed, 401
  energy, 259, 261
  reconstruction, 408, 409
  structure, 395
    easy core, 415
    edge in BCC metal, 414
    hard core, 415
    in DC crystal, 408, 409
    screw in BCC metal, 414
displacement shift complete (DSC) lattice, 446
  DSC lattice vector, 473, 475
  for Σ5 CSL
    simple cubic lattice, 447
  for Σ11 CSL
    FCC lattice, 451
  for Σ5 CSL
    FCC lattice, 449
disregistry, 397
distinguishable configurations, 106
domain boundary
  ferroelectric, 14
  magnetic, 14
double cross-slip, 235
dynamic strain aging, 423

Einstein model, 114
Einstein notation, 24
elastic field
  edge dislocation, 251, 253
    array, 460
    in finite radius cylinder, 255
  misfitting inclusion, 88
  mixed dislocation, 257
  point defect, 85
  screw dislocation, 245, 246
    array, 463
    in finite cylinder, 247
elastic strain energy, 33
  edge dislocation, 261
  screw dislocation, 259
electronic defect, 140
energy, 48, 102, 103
  bond energy, 103
  interaction energy, 104
  strain energy, 104
enthalpy, 50, 53, 102
entropy, 49, 50, 57
  configurational, 105, 106
  mixing, 101, 105
  vibrational, 105, 107, 114, 115
entropy expression
  Boltzmann's, 50, 57, 63
  Shannon's, 58
epitaxial film, 312
equation of state, 51
equilibrium concentration, *see* point defect
equilibrium condition, *see* stress
equilibrium fraction
  interstitial point defect, 110, 111, 125
  substitutional point defect, 109, 127
  vacancy, 113, 129
Escaig stress, 357
Eshelby twist, 248, 250
Euler's theorem, 52

F-center, 10
Fermi–Dirac distribution, 419, 423
Fermi–Dirac relation, 110, 127
Fick's first law, 184
Fick's second law, 186
Fleischer mechanism, 369
force dipole, 41, 84
Frank partial, 361, 373
  loop, 373
Frank's rule, 221
Frank–Bilby equation, 470
Frank–Read source, 233, 234
Frenkel defect, 142, 144, 154, 159
Frenkel disorder, 142
Frenkel model of slip, 210
Friedel–Escaig mechanism, 369

GaAs, 6, 9, 232
gas constant, 55
generalized stacking fault, 403, 404
Gibbs free energy, 50, 54, 102
Gibbs–Duhem relation, 52
grain boundary, 14
  axis–angle representation, 434
  GB dislocation, 450, 470, 473, 478
  GB engineering, 431
  general grain boundary, 455
  misorientation, 433
  orientation variables, 433
  singular grain boundary, 438
  sliding, 199
  tilt boundary, 437
    $\Sigma 11$, 444
    $\Sigma 5$, 443
    asymmetric, 469
    low angle, 297, 456, 460
    symmetric, 445, 479
  twist boundary, 437
    $\Sigma 5$, 442
    low angle, 293, 463
  vicinal grain boundary, 455
    tilt, 471
    twist, 474
grains, 1
Green's function, 39

halogen vacancy, 152
hard sphere radius, 72, 80
hardening mechanism
  dispersion hardening, 303
  solid solution hardening, 77
  strain hardening, 290, 302, 306, 311, 312, 337, 344
  Taylor hardening, 302
harmonic function, 38
harmonic oscillator, 60, 115
Head's problem, 328
Heisenberg's uncertainty principle, 61
Helmholtz free energy, 50, 53, 56
Hooke's law, 30
hydrogen interstitial, 125
hydrostatic pressure, 32, 190

ideal shear strength, *see* strength
ideal solution, 81
image dislocation
  edge
    free surface, 334
  screw
    curved interfaces, 331
    free surface, 320
    general image problem, 325
    inside cylindrical inhomogeneity, 332
    multiple images, 324
    multiple phase boundaries, 328
    near cylindrical inhomogeneity, 331
    rigid surface, 322
image pressure, 420
image stress, 90, 256, 420
  hydrostatic, 92
image torque, 249
InAs, 9
inclusion, 419
intermetallic alloy, 161
internal energy, *see* energy
interstitial, *see* point defect
isotropic elasticity, 28

KCl, 10
Kear–Wilsdorf lock, 385
Kelvin's problem, 40
King graph, 80
King table, 82, 93, 97, 488
Kirchoff's law, 222

Lamé coefficient, 32, 89
Laplace's equation, 38, 245
lattice, 5, 6
  BCC/body-centered cubic, 4
  FCC/face-centered cubic, 4
  hexagonal, 7, 375
  primitive tetragonal, 440
  SC/simple cubic, 4
lattice constant, 3, 71, 72, 80
lattice expansion, 121
lattice parameter, *see* lattice constant
lattice points, 4
Legendre transform, 52
LiF, 308
line tension, *see* dislocation
Lomer dislocation, 370
Lomer–Cottrell dislocation, 369
Lomer–Cottrell reaction, 369
low symmetry site, 75
meta-stable state, 178
metal vacancy, 152
method
  chemical equilibrium, 146, 165
  Lagrange multiplier, 145
Miller indices, 3
Miller–Bravais indices, 375
misfit
  dislocation, 313, 315–317
  strain, 232, 313
misfitting
  defect, 91
  inclusion, 86, 87, 94
  sphere, 93

Nabarro–Herring creep, 193
NaCl, 10
Nernst's theorem, 50
neutral vacancy, 149
Ni_3Al, 382
non-hydrostatic stress, 131
non-shearable particles, 304
nonstoichiometric ionic solid, 148
normal occupancy, 146
nucleation theory, 178

octahedral site, 74, 75, 181
orientation matrix, 434
Orowan bowing, 304
  formula, 305
Orowan equation, 307
Orowan loop, 305
oxidation reaction, 155

partition function, 57, 60
PbSe, 250
Peach–Koehler force, 268, 270, 273, 424
Peach–Koehler formula, 270, 271, 284
Peierls–Nabarro dislocation, 402
Peierls–Nabarro model, 351, 395
phase boundary, 14, 325
pinning point, 235, 372
  self, 418
Planck's constant, 61
plane strain, 38, 251, 476
point defect
  charged, 148
  chemical potential, 122
  electronic, 11

point defect (*cont.*)
  equilibrium concentration, 102, 108–110, 116
  equilibrium fraction, 109, 110, 112, 125
  formation energy, 103, 109
  formation entropy, 108, 109
  formation volume, 105, 109
  interaction with dislocation, 77, 418
  interstitial, 11, 72, 73, 109, 123
  neutral, 148
  non-symmetric, 84, 86
  substitutional, 11, 72, 108, 126
  symmetric, 84
  vacancies, 11
point force, 40
Poisson's ratio, 29, 252, 262
polycrystal, 1
polygonization, 458–460
positron annihilation, 118
precipitate, 93, 94
primitive cell, 4, 439
prismatic plane, 377
pyramidal plane, 377

quasi-chemical approach, 101
quench-resistivity, 119

reaction rate constant, *see* chemical equilibrium
Read–Shockley model, 456
reduction reaction, 155
regular solution, 101
resistivity, 119
right hand, start-finish (RH/SF) convention, *see* Burgers circuit

saddle point, 177
sapphire, 10
Schmid factor, 45
Schottky defect, 142, 154
Schottky disorder, 142
Seitz radius, 79, 488
self-interstitial, 140, 141
semi-coherent
  film, 314
  interface, 318
  particle, 319
shear modulus, 29, 30, 212, 497
Shockley partial, 350, 371
  loop, 374
Si, 140
Sievert's law, 126
single crystal, 1
size factor
  linear, 82
  volume, 81
slip, 225
slip direction, 349
slip plane, 210, 349
  active, 349
  glide-set, 406
  primary, 235
  shuffle-set, 406, 410
slip pyramid, 362
slip system, 350
slip traces, 209
solid solution hardening, *see* hardening mechanism, 78
solubility, 112
solute, 72
  as inclusion, 419
  atmosphere, 425
  distribution around edge dislocation, 422
  equilibrium concentration, 420
  interstitial, 101, 181, 187
  shear stress field, 424
  substitutional, 101, 182, 187
solvent, 73
solvus line, 102, 112
spacing defect, 482
spherically symmetric distortion, 73, 76
stacking fault, 12, 14, 351, 352, 354, 358, 413
  complex, 384
  extrinsic, 354, 378
  generalized, 403, 404
  in HCP crystal, 377
  intrinsic, 354, 378
stacking fault tetrahedron (SFT), 374
stacking sequence, 352, 354, 383, 385
Stirling's approximation, 58, 107
Stirling's formula, 58
stoichiometry, 161, 162
strain
  axial, 25
  compatibility condition, 36
  engineering, 26
  shear, 25
  tensor, 26
  transformation, 27
strain hardening, *see* hardening mechanism
strength, 77
  theoretical shear, 211, 212
strengthening, *see* hardening mechanism
stress
  axial, 21
  dimension, 21
  equilibrium condition, 35
  shear, 21
  tensor, 21, 23
  transformation, 24
  unit, 21
stress and strain fields, *see* elastic field
stress function, *see* Airy stress function
substrate, 232
superalloy, 1, 388
surface
  energy, 118
  step, 209, 225, 480
  stress, 66, 118

Taylor hardening, *see* hardening mechanism
Taylor relation, 303
tetragonal defect, 77
tetrahedral site, 74, 75, 181
thermal expansion, 46, 53, 65, 83, 121, 497, 498
thermodynamic potentials, 50
thermodynamics
  first law, 48
  second law, 49
  third law, 50
Thompson's notation, 360
Thompson's tetrahedron, 360, 372
triple junction, 460
twin boundary, 355, 413
twinning partial, 413

unit cell, 4

vacancy
  activation free energy, 178
  activation volume, 178
  annihilation, 176

chemical potential, 114, 128–131
condensation, 176, 177
constrained equilibrium, 167
divacancies, 164
equilibrium concentration, 112, 116, 164
equilibrium fraction, 116
formation energy, 113, 116
formation free energy, 132
in Si, 140
jump rate, 177, 178
loop, 177, 295
monovacancies, 164
motion, 177
quadvacancies, 167
relaxation volume, 114
sinks, 176
sources, 176
trivacancies, 166

vacancy diffusion, 230
inhomogeneous stress, 190
uniform hydrostatic stress, 190

valence band, 150, 153
Vegard's law, 81
Voigt notation, 27, 34
Volterra dislocation, 214, 215, 402

X-ray measurement, 121

yield point drop, 308
Young's modulus, 29

ZnO, 159

Printed in the United States
By Bookmasters